23rd Association for the Advancement of Animal Breeding and Genetics Conference 2019

Armidale, Australia
27 October – 1 November 2019

ISBN: 978-1-7138-0335-5

Printed from e-media with permission by:

Curran Associates, Inc.
57 Morehouse Lane
Red Hook, NY 12571

Some format issues inherent in the e-media version may also appear in this print version.

Printed with permission by Curran Associates, Inc. (2020)

AAABG – Association for the Advancement of Animal Breeding and Genetics
University of New England
Armidale NSW 2351
Australia

Phone: 61 2 677 320 55
Fax: 61 2 677 332 66

AGBU@une.edu.au

Additional copies of this publication are available from:

Curran Associates, Inc.
57 Morehouse Lane
Red Hook, NY 12571 USA
Phone: 845-758-0400
Fax: 845-758-2633
Email: curran@proceedings.com
Web: www.proceedings.com

Association for the Advancement of Animal Breeding and Genetics

AAABG

Association for the Advancement of Animal Breeding and Genetics

Proceedings of the Twenty-third (23rd) Conference

Armidale, New South Wales, Australia
27th October – 1st November 2019

PRESIDENT'S MESSAGE

I wish you a very warm welcome to AAABG's Ruby Anniversary meeting, the 23rd conference in Armidale, NSW. When the steering committee behind the AAABG held the very first conference in Armidale in 1979, they might not have expected the organisation to still be going strong 40 years later. Since then, 22 conferences have been held in all Australian states and both the North and South Islands of New Zealand. While other organisations have sometimes struggled with memberships, all AAABG conferences have attracted substantial numbers of contributors and delegates, with regular participants from around the globe as well. This highlights how important it is to provide a great forum for communication amongst scientists, educators, students and service providers, who traditionally make up the bulk of attendees, ultimately to increase knowledge and foster ideas and collaboration.

This year, we also introduce an extended program which starts with a student workshop and ends with a program which will contain some talks of interest to breeders. Allowing students to meet each other before the conference, and obtain some wise words from educators and extension specialists, should improve their conference experience and provide some valuable insight for their future progression. Additionally, the attendance of breeders at AAABG has dropped off compared to early years. This is in part due to the increasing complexity of livestock breeding, making many talks less accessible to a general audience and leading to an increasing distance between many researchers and those who benefit from their work. We hope to encourage more breeder participation this year and that there will be plenty of mix and mingle during the last 1.5 days of the program.

This year also marks one of the most extensive droughts across large areas of Eastern Australia, with rainfall in the 2019 year to date the lowest on record for the New England-Northern Tablelands area. So, while we were hoping to dazzle you with some beautiful spring green at this conference, the reality is that conditions may still be very poor by the time delegates arrive, and high level water restrictions will be in place for Armidale. This is a timely reminder of the difficult conditions under which our livestock breeders and producers function, and I take off my hat to their resilience under these circumstances. In particular, I thank those breeders who are still able to welcome delegates to their properties on tours, and trust that delegates recognise the courage this must take.

The scientific program again reports a wide variety of research. The implementation of genomic selection is not without its challenges, but is becoming a more mature part of modern breeding programs. New technologies offer future opportunities, both in terms of novel phenotyping and techniques such as gene editing. These developments are combined with talks which touch on aspects important to effective implementation of breeding programs today in the livestock industries.

I wish to thank the sponsors who have supported the conference, and the willingness of both our local and overseas speakers to contribute to the conference program. I also thank staff at ASN (the event organisers), the committee who have helped organise the Armidale meeting, and Kathy Dobos for preparing this booklet. Thanks also go to reviewers of papers and the session chairs, as well as those involved in organising tours and assisting with the student program.

I trust you have an enjoyable conference.

Kim Bunter

ASSOCIATION FOR THE ADVANCEMENT OF ANIMAL BREEDING AND GENETICS

2019

TWENTY THIRD CONFERENCE

COMMITTEE

President	Kim Bunter
President Elect	Forbes Brien
Vice President	Sonja Dominik
Secretary	Kath Donoghue
Treasurer	Sam Clark
Editors	Sue Hatcher Susanne Hermesch Kim Bunter
Ordinary Committee Members	Luke Stephen Sarita Guy Matias Suarez
Handbook Production	Kathy Dobos
Professional Conference Organiser	ASN Events

CITATION OF PAPERS

Papers in this publication should be cited as appearing in the Proceedings of the Association for the Advancement of Animal Breeding and Genetics (Abbreviation: Proc. Assoc. Advmt. Anim. Breed. Genet.)

For example:
Bowley F.E., Amer P.R. and Meier S. (2013) New approaches to genetic analysis of fertility traits in New Zealand dairy cattle. *Proc. Assoc. Advmt. Anim. Breed. Genet.* **20**: 37-40.

REVIEWERS and SECTION EDITORS

All papers, invited and contributed, were subjected to peer review by two referees. We acknowledge and thank those people listed below for their work in reviewing the papers (and apologise if we have inadvertently omitted any reviewer from the list).

Reviewers:

Hassan Aliloo
Michael Aldridge
Peter Amer
Jason Archer
Paul Arthur
Robert Banks
Stephen Barwick
Amy Bell
Tracie Bird-Gardiner
Hugh Blair
Vinzent Boerner
Clara Bradford
Forbes Brien
Daniel Brown
Kim Bunter
Annika Bunz
Sam Clark
Bronwyn Clarke
Schalk Cloete
Natalie Connors
Brad Crook
Neil Cullen
Hans Daetwyler
Lino De La Cruz Colin
Sara De las Heras Saldana
Dion Detterer
Ken Dodds
Sonja Dominik
Kath Donoghue
Christian Duff
Naomi Duijvesteijn
Katherine Edgerton-Warburton
Mohammad Ferdosi
Marina Fortes
Dorian Garrick
Daniel Garrick
Arthur Gilmour
Tom Granleese
Johan Greeff
Boyd Gudex
Phillip Gurman
Sarita Guy
Mekonnen Haile-Mariam
Bruce Hancock
Rhiannon Handcock
Mehedi Hasan
Sue Hatcher
Ben Hayes
Michelle Hebart
Fiona Hely
Mark Henryon
John Henshall
Susanne Hermesch
Melanie Hess
John Holmes
Dean Jerry
Gilbert Jeyaruban
Patricia Johnson
Matthew Kelly
Majid Khansefid
Mehar Khatkar
Brian Kinghorn
Stephen Lee
Andres Legarra
Sigrid Lehnert
Craig Lewis
Yutao Li
Li Li
Daniela Lourenco
Russell Lyons
Iona MacLeod
Karen Marshall
Rudi McEwin
Aaron McMillan
Karin Meyer
Catriona Millen
Nasir Moghaddar
Kirsty Moore
Stephen Moore
Suzanne Mortimer
Carel Muller
Anieka Muller
Cornelius Nel
Loan Nguyen
Bang Nguyen
Thuy Nguyen
Hugh Nivison
Peter Parnell
Gertje Petersen
Wayne Pitchford
Raul Ponzoni
Greg Popplewell
Laercio Porto-Neto
Jennie Pryce
Ian Purvis
Cheryl Quinton
Anne Ramsay
Imtiaz Randhawa
Antonio Reverter
Matt Reynolds
Malshani Samaraweera
Bruno Santos
Penny Schulz
Ric Sherlock
Jen Smith
Luke Stephen
Eva Strucken
Matias Suarez
Andrew Swan
Imke Tammen
Peter Thomson
Bruce Tier
Irene Van Den Berg
Julius van der Werf
Alison Van Eenennaam
Laura Vargovic
Sam Walkom
Bradley Walmsley
Christie Warburton
Matt Wolcott
Rob Woolaston
Shernae Woolley
Ruidong Xiang
Yuandan Zhang

SPONSORS of the 23rd AAABG Conference 2019

The financial assistance of the following organisations is gratefully acknowledged.

Angus Australia — Celebrating 100 years · 1919-2019 ·

Weatherbys Scientific Australia

Humanizing Genomics macrogen OCEANIA

NEOGEN Australasia

mla MEAT & LIVESTOCK AUSTRALIA

awi Australian Wool Innovation Limited

agbu — A joint unit of the NSW Department of Primary Industries and the University of New England

CSIRO

une University of New England

NSW GOVERNMENT Department of Primary Industries

AAABG was formerly known as the Australian Association for Animal Breeding and Genetics. Following the 1995 OGM the name was changed when it became an organisation with a joint Australian and New Zealand membership. The Association for the Advancement of Animal Breeding and Genetics is incorporated in South Australia.

THE ASSOCIATION FOR THE ADVANCEMENT OF ANIMAL BREEDING AND GENETICS INCORPORATED

OBJECTIVES

(i) to promote scientific research on the genetics of animals;
(ii) to foster the application of genetics in animal production;
(iii) to promote communication among all those interested in the application of genetics to animal production, particularly breeders and their organisations, consultants, extension workers, educators and geneticists.

To meet these objectives, the Association will:

(i) hold regular conferences to provide a forum for:
  (a) presentation of papers and in-depth discussions of general and industry-specific topics concerning the application of genetics in commercial animal production;
  (b) scientific discussions and presentation of papers on completed research and on proposed research projects;
(ii) publish the proceedings of each Regular Conference and circulate them to all financial members;
(iii) use any such other means as may from time to time be deemed appropriate.

MEMBERSHIP

Any person interested in the application of genetics to animal production may apply for membership of the Association and, at the discretion of the Committee, be admitted to membership as an Ordinary Member.

Any organisations interested in the application of genetics to animal production may apply for membership and, at the discretion of the Committee, be admitted to membership as a Corporate member. Each such Corporate Member shall have the privilege of being represented at any meeting of the Association by one delegate appointed by the Corporate Member.

Benefits to Individual Members

- While it is not possible to produce specific recommendations or "recipes" for breeding plans that are applicable for all herd/flock sizes and management systems, principles for the development of breeding plans can be specified. Discussion of these principles, consideration of particular case studies, and demonstration of breeding programs that are in use will all be of benefit to breeders.
- Geneticists will benefit from the continuing contact with other research workers in refreshing and updating their knowledge.
- The opportunity for contact and discussions between breeders and geneticists in individual members' programs, and for geneticists in allowing for detailed discussion and appreciation of the practical management factors that often restrict application of optimum breeding programs.

Benefits to Member Organisations

- Many of the benefits to individual breeders will also apply to breeding organisations. In addition, there are benefits to be gained through coordination and integration of their efforts. Recognition of this should follow from understanding of common problems, and would lead to increased effectiveness of action and initiatives.
- Corporate members can use the Association as a forum to float ideas aimed at improving and/or increasing service to their members.

General Benefits

- Membership of the Association may be expected to provide a variety of benefits and, through the members, indirect benefits to all the animal industries.
- All members should benefit through increased recognition of problems, both at the level of research and of application, and increased understanding of current approaches to their solution.
- Well-documented communication of gains to be realised through effective breeding programs will stimulate breeders and breeding organisations, allowing increased effectiveness of application and, consequently, increased efficiency of operation.
- Increased recognition of practical problems and specific areas of major concern to individual industries should lead to increased relevance of applied research.
- All breeders will benefit indirectly because of improved services offered by the organisations which service them.
- The existence of the Association will increase appreciably the amount and use of factual information in public relations in the animal industries.
- Association members will comprise a pool of expertise – at both the applied and research levels – and, as such, individual members and the Association itself must have an impact on administrators at all levels of the animal industries and on Government organisations, leading to wiser decisions on all aspects of livestock improvement, and increased efficiency of animal production.

CONFERENCES

One of the main activities of the Association is the Conference. These Conferences will be structured to provide a forum for discussion of research problems and for breeders to discuss their problems with each other, with extension specialists and with geneticists.

ASSOCIATION FOR THE ADVANCEMENT OF ANIMAL BREEDING AND GENETICS
FELLOWS OF THE ASSOCIATION

"Persons who have rendered eminent service to animal breeding in Australia and/or New Zealand or elsewhere in the world, may be elected to Fellowship of the Association…"

Elected February 1990 R.B.M. Dun *Elected September 1992* K. Hammond *Elected July 1995* C.H.S. Dolling J.R. Hawker J. Litchfield *Elected February 1997* J.S.F. Barker R.E. Freer *Elected June 1999* J. Gough J.W. James *Elected July 2001* J.N. Clarke A.R. Gilmour L.R. Piper *Elected September 2005* B.M. Bindon M.E. Goddard H.-U. Graser F.W. Nicholas *Elected September 2007* K.D. Atkins R.G. Banks G.H. Davis	*Elected September 2009* N. Fogarty A. Fyfe J. McEwan R. Mortimer R. Ponzoni *Elected September 2011* B.P. Kinghorn A. McDonald *Elected October 2013* H. Burrow P. Fennessy G. Nicoll P. Parnell *Elected October 2015* P. Arthur D. Johnson K. Meyer B. Tier R. Woolaston *Elected October 2019* S.A. Barwick H.T. Blair S.W.P. Cloete I.W. Purvis

HONORARY MEMBERS OF THE ASSOCIATION

"Members who have rendered eminent service to the Association may be elected to Honorary Membership…"

Elected September 2009
W.A. Pattie J. Walkley

HELEN NEWTON TURNER MEDAL TRUST

The Helen Newton Turner Medal Trust was established in 1993 following an anonymous donation to the Animal Genetics and Breeding Unit. The Helen Newton Turner Medal is awarded to provide encouragement and inspiration to those engaged in animal genetics. The Medal is named after Dr Helen Newton Turner whose career with CSIRO was dedicated to research into the genetic improvement of sheep for wool production. The Medallist is chosen by Trustees from the ranks of those persons who have made an outstanding contribution to genetic improvement of Australian livestock.

The Helen Newton Turner Medal was first awarded in 1994 to Associate Professor John James and a list of all recipients to date is given below. The recipient of the Medal is invited to deliver an Oration on a topical subject of their choice. The Oration of the 2015 Medal recipient, Dr. Arthur Gilmour, is reproduced in these proceedings.

MEDALLISTS

1994 J.W. James
1995 L.R. Piper
1997 J. Litchfield
1998 J.S.F. Barker
1999 C.W. Sandilands
2001 G.A. Carnaby
2003 F.W. Nicholas
2005 K. Hammond
2007 L. Corrigan
2009 R. Hawker
2011 R. Banks
2013 M. Goddard
2015 A. Gilmour
2017 A. Collins
2019 K. Atkins

HELEN NEWTON TURNER AO

HELEN NEWTON TURNER MEDALIST ORATION 2017

Alf Collins

Alf Collins snr is one of the most innovative beef cattle breeders in the world. Building on the foundations established by his father, he has applied enormous dedication, careful recording and rigorous focus on breeding for profitability, to the continuous improvement of Brahman cattle.

Brahman cattle have to perform in very challenging environments, and breeding programs to deliver genetic improvement in those environments are challenging too – reflecting large scale of operations and variable climatic conditions.

Alf has met these challenges head on and collected performance records underpinning reliable EBVs, and used the information backed by hard-nosed practical understanding of functionality and survival ability, to generate very impressive genetic progress over several decades. Perhaps the most outstanding aspect of that genetic progress is that it includes very substantial progress in female fertility – something that has almost been treated as "too hard" by most breeders of tropically adapted cattle. CBV has actively participated in industry R&D, including significant contributions to Beef CRC I, II and III.

The breeding program includes several fertility traits within overall selection for profit: recording includes speed of re-breed, puberty threshold, calving interval, age at first calving, number of calves, speed of growth, dry season gain, wet season acceleration, as well as good temperament, and fleshiness.

Alf Collins is a deep thinker about what cattle need to do in the tropical environment, and has never been afraid to try novel approaches or include new traits if they will help breeding cattle better and better suited to the environment and to improving profit:

"At CBV, from 1981 to the present day, our management has been relentless in the development and multiplication of the traits that have greatest commercial significance. In total, this represents over 50 years of development, using steadily improving tools of analysis and selection.

We have absolutely no tolerance of cattle that do not earn every single year. We get our share of non-performing stock and have management strategies to convert them to beef carcases immediately when they fail.

The genetic trends reflect this strategy at CBV.

Reproduction and survival are paramount, coupled with gentle temperament, fleshy bodies and thrift at grazing. CBV cattle are true examples of a highly adapted breed. This equates to a high speed beef machine at minimal cost.

We have received very high levels of support from researchers, scientists, clients, family and friends. Intellectual inputs have been considerable, along with personal effort. CBV has an ongoing involvement in research and analysis every year.

Our matings commence in the dry season on October 1, to identify the most efficient adapted females, by their ability to conceive whilst lactating in very dry grazing and to hold that pregnancy, calve un-assisted, raise a sound calf and to rebreed within our low cost management. Our stocking rate of kilograms per hectare per 100mm of rainfall is high but ecologically responsible.

Consequently earnings per hectare per financial year are optimised."

Alf Collins continues to be an outstanding pioneer and innovator in real-world application of genetics technology, and the demonstration that it is possible to breed genetically fertile, productive and profitable tropically adapted cattle is an inspiration.

TABLE OF CONTENTS

Genomic Selection 1

Efficiency and Product Quality

Posters

Proc. Assoc. Advmt. Anim. Breed. Genet. 23:1-6

INVESTMENTS IN BREEDING TECHNOLOGIES AND ORGANIZATION TO MEET GLOBAL NEEDS

J.A.M. van Arendonk, M.C.A.M Bink, K. Peeters, B. Visser, N. Duijvesteijn and P. van As

Hendrix Genetics, PO Box 114, 5830 AC Boxmeer, The Netherlands

SUMMARY

Animal breeding has a vital role to play in solving the global food challenge. This paper will concentrate on investments that are needed for animal breeding to meet the challenges of the future and begins with describing the global challenge. There is not a single solution that will work in all species in all regions, so solutions need to be tailored to the local conditions. There is a clear need for both more sustainable production of animal proteins and a reduction of waste in the food chain. There is regional diversity in emphasis on the different components of sustainability, but the general trend is towards animal protein production with a lower ecological impact, with a minimum use of antibiotics and with good animal welfare. This requires not only investments in genetic technologies like genomic selection but also in methods for phenotyping individual animals under commercial conditions.

INTRODUCTION

Animal breeding is a powerful tool to improve many aspects of animal production. In this paper, we describe the contributions of animal breeding to solving the global challenges when it comes to feeding the growing world population sustainably.

Hendrix Genetics is a multi-species animal breeding company with breeding programs in turkeys, layers, swine, salmon, trout, shrimp and coloured broilers. To be a competitive animal breeding company in any species requires substantial investments in research and development. By working in multiple species, these investments can be more cost effective as there are many similarities between species. For example, the IT infrastructure for collecting and storing information on individual animals and the methods for performing genomic evaluations are very similar for different species.

After a brief description of the global challenges and the expected changes in our value chains, we will describe in more detail the role of animal breeding and how new technologies can help to better meet the challenges.

GLOBAL CHALLENGE

We face major global challenges when it comes to feeding the growing world population sustainably. Rabobank has predicted that the animal protein market will grow by 45% in the next two decades and this global growth will be largely in Asia and to a lesser extent in Africa. We see more and more developing countries reaching middle income status, the inflection point for protein consumption, leading to an increased need for locally produced animal protein. The contribution of species to animal protein production differs between regions. For example, currently close to 90% of aquaculture production takes place in Asia, which is also the biggest growth market for layers and swine. In contrast, North America remains a high value and volume market for poultry, pigs and cattle, whereas aquaculture is expected to remain limited.

There is a clear need for more sustainable methods of producing all animal proteins. There is regional diversity in emphasis of the different components of sustainability, but the general trend is towards animal protein production with a lower ecological impact, with a minimum use of antibiotics and with good animal welfare.

At all levels in our value chains we see scale increasing. The number of people working in animal production is declining, the farms are getting bigger, and value chains are getting shorter and increasingly coordinated. Innovative farming methods using robotics and data driven management support will help not only to meet the labour challenge but also to improve sustainability.

Worldwide, the use of technology and software is rapidly increasing. Already thousands of companies offer data-based services to support farm management, increasingly making use of sensors, machine learning, and other decision-support tools. We also see increasing societal pressure in the developed world regarding environmental impact, livestock treatment and biotechnology. Also, large food companies and supermarket chains are forcing changes to production practices.

We anticipate the following changes in our value chains:

- Increased use of digital technology and software for managing farming operations, with large companies fulfilling this demand
- Increased mechanisation and automation, driving standardization
- Stronger presence of alternative sources of protein, including insects
- More and more varied animal protein "brands" differentiated by farming system, animal type, and product quality.
- More ready-to-eat providers, such as food delivery companies, and ready meals.

Animal protein production. There are many individuals on this planet who live relatively healthy lives consuming little or no animal protein, and many would argue that the challenge of feeding the human population could be met by reducing the amount of livestock products in our diet. However, the demand for animal protein, especially in developing countries, is expected to grow as they become more affluent. Part of the animals' proteins are produced from feed, such as grain, that could be directly consumed by humans, while another part is produced from feed resources that would not feed humans directly, such as grass and by-products from the human food industry.

According to the FAO, an estimated one third of all food produced globally is either lost or wasted. This represents a large inefficiency in the food system. Food loss refers to any food that is lost in the supply chain between the producer and the market. Food waste, on the other hand, refers to the discarding or alternative (non-food) use of food that is otherwise safe and nutritious for human consumption. Meeting the food challenge is not only about more sustainable production but also about reducing food loss and waste.

The challenge for livestock production is to meet the growing demand for animal protein while at the same time reducing the environmental impact. This implies that livestock production needs to improve the efficiency of production, robustness of animals and quality of animal products. Improvement of efficiency of animal production needs to focus on improving lifetime productivity, which can be achieved by improving not only individual productivity but also by reducing losses through improved health and reproductive performance. Robustness of animals refers to the ability of animals to handle variation in the environment, in particular feed quality and climate. The quality of animal products refers not only to the food safety and taste but also to animal welfare.

THE ROLE OF ANIMAL BREEDING

Animal breeding has a vital role to play in solving the global food challenge. In the last 4 decades, animal breeding has halved the amount of feed required to produce animal proteins in poultry and pigs. Reducing the ecological food print is an important contribution to improved sustainability. Improving sustainability also requires reducing the feed-food competition, reducing the use of antibiotics, and improving animal well-being.

Breeding goal. The breeding goal summarizes the direction of change of a population. Over the years, the breeding goal has changed in response to the changes in production circumstances and the increased attention to sustainability. Commercial poultry and pig breeding goals have broadened widely since the 1970s (Neeteson-van Nieuwenhoven *et al.* 2013). Over time, the relative focus on productivity has decreased and objectives such as efficiency, welfare, robustness and product quality have increased. Production circumstances and consumer demands will continue to change and impact the breeding goal not only in terms of the number of traits but also in terms of the relative emphasis.

Sustainability program: As a breeding company, we also keep many animals ourselves. That is why our efforts to achieving sustainability are not only directed towards our breeding program but also improving our own performance. For improving our own performance, we have established in 2013 a sustainability program comprising of three building blocks: animals, people and planet.

- Animal welfare, biosecurity and genetic resources are the key priorities within the building block animals. Ensuring animals are treated with care and respect and are kept under the highest standards of welfare is essential. We ensure that taking good care of animals is embedded in our company culture. As global suppliers of breeding stock, we have a responsibility for ensuring biosecurity and animal health. In addition, we also have an obligation to protect our genetic resources.
- People make our business and deliver our products and service to our customers. We started off with setting KPI's for health and safety including illness percentage, accidents and time lost time due to accidents. More recently, we have added employee engagement and expertise.
- Minimizing the environmental impact of livestock through improving input efficiency and helping to reduce the use of antibiotics are key parts of the building block planet. In addition, the company is investing in minimizing its own ecological footprint to preserve and improve the environment that its activities impact.

We have implemented a sustainability reporting cycle, which includes a regular program of data collection, target setting and evaluation which is aimed at making improvements year after year. In addition, we will publish a CSR report to increase the awareness on our activities both internally and externally.

DISSEMINATION

Not only generation but also dissemination of genetic progress plays an important role in an animal breeding organisation. In cattle, frozen semen is the most commonly used method of distributing genetic progress. In poultry and swine, frozen semen is not an option. In swine fresh semen and live animals are used for dissemination. In poultry hatching eggs and one-day old animals are used for dissemination. The use of live animals rather than frozen semen comes with logistic and biosecurity challenges.

In poultry and swine, a multi-tier crossbreeding system is used. In a typical laying-hen program, pure-line birds are used to produce grandparents which are crossbred to produce the parent stock males and parent stock females. The parent stock is used to produce the commercial birds as illustrated in Figure 1. The genetic progress is generated in the pure lines under bio secure conditions. Subsequently this progress is disseminated from the pure line to the commercial offspring through several multiplication steps. The system also allows capturing the benefits of crossbreeding. Furthermore, it allows making the best combination of different pure lines to meet the needs of farmers operating in different countries and markets. This system also offers the breeding organisation two options to react to a change in product demand and to a change in production environment. First, there is the option to change the combination of lines to produce the commercial product. Second, there is the option to change the breeding goal in one or more pure lines. By changing the combination of lines, we can react more rapidly to changes compared to changing the breeding goal of a line. We continuously evaluate the expected developments to ensure that the product portfolio not only meets the current needs but also the expected needs in the years to come.

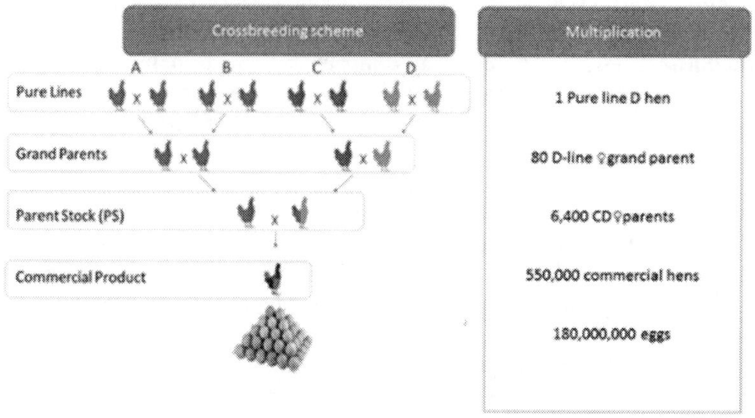

Figure 1. Schematic diagram of the poultry production pyramid in which genetics of 4 pure lines (A, B, C and D) is used in a crossbreeding scheme to produce parent stock (PS) males (AB) and females (CD) and commercial products (ABCD). The relative size (multiplication) of each layer in the production pyramid is given for the female lineages (from pure line D hen through grandparents and parents to commercial hens and eggs produced by these hens)

TECHNOLOGIES

Our future is tied directly to product superiority, which requires the implementation of state-of-the-art breeding technology for all our products. This implies that we invest in tools for collection of information on individual animals, in genomic selection to ensure that we make best use of the collected information and breeding scheme design. Investments in technology should also provide solutions for labour shortages on our breeding farms and on the farms of our customers.

We see many promising developments in the domains of phenotyping, digitalization, and genetics technologies. We will continue to make targeted investments in the most promising technologies starting from a business needs perspective. In the following sections, more background is given on activities in the domain on phenotyping and gene editing.

Phenotyping. We invest in phenotyping methods not only to collect novel traits in the domain of animal behaviour but also to measure performance of animals under commercial conditions. Remote sensors such as cameras, microphones, thermometers and accelerometers offer the opportunity to capture data from groups or individual animals. Data from remote monitoring sensors combined with individual animal identification can provide information regarding pig welfare, health and productivity (Benjamin and Yik 2019).

Livestock are nowadays more frequently kept in larger groups, resulting in an increase in social interactions between individuals. Moreover, treatments to limit the consequences of adverse social interactions, such as beak trimming in poultry and tail docking in pigs, will probably be banned in the future (at least in EU countries), so that the negative effects of social interactions will likely increase unless action is taken to avoid that. Actions are needed to prevent or diminish the negative effects of social interactions. Bijma (2007) demonstrated that pecking in laying hens is a socially affected trait which not only depends on the hen's ability to avoid being pecked (direct genetic effect) but also on the pecking behaviour of her group mates (indirect genetic effect). Using this knowledge, we have demonstrated that we can select animals that are less likely to perform damaging behaviour. Selection can be further improved using sensor technologies that allow the identification of laying hens in large groups that show less pecking behaviour (Ellen *et al.* 2019).

Traditionally, egg production on laying hens is measured in single bird or family group cages. This housing system is needed to link the egg production to a single individual or parent. The housing system, however, does not reflect the commercial conditions for laying hens which are increasingly kept in cage-free conditions. The difference between selection and commercial environment might lead to genotype by environment interaction which would make selection less effective. To overcome this, we are investing in automatic nests for laying hens which allows the recording of individual egg production of animals kept in a group. These automatic nests are not available on the market and need to be developed internally.

Gene editing is a rapidly developing technology with many potential applications, including in animal breeding. Hendrix Genetics is committed to responsible farm animal breeding. We strive to meet growing global demands for food by supporting animal protein producers worldwide with innovative and sustainable genetic solutions. New technologies like gene editing can be part of our future solutions. Alongside delivering benefits to producers, our solutions must also meet the rigorous needs of consumers and society.

While we rely on genomic selection in our breeding programs, Hendrix Genetics does not currently use any form of gene modification. We, however, continue to closely monitor the rapid developments in gene editing and invest in research in this new technology to evaluate its potential application. Gene editing will help us to get a better understanding of genes and mutations in genes that contribute to genetic variation in traits. That knowledge can be used to improve genomic selection schemes provided that the desired variants are present in the population. When the desired variant is not present, genetic improvement via gene editing is an innovative solution.

Investment in research into gene editing does not imply that Hendrix Genetics will necessarily use this technology in the future. Before using a new technology, we need to understand the full impact of it on animals, animal products and humans. We must be convinced of the added value of gene editing before entering any discussion on commercial application. Such discussion will not only cover technical issues but more important ethical and regulatory issues. Now, Hendrix Genetics sees several critical challenges ahead for gene editing that must be resolved before commercial application can even be considered.

Even with satisfactory results from research, Hendrix Genetics would only ever consider gene editing for applications when it clearly outperforms any alternatives. The most likely application of gene editing appears to be to improve the health and welfare of farm animals (including fish). It is very unlikely that we will use gene editing for realizing higher production efficiency directly. We are, for example, involved in research on the opportunity to use gene editing to stop surgical castration of male pigs.

POULTRY BREEDING FOR AFRICAN SMALLHOLDER FARMERS

There is a wide variation in climate, production circumstances and consumer preferences around the world. This implies that when it comes to animal breeding, one size does not fit all. As an international breeding organisation, we need to have a product portfolio to meet that diversity. This can be illustrated when looking at smallholder farmers in Africa. To also meet their needs, we not only breed birds that are specialized in egg production but also dual-purpose birds, intended to produce both eggs and meat.

Poultry constitutes an important economic activity for the rural poor in many African countries. Several researchers have shown that the performance of smallholder poultry production can be greatly improved by using improved genetics. The local indigenous breeds are inefficient and unproductive compared to other alternative breed options, such as Sasso and Kuroiler. In many instances the smallholder farmers in rural areas do not have access to improved genetics and are forced to use birds that have low levels of productivity and high mortality rates. The access to an improved low-input and dual-purpose chicken to supplement the local indigenous breeds has the potential to transform the rural poultry enterprise.

This situation can be changed as demonstrated by the African Poultry Multiplication Initiative (APMI) led by the World Poultry Foundation (WPF), with investments in Uganda, Ethiopia, Tanzania, and Nigeria as well as other poultry initiatives in Burkina Faso. The APMI model operates through capable local private companies to establish a parent stock and hatchery operation for the supply of improved genetics of low-input, dual purpose chicken breeds to farmers in their communities. These initiatives are dependent on access to poultry parent stock for the improved breeds. We have partnered with WPF to ensure reliable access to improved parent stock genetics. The supply of parent stock is frequently disrupted by outbreaks of diseases such as avian influenza. An outbreak of avian influenza in the source country leads to a ban on export of parent stock. A long-term sustainable solution to mitigate this risk is duplication of the germplasm at multiple locations.

Although breeds such as Kuroiler and Sasso perform better than most local ecotypes, the productivity and feed utilization efficiency of these breeds is far lower than current commercial breeds. Results from ILRI's African Chicken Genetic Gain project shows that there is a wide variability in the performance of Kuroiler and Sasso in different agro ecologies. We have, therefore, implemented a genetic improvement program to further improve the productivity, adaptability, and resilience of the lines that are used to produce the dual-purpose breed. The genetic gain of the lines may be further accelerated by the application of genomics selection. However, implementation of this technology for the benefit of smallholder farmers in Africa has failed due a combination of two factors. First, the lack of support for such genetic improvement schemes to develop proper infrastructure (such as performance recording and genetic evaluation schemes). Second, the lack of a system to sustainably multiply and distribute the improved genetic material to the smallholders. We aim to overcome these factors due to our experience and knowledge and more importantly our access to a larger international market. The ability to sell genetic material in multiple countries is crucial for offsetting the cost of a breeding program to improve the dual-purpose chicken. With these improved breeds, smallholder farmers in Africa are not only able to increase their income but also to contribute to feeding the growing population with nutritious protein.

COLLABORATION

In order to find sustainable solutions for the global food challenge, we are continuously exploring innovations in the domain of measuring health, welfare and productivity of animals. These innovations need to be based not only on a solid understanding of the underlying biology but also on an overall view on the issue at stake. Developing a solid understanding is an important but not the only driver to be involved in research collaboration with knowledge institutes. Equally important drivers for participation in a research project are creating awareness in the scientific community for the issues involved in improving sustainability and training a new generation of researchers. Solving sustainability issues often requires collaboration in multidisciplinary teams. Industry participation in research projects is expected to speed-up innovations and contribute to training of new talents that are focussed on generating solutions. Collaboration is therefore crucial for realizing sustainable solutions for the global food challenge.

REFERENCES

Benjamin M. and Yik S. (2019) *Animals* **9**: 133.

Bijma P., Muir W.M. and van Arendonk, J.A.M. (2007) Genetics **175**: 277.

Ellen E.D., van der Sluis M., Siegford J., ….and Rodenburg, T.B. (2019) *Animals* **9**: 108.

Neeteson-van Nieuwenhoven,A.-M., P. Knap S. and Avendaño S. (2013) *Animal Frontiers* **3**: 52.

Proc. Assoc. Advmt. Anim. Breed. Genet. 23:7-10

IMPROVED RATE OF TARGETED GENE KNOCK-IN OF *IN-VITRO* FERTILIZED BOVINE EMBRYOS

J.R. Owen[1], S.L. Hennig[1], E.E. Paulson[1], J.L. Lin[1], P.J. Ross[1] and A.L. Van Eenennaam[1]

[1]University of California, Davis, CA, 95616

SUMMARY

Variables for achieving targeted gene knock-ins using CRISPR/Cas9 mediated gene insertion in bovine embryos following in-vitro maturation were tested to evaluate the rate of integration at a target genomic location, and the level of mosaicism. Guide-RNAs (gRNA) were developed targeting downstream of the Zinc Finger X-linked (ZFX) gene located on the bovine X-chromosome. One gRNA (ZFXg3) was found to cut with high frequency in-vivo (82%). Donor vectors utilizing different endogenous repair pathways: homologous recombination (HR) or homology-mediated end joining (HMEJ), were then designed to insert the sex determining region on the Y-chromosome (SRY) gene into the target cut-site of ZFXg3 to produce bulls that would sire all male offspring (XY males, and $X_{SRY}X$ males). CRISPR/Cas9 reagents were introduced into either MII oocytes, or six hours after in-vitro insemination (hpi). The HMEJ donor vector (hmejSRYp) showed a significantly higher insertion rate compared to the HR donor vector (hrSRYp) (32.5% vs. 0%; $p < 0.0001$). Additionally, of those that were positive for the insert, 23.4% were non-mosaic hemizygous (males) or homozygous (female) knock-ins There was no significant difference in the level of mosaicism when injecting hmejSRYp in mature oocytes as compared to six hours post in-vitro insemination (hpi), although to date a limited number of blastocysts injected 6hpi have been analyzed. Finally, there was no significant difference between the knock-in efficiency, or the level of mosaicism when comparing XX and XY embryos ($p > 0.05$). Utilizing the HMEJ pathway in bovine embryos resulted in a significantly higher rate of CRISPR-mediated gene knock-in as compared to HR, and approximately a quarter of these X chromosome knock-ins were non-mosaic (hemizygous males or homozygous females) by PCR.

INTRODUCTION

Genome editing technologies have the potential to have a positive impact on livestock genetic improvement (Van Eenennaam and Young 2019). However, for these tools to be implemented, they must seamlessly integrate into existing breeding program designs to maintain or accelerate the rate of genetic gain. Obtaining high rates of targeted gene knock-ins through homology-directed repair (HDR) using site-directed nucleases in the presence of a repair template has proven difficult in livestock embryos, often resulting in a low integration rate and/or mosaic individuals (Georges *et al.* 2018). The primary method that has been trialed for HDR-mediated knock-ins in bovine embryos is the homologous recombination (HR) pathway. However, the primary method for double-strand break (DSB) repair in gametes and the early zygote is the end-joining pathway (Rothkamm *et al.* 2003). The HDR pathway is primarily restricted to actively dividing cells (S/G2-phase) and only becomes highly active towards the end of the first round of DNA replication in the one-cell zygote (Hustedt and Durocher 2017). Consequently, gene knock-ins in livestock in livestock have typically been achieved by HR in cell culture, followed by somatic cell nuclear transfer (SCNT) cloning of the edited cell line. However, this method can be costly and inefficient (Tan *et al.* 2016). We describe an approach to achieve improved rates of knock-ins in developing bovine embryos using the alternative homology-mediated end joining (HMEJ) DSB repair pathway, and a method to screen for non-mosaic founder individuals prior to embryo transfer, thereby avoiding the need for SCNT to obtain knock-in founders, and allowing the opportunity to edit the next generation of animals in a breeding program in a single step.

MATERIALS AND METHODS

Four single–guide RNAs (sgRNAs) were designed for high specificity and limited off-target potential using the online tools sgRNA Scorer 2.0 (Chari *et al.* 2017) and Cas-OFFinder (Bae *et al.* 2014), respectively. *In-vitro* fertilized bovine embryos were produced using methods previously described (Bakhtari and Ross 2014). The sgRNAs (ZFXg1-4) Cas9 individually injected by laser-assisted cytoplasmic injection (Bogliotti *et al.* 2016) of a solution containing 67ng/µL of each sg-RNA alongside 167ng/µL of Cas9 protein (PNA Bio, Inc., Newbury Park, CA) as ribonucleoprotein complexes (RNP) in three replicates of 30 embryos per guide. Embryos that reached blastocyst stage were collected, lysed, and analyzed using PCR (Table 1), followed by Sanger sequencing.

Table 1. Sequence of primers used for PCR evaluation and confirmation of SRY knock-in and sex, and guide-RNA sequences (*sequences developed by Gokulakrishnan *et al.* 2012)

	Name	Sequence 5'- 3'	T_m (°C)
PCR primers	ZFXgF	TCCAAGGAGCTATGTCACAGAA	60.8
	ZFXgR	CACTAGCTTTGGGCGATATGA	60.8
	ecZFXknF	CCGCTTCAAATCAGTTTAATCC	58.9
	ecZFXknR	CCCCACCAGGAAAGTACAAA	60.4
	srnckF	TGGTCCTCTGTTAATCAGTTCTTTC	61.3
	srnckR	GGAACTGCTTGGGTACCAAG	62.4
	DDX3-1F*	AGGAAGCCAGGAAAGTAA	55.3
	DDX3-1R*	CATCCACGTTCTAAGTCTC	58.0
Guide RNA	ZFXg1	ACAACCCAAAATGAAGGGGG	-
	ZFXg2	AATACAACCCAAAATGAAGG	-
	ZFXg3	CTCCCATGTCATAACTTCTG	-
	ZFXg4	GATATGAAATTACACTGGAC	-

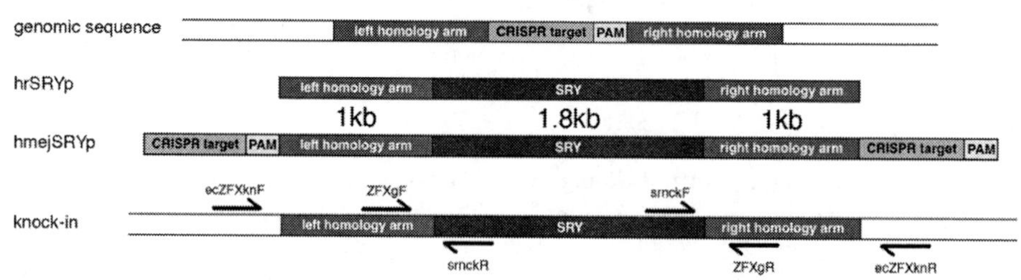

Figure 1. Schematic representation of donor vectors used to test knock-in efficiency in *in* bovine embryos

Donor vectors contained the 1.8kb *Bos taurus* SRY promoter and coding sequence (Accession: U145569), 1kb homology arms flanking each side of the Cas9 cut site, with (hmejSRYp) or without (hrSRYp) the CRISPR target site flanking each homology arm (Figure 1).

Oocytes were collected and *in-vitro* matured for 18 hours prior to injection or *in-vitro* fertilization (Bakhtari and Ross 2014). CRISPR/Cas9 reagents for each donor were introduced by laser-assisted cytoplasmic injection (Bogliotti *et al.* 2016) of a solution containing 67ng/µL of guide-RNA, 167ng/µL of Cas9 protein (PNA Bio, Inc., Newbury Park, CA) and 133 ng/µL of circular plasmid after stripping of cumulus cells from mature oocytes. Injected mature oocytes were in-vitro fertilized and co-cultured with cumulus-oocyte complexes (COCs) for 16 hours. Un-injected in-vitro fertilized embryos were stripped of cumulus cells six hours after fertilization and injected as described above. Injected embryos were scored to developmental stage reached. Embryos that reached blastocyst stage were collected, lysed and underwent whole-genome amplification using the REPLIg Mini Kit (Qiagen, Valencia, CA), PCR and Sanger sequencing. Data were analyzed with GLM in R to test which variables were statistically different. A $\chi 2$ test was used to test whether total knock-in and mosaicism rates differed between donor vector types.

RESULTS AND DISCUSSION

Four sgRNAs (ZFXg1-4) Cas9 ribonucleoprotein complexes (RNP) were individually injected into 90 embryos resulting in *in-vivo* mutation rates of 38%, 57%, 82% and 40%, respectively. Based on these results, we selected sgRNA ZFX3 for the knock-in experiments. Treatment group did not affect overall mutation rate ($P > 0.05$), however embryos injected with ZFX3 RNP and donor hmejSRYp showed a significantly higher rate of total knock-ins (targeted SRY integration) compared to hrSRYp, which showed zero knock-ins (Table 2; P-value < 0.01). When comparing the effect of sex of the embryo, and the time of injection between MII injected oocytes and 6hpi, there was no significant difference on the knock-in efficiency or the level of mosaicism (Table 2; $P > 0.05$). Because we were targeting the X-chromosome, PCR-analysis of embryo biopsies limited our ability to differentiate between heterozygous and mosaic female embryos.

Table 2. Mutation, knock-in, and mosaicism rate of blastocysts after cytoplasmic injection of ZFX3 RNP hmejSRYp or hrSRYp at the MII oocyte, or Embryo (6 hpi) development stage

						Knocked-in subset	
Sex	n	Donor	Time of Injection	%Mutation Rate (n)	%Total Knock-In (n)	%Hemi/ Homo (n)	%Hetero/ Mosaic (n)
Female	78	hmejSRYp	MII oocyte	83^a (65)	**40^a** (31)	19^a (6)	81^a (25)
	8		Embryo	88^a (7)	**25^a** (2)	0^a (0)	100^a (2)
	6	hrSRYp	MII oocyte	83^a (5)	0^b (0)	n/a	n/a
	6		Embryo	67^a (4)	0^b (0)	n/a	n/a
Male	97	hmejSRYp	MII oocyte	70^a (68)	**29^a** (28)	29^a (8)	71^a (20)
	14		Embryo	86^a (12)	**21^a** (3)	33^a (1)	67^a (2)
	10	hrSRYp	MII oocyte	70^a (7)	0^b (0)	n/a	n/a
	8		Embryo	75^a (6)	0^b (0)	n/a	n/a
Total	175	hmejSRYp	MII oocyte	76^a (133)	**34^a** (59)	24^a (14)	76^a (45)
	22		Embryo	86^a (19)	**23^a** (5)	20^a (1)	80^a (4)
	16	hrSRYp	MII oocyte	75^a (12)	0^b (0)	n/a	n/a
	14		Embryo	71^a (10)	0^b (0)	n/a	n/a

Letters that differ in the same column are statistically different (P-value < 0.05)

This increased rate of knock-ins with donor hmejSRYp is likely the result of the DSB repair pathway triggered by the different donor vectors. The hrSRYp donor vector required initiation of the homologous recombination (HR) pathway for integration, which has been shown to have a low activity in early embryos. In contrast, hmejSRYp utilizes the homology-mediated end-joining (HMEJ) pathway (Yao *et al.* 2017). In mice zygotes, this pathway was found to have a significantly higher efficiency of targeted knock-ins as compared to HR, which is consistent with the end-joining pathway being the primary DSB repair mechanism in gametes and pre-S-phase zygotes (Rothkamm *et al.* 2003). It should be noted that the MII injected oocytes were observed to have lower post-fertilization development rates compared to zygotes injected after insemination (12.1% (n=1,584) versus 18.4% (n = 163), respectively), perhaps due to increased rates of polyspermy in the stripped oocytes. Targeting the HMEJ pathway in developing embryos, alongside a method to screen for non-mosaic founder individuals prior to embryo transfer (Figure 2), has the potential to be an alternative to SCNT cloning of genome-edited knock-in cells. The implementation of a gene editing approach such as this alongside genetic breeding programs could enable the introduction of useful genetic variants such as polled (hornlessness), while maintaining the rate of genetic gain without increasing inbreeding above acceptable levels (Mueller *et al.* 2019). Recent Australian regulation would categorize the use of a donor template to guide the DSB repair to produce a cisgenic knock-in, as detailed in this paper, as resulting in a genetically modified organism (GMO) which may limit the use of this approach in animal breeding programs.

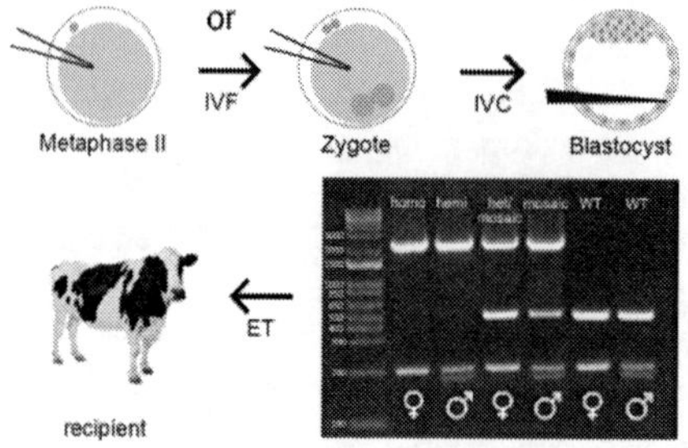

Figure 2. Schematic representation of CRISPR-mediated development of SRY knock-in bovine offspring by cytoplasmic injection (CPI)

Biopsies taken at day 7 and are analyzed via PCR to simultaneously detect sex, success of knock-in, and mosaicism prior to embryo transfer (ET) to synchronized recipients. Upper bands using ZFXgF/R PCR primers: wild type (WT) 520bp, knock-in 2349bp. Lower bands using DDX3-1F/R PCR primers: female 208bp, male 189bp and 208bp. IVF: in-vitro fertilization, IVC: in-vitro culture, het: heterozygous, hemi: hemizygous male, homo: homozygous knock-in female.

CONCLUSION

In-vitro production of bovine embryos combined with CPI of CRISPR Cas9 RNP in MII oocytes or 6 hpi bovine embryos, along with a donor vector designed to target the HMEJ repair pathway, yielded a 32.5% knock-in rate of the 1.8 kb SRY target gene of which 23.4% were non-mosaic, hemizygous (males) or homozygous (females), targeted X-chromosome knock-ins.

ACKNOWLEDGEMENTS

This project was supported by Biotechnology Risk Assessment Grant Program competitive grant no. 2015-33522-24106 from the U.S. Department of Agriculture.

REFERENCES

Bae S., Park J. and Kim J. (2014) *Bioinformatics* **30:** 1473.
Bakhtari A. and Ross P.J. (2014) *Epigenetics* **9:** 1271.
Bogliotti Y.S., Vilarino M. and Ross P.J. (2016) *J. Vis. Exp.* **116**: doi:10.3791/54465
Chari R., Yeo N.C., Chavez A. and Church G.M. (2017) *ACS Synth. Biol.* **6**: 902.
Georges, M., Charlier, C. and Hayes, B. (2018). *Nat. Rev. Genet.* **20**: 135 .
Gokulakrishnan P., Kumar R.R., Sharma B.D., Mendiratta S.K. et al. (2012) *J. Anim. Sci.* **5**: 733.
Hustedt N and Durocher D. (2017) *Nat. Cell Biol.* **19**: 1.
Mueller M.L., Cole JB., Sonstegard T.S. and Van Eenennaam A.L. (2019) *J. Dairy Sci.* **102** 4215.
Rothkamm K., Krüger I., Thompson L.H. and Löbrich M. (2003) *Mol. Cell. Biol.* **23**: 5706.
Tan W., Proudfoot C., Lillico S.G. and Whitelaw C.B. (2016). *Transgenic Res.* **25**: 273-87
Van Eenennaam, AL. and Young, AE. (2019). *Encycl. Food Secur. Sustainability* **3**: 171.
Yao X, Liu Z, Liu J, Zhou H, Shen X, *et al.* (2017) *Cell Res.* **27**: 801.

Proc. Assoc. Advmt. Anim. Breed. Genet. 23:11-14

INTEGRATION OF FUNCTIONAL GENOMICS AND PHENOMICS INTO GENOMIC PREDICTION RAISES ITS ACCURACY IN SHEEP AND DAIRY CATTLE

H.D. Daetwyler[1,2], R. Xiang[1,3], Z. Yuan[1,4], S. Bolormaa[1], C.J. Vander Jagt[1], B.J. Hayes[5], J.H.J. van der Werf[6], J.E. Pryce[1,2], A.J. Chamberlain[1], I.M. MacLeod[1] and M.E. Goddard[1,3]

[1]AgriBio, Centre for AgriBioscience, Agriculture Victoria, Bundoora, VIC, 3083 Australia
[2]School of Applied Systems Biology, La Trobe University, Bundoora, VIC, 3083 Australia
[3]Faculty of Veterinary and Agricultural Sciences, University of Melbourne, Parkville, VIC, 3052 Australia
[4]Lanzhou University, Lanzhou, China
[5]QUAAFI, University of Queensland, St. Lucia, QLD, 4067 Australia
[6]University of New England, Armidale, NSW, 2351 Australia

SUMMARY

In the animal breeding there is debate on whether knowledge of functional genomics is useful for genomic prediction. Black box approaches have worked well but technological change now allows for the generation of functional genomic and phenomic information at high resolution. This will allow us to come closer to actual functional variants, thereby increasing genomic prediction accuracy in animals less related to the reference population, such as across breeds and across generations. Here we demonstrate that even with current imperfect knowledge the use of functional information in genomic prediction results in immediate benefits to prediction accuracy and industry breeding decisions.

INTRODUCTION

Currently implemented industry genomic evaluations usually use single nucleotide polymorphisms (SNP) that are neutral and of medium density (e.g. 50k SNP chips in sheep and cattle). The evaluations rely on SNP being in linkage disequilibrium (LD) with causative mutations. This has been effective and has resulted in good prediction accuracy when reference populations are of sufficient size and when predictions are for animals that are relatively closely related to the reference. However, large LD blocks break down quite quickly across generations and LD is also only consistent across breeds at short distances that are not captured by medium density genotyping platforms. This reduces genomic prediction accuracy in these animal groups and imposes a shelf-life on reference populations. A solution is to find SNP that are not neutral but that are more closely linked to, or, are causative mutations. Purely statistical methods can do that with some success, but they are often limited in their ability to fine map causal variants and are susceptible to biases because it is difficult to keep association discovery and prediction reference populations independent. This is where additional independent functional information from other "omics" is helpful to prioritise SNP at finer scale. The overall idea is to reduce the millions of sequence SNP in whole genome sequence data to thousands, such that they can be routinely genotyped by industry and used in genetic evaluations without great computational challenges.

A plethora of high-resolution "omics" data can now be collected in relatively large numbers of animals providing newly defined intermediate phenotypes. Genome sequencing technologies have enabled several approaches to investigate regions of the genome that are associated with phenotypes as well as gene expression and regulation. Large global collaborative projects have created inventories of sequence variants in cattle (1000 Bull Genomes Project) and sheep (SheepGenomesDB) (Daetwyler *et al.* 2014; Daetwyler *et al.* 2017; Bouwman *et al.* 2018). The advantage of sequence data is that the vast majority of SNP and short insertions/deletions (Indels) will be contained in the dataset, thereby enabling quicker discovery of causative mutations or variants that are very closely linked to these mutations (Hayes and Daetwyler 2019). Next-generation genome sequencing also underpins most assays that aim to interrogate gene expression and regulation, for example RNA and chromatin immunoprecipitation (ChIP) sequencing. Regulators of gene expression have been found to be important and enriched in regions that have been associated with phenotypes (Wang *et al.* 2018). Regulatory regions can be identified with expression quantitative trait loci (eQTL) mapping, where variants are associated with gene and exon expression as well as with splice variants (Chamberlain *et al.* 2018; Xiang *et al.* 2018). Similarly, SNP in highly expressed genes in relevant tissues can be identified and such information can be utilized directly in genomic prediction (MacLeod *et al.* 2019). Another functional assay that provides insight into regulatory regions is ChIP sequencing, which can provide information on histones with specific modifications that indicate regions that are likely to be enhancers, promotors or repressors of gene expression. Finally, molecular phenomics (e.g. metabolite levels) can reveal the abundance of compounds in the pathway between gene expression signals and phenotypes and can also be genetically mapped.

Our aim was to combine information from several omics-derived datasets to prioritize variants to increase the accuracy of genomic prediction. We demonstrate the advantage of using this additional information to raise the accuracy of genomic prediction with examples in sheep and dairy cattle.

MATERIALS AND METHODS

Sheep. 42 million sequence variants discovered by SheepGenomesDB Run2 (Daetwyler *et al.* 2017) were imputed into 46,000 sheep (Bolormaa *et al.* 2019). Only the 31 million sequence variants with a Minimac R2 >0.4 were used for downstream analyses. RNA sequencing was carried out on 150 wethers for muscle and liver tissues (Bolormaa *et al.* 2015). All data was aligned with the program STAR, counts were generated with the R package feature Counts, normalised for read depth. Expression QTL (eQTL) mapping was performed with gene and exon counts, as well as with splice variants at SNP 1 megabase (Mb) up and downstream of genes. A false discovery rate (FDR) threshold of 0.05 was used to determine significant SNP, which were then overlapped with significant QTL regions from a genome-wide association study on meat and carcass traits (individual animal phenotypes) also imposing a FDR of 0.05 (Bolormaa *et al.* 2016) and pruned for LD > 0.9. The same multi-breed reference population and traits as Khansefid et al (2018) were used to test two SNP sets: i) the 50k Ovine SNP chip and ii) the 50k Ovine SNP chip with the 10,000 significant eQTL sequence SNP added. Genomic prediction accuracy was validated in approximately 1000 Merino and 500 Border Leicester/Merino cross sheep for 6 meat traits (individual animal phenotypes). Validation animals were chosen to not have half-sibs in the training set to restrict relationships (Khansefid *et al.* 2018).

Dairy Cattle. 17 million sequence variants identified in the 1000 Bull Genomes Project Run6 were imputed into 44,260 animals (about 75% Holstein, 20% Jersey and 5% Australian Red breeds). Sequence variants associated with gene expression (eQTLs) and concentration of milk metabolites (mQTLs, phospholipids), and under histone modification marks (providing information on protein – DNA interactions) were discovered from multi-omics data in several tissues of over 400 cattle. Variants were also identified from 1000 Bull Genomes database (N=2,330) beef-dairy selection signatures. These analyses defined 30 variant sets and for each set we estimated the genetic variance it explained across 34 complex traits in 11,923 bulls and 32,347 cows. Only sets that explained more variance than a random set were carried forward in the analysis leaving approximately one million variants. We defined a Functional-And-Evolutionary Trait Heritability (FAETH) score indicating the proportion of the variance explained by each variant (Xiang *et al.* 2019). Further LD pruning and variant classification reduced the set to 40,000 variants that were included on a new Illumina XT SNP chip design. Finally, we tested whether this new variant set increased genomic prediction accuracy using Bayesian genomic prediction method BayesR across milk, fat and protein yield, somatic cell count and fertility, when compared to the standard Illumina 50k SNP chip in an independent cow dataset (N range 538 (Crossbreds) to 2740 (Holstein)). Similarly to sheep, validation animals were not allowed to have sires or half-sibs in the training set.

RESULTS AND DISCUSSION

Sheep. One million eQTL were detected with significant overlap of eQTL between gene, exon expression and splice variation. Overlapping the eQTL with significant GWAS peaks resulted in 10,000 selected SNP that were added to the 50k Ovine SNP chip for genomic prediction. The increase in prediction accuracy from adding the 10,000 functional SNP was approximately 2 to 3% and varied between traits (Figure 1). In most traits Bayesian methods attained higher prediction accuracy than GBLUP as they are better at accommodating SNP with large effects (data not shown). Bias of genomic breeding values (slope of phenotypes on genomic breeding values) was unaffected compared to Ovine 50k results.

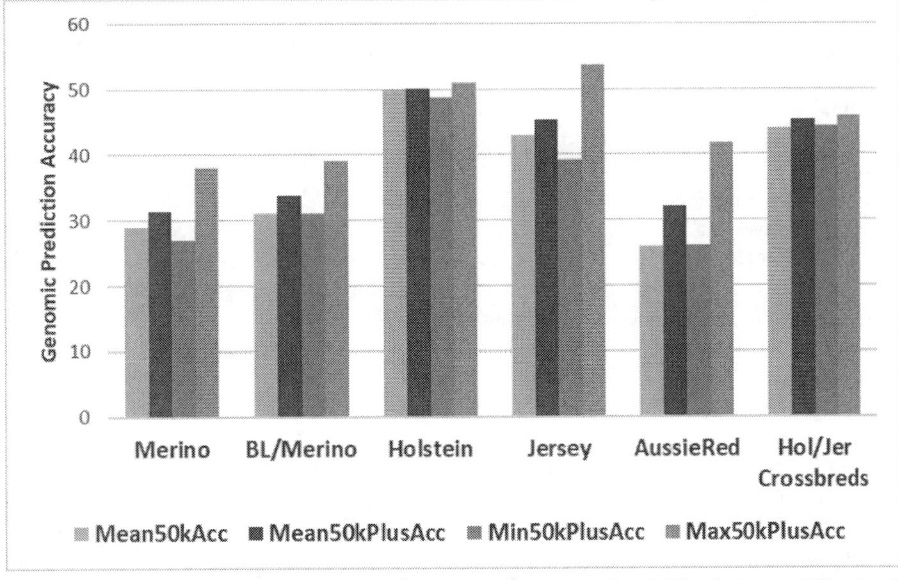

Figure 1. Genomic prediction accuracy when comparing standard 50k Ovine and Bovine SNP chips (50k) to SNP sets that include prioritised markers using functional information (50kPLus) in Merino and Border Leicester/ Merino cross sheep, as well as Holstein, Jersey, Aussie Red, and Holstein/Jersey crossbred cattle

Dairy Cattle. In the variant prioritisation work, the per-variant trait variance explained was highly consistent ($r > 0.98$) between bulls and cows across traits. Based on the per-variant heritability, the sets of mQTL, eQTL and variants associated with non-coding RNAs ranked the highest, followed by more recent mutations, those under histone modification marks, and selection signatures. A XT SNP chip with 40,000 variants from the prioritisation (as well as 8,000 markers overlapping with the Low-Density Dairy SNP chip) is currently in use for genotyping these variants directly (to avoid imputation errors). An early validation in cows not used in the prioritisation and using the imputed high-value variants has increased prediction accuracy on average by 2.5% across all pure breed groups and traits (Figure 1). The increase in accuracy was more pronounced in crossbred, Jersey and Australian Red cattle, which is encouraging for these smaller breed groups, but could also be partly due to lower reference population sizes in those groups. Additional XT SNP chip results can be found in van den Berg *et al.* (2019).

CONCLUSIONS

A strategy to prioritize variants from whole-genome sequence using functional genomic, annotation, and phenomic information combined with target trait phenotypes has increased genomic prediction accuracy in animals that are less related to the reference population in both sheep and dairy cattle. This results in genomic breeding values that are more widely applicable across breeds (shown) and more robust across generations (not shown). The prioritized SNP sets can be utilized by industry immediately to increase prediction accuracy and genetic gain.

ACKNOWLEDGEMENTS

The authors acknowledge financial support from DairyBio, a joint venture between Agriculture Victoria, Dairy Australia and the Gardiner Foundation. We are grateful to have access to data from the 1000 Bull Genomes Project, SheepGenomesDB, SheepCRC, SheepGenetics, and DataGene.

REFERENCES

Bolormaa S., Behrendt R., Knight M.I., … Daetwyler H.D. (2015) *Proc. Assoc. Advmt. Anim. Breed. Genet.* **21**: 489.

Bolormaa S., Chamberlain A.J., Khansefid M., … MacLeod I.M. (2019) *Genet. Sel. Evol.* **51**: 1.

Bolormaa S., Hayes B.J., van der Werf J.H.J., … Daetwyler H.D. (2016) *BMC Genomics* **17**: 224.

Bouwman A.C., Daetwyler H.D., Chamberlain A.J., … Hayes B.J. (2018) *Nat. Genet.* **50**: 362-7.

Chamberlain A.J., Xiang R., Vander Jagt C.J., … Goddard M.E. (2018) *11th World Congress in Genetics Applied to Livestock Production*, Auckland.

Daetwyler H.D., Brauning R., Chamberlain A.J., … Kijas J.W. (2017) *Proc. Assoc. Advmt. Anim. Breed. Genet.* **22**: .

Daetwyler H.D., Capitan A., Pausch H., … Hayes B.J. R.F. (2014) *Nat. Genet.* **46**.

Hayes B.J. and Daetwyler H.D. (2019) *Ann. Rev. Anim. Biosci.* **7**.

Khansefid M., Bolormaa S., Swan A.A., … MacLeod I.M. (2018) *11th World Congress of Genetics Applied to Livestock Production*, Auckland, NZ.

MacLeod I.M., Bowman P.J., Chamberlain A.J., … Goddard M.E. (2019) *Proc. Assoc. Advmt. Anim. Breed. Genet.* **23**:

van den Berg I., MacLeod I.M. & Pryce J.E. (2019) *Proc. Assoc. Advmt. Anim. Breed. Genet.* **23**.

Wang M., Hancock T.P., Chamberlain A.J., Vander Jagt C.J., … Hayes B.J. (2018) *BMC Genomics* **19**: 395.

Xiang R., Hayes B.J., Vander Jagt C.J., MacLeod I.M., … Goddard M.E. (2018) *BMC Genomics* **19**: 521.

Xiang R., Van Den Berg I., MacLeod I.M., Hayes B.J., … Goddard M.E. (2019) *PNAS (online)*.

Proc. Assoc. Advmt. Anim. Breed. Genet. 23:15-18

MORE GENOTYPES THAN MARKERS: THE SS-T-BLUP MODEL IN ACTION. AN APPLICATION STUDY IN MULTI-TRAIT AUSTRALIAN ANGUS BREEDPLAN GENETIC EVALUATION

V. Boerner and D.J. Johnston

Animal Genetics & Breeding Unit*, University of New England, Armidale, NSW, 2351 Australia

SUMMARY

Multi-trait single step genetic evaluation is increasingly facing the situation of having more individuals with genotypes than an individuals' genotype has markers. This leads to an algebraically impossible inversion of the genomic relationship matrix (G). Recent derivations in single step equations called SS-T-BLUP have provided an elegant way to circumvent the inversion of the G and therefore accommodate the described situation. In this paper we examine the applicability of the SS-T-BLUP model to the multi-trait Australian Angus BREEDPLAN genetic evaluation and compare the results to applying two different ways of using G in a single step model. Results clearly show that SS-T-BLUP outperforms other single step formulations and allows users to avoid approximating the inverse of G.

INTRODUCTION

Within the last decade genotyping thousands of individuals with Single Nucleotide Polymorphism (SNP) chips at the commercial level has become common practice in many species of economic relevance. However, due to cost effectiveness these individuals are being genotyped with low to medium density SNP chips, with usually not more than 50,000 markers. To date, genetic evalua-tion systems allow for SNP marker genotypes via the so-called Single Step model (Christensen and Lund 2010). In this model most often markers are used to pre-calculate a marker based relation-ship matrix which subsequently combined with the usual pedigree derived relationship matrix to a so-called *H* matrix (SS-H-BLUP). This requires the inverse of G as well. The described situation of having thousands of individuals genotyped at medium to low density has led to the situation where G is algebraically no longer invertible due to rank deficiencies. A possible solution is to abandon G and move to a model which incorporates the markers directly (SS-SNP-BLUP). While SS-SNP-BLUP is generally equivalent to SS-H-BLUP many of its final implementations suffer from convergence problems with regard to iterative solving or demanding pre-conditioner computation. Recently an elegant intermediate model has been formulated which may be seen as a mix of SS-H-BLUP and SS-SNP-BLUP called SS-T-BLUP (Mäntysaari *et al.* 2017). SS-T-BLUP does not need G nor its inverse and fits the markers directly. As it fits G implicitly, it is algebraically equivalent to SS-H-BLUP under certain assumptions. In addition, it provides EBVs at the individual level which can be readily transformed into marker solutions. In this paper we will examine the effect of SS-T-BLUP on the computational load to a Single Step genetic evaluation of Australian Angus. We will compare the results relative to the ordinary SS-H-BLUP approach.

METHODS

Model. The "H" matrix (Christensen and Lund 2010) required for SS-H-BLUP can be written as

$$\begin{array}{c|c} A_{1,1}-A_iA_{2,2}A_i'+A_iG_wA_i' & A_iG_w \\ \hline G_wA_i' & G_w \end{array} \quad (1)$$

* A joint venture of NSW Department of Primary Industries and the University of New England

where 1 is a vector indexing the subset of n_{ng} non-genotyped of individuals, 2 is a vector indexing the subset of n_g genotyped individuals, A is the pedigree-based relationship matrix, $A_i = A_{1,2}A_{2,2}^{-1}$, and G_w is a genomic relationship matrix dimension $n_g \times n_g$ which is constructed from a centred and scaled marker genotypes matrix M of dimension $n_g \times n_m$ and subsequently blended. Thus $G_w = \gamma MM' + \lambda C$, where C is an arbitrary but symmetric matrix and γ and λ are arbitrary non-zero weights. For the sake of simplicity we will set $C = A_{22}$ and $1 = \gamma + \lambda, \gamma > 0, \lambda > 0$. H^{-1} can be written as

$$\left(\begin{array}{c|c} A^{1,1} & A^{1,2} \\ \hline A^{2,1} & A^{2,2} \end{array}\right) + \left(\begin{array}{c|c} 0 & 0 \\ \hline 0 & G_w^{-1} - A_{2,2}^{-1} \end{array}\right) \quad (2)$$

(Christensen and Lund 2010) or as $\widetilde{H}^{-1}$ (Strandén *et al.* 2017)

$$\left(\begin{array}{c|c} A^{1,1} & A^{1,2} \\ \hline A^{2,1} & A^{2,2} \end{array}\right) + \left(\begin{array}{c|c} 0 & 0 \\ \hline 0 & G_w^{-1} - (A^{2,2} - A^{2,1}(A^{1,1})^{-1}A^{2,1}) \end{array}\right), \quad (3)$$

where $A^{\cdot,\cdot}$ is a respective block of the inverse of A. However, replacing G_w with $\gamma MM' + \lambda C$ in equation 1 and inverting the resulting matrix yields matrix Ψ^{-1}

$$\left(\begin{array}{c|c} A^{1,1} & A^{1,2} \\ \hline A^{2,1} & A^{2,2} \end{array}\right) + \left(\begin{array}{c|c} 0 & 0 \\ \hline 0 & \lambda^{-1}(A^{2,2} - A^{2,1}(A^{1,1})^{-1}A^{2,1}) \end{array}\right) - \left(\begin{array}{c|c} 0 & 0 \\ \hline 0 & M^*M^{*'} \end{array}\right) \quad (4)$$

where $M^* = M^\dagger(K_u)^{-1}$, $M^\dagger = (\lambda^{-1}(A^{2,2} - A^{2,1}(A^{1,1})^{-1}A^{1,2}))M$, $(K_u)^{-1}$ is an upper triangular matrix derived from $K^{-1} = (K_u)^{-1}(K_u')^{-1}$, $K = (\gamma^{-1}D^{-1} + M'M^\dagger)$ and D^{-1} is the inverse of D which is an arbitrary but symmetric and positive definite matrix of dimension $n_m \times n_m$ (Mäntysaari *et al.* 2017). Further D may contain marker specific weights, or allele frequencies if M is not scaled. Given matrices H^{-1}, $\widetilde{H}^{-1}$ and Ψ^{-1} one can define three different BLUP models, SS-H-BLUP, SS-$\widetilde{\text{H}}$-BLUP, and SS-T-BLUP, which differ solely in which formulation of the inverse of H is used (H^{-1}, $\widetilde{H}^{-1}$ or Ψ^{-1}). However, the different formulations will have consequences for solver preparation and iteration time.

Data. The SS-H-BLUP, SS-$\widetilde{\text{H}}$-BLUP and SS-T-BLUP models were applied to an Australian Angus data set currently used in commercial genetic evaluation. The data set comprised of 35 traits with a total of 9,565,814 records across all traits, and 2,621,403 individuals in the pedigree which allowed for multiple sire mating. The number of animals with genotypes was 58,705 comprising of SNP marker genotypes of various densities and panel manufacturers imputed to a common set of 56009 SNPs. To increase the computational load additional 91,295 genotypes (data set 150k) and 341,295 genotypes (data set 400k) were artificially imputed in a combined regression-sampling approach. The 400k data set was only used for SS-T-BLUP because the other models were computationally infeasible.

The multi-trait model included a single fixed factor per trait, 27 correlated genetic factors, 27 correlated genetic groups factors with 19 genetic groups each, 3 correlated maternal permanent environmental factors and 22 correlated sire-by-herd factors. The total number of equations was 76,823,378. λ and γ were set 0.05 and 0.95, respectively.

Software. The system of equations was solved with AGBU's current large scale linear mixed model library solver which uses the preconditioned gradient algorithm (PCG) for iteratively solving linear mixed models and integrates Intel(R) MKL(R), version 2017 update 8. Convergence was achieved when the L2 norm of PCG residuals scaled by the L2 norm of the mixed model equations' right hand side was $\leq 2.68e^{-9}$. All computationally relevant integer and all real numbers were represented in a 64 bit. Computations for the 150k data set were carried out on a computer with two sockets each carrying an Intel(R) Xeon(R) CPU E5-2697 v3 with 2.60GHz, a total of 28 cores, and 528GB of random access memory (RAM). Computations for the 400k data set were carried out on a computer with two sockets each carrying an Intel(R) Xeon(R) CPU E5-2697 v4 with 2.30GHz, a total of 36 cores, and 256GB of RAM.

RESULTS

Table 1: Processing time in real time seconds (hours) for various steps when iteratively solving a SS-T-BLUP, SS-H-BLUP and SS-$\tilde{H}$-BLUP model using an Australian Angus BREEDPLAN dataset

task	SS-H-BLUP$^1_{150}$	SS-$\tilde{H}$-BLUP$_{150}$	SS-T-BLUP$_{150}$	SS-T-BLUP$^2_{400}$
G	1,756	1,756	-	-
$A_{2,2}$	250	250	-	-
G^{-1}	9,150	9,150	-	-
${A_{2,2}}^{-1}$	3,500	-	-	-
$M^{\dagger}$ and K	-	-	3,422	4,210
K_L	-	-	352	320
M^{*}	-	-	629	1170
$A^{-1}_{2,2}$ diag^3	-	262	262	219
preparation	14,656 (4)	11,418 (3.2)	4,665 (1.3)	5,919(1.6)
iteration	7.5	11.2	8.6	12
$\sum$ iteration	19,123 (5.3)	28,716 (7.9)	22,134 (6.1)	30,809 (8.5)
run time	33,779 (9.4)	40,134 (11.1)	26,799 (7.4)	36,728 (10.2)

1: 150,000 individuals with genotypes. 2: 400,000 individuals with genotypes. 3: sampling of diagonal elements of $A^{-1}_{2,2}$ using 10,000 samples.

Results for the different parts of the setup and solving steps are provided in Table 1. SS-H-BLUP$_{150}$, SS-$\tilde{H}$-BLUP$_{150}$, SS-T-BLUP$_{150}$ and SS-T-BLUP$_{400}$ converged in equal number of rounds which was $\simeq$2,560. The major differences between SS-H-BLUP$_{150}$, SS-$\tilde{H}$-BLUP$_{150}$ and SS-T-BLUP$_{150}$ are the computation time for run preparation and the computation time per round of iteration. The preparation time for model specific parts for SS-T-BLUP$_{150}$ was 1.3 hours, for SS-H-BLUP$_{150}$ 4 hours and for SS-$\tilde{H}$-BLUP$_{150}$ 3.2 hours. Thus, compared to SS-T-BLUP, SS-H-BLUP needed 3 times and SS-$\tilde{H}$-BLUP 2.5 times more real time for all necessary pre-calculations. In terms of time per iteration SS-H-BLUP$_{150}$ took 7.5 real time seconds for a single round of the preconditioned gradient solver, followed by SS-T-BLUP$_{150}$ with 8.5 real time seconds. With 11.2 seconds per iteration SS-$\tilde{H}$-BLUP was slowest. Due to the huge time savings for run preparation and only a slightly longer time while iterating SS-T-BLUP$_{150}$ needed only 80 % of the total processing time required by SS-H-BLUP$_{150}$ and only 66 % of SS-$\tilde{H}$-BLUP$_{150}$. The last column in Table 1 shows the computing time for SS-T-BLUP$_{400}$.

DISCUSSION

SS-T-BLUP has been proposed as a single step model which can facilitate data sets where the number of genotyped individuals exceeds the number of markers and the G matrix is algebraically not invertible. These situations become more common in commercial plant and livestock species where individuals are genotyped with low to medium density SNP chips (Mäntysaari *et al.* 2017). This is achieved by reformulating the "H" matrix representation such that neither the G or $A_{2,2}$ matrices nor their inverses need to be built or approximated. As shown by the results, SS-T-BLUP clearly outperforms SS-H-BLUP in terms of total processing time which is mainly due to the huge computational cost for setting up G, $A_{2,2}$ and inverting both as the inversion cost grows cubicly with n_g, whereas at a constant n_m the cost for generating $M^{\dagger}$ grows less than linearly and the cost for K grow $(n_m \times n_m + 1)/2 \times n_g$. In terms of seconds per iteration the main difference between SS-T-BLUP, SS-H-BLUP and SS-$\widetilde{\text{H}}$-BLUP is caused by the operations of Ψ^{-1}, H^{-1} and $\widetilde{H^{-1}}$ times a vector y. This can be narrowed down further to a single matrix vector operation $\Delta H_{2,2}^{-1}y = (G_w^{-1} - A_{2,2}^{-1})y$ in SS-H-BLUP, or one matrix vector operation $\Delta H_{2,2}^{-1}y = G_w^{-1}y$ and one solver operation $y = (A^{2,2} - A^{2,1}(A^{1,1})^{-1}A^{1,2})x$ in SS-$\widetilde{\text{H}}$-BLUP, or two matrix vector operations $M^{\star\prime}M^{\star}y$ and one solver operation $y = (A^{2,2} - A^{2,1}(A^{1,1})^{-1}A^{1,2})x$ in SS-T-BLUP. In the example given here operations $\Delta H_{2,2}^{-1}y$ and $G_w^{-1}y$ required $\approx 2.25e10$ floating point operations (FLOPs), whereas operation $M^{\dagger\prime}M^{\dagger}y$ required $\approx 1.5e10$ FLOPs. SS-T-BLUP and SS-$\widetilde{\text{H}}$-BLUP have additional cost for solving $y = (A^{2,2} - A^{2,1}(A^{1,1})^{-1}A^{1,2})x$ which offsets the FLOP advantage of SS-T-BLUP and produce an additional overhead for SS-$\widetilde{\text{H}}$-BLUP. For SS-$\widetilde{\text{H}}$-BLUP these disadvantages whilst iterating are not balanced due to not inverting $A_{2,2}$, because its inverse can be calculated much quicker than the inverse of G, resulting in almost 20% more total processing time compared to SS-H-BLUP. For SS-T-BLUP the combination of an advantage in terms of FLOPs, extra burden for solving and huge saving in preparation time resulted in a 20% and 33% decrease in processing time compared to SS-H-BLUP and SS-$\widetilde{\text{H}}$-BLUP, respectively.

CONCLUSION

These results support the conclusion that SS-T-BLUP provides a feasible algorithm to calculate exact solutions for estimated breeding values when the number of genotyped individuals exceeds the number of markers. A limitation to the number of genotyped individuals is solely set by the available RAM. Therefore SS-T-BLUP allows solving Single Step equation systems iteratively without generating G or $A_{2,2}$ or their inverse matrices or any approximation of these matrices.

ACKNOWLEDGEMENTS

The authors thank Angus Australia for providing access to their data. This work was funded by Meat and Livestock Australia (Project L.GEN.0174).

REFERENCES

Christensen O. F. and Lund M. S. (2010) *Genet. Sel. Evol.* **42**:2.
Mäntysaari E. A., Evans R. and Strandén I. (2017) *J. Anim. Sci.* **95**(11):4728.
Strandén I., Matilainen K., Aamand G. P. and Mäntysaari E. A. (2017) *J. Anim. Breed. Genet.* **134**:264.

Proc. Assoc. Advmt. Anim. Breed. Genet. 23:19-22

SIMPLE EXAMPLE TO DEMONSTRATE THE EFFECT OF ALLELE FREQUENCIES ON THE GENOMIC RELATIONSHIP MATRIX VALUES

M.H. Ferdosi, N.K. Connors, V. Boerner and D.J. Johnston

Animal Genetics & Breeding Unit*, University of New England, Armidale, NSW, 2351 Australia

SUMMARY

Genomic evaluations using single-step genomic best linear unbiased prediction (ssGBLUP) combine the genomic relationship matrix (GRM) and numerator relationship matrix (NRM) together, to form the H matrix. The GRM values represent relationships between individuals and are dependent on allele frequencies. In this study, a simple example is used to demonstrate how the change in allele frequency can effect the values in the GRM, while also exploring the possible range of GRM values.

INTRODUCTION

In the pre-genomic era, pedigree was used to build the Numerator Relationship Matrix (NRM) that shows the relationship among individuals. The NRM is double the coancestry and can only show the relatedness between individuals, so the NRM values are always positive and range between 0 to 2. The NRM is a key component in Mixed Model Equations (MME) to calculate variance components and Estimated Breeding Values (EBVs). Genomics is used routinely in genetic evaluations nowadays, such as Australia's national beef recording and genetic evaluation system (BREEDPLAN), and with decreasing prices of genotyping, large numbers of individuals are genotyped. VanRaden (2008) showed that a Genomic Relationship Matrix (GRM) can replace the NRM in MME. The GRM is a variance and covariance matrix that can not only show relatedness among individuals but can also show the unrelatedness among individuals through negative values. The GRM values are dependent on allele frequencies and coding (Strandén and Christensen 2011; Tier *et al.* 2015). In the situation that both genotype (GRM) and pedigree (NRM) are available as current and historical information, a new method is required to make best use of both information sources appropriately. Single-Step genomic best linear unbiased prediction (ssGBLUP) was suggested by (Aguilar *et al.* 2010) to address this issue by building the new matrix H, combining both NRM and GRM information. Currently, ssGBLUP used in BREEDPLAN uses realised population allele frequencies to build the GRM. In this study, a simple example is used to demonstrate how the change in allele frequency can change the GRM values, whilst also exploring the possible range of GRM values. A better understanding of effects of allele frequency on GRM values will lead to a better understanding of the H matrix.

MATERIAL AND METHODS

Theory. This study considers a very simple situation where we have three animals, each with one marker (alleles AA, AB and BB). Summarising the GRM value (r) for one locus and two individuals using VanRaden first method (VanRaden 2008):

$$r = \frac{(b-2p+1)(c-2p+1)}{2p(1-p)} = \frac{bc+b+c+1}{2p(1-p)} - \frac{b+c+2-2p}{1-p} \qquad (1)$$

where 'b' and 'c' were genotypes (only one marker) for two individuals and 'p' was the allele frequency. The 'b' and 'c' are coded -1, 0 and 1 for AA, AB and BB. The $(2p-1)$ that is subtracted from 'b' and 'c' is the mean genotype score. The $2p(1-p)$ is a scaling factor in order to make the GRM values comparable to NRM. For the case where 'b' and 'c' are opposing homozygotes i.e. $b =$

---

* A joint venture of NSW Department of Primary Industries and the University of New England

-1 and $c = 1$ then 'r' is

$$r = \frac{-1-1+1+1}{2p(1-p)} - \frac{-1+1+2-2p}{1-p} = 0 - \frac{2-2p}{1-p} = -2. \quad (2)$$

For the case where 'b' and 'c' are both heterozygote i.e. $b = 0$ and $c = 0$ then

$$r = \frac{1}{2p(1-p)} - \frac{2-2p}{1-p} = \frac{1}{2p(1-p)} - 2. \quad (3)$$

Table 1 - (A) shows the formulas for all genotype pairs and Table 1 - (B) shows similar formulas prior to dividing the GRM values by the scaling factor $2p(1-p)$. The determinant for both matrices were equal to 0, i.e. this matrix is singular and cannot be inverted as mentioned in Strandén and Christensen (2011). Table 2 shows the formula for which allele frequency can be calculated if wanting to obtain a specific relationship value. A relationship cannot be calculated for opposing homozygotes by using Table 1 - (A) when scaling factor is used, and as such there is no formula for this combination in Table 2. However, without the scaling factor (Table 1 - (B)) or changing the scaling factor the relationship can be calculated.

Table 1. Formula to calculate GRM value (r) for all possible genotype pairs - single marker only

Formula	(A) - with division by $2p(1-p)$			(B) - without division by $2p(1-p)$		
Allele	-1	0	1	-1	0	1
-1	$\frac{2p}{1-p}$	$\frac{2p-1}{1-p}$	-2	$4p^2$	$4p^2-2p$	$4p^2-4p$
0		$\frac{1}{2p(1-p)}-2$	$\frac{-2p+1}{p}$		$4p^2-4p+1$ or $(1-2p)^2$	$4p^2-6p+2$
1			$\frac{2}{p}-2$			$4p^2-8p+4$ or $(2-2p)^2$

p is the allele frequency

Table 2. Formula to calculate allele frequency (p) based on the specific relationship (r) in GRM - single marker only

Allele	-1	0	1
-1	$\frac{r}{r+2}$	$\frac{r+1}{r+2}$	?
0		$\frac{\sqrt{r^2+2r}+r+2}{2(r+2)}-2$	$\frac{1}{r+2}$
1			$\frac{2}{r+2}$

r is the relationship

RESULTS AND DISCUSSION

The formulas shown in Table 1 - (A) were used to calculate the GRM values that would be generated when p is 0.5 and 1. Table 3 - (A) shows the GRM values when p is 0.5, and Table 3 - (B) shows the GRM values when p is 1. Since the $2p(1-p)$ becomes 0 when the p value is 1, the limit was used when p approaches 1 (or 0 - Table 3 - (B)). Table 2 can be useful for simulation purposes. For example, Tables 4 (A) and (B) show the allele frequencies required to get a GRM values for important relationships of 0.5 (expected value for parent and offspring relationships or full-sib relationships) and 0.25 (expected value for half-sibs relationships) respectively. Figure 1 summarises the results shown in Tables 3 and 4.

Table 3. Table shows GRM values when $p = 0.5$ (A) and when p approaches 1 (B) - by using the formula in Table 1 - A

Formula	(A) - $p = 0.5$ and $2p(1-p) = 0.5$			(B) - $lim_{p\to 1^+}$ and $2p(1-p) = 0$		
Allele	-1	0	1	-1	0	1
-1	1/0.5	0	−1/0.5	0	−1	−2
0		0	0		∞	∞
1			1/0.5			∞

Table 4. For different relationships (r) using formula in Table 2 the p would be

Formula	(A) - for 0.5 relationships			(B) - for 0.25 relationships		
Allele	-1	0	1	-1	0	1
-1	1/5	3/5	−	1/9	5/9	−
0		$-\frac{\sqrt{5}-5}{10}, \frac{\sqrt{5}+5}{10}$	2/5		1/3,2/3	4/9
1			4/5			8/9

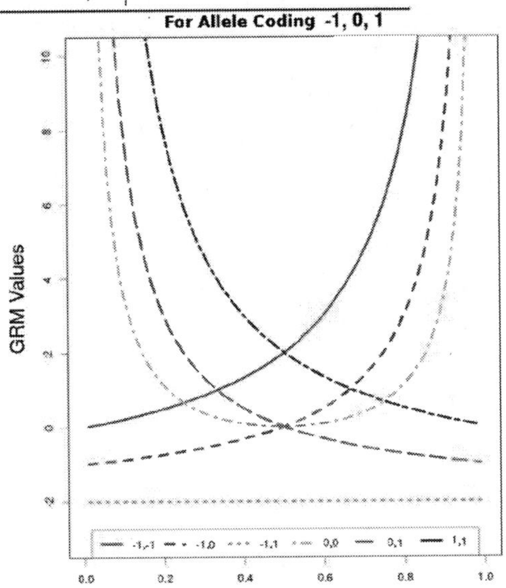

Figure 1. Effect of different allele frequencies on the GRM values using three individuals and one locus. The legend shows the genotypes pairs.

For a single marker only GRM, as discussed in this article, allele frequencies have significan effects on the GRM values. As shown in Figure 1, the more extreme the allele frequency (i.e. (or 1) the more extreme the GRM value. Table 3 - (B) shows that allele frequencies of 0 and 1 cai result in infinite GRM values, demonstrated also in Figure 1. The lower limit of GRM for opposin homozygote is always -2, regardless of allele frequency. Figure 1 demonstrates how rare alleles anc extreme allele frequencies can cause very large numbers in the GRM. This is amplified here due tc only using a single marker. It should be noted that in practice, usually using thousands of markers the effect of extreme allele frequencies will be minimized. This is dependent on SNP selection and whether the population is multi-breed for example. This simple example shows the importance of choosing the appropriate allele frequency (e.g. base population allele frequency – VanRaden (2008)) in order to reflect the true relationship among individuals in a GRM. Removing SNPs with very high or low allele frequencies or replacing their allele frequencies with pre-set allele frequencies may lead to more compatible values in GRM (in comparison to NRM), with no or negligible effect on estimated breeding values kings (Tier *et al.* 2015).

CONCLUSIONS

In this article a simplified version of the GRM was presented to demonstrate the effect of allele frequency on GRM values. In addition, simple formula were presented to calculate GRM values based on the specific allele frequency, or what allele frequency to use to obtain a specific GRM relationship value. These formulas can further be used for simulation purposes and development of methods to build the GRM efficiently.

ACKNOWLEDGMENTS

This research is supported by Meat and Livestock Australia (MLA) project L.GEN.0174. . The authors also wish to thank the reviewers for comments leading to the improvement of this article.

REFERENCES

Aguilar I., Misztal I., Johnson D., Legarra A., Tsuruta S. and Lawlor T. (2010) *J Dairy Sci* **93**:743.
Strandén I. and Christensen O.F. (2011) *Genet Sel Evol* **43**:25.
Tier B., Meyer K. and Ferdosi M. (2015) In *Proc. Assoc. Advmt. Anim. Breed. Genet*, 22. pp. 28–30
VanRaden P.M. (2008) *J Dairy Sci* **91**:4414.

Proc. Assoc. Advmt. Anim. Breed. Genet. 23:23-26

DEEP LEARNING FOR GENOTYPE QUALITY CONTROL

D.P. Garrick[1,2]

[1]Theta Solutions, LLC, Atascadero CA, USA 93422
[2]A.L. Rae Centre for Genetics and Breeding, School of Agriculture and Environment, Massey University, New Zealand

SUMMARY

SNP genotype data are increasingly employed across a range of species for routine use in parentage verification and identification, and single-step evaluations. Robust and automated quality control processes are a critical step in maximizing the value of genotype data in these, and other, applications. Prediction of "genotype sex" is a common quality control metric but can be problematic for example on mammalian chips that do not contain Y chromosome markers because methods based on heterozygosity of X chromosome markers can incorrectly flag inbred females as male. A deep learning model is trained to predict "genotype sex" and validated and tested using real-world data routinely used in the American Hereford Association's single-step evaluation.

INTRODUCTION

A major challenge that comes with the advent of low-cost SNP genotyping is curation and management of the vast quantities of data that are produced. Take the case when the genotype sample for a particular animal fails to verify against its genotyped parents in a SNP based parentage verification. If this was to occur, an ideal system would automatically initiate a search against other relevant genotype samples to try and find the true parent without any extra input from the user. If such functionality is not available, or if such a search fails to find a match, then there is the question of a) is the true parent not genotyped, or b) is one or more of the relevant genotypes involved in the parent verification a bad or mismatched sample. In either case this typically requires the breed society and/or breeder to be contacted in order to generate a list of potential parents or to query any potential issues with the sample. This can be complicated by the use of non-standard or otherwise inconsistent animal, sample, and genotype identifiers. The length of time for this process can be significantly shortened by gleaning various information from the genotype sample(s) in question such as potential relatives or phenotypic characteristics. For example, if a genotype is clearly from a female and the animal in question is definitively male (or animal is black and horned and genotype indicates red and polled), it is reasonable to assume the sample in question is incorrect and the animal should have a new sample taken for regenotyping.

Prediction of "genotype sex" is an important quality control metric for genotype samples and is predicted from the sex chromosomes, i.e. in mammals the X and Y chromosomes for males and two X chromosomes for females. Females inherit one X chromosome from their mother, and one X chromosome from their father. With no inbreeding, the copy of each allele from each chromosome will not always be the same and the resulting SNPs will exhibit heterozygosity. As males only have one copy of the X and Y chromosomes, any alleles called from the unmatched parts of those chromosomes should always be the same, resulting in homozygosity within that region.

Deep learning is a subset of machine learning algorithms that passes an input training dataset through multiple layers of neurons in a neural network to successively transform and extract features from the output of the previous layer (Deng and Yu 2014). Leveraging the unique computational capabilities of Graphics Processing Units (GPUs) developed to render modern video games, deep learning approaches have gained significant media attention recently due to associated large technological advances in applications such as self-driving cars, image recognition and classification, medical diagnostics, and many others.

Certain X chromosome SNPs, even those outside the pseudo autosomal region (PAR), can be heterozygous in males if they are located in regions exhibiting copy number variation. Further, X chromosome SNPs can be homozygous in females, especially inbred females who may have inherited the same X chromosome from both her sire and dam, e.g. if her sire is also her maternal grandsire. Thus in rules-based approaches selecting an appropriate subset of SNPs and male/female heterozygosity cut-offs can greatly affect the subsequent genotype sex prediction and without Y chromosome SNPs inbred females can be misclassified. On the other hand, given a suitable training dataset with realistic data and known true sex of the associated samples, a deep learning model can in theory account for the nuances and variation of specific SNPs in the given training dataset to generate accurate predictions. This is possible using a table containing the relevant sex SNPs and utilizing approaches for deep learning on tabular data via the fast.ai toolbox (Rachel Thomas 2018). The objective of this study was to determine if a deep learning approach can accurately predict the genotype sex of an animal and to assess the value of such a tool as a routine automated quality control step within a genomic database information system.

MATERIALS AND METHODS

The genotype data employed for the study consisted of a subset of those SNP genotypes from 67,304 animals used in a recent single-step evaluation from the full American Hereford Association genomic database of >110,000 genotyped animals. The samples originate from several platforms, genotyping laboratories, and chips across a number of years but consist predominantly of GeneSeek 50K and 30K genotypes. Of these, a subset of 15,619 "pedigree verified true" male and female genotypes was determined by taking samples from only those animals who were recorded in the current pedigree as a sire or dam and who subsequently passed SNP-based pedigree verification with at least 1 genotyped offspring. For pedigree verification, no samples used in this study had less than 5,000 called SNPs in common. Pedigree verified animals recorded as a sire in the pedigree were then considered a "true" male while those recorded as a dam were considered a "true" female totalling 5,058 and 10,561 for males and females respectively. As the American Hereford Association has utilized the international ICAR ID format for many years, the pedigree recorded sex for each animal is recorded as the 7th character of the ID, e.g., HERUSAM000000000001 is recorded as a male and HERUSAF000000000002 is recorded as a female. Comparing a predicted genotype sex to its pedigree recorded sex is straightforward as a result.

Three approaches for computing "genotype sex" were examined. The first consists of a simple rule-based non PAR (nPAR) X-chromosome heterozygosity check using all available called nPAR X SNPs from a list of 3,035 SNPs which exist across a variety of genotyping chips and platforms. No sample used in this study had less than 700 called nPAR X SNPs. Samples with ≤5% heterozygosity amongst their called nPAR X SNPs were classified as males while samples with >5% were classified as females. The second approach tested is the rule-based protocol developed by ICBF and is as follows using only a specific small subset of 280 nPAR X chromosome SNPs as described by McClure *et al.* 2018: 1) Determine heterozygosity rate (#AB/ (#AA+#AB+#BB)) for nPAR SNP; 2) If ≤5% het rate = male; 3) If ≥15% female; 4) If between 5 and 15%= ambiguous sex. Additionally, ICBF employs a subset of 7 Y chromosome SNPs: 1) Count nPAR chrY genotypes; 2) If 0–1 genotypes = female; 3) If 6–7 =male; 4) If 2–5 = ambiguous sex. Between the X and Y chromosome predictions any non-conflicting unambiguous sex is reported, otherwise an ambiguous or conflicting sex is reported. The Y sex prediction is dependent on samples having been genotyped on a chip where Y SNPs are available and several thousand samples used in this study did not have Y SNPs available. Instead of excluding those samples a two-step ICBF (X+Y) sex prediction was utilised instead of the fully joint

ICBF(X+Y) sex prediction described above and by McClure *et al.* 2018, such that Y chromosome predictions were used only if the X chromosome predictions were ambiguous.

Finally, a deep neural network (DNN) genotype sex predictor was built utilizing the fast.ai deep learning tabular toolbox (Howard & others 2018) in conjunction with a dataset consisting of just the 280 ICBF X chromosome sex SNPs. Some 2,500 male and 5,000 female genotypes chosen at random from the "pedigree verified true" samples were used as the training data for the DNN while the remaining of the 15,619 samples were used as the validation data. The only dependent variable is the sex prediction while each called SNP was treated as an input categorical variable with values -1, 0, 1, or 5 (no call). Prediction accuracy was used as the training metric and neural networks with various numbers of hidden layers and neurons per layer were tested for training over 25 epochs which took ~5-6 minutes each. The sex prediction is output as a probability of being male and a probability of being female. Sex predictions with ≥ 80% probability were taken as the predicted sex with the remaining assumed to be ambiguous.

RESULTS AND DISCUSSION

Table 1 summarises the number of predicted male, female, and ambiguous sex animals from each approach. An ambiguous male or female means the sex prediction was ambiguous and the pedigree recorded sex was male or female respectively. A conflicting male or female refers to the pedigree recorded sex being male or female respectively and the genotype sex predicted as female or male, respectively. The DNN results were reported from a network with 600 hidden layers and 300 neurons per layer which was found to have the most accurate results of those tested. However, other network sizes with neurons on the order of the number of SNPs (280) achieved very similar results. Perhaps unsurprisingly, the DNN achieves the highest accuracy on this "pedigree verified true" dataset as it is the same dataset that was used for training and validation of the neural network.

Table 2 summarises the differences between the sex predictions from each approach compared to the pedigree recorded sex of each animal in the larger genotype database not including samples otherwise used in the training and validation set for the DNN consisting of 67,304-15,619=51,685 samples. In both the "training" and "test" datasets, use of the ICBF Y chromosome data to augment otherwise ambiguous predictions using only the ICBF X chromosome results does appear to improve prediction accuracy. The nPAR X approach with the hard cut-off between male and female means no "ambiguous" sex samples are flagged, however, the overall percentage of animals matching their pedigree recorded sex is roughly the same as the ICBF approach. The DNN achieves a similar percentage of predictions matching the pedigree recorded sex in the test dataset as the rules-based approaches while using only the 280 ICBF X SNPs and after training with a dataset of only 2,500 male and 5,000 female genotypes randomly selected from the 15,619 "true" sexed samples. The remainder of the 15,619 samples were used for cross-validation during training.

Table 1. Results summary against the "pedigree verified true" sex of 15,619 individuals used for training and validation of the DNN

	nPAR (X)	ICBF(X)	ICBF(X+Y)	DNN(X)
% Correctly Predicted	99.76	99.23	99.86	99.88
Total Predicted Female	10,525	10,444	10,542	10,549
Total Predicted Male	5,094	5,074	5,076	5,065
Ambiguous Female	N/A	97	0	5
Ambiguous Male	N/A	4	1	0
Conflicting Female	37	20	20	10
Conflicting Male	1	0	1	3

Table 2. Genotype sex prediction results summary against the pedigree recorded sex of the animals in the 51,685 samples test dataset, which does not include any animals used in the training and validation set

	nPAR (X)	ICBF(X)	ICBF(X+Y)	DNN(X)
% Matching pedigree	99.79	99.40	99.82	99.70
Ambiguous Female	N/A	197	2	10
Ambiguous Male	N/A	37	14	46
Conflicting Female	57	53	53	61
Conflicting Male	51	23	25	82

CONCLUSIONS

This study shows deep learning approaches have potential as an accurate genotype sex prediction tool in routine and automated genotype sample quality control processes. The accuracy of a deep learning tool trained on a random subset of "pedigree verified true" gendered samples is found to be comparable to that of existing rules-based approaches. A purely X chromosome heterozygosity rules-based approach can benefit from using Y chromosome data to improve otherwise ambiguous predictions.

The benefits of a deep learning tool are that it can be integrated and automated with an existing suite of quality control protocols. In a production system the tool could be routinely tuned and further trained against new and verified data as it arrives. This in theory should allow it to better account for the nuances in the specific datasets of interest.

There are a significant number of avenues for further investigation with regards to the deep learning approach. These include greater exploration of the effect of the deep learning parameters on prediction results, e.g. number of hidden layers and neurons per layer, as well as the size of the dataset used for training and validation both in terms of the SNPs included and the particular individuals that comprise the training and validation sets. Other avenues include incorporation of other data features into the deep learning model such as genotyping platform or chip, Y chromosome SNPs, recorded breed, sample call rate or individual SNP GC scores, inbreeding coefficients, and/or other pedigree information. If genotype data on individuals exhibiting sex chromosome defects or being intersex are available, these could also be incorporated. Extension of the model to additional prediction outputs (e.g. breed) would also be valuable.

Some drawbacks of the deep learning approach are that it does require a suitable training dataset, finding the optimal DNN architecture (e.g. number of layers and neurons per layer) and training parameters is unclear, it requires GPU-based hardware and expertise to run. Finally, even though the deep learning model returns the probability a given sample is male or female unlike the rules-based approaches, the abstract nature of the deep learning model can create extra challenges in communicating prediction results back to breeders or other stakeholders.

REFERENCES

Deng, L., & Yu, D. (2014). *Foundations and Trends® in Signal Processing*, *7***(3–4)**:197. https://doi.org/10.1561/2000000039

Howard, J., & others. (2018). fastai. GitHub.

McClure, M. C., McCarthy, J., Flynn, P., McClure, J. C., Dair, E., O'Connell, D. K., & Kearney, J. F. (2018). *Frontiers in Genetics*, *9***(84)**. https://doi.org/10.3389/fgene.2018.00084

Thomas, R. (2018). Retrieved from https://www.fast.ai/2018/04/29/categorical-embeddings/

Proc. Assoc. Advmt. Anim. Breed. Genet. 23:27-30

'METAFOUNDERS' TO MODEL BASE POPULATIONS IN GENOMIC EVALUATION FOR MULTI-BREED SHEEP DATA

K. Meyer and A.A. Swan

Animal Genetics & Breeding Unit*, University of New England, Armidale, NSW, 2351 Australia

SUMMARY

Models for genetic evaluation of animals from different base populations need to account for systematic differences not explained by genetic relationships considered. These include differences between breeds, animals with unknown parentage born in different time periods or, for single-step evaluation, founders for animals with or without genotype information. A standard method to achieve this, is to define appropriate genetic groups and fit these as additional effects in the model of analysis. Recently, so-called meta-founders have been proposed as an alternative which accounts for ancestral inbreeding and relationships, estimated from genomic information. We examine estimates of ancestral relationships and their impact on predicted breeding values for a practical data set from a multi-breed sheep population. While estimates were afflicted by insufficient genomic information for some groups, results correctly identified some known breed or strain differences and patterns of introgression. Correlations between predicted breeding values from respective analyses fitting genetic groups and meta-founders were high, suggesting that there is scope for meta-founders to replace genetic groups. However, fitting meta-founders reduced variances of predicted breeding values. Further investigations when more genotype information becomes available are warranted.

INTRODUCTION

The single-step procedure for joint genetic evaluation of genotyped and non-genotyped animals using both pedigree and genomic information has become routine for many livestock improvement schemes. Commonly, this is implemented as the so-called ssGBLUP which replaces the classic, pedigree based relationship matrix, **A**, with its counterpart, **H**, which combines the genomic relationship matrix, **G**, with **A**. An inherent problem with this approach is that **A** and **G** imply conceptually different base population: For **A**, parents of animals at the time when pedigree recording began are considered to be the unrelated, non-inbred founders. In contrast, genomic relationships reference an ancestral base population in the distant past. Several methods have been described to align the two matrices; see Meyer *et al.* (2018) for a recent review. Some proposals involve scaling **G** to 'match' **A** while others suggest to modify **A** to account for ancestral inbreeding (Christensen 2012). Specifically, the latter can be achieved by replacing unknown parents in the pedigree with 'meta-founders' (MF), allowing for ancestral inbreeding and relationships between them, estimated from genomic information (Legarra *et al.* 2015; Garcia-Baccino *et al.* 2017). MF are conceptually similar to the 'phantom' parents (Westell *et al.* 1988) used routinely to account for unknown parent groups. Thus, in addition to aligning **G** and **A**, they may provide an alternative to modeling genetic groups. Moreover, MF are treated as correlated and may model genetic relationships between different base populations more appropriately. This paper examines estimates of ancestral relationships and their effects on estimates of breeding values for a practical data set recorded for Australian sheep.

MATERIAL AND METHODS

Data consisted of 1,206,908 records for eye muscle depth, recorded for Australian terminal sire sheep breeds between 1990 and 2018. These included 5 main breeds, namely Poll Dorset, Suffolk, White Suffolk, Merino and Texel, and 17 minor breeds with breed differences modeled by appropriately defined genetic group effects. After eliminating individuals not connected to the data or

* A joint venture of NSW Department of Primary Industries and the University of New England

genotyped animals, there were 1,698,838 animals in the pedigree. Genotype information, consisting of marker counts for 48,599 SNPs, was available for 23,040 animals, of which 18,396 had phenotypes. Data were pre-corrected for fixed effects of birth and rearing type, age, dam age and body weight.

Routine analyses currently classify animals of unknown parentage into 93 genetic groups (GG), based on flock and year of birth. These animals were assigned MF 'parents' based on GG memberships. A total of 10.6% of animals had both parents unknown and 7.8% had no sire identified. All animals with both parents unknown belonged to a single GG thus had the same MF as 'sire' and 'dam'.

Estimates of the matrix of ancestral relationships, $\mathbf{\Gamma}$, were obtained from marker information using a pseudo-EM algorithm (Garcia-Baccino *et al.* 2017; Legarra and Astruc 2018). For two of the GG, no genotypes were available. For these, diagonal elements of $\mathbf{\Gamma}$ were set to the minimum value found for the other groups. Similarly, off-diagonal elements were replaced by values reflecting the minimum correlation encountered. In addition, the resulting estimate of $\mathbf{\Gamma}$ was regularised by shrinking its eigenvalues towards their mean, so that the smallest value exceeded 0.01. The inverse numerator relationship matrix including MF, $\mathbf{A}^{-\Gamma}$, and the corresponding submatrix of $\mathbf{A}$ for genotyped animals, $\mathbf{A}_{22}^{\Gamma}$, were obtained as outlined by Legarra *et al.* (2015).

A 'raw' genomic relationship matrix, $\mathbf{G}_M$, was build from marker counts using method 1 of Van Raden (2008). This was transformed into $\mathbf{G} = \lambda(\mathbf{G}_M + \alpha\mathbf{J}) + (1-\lambda)\mathbf{A}_{22}$ with $\lambda = 0.95$ and $\mathbf{A}_{22}$ the submatrix of $\mathbf{A}$ for genotyped animals. To build the 'standard' $\mathbf{H}^{-1}$ (no MF) markers were centered using observed frequencies and $\alpha = 0.025$ was estimated following Vitezica *et al.* (2011). To build $\mathbf{H}^{-\Gamma}$ (including MF), markers were centered assuming allele frequencies of 0.5 and $\alpha = 0$, and $\mathbf{A}^{-1}$ and $\mathbf{A}_{22}$ were replaced by $\mathbf{A}^{-\Gamma}$ and $\mathbf{A}_{22}^{\Gamma}$, respectively. In addition, $\mathbf{H}^{-\Gamma}$ was scaled (see Legarra *et al.* 2015) so that the same variance components were appropriate for analyses with and without MF.

The model for ssGBLUP analyses fitted animals' additive genetic effects, 54,094 contemporary groups (fixed) and 56,212 sire × flock-year (random) effects throughout. A standard analysis (no MF) fitted 93 GG as additional random effects. For analyses with MF, $\mathbf{H}^{-1}$ was replaced with $\mathbf{H}^{-\Gamma}$ either including or excluding GG. Mixed model equations were solved iteratively using a preconditioned conjugate gradient algorithm with diagonal preconditioner. All calculations were performed using WOMBAT (Meyer 2007).

RESULTS AND DISCUSSION

Means and ranges for estimates of ancestral inbreeding or 'self-relationships', i.e. the diagonal elements of $\mathbf{\Gamma}$, and correlations between MF (derived from $\mathbf{\Gamma}$) are summarised in Table 1. Mean across breed group correlations ranged from 0.48 (Suffolk × Texel) to 0.71 (Poll Dorset × White Suffolk), with the range of individual values similar to that within the minor breeds and Merinos (0.14 to 0.95). Some correlations close to unity suggest scope for merging selected GG. Overall, however, estimates fluctuated considerably and no consistent breed group differences or time trends were evident. To some extent, this can be attributed to definitions of GG, e.g. multiple groups for the same breeds in different flocks and overlapping years of birth. In contrast, Legarra and Astruc (2018) found increasing inbreeding and covariances between MF with time for a breed of French dairy sheep. Accurate estimation of $\mathbf{\Gamma}$ requires sufficient genomic information for all MF. Hence, in part at least, this variability

Table 1. Estimates of self-relationships and ancestral correlations between meta-founders

	n[a]	Self-relationship		Correlation	
		$\bar{x}$[b]	range	$\bar{x}$	range
P. Dorset	15	0.68	0.57–0.89	0.83	0.53–0.98
Suffolk	14	0.88	0.57–1.07	0.61	0.37–0.87
W.Suffolk	14	0.62	0.50–0.81	0.75	0.58–0.98
Merino	23	0.66	0.48–1.00	0.67	0.14–0.95
Texel	10	0.80	0.59–0.96	0.62	0.40–0.96
Other	17	0.70	0.45–0.99	0.51	0.14–0.95

[a] No. of MFs per breed group [b] Mean

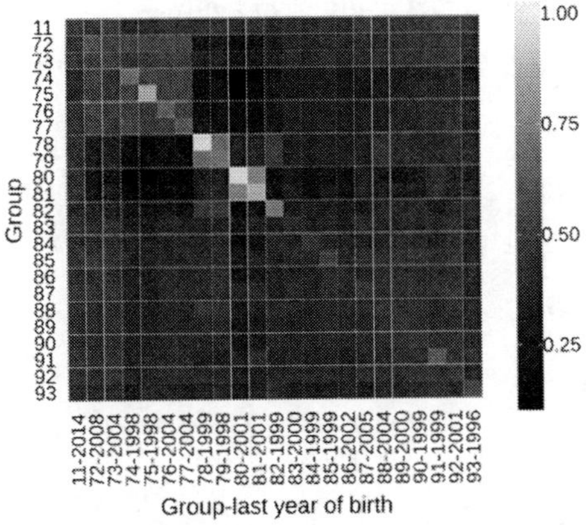

Figure 1. Heatmap plot of diagonal block of Γ for Merino groups

may be attributed to a rather uneven distribution of genotypes over GG and MF. Genotyped animals mostly contributed to numerous GG. Sums of relative contributions (0 to 1; summed over genotyped animals) were less than 5 for 23 GG and exceeded 500 for 7 GG.

Nevertheless, estimates of ancestral relationships correctly identified some known strain differences or patterns of introgression. Figure 1 shows the diagonal block of Γ for Merino GG. Groups 72–81 originate from a resource flock (e.g. Taylor and Atkins 1997), with 72–77, 78–79 and 80–81 representing medium, strong and fine wool selection lines, respectively, while 82 is another strong wool flock. The pattern of covariances between these GG reflects the divergence between strains. Similarly, high ancestral correlations between breed groups shown in Figure 2 highlight the role of Poll Dorset sheep in the formation of the White Suffolk breed.

Statistics comparing predicted breeding values (EBV) from analyses fitting GG or MF are summarised in Table 2 for different categories of animals. Overall, correlations were high suggesting that there is scope for MF to replace explicit GG in the model of analysis. Variances of EBV fitting MF only were considerably lower than those obtained fitting GG as an additional random effect. This implies somewhat stronger shrinkage of predictions when fitting MF, inspite of assuming the same variances for GG and animals' additive genetic effects, or, on average, higher error variances for MF than GG effects. The correlation between predicted GG and MF was 0.87. Negative intercepts highlight the change in alignment of conceptual base populations due to MF. As to be expected, fitting both MF and GG tended to increase these variances but had little effect on correlations. It also increased some of the regressions coefficients for EBV fitting MF on EBV fitting GG, presumably by accounting for group differences which were not quite modeled adequately, possibly due to lack of genomic information and thus less reliable estimates of ancestral relationships.

For routine implementation of ssGBLUP, convergence behaviour of iterative schemes to solve the mixed model equations is important. Fitting GG is known to increase the number of iterates required considerably. Replacing GG by MF did not prove advantageous in this respect: For our analysis fitting MF increased the number of iterates substantially, from 619 (GG) to 1,014 (MF). This

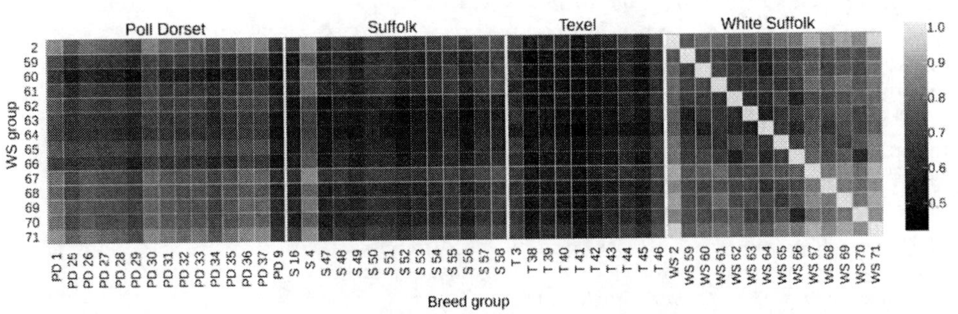

Figure 2. Heatmap plot of estimates of ancestral correlations for White Suffolk groups

Table 2. Summary statistics for predicted breeding values from different analyses

Gen.[a]	Ph.[b]	MF[c]	N[d]	Fit MF only				Fit MF and GG			
				V[e]	r[f]	b[g]	a[h]	V	r	b	a
No	No	0	204,688	0.724	0.991	0.887	-0.317	0.897	0.993	0.940	-0.229
No	Yes	0	1,159,639	0.816	0.996	0.912	-0.337	0.850	0.997	0.919	-0.220
No	No	1	103,126	0.517	0.978	0.836	-0.292	0.930	0.976	0.941	-0.227
No	Yes	1	28,873	0.554	0.992	0.824	-0.294	0.740	0.992	0.853	-0.174
No	No	2	179,472	0.421	0.976	0.840	-0.291	0.940	0.980	0.951	-0.218
Yes	No	0	4,562	0.916	0.989	0.937	-0.423	1.009	0.994	0.999	-0.459
Yes	Yes	0	18,341	0.900	0.995	0.937	-0.398	0.989	0.996	0.990	-0.436
Yes	No	1	82	0.815	0.981	0.898	-0.945	0.768	0.973	0.853	-0.570
Yes	Yes	1	55	0.838	0.970	0.884	-0.737	0.837	0.972	0.889	-0.582

[a] Genotype [b] Phenotype [c] Number of MF parents [d] Number of animals [e] Variance of predicted breeding values as proportion of variance fitting GG only [f] Correlation with breeding values fitting GG only [g] Regression on breeding values fitting GG only [h] Intercept

can be attributed to allowing for correlations between groups when fitting MF. In particular, some correlation estimates were close to unity (see Table 1). Hence, convergence is likely to be improved if groups can be redefined so as to avoid small eigenvalues in $\boldsymbol{\Gamma}$. Furthermore, additional analyses using a 'deflated' preconditioner (see Meyer and Swan 2019) reduced iterates needed when fitting MF to 594, suggesting that there is scope to compensate for any increases in numbers of iterates required due to fitting MF rather than GG.

CONCLUSIONS

Meta-founders have been proposed to align base populations for pedigree based and genomic relationship matrices in ssGBLUP and as an alternative to modeling breeds or genetic groups. Results demonstrate that estimating ancestral relationships and fitting MF is feasible for practical data with many genetic groups. However, optimal performance requires careful definition of groups and sufficient genomic data for all groups to ensure reliable estimates of ancestral relationships.

ACKNOWLEDGEMENTS

This work was supported by Meat and Livestock Australia grant L.GEN.1704.

REFERENCES

Christensen O.F. (2012) *Genet. Sel. Evol.* **44**:37.

Garcia-Baccino C.A., Legarra A., Christensen O.F., Misztal I., Pocrnic I., Vitezica Z.G. and Cantet R.J.C. (2017) *Genet. Sel. Evol.* **49**:34.

Legarra A. and Astruc J.M. (2018) In *Book of Abstracts of the 69th Annual Meeting of the European Federation of Animal Science*. Dubrovnik, Croatia, 27-31 August, pp. 616, Paper no. 66.47.

Legarra A., Christensen O.F., Vitezica Z.G., Aguilar I. and Misztal I. (2015) *Genetics* **200**:455.

Meyer K. (2007) *J. Zhejiang Univ. SCIENCE B* **8**:815.

Meyer K. and Swan A. (2019) *Proc. Ass. Advan. Anim. Breed. Genet.* **23**:000.

Meyer K., Tier B. and Swan A. (2018) *Genet. Sel. Evol.* **50**:39.

Taylor P.J. and Atkins K.D. (1997) *Int. J. Sheep Wool Sci.* **45**:92.

Van Raden P.M. (2008) *J. Dairy Sci.* **91**:4414.

Vitezica Z.G., Aguilar I., Misztal I. and Legarra A. (2011) *Genet. Res.* **93**:357.

Westell R.A., Quaas R.L. and Van Vleck L.D. (1988) *J. Dairy Sci.* **71**:1310.

Proc. Assoc. Advmt. Anim. Breed. Genet. 23:31-34

USING RANDOM FOREST TO IDENTIFY SNPS THAT DECREASE ACCURACY OF GENOMIC PREDICTION – BEHAVIOUR OF SNPS WITH NEGATIVE VIM VALUES

Y. Li[1], F.S.S. Raidan[2], M. Naval Sanchez[1], A.W. George[3] and A. Reverter[1]

[1]CSIRO Agriculture & Food, St Lucia, QLD, 4067 Australia
[2]CSIRO Agriculture & Food, Hobart, TAS, 7004 Australia
[3]CSIRO Data61, Dutton Park, QLD, 4102 Australia

SUMMARY

Random Forest (RF) is one of the most popular machine learning methods for large genomics data analysis. It produces the variable important measures (VIMs) for individual features, which can be positive, zero or negative, indicating a positive or negative contribution of the feature. It is easy to interpret single nucleotide polymorphisms (SNPs) with positive or zero VIM values when applying RF for genomic prediction. However, little is known about the interpretation of SNPs with negative VIM values. Most importantly, what impact of these SNPs have on the genomic prediction accuracy of breeding values? In this study, using genotype information from 651,253 SNPs for 2,109 Brahman cattle with yearling weight phenotype, we applied the RF to identify 8,195 SNPs with negative VIM values and investigated their impact on genomic prediction. Specifically, we addressed the questions: 1) How did these SNPs differ from the top SNPs chosen from the RF with positive VIM values or the SNPs randomly selected but evenly spaced along a genome? 2) Did these SNPs have any biological relevance? Our results show that 1) including the SNPs with negative VIM values in the genomic prediction would result in the increase in error variance and decrease in the accuracy of genomic prediction; 2) these SNPs had no biological functions.

INTRODUCTION

Random Forest (RF, Breiman 2001) is one of the most commonly used machine learning methods for large genomics data analysis (Chen and Ishwaran 2012). One of its analysis output parameters is the variable importance measure (VIM). When applied to a continuous phenotype, RF generates the VIM - %IncMSE (percentage increase in Mean Squared Error). It measures an individual feature's contribution to the prediction accuracy of decision trees, via the change of MSE when the data for a feature (here a SNP) is permuted while all others are kept constant, with valid VIM values being positive, zero or negative. The larger the value (i.e., more positive), the more important the feature is. When applying this method to a high-density SNP panel for genomic prediction of a quantitative trait with a moderate heritability, the questions are: 1) how do SNPs with negative VIM values behave? 2) Do they have any biological relevance? In this study, we investigated the impact of SNPS with negative VIM values on the accuracy of genomic prediction and their possible molecular functions.

MATERIALS AND METHODS

Data. A Brahman cattle dataset, consisting of 2,109 genotyped animals with 651,253 SNPs per animal from the CRC for Beef Genetic Technologies (Porto-Neto *et al.* 2014), was used for this study. The animals were measured for yearling weight (YWT), which ranged from 115 to 353 kg with an average of 227.7 kg (±34.32kg). Since RF does not fit fixed effects into the process, prior to the RF analysis, the phenotypic values were adjusted for the fixed effects. These include contemporary group (combination of sex, year and location and 41 levels) and age (302-416 days). The residuals from the linear model of analysis of variance were then combined with the SNP information for the RF analysis.

Identification of SNPs with negative VIM values (SNP_{negvim}) using RF. The detailed RF method can be found in Li *et al.* (2018). The algorithm incorporates both training and validation procedures in its process to build decision trees to examine individual SNP contributions to prediction accuracy. We carried out an initial hyper-parameter fine-tuning for tree size (NTree) from 10,000, 12,000, … 20,000 using all SNPs, while the Mtry value was set as two times of the squared root of total number of SNPs. A CSIRO high performance cluster computer with the R program (version 3.4.0) and the library randomForest was used for the analyses.

Genomic prediction accuracy with and without SNP_{negvim}. A five-fold cross-validation scheme was applied to the RF and genomic prediction. The population was partitioned into 5 subsets and each time 4 subsets was used for training and the remaining subset was used for validating. In addition to the genomic prediction accuracy comparison between all SNPs with and without SNP_{negvim}, we also examined the results from the subsets of the top 1,000, 5,000, 10,000 and 50,000 SNPs with positive VIM values from the RF, and those of the same size but evenly spaced SNPs along the genome (denoted "Even"). A GBLUP model (VanRaden 2008) was used to estimate variance components and genomic breeding values (gEBVs), where the fixed effects in the model included the contemporary group and age. The accuracy of genomic prediction was calculated as the correlation between gEBVs and the adjusted phenotypic values, and then divided by the square root of heritability. The final estimates of genetic parameters were the average values from five validation analyses. The program AIREMLF90 (Misztal *et al.* 2002) was used in the GBLUP analyses.

Gene Ontology (GO) Enrichment Analysis. A locus-based gene ontology enrichment analysis using GREAT v3.00 (McLean *et al.* 2010) was undertaken. SNPs (±10 bp) were translated to human coordinates (GRC37/hg19) using UCSC's liftOver tool (minMatch=0.1) (Hinrichs *et al.* 2006). A binomial and a hypergeometric test were used to assess the enrichment of molecular function terms and biological process terms.

Functional Enrichment Analysis. Cattle functional annotation was derived from i) histone chromatin marks in liver H3K27ac, and H3K4me3 (Villar *et al.* 2015); ii) ATAC-seq information from CD4+ and CD8+ from the Fr-AgENCODE (Foissac *et al.* 2018); iii) experimental in-house ATAC-seq in liver and muscle tissues; and iv) derived from current UMD3.1 annotation. To assess the significance of overlap between SNP datasets and functional genomic features we performed a Fisher's exact test with false discovery rate correction using the R package LOLA (Sheffield and Bock 2016).

RESULTS AND DISCUSSION

Characteristics of the SNPs with negative %IncMSE values. The distribution of average VIM (%IncMSE) values (from 5-fold training datasets) for ranked SNPs (from the most important to the least important) is shown in Figure 1. Surprisingly, of the 651,253 SNPs, 180,056 (27.7%) were found to have a negative average VIM value. However, when investigated further, we found that only 8,195 of these SNPs had the negative VIM values in all 5-fold datasets, and the remaining 171,861 SNPs varied between the datasets used. This clearly indicates that extreme caution needs to be taken when using the average of the VIM values from a cross-validation scheme as the criteria to identify the SNPs with negative VIM values. An extra step is required to validate the SNPs, because the SNPs with negative VIM values in one population could have positive VIM values in another population.

For these 8,195 SNP_{negvim}, the average MAF was 0.21 (with the range 0.01-0.50). We also checked the allele substitution effects from the previous GWAS study on this population (Porto-Neto *et al.* 2014) and found that these SNPs distributed along the whole genome, whereby 4,143 had positive effects and the remaining 4,052 had negative effects. However, the genotypes of these SNP_{negvim} were in fact imputed from an initial low-density panel of cattle 60k. These may reflect the quality of imputation.

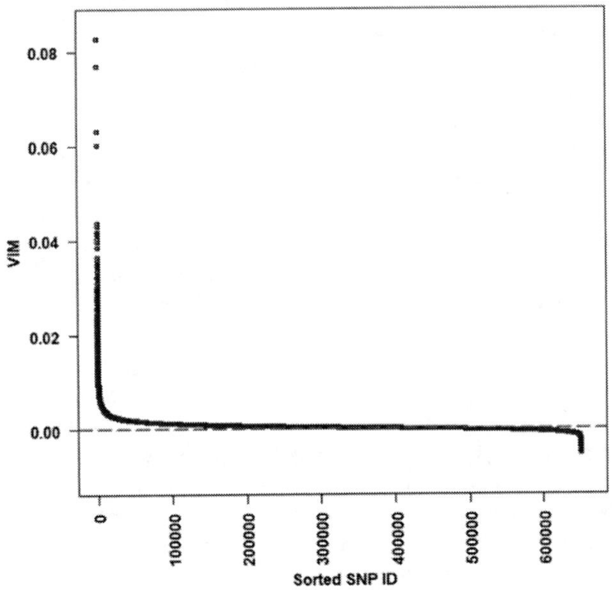

Figure 1. Distribution of average variable importance measures of ranked SNPs

Table 1. Average estimates of variance components and genomic prediction accuracy for different subsets of SNPs

Marker	Additive Model			
	h_a^2	σ_a^2	σ_p^2	†ACC
RF1,000	0.26±0.03	171.6±25.0	658.8±26.1	0.47
RF5,000	0.39±0.04	254.9±32.7	658.2±26.3	0.53
RF10,000	0.42±0.04	278.5±35.2	659.1±26.4	0.55
RF50,000	0.45±0.04	299.0±38.7	669.2±26.7	0.58
Even1,000	0.18±0.03	124.1±22.2	682.8±25.2	0.28
Even5,000	0.30±0.04	218.9±32.2	680.0±26.0	0.47
Even10,000	0.36±0.04	245.4±35.2	681.3±26.3	0.47
Even50,000	0.40±0.04	275.9±38.7	681.4±26.3	0.48
§643,058	0.41±0.05	281.4±39.4	679.5± 26.7	0.59
All SNPs (651,253)	0.41±0.05	281.0±39.6	679.6±26.7	0.55

§ All SNPs without 8,195 VIM negative SNPs; † Accuracy of genomic prediction

Genomic prediction accuracy with and without the negative VIM SNPs. Table 1 presents the estimates of variance components and the genomic prediction accuracies from using different sources of SNPs. In comparison to the accuracy results from using the whole panel (All SNPs, last row in Table 1, ACC = 0.55), the top SNPs from the RF (i.e. RF5,000 and RF10,000) showed very similar or higher (RF50,000) genomic prediction accuracy values. They significantly outperformed the same-size SNPs randomly selected but evenly distributed along the genome (Even-). Interestingly, after removing 8,195 SNP_{negvim}, the genomic prediction with the remaining 643,058 SNPs (Table 1) resulted in an improved accuracy value (0.59) compared to the whole panel (0.55). This value was similar to that of using RF50,000. In addition, we discovered that all the evenly distributed SNP datasets contained about 20% SNP_{negvim}. These results suggest that including SNP_{negvim} in the whole panel would have caused the reduction in accuracy estimates.

Gene Enrichment Analysis. When comparing the biological functions of the genes near 8,195 SNP_{negvim} with those of RF5,000 or Even5,000, there was no significant enrichment found for 8,195 SNP_{negvim}, nor for Even5,000. However, for RF5,000, were enriched for "RNA polymerase II core promoter sequence-specific DNA binding", consisting of several transcription factors such as EGRF1, GATA3, GATA6, NFIL3, PAX6, PAX8 or SOX11. The latter, renowned for its role in embryonic development and determination of cell fate (Jiang *et al.* 2013). Finally, at the functional level, RF 5,000 showed significant enrichment for experimental promoters and muscle regulatory regions.

CONCLUSIONS

In low commodity livestock or aquaculture species, a common practice in applying genomic selection is to genotype parents with a high-density SNP panel, genotype young progeny with a low-density panel and then impute the low-density panel to the high-density panel for genomic prediction. This study demonstrates that it is important to identify and remove the problematic SNPs (with negative VIM values) that increase the error variance and decrease accuracy of genomic prediction. The machine learning method – Random Forest has merit in use as a pre-screening tool for i) identifying problematic SNPs; and ii) identifying subsets of SNPs that have biological functions for low-density panels.

REFERENCES

Breiman L. (2001) *Machine Learning*. **45**: 5.

Chen X. and Ishwaran H. (2012) *Genomics*. **99**: 323.

Foissac S, Djebali S, Munyard K, Vialaneix N, Rau A, Muret K, Esquerre D, et al. (2018) *bioRxiv* 316091.

HinrichsA .S., Karolchik D., Baertsch R., Barber G.P,. Bejerano G., Clawson H., Diekhans M., Furey T.S., Harte R.A., Hsu F., Hillman-Jackson J., Kuhn R.M., Pedersen J.S., Pohl A., Raney B.J., Rosenbloom K.R., et al. (2006) *Nucleic Acids Res.* **34**: D590.

Jiang Y., Qing Q., Xie X., Libby R.T., Lefebvre V. and Gan L. (2013). *J Biol Chem*. **288**: 18429.

Li B., Zhang N., Wang Y.-G., George A., Reverter A. and Li Y. (2018) *Front. Genet. 9:237.*

Misztal I., Tsuruta S., Strabel T., Auvray B., Druet T., and Lee D.H. (2002) *Proc. 7th World Congr. Genet. Appl. Livest. Prod.* Communication N° 28-07.

McLean C.Y., Bristor D., Hiller, M., Clarke S.L., Schaar B.T., Lowe C.B., Wenger A.M., and Bejerano G. (2016) *Nat. Biotechnol.* **28**:495.

Porto-Neto L.R., Reverter A., Prayaga K.C., Chan E.K.F., Johnston D.J, Hawken R.J., Fordyce G, Garcia J.F., Sonstegard T.S., Bolormaa S., Goddard M.E., Burrow H.M., Henshall J.M., Lehnert S.A. and Barendse W. (2014). *PLOS One*. **9**: e113284.

Sheffield N.C. and Bock, C. (2016) *Bioinformatics* **32:** 587.

VanRaden P.M. (2008) *J Dairy Sci.* **91**: 4414.

Villar D., Berthelot C., Aldridge S., Rayner T.F., Lukk M., Pignatelli M., Park T.J., Deaville R., Erichsen J.T., Jasinska A.J., Turner J.M., Bertelsen M.F., Murchison E.P., Flicek P., Odom D.T. (2015). *Cell* **160**: 554.

Proc. Assoc. Advmt. Anim. Breed. Genet. 23:35-38

IMPORTANCE OF HEAT STRESS ADAPTATION FOR NEW ZEALAND DAIRY CATTLE

S. Harburg, P.R. Amer, J. Duckles, G.M. Jenkins and J. Sise

AbacusBio Limited, PO Box 5585, Dunedin 9058, New Zealand

SUMMARY

An analysis was undertaken to explore the potential impacts of increased frequency of heat stress events on New Zealand dairy production systems, with subsequent consideration of the implications for current breeding strategies. Based on current forecasts, the expected impact of climate change will increase the frequency of heat stress events. However, it is unlikely that the expected impacts of heat stress require major deviations from current practices and breeding objectives based on unmitigated impacts on milk production and the trade-offs associated with mitigation.

INTRODUCTION

Anthropogenic climate change represents a key threat to global agricultural industries and food production systems via increased temperatures, changes in rainfall patterns, more frequent extreme weather events, and exposure to new pests and diseases. Given the importance of the dairy industry to the New Zealand economy, understanding the impacts of climate change on domestic dairy production is of national significance.

Increased frequency of hot weather could adversely affect the dairy industry via increased milk production losses due to heat stress. When exposed to hot conditions, cattle reduce dry matter intake to reduce production of metabolic heat, and partition energy into heat dissipation behaviours at the expense of production (Gaughan, Sejian, Mader, & Dunshea, 2019). Consequently, hot and humid weather is frequently associated with reductions in milk production because of heat stress.

This paper explores the long-term climate change forecasts across key New Zealand dairy regions to estimate the potential impact of increased heat stress and implications for current breeding objectives.

MATERIALS AND METHODS

Dairy production occurs across all New Zealand regions, albeit with the largest concentrations of dairy cow numbers occurring in Waikato (23%) and North Canterbury (14%) (LIC and DairyNZ 2018). With Waikato located in the north-western section of the North Island, and North Canterbury on the eastern coast of the South Island, these locations were selected as case studies in order to represent geographically diverse locations.

NIWA, the National Institute of Water and Atmospheric Research, produces long range climate change forecasts for key New Zealand locations. Changes in the frequency of heat stress events for both Waikato and North Canterbury were obtained using NIWA datasets. Climate comparisons occurred between a historical average from 1970 to 2015 as a baseline and forecast future climate in 2090.

NIWA climate change forecasts were configured using three Representative Concentration Pathways scenarios (RCPs) – RCP2.6 (low), RCP4.5 (low-mid), and RCP8.5 (high)- representing hypothetical pathways for the accumulation of greenhouse gases within the earth's atmosphere. These pathways broadly represent conservative (RCP2.6) through to extreme (RCP8.5) levels of climate change impacts on temperature and rainfall. Across each RCP scenario an average of six different global climate models was used to forecast changes in the number of annual 'hot days' above 25C (NIWA, 2019).

New Zealand dairy cattle have been reported to possess a threshold associated with the onset of heat stress over a Temperature Humidity Index (THI) range of 68 to 74 (Bryant *et al.* 2007). Based on prevailing levels of relative humidity in both regions, this heat stress threshold overlaps neatly with a temperature of 25C whereby the THI value at 50% relative humidity is 72, and at 80% relative humidity the THI value is 75. Consequently, the forecast annual 'hot days' frequency was used as a proxy for the expected annual frequency of days exceeding heat stress thresholds.

Regional milk solid production data was sourced for Waikato (358kg per cow per annum) and North Canterbury (413kg per cow per annum) from LIC and DairyNZ (2018). Future 2090 production levels were forecast by adjusting these baseline production levels to account for current genetic trends in milk solid production (National genetic progress of 2.15kg per year for milk solids). Consequently, future production was estimated to be 504kg per cow per year in Waikato, and 582kg per cow per year in North Canterbury.

Berry *et al.* (1964) established a formula for the prediction of milk production impacts due to heat stress: Decline in milk production (kg/d) = -1.075 - 1.736 x NL + 0.02474 x NL x THI, where NL is normal milk production (kg/d) during exposure to temperatures between 10 to 18 °C. NL was derived from the previously reported regional milk solid production forecasts.

Forecasts of current and future levels of milk loss attributable to heat stress were estimated using the above formula to determine daily losses at indicative THI values of 75 and 80. Annual losses were derived by multiplying these daily losses by the expected 'hot day' frequency for each RCP scenario. Due to the uncertainty surrounding average THI values across future 'hot days', a conservative average THI value (THI 75) and extreme average THI value (THI 80) were adopted.

RESULTS AND DISCUSSION

Table 1. displays forecast changes in the forecast frequency of hot days (days exceeding heat stress thresholds) for each location under the three climate change RCP scenarios.

Table 1. Forecast change in annual hot days under climate change

	Current annual hot days	Forecast hot day frequency		
		RCP2.6 (low)	RCP4.5 (mid)	RCP8.5 (high)
Waikato	30	40	60	100
North Canterbury	35	40	50	70

Table 2. displays estimated milk production losses associated with the increased frequency of hot days and subsequent heat stress effects.

Table 2. Forecast annual milk solid production losses in year 2090 attributable to heat stress

	Average THI on 'Hot Days'	Current annual milk solid loss	Forecast annual losses in milk solid production (2090)		
			RCP2.6 (low)	RCP4.5 (low-mid)	RCP8.5 (high)
Waikato	75	2.1kg (0.6%)	5.3kg (1.0%)	7.9kg (1.6%)	13.2kg (2.6%)
Waikato	80	7.2kg (2.0%)	14.9kg (3.0%)	22.3kg (4.4%)	37.2kg (7.4%)
North Canterbury	75	4.2kg (1.0%)	8.2kg (1.4%)	10.2kg (1.8%)	14.3kg (2.5%)
North Canterbury	80	12.0kg (2.9%)	20.7kg (3.6%)	25.9kg (4.5%)	36.3kg (6.2%)

Current heat stress losses are approximately 0.5% to 2.0% of annual production in Waikato and 1% to 2.9% in Canterbury. Under the more moderate RCP scenarios, expected milk solid loss attributable to heat stress is proportionally similar to current losses after accounting for expected genetic progress in milk solid production (2.15kg per year) to 2090. Under the most extreme RCP scenario (RCP8.5), losses increase up to 7.4% of expected 2090 milk solid production in Waikato and 6.2% in Canterbury.

To provide perspective, under the most extreme THI and RCP scenario (RCP8.5 & THI80), additional heat stress losses will amount to 14% of expected genetic progress (at current genetic trends) for milk solid production for North Canterbury farmers, 21% of expected genetic progress for Waikato farmers.

Mitigation of expected heat stress impacts on milk production could be undertaken via selection for heat tolerance. Research undertaken by Garner *et al.* (2016) and Nguyen *et al.* (2016) has led to the development of a genomic-based heat tolerance ABV for Australian dairy cattle to facilitate selection for improved heat tolerance.

The Australian heat tolerance ABV is moderately to strongly antagonistically correlated to milk production traits (r_g = -0.75 to the milk production index). In the absence of a very strong economic signal for improved heat tolerance it is likely that limited genetic progress will be made due to the relationship between heat tolerance and current key production traits. Diversion of index selection emphasis toward heat tolerance could also affect future genetic progress for production traits to an extent that is equivalent to the expected heat stress impacts.

Based on our analysis of forecast heat stress impacts it is likely that insufficient economic incentive will exist to warrant the inclusion of a heat tolerance trait within the New Zealand dairy breeding objective.

Some genetic mitigation of heat stress could be justified to mitigate potential impacts on cow fertility. The scale of potential impacts was on conception rates was not explored within this study and is more difficult to quantify and predict. Mitigation could be achieved by revising the index economic values for fertility based on potential conception rates under future climatic conditions as opposed to the development of a new trait. This would increase selection emphasis on fertility as a means of offsetting expected adverse heat stress impacts.

Further options for genetic mitigation could include development of homozygous 'slick' sires. The 'slick gene' represents an adaptive mechanism utilised by Senepol beef cattle, a tropically adapted Bos Taurus beef breed originating from Central America. The 'slick gene' represents a single gene haplotype located on chromosome 20 that produces a short, sleek coat and enhanced sweating capacity (Dikmen *et al.* 2014). However, validation is required of the heat tolerance benefits within a humid, pastoral environment with low evaporative cooling potential.

CONCLUSIONS

The forecast impacts of climate change on the frequency of heat stress events do not warrant significant genetic adaptation strategies for New Zealand dairy farmers. Farmers are encouraged to understand the expected level of adaptation challenge they will face into the future and make rational and objective decisions about the relative importance of adaptation within a genetic context. Trading off significant differences in production for greater heat tolerance would be unwarranted in most New Zealand dairy regions under the climate change forecasts contained within this paper.

ACKNOWLEDGEMENTS

Funded by the New Zealand Government to support the objectives of the Livestock Research Group of the Global Research Alliance on Agricultural Greenhouse Gases.

REFERENCES

Berry I., Shanklin M. and Johnson H. (1964) *Trans. Am. Soc. Agric. Eng.* **7**:329.

Bryant J., López-Villalobos N., Pryce J., Holmes C. and Johnson D. (2007) *NZ J. Agric. Res.* **50**: 327.

DairyNZ. (2019, April 3) All about BW. Retrieved from Dairy NZ: https://www.dairynz.co.nz/animal/animal-evaluation/interpreting-the-info/all-about-bw/

Dikmen S., Khan F., Huson H., Sonstegard T., Moss J., Dahl G. and Hansen P. (2014) *J. Dairy Sci.* **97**: 5508.

Datagene. (2019, April 29th) ABVs for NASIS bulls, April 2019. Retrieved from Datagene: https://datagene.com.au/v2/sitev2.nsf/0/d404c78a0229cee2ca2580f40020cca1

Garner J., Douglas M., Williams S., Wales W., Marrett L., Ngyuen T. and Hayes, B. (2016) *Sci. Rep.* **6**:. doi:10.1038/srep34114

Gaughan J., Sejian V., Mader T. and Dunshea F. (2019) *Anim. Fronteirs.* **9**: 47.

LIC & DairyNZ . (2018) New Zealand Dairy Statistics 2017/18. Hamilton, New Zealand: LIC & DairyNZ.

Nguyen T., Bowman P., Haile-Mariam M., Pryce J. and Hayes, B. (2016) *J. Dairy Sci.* **99**: 2849.

NIWA. (2019, May 15). Our future climate New Zealand. Retrieved from NIWA: https://ofcnz.niwa.co.nz/#/localCharts

Proc. Assoc. Advmt. Anim. Breed. Genet. 23:39-42

GENOTYPE BY ENVIRONMENT INTERACTION FOR HEAT TOLERANCE IN AUSTRALIAN HOLSTEIN DAIRY CATTLE

E.K. Cheruiyot[1,2], M. Haile-Mariam[1], T.T.T. Nguyen[1], B.G. Cocks[1,2] and J.E. Pryce[1,2]

[1]Agriculture Victoria, Department of Jobs, Precincts and Regions, Bundoora, VIC 3083, Australia
[2]School of Applied Systems Biology, La Trobe University, Bundoora, VIC 3083, Australia

SUMMARY

Genomic breeding values for heat tolerance in dairy cattle were first released in Australia in December 2017 to select animals with better tolerance to heat stress. It is also important to identify animals which perform well in a wide range of temperature and humidities, given the large seasonal and geographical variation in Australia. The aim of this study was to investigate the magnitude of genotype by environment interactions for heat tolerance in Australian Holsteins. A total of 2.5 million test-day milk yield records from 823,055 cows and 6,615 sires were included in the analysis. The heritability estimates at 5th and 95th percentile of temperature-humidity index (THI) were: 0.27 and 0.21, 0.21 and 0.14, and 0.19 and 0.14 for milk, protein and fat yield, respectively. The genetic correlations at the extreme THI values, that is THI = 60 and THI = 75 (equivalent to the temperature and relative humidity of around 20 °C and 45 and, 31 °C and 50, respectively) were: 0.87, 0.84, and 0.86 for milk, protein and fat, respectively. A re-ranking among sires was observed in different environments. These results could allow farmers to make decisions on whether to select sires which are best suited to specific environments, or those that are consistent across a range of environments.

INTRODUCTION

The desire to breed for robustness in the dairy industry is intensifying, driven in part by climate change. One of the key components of robustness is genotype by environment interactions (G × E), which refers to the change in performance or a change in the ranking of animals in different environments. In Australia, dairying is carried out in a wide range of production systems and climatic conditions suggesting that reranking of genotypes may occur.

Various studies have demonstrated the presence of G × E due to heat stress in dairy cattle as reviewed by Carabaño *et al.* (2017). Previous studies in Australia using test-day records reported evidence of G × E for production traits due to heat stress for Australian Holsteins (Hayes *et al.* 2003; Haile-Mariam *et al.* 2008). These studies used first parity or whole lactation data.

Genetic selection for production traits in Australian dairy cattle has resulted in considerable genetic gains. However, this may have led to increased sensitivity to heat stress in dairy animals (Carabaño *et al.* 2017) and possibly increased G × E because of an unfavourable genetic correlation between heat tolerance and milk production traits (Ravagnolo *et al.* 2000). Nguyen *et al.* (2017) noted a declining genetic trend for heat tolerance in Australian Holstein and Jersey dairy cattle at a rate of 0.3%/year. This declining trend coupled with increasing temperature and frequency of heat events suggests the importance of revisiting the magnitude of changes in animal performance at different environmental temperature and humidities. The objective of this study was to investigate G × E for heat tolerance using test-day milk yield records in combination with temperature and humidity data from publicly available weather stations over a 15-year period.

MATERIALS AND METHODS

Test-day data. First lactation milk, protein and fat yield data (consisting of 6.6 million records for Holstein cows between 2003 to 2017) were obtained from DataGene (DataGene Ltd., Melbourne,

Australia). Data editing was as follows: (1) tests < 5d or > 305d days in milk (DIM) and herd test days with less than 10 cows were removed; (2) sires with daughters in less than 2 herds and herds using fewer than 2 sires were excluded and (3) only cows with at least 4 records were retained for analyses. The final dataset comprised 5.2 million records for 823,055 cows and 6,615 sires from 3,732 herds. The pedigree for these data included up to 15 generations.

Climate data. Climate data included hourly dry bulb and dew point temperature and relative humidity obtained from the Bureau of Meteorology (Melbourne, Australia) for 163 weather stations in Australia from 2003 to 2017. The pairwise distances between herds were calculated from geographical coordinates and assigned to the nearest weather station. Hourly temperature-humidity indexes (THI) for each weather station were calculated as follows (Yousef 1985): $THI = T_{db} + \left(0.36\ T_{dp}\right) + 41.2\big)$, where T_{db} = hourly dry bulb temperature (°C); T_{dp} is dew point temperature (°C) and T_{dp} = (237.3b)/(1.0 − b), where b = [log (RH/100.0) + (17.27T_{db})/(237.3 + T_{db})]/17.27, and RH = relative humidity. The THI values were then averaged for 24 hours to get the daily THI. The daily THI on the test day, 1, 2, 3, and 4th day before test day were then averaged and assigned to the test-day records.

Milk yield traits in Australia have been reported to begin declining at THI > 60 (Hayes *et al.* 2003; Nguyen *et al.* 2016). Therefore, the THI threshold was set at 60 in this study (i.e., if THI < 60 then THI = 60). A small proportion (0.004%) of tests obtained at THI ≥ 75 were given a value of 75. This was to avoid unexpected trajectories as possible artefacts, which are often related to fitting polynomials with few extreme data points.

Statistical analysis. A univariate random regression sire model was applied to the data as follows: $y_{ijk} = \mu + HTD_i + YS_j + \sum_{n=1}^{3} A_n X_n + \sum_{n=1}^{8} D_n Z_n + \sum_{n-1}^{2} P_n T_n + \sum_{n=0}^{1} S_{kn} W_n + e_{ijk}$, where y_{ijk} is yield of milk in litres, fat or protein in kg from the *i*th herd test day, *j*th year season of calving, and daughter of the *k*th sire; μ is the intercept; HTD_i is the effect of the *i*th herd test day; YS_j is the effect of the *j*th year season of calving; X_n, Z_n and P_n are the *n*th-order Legendre polynomials corresponding to age on day of test, DIM at test, and THI, respectively; A_n, D_n and T_n are the fixed regression coefficient of traits on age at test, DIM at test, and THI, respectively; S_{kn} is a random regression coefficient on THI for the *k*th sire; W_n is either the intercept ($n = 0$) or slope solution ($n = 1$) for heat load index (THI) for cows and sires; and e_{ijk} is the vector of residual effects. The following (co)variance structure was assumed: $Var\ (S) = \begin{bmatrix} s_0 \\ s_1 \end{bmatrix} = \begin{bmatrix} \mathbf{A}\sigma^2_{s_0} & \mathbf{A}\sigma_{s_0s_1} \\ \mathbf{A}\sigma_{s_0s_1} & \mathbf{A}\sigma^2_{s_1} \end{bmatrix}$, where **A** is the relationship matrix for sires constructed from pedigree data; s_0, s_1 are the intercept and slope for sires; $\sigma^2_{s_0}$, $\sigma_{s_0s_1}$, $\sigma^2_{s_1}$ are (co)variance for sire effects of THI. Heterogeneous error variance was modelled for 10 DIM intervals over a lactation (DIM = 5–30, 31–60, 61–90, 91–120, 121–150, 151–180, 181–210, 211–240, 241–270, and 271–300) as follows: $Var\ (e) = \mathbf{R} = \text{diag}\{\mathbf{I}\sigma^2_{e_1}, \mathbf{I}\sigma^2_{e_2} \ldots \mathbf{I}\sigma^2_{e_{10}}\}$, where $\sigma^2_{e_1}, \sigma^2_{e_2} \ldots \sigma^2_{e_{10}}$ represents error variances and **I** is the identity matrix. (Co)variance components were estimated using ASREML version 4.2 (Gilmour *et al.* 2015).

Calculation of genetic parameters. Additive genetic variances for sires were extracted from the diagonal elements of the covariance $\hat{\mathbf{G}}$ matrix calculated as $\hat{\mathbf{G}} = 4 * \mathbf{\Phi} Var(\hat{\mathbf{S}})\mathbf{\Phi}'$, where $\mathbf{\Phi}$ is the matrix of Legendre polynomial functions for THI; $\hat{\mathbf{S}}$ is the sire (co)variance matrix. The genetic correlations were obtained from transforming the covariance $\hat{\mathbf{G}}$ matrix to a correlation matrix. The heritability as a function of THI was calculated as $h^2_i = \frac{4 * \hat{\sigma}^2_{s(i)}}{\hat{\sigma}^2_{s(i)} + \hat{\sigma}^2_e}$, where $\hat{\sigma}^2_{s(i)}$ is sire variance at *i* THI and $\hat{\sigma}^2_e$ is the average residual variance over the lactation. The estimated breeding value (EBV) for the sire *i* along the THI trajectory was calculated as $EBV_i = ø_j * \hat{a}'_i$, where $\hat{a}'_i$ is the vector of estimated random regression coefficients for the slope and intercept for sire *i*; $ø_j$ is the vector of Legendre polynomials evaluated at THI *j*. To examine the changes in performance along the THI trajectory, we estimated EBVs for sires with more than 1000 daughters with yield records.

RESULTS AND DISCUSSION

Table 1 shows genetic variances and heritability estimates at the 5th, 50th and 95th percentiles of THI. The genetic variance and heritability estimates decrease with increasing THI values. The heritability was greater for milk yield at the 5th and 95th percentiles (0.27 and 0.21) compared to protein yield (0.21 and 0.14) and fat yield (0.19 and 0.14).

Table 1. Additive genetic variances and heritabilities for milk, fat and protein yields at the 5th, 50th and 95th percentiles of the temperature-humidity index (THI)

	Additive genetic variance			Heritability		
	5th	50th	95th	5th	50th	95th
Milk (kg)	4.55	3.86	3.54	0.27	0.23	0.21
Fat (kg)	0.005	0.004	0.003	0.19	0.17	0.14
Protein (kg)	0.004	0.003	0.002	0.21	0.17	0.14

At the extremes of the trajectory of THI (i.e., THI 60 vs 75), the genetic correlations were 0.87, 0.84, 0.86 for milk, protein and fat, respectively (Figure 1). In the previous study, Hayes *et al.* (2003) reported smaller G × E estimates for milk (0.94), protein (0.92) and fat (0.90). Greater G × E in our study is likely in part due to increased sensitivity to heat stress in study population following continued selection for production traits over the years or a slight difference between the analyses; Hayes *et al.* (2003) included a random regression coefficient on THI for cows in their models.

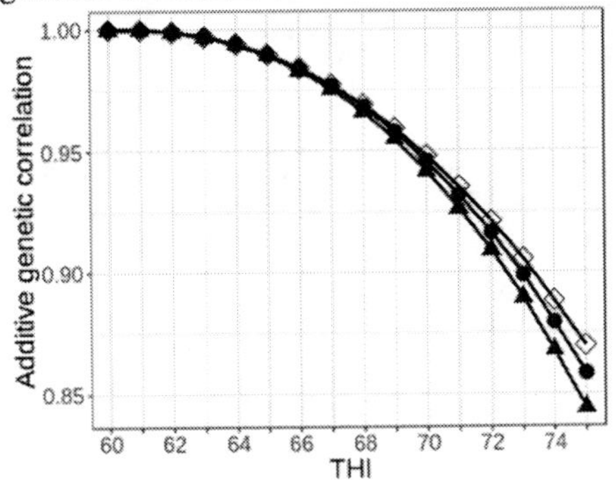

Figure 1. Additive genetic correlations for milk (□), protein (▲) and fat (●) yields at temperature-humidity index (THI) = 60 and those at THI up to 75

Reranking exists among sires, as seen from the differences in the reaction norms of EBVs for fat yield (Figure 2). Two groups of sires were identified based their EBVs at thermoneutral (THI = 60) and heat stress (THI = 75) conditions. The first group (shown in gray) are sires with above-average EBVs at THI = 60 and smaller EBVs at THI = 75 (i.e., environmentally sensitive sires). Daughters of these sires will likely produce less under heat stress conditions and therefore can be used in regions with the consistently low heat load. A more controlled environment, such as the provision of shade and diets designed to reduce core body temperature will be necessary if their daughters are to perform optimally under high heat load conditions.

The second group (Figure 2; shown in black) are sires with above-average and stable EBVs (i.e., resilient or robust sires); their performances are comparatively consistent and are well suited for variable environments. If the objective is to breed for robustness or resilience, then these sires are ideal candidates for selection. Australian dairying is predominantly pasture-based characterised by an array of factors including weather conditions which vary considerably between years as well as seasonal variability in feed quantity and quality feeds. Under these conditions and considering current trends towards extensive exchange of sires between regions or export of sires to other countries, it would be more beneficial to select for robust sires.

This study only considered first lactation data. Greater reranking is expected with later lactations due to relatively higher sensitivity to heat stress associated with greater milk yield in multiparous cows (Carabaño *et al.* 2017). This will be investigated in further studies.

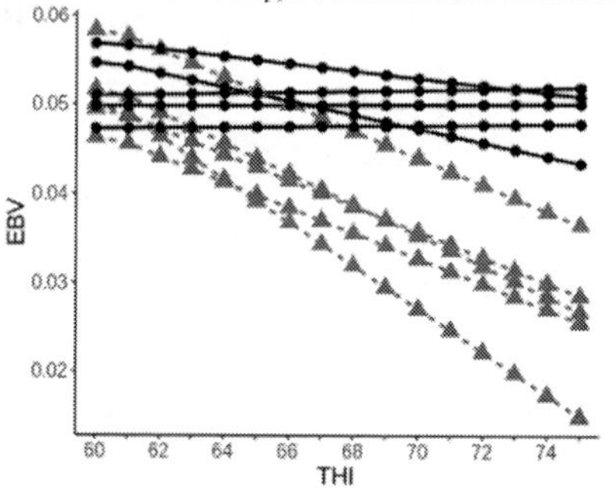

Figure 2. Estimated breeding values (reaction norms) along the THI for a sample of 10 sires with over 1000 daughters with fat yield records; the gray lines (▲) represent sires with above-average EBV at the thermoneutral conditions (THI = 60) and smaller EBV at heat stress conditions (THI = 75) whereas the black lines (•) are sires with above-average and stable EBVs

CONCLUSION

The results from this study indicate G × E due to heat stress exists at extreme THI for all the milk traits studied. The differences observed in the reaction norms (i.e., EBVs along the trajectory of THI) among the sires suggest that genetic variation in sire sensitivity to heat stress exist, which can be used to select animals that perform optimally in different environments.

ACKNOWLEDGEMENTS

This study was thankfully supported by DairyBio, funded by Dairy Australia, the Gardiner Foundation and Agriculture Victoria. The first author acknowledges a PhD scholarship from La Trobe University and DairyBio. Special thanks to all farmers involved in the study and DataGene Ltd., for providing the data.

REFERENCES

Carabaño M.J., Ramón M., Díaz C., Molina A., Pérez-Guzmán M.D. and Serradilla J.M. (2017) *J. Anim. Sci.* **95**: 1813.

Gilmour, A.R., Gogel, B.J., Cullis, B.R., Welham, S. and Thompson R. (2015) ASReml user guide release 4.1 structural specification. Hemel hempstead, VSN international ltd.

Haile-Mariam M., Carrick M.J. and Goddard M.E. (2008) *J. Dairy Sci.* **91**: 4840.

Hayes B.J., Carrick M., Bowman P. and Goddard M.E. (2003) *J. Dairy Sci.* **86**: 3736.

Nguyen T.T.T., Bowman P.J., Haile-Mariam M., Nieuwhof G.J., Hayes B.J. and Pryce J.E. (2017) *J. Dairy Sci.* **100**: 7362.

Nguyen T.T.T., Bowman P. J., Haile-Mariam M., Pryce J.E. and Hayes B.J. (2016) *J. Dairy Sci.* **99**: 2849.

Ravagnolo O., Misztal I. and Hoogenboom G. (2000) *J Dairy Sci* **83**: 2120.

Yousef, M.K. (1985) Stress physiology in livestock. Volume I. Basic principles. CRC press.

Proc. Assoc. Advmt. Anim. Breed. Genet. 23:43-46

NOVEL SELECTION CRITERIA WILL BE REQUIRED FOR REDUCTION OF NEW ZEALAND'S NATIONAL GREENHOUSE GAS EMISSIONS INVENTORY THROUGH DAIRY GENETICS

X. Zhang, G.M. Jenkins, J.A. Sise, B. Santos, C.D. Quinton and P.R. Amer

AbacusBio Limited, PO Box 5585, Dunedin 9058, New Zealand

SUMMARY

The objective of this study was to estimate the reductions in national methane emissions from the New Zealand dairy industry arising though current genetic trends. Based on recent genetic trends, the emissions intensity per milk protein equivalents was calculated to be reducing by 0.43% per year reflecting production efficiency gains. In contrast, emissions per hectare was calculated to be reducing by only 0.03% per year, and this reduction is critically dependent on the assumption that genetic gain in milk yield potential is not exacerbating intensification of dairy farming systems. Novel selection criteria will be required to achieve national reductions in methane emissions from the New Zealand dairy industry.

INTRODUCTION

The Productivity Commission of New Zealand estimated in 2018 that the methane emissions from livestock need to be reduced by 10-22% of the amount in 2016, i.e. 2.8-6.1 million tonnes by 2050. Along with efforts from other sectors, New Zealand would therefore contribute a fair share towards maintaining the current global warming levels. Genetic improvement is one possible tool that could assist the New Zealand dairy industry to achieve this goal while still maintaining the critical role of the industry in export revenue and rural livelihoods.

Previous studies have concluded that methane emissions in dairy cattle were strongly correlated with dry matter intake (DMI) (Pickering *et al.* 2015). Therefore, we applied in this study a methodology which quantifies methane emissions from changes in DMI due to unit genetic changes. This method was applied to traits in the national breeding goal for the New Zealand dairy industry, Breeding Worth (BW).

The objective of this study was to compare how current genetic trends in key dairy production traits are impacting on a range of emission metrics so as to evaluate whether the current breeding strategy would need to be modified in order to help meet the national methane emissions reduction policy.

MATERIALS AND METHODS

The methane emissions were estimated as their carbon dioxide equivalents (CO_2-eq) as a direct conversion from feed intake energy, i.e. kg DMI × 0.583 kg CO_2-eq/kg DM (Fennessy *et al.* 2015). Feed energy consumed by a breeding cow, and her replacement both on and off the milking platform were estimated. We proposed 3 measurement definitions to describe the impact of genetic trait changes on methane emissions as follows:

Gross methane emissions. The gross methane emissions as CO_2-eq emitted by a breeding cow in a year prior to genetic change (E) was estimated as a product of number of animals, feed intake, and the conversion coefficient described above.

Methane per hectare (ha). The gross CO_2-eq emissions per ha of grazing land (EH) was expressed as a ratio of E and the total number of ha for grazing land required per cow per annum (H).

Methane intensity on an animal product basis. The emission intensity (EI) was calculated as a ratio of the E and total number of product outputs per cow. Here all types of animal product outputs were converted to milk protein equivalents (milk protein-eq) using a revenue ratio.

The changes of *E*, *EH* and *EI* due to genetic improvement were denoted as gross value (*GV*), emission value on a ha basis (EV^h) and emission value on an animal product basis (EV^m) and were calculated by obtaining the first derivative of each of the 3 equations with respect to an unit change in one genetic trait (*g*) at a time following Amer *et al.* (2017).

Response to index selection. Genetic trends averaged over the past 5 years were accessed from New Zealand Animal Evaluation Ltd (NZAEL). Trait-wise annual responses in *E*, *EH* and *EI* from index selection were calculated as a product of *GV*, EV^h or EV^m and genetic trend and aggregated over all breeding objective traits.

RESULTS AND DISCUSSION

The emission values for each of the traits within the breeding objective are listed in Table 1. By achieving a 1-unit increase in trait genetic merit, the associated annual gross emissions per breeding cow were estimated to increase for all traits except Residual Survival and Fertility. Similar patterns were observed for emission per ha. In contrast, emission intensity values per unit of milk protein-eq were estimated to decrease for Milk Fat, Milk Protein, Residual Survival, Fertility and Body Condition Score (BCS) as genetic merit improves. Liveweight and Milk Volume emission intensity values were estimated to be positive but on a much smaller scale compared to other traits. A negative emission intensity value for any trait indicates that the increase in gross emissions associated with that trait is proportionally smaller than the increase in either ha or animal product output.

The Productivity Commission (2018) suggested a 10-22% target reduction of gross methane emissions by 2050 of that in 2016, equals to 2.8-6.1 million tonnes (Ministry for the Environment 2018). However, direct selection for reductions in gross emissions per animal would result in direct selection against efficiency improving traits (i.e. against Milk Fat and Milk Protein yield). A better overall outcome than direct selection for inefficiency would be to continue selecting for animal efficiency, but then use other policy mechanisms to reduce the total number of animals or hectare areas farmed (Quinton *et al.* 2017).

Table 1. Estimated effects of a 1-unit trait change in gross methane emissions (kg CO_2-eq emission/breeding cow/year, *GV*), emissions per hectare (kg CO_2-eq emission/ha, EV^h) and emission intensity (kg CO_2-eq emission/kg milk protein-eq, EV^m)

Trait	Unit	*GV*	EV^h	EV^m
Milk Fat	kg	3.57	0.04	-0.02
Milk Protein	kg	2.19	0.02	-0.02
Milk Volume	L	0.07	0.001	0.00004
Liveweight	kg	2.40	0.12	0.005
Residual Survival	day	-0.24	-0.32	-0.0007
Somatic Cell Score	score	0	0	0.04
Fertility	%	-6.28	-8.80	-0.04
Body Condition Score	score	22	26	-0.29

Table 2 shows the current (2019) values for gross emissions, emission per ha and emission intensity for all traits within the current breeding objective. The annual and 20-year change estimates for the aggregated genetic trend are also listed. On average, one breeding cow in New Zealand was estimated to emit 3.087 tonnes of CO_2-eq in year 2019. Over the years, the gross emissions are estimated to increase but emission per ha and emission intensity would reduce, and *EI* was estimated to reduce proportionally faster than the changes of *E* and *EH*.

Given there are 4.8 M dairy cattle in total across New Zealand (DairyNZ 2017), the country-wise gross CO_2-eq by 2050 would increase by 4.8 M animals × 9.95 kg/year/animal × (2050 - 2019) = 1.5 M tonnes, if there was no reduction in the number of dairy cattle. If the land area remained the same from 2017 with 2.4 M ha in dairy sector (Beef + Lamb NZ Economic Service statistics 2017), the country-wise gross CO_2-eq by 2050 would change by 2.4 M ha × (-2.31 kg/ha) × (2050 - 2019) = -171-k tonnes. This is less than 6% of the Productivity Commission 2050 target of 2.8 M tonnes.

Table 2. Aggregated genetic trend predictions for gross CO_2-eq emission (kg CO_2-eq/cow/year, *E*), emission per hectare (kg CO_2-eq/ha, *EH*) and emission intensity (kg CO_2-eq/kg milk protein-eq, *EI*)

	Total value at 2019 (kg)	Annual change (kg)	Annual change percentage (%)[1]	20-year change (kg)	20-year change percentage (%)[1]
E	3,087	9.95	0.32	199	6.45
EH	6,915	-2.31	-0.03	-46	-0.67
EI	9.27	-0.04	-0.43	-0.80	-8.63

[1]percentage compared to 2019.

In emissions per ha measurements, we have assumed that stocking rate gets adjusted as feed requirements per cow increases hence these measurements could adapt to intensive farming system. In another scenario, often dairy farmers in New Zealand increase supplements, e.g. concentrates, for higher genetic merit cows to milk more. This part could be assessed by sensitivity tests.

CONCLUSIONS

This study shows that under the current breeding objective, each New Zealand dairy cow was estimated to produce more gross methane emissions, but also to become more production efficient. Gains in emissions per ha are at best very modest and critically dependent on the assumption that future genetic gain in milk production potential will not encourage further trends towards intensification of New Zealand's dairy production systems. To reach the 2050 methane reduction goal, new selection criteria and a changed emphasis of selection beyond the current tightly defined goal of increasing farm profitability will be required.

ACKNOWLEDGEMENTS

Funded by the New Zealand Government to support the objectives of the Livestock Research Group of the Global Research Alliance on Agricultural Greenhouse Gases. The authors also wish to acknowledge NZAEL for use of the National Breeding Objective model.

REFERENCES

Amer, P., F. Hely, C. Quinton, and A. Cromie. (2017) *Animal* **12**: 5.

Fennessy, P., T. Byrne, and L. Proctor. (2015) Meat and Livestock Australia Ltd, North Sydney, Australia.

Ministry for the Environment. (2018) New Zealand's Greenhouse Gas Inventory 1990-2016. PO Box 10-362, Wellington, New Zealand.

New Zealand dairy industry. (2017) New Zealand Dairy Statistics 2016-2017. Accessed October

31, 2018.
New Zealand Productivity Commission. (2018) Low-emissions economy: Final report. Accessed October 31, 2018.
Pickering, N., M. Chagunda, G. Banos, R. Mrode, J. McEwan, and E. Wall. (2015) *J. Anim. Sci.* **93**: 11
Quinton, C., F. Hely, P. Amer, T. Byrne, and A. Cromie. (2018) *Animal* **12**: 889.

Proc. Assoc. Advmt. Anim. Breed. Genet. 23:47-50

EXPERIENCES WITH NON-LINEAR ECONOMIC VALUES IN SELECTION INDEX DESIGN

C.D. Quinton[1], P.R. Amer[1], T.J. Byrne[2], J.A. Archer[1], B. Santos[1], F. Hely[1]

[1]AbacusBio Limited, 442 Moray Place, PO Box 5585, Dunedin 9058, New Zealand
[2]AbacusBio International, Roslin Innovation Centre, Midlothian EH25 9RG, United Kingdom

SUMMARY

In breeding objectives, linear economic values (LEV) are typically applied because they are effective and easy to implement. However, LEVs can be over-simplifications for some traits in diverse populations that span a wide range of economic and biological conditions. We have been helping an increasing number of breeding programs by applying non-linear economic value functions (NLEV). Although NLEV are more complex to implement in breeding objectives, they can provide more specific and robust trait and therefore overall index valuation. We describe experiences applying NLEV for prolificacy, wool quality, dystocia, and maternal ability in sheep and cattle breeding objectives.

INTRODUCTION

Most animal breeding objectives and selection indexes are built as linear functions. For example, a linear selection index that estimates individuals' total merit in units of currency is defined as $I = \Sigma(b_i \times \hat{g}_i)$, where, for each trait i, the individual's trait value in units of currency is the trait weighting common to all individuals (b_i, index weight) multiplied by the individual's estimated genetic value for that trait ($\hat{g}_i$, e.g. EBV). The individual's index value I is then the sum of all trait values.

However, many traits have non-linear relationships between genetic values and trait values caused by complex market signals or biological limits. A classic example is where carcass sale price ($/kg) has an intermediate optimum relationship with fat cover: below- and above-optimum levels earn reduced prices. Non-linear economic value functions (NLEV) and selection indexes have been discussed in scientific literature (see Martin-Collado *et al*. 2016), but rarely implemented in practice. Commonly, breeding objectives apply a linear economic value (LEV) and index weighting that reflects the population mean genotype; i.e. b_i = partial derivative of a non-linear function at the population mean. When the breeding objective is periodically reviewed, the LEV is updated in accordance with the population mean. This approach is effective for selection and genetic change on a large population scale (e.g., Goddard 1983) and furthermore is simple to configure in genetic evaluation systems, and straightforward to report to users.

A crucial limitation to LEVs as approximations of non-linear value functions is that for diverse populations that span a wide range of economic or biological conditions, LEVs can result in genotypes at the extremes of the distributions being severely over- or under-valued for that trait. This has further implications for multi-trait breeding objectives if it causes individuals to rank highly only because of that trait while being merely average for others. For these reasons, an increasing number of genetic evaluation systems are applying NLEVs in breeding objectives and selection indexes.

NON-LINEAR ECONOMIC VALUE FUNCTIONS

For NLEV, a full function is defined that describes the relationship between individuals' genotype and profitability. The function form may be a simple quadratic or exponential, or more complex combined function. The full range of available genotypes need to be considered to ensure that the function properly values extreme genotypes. Ideally, the function should represent industry conditions, yet be robust and easy to code into genetic evaluation systems.

A primary outcome of implementing NLEV in a selection index is that relative trait weightings within the index depend on the individual's genotype and its location on the function. This is described further in examples below.

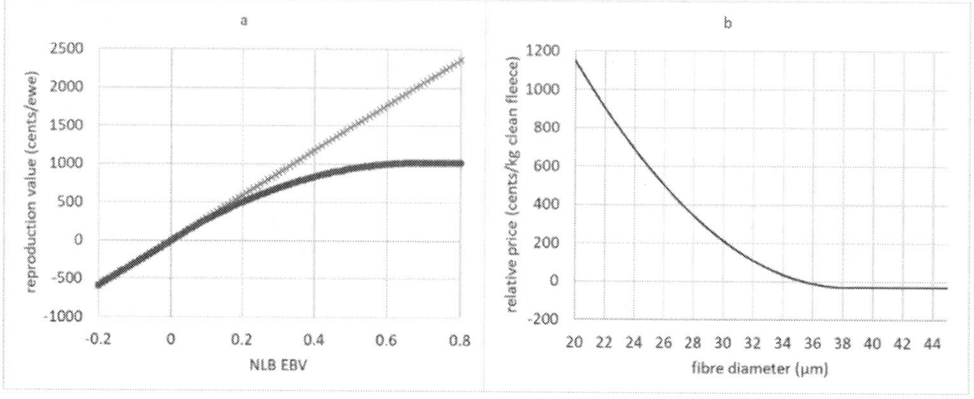

Figure 1. Illustrations of economic value functions. (a) Sheep reproduction value linear (×) and non-linear (●) functions. (b) Wool adult ewe fibre diameter non-linear relationship with relative price

Sheep prolificacy. In 2017, the New Zealand national sheep evaluation system implemented a NLEV for number of lambs born per litter (NLB) in the NZMW maternal index (https://www.sil.co.nz/files/151191893412.pdf) which includes reproduction, growth, survival, and wool sub-indexes. Previously, the index applied a LEV for NLB which was based on the national population mean. Although the population mean is below optimum, there is a wide diversity of prolificacy genotypes in the evaluation so that many individuals have substantially above-optimum genotypes. These individuals were over-valued for reproduction under the linear system, with the outcome that many high-prolificacy rams would achieve top index ranking due to their NLB EBV while having only average EBVs for other index traits such as growth.

A NLEV was developed to better value high prolificacy genetics (preliminary function described by Quinton *et al.* 2017). The function (Figure 1a) is composed of 3 parts: at low prolificacy individuals' value (cents) increases linearly up to the population mean NLB EBV; from mean to optimum NLB, value increases in quadratic fashion with diminishing gains; then above the optimum, a flat "capped" value is imposed so all genotypes receive the same value. Therefore, average rams' reproduction values remained similar, but very high prolificacy rams' values were capped and therefore full NZMW index ranking differences amongst these became due to their genotypes for other traits. Thus, NLB has less influence on the full index value at high prolificacy levels.

This non-linear then flat function has been demonstrated to be the most efficient approach to value an intermediate optimum trait in a multi-trait selection index, when the population mean is below and close to optimum (Martin-Collado *et al.* 2016). From a full index perspective, this approach is predicted to mitigate the risk of highly prolific genetics badly overshooting optimum NLB, while improving selection response in other traits.

Wool fibre diameter. The NZMW index also includes a wool sub-index, which currently values fleece weights, but a recent industry survey revealed substantial interest in valuing crossbred wool quality traits including fibre diameter. A NZ wool sale price analysis (unpublished) quantified the well-known non-linear relationship of fibre diameter with price (c/kg). At stronger micron range (35µm+), micron has little effect on price. However, at finer microns (33-35μm), some price premiums are awarded. The premiums become greater as fleeces move to the mid-micron and finer ranges. Because of these differing relationships, a single LEV for fibre diameter is not suited for the diversity of wool in NZ. The conventional approach of calculating separate LEVs and therefore separate breeding objectives for categories of sheep based on typical fibre diameter ranges has drawbacks: multiple ranking systems are confusing to users who will be considering ram purchases across a wide fibre diameter range; and also incorrectly values individuals that are at the borders of these categories.

A NLEV for fibre diameter (Figure 1b) has been proposed featuring high values for finer micron, with a quadratic curve of decreasing values over medium and stronger microns (<38μm). The lowest (base) wool price occurs. At ≥38μm, all are assigned the base price. This approach is suited to the greater price premiums (c/kg fleece) awarded to mid-micron and finer wool types, compared to cross-bred and strong wool types. Therefore, the same function can be used to value fibre diameter in all NZ crossbred and mid-micron breeds and separate breeding objectives are not required for each type.

Dystocia. Dystocia is typically a categorically observed phenotype with an underlying normal distribution of birthing ease genotypes that results in proportions of a population falling into observed categories. With an economic value defined as the change in profit per unit change in population EBV, then a non-linear relationship between profit (costs) and genotype emerges as the population mean shifts. Distinct category costs (e.g. labour, veterinary, and potential replacement costs) may also contribute to non-linearity.

A survey of Irish beef and dairy farmers (Martin-Collado *et al.* 2017) and a recent American Angus industry survey (unpublished) showed that farmers are prepared to tolerate a small amount of dystocia, but as herd dystocia levels rise this trait is considered to be increasingly problematic. The American Angus trait preference survey also revealed that farmers' opinions of the relative importance of calving ease within the full breeding objective depends on their herd's current levels.

We have helped develop NLEV for dystocia in breeding objectives for American Angus and for an Irish index aimed at selecting beef bulls to mate to dairy cows. In both cases, the NLEV implements a high cost of differentiation at high levels of dystocia, with diminishing marginal benefits as genetic values for dystocia improve. Therefore, bulls with poor dystocia have a larger penalty applied, meaning that fewer of these will appear on leading index lists; conversely, bulls with exceptionally low dystocia (i.e. less than required by most producers) are unlikely to appear on leading lists based on this trait alone.

Maternal ability. In the American Angus beef industry survey mentioned above, respondents judged that the trait weaning weight maternal (WWM, aka maternal ability or "milk") was over-valued at the higher range. Similar to NLB, farmers opinion was that increased WWM is desirable up to a point, but then in environments where feed has high availability to cows or supplements can be provided increased WWM has no further value. In harder environments with low feed availability, over-optimum WWM is considered a liability as high milk cows lose condition and subsequent fertility. For this trait, an intermediate optimum NLEV was built that incorporated survey results of farmers reported lower and upper thresholds of accepted WWM breeding values.

PRACTICAL CONSIDERATIONS FOR IMPLEMENTION

NLEV are more complex than linear EV and therefore do present some challenges for implementation in large-scale breeding objectives.

First, the genetic evaluation program software needs to be adapted to incorporate the NLEV and calculate individual trait values. Most evaluation software code is designed to apply a single linear index weighting coefficient per trait; therefore, experts are required to program NLEV and test index value calculations.

Genetic evaluation systems also must recognize that individual trait values calculated with NLEV are more sensitive to changes in the genetic base definition. A change in the EBV will change any individual's location on the NLEV which may also cause re-ranking.

We have found NLEV most practical if incorporated into modular breeding objective where each trait economic value fully and independently quantifies revenues and costs associated with the trait. E.g. a three-trait index containing a non-linear trait weighting may be described as follows: $I = (b_1 \times \widehat{g}_1) + (b_2 \times \widehat{g}_2) + f(\widehat{g}_3)$, where the individual's trait values for traits 1 and 2 are calculated in the usual linear approach, but where the trait 3 value is calculated according to NLEV. With this modular perspective, NLEV can be substituted for LEV or added on to conventional linear breeding objectives. This modular approach is increasingly useful as breeding programs add new traits (e.g., health and welfare, environmental, novel genomics).

Predicting selection response with NLEV requires different approaches than conventional linear indexes. Most breeding methodologies and software are built around linear breeding objectives and prediction methods use linear regressions, assuming normal distributions. However, NLEVs can skew distributions, especially if values are capped as in the sheep prolificacy function. In these cases, it is preferable to evaluate potential selection intensity and response by analysing real genetic evaluation data sets and calculating trait mean EBVs of selected individuals. For longer-term predictions, stochastic simulations could be employed.

Our experiences with NLEV are that users (breeders, farmers using GE to select animals) are generally very receptive to the concept because the resultant individual animal trait values and rankings tend to better reflect industry realities and their preferences for selection candidates. However, additional education is required for extension services and users who are familiar with reports formatted for simple linear index coefficients. Similarly, for users who are used to pie or bar charts to illustrate relative trait emphases within an index, education is needed to understand how NLEV can shift relative importance of traits.

CONCLUSIONS

Non-linear economic value functions and selection indexes have been well discussed in breeding objective theory, but until recently rarely implemented genetic evaluation systems. Although NLEVs are more complex to apply, these functions are flexible solutions for valuing genetics in diverse populations and our experience is that they are typically very well received by industry stakeholders.

ACKNOWLEDGEMENTS

The authors thank Angus Genetics Inc. and the American Angus Association, Beef+Lamb New Zealand Genetics, and the Irish Cattle Breeding Federation for their contributions.

REFERENCES

Goddard M.E. (1983) *Theor. Appl. Genet.* **64:**339.

Martin-Collado D., Byrne T.J., Visser B., and Amer P.R. (2016) *J. Anim. Breed. Genet.* **133:**476.

Martin-Collado D., Hely F., Byrne T.J., Evans R., Cromie A.R, and Amer P.R. (2017) *Animal* **11:**318

Quinton C., Byrne T., and Amer P. (2017) *Proc. Assoc. Advmt. Anim. Breed. Genet.* **22:**94.

Proc. Assoc. Advmt. Anim. Breed. Genet. 23:51-54

CURRENT PROGRESS ON DEVELOPING A SELECTION INDEX FOR AUSTRALIAN MEAT GOATS

M.N. Aldridge[1,2], W.S. Pitchford[1] and D.J. Brown[3]

[1]Animal and Veterinary Science, University of Adelaide, Roseworthy, SA, 5371, Australia
[2]Animal Breeding and Genomics, Wageningen University and Research, Wageningen, Gelderland, 6708PB, the Netherlands
[3]Animal Genetics & Breeding Unit*, University of New England, Armidale, NSW, 2351 Australia

SUMMARY

Previously meat goat breeders in Australia have used the Carcase Plus (CPLUS) index to make genetic selections. CPLUS is an index focused on lean meat production which used sheep parameter estimates and economic values. It was recommended that a new dual purpose index be developed for increased weaning rate and meat production of goats. The new index "Kid Plus" (K+) uses parameter estimates and economic values calculated for goats and places an economic value on reproductive traits, including kid survival. The dollar value response for each doe joined was higher for K+ ($16.56) compared to CPLUS ($9.53).

INTRODUCTION

Australian goat breeders using the national performance recording scheme (KIDPLAN) use the Carcase Plus selection index which was designed for Australian terminal sire sheep (Sheep Genetics 2016). The CPLUS index puts a large emphasis on increasing growth and eye muscle depth while maintaining leanness. There are several issues with this index when applied to KIDPLAN. Currently there are insufficient breeders consistently recording and submitting data for eye muscle depth or fat depth to justify the emphasis placed on these traits. The CPLUS index places a negative economic value on fat depth, but goats are already very lean and have a small amount of variation in fat depth. Another issue is the economic values used in CPLUS are based on lamb and not representative of the Australian meat goat market. Lastly, the genetic and phenotypic covariance matrices rely on values estimated from Terminal sheep breeds, which have been somewhat modified to suit the KIDPLAN dataset. Australian goat producers have a growing demand for an index built specifically for Australian meat goats (BCS Agribusiness 2012). The aim of this project was to develop the first Australian meat goat specific index.

MATERIALS AND METHODS

There were nine traits of interest used in the analysis; birth weight (BWT), weaning weight (WWT), post-weaning weight (PWT), maternal weaning weight (MWWT), number of kids born (NKB), number of kids weaned (NKW), kid survival (KSV), eye muscle depth (EMD), fat depth (FAT), and worm egg count (WEC). Parameter estimates were made with bivariate animal models in ASReml (Gilmour *et al.* 2009) using KIDPLAN data (Table 1). Body weight was defined as 50% emphasis of WWT and PWT. Kid survival was defined as a trait of the kid, between birth and weaning, it was corrected for birth weight and number of kids born. For EMD and FAT parameter estimates were combined post-weaning and yearling traits, due to limited records, and the low phenotypic variation of fat traits. There was insufficient data in KIDPLAN or published literature for genetic and phenotypic correlations of maternal weaning weight or worm egg count, any analysis that included these traits used the previous covariance estimates from CPLUS (these traits are only included in CPLUS to

* A joint venture of NSW Department of Primary Industries and the University of New England

monitor trait changes and are not included in selection).

Table 1. Summary of parameter estimates. Genetic variance (σ_A^2), residual variance (σ_e^2) and maternal permanent environmental variance (MPE). The heritabilities are on the diagonal, genetic correlations are below the diagonal, and the phenotypic correlations above

	BWT	WWT	PWT	MWWT	EMD	FAT	WEC	NKB	NKW	KSV
σ_A^2	0.21	1.17	2.45	1.00	0.25	0.014	1.40	0.012	0.013	0.013
σ_e^2	0.12	8.28	15.56	9.20	2.01	0.206	5.39	0.300	0.307	0.133
MPE	0.07	1.56	2.42	1.00	0.09	0.005	7.00	0.030	*0.321	0.007
BWT	0.53	0.35	0.32	0.20	0.01	-0.02	-0.03	0.00	0.00	0.01
WWT	0.53	**0.11**	0.81	0.11	0.03	-0.06	0.00	0.00	0.00	0.00
PWT	0.50	0.88	**0.12**	0.08	0.06	-0.04	0.03	0.08	0.06	0.00
MWWT	0.48	0.50	0.50	**0.09**	0.00	0.00	0.00	0.00	0.00	0.00
EMD	-0.22	-0.21	-0.26	-0.38	**0.11**	0.27	-0.06	-0.07	0.01	0.00
FAT	-0.27	-0.24	-0.19	-0.27	0.26	**0.06**	-0.11	-0.29	0.01	0.00
WEC	0.11	-0.03	-0.24	-0.12	<0.01	<0.01	**0.10**	-0.02	0.04	0.00
NKB	0.10	0.08	0.12	0.15	<0.01	<0.01	<0.01	**0.04**	0.41	0.00
NKW	0.01	0.18	0.29	0.33	<0.01	<0.01	<0.01	0.90	**0.04**	0.00
KSV	0.19	0.05	0.03	-0.06	0.05	0.05	<0.01	0.57	0.63	**0.08**

*Animal permanent environmental variance

Table 2. Summary of economic values used for each index based on survey results and Sheep-Object2 (values in $AUD per trait unit)

Trait	Units	CPLUS	LP2020	SRC	LMG	MMG	K+
BWT	kg	0.00	-0.21	-0.21	0.00	0.00	0.00
WWT	kg	2.33	0.32	0.40	2.53	2.53	2.53
PWT	kg	3.50	0.47	1.48	2.53	2.53	2.53
MWWT	kg	0.00	0.00	1.88	0.00	0.00	0.00
EMD	mm	11.40	1.54	2.40	11.40	11.40	11.40
FAT	mm	-4.07	-0.55	0.00	-4.07	-4.07	-4.07
WEC	%	0.00	-1.71	-1.71	-1.71	-1.71	-1.71
NKB	Number	0.00	0.00	0.00	0.00	11.00	11.00
NKW	Number	0.00	0.00	75.00	0.00	30.00	30.00
KSV	Number	0.00	0.00	0.00	0.00	0.00	87.00

Surveys from key industry stakeholders were used to determine breeding objectives, herd structures and economic values were calculated with SheepObject2, a breeding objective software program developed by Andrew Swan (AGBU). There were six indexes of interest; including the CPLUS index. The Lamb 2020 (LP2020) index, designed to increase worm resistance as producers identified internal parasites as an industry issue. The maternal sheep index, Self-replacing Carcase (SRC). The first new KIDPLAN index is a Lean Meat Goat index (LMG) that included economic weights for the body weights and carcase traits. The second KIDPLAN index was a Maternal Meat Goat index (MMG), which added values for NKB and NKW. The final KIDPLAN index Kid Plus (K+), was a dual purpose index for lean meat production and reproduction which included a weight for KSV. The economic values are summarised in Table 2.

A herd of 280 does was used to model the indexes as per the calculations for the average herd size of commercial and seedstock producers. The proportion of males selected was 5%, and 50% for females. Generation intervals of 3 and 4 years were used for males and females respectively. The selection emphasis for EBVs was 65%. To address the Bulmer effect (Bulmer 1971) for a reduction in genetic variance caused by genetic selection, an adjustment for males and females was calculated using Normal distribution theory.

The index calculations were done using R (R Core Team 2016). The index selection theory of Hazel (1943) was used with the variances and covariances in Table 1. The economic values of Table 2 were used for an economic weights vector (a). The index weights ($\boldsymbol{b} = \boldsymbol{P}^{-1}\boldsymbol{Ga}$) were then calculated. The genetic gain ($\boldsymbol{R} = \boldsymbol{b}'\boldsymbol{G}(\boldsymbol{b}'\boldsymbol{Pb})^{-0.5}$) and the total economic gain ($\boldsymbol{\sigma_I} = (\boldsymbol{b}'\boldsymbol{Pb})^{0.5}$) of the index response for one standard deviation of selection was calculated for each of the indexes under different recording scenarios. The recording scenarios were for growth (only BWT, WWT, and YWT recorded), carcass (adds EMD and FAT records), reproduction (no carcass traits but NLB, NLW, and KSV added), standard practice (includes growth traits and reproductive traits but limited carcass traits recorded), best practice (standard practice with full carcass trait records), and gold standard (best practice with WEC recorded).

RESULTS AND DISCUSSION

The index dollar value is the $AUD of additional income per doe joined, per generation, with 5% of males selected and 50% of females, and using the index for the Australian market (Figure 1). The CPLUS index had an index dollar value of between $6.86 and $9.53 across recording scenarios, and was similar to the LMG, which was between $5.67 and $8.84. Both indexes had an increasing value under the following recording scenarios; Growth, Reproduction, Standard practice, Carcase, Best practice, and Gold standard. The maternal index SRC had index dollar values of between $5.99 and $8.33. In comparison, MMG had a value of between $6.64 and $9.86 and K+ had the highest values of between $9.39 and $16.27. Indexes SRC, MMG and K+ increased for the recording scenarios from Growth, Carcase, Standard practice, Reproduction, Best practice to Gold standard. LP2020 had the lowest index dollar values of $2.34 for the recording scenario Gold standard and between $1.25 and $1.35 for the remaining recording scenarios.

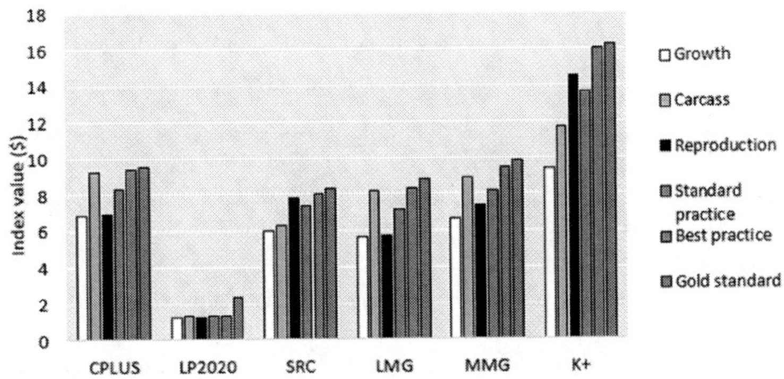

Figure 1. Summary of index response values ($ / doe joined / generation) for each index type and under Growth (white), Carcase (grey), Reproduction (black), Standard practice (green), Best practice (blue) and Gold standard (red) recording scenarios

There are a number of reasons why the index K+ should replace the current CPLUS for KIDPLAN users. Most importantly, it better described profit for the breeding objective of Australian meat goat producers. This was illustrated by the higher index dollar value responses for all recording scenarios. This was primarily due to the inclusion of KSV and the high economic value calculated with SheepObject2. Even under the Growth and Carcase recording scenarios, K+ was similar to CPLUS due to the high economic value placed on body weight and the positive genetic correlations those traits have between each other and KSV. The higher heritability and variation of survival compared to sheep was another reason why KSV is a suitable trait to be included in a KIDPLAN index. The fact that producers must submit the required birth type and rearing types for the KSV calculation improves the accuracy of estimates. Both NKB and NKW are traits of the doe, including both in the index could encourage breeders to better record birth and rearing type which has historically been an issue with the CPLUS index. The high genetic correlation between NKB and NKW could make reducing the index to NKW beneficial as it is easier to record. However, it is also important to monitor the direction of changes for both traits as larger litters resulting from increasing NKB, could result in higher rates of dystocia. Most importantly producers need to have further education on the importance of accurate pedigree and birth type recording.

CONCLUSIONS

Goats differ to sheep in higher heritabilities for kid survival, even with similar trait definitions. These differences include a higher genetic correlated between kid survival and birth weight, greater variation in number of kids born and weaned, less variation for eye muscle and fat depth, and genetic correlations between production traits were significantly different from sheep. The differences in genetic and phenotypic parameters, recording practices, economic values, and breeding objectives of goat breeders led to the creation of new Australian meat goat indexes for KIDPLAN users. The K+ index is based on the best defined breeding objective. This places selection pressure on growth and reproductive traits, especially kid survival calculated from existing birth and rearing type data. Before the K+ index is adopted by KIDPLAN users, further investigation is needed, including; predicted trait changes, differences in economic selection emphasis, selection differential of sires selected between different indexes, and a sensitivity analysis of the economic values used. Future testing of the indexes is recommended to compare the theoretical response to the real world and to demonstrate to producers that a index designed for meat goats is better than the current CPLUS index. Producers are also strongly recommended to record key traits for WEC and carcase traits.

REFERENCES

Gilmour, A. R., Gogel, B. J., Cullis, B. R., Thompson, R., & Butler, D. (2009). ASReml user guide release 3.0. *VSN International Ltd, Hemel Hempstead, UK.*

BCS Agribusiness (2012) Goat Seedstock Industry Stocktake. *Meat and Livestock Australia*

Bulmer, M (1971) *Am. Nat.* **105:** 201.

Hazel, LN (1943) *Genetics* **28:** 476.

R Core Team (2016) R: A Language and Environment for Statistical Computing. *R Foundation for Statistical Computing*

Sheep Genetics (2016) 'ASBVS and Indexes explained.' Available at http://www.sheepgenetics.org.au/Getting-started/ASBVs-and-Indexes [Accessed 09/04/2019]

Proc. Assoc. Advmt. Anim. Breed. Genet. 23:55-58

INDUSTRY CONSULTATION SURVEY FOR THE AMERICAN ANGUS $VALUE INDEXES REVIEW

B. Santos[1], J.A. Archer[1], D. Martin-Collado[2], C.D. Quinton[1], J. Crowley[1], P.R. Amer[1] and S. Miller[3]

[1]AbacusBio Limited, PO Box 5585, Dunedin 9058, New Zealand
[2]CITA Aragón-IA2, Zaragoza, Spain
[3]Angus Genetics Inc., Saint Joseph, MO 64506, USA

SUMMARY

An industry consultation which included an on-line survey distributed to breeders, commercial cattle producers and finishers was carried out as part of a larger revision of the American Angus Association's $Value indexes. A total of 3,174 responses were received. Survey outcomes were used to cluster respondents according to their farming systems and demographic profiles, understand their preferences for traits and to gain insight on whether there are different trait priorities within and between respondents. The survey provided a mandate from industry to review and propose changes to current $Value indexes. It also provided insight to modify bio-economic models that calculate trait economic values to accommodate non-economic factors that systematically influence preferences. The trait preference survey revealed that cow survival, docility, foot score, heifer pregnancy and weaning weight ranked higher on average than what we would have expected based on provisional bioeconomic model calculations. There are differences in trait preferences caused by intrinsic views and beliefs between groups of respondents across and within business activities. These differences reach beyond typical characteristics that can be readily described, such as production system or location. The survey has provided important information for development of indexes which are well aligned with requirements of stakeholders in Angus beef production.

INTRODUCTION

Selection indexes are often developed by bio-economic modelling of production systems. These models do not fully account for the large heterogeneity of trait preferences that is usually found within livestock industries (Paakala *et al.* 2018), for instance when beef cattle farmers choose bulls or select replacements for their herds. Experience has shown that indexes have greater uptake when they are aligned with farmer views and preferences. Industry consultation through survey methods provide a significant and valuable resource to analyse views of farmer trait preferences.

The American Angus Association (AAA) has recently reviewed its current multi-trait economic selection indexes, also known as Angus $Value Indexes. The aim was to update breeding objectives and economic selection indexes based on sound scientific methods, and in line with the preferences of American Angus breeders, cow-calf and feedlot producers and other industry stakeholders.

An on-line survey was designed to describe farming systems and demographic profiles. This stage is hereafter referred to as industry consultation and it aimed to understand drivers of selection decisions when breeders and ranchers choose bulls and replacement candidates. The industry consultation also sought to facilitate understanding of stakeholders' perceptions of the impact that breeding decisions have on their businesses, with a goal of understanding the factors that drive industry engagement.

The objective of this paper is to provide an overview of the industry consultation survey and its key findings. We also provide some perspective on how results of the survey were used to inform subsequent bioeconomic model calculations.

MATERIALS AND METHODS

The survey was conducted from July to early October 2018 and was distributed to all AAA members and made available widely to commercial cattle producers, retained owners and finishers.

The on-line survey was hosted at the American Angus Association; a link directed respondents to the demographics survey which then conducted respondents to the trait preference survey through a seamless process. Respondents had to complete the survey once it was initiated, with no option to pause and return later. The expected time to complete the survey was around 20 to 30 minutes per respondent with a target of 500 to 600 responses. Respondents had the option to either complete the process under total anonymity, or to provide their AAA membership number.

Demographics survey. This survey consisted of 53 questions on farmer and farm systems' to provide details of the farm operation, such as farm and herd size, location, feeding system, etc. Further questions were presented to farmers to determine their views on $Value index and EPDs, and to understand the importance placed on a range of selection criteria when buying or selecting bulls and heifers. The demographic survey asked 53 questions.

Demographic data were used to form a priori groups or, where appropriate, to define farmer typologies which are points of commonality and/or heterogeneity in trait preferences among respondents. Typologies might be associated with respondents' farming system, location, age or any other demographic factor.

Trait preference survey. We used the PAPRIKA pairwise comparison methodology which successively presented two options at a time for respondents to choose between. This approach is practical and requires less intellectual effort from participants when compared to other methods, such as choice experiments. The pairwise comparison makes choice decisions simpler and therefore may be nearer to "true" preferences of respondents. We used the on-line tool 1000Minds® (Hansen and Ombler, 2009) to prioritize choice alternatives. Fourteen traits of interest for farmers were included in the preference survey, and the list of traits and extent of trade-offs between them is presented in Table 1. Trait trade-offs were quantified based on industry data and market prices such that each trade-off produces a similar economic impact, assuming they make sense from a respondent point of view.

Table 1. Trait preference survey questions for the $Value indexes review

Trait Name	Unit of trade-off, comparison and clear trade-off
Weaning Weight	15 lbs more weaning weight because of growth potential
Milk	15 lbs more weaning weight because of cow milking ability
Heifer pregnancy	4 more heifers calve per 100 mated per year
Calving ease	3 less assisted calvings per 100 heifers
Cow survival	6 more cows per 100 live past 5 calvings
Cow mature weight	60 lbs less cow mature weight
Cow frame score	1 less unit (2 inches) of frame score
Body condition score	1 more unit of cow condition score under nutritional stress
Foot score	8 more heifers per 100 suitable as replacements because of good feet
Docility	8 more heifers per 100 suitable as replacements because of good temperament
Feedlot gain	14 less days to commercial endpoint due to feedlot growth performance
Feedlot efficiency	0.5 lb less feed per lb of live weight gain
Yield grade	5 less carcasses per 100 grading Yield Grade 4+
Marbling grade	30 more carcasses per 100 exceeding Mid-Choice grade or better for marbling

Farmers' preferences for traits are known to be heterogeneous, and accounting for this heterogeneity is an attempt to reflect the preferences of a large proportion of farmers. The objective of incorporating farmer's trait preferences is to account for intangible non-economic factors when formulating economic selection indexes.

Survey result analysis. The demographic and trait preference surveys were analysed both separately and jointly to allow a better understanding of the heterogeneity of responses. Three analyses were undertaken; an a priori analysis based on demographic information; a principle component analysis (PCA) to reduce the dimensionality of the data; and a cluster analysis (CA) of the resultant principle components.

The PCA procedure explores the correlation and the variation in trait preferences from which the principal components of the preferences are calculated. For CA, the K-means clustering method was used to measure the distance between preference means for each variable (i.e. trait preference). K-means clustering aims to group *n* observations into *k* clusters in such a way that each observation belongs to the cluster with the nearest mean.

The combination of these analyses enables application of typologies, or drivers of preferences, into clustered groups of factors with statistically different patterns of trait preferences. These patterns can assist in designing selection indexes and tailoring extension efforts.

RESULTS AND DISCUSSION

A total of 3,174 responses were received, including 1,709 full completions of both demographic and trait preference survey sections. Results indicated a general positive perception about AAA's EPDs and $Value indexes. Over 70% of respondents use $Value indexes; there was 50-75% total agreement regarding the importance and usefulness of the $Value indexes; and over 80% of Breeders offer $Value figures to their clients. Of commercial cow-calf producers who responded to the survey, 68% ask for $Values when purchasing bulls.

Table 2. Mean preference ranks (lower ranks mean higher preference) for traits across business activities

Trait Name	Commercial cow-calf		Retained owner		Seedstock breeder		K.W. P value
	Mean	Sd	Mean	Sd	Mean	Sd	
Cow survival	3.9	2.9	5.3	3.5	4.6	3.1	0.642
Docility	5.4	3.3	5.4	3.0	5.1	3.1	0.176
Foot score	6.2	3.4	6.1	3.5	5.0	3.2	<0.001
Heifer pregnancy	5.8	3.1	6.5	3.4	5.4	3.1	0.046
Weaning weight	6.3	3.3	7.7	3.3	6.7	3.3	0.597
Calving ease	6.3	3.7	7.2	3.9	6.7	3.4	0.084
Body condition score	7.4	3.7	8.1	3.7	7.5	3.5	0.433
Marbling grade	8.1	4.0	5.4	3.5	7.7	3.9	0.308
Feedlot efficiency	8.0	3.4	6.8	3.4	7.8	3.3	0.877
Milk	7.6	4.0	9.5	3.9	7.9	3.8	0.531
Feedlot gain	9.4	3.4	7.9	3.6	9.1	3.3	0.049
Cow mature weight	9.1	3.5	9.7	3.6	10.1	3.4	<0.001
Cow frame score	10.2	3.3	10.6	3.4	10.8	3.2	0.203
Yield grade	11.1	2.8	8.8	3.6	10.7	3.0	0.396

There was also support to review and refine $Values, with 75% of respondents at least somewhat agreeing that there would be value in revised indexes that weight traits differently. Also, about 70% of respondents agreed there was need for a specific maternal index, which includes fertility and functional traits such as foot score and docility.

The trait preference survey revealed that the specified changes (Table 1) in cow survival, docility, foot score, heifer pregnancy and weaning weight ranked the highest on average (Table 2). There were differences in trait preferences between groups of respondents across and within business activities. These differences are caused by intrinsic views and beliefs and reach beyond typical characteristics that can be readily described, such as production system or location.

The PCA and CA analyses resulted in three distinct groups (or clusters) of respondents, named Maternal, Production and Cow Hard Environment, according to their pattern of trait preferences across regional or climatic attributes, and in all production or feeding systems (Table 3). These groups were distributed among cow-calf producers, seedstock breeders and retained owners. No difference was found between pattern of preference and business activity. The largest variation in preferences among respondents were on milk, MW, BCS, feedlot gain and marbling.

Table 3. Definition of preferences clusters with average trait rankings across clusters1

Group	WWT	Milk	HP	CE	Cow survival	MW	Cow frame	BCS	Foot	Docy	feedlot gain	Feedlot efficiency	Yield grade	Marb
Maternal	5.7	5.7	4.8	5.5	3.9	10.4	11.2	7.9	4.5	4.3	10.3	8.9	11.8	10.1
Production	6.2	7.2	6.5	8.1	5.6	11.3	11.8	8.5	6.3	5.9	7.1	6.7	9.4	4.4
Cow hard	8.5	11.9	5.6	6.6	3.8	7.1	8.4	5.7	5.8	5.6	9.8	7.4	10.6	8.2

The survey has provided important information for development of indexes which are well aligned with requirements of stakeholders in Angus beef production. Differing trait priorities related to cow feed requirements (e.g. mature weight, milk, condition score) were identified, but ultimately were not deemed enough to justify presentation of multiple indexes. Consequently, the current maternal sub-index was updated targeting the most common feeding systems, with downward pressure on cow maintenance requirements based on the cost of providing additional feed, and a non-linear emphasis on maternal weaning weight. The non-linear milk function (Quinton et al. 2019) was constructed to reward bulls with milk EPDs in the range desired by most breeders, while ensuring that bulls with very high milk do not rise to the top of the index without being exceptional for other traits. Modifications were also made to existing terminal sub-indexes (focused on growth, yield and marbling traits), and a new overall index combining maternal and terminal traits will be implemented based on the industry consultation survey results.

CONCLUSIONS

An on-line industry consultation survey was used to inform economic modelling, and selection index theory principles to propose revised options for $Value indexes. Different groups of farmers were identified according to their pattern of trait preferences. The resulting indexes and sub-indexes are therefore more closely aligned to the requirements of stakeholders in Angus beef production than those being replaced.

ACKNOWLEDGEMENTS

This project has been funded by the American Angus Association Trust. The authors wish to acknowledge American Angus Association and Angus Genetics Inc. for the valuable contribution during the development of the project.

REFERENCES

Hansen P. and F. Ombler (2009) *Journal of Multi-Criteria Decision Analyses*. **15**: 87.

Paakala E., Martín-Collado D., Mäki-Tanila A. and Juga J. (2018) *J. Anim. Breed. Genet.* **135**: 410.

Quinton C. D., Amer P.R., Byrne T.J., Archer J.A., Santos B. and Hely F. (2019) *Proc. Assoc. Advmt. Anim. Breed. Genet.* **23**: *These proceedings*.

Proc. Assoc. Advmt. Anim. Breed. Genet. 23:59-62

COMPARISON OF GENE EDITING VERSUS CONVENTIONAL BREEDING TO INTROGRESS THE *POLLED* ALLELE INTO THE TROPICALLY ADAPTED AUSTRALIAN BEEF CATTLE POPULATION

M.L. Mueller[1], J.B. Cole[2], N.K. Connors [3], D.J. Johnston[3], I.A.S. Randhawa[4], A.L. Van Eenennaam[1]

[1]University of California, Davis, CA, 95616 United States
[2]U.S. Department of Agriculture, Beltsville, MD, 20705 United States
[3]Animal Genetics & Breeding Unit*, University of New England, Armidale, NSW, 2351 Australia
[4]University of Queensland, Gatton, QLD 4343 Australia

SUMMARY

Breeding polled (hornless) cattle is a long-term solution to the costly and increasingly unacceptable cattle management practice of dehorning. This study simulated introgression of the *POLLED* allele into a tropically adapted Australian beef cattle population via conventional breeding or gene editing for multiple polled mating schemes and compared results to baseline selection on genetic merit using the Japan Ox Economic Index ($JapOx) alone, over the course of 20 years. Overall, our simulations show that given the limited number of polled Brahman sires, conventional breeding to increase the *POLLED* allele frequency will have to occur gradually to prevent major impacts on the rate of genetic gain. Furthermore, this study demonstrates how gene editing could help to ameliorate these impacts if a rapid decrease in *HORNED* allele frequency is required due to public pressure or legislation requiring the immediate cessation of dehorning practices.

INTRODUCTION

Dehorning is a standard cattle management practice to protect animals and humans from injury. It is an unpleasant, costly process subject to public scrutiny. Horns are inherited as an autosomal recessive trait (Long and Gregory 1978). However, the Brahman breed, which is most commonly used in extensive grazing systems in Northern Australia (Bunter *et al.* 2013), is predominantly horned. Therefore, decreasing *HORNED* allele frequency through conventional breeding strategies has been challenging (Prayaga 2007). Alternatively, the use of gene editing to produce high-genetic-merit polled sires has been proposed (Carlson *et al.* 2016). Although other genetic factors (i.e., scur and African horn) have been associated with the presence/absence of horns, these factors are believed to segregate independently so this study only modeled *HORNED* and *POLLED* alleles. The objective of this study was to simulate introgression of *POLLED* into a tropically adapted Australian beef cattle population via conventional breeding or gene editing for multiple polled mating schemes and compare to baseline selection on genetic merit, using the Japan Ox Economic Index ($JapOx) alone, over the course of 20 years (yr).

MATERIALS AND METHODS

Simulation. Geneedit.py (Cole and Mueller 2019) was used to simulate introgression of *POLLED* into the Australian Brahman population via conventional breeding or gene editing. Ten nucleus (seedstock) herds supplied bulls to 200 multiplier (commercial) herds. The seedstock base population was 15,000 cows and 40 bulls. The commercial base population was 35,000 cows and 800 bulls. True breeding values for $JapOx were determined by randomly sampling from a normal distribution, with a standard deviation (SD) of $34 for both the seedstock and commercial populations, and a mean of

* A joint venture of NSW Department of Primary Industries and the University of New England

$34 for seedstock cows and $0 for commercial cows (Johnston and Graser, 2009). Base population bulls averaged 1 genetic SD higher than cows. The proportion of polled bulls was set to 30% heterozygous (Pp) and 2.6% homozygous (PP). PP bulls averaged 0.16 SD lower $JapOx than horned bulls, and the *HORNED* frequency for both base populations was set to 80% (Connors *et al.* 2018). Pre-weaning calf loss was set to 8% (seedstock) and 13% (commercial), and the dehorning mortality for both populations was 2% (Bunter *et al.* 2013).

To maintain a maximum population size of 3,000 (~1,800 breeding age) seedstock and 100,000 (~61,000 breeding age) commercial cows, cows were culled first by age (≥ 10 yr) and then at random. Both seedstock and commercial females had their first calf at age 3 and seedstock bulls were eligible for breeding at age 2. The seedstock population kept the top 5% of $JapOx 2-yr-old bulls for breeding to seedstock cows and the remainder were mated to commercial cows. To maintain a population size of 60 seedstock and 1,800 commercial bulls, bulls were culled first by age (≥ 5 yr) and then by $JapOx ranking. Ten replicates of each scenario were simulated for 20 yr, with overlapping generations as described previously (Cole 2015; Mueller *et al.* 2019).

Mating schemes. Each herd used a unique portfolio of sires and the maximum sire portfolio sizes were 6 and 10 bulls for seedstock and commercial herds, respectively. To model mating via natural service, each bull was limited to 35 matings per year and bulls within a sire portfolio were mated randomly to cows in all scenarios. Three mating schemes, 1 baseline (A) and 2 polled (B, C) were modeled. To establish a baseline and model current practice, scheme A used $JapOx as the sole sire selection criterion. In scheme B, PP bulls were preferentially selected for sire portfolios, and then both Pp and horned sires were used for the remaining sire portfolios. In contrast, in scheme C only PP bulls could be included in the sire portfolios and if the mating limit was reached then cows were left open. Scheme C models a potential situation if producers are prohibited from using sires that result in horned offspring.

Gene editing. Polled mating scheme C described above was also simulated with the addition of gene editing for polled. In these scenarios, gene editing was modeled as an added step to the elite sire production system proposed by Kasinathan *et al.* (2015), which combines the use of advanced reproductive technologies and somatic cell nuclear transfer cloning with embryo transfer. In the C-1% and C-10% scenarios, seedstock bull calves were sorted yearly on $JapOx and the top 1% or 10%, respectively, of Pp and horned bulls were cloned and then gene edited to be PP.

RESULTS AND DISCUSSION

***HORNED* frequency.** The baseline scenario A did not result in a significant decrease of *HORNED* frequency in the Australian Brahman population after 20 yr (Figure 1), which is consistent with US dairy simulation results (Cole 2015; Mueller *et al.* 2019). The preferential selection of PP sires in scheme B, resulted in a significant decrease ($P \leq 0.05$) in *HORNED* frequency after 20 yr compared to baseline scheme A. However, after only 5 yr scheme C resulted in a significantly lower ($P \leq 0.05$) *HORNED* frequency (66%), than scheme B (74%). Both scenario C and C-1%, which included gene editing only the top 1% of seedstock bull calves per year, resulted in a similar ($P = 0.81$) rapid decrease in *HORNED* frequency to 10.2% after 20 yr. Additionally, scenario C-10% resulted in a slightly lower *HORNED* frequency (9.8%; $P \leq 0.05$) after 20 years than either scenario C or C-1%.

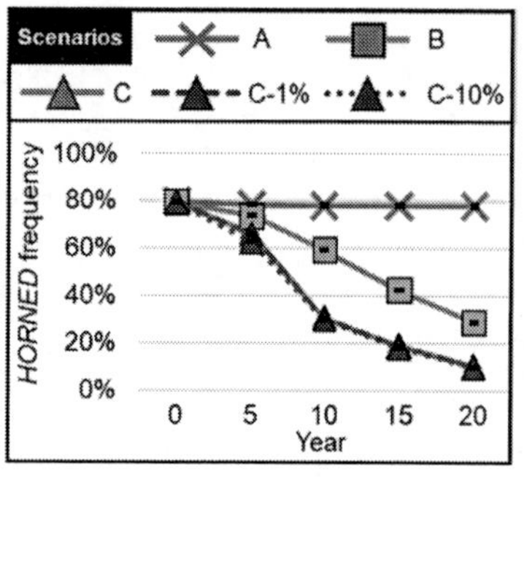

Figure 1. Average effect of each mating scenario on *HORNED* frequency. Error bars (black bars) represent SEM

Inbreeding. In all scenarios inbreeding increased less than 1% per generation. This level of inbreeding has been found to have relatively minor effects on traits of economic or biological significance in tropical beef cattle (Burrow, 1998). A limitation of the simulation is the assumption that all base population animals were initially unrelated, which is unlikely to be valid in a commercial setting.

Genetic gain. The greatest genetic gain ($JapOx) after 20 yr was achieved in baseline scheme A ($160). Selection of polled sires resulted in significantly slower ($P \leq 0.05$) rates of genetic gain ($JapOx) compared to baseline scheme A (Figure 2), which is consistent with previous findings in dairy (Spurlock et al., 2014; Mueller et al., 2019). However, the addition of gene editing to scheme C, scenarios C-1% and C-10% both resulted in significantly greater ($P \leq 0.05$) genetic gain than the polled conventional breeding scenarios B and C. Of the polled scenarios, C-10% resulted in the greatest genetic gain after 20 yr ($154), which was significantly higher ($P \leq 0.05$) than C-1% ($144). A limitation of the simulation is the assumption that true breeding values for $JapOx are known (i.e., breeding value accuracy = 1). Accuracies for cattle in commercial populations with little performance or pedigree information are likely considerably lower, decreasing the rate of gain.

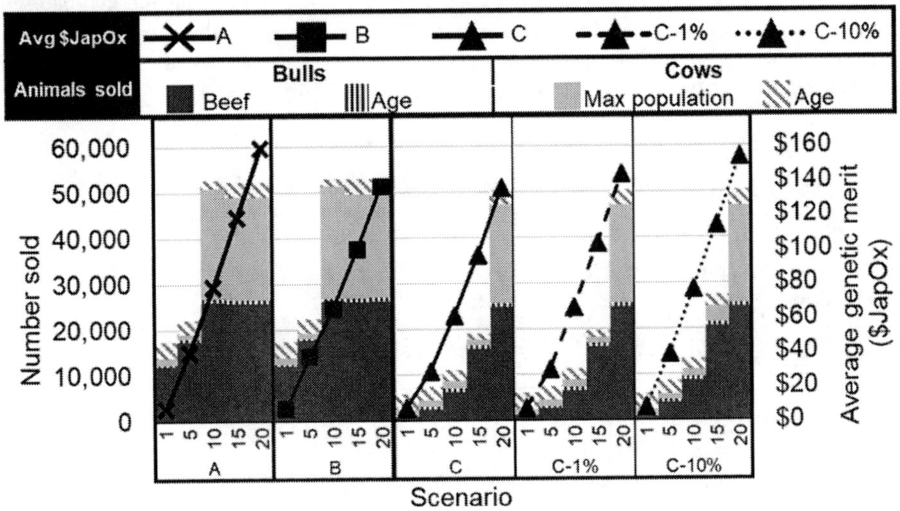

Figure 2. Average effect of each mating scenario on the number of animals sold per year by category on the primary y-axis and the average $JapOx per scenario on the secondary y-axis

Number of cows bred and animals sold for beef. After 10 yr of both scheme A (baseline) and B (preferential PP), the maximum multiplier cow population size was reached. Due to the delayed mating age (3 yr) there were ~61,000 cows bred in yr 10 and thereafter, and there were no cows left open in these mating schemes. Therefore, at maximum population size, scheme A and B resulted in ~26,000 steers sold for beef per year (Figure 2). In contrast, due to the limited number of PP sires available a significantly greater ($P \leq 0.05$) number of cows were left open yr 1 to 7 in scheme C, which resulted in a significantly smaller ($P \leq 0.05$) cow population size until yr 18. Consequently, scheme C resulted in significantly less ($P \leq 0.05$) total animals sold per year until yr 18 (Figure 2). The addition of gene editing only the top 1% of seedstock bull calves per year to this mating scheme (C-1%) resulted in similar numbers of total animals sold for beef per year. However, scenario C-10% resulted in significantly more ($P \leq 0.05$) total animals sold per year from yr 3-18 than either scenario C or C-1%. All 4 polled mating scenarios (B, C, C-1% and C-10%) resulted in significantly more (*P*

≤ 0.05) total animals sold for beef in year 20 than baseline scheme A, as a result of decreased calf loss due to less calves needing to be dehorned.

Scenarios. Preferential selection of PP sires (B) decreased *HORNED* frequency to 29% after 20 yr, whereas the obligatory use of only PP sires (C) decreased the frequency to 10% after 20 yr. The C-1% scenario, which added gene editing only the top 1% of seedstock bull calves per year to mating scheme C, resulted in similar *HORNED* frequency, genetic gain and number of total animals sold for beef per year to scenario C. However, gene editing the top 10% of seedstock bull calves per year (C-10%) resulted in significantly higher *POLLED* frequency, genetic gain and number of total animals sold for beef per year to scenario C.

Scheme C models a situation that could arise if producers are prohibited from using genetics that result in horned offspring. In this simulation cows were left open if no suitable PP sire was available. A more realistic alternative would be to use PP bulls from other less tropically adapted breeds, which could result in higher levels of mortality due to ill-adapted sires and progeny.

Regulatory considerations. Given recent developments outlined by the Australian Office of the Gene Technology Regulator (OGTR) it appears that animals modified using template-guided techniques, like the *POLLED* allele, will be regulated as genetically modified organisms (GMO) in Australia (Mallapaty, 2019). This is not the case in other countries (e.g., Brazil) and may effectively preclude the use of gene editing to introduce the *POLLED* into Australian cattle breeding programs.

CONCLUSIONS

Overall, our simulations show that given the limited number of polled Brahman sires, conventional breeding to increase *POLLED* frequency will have to occur gradually to prevent a major impact on the rate of genetic gain ($JapOx). Furthermore, this study demonstrates how gene editing could help reduce this loss if a rapid decrease in *HORNED* frequency is necessary due to public pressure or legislation ceasing dehorning practices immediately.

ACKNOWLEDGEMENTS

This project was supported in part by funds from the USDA National Institute of Food and Agriculture National Research Initiative Competitive Grant no. 2015-67015-23316 (T. S. Sonstegard, PD). The authors would like to acknowledge GeneSeek®, a Neogen Corporation Company, and the Australian Brahman Breeders Association Limited for providing genotype and estimated breeding value data.

REFERENCES

Bunter K.L., Johnston D.J., Wolcott M.L. and Fordyce G. (2013) *Anim. Prod. Sci.* **54**: 25.

Burrow, H.M. (1998) *Livestock Production Science* **55**:227.

Carlson D.F., Lancto C.A., Zang B., Kim E.-S., Walton M., Oldeschulte D., Seabury C., Sonstegard T.S. and Fahrenkrug S.C. (2016) *Nat. Biotechnol.* **34**:479.

Cole J.B. (2015) *Genetics, selection, evolution : GSE* **47**: 94.

Cole J.B., and Mueller M.L. (2019) Geneedit.py. https://github.com/wintermind/gene-editing.

Connors N.K., Randhawa I.A.S. and Johnston D.J.. (2018) Brahman polled data. unpublished data.

Johnston D.J. and Graser H. (2009) Meat and Livestock Australia Ltd. ISBN: 9781741919202.

Kasinathan P., H. Wei T., Xiang, J.A., Molina J., Metzger D., Broek S., Kasinathan D.C., Faber and Allan M.F. (2015) *Scientific Reports* **5**: 8674.

Mallapaty S. (2019) *Nature*

Mueller,M.L., Cole J.B., Sonstegard T.S. and Van Eenennaam A.L. (2019) *J. Dairy Sci.* **102**: 4215.

Prayaga K.C. (2007) *Aust. J. Agric. Res.* **58**: 1.

Spurlock D.M., Stock M.L. and Coetzee J.F. (2014) *J. Dairy Sci.* **97**: 5265.

Proc. Assoc. Advmt. Anim. Breed. Genet. 23:63-66

PRE- AND POST-PUBERTY CO-EXPRESSION GENE NETWORKS FROM RNA-SEQUENCING OF BRAHMAN HEIFERS

L.T. Nguyen[1,2,3], A. Reverter[4], A. Cánovas[5], L.R. Porto-Neto[4], B. Venus[3], A. Islas-Trejo[6], S.A. Lehnert[4], J.F. Medrano[6], M.G. Thomas[7], S.S. Moore[3] and M.R.S. Fortes[1]

[1]School of Chemistry and Molecular Biosciences, The University of Queensland, Australia
[2]Faculty of Biotechnology, Vietnam National University of Agriculture, Vietnam
[3]Queensland Alliance for Agriculture and Food Innovation, The University of Queensland, Australia
[4]CSIRO Agriculture and Food, Queensland Bioscience Precinct, Australia
[5]Centre for Genetic Improvement of Livestock, Department of Animal Biosciences, University of Guelph, N1G 2W1, Canada
[6]Department of Animal Science, University of California-Davis, CA, USA
[7]Department of Animal Science, Colorado State University, Fort Collins, CO 80523, USA

SUMMARY

Brahman cattle, a *Bos indicus* breed, are well adapted to the harsh environment of northern Australia but reach puberty at an older age compared to *Bos taurus* breeds. Samples from hypothalamus (HYP), pituitary gland (PIT), both ovaries (OVA), liver (LIV), adipose tissue (AT), uterus (UTE) and *longissimus dorsi* muscle (MUS) from pre- and post-pubertal heifers were harvested for RNA sequencing (RNA-Seq). Four gene categories, including differentially expressed (DE) genes, tissue specific (TS) genes, key transcription factors (TF) and genes harbouring SNP associated with heifer fertility, were utilized as nodes of the gene co-expression networks. Significant network connections were identified using an algorithm that exploits the dual concepts of partial correlation and information theory (PCIT). Significance analysis ($P < 0.01$) of RNA-Seq data revealed 2,116 DE genes, 624 TS genes, 186 TF and 179 genes having SNP associated with heifer fertility within the 14,437 expressed genes (genes with reads per kilobase of exon per million mapped reads (RPKM) > 0.2). PCIT analysis pinpoints *ZEB1*, *TEF* and *NFATC2* as the best trio of TF in terms of their ability to span the majority of the topology of the pre- and post-puberty networks. A new role for *SEMA7A* in bovine pubertal development is also postulated. Taken together, our multi-tissue omics analysis revealed candidate genes that could lead to improved understanding of the mechanisms that guide pubertal development.

INTRODUCTION

Fertility traits are economically important for beef cattle operations. Improvements in reproductive efficiency can increase profitability and reproduction rate of beef cattle. Although events involved in the puberty process are similar in *Bos indicus* and *Bos taurus* cattle, they are initiated earlier in *Bos taurus* (Johnston *et al.*, 2009). Selection programs for early pubertal cattle based on phenotype require additional expenditure and labour. As the precise mechanisms inhibiting or stimulating bovine puberty are not entirely clear, identification of molecular regulatory networks modulating puberty in *Bos indicus* cattle is required to better manage heifer development, support development of new biotechnologies, and perhaps develop genetic selection tools of early pubertal cattle.

Our study aimed to identify DE genes, TF, metabolic pathways and networks involved in Brahman cattle puberty. Key tissues for puberty (HYP, PIT, OVA and UTE) and for growth and metabolism (LIV, MUS and AT) were collected from six pre- and six post-pubertal Brahman heifers for RNA-Seq analyses. Gene expression values were obtained and used to construct pre- and post-puberty co-expression gene networks using an algorithm based on PCIT. The predicted co-expression networks were linked by DE genes, TS genes, known TF and genes harbouring SNP associated with heifer fertility traits. These analyses provide new insights into candidate regulatory genes and gene expression pathways involved in bovine puberty.

MATERIALS AND METHODS

Twelve heifers of similar age were managed, handled and euthanized under protocols approved by the Animal Ethics Committee of the University of Queensland (UQ), Production and Companion Animal group (certificate number QAAFI/279/12). Heifers were examined every two weeks for observation of the pubertal development. Post-puberty heifers were in the luteal phase of their second cycle. There was no statistical difference in either BW (338 ± 54 and 363 ± 39 kg, $P = 0.38$) or CS (3.5 ± 0.4 and 3.8 ± 0.4, $P = 0.18$) between pre- and post-pubertal heifers.

Tissue samples (HYP, PIT, OVA, UTE, LIV, MUS and AT) were harvested as fast as possible after slaughter to preserve quality of RNA. In total, 96 tissue samples were available for RNA extraction (12 per tissue, except for OVA which had 24 samples available corresponding to the left and right ovaries). Total RNA was purified using a combination of RNeasy (QIAGEN, Australia) and TRIzol methods as previously described (Fortes *et al.* 2016; Nguyen *et al.* 2017a; Nguyen *et al.* 2018). All samples were passed quality control with RNA integrity numbers higher than 6.9.

The Illumina TruSeq sample preparation kit (Illumina, San Diego, CA) was utilized to construct cDNA libraries for each sample. Standard HiSeq 2000 sequencer analyser (Illumina, San Diego, CA) protocols were used to conduct RNA sequencing. Sequence reads were assembled and mapped to the annotated bovine genome (UMD3.1). Quality control and RNA-Seq expression analyses were performed using CLC Bio Genomic workbench software (CLC Bio, Aarhus, Denmark), with procedures described previously (Nguyen *et al.* 2017a; Nguyen *et al.* 2018). A threshold of the gene expression value (RPKM) ≥ 0.2 was utilized to annotated expressed genes (Mortazavi *et al.* 2008).

We applied "omics" pipeline developed by Nguyen *et al.* (2017b) to identify DE genes, TS genes, genes harbouring SNP associated with female fertility (heifer pregnancy, first service conception and age at first corpus luteum). From the predicted pre-pubertal and post-pubertal networks using PCIT which comprised DE, TS, TF and genes harbouring associated SNP (Reverter and Chan 2008), we applied an information lossless approach (Reverter and Fortes 2013) to explore the connectivity degree of all TF in the network. This approach allowed identification of the best trio of TF that, through their first neighbours, span most of the network topology. Finally, the list of DE genes (n = 2,116) was used as target list for functional enrichment analysis using Database for Annotation, Visualization, and Integrated Discovery (DAVID, Dennis 2013).

RESULTS AND DISCUSSION

An average of 60 million sequence reads were obtained for each individual sample. Previous studies demonstrated that approximately 30 million reads are sufficient to detect more than 90% of annotated genes in mammalian genomes (Lee *et al.* 2013; Wang *et al.* 2011). Despite the absence of a *Bos indicus* reference genome, our transcriptome data provided 60 to 70 % mapped reads. The relatively high number of sequence reads and mapped reads indicates that our data are adequate for differential expression studies.

A total of 2,116 DE genes, 624 TS genes, 186 TF and 179 genes harbouring SNP associated with heifer fertility traits were identified by comparing the pubertal status. Compared to a study by Cánovas *et al.* (2014) which used similar methods to identify genes in pre- and post-pubertal Brangus heifers, we found a higher number of DE genes, but lower numbers of TS genes, TF and genes harbouring associated SNP. The genetic makeup of Brangus heifers is 3/8 Brahman and 5/8 Angus. Differences in the breed type, the experimental design and sample size need to be considered when comparing the results of these two studies. Despite these discrepancies, comparing data from these two studies could be useful to elucidate genes relevant for pubertal development in cattle, regardless of breed. Alternatively, specific genes delaying the pubertal process in Brahman heifers may be identified.

Based on gene ontology (GO) analysis of the 2,116 DE genes, we found enriched GO terms "G-protein coupled receptor protein signalling pathway", "regulation of hormone levels" and "steroid metabolic process". Metabolites and hormones are integrating peripheral signals for reproduction. Moreover, we also identified the most enriched biological process GO term: "immune response" (adjusted $P = 8.3 \times 10^{-13}$). Reproduction is intimately connected to the immune function in women (Abrams and Miller 2011). The enrichment we found in cattle for the DE genes supports the idea of a relationship between reproduction and the immune system in cattle. The KEGG pathway neuroactive ligand–receptor interaction (adjusted $P = 2.5 \times 10^{-06}$) has well known roles in puberty. This pathway comprises ligands and receptors noted to be involved in pubertal signalling such as glycoprotein hormones, alpha polypeptide, GABA receptor, OB-R, prolactin, prolactin receptor and growth hormone receptor (Ainu Husna *et al.* 2012).

The hub nodes of pre- and post-pubertal Brahman heifers sub-networks were *ZEB1*, *TEF* and *NFATC2* (Figure 1). Of note, *ZEB1* may control GnRH expression directly as well as indirectly (Messina *et al.* 2016), and was suggested as a candidate gene in a quantitative trait locus (QTL) study with pleiotropic effects on fatness, stature and reproduction in beef cattle (Bolormaa *et al.* 2014). Both our present study and the Brangus study (Cánovas *et al.* 2014) identified *ZEB1* as a key regulatory factor for bovine puberty. The gene *TEF* was reported as a transcription factor expressed in the pituitary gland during embryogenesis (Droplet *et al.* 1991). The initiation of *TEF* gene expression coincides with that of thyroid stimulating hormone beta (TSHβ). Droplet *et al.* (1991) reported that *TEF* can bind to and lead to effective transactivation of the TSHβ promoter. Thyroid hormones have a role in normal growth and reproductive function (Weber *et al.* 2013). The third TF of the best trio, *NFATC2*, belongs to the nuclear factor of activated T cells family that has been suggested to mediate GnRH action (Armstrong *et al.* 2009). These nuclear factors often generate signals in coordination with MAPKs (Macian 2005), which also play a role in GnRH regulation (Armstrong *et al.* 2009). In summary, our results amount to a growing body of evidence that supports these TF as important in the complex modulation of GnRH signaling and pubertal development.

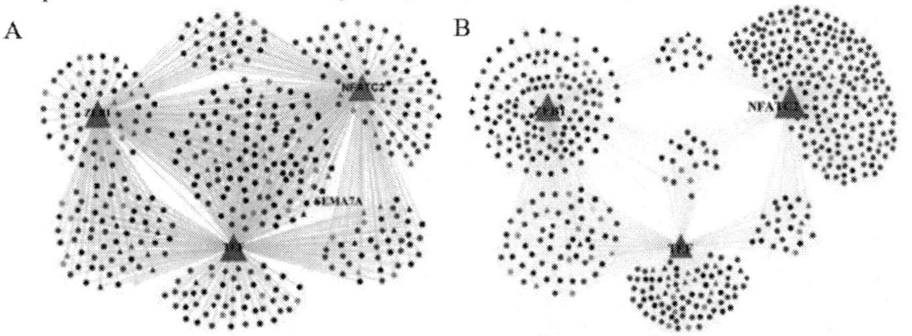

Figure 1. Sub-networks created with the best trio of transcription factors that span most of the network topology. A: pre-puberty network, B: post-puberty network. Genes are coloured according to their categories as follows: red = DE genes; pink = TF; blue = TS; dark brown = genes pertaining to two categories; and yellow = genes pertaining to three categories

Furthermore, examining the interaction between the best TF trio and other nodes in our sub-networks, we found that *SEMA7A* only interacted with the three TF in the pre-puberty network. In mice, during early development, loss of SEMA7A signaling can alter GnRH neuron migration and therefore lead to abnormal gonadal development and altered fertility (Messina *et al.* 2011). Protein and mRNA expression of *SEMA7A* were observed in multiple neuronal systems (Pasterkamp *et al.* 2007). A study of the adult female rat brain suggested that *SEMA7A* was required for the neuroendocrine control of ovarian cycle (Parkash *et al.* 2015). Our result revealed only a slight and insignificant increase in the expression level of *SEMA7A* after puberty in HYP (FC = 0.2). However, significant DE *SEMA7A* (P < 0.01) was observed in the UTE (FC = -1.3) and PIT (FC = -0.9), representing a decrease in expression when progesterone signaling was present. We hypothesize that *SEMA7A* is regulated by the best trio of TF and could contribute to events leading to GnRH release in pre-pubertal Brahman heifers.

CONCLUSIONS

Our results provided potential candidate genes, pathways and networks related to pubertal development. Gene ontology terms and pathways identified from our target gene list might be informative to explain the molecular mechanisms involving in the onset of puberty in Brahman heifers. However, our current work was relying only on gene expression data and bioinformatics tools. Therefore, extensive functional experimental validation for these candidate genes is warranted.

ACKNOWLEDGEMENTS

The authors acknowledge the financial support provided by UQ to this research under a start-up grant awarded to Professor Stephen Moore.

REFERENCES

Abrams E.T and Miller E.M. (2011) *Am. J. Phys. Anthropol.* **146** Suppl 53: 134.

Ainu Husna M.S.S., Flatscher-Bader T., Lehnert S.A., Reverter A., Chan E.K., *et al.* (2012) *J. Trop. Agric. Food Sci.* **40**: 245.

Armstrong S.P., Caunt C.J., Fowkes R.C., Tsaneva-Atanasova K. and McArdle C.A. (2009) *J. Biol. Chem.* **284**: 35746.

Bolormaa S., Pryce J.E., Reverter A., Zhang Y., Barendse W., Kemper K., *et al.* (2014) *PLoS Genet.* **10**: e1004198

Cánovas A., Reverter A., DeAtley K.L., Ashley R.L., Colgrave M.L., *et al.* (2014) *PloS one.* **9**: e102551.

Dennis G.Jr., Sherman B.T., Hosack D.A., Yang J., Gao W., *et al.* (2003). *Genome Biol.* **4:** P3.

Drolet D.W., Scully K.M., Simmons D.M., Wegner M., Chu K.T., *et al.* (1991). *Genes Dev.* **5**: 1739.

Fortes M.R.S., Nguyen L.T., Weller M.M.D.C.A, Cánovas A, Islas-Trejo A., *et al.* (2016) *J. Anim. Sci.* **94**: 3693.

Lee H-J., Jang M., Kim H., Kwak W., Park W., Hwang J.Y., *et al.* (2013) *PloS one.* **8** :e66267.

Macian F. (2005) *Nat. Rev. Immunol.* **5**: 472.

Messina A., Langlet F., Chachlaki K., Roa J., Rasika S., *et al.* (2016) *Nat. Neurosci.* **19**: 835.

Messina A., Ferraris N., Wray S., Cagnoni G., Donohue D.E., *et al.* (2011) *Hum. Mol. Genet.* **20**: 4759.

Mortazavi A., Williams B.A., McCue K., Schaeffer L., Wold B. (2008) *Nature methods.* **5**: 621.

Nguyen L.T., Reverter A., Cánovas A., Venus B.K., Islas-Trejo A., *et al.* (2017a) *J. Anim. Sci.* **95**: 599.

Nguyen L.T., Reverter A., Cánovas A., Venus B.K., Anderson S.T., *et al.* (2018) *Front. Genet.* **9** (87).

Nguyen L.T., Fortes M.R.S. and Reverter A. (2017b). *Proc. Assoc. Advmt. Anim. Breed. Genet.* **22**: 289.

Parkash J., Messina A., Langlet F., Cimino I., Loyens A., Mazur D., *et al.*, (2015) *Nat. Commun.* **6**: 6385.

Pasterkamp R.J., Kolk S.M., Hellemons A.J. and Kolodkin A.L. (2007) *BMC Dev. Biol.* **7**: 98.

Reverter A and Chan E.K. (2008) *Bioinformatics.* **24**: 2491.

Reverter A and Fortes M.R. (2013) *J. Anim. Sci.* **91**: 530.

Wang Y., Ghaffari N., Johnson C.D., Braga-Neto U.M., Wang H., Chen R., *et al.* (2011) *BMC Bioinformatics.* **12**: S5.

Weber G., Vigone M.C., Stroppa L., Chiumello G., (2003). *J. Pediatr. Endocrinol. Metab.* **16** (Suppl 2): 253.

Proc. Assoc. Advmt. Anim. Breed. Genet. 23:67-70

GENETICALLY ENGINEERED AND GENOME EDITED LARGE ANIMAL MODELS FOR NEURONAL CEROID LIPOFUSCINOSES – A REVIEW

I. Tammen[1,2], C.G. Grupen[1], C.L. Pollard, N. Morey and F. Delerue[3]

[1]The University of Sydney, Faculty of Science, Sydney School of Veterinary Science, Camden, NSW, 2570 Australia
[2]Batten Animal Research Network (BARN)
[3]Macquarie University, Faculty of Medicine and Health Sciences, Genome Editing at Macquarie (GEM), Sydney, NSW, 2109 Australia

SUMMARY

The neuronal ceroid lipofuscinoses (NCL) are a group of fatal neurodegenerative inherited diseases. Ovine models have been instrumental to advance the understanding of the genetics and the underlying disease mechanism, but most importantly are crucial for the development of therapeutic interventions. We have commenced to use CRISPR/Cas9 technology to generate an ovine model for the so-called Turkish variant of late-infantile neuronal ceroid lipofuscinosis (CLN7), a relatively common disease variant in humans for which currently no ovine model exists. Other groups have created genome edited and genetically engineered models for CLN1 and CLN3 variants, respectively. We summarise information about naturally occurring variants of NCL in animals and review the limited information about genome edited and genetically engineered non-laboratory animal models for NCL.

INTRODUCTION

Neuronal ceroid lipofuscinoses (NCLs/Batten disease) are a group of lysosomal storage disorders affecting humans and animals. Common characteristics of these diseases include distinctive auto-fluorescent storage bodies in neurons and many other cells and progressive brain and retinal atrophy leading to loss of vision, mental and motor deterioration, epileptic seizures and premature death. In humans, NCL variants have been categorized based on the disease causing genes, i.e. CLN1/*PPT1*, CLN2/*TPP1*, CLN3/*CLN3*, CLN4/*DNAJC5*, CLN5/*CLN5*, CLN6/*CLN6*, CLN7/*MFSD8*, CLN8/*CLN8*, CLN10/*CTSD*, CLN11/*GRN*, CLN12/*ATP13A2*, CLN13/*CTSF*, CLN14/*KCTD7* (Warrier *et al.* 2013). Despite the identification of the disease-causing genes, the links between protein defects, lysosomal storage and pathogenesis are not well understood (Cooper *et al.* 2015). There is no cure, but enzyme replacement therapy (ERT) has shown to attenuate the progression of the CLN2 variant of disease; and research in animal models and human clinical trials suggest that promising results can be achieved with both ERT and gene therapy for variants that are caused by mutations in genes coding for the soluble proteins PPT1, TPP1, CLN5, CTSD, GRN, CTSF (Kohlschütter *et al.* 2019; Mole *et al.* 2019). However, effective therapeutic interventions for variants that are caused by mutations in genes coding for the membrane proteins CLN3, DNAJC5, CLN6, MFSD8, CLN8, ATP13A2 and KCTD7 are lacking.

NON-LABORATORY ANIMAL MODELS FOR NCL

Naturally occurring NCL diseases have been described in many animal species (Table 1) and both naturally occurring, and genetically engineered animal models have been crucial in research efforts to improve our understanding of the genetics and the underlying disease mechanism. Such animal models of NCL disease are required for safety and proof of concept studies for therapeutic interventions (Bond *et al.* 2013). Non-laboratory animal models, such as dogs and sheep, are of specific interests due to their comparatively large and complex brains, long lifespan and the spectrum of clinical signs with which they present. Considerable progress has come from studying sheep with naturally occurring CLN5 and CLN6 forms of disease by the Batten Animal Research Network (BARN) (Palmer *et al.* 2015; Mitchell *et al.* 2018). However, naturally occurring models are not available for all variants of NCL disease (Table 1) and very few non-laboratory animal models have been maintained as research populations. Recently ovine and porcine models for the NCL variants CLN1 and CLN3 have been developed using homologous recombination followed by somatic cell nuclear transfer as well as CRISPR/Cas9 genome editing methods (Table 2; Beraldi *et al.* 2016; Eaton *et al.* 2019).

Table 1. Natural occurring NCLs in animals. NCL variants, genes, species, OMIA/MGI ID, and breed are shown (OMIA: https://omia.org/home/; MGI: http://www.informatics.jax.org)

NCL variant/*gene*	Species (OMIA or MGI ID: breed)
CLN1/*PPT1*	• *Canis lupus familiaris* (001504-9615: Miniature Dachshund; Italian Cane Corso)
CLN2/*TPP1*	• *Canis lupus familiaris* (001472-9615: Longhaired Dachshund)
CLN5/*CLN5*	• *Bos taurus* (001482-9913: Devon) • *Canis lupus familiaris* (001482-9615: Border Collie, Australian Cattle Dog; Golden Retriever) • *Ovis aries* (001482-9940: Borderdale)
CLN6/*CLN6*	• *Canis lupus familiaris* (001443-9615: Australian Shepherd) • *Mus musculus* (MGI:2159328) • *Ovis aries* (001443-9940: Merino)
CLN7/*MFSD8*	• *Canis lupus familiaris* (001962-9615: Chinese Crested Dog, Chihuahua) • *Macaca fuscata* (001962-9542: Japanese macaque)
CLN8/*CLN8*	• *Canis lupus familiaris* (001506-9615: English Setter, Australian Shepherd, Alpenlaendische Dachsbracke, Saluki) • *Mus musculus* (MGI:1856959)
CLN10/*CTSD*	• *Canis lupus familiaris* (001505-9615: American Bulldog) • *Ovis aries* (001505-9940: Swedish Landrace)
CLN12/*ATP13A2*	• *Canis lupus familiaris* (001552-9615: Tibetan Terrier)
n.d./*ARSG*	• *Canis lupus familiaris* (001503-9615: American Staffordshire Terrier)
n.d./n.d.	• *Agapornis roseicollis* (000181-60468) • *Anas platyrhynchos* (000181-8839) • *Bos taurus* (000181-9913: Holstein, Beefmaster) • *Canis lupus familiaris* (000181-9615: American Pit Bull Terrier, Cocker Spaniel, Dalmatian, Japanese Retriever, Labrador Retriever, Minature Schnauzer, Polish Owczarek Nizinny, Saluki, Welsh Corgi) • *Capra hircus* (000181-9925: Nubian) • *Equus caballus* (000181-9796: Aegidienberger) • *Felis catus* (000181-9685: domestic short-haired, Siamese) • *Macaca fascicularis* (000181-9541) • *Mustela putorius furo* (000181-9669) • *Ovis aries* (000181-9940: Rambouillet) • *Sus scrofa* (000181-9823:Vietnamese pot-bellied)

Due to the large amount of research conducted on naturally occurring ovine CLN5 and CLN6 variants, creation of additional ovine models of NCL disease is of particular interest. Direct comparison of natural disease history across these different ovine models would be possible. Standardised assessments of the disease progression as well as gene therapy methods that have been developed for the ovine CLN5 and CLN6 research flocks in Australia and New Zealand (Palmer *et al.* 2015;

Mitchell *et al.* 2018) could be directly transferred to newly developed ovine models for NCL variants for which there is currently no non-laboratory research population.

OVINE CRISPR/CAS9 CLN7 MODEL

Until recently there were no non-laboratory animals diagnosed with CLN7 disease (MIM # 610951), which is the 5th most common variant of NCL disease in humans (NCL-Resource https://www.ucl.ac.uk/ncl-disease/mutation-and-patient-database). It is unclear if CLN7 research populations can be established from the recently reported Chihuahua (Ashwini *et al.* 2016) and macaque (McBride *et al.* 2018) cases. We have therefore commenced to develop a CRISPR/Cas9 genome edited CLN7 sheep model (Table 2; Tammen *et al.* 2019) that mimics one of the 39 known human *MFSD8* mutations and will allow direct comparison to the existing natural occurring ovine variants of NCL disease. We have confirmed that our chosen electroporation approach modified from Kaneko *et al.* (2013) is an efficient way to deliver CRISPR/Cas9 components to *in vitro* produced embryos. We identified sgRNAs and donor template that create the desired genome edit. However, regulatory uncertainties have delayed this work as the current requirement to maintain CRISPR/Cas9 genome edited sheep as genetically modified organisms (GMO) substantially increases the costs for the planned research. However, amended regulations, which consider animals that are created using CRISP/Cas9 and Cas9-induced non-homologous end joining (NHEJ) as non-GMO, will take effect in October 2019 in Australia and will allow us to proceed with this research.

Table 2. Genetically engineered and genome edited non-laboratory animal models for CLN3, CLN1 and CLN7 variants of NCL disease

NCL variant	CLN3/*CLN3*	CLN1/*PPT1*	CLN7/*MFSD8*
Number of human patients / families with disease variant*	432 / 401	230 / 177	104 / 88
Protein location**	late endosomal/ lysosomal membrane, presynaptic vesicles	lysosomal matrix	lysosomal membrane
Protein function**	unknown	palmitoylthioesterase	predicted transporter
GE model species	*Sus scrofa*	*Ovis aries*	*Ovis aries*
Targeted gene / mutation	*CLN3* $^{\Delta ex7-8/\Delta ex7-8}$	*PPT1* p.Arg151Ter	*MFSD8* c.103C>T
Methodology	homologous recombination in fetal fibroblasts & somatic cell nuclear transfer	CRISPR/Cas9 HDR via microinjection of *in vitro* derived embryos	CRISPR/Cas9 HDR & NHEJ via electroporation of *in vitro* derived embryo
Animals with targeted mutation	yes	yes (3 Indel, 6 heterozygous HDR and 3 homozygous HDR)	embryos only
Clinical signs/ histopathology characteristic of NCL disease	yes	yes	unknown
Reference	Beraldi *et al.* 2016; Johnson *et al.* 2019	Eaton *et al.* 2019	Tammen *et al.* 2019

* NCL-Resource: https://www.ucl.ac.uk/ncl-disease/mutation-and-patient-database

** Kollman *et al.* (2013)

CONCLUSIONS

Variants of NCL have been described in many animal species and the identification of disease-causing mutations and development of DNA diagnostics allows for effective management of these diseases in companion animals and livestock. Non-laboratory animal models for NCL have been instrumental in increasing our understanding of this devastating group of diseases in humans and are of particular importance for safety and proof of concept studies for therapeutic interventions. CRISPR/Cas9 technology is an efficient method to develop new animal models for human disease and can be used to validate the effect of predicted disease-causing mutations in animals. Changes to the regulation relating to the use of CRISPR/Cas9 technology will make it easier to create animal models for human disease.

ACKNOWLEDGEMENTS

The development of the ovine CRISPR/Cas9 CLN7 model is funded by the Batten Disease Support and Research Association (BDSRA).

REFERENCES

Ashwini A., D'Angelo A., Yamato O., Giordano C., Cagnotti G., Harcourt-Brown T., Mhlanga-Mutangadura T., Guo J., Johnson G.S. and Katz M.L. (2016) *Mol. Genet. Metab.* **118**: 326.

Beraldi R., Mdaki K., Kovács A.D. and Pearce D.A. (2106) *15th International Conference on Neuronal Ceroid Lipofuscinosis (Batten Disease), Boston USA:* p26

Bond M., Holthaus S.M., Tammen I., Tear G. and Russell C. (2013) *Biochim. Biophys. Acta.* **1832**: 1842.

Cooper J.D., Tarczyluk M.A. and Nelvagal H.R. (2015) *Biochim. Biophys. Acta.* **1852:** 2256.

Eaton, S.L., Proudfoot, C., Lillico, S.G., Skehel, P., Kline, R.A., Hamer, K., Rzechorzek, N.M., Clutton, E., Gregson, R., King, T., O'Neill, C.A., Cooper, J.D., Thompson, G., Whitelaw, C.B. and Wishart, T. M. (2019) *Sci. Rep.* **9**: 9891

Johnson T.B., Sturdevant D.A., White K.A., Drack A.V., Bhattarai S., Rogers C., Cooper J.D., Pearce D.A. and Weimer J.M. (2019) *Mol. Genet. Metab.* **126**: S81.

Kaneko T., Sakuma T., Yamamoto T. and Mashimo T. (2014) *Sci. Rep.* **4**: 6382.

Kohlschütter A., Schulz A., Bartsch U. and Storch S. (2019) *CNS Drugs* **33:** 315.

Kollmann K., Uusi-Rauva K., Scifo E., Tyynelä J., Jalanko A., and Braulke T. (2013) *Biochim. Biophys. Acta.* **1832**: 1866.

McBride J.L., Neuringer M., Ferguson B., Kohama S.G., Tagge I.J., Zweig R.C., Renner L.M., McGill T.J., Stoddard J., Peterson S., Su W., Sherman L.S., Domire J.S., Ducore R.M., Colgin L.M. and Lewis A.D. (2018) *Neurobiol. Dis.* **119:** 65.

Mitchell N.L., Russell K.N., Wellby M.P., Wicky H.E., Schoderboeck L., Barrell G.K.,Melzer T.R., Gray S.J., Hughes S.M. and Palmer D.N. (2018) *Mol. Ther.* **26:** 2366.

Mole S.E., Anderson G., Band H.A., Berkovic S.F., Cooper J.D., Kleine Holthaus S.M.,

Palmer D.N., Neverman N.J., Chen J.Z., Chang C.T., Houweling P.J., Barry L.A., Tammen I., Hughes S.M. and Mitchell N.L. (2013) *Biochim. Biophys. Acta.* **1852:** 2279.

Tammen I., Grupen C.G., Pollard C.L., Morey N., Delerue F. (2019) *2nd Asia Pacific Lysosomal Conference, Auckland, New Zealand*: 23.

Warrier V., Vieira M. and Mole S.E. (2013). *Biochim. Biophys. Acta.* **1832**: 1827.

Proc. Assoc. Advmt. Anim. Breed. Genet. 23:71-74

GENOMIC PREDICTION IN A NUMERICALLY SMALL SHEEP BREED POPULATION USING IMPUTED SEQUENCE VARIANTS

N. Moghaddar[1,2], D.J. Brown[2,3], A.A. Swan[2,3], I.M. MacLeod[4] and J.H.J. van der Werf[1,2]

[1]Animal Science, ERS, University of New England, Armidale, NSW 2351, Australia
[2]CRC for Sheep Industry Innovation, Armidale, NSW 2351, Australia
[3]Animal Genetics & Breeding Unit*, University of New England, Armidale, NSW, 2351 Australia
[4]School of Applied Systems Biology, La Trobe University, Bundoora, VIC 3083, Australia

SUMMARY

The accuracy of genomic prediction for a numerically small sheep breed was investigated based on a large multi-breed admixed reference set using moderate or high density SNP genotypes, imputed whole genome sequence genotypes or selected sequence variants based on a genome wide association study (GWAS). Reference set with weight and eating quality phenotypes was divided into a GWAS sub set (n=4,000), a training set (n=13,466 to 38,098) and a validation set with data of 143 to 169 purebred Dorper sheep. Genomic BLUP was used to estimate genomic breeding values and prediction accuracy was evaluated in the validation set based on the correlation between GBV and corrected phenotypes. Results showed a prediction accuracy between 20% and 30% based on 50k genotypes across different trait, which increased on average by 2.5% to 7.0% by using HD genotypes or selected sequence variants derived from an independent GWAS.

INTRODUCTION

Genomic prediction has been successfully implemented in breeding programs of the main livestock species. In numerically small breeds, it is difficult to establish a reasonably large reference population and prediction based on other main breeds was shown to be of limited value, (Kachman *et al.* 2013; Moghaddar *et al.* 2014). Low GBV predictability from other breeds would be partly because of low linkage disequilibrium (LD) across breeds between genetic markers and the causative mutation, a different distribution of QTL effect and QTL frequency between breeds, or due to genotype by background genotypes interaction. The problem of low LD maybe overcome when using denser marker sets or whole genome sequence (WGS) variants in genomic prediction. This study evaluated the accuracy of genomic prediction for growth and eating quality traits in purebred Dorper sheep based on a large multi-breed admixed sheep reference population, and to compare predictions based on common 50k or HD SNP genotypes, imputed WGS genotypes or using selected sequence variants based on an association study.

MATERIALS AND METHODS

Phenotypes and Animals. Data on post weaning weight (PWT), carcass scanned fat (CCFAT)and eye muscle depth (CEMD), intramuscular fat (IMF) and shear force at 5 days aging (SF5) recorded in research and industry flocks between 1999 and 2017 were used in this study. Figure 1 shows the genetic diversity of the sheep breeds used in this study as a plot of the first versus the second principal component derived from a genomic relationship matrix (GRM). Phenotypes were corrected for fixed environmental effects separately for research and industry animals. The fixed effects of the model were flock, year, sex, management groups, birth and rearing type, age of dam, age at and weight at measurement (for scanned traits). Random maternal effects were fitted for post weaning weight. Corrected phenotypes from research and industry data were combined and then corrected for source

* A joint venture of NSW Department of Primary Industries and the University of New England

of data (research/industry) and random effect of breed proportion derived from a multi generation pedigree using ASReml 3.0 (Gilmour *et al.* 2009). Between 143 and 169 purebred Dorper sheep with phenotypes and genotypes were used as validation set to represent a numerically small breed. Two data subsets were formed for a genome wide association study (GWAS); n=4000, either randomly assigned or selected based on possible higher relationship to the validation set. The rest of population (between 17,466 and 42,098 across different traits) was used as genomic prediction training set.

Genotypes. Animals were genotyped with the Illumina 50k-ovine (~70%) or 12k-ovine SNP panel (~30%), which yielded a final 44,101 and 11,377 SNP per animal respectively. Genotypes were imputed to HD genotypes based on 2,266 animals as reference set and then to WGS based on 726 animals as reference set. The final set was comprised of 31,154,249 SNP and InDels. Selection of sequence variants was based on significant SNP (–Log Pvalue ≥ 3.5) in GWAS performed on sequence data and then pruned locally for high LD (≥0.95). Association analysis was based on regression of corrected phenotypes on single sequence variant in linear mixed model (LMM) using Gemma V0.96 (Zhou and Stephens 2012).

Genomic prediction. GBV were calculated based on GBLUP with MTG2 2.02 (Lee *et al.* 2016) using the following SNP arrays: 1) 50k (44,101) genotypes, 2) HD (452,998) genotypes, 3) WGS (30,724,780) and 4) 50k and selected sequence variants (2,583-2,865). The following model was used to estimate variance components and genomic breeding values in scenarios 1, 2 and 3: $y=Xb + Za + e$, where y is a vector of corrected phenotypes, b is a vector of fixed effect (only mean), a is a vector of random additive genetic effects and e is a vector of random residual effects. X and Z are incidence matrices that relate fixed and additive genetic effects to phenotypes respectively. The additive genetic effects were assumed to be normally distributed with a covariance structure based on the GRM derived from the respective SNP panels. The genomic prediction model in scenario 4 was based on fitting two genetic component simultaneously, with covariance structure based on a GRM from 50k genotypes and selected variants, respectively. Accuracy of genomic prediction in purebred Dorper sheep was evaluated based on Pearson correlation coefficient between GBV and corrected phenotypes in the validation set divided by the square root of the trait's heritability.

RESULTS AND DISCUSSION

Slightly higher heritability, but consistent across different traits, was observed based on imputed HD genotypes and imputed sequence data compared to 50k genotypes (Table 1). Higher heritability is related to stronger LD between markers and QTLs and better estimation of realized genetic relationship.

The sum of the heritability based on fitting two random components simultaneously was on average similar to heritability estimates based on 50k or HD genotypes. Figures 2 and 3 compare the accuracy of genomic prediction for Dorper sheep according to using 50k or imputed HD genotypes, imputed WGS variants and 50k SNPs plus selected imputed WGS variants, respectively. Results show a higher accuracy of genomic evaluation by including the effect of selected sequence variants in the prediction model as an additional random effect. The extra accuracy was on average 0.065 and 0.077 higher when fitting selected sequence variants from a random or selected GWAS population, respectively. SF5 and IMF showed the highest increase in prediction accuracy; 0.11 and 0.09 when using selected variants derived from random or selected GWAS populations, respectively. Accuracy of genomic evaluation from using all called sequence variants (~$31x10^6$ variants) was not consistently higher than 50k genotypes. SF5 showed an increase of 0.05 and the prediction accuracy was equal or even lower than 50k genotypes. Prediction from imputed HD genotypes was more accurate (2.4%) compared to prediction using 50k genotypes in most cases except for PWT and IMF. Results show a base of between 20% and 32% genomic prediction accuracy on growth and eating quality traits using 50k genotype data for Dorper sheep based on the use of a large multi-breed reference population (13,466 to 38,098). This base prediction accuracy was expected and would be related to the use of the large multi-breed reference set which includes breeds that are genetically close to Dorper sheep (Figure.1).

Table 1. Heritability (h^2) estimates based on 50k, HD, WGS and 50k and Selected Sequence variants for different traits

Trait	No of Records	h^2,50k	h^2,HD	h^2,WGS	h^2(50k,Sel_SNPs)
Post Weaning Weight (PWT)	38,098	0.182	0.182	0.184	0.174, 0.04
Carcass Scanned Fat (CCFAT)	14,369	0.185	0.214	0.229	0.163,0.06
Carcass Eye Muscle Depth (EMD)	14,507	0.148	0.151	0.149	0.135,0.02
Intra Muscular Fat (IMF)	13,466	0.404	0.434	0.455	0.412,0.03
Shear Force day5 Aging (SF5)	14,394	0.172	0.178	0.196	0.146,0.03

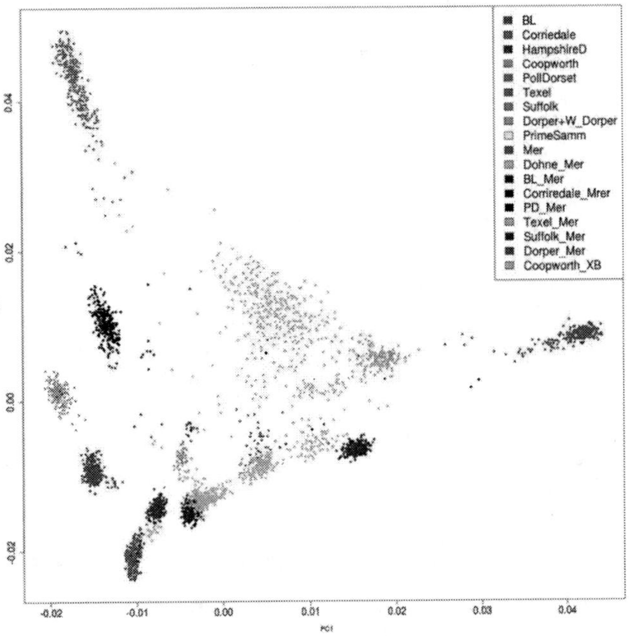

Figure 1. Genetic diversity of the sheep breeds as a plot of the first vs second principal components

Improvement in prediction accuracy by using selected sequence variants in the current study is in similar range to previous study in main sheep breeds (Moghaddar *et al.* 2018) and is in line with the results of studies on multi-breed dairy cattle. In dairy cattle, Van den Berg *et al.* (2016) showed on average up to 7% higher genomic prediction reliabilities (R^2) across milk yield, protein and fat from a multi-breed reference population. Brøndum *et al.* (2015) reported up to 5% improvement in genomic prediction reliability on a range of production traits in multi-breed dairy cattle based on including selected sequence data from GWAS in GBLUP. Using a complete set of imputed WGS a marginal, zero or even some drop in GBV accuracy observed. This is because WGS provide a very large amount of genetic markers of which a small subset would be at or in high LD with causative mutations. Majority of these imputed sequence variants would not be able to capture genetic variance and their contribution would be limited to capturing the family relationships between animals, which would be similar or slightly higher to the relationship captured by 50k genotypes. Similar results of no improvement in prediction accuracy from using all the sequence variants data have been reported in Holstein-Friesian dairy cattle (VanRaden *et al.* 2015).

The extra prediction accuracy based on selected variants derived from a GWAS subset that used data from animals closely related to the target breed appears to be slightly higher (2% on average) than using a random GWAS subset. The differences may be not statistically significant and requires more verification in further studies, particularly based on larger GWAS populations. However, higher accuracy would be related to probably larger proportion of SNPs derived from a more related GWAS subset in association with gene that segregate in target breed. This indicates that while multi-breed GWAS population is more powerful to find larger numbers of causal genomic regions (Duijvesteijn *et al.* 2018; van der Berg *et al.* 2016), our study showed more genetically related GWAS population to target population is preferable to obtain more accurate genomic breeding values. The GWAS results, which showed there are some significant genomic regions limited to a random or a selected GWAS subsets, support these results.

CONCLUSIONS

Genomic prediction accuracy for a numerically small breed population increased by 2.5% and 7% based on using imputed high-density marker genotypes and imputed sequence variants derived in an independent population respectively. Selection of sequence variants from a genetically more related population was in favour of higher genomic prediction accuracy in small breed populations.

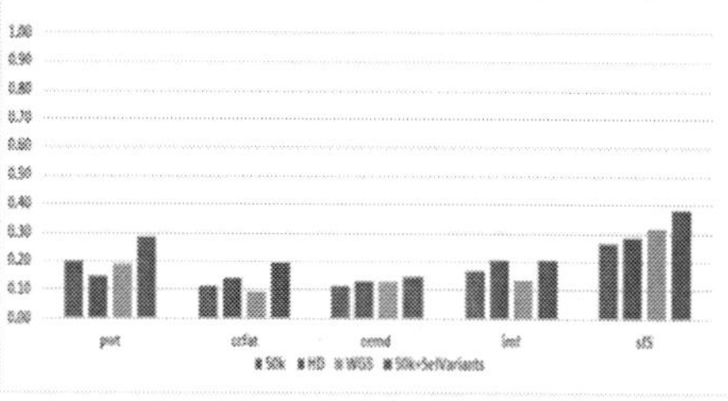

Figure 2. Accuracy of genomic prediction from 50k, HD and using selected SNPs from random GWAS set

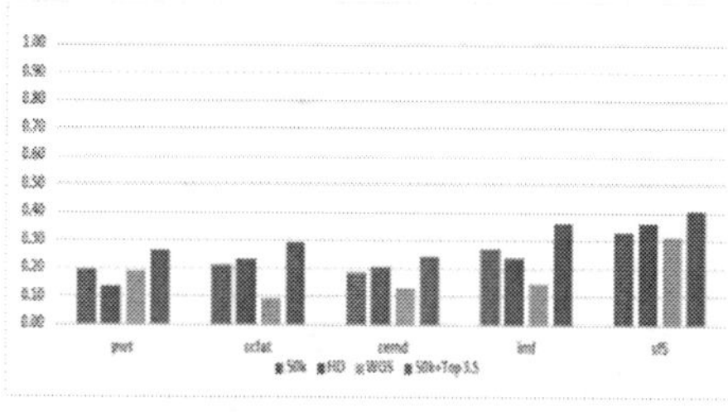

Figure 3. Accuracy of genomic prediction using 50k, HD, WGS and selected SNPs from selected GWAS set

REFERENCES

Brøndum R.F., Su G., Janss L. and Sahana G. (2015) *J. Dairy Sci.* **98**: 4107.

Kachman S., Spangler M., Bennett G., Hanford K., Kuehn L., Snelling W., Thallman R., Saatchi M., Garrick D., Schnabel R., Taylor J. and Pollak E. (2013) *Genet. Sel. Evol.* **45**: 30.

Gilmour A.R., Gogel B.J., Cullis B.R. and Thompson R (2009). VSN International Ltd; 2009.

Lee, S.H. and van der Werf, J.H.J (2016). Bioinformatics 32, 1420.

Moghaddar N., Swan A.A. and van der Werf J.H.J. (2014) *Genet. Sel. Evol.* **46**: 58.

Van den Berg I., Boichard D. and Lund M.S. (2016) *Genet. Sel. Evol.* **48**: 83.

VanRaden P.M., Tooker M.E., O'Conell J.R., Cole J.B. and Bickhart D.M. (2017) *Genet. Sel. Evol.* **49**:32.

Zhou X. and Stephens M. (2012) *Nature Genetics.* **44**: 82.

Proc. Assoc. Advmt. Anim. Breed. Genet. 23:75-78

GENETIC DIVERSITY IN AUSTRALIAN ANGUS BEEF CATTLE

S.A. Clark[1], T. Granleese[1] and P.F. Parnell[2]

[1]School of Environmental and Rural Science, University of New England, Armidale, Australia
[2]Angus Australia, 86 Glen Innes Road, Armidale, New South Wales, Australia

SUMMARY

This study examined trends in the genetic diversity in the Australian Angus cattle population through the calculation of inbreeding, effective population size, effective number of ancestors and effective number of founders over time. The effective population size ranged from 68 to 122 depending on the assumed generation interval. For animals born in 2018, 10 key ancestors explained ~42% of the genetic diversity within the population. Knowledge of overall genetic diversity will help manage the population to maintain long-term rates of genetic gain.

INTRODUCTION

The practice of selection in livestock breeding programs has been shown to lead to increases in inbreeding over time. This has become particularly evident in populations where there is widespread use of artificial breeding technologies (Bijma 2000). Inbreeding is essentially an increase in the number of homozygous individuals within a population. With this increase in homozygosity (or the subsequent reduction is heterozygosity) genetic variation is reduced, which can cause a depression in fitness (inbreeding depression) and a decrease in future selection response (Falconer and Mackay 1996). Selection based on estimated breeding values (EBVs) that incorporate family information (genomic or pedigree based) can lead to increased rates of inbreeding due to the high correlation between EBVs within family, especially when animals are selected at a young age and EBVs are based on ancestral information.

A number of measures have been used to describe genetic diversity in a selected population, including the rate of inbreeding, the effective number of founder individuals, ancestral contributions and effective population size (Boichard *et al.* 1997). Such measures can give a useful insight into whether potential reductions in future response to selection may be expected. Knowledge about the ancestral make up of a population can also have important application in genomic selection where key ancestors are ideal candidates for genotyping at higher marker densities or whole genome sequencing.

The Australian Angus cattle population has achieved substantial genetic progress in the last several decades (Parnell 2015). This genetic progress may have impacted the amount of genetic diversity within the population. The aim of this study was to examine the past and current genetic diversity present in the Angus Australia population.

MATERIALS AND METHODS

This study used data provided by the Angus Society of Australia. The analysis focused on pedigree information on animals born between 1990 to 2018. In total, the pedigree data consisted of 1,551,078 animals, including 42,476 unique sires and 447,000 unique dams.

Measures of diversity. *Inbreeding.* Inbreeding was estimated using the algorithm suggested by Meuwissen and Luo (1992) for the entire pedigree. The rate of inbreeding per year was estimated as the regression of year on inbreeding. As stated in Falconer and Mackay (1997) the effective population size (Ne) is a function of the rate of inbreeding (ΔF) observed per generation. Therefore, the rate of inbreeding per year was estimated for assumed generation intervals ranging from 5 years to 9 years. Effective population size can be described as:

$$Ne = \frac{1}{2\Delta F}$$

Effective number of ancestors. The effective number of ancestors (f_a) accounts for bottlenecks since the population formation, adjusting for losses of allelic diversity since the founder generation. It is estimated by:

$$f_a = \left[\sum_{i=1}^{N_t} p_i^{\,2}\right]^{-1}$$

where p_i is the marginal genetic contribution of ancestor i as defined by Boichard *et al.*, (1997). The marginal contribution was generated for a given number of ancestors such that the upper and lower limits to the effective number of ancestors were zero (N_t=1000).

Effective number of founders. The effective number of founders (f_e) is an alternative measure to estimating the total number of ancestors in the population, accounting for the fact that some ancestors contributed more descendants than others. It is calculated as the number of equally contributing founders it would take to achieve a similar amount of genetic diversity observed in the current population, i.e.

$$f_e = \left[\sum_{i=1}^{N_t} q^2\right]^{-1}$$

where q is the genetic contribution of founder i as defined by Lacey (1989)).

As noted by Sorensen *et al.* (2005), the effective number of founders is a useful historical observation of changes in population structure. It can be used in conjunction with the effective number of ancestors such that if the ratio between the two measures is less than 1 then some bottlenecks have occurred since the foundation generation in the population.

RESULTS AND DISCUSSION

Inbreeding. The rate of inbreeding since 1990, shown in Figure 1, was estimated as 0.0082 per year. The total inbreeding level was on average of 0.03 in 2018. Inbreeding was steadily accumulating until 2011, after which it has remained steady or slightly reduced. The reduction in inbreeding is most likely a reaction from breeders to greater efforts to utilise "outcross" genetics, partially in response to avoidance of sires known to be carriers of recessive genetic disorders identified in the Angus population (Beever 2009).

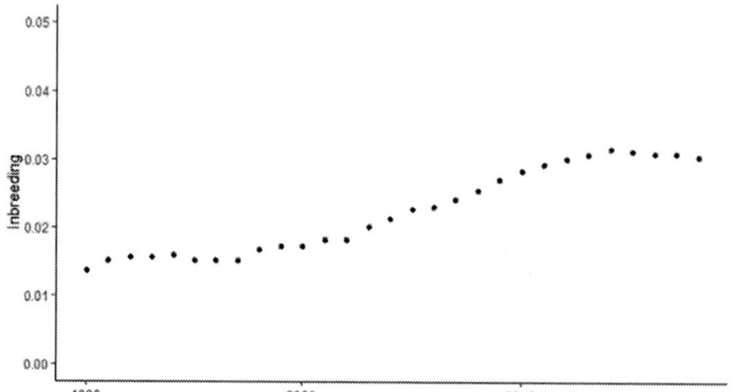

Figure 1. Average inbreeding since 1990 in the Angus Australia population

Table 1 shows the rate of inbreeding per generation and the corresponding effective population size for different assumed generation intervals. The effective population size is higher than those estimated in dairy populations, were values of below 50 are regularly observed (Sorensen *et al.,* 2005). It is often recommended in animal breeding that it is important to maintain an effective population size of at least 50 to 100 (Bijma 2000). Such values have been derived from theoretical expectations, where natural selection counteracts inbreeding depression. Although this is usually not the case in livestock breeding it gives a useful guide for the management of diversity. The maintenance of the current level of diversity will ensure that long-term response to selection can be maintained.

It is likely that the estimates of Ne presented are an overestimate of true genetic diversity, given that the pedigree of the population is relative to a given base. Although, recent estimates of N_e from genomic data (N_e=93) (results not shown) agree with the current estimates from pedigree data.

Table 1. The rate of inbreeding and effective population for alternative generation intervals

	Assumed Generation Interval				
	5	6	7	8	9
ΔF	0.0041	0.0049	0.0057	0.0067	0.0074
Ne	122	102	87	76	68

The effective number of founders (f_e) and ancestors (f_a) rapidly declined until 2008, where both measures plateau (Figure 2). In 2018, the ratio between f_e and f_a was 0.32 indicating that a genetic bottleneck has occurred since the founder generation as a result of selection applied in the population.

Individual marginal contributions of founders to the population were required for the estimation of the effective number of ancestors. This gave the opportunity to observe the importance of key ancestors to the population. Table 2 shows the top 10 ancestors based on their marginal contribution to the population. The top sire explained ~12% of the genetic contributions to the population. The top 10 ancestors collectively accounted for 42% of the total genetic diversity, with the top 50 ancestors explaining 70% (results not shown).

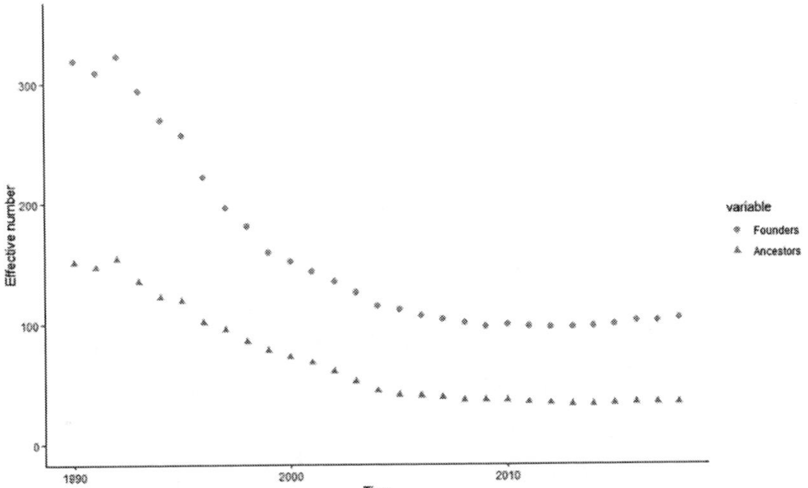

Figure 2. Effective number of founders and ancestors since 1990

Table 2. Summary of marginal contributions for the top 10 individuals

Sire	Birth Year	Total	Marginal	Cumulative	Offspring
1	1990	0.117	0.117	0.117	7772
2	1995	0.0619	0.0619	0.1789	9862
3	1982	0.0609	0.0609	0.2398	2969
4	1978	0.0548	0.051	0.2908	1272
5	1986	0.0322	0.0322	0.323	1265
6	2006	0.0459	0.027	0.35	5356
7	1988	0.0803	0.0204	0.3704	3708
8	1980	0.0242	0.0196	0.39	117
9	1990	0.0169	0.0169	0.4069	3774
10	1992	0.0205	0.0153	0.4222	703

CONCLUSIONS

This study shows that while the diversity of the Angus cattle population in Australia reduced until 2008, the amount of diversity has been maintained since this time. The Angus population has been founded by a relatively small number of ancestors, with the top 10 ancestors explaining 42% of the genetic diversity. Current levels of diversity need to be maintained to ensure losses in response due to inbreeding are not observed.

REFERENCES

Beever J. (2009) An update on neuropathic hydrocephalus in cattle.

Bijma P. (2000) Ph.D. *Thesis*, Wageningen University, the Netherlands.

Boichard D., Maignel L. and Verrier E. (1997) *Gen. Sel. Evol.* **29**: 5.

Falconer D.S. and MacKay T.F.C. (1996) *Harlow*: Longman Scientific & Technical, 4.

Lacy R.C. (1989) *Zoo Biology* **8**: 111.

Meuwissen T.H.E. and Luo Z. (1992) *Gen. Sel. Evol.* **24**: 305.

Parnell P. (2015) *Proc. Assoc. Advmt. Anim. Breed. Genet.* **21**: 221.

Sørensen A,C., Sørensen M,K., and Berg P. (2005) *J. Dairy Sci.* **88**: 1865.

Proc. Assoc. Advmt. Anim. Breed. Genet. 23:79-82

VALIDATION OF SINGLE STEP GENOMIC BEST LINEAR UNBIASED PREDICTION IN BEEF CATTLE

M.G. Jeyaruban, P.M. Gurman, D.J. Johnston, A.A. Swan, R.G. Banks and C.J. Girard

Animal Genetics & Breeding Unit*, University of New England, Armidale, NSW, 2351 Australia

SUMMARY

This study investigated the accuracy of predicting future phenotypes of young Angus and Hereford cattle using Single-step Genomic BLUP (SSGBLUP) compared to the traditional pedigree-based BLUP evaluation (NRMBLUP). Forward cross-validation, using two comparison methods, was used to quantify the predictability of the two evaluations. For each breed, two data sets named 'full' and 'partial' were generated. The 'full' data set included all relationships, all genotypes and phenotypes of animals born up to November 2018. For 'partial' data sets, phenotypes of animals born after December 2014 were removed and the data for animals removed after December 2014 were used as the 'validation data set'. SSGBLUP and NRMBLUP analyses were performed separately for the full and partial data sets and EBVs were predicted for animals in the validation data set. In Method 1, R squared values (R^2), regression coefficients (REG) and adjusted correlation (ACOR), between pre-corrected phenotypes and predicted EBVs were compared. In Method 2, correlation ratios between EBVs from full and partial evaluations were estimated to calculate the increase in predictability between the SSGBLUP and NRMBLUP. The estimated R^2, REG and ACOR using SSGBLUP were higher than those from NRMBLUP. A similar pattern was observed for correlation ratios from Method 2. The increase in ability to predict future phenotypes using Method 1 ranged from 30 to 50% and 10 to 36% for genotyped and 2 to 4% and 1 to 2 % for non-genotyped Angus and Hereford cattle, respectively. Using Method 2, the ability to predict future phenotypes ranged from 22 to 40% and 6 to 28% for genotyped and 1 to 2% and 0.5 to 1 % for non-genotyped Angus and Hereford cattle in the validation set, respectively. This study showed that there was an increase in the accuracy to predict future performance from SSGBLUP compared to NRMBLUP in Angus and Hereford cattle. The increase in predictive ability varied according to the heritability of a trait, the number of phenotypes and genotypes included in the evaluation and whether the animals were genotyped or not in the evaluation.

INTRODUCTION

BREEDPLAN analytical software developed by the Animal Genetics and Breeding Unit (AGBU) is used for genetic evaluation of beef cattle using best linear unbiased prediction (BLUP) (Graser *et al.* 2005). Prior to 2012, EBVs were predicted using pedigree based BLUP models (NRMBLUP). Since 2012, the BREEDPLAN software has been upgraded to include a range of DNA marker-based predictions. With the development of 50K micro arrays in 2008, genome wide SNP based prediction called Molecular Breeding Values 'MBVs' were included using a post-BLUP blending method. This meant that genotype information did not influence EBVs of pedigree-only animals. Furthermore, blending of MBVs into existing EBVs is sensitive to various biases which can be complicated to eliminate. These biases are mostly overcome by implementing Single-step Genomic BLUP (SSGBLUP). In SSGBLUP, information from pedigree, phenotypes and genotypes are jointly used. SSGBLUP combines the genomic relationship matrix (G) for genotyped animals with the pedigree-based relationship (A) for non-genotyped animals (Christensen and Lund 2010). Therefore, SSGBLUP is expect to produce more accurate EBVs for animals with genotypes than NRMBLUP.

Since 2017, SSGBLUP has been implemented for the genetic evaluation and use in Angus, Brahman,

Hereford and Wagyu breeds in Australia (Johnston *et al.* 2018). An important implementation step is to quantify the extent of increase in predictability of SSGBLUP over NRMBLUP. A forward cross validation method proposed by Legarra and Reverter (2018) was used in this study to compare the predictability of SSGBLUP and NRMBLUP. Predictability is defined as how well the EBVs predict observed performance.

MATERIALS AND METHODS

Data used in this study were submitted by Angus and Hereford breeders and their breed societies for use in the November 2018 BREEDPLAN evaluation. Data included 600 day weight (FWT), scan eye muscle area in heifers (HEMA) and bulls (BEMA), and scrotal circumference (SC). Univariate analyses were performed for each trait using models described by Graser *et al.* (2005). Table 1 summarises the number of animals with phenotypes and genotypes for each trait across the two breeds.

Forward cross-validation described by Legarra and Reverter (2018) was used to compare the predictability of SSGBLUP and NRMBLUP. For each breed, two data sets named 'full' and 'partial' were generated. The Full data set included all relationships, genotypes and phenotypes of animals born up to November 2018. For the 'partial' data set, phenotypes of animals born after December 2014 were removed and the data for animals removed were used as the 'validation data set'. The SSGBLUP and NRMBLUP analysis were performed separately for the full and partial data sets and EBVs were predicted for animals in the validation data set. A strict criteria was implemented to ensure good convergence.

Two approaches were used to assess the ability to predict the future phenotypes in the validation data set using EBVs estimated from the partial data. In approach 1, adjusted phenotypes in the 'validation set' were regressed against the EBVs from partial analyses of SSGBLUP (SEBVp) and NRMBLUP (NEBVp) within their respective contemporary group. R-squared values (R^2) and regression coefficients (REG) were estimated. Accuracy of prediction was calculated as a correlation between adjusted phenotypes and SEBVp or NEBVp and the correlations were adjusted for by dividing by the square root of the heritability (ACOR). The increase in ability to predict future genotypes (PRED1) of young Angus and Hereford cattle was assessed as a ratio between ACOR of SSGBLUP and NRMBLUP.

In approach 2, the Pearson correlations between EBVs using full ($\hat{U}_f$) and partial ($\hat{U}_p$) for animals in the validation data set were computed as per the formula given below from Legarra and Reverter (2018),

$$\hat{\rho}_{f,p} = \frac{\frac{1}{n}\left(\hat{U}_p - \bar{\hat{U}}_p\right)'\left(\hat{U}_f - \bar{\hat{U}}_f\right)}{\sqrt{\frac{1}{n}\left(\hat{U}_f - \bar{\hat{U}}_f\right)'\left(\hat{U}_f - \bar{\hat{U}}_f\right)\frac{1}{n}\left(\hat{U}_p - \bar{\hat{U}}_p\right)'\left(\hat{U}_p - \bar{\hat{U}}_p\right)}}$$

Where n is the number of animals in validation set, $\hat{U}_f$ are the full EBVs, $\bar{\hat{U}}_f$ the mean of the full EBVs, $\hat{U}_p$ are the partial EBVs, $\bar{\hat{U}}_p$ the mean of the partial EBVs. Legarra and Reverter (2018) showed that $\hat{\rho}_{f,p}$ was equal to the ratio of accuracy of partial (acc_p) and accuracy of full (acc_f) of SSGBLUP or NRMBLUP. This was modified to get the increase in predictive ability (PRED2) of SSGBLUP by calculating the ratio between acc_p of SSGBLUP and acc_p of NRMBLUP as per the equation given below,

$$PRED2 = ((corr\ (SEBV_p, SEBV_f) / corr\ (NEBV_p, SEBV_f)) - 1)*100$$

RESULTS AND DISCUSSION

The data used in the methods is summarised in Table 1. In addition to the number of records given in Table 1, Angus and Hereford had 55999 and 10,971 genotyped animals, respectively, in the full and partial analyses. The number of animals with phenotypes and genotypes in the validation data for each trait ranged from 11,455 to 14,162 for Angus and 1,507 and 3,908 for Hereford. Heritabilities used in the prediction for FWT, HEMA, BEMA and SC for Angus were 0.38, 0.26, 0.24 and 0.39, respectively and for Hereford were 0.31, 0.24, 0.23 and 0.44, respectively.

Table 1. Summary of data used in the prediction

Trait	Angus				Hereford			
	Number of records		[1] Number in validation set		Number of records		Number in validation set	
	Full	Partial	Geno	Non	Full	Partial	Geno	Non
FWT	801,991	673,969	14,162	100,076	514,345	464,703	3,569	40,959
HEMA	368,832	289,344	11,455	68,033	128,810	104,557	1,507	22,746
BEMA	406,378	316,707	13,546	76,125	177,311	148,585	3,908	24,818
SC	335,437	256,152	12,404	66,881	133,276	108,026	3,432	21,818

[1] 'Geno': genotyped animals; 'Non': non-genotyped animals.

Genotyped animals. For genotyped animals in the validation set, estimated R^2, REG, ACOR and PRED1 from Method 1 and the PRED2 from Method 2 are given in Table 2. Using Method 1 for Angus, estimated R^2 values ranged from 0.11 to 0.22 for SSGBLUP and from 0.06 to 0.12 for NRMBLUP. Estimated R^2 values were higher for SSGBLUP than NRMBLUP for all traits. The estimated REG using SSGBLUP were also higher than those using NRMBLUP. However, the estimated REG was higher than 1 for SSGBLUP indicating that EBVs were under-predicted for SSGBLUP. The ACOR ranged from 0.67 to 0.79 for SSGBLUP and 0.48 to 0.59 using NRMBLUP. Adjusted correlations were higher for SSGBLUP than for NRMBLUP for all traits. The PRED1 ranged from 30 to 53%.

For Hereford, estimated R^2 values ranged from 0.08 to 0.17 for SSGBLUP and from 0.05 to 0.13 for NRMBLUP. As observed for Angus, estimated R^2 values were higher for SSGBLUP than NRMBLUP for all traits. Estimated REG using SSGBLUP were also higher than those using NRMBLUP. The ACOR ranged from 0.56 to 0.62 for SSGBLUP and 0.41 to 0.54 using NRMBLUP. The ACOR were higher for SSGBLUP than for NRMBLUP for all traits. PRED1 ranged from 10 to 36%.

Using Method 2 for Angus, PRED2 ranged from 23 to 50%, respectively. For Hereford PRED2 ranged from 6 and 28%, respectively.

Non-genotyped animals. Using Method 1 for Angus, changes in the estimated R^2, REG, ACOR and PRED1 between SSGBLUP and NRMBLUP were similar to those observed for genotyped animals. However, increases were lower than the values observed for genotyped animals, with results for PRED1 ranging from 3 to 6%. A similar pattern was observed for Hereford where PRED1 ranged from 1to 3%.

Using Method 2 for Angus, similar to genotyped animals, the predictability of SSGBLUP was higher than for NRMBLUP for all traits. The PRED2 ranged from 2 to 5%. For Hereford, PRED2 ranged from 1 to 2%.

Results for both procedures showed higher predictability for SSGBLUP as compared to NRMBLUP. However, estimated regression slopes greater than one indicate that cross-validation using Method 1 may be biased due to errors in adjusting the fixed effects, selection and the heritability used in the evaluation (Legarra and Reverter 2018). As expected, the advantage in predictability of both procedures using SSGBLUP (compared to NRMBLUP) was higher for genotyped animals than non-genotyped animals. Furthermore, Angus, with a higher number of phenotypes and genotypes animals gave higher PRED1 and PRED2 for all traits than in Hereford. When the genotyped and non-genotyped animals were combined, the increase in predictability estimated for SSGBLUP in this study was lower than the range (25 to 36%) published by Lourenco et al (2018) for Angus cattle in USA. Lourenco et al (2018) had more animals with records and genotypes than the numbers available in this study.

Table 2. Estimated R squared (R^2), regression coefficient (REG) and adjusted correlations (ACOR) from Method 1and increase in predictability from Method 1 (PRED 1 %) and Method 2 (PRED 2 %) for SSGBLUP over NRMBLUP for genotyped animals

Trait	Method 1						Methods	
	SSGBLUP			NRMBLUP			1	2
	R^2	REG	ACOR	R^2	REG	ACOR	PRED1	PRED2
				Angus				
FWT	0.22	1.16±0.02	0.79	0.12	1.07±0.02	0.58	36	28
HEMA	0.15	1.07±0.02	0.77	0.09	1.08±0.03	0.59	30	23
BEMA	0.11	1.04±0.03	0.67	0.06	0.93±0.03	0.48	39	26
SC	0.22	1.22±0.02	0.75	0.09	1.09±0.03	0.49	53	50
				Hereford				
FWT	0.10	1.11±0.05	0.56	0.05	0.93±0.06	0.41	36	28
HEMA	0.08	0.99±0.09	0.56	0.06	0.92±0.09	0.51	10	6
BEMA	0.08	1.17±0.06	0.61	0.06	1.05±0.06	0.51	19	13
SC	0.17	1.07±0.04	0.62	0.13	0.99±0.04	0.54	15	12

CONCLUSIONS

Ability to predict the future phenotypes of both genotyped and non-genotyped animals was higher for SSGBLUP compared to NRMBLUP. Both methods of comparisons yielded very similar results. Furthermore, ability to predict the future phenotypes was influenced by the number of genotyped animals in the evaluation and the heritability of the trait used. Higher numbers of genotyped animals and higher heritability resulted in increased predictability for SSGBLUP.

ACKNOWLEDGEMENT

The authors would like to thank the Meat and Livestock Australia (MLA) for their financial support through L.GEN.0174 and Angus Australia, Hereford Australia and Agricultural Business Research Institute, UNE for providing data for this study.

REFERENCES

Christensen O.F. and Lund M.S. (2010). *Genet. Sel. Evol.* **42:** 2.

Graser H-U., Tier B., Johnston D.J. and Barwick S.A. (2005) *Aust. J. Exp. Agric.* **45**:913.

Johnston D.J, Ferdosi M.H, Connors N.K, Boerner V, Cook J, Girard C.J, Swan A.A, Tier B. (2018). In *Proc. 11th World Cong. Genet. Appl. Livest. Prod.* **11**:.269.

Legarra A. and Reverter A. (2017). *Proc. Assoc. Advmt. Anim. Breed. Genet.* **22**:73

Legarra A. and Reverter A. (2018) *Genet. Sel. Evol.* **50:** 53.

Lourenco D.A.L., Tsuruta S., Fragomeni B.O., Masuda Y., Aguilar I., Legarra A., Millier S., Moser D. and Misztal I. (2018). In *Proc. 11th World Cong. Genet. Appl. Livest. Prod.* **11:** 495.

Proc. Assoc. Advmt. Anim. Breed. Genet. 23:83-86

FEASIBILITY OF USING IMAGING CARCASS TRAITS IN GENETIC EVALUATION FOR AUSTRALIAN WAGYU BEEF CATTLE

Y. Zhang and R.G. Banks

Animal Genetics & Breeding Unit*, University of New England, Armidale, NSW, 2351 Australia

SUMMARY

This study estimated genetic parameters for AusMeat and camera image analysis carcass traits. Most carcass traits were moderately to highly heritable. The genetic correlation between AusMeat marble score and the image analysis marbling percentage traits was close to unity, as was the estimate between the two eye muscle area traits. Accuracies of genomic breeding values from single step genomic BLUP (ssGBLUP) were up to 4% higher than those from pedigree based BLUP (PBLUP) evaluations. The highest increase in EBV accuracies from ssGBLUP over those from PBLUP was for animals with a genotype but no phenotype. The use of image carcass traits for selection is feasible for genetic evaluation.

INTRODUCTION

Wagyu is a collective term for Japanese beef cattle breeds (Japanese Black, Japanese Brown, Japanese Shorthorn and Japanese Polled). Australian Wagyu production started in the 1990s. Genetic analysis of Wagyu cattle has been reported in a number of studies, with most from Japan and the USA. As summarized by Oyama (2011), heritability estimates for carcass traits were moderate to high, for instance, 0.23 to 0.78 for carcass weight, 0.28 to 0.61 for rib eye area, 0.24 to 0.50 for fatness and 0.16 to 0.74 for marble score. Recently applied imaging technology for assessing carcass characteristics has the potential to accurately and objectively capture carcass characteristics. Those carcass image analysis traits have been tested in Australian Wagyu cattle on a small scale (Maeda *et al.* 2014). Application of genomic selection in livestock could improve the accuracy of selection and enhance genetic gain. The aims of this study were 1) to estimate genetic parameters for carcass AUSMEAT and image analysis traits Australian Wagyu cattle, in AusMeat and image analysis traits, and 2) to test the accuracies of the Estimated Breeding Values (EBV).

MATERIALS AND METHODS

Phenotypes. Phenotypes were extracted from the Australia Wagyu Association BREEDPLAN database (Aug 2018). Animals used were progeny of 462 sires, with the number of progeny per sire ranging from 1 to 271. Amongst sires whose progeny had carcass records, 168 had only one progeny (3% of total carcass records), 207 sires had more than 5 progeny and 12 sires had more than 100 progeny. The average number of progeny per sire was 12. After editing, 6068 carcass records were used in the analysis. Carcass traits were measured using the AusMeat grading system (AusMeat Limited 2005), including hot carcass weight (CWT, kg), marble score (CMAU, on a scale of 0 to 12), P8 fat (CP8, mm) depth and carcass eye muscle area (CEMA, cm^2). Image analysis traits were obtained in two steps, 1) colour images of carcass cross-sections between the 5th and 6th ribs were collected using the digital camera (HK-333, Hayasaka Rikoh, Sapporo Japan, as described by Kuchida *et al.* 2001); 2) images were analysed using the image analysis software, BeefAnalyserII (Hayasaka Rikoh, Sapporo Japan) to generate carcass traits. Details of the processes have been reported previously (Maeda *et al.* 2014). The traits generated by the image analysis software are marbling percentage (CCMP, %), eye muscle area (CCRA, cm^2), fineness index or fine marbling particles per cm^2 (CCFI, count/cm^2),

* A joint venture of NSW Department of Primary Industries and the University of New England

percentages of coarse fat particles (all > 1 pixel, CCCI; the 5 largest, CCCJ; the 10 largest, CCCK, or the largest one, CCMX, %), number of fat particles (CCNM, count/cm^2) and brightness of the eye muscle area (CCLL).

Genotype and genomic relationship matrix. Animals were genotyped for various sizes of Illumina Bovine chips (Illumina Inc., San Diego, CA, USA), ranging from 6K to 800K, with most genotyped with 50K or 150K panels. Genomic data were subjected to quality control (Connors *et al.* 2017) and imputed to 150K using Fimpute 2.2 (Sargolzaei *et al.* 2014). SNP genotypes for 12956 animals were used to calculate the genomic relationship matrix $\boldsymbol{G}_m$ (VanRaden 2008)967 bulls and 50,000 markers distributed randomly across 30 chromosomes. Estimation of genomic inbreeding coefficients required accurate estimates of allele frequencies in the base population. Linear model predictions of breeding values were computed by 3 equivalent methods: 1. The numerical relationship matrix $\boldsymbol{H}$, that combines the pedigree relationship matrix, $\boldsymbol{A}$, and a modified genomic relationship matrix, $\boldsymbol{G}$, was used in ssGBLUP analyses. $\boldsymbol{G}$ was manipulated as, $\boldsymbol{G} = \boldsymbol{\lambda G}_m + (\mathbf{1} - \boldsymbol{\lambda})\boldsymbol{A}_{22}$, where $\boldsymbol{\lambda}$ is the fraction of additive genetic variance explained by markers, ranging between 0 and 1.

Statistical Models. Data were analysed using an animal model fitted with fixed effects and covariates to estimate breeding values, genetic variances and heritabilities. For CWT, the fixed effects were contemporary group (defined by herd, original owner, sex, management group and killing dates), and age (days) as linear and quadratic regressors. For other carcass traits linear and quadratic forms of carcass weight (kg) were fitted instead of age. The same model was implemented for each trait in PBLUP and ssGBLUP using Wombat (Meyer 2007). ssGBLUP analyses were performed with the $\boldsymbol{H}^{-1}$ matrix calculated for four levels of λ: 0.25, 0.50, 0.75 and 0.95 to identify the optimal λ. The prediction accuracies of EBVs were calculated as $Acc = \sqrt{1 - \left(\frac{PEV}{(1+f)\sigma_a^2}\right)}$, where PEV is the prediction error variance of the EBV, f is the inbreeding coefficient and σ_a^2 is the additive genetic variance. The average EBV accuracies were calculated for all animals or subsets of animals in each of PBLUP or ssGBLUP analyses. The EBV accuracies were compared amongst subsets of animals which were identified as animals that were phenotyped, genotyped or both. The comparisons were conducted by firstly, identifying the highest average EBV accuracy from the 4 ssGBLUP analyses for each trait. The difference between this accuracy and the PBLUP accuracy was obtained for this subset of animals.

Bivariate analyses were performed for pairs of marbling traits (CCMP, CCFI or CCCI vs CMAU) and eye muscle area traits (CCRA vs CEMA).

RESULTS AND DISCUSSION

Trait summary. The average slaughter age was 1003 days and the average carcass weight was 419 kg with an AusMeat marble score of 7.2 and an eye muscle area of 66 cm^2 (Table 1). Image carcass traits showed the same eye muscle area with higher variation. An average of 27% of the eye muscle was intramuscular fat (CCMP), 7.3% was coarse fat flecks (CCCI) and 2.7 was fine fat particles per cm^2 of rib eye area (CCFI).

Variance and genetic parameters. Heritability estimates for most traits were moderate, with relatively low standard errors (Table 2). Heritability estimates for AusMeat traits ranged from 0.42 for CEMA to 0.60 for CWT. This is in line with estimates reported previously. For image analysis traits, CCMP, CCFI and CCRA were moderately heritable and those estimates tended to be significantly different from zero. The heritability estimates for coarseness (CCCI, CCCJ, CCCK) were moderate but with large standard errors. The brightness of eye muscle (CCLL) was also moderately heritable. The relative proportion of the largest marbling particle (CCMX) had low heritability which was not significantly different from zero. The current estimates were similar to those by Maeda *et al.* (2014), but with lower standard errors. Heritabilities from ssGBLUP (not shown) for different λ values varied. The estimates at a λ of 0.25 were the highest and were higher than those estimated from PBLUP.

EBV Accuracies from different analyses. The EBV accuracies from PBLUP and ssGBLUP at 4 levels of λ: 0.25, 0.50, 0.75 and 0.95, are shown in Table 3. The highest EBV accuracies from ssGBLUP were higher than those from PBLUP. Accuracy increases ranged from 0.02 for CCMP to 0.04 for CCMX. The maximum eye muscle area (CEMA or CCRA) EBV accuracies from ssGBLUP were the same or slightly lower than those from PBLUP. The highest ssGBLUP EBV accuracies were found at a λ of 0.95 and mostly for the fatness traits. The highest ssGBLUP EBV accuracy for CCLL was identified at a λ of 0.25, which was slightly lower than that from PBLUP (-0.02).

The highest increase in EBV accuracies was 0.04 and found in the subset of genotyped animals, either with (0.03) or without (0.04) phenotypes. For non-genotyped animals, the maximum EBV accuracies from ssGBLUP were almost identical to those from PBLUP (with an average increase of 0.01).

Table 1. Descriptive statistics for carcass traits of Australian Wagyu cattle

Trait	Count					Trait			
	Animal	Sire	Dam	cg	Genotype	Mean	Std	Min	Max
CWT	6068	462	4007	1543	1380	418.7	54.28	256	580
CMAU	5634	422	3744	1296	1368	7.23	1.77	2	12
CP8	3496	242	2169	851	1100	22.04	8.34	4	46
CEMA	3374	305	2374	993	1116	65.83	23.17	41	128
CCMP	2109	281	1867	587	727	27.02	7.31	9	49
CCRA	1849	250	1700	386	750	65.92	26.51	0	129
CCFI	1942	263	1743	515	692	2.70	0.60	0	4
CCCI	1838	247	1689	383	750	7.26	5.52	0	42
CCCJ	608	88	601	113	45	4.66	3.02	0	25
CCCK	608	88	601	113	45	5.68	3.42	0	26
CCLL	1113	156	1001	246	656	79.57	10.89	0	118
CCMX	1766	233	1633	371	728	2.47	2.19	0	24
CCNM	608	88	601	113	45	1099.10	688.95	49	4389

Table 2. Variance components and heritability for carcass traits of Australian Wagyu cattle

Trait	σ_a^2±se	σ_e^2±se	h^2±se
CWT	832.64±105.65	552.49±72.55	0.60±0.06
CMAU	1.01±0.15	1.38±0.11	0.42±0.05
CP8	17.95±3.09	17.6±2.11	0.51±0.07
CEMA	32.09±6.45	23.32±4.44	0.58±0.09
CCMP	11.97±3.66	20.98±2.94	0.36±0.10
CCRA	36.04±9.31	38.50±7.25	0.48±0.11
CCFI	0.10±0.03	0.16±0.02	0.39±0.11
CCCI	1.39±0.71	8.92±0.67	0.14±0.07
CCCJ	1.97±1.61	4.48±1.43	0.30±0.24
CCCK	2.57±1.95	5.27±1.73	0.33±0.24
CCLL	24.14±9.23	32.98±7.07	0.42±0.14
CCMX	0.11±0.16	3.65±0.20	0.03±0.04

Bivariate analyses. The genetic correlations between eye muscle area (CCRA vs CEMA) and marbling traits (CMAU vs CCMP) were close to unity (not shown). Genetic correlations among the three measures of fat particle coarseness (CCCI, CCCJ and CCCK) were high and positive, suggesting they were measurements of the same trait. The fat fineness index (CCFI) was negatively correlated with the coarseness measures CCCJ and CCCK, but not with CCCI.

Phenotypically, the two measurements of marbling were highly correlated, but were on different scales. An increase of one unit of AusMeat marble score (CMAU) is equivalent to a 3.5% increase in marbling percentage (CCMP). Two eye muscle area measures were almost identical (1.0 cm^2 of CEAM is equivalent to 0.99 cm^2 of CCRA). All coarseness indices represent essentially the same marbling trait, being similarly correlated to the marbling percentage (CCMP) trait. The regression of CCMP on CCFI showed that increasing fine fat particle per cm^2 of eye-muscle area by 1.0 increases CCMP by 9.8%.

Table 3. Accuracies of EBVs from PBLUP (Ped) analyses and ssGBLUP analyses with 4 levels of λ, 0.25, 0.50, 0.75 and 0.95 (H25, H50, H75 and H95)

Trait	Ped	H25	H50	H75	H95	Increment
CWT	0.53	0.53	0.53	0.55	0.56	0.03
CMAU	0.50	0.50	0.50	0.51	0.52	0.03
CP8	0.48	0.48	0.48	0.49	0.51	0.03
CEMA	0.49	0.48	0.47	0.46	0.47	0.00
CCMP	0.44	0.44	0.43	0.44	0.45	0.02
CCRA	0.45	0.45	0.44	0.44	0.44	0.00
CCFI	0.42	0.41	0.40	0.41	0.42	0.00
CCCI	0.36	0.37	0.32	0.39	0.41	0.05
CCCJ	0.29	0.27	0.27	0.29	0.31	0.02
CCCK	0.30	0.28	0.27	0.29	0.31	0.02
CCLL	0.42	0.40	0.36	0.36	0.37	-0.02
CCMX	0.22	0.21	0.21	0.23	0.26	0.04

CONCLUSIONS

Most of the AusMeat and image analysis carcass traits were moderately to highly heritable with moderate standard errors. The genetic correlation between AusMeat marble score and the image analysis marbling percentage was close to unity. A similarly high genetic correlation was estimated between the eye muscle area traits. Image analysis of carcass characters is feasible for use during selection in Australian Wagyu cattle. Accuracies of genomic breeding values at optimal levels of λ could be increased by 4% across traits. For traits from reasonably sized datasets, an increase of 6% in EBV accuracies could be achievable.

ACKNOWLEDGEMENTS

We gratefully acknowledge the Australian Wagyu Association (AWA, http://www.wagyu.org.au/) for making the data available and Gilbert Jeyaruban and Christian Girard for helping in data preparations.

REFERENCES

AusMeat Limited. (2005) Handbook of Australian Meat. 7th ed. (AusMeat: Brisbane).

Connors N.K., Cook J., Girard C.J., Tier B., Gore K.P., Johnston D.J. and Ferdosi M.H. (2017) *Proc. Assoc. Advmt. Anim. Breed. Genet.* **22**:317.

Kuchida K., Suzuki M. and Miyoshi S. (2001) *Anim. Sci. J.* **72**:J224.

Maeda S., Grose J., Kato K. and Kuchida K. (2014) *Anim. Prod. Sci.* **54**:557.

Meyer K. (2007) *Journal of Zhejiang University. Science. B* **8**:815.

Oyama K. (2011) *Anim. Sci. .J* **82**:367.

Sargolzaei M., Chesnais J.P. and Schenkel F.S. (2014) *BMC Genomics*, **15**:478.

VanRaden P.M. (2008) *J. Dairy Sci.* **91**: 4414.0.

Proc. Assoc. Advmt. Anim. Breed. Genet. 23:87-90

GBLUP ANALYSIS PREDICTED FERTILITY PHENOTYPES OF CROSSBRED BULLS USING DATA FROM BRAHMAN AND TROPICAL COMPOSITE

L.R. Porto-Neto[1], M.R.S. Fortes[2], A. Reverter[1]

[1]CSIRO Agriculture and Food, Queensland Bioscience Precinct South, St Lucia, Qld 4067
[2]University of Queensland, School of Chemistry and Molecular Biosciences, St Lucia, Qld 4072

SUMMARY

Bull fertility directly impacts the outcome of cow herds under a natural mating system. Blood concentration of the hormone inhibin (INH), measurements of the scrotal circumference (SC18 and SC24) and the percentage of normal sperm (PNS) in an ejaculate are heritable indicators of bull fertility. We analyzed bulls from the CRC for Beef Genetic Technologies consisting of three breed types (Brahman, Tropical Composite, and Crossbreds) to which those four fertility-related traits were observed. We used 9,012 SNP markers to generate a genomic relationship matrix and to run a GBLUP analysis. We adjusted the model for the population substructure using the first two principal components derived from all genotypes. The GBLUP analyses were run twice, one with the whole dataset and another setting the phenotypes of the Crossbred animals to missing. The accuracy and bias of genomic estimated breeding values (GEBV) was estimated using the Method LR. Heritability estimates ranged from 0.17 (PNS) to 0.43 (SC24), and GEBV accuracies from 0.54 (PNS) to 0.81 (SC24). No bias was observed for any trait. Also, there is no evidence of over- or under-dispersion for INH. However, the GEBVs for PNS seems to be over-dispersed, and the ones of SCs (both SC18 and SC24) seem to be under-dispersed. The use of large enough multi-breed reference populations can lead to accurate GEBV for bull fertility traits.

INTRODUCTION

The vast majority of Australian beef cows are bull mated, especially in the north where artificial insemination is virtually inexistent. Therefore, the bull's ability to reach puberty, produce good quality sperm and effectiveness in serving cows are of fundamental importance with a direct impact on herd productivity. There are several indicators of bull fertility that are polygenic and heritable traits (Corbet *et al.* 2013). Serum levels of Inhibin (INH) measured at approx. four months of age is an early indicator of puberty (Burns *et al.* 2013). Scrotal circumference (SC) is related to bull fertility and correlated to heifer puberty (Fortes *et al.* 2012, 2013). Percentage of normal sperm (PNS) is an indicator of calf-output (Holroyd *et al.* 2002).

The application of genomic selection approaches for fertility-related traits is of interest. However, the collection of fertility-related phenotypes is expensive and the number of available animals with phenotypes and genotypes of any particular breed is too small to generate accurate estimates of breeding value. Therefore, the use of a multi-breed reference population is a valid alternative approach. The use of multi-breed genomic selection is a current hot topic of research, with some promissing results in hard to measure traits, as female fertility (Hayes *et al.* 2019).

Here we analyzed data on four traits related to bull fertility, and built a multi-breed reference population that included Brahman and Tropical Composite, to estimate GEBVs of crossbred animals. It should be noted that the resutls presented here are part of a work in progress towards a multi-breed evaluation, and are not final.

MATERIALS AND METHODS

Animals and phenotypes. There were 2,979 bulls of three breed types: Brahman, Tropical

Composites and crossbreds. They were the progeny of cows from the Beef CRC Lifetime Performance Population previously described (Barwick *et al.* 2009; Johnston *et al.* 2009). The crossbred bulls were the product of Brahman crosses with Tropical Composites. Four indicators of bull fertility (INH, SC18, SC24 and PNS) were considered; Descriptive statistics in Table 1.

SNP genotypes. Two SNP genotyping arrays were used, the BovineSNP50 (Illumina Inc., San Diego, CA) and the Indicus 74K array (Neogen). Initial quality control (QC) for genotypes were performed within breed and specifically to each SNP chip. After initial QC, SNP were remapped to the new bovine reference genome ARS-UCD1.2. Only SNP that were genotyped in both platforms and had a call rate greater than 95% were kept for analyses (n = 9,012 SNP). This SNP set was distributed across the genome, including the X chromosome.

Statistical Analyses. Principal components analysis on SNP genotypes was conducted using PLINK 1.09 (Chang et al, 2015; www.cog-genomics.org/plink/1.9/). Following recent approaches of multibreed datasets (Hayes *et al.* 2019), our GBLUP was performed using the software Golden Helix, fitting a mixed linear model with cohort (year and contemporary group) as fixed effect, and the covariates of age at measurement and PC1 and PC2 that accounts for the different breed composition. Two GBLUP runs were performed for each trait, one using the full dataset and a second setting the phenotypes of the Crossbred animals as missing. The accuracy, dispersion and bias were calculated using the Method LR (Legarra and Reverter 2018). In brief, bias was computed from the difference between the GEBV using the full data minus the GEBV setting the crossbred data as missing. Dispersion was computed from the slope of the regression of the GEBV using the full data on the GEBV with the crossbred data as missing. Finally, accuracy was computed from the covariance between the two GEBV divided by the genetic variance weighted by the average inbreeding coefficient and the average relationship between individuals.

Table 1. Descriptive statistics of samples* and phenotypes used for analysis**

	PNS			Inhibin			SC18			SC24		
Breed	n	Mean	SD	n	Mean	SD	n	Mean	SD	n	Mean	SD
BRM	1023	0.70	0.22	806	7.41	1.89	1098	26.70	2.71	1098	29.89	2.86
Cross	159	0.60	0.24	161	8.34	2.05	161	30.18	2.96	161	33.07	3.00
TCO	1648	0.72	0.19	1329	7.76	1.88	1719	29.82	2.82	1719	31.43	2.80

*BRM – Brahman, Cross – Crossbred, TCO – Tropical Composite.

** PNS – Percent of normal sperm at 24 month of age, Inhibin – Blood level of inhibin at around 4 months of age, SC18 and SC24 scrotal circumference at 18 and 24 months of age. In table, n is the number of animals, and SD is the standard deviation.

RESULTS AND DISCUSSION

Using the principal components analysis, we captured the expected sub-structure of our population. Three main clusters were observed corresponding to each of the breed types included in the study (Figure 1). Also as expected, the Tropical Composite designation showed the highest variation within each of the breed types. While PC1 captured the differences between the three main populations (Brahman, Tropical Composites and Crossbreeds), it is the combination of PC1 and PC2 that allows the separation of substructures within populations. This is particularly the case for the two sub-populations within the crossbreds (Figure 1).

The estimates of heritability were similar to previously described for Brahman or Tropical Composite (Corbet *et al.* 2013), apart from INH that was lower in both cases (0.42, opposed to 0.72-0.74). SC have higher heritabilities in Brahman (~0.75) compared to Tropical Composite (~0.43), and in this study was 0.42. PNS on the other hand have higher estimates of heritability in Tropical Composite

(0.27) than Brahman (0.15), which was closer to the estimate of this study (0.18). Often it is observed variation in heritability estimates that can be mostly attributed to sample variation. As SC is easy to measure, relatively inexpensive and highly heritable, it is likely this will remain the reference trait for bull fertility. Considering the GEBV, there is no evidence of bias for any of the observed traits. Also, there is no evidence of over- or under-dispersion for INH (Table 2). However, the GEBV for PNS seems to be over-dispersed, and the ones for SC18 and SC24 seem to be under-dispersed. The population accuracies estimated using method LR are strong, especially for SC measurements.

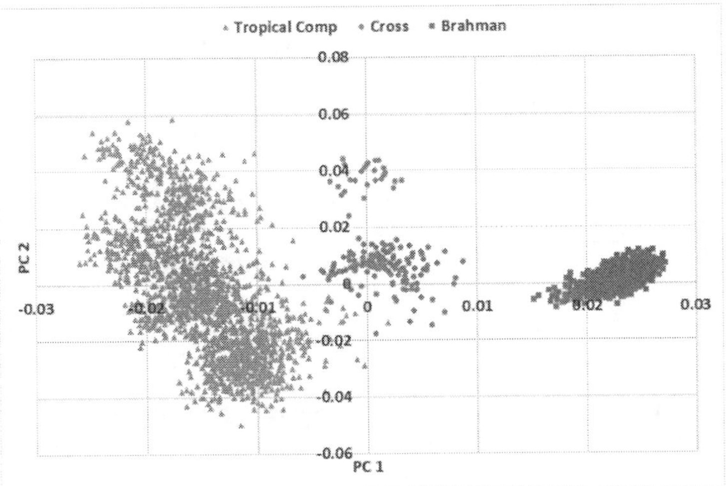

Figure 1. Principal component analysis on SNP genotypes for 2,979 bulls of three breed types: Brahman (blue), Crossbreds (orange) and Tropical Composites (grey)

The correlation between the GEBV estimated using all dataset, including the crossbred data, and those estimated setting the crossbred data to missing values varied between traits (Figure 2), from moderate (0.35) for PNS to high (0.77) for SC18.

Table 2. Estimates of heritability, accuracy, bias and dispersion for GEBV of fertility-related traits in bulls

Trait	Heritability	GEBV accuracy	Bias	Slope
PNS	0.176	0.544	-4.36×10^{-10}	0.970
Inhibin	0.419	0.685	-3.50×10^{-9}	1.006
SC18	0.423	0.799	8.15×10^{-9}	1.033
SC24	0.428	0.811	1.01×10^{-8}	1.018

* PNS – Percent of normal sperm at 24 month of age, Inhibin – Blood level of inhibin at around 4 months of age, SC18 and SC24 – Scrotal circumference at 18 and 24 months of age.

CONCLUSIONS

There are still some improvements that could be done before implementation of multi-breed genomic selection for bull fertility-related traits e.g. better understanding how to model different populations in different environments, and consistency in trait measurement. The lack of bias and the high accuracy of the estimates are encouraging and warrant further research.

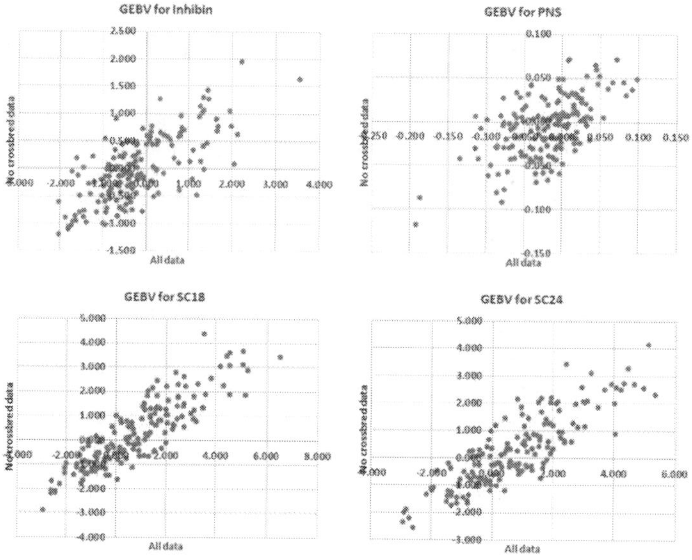

Figure 2. GEBV of Crossbred using all data, including own record (x-axis) and without own records (y-axis)

ACKNOWLEDGEMENTS

This work was performed using the legacy database from the CRC for Beef Genetics Technologies. Two MLA projects provided financial support for genotyping the bulls, they are project B.NBP.0786 and project B.NBP.0604.

REFERENCES

Barwick S.A., Johnston D.J., Burrow H.M., Holroyd R.G., Fordyce G., Wolcott M.L., Sim W.D. and Sullivan M.T. (2009) *Anim. Prod. Sci.* **49**: 367.

Burns B.M., Corbet N.J., Corbet D.H., Crisp J.M., Venus B.K., Johnston D.J., Li Y., McGowan M.R., Holroyd R.G. (2013) *Anim Prod Sci.* **53**: 87.

Chang C.C., Chow C.C., Tellier L.C.A.M., Vattikuti S., Purcell S.M. and Lee J.J. (2015) *GigaScience* **4**.

Corbet N.J., Burns B.M., Johnston D.J., Wolcott M.L., Corbet D.H., Venus B.K., Li Y., McGowan M.R.and Holroyd R.G. (2013) *Anim. Prod. Sci.* **53**: 101.

Fortes M.R.S., Lehnert S.A., Bolormaa S., Reich C., Fordyce G., Corbet N.J., Whan V., Hawken R.J. and Reverter A. (2012) *Anim Prod Sci.* **52**: 143.

Fortes M.R.S., Reverter A., Kelly M., McCulloch R. and Lehnert S.A. (2013) *Androl.* **1**: 644.

Hayes B.J., Corbet N.J., Allen J.M., Laing A.R., Fordyce G., Lyons R., McGowan M.R. and Burns B.M. (2019) *J Anim Sci.* **97**: 55.

Holroyd R.G., Doogan V.J., De Faveri J., Fordyce G., McGowan M.R., Bertram J.D., Vankan D.M., Fitzpatrick L.A., Jayawardhana G.A. and Miller R.G. (2002) *Anim Reprod Sci.* **71** :67.

Johnston D.J., Barwick S.A., Corbet N.J., Fordyce G., Holroyd R.G., Williams P.J. and Burrow H.M. (2009) *Anim. Prod. Sci.* **49**: 399.

Legarra A. and Reverter A. (2018) *Genet Sel Evol.* **50**: 53.

Proc. Assoc. Advmt. Anim. Breed. Genet. 23:91-94

GENETICS OF HEIFER AGE AT PUBERTY IN AUSTRALIAN HEREFORD CATTLE

M.L. Wolcott[1], R.G. Banks[1], M. Tweedie[2] and G. Alder[2]

[1]Animal Genetics & Breeding Unit*, University of New England, Armidale, NSW, 2351 Australia
[2]Beef and Lamb New Zealand Genetics, Dunedin 9058, New Zealand

SUMMARY

Age at puberty has become a key trait in the genetic evaluation of female reproduction traits for tropically adapted beef breeds in northern Australia. This study aimed to characterise the trait in Australian Hereford seedstock heifers and to determine the degree to which it, and associated traits, were under genetic control. Hereford heifers (n = 922) from three seedstock herds were serially ultrasound scanned to detect their first *corpus luteum* (indicative of age at puberty) at 4 - 6 week intervals from 10.6 to 13.2 months of age, at which time heifers were synchronised for artificial insemination. Results showed that only 52% of heifers were pubertal at synchronisation, and for these heifers, age at puberty had a heritability of 0.26. When a penalised record (equal to the maximum age at puberty for their contemporary group plus 21 days) was included for heifers which were not pubertal into mating, heritability increased to 0.38. For sires with at least 10 progeny, EBVs for age at puberty ranged from -42 to 28 days. The ability of heifers to conceive early in their first mating season is linked to lifetime reproductive performance. These results suggest that the proportion which have reached sexual maturity as they enter their first mating is significantly less than 100% and that opportunities exist, if the trait were included in the genetic evaluation for the breed, to monitor and apply selection to improve age at puberty in Hereford heifers.

INTRODUCTION

Results from the Co-operative Research Centre for Beef Genetic Technologies' Northern Breeding Project (Beef CRC) showed that age at puberty, identified by serial ultrasound scanning to determine date at first ovulation, was heritable in tropically adapted beef genotypes (Johnston *et al.* 2009). These results have been supported by subsequent research in the Repronomics™ project (Johnston *et al.* 2019) (h^2 = 0.32 to 0.56). Associated research also demonstrated that lower age at puberty was favourably genetically correlated with lifetime reproductive outcomes (r_g = -0.29 to -0.40), and that selection to improve (reduce) age at puberty would have favourable consequences for lifetime reproductive performance (Johnston *et al.* 2014). Morris *et al.* (2000) showed moderate heritability for age at puberty in Angus heifers when the trait was based on observed first oestrus (h^2 = 0.31), and a high genetic correlation with first mating pregnancy rate (r_g = -0.89). The current study aimed to exploit methods developed in the Beef CRC to characterise age at puberty in Hereford heifers, to determine the heritability of the trait and its potential to provide a means to improve and monitoring female reproduction in the genetic evaluation for the breed.

MATERIALS AND METHODS

Animals and management. Heifers used for this study were made available by three Hereford seedstock breeders, and represented the entire cohort of females weaned in 2017 and 2018 from each herd. Herds were selected for inclusion based on a history of high quality pedigree and performance recording, and a willingness to endure the significant imposition associated with serial ultrasound scanning required to identify first oestrous. Heifers were managed in accordance with standard practices for the three seedstock herds, one of which was located in the Southeast of New South Wales (n =

* A joint venture of NSW Department of Primary Industries and the University of New England

534) and the other two in the New England region (n = 149 and 239).

Heifers were born over a 2-month spring calving period at the Southern New South Wales property and over three months for the New England herds. The animals evaluated for this study were the progeny of 99 sires, with 71% from sires with at least 10 progeny, and 20% of heifers from sires used in at least two herds. Heifers were weaned at an average of 5.4 months, with the two New England properties weaning at 6.6 months and the remaining herd weaning earlier (averaging 4.5 months old). Heifers weaned in 2018 were reared under significantly dryer conditions than those in 2017. This meant that more supplementary feeding was provided for heifers in 2018, but within herd and year, all animals received the same nutritional interventions. This was also the case for routine management practices (animals identification and branding, vaccination, parasite control treatments, etc.) as well as culling for conformation related traits between weaning and syncronisation for artificial insemination. All herds routinely submit data to BREEDPLAN for genetic evaluation. For the heifers involved in this study, this included pedigree information, date of birth and weaning weight, and these data were extracted from the Hereford Australia Ltd. database for these analyses.

Scanning for ovarian function. Ultrasound scanning to detect first oestrous followed the protocols described by Johnston *et al.* (2009) for tropical beef females in the Beef CRC. Within herd and year, scanning was performed by one of three technicians using a Mindray M7Vet real-time ultrasound unit equipped with a variable frequency 6LE5Vs intra-rectal transducer, set at 8MHz. The timing of first scans to detect the presence of a *corpus luteum* (CL), was undertaken when managers at each location observed the first signs of heat in the heifer cohorts examined for this study (post-weaning). Subsequent scans were undertaken at 4 - 6 week intervals, until the first progesterone based synchronisation treatment occurred in each herd, prior to artificial insemination (into-mating). All heifers in the cohort were scanned at post-weaning and at mating synchronization, with interim scans performed on heifers which had not displayed a CL. This resulted in the majority of heifers scanned three times up to synchronisation, with average number of scans per animal, within herd and year, between 2.3 and 2.8. Based on ovarian scanning results, the following traits were defined:

- **Age at puberty (AP)** was a trait in females which displayed a CL prior to mating, calculated as the scanning date at which the first CL was detected minus date of birth.
- **Penalised AP (APP)** generated an age at puberty record for heifers which had failed to display a corpus luteum prior to mating. APP was calculated for these animals as the maximum AP for their contemporary group plus 21 days. For a small number of heifers which failed to display a CL prior to mating and were in small contemporary groups (for which the maximum AP was based on too few records ($N \leq 3$) to be reliable) no APP was analysed (N = 15 heifers).
- **Pubertal into mating (PUB)** was a binary trait which identified heifers which had cycled at any time up to mating (1) or not (0).
- **Antral follicle count (FC)** was the total number of follicles greater than 2mm, visible by ultrasound examination of both ovaries at the first scan in heifers which did not have a CL.

Growth and body composition traits. At each scan, records of liveweight weight (LWT), hip height (HH) and body condition score (BCS) were collected for each heifer following the protocols for growth and body composition traits described by Johnston *et al.* (2009). P8 fat depth (P8) was also measured at each scan using the scanner's inbuilt callipers, with the exception of the first scan for heifers from one herd where the records could not be collected.

Modelling, variance component and EBV estimation. Descriptive statistics were generated using PROC MEANS in SAS. Contemporary group information was extracted from the Hereford Australia Ltd. database, and was built based on information supplied by participating breeders as described by Graser *et al.* (2005).

The contemporary group for 200 day weight was used to analyse heifer growth, body composition and the descriptors of ovarian function evaluated for this study. For growth and body composition traits, dam age and linear animal age were fitted as covariates. Consistent with the protocols established by Johnston *et al.* (2009) heifer age was modelled for ovarian scanned traits as month of birth nested within herd and year. Variance components for each trait were estimated in univariate analyses in ASReml (Gilmour *et al.* 2009), with EBVs for all animals in the three generation pedigree estimated as the solution for the random animals effect. For this study genetic parameters for the binary PUB trait were estimated on the observed scale.

RESULTS AND DISCUSSION

Growth and body composition traits. Summary statistics, additive variances and heritabilities for post-weaning growth and body composition traits are presented in Table 1. On average, heifers were 10.6 months of age at their post-weaning scan, with mean ages at first scan consistent across herds. Additive variances and heirtabilities for post-weaning LWT and HH were consistent with those reported by Donoghue *et al.* (2018) for Angus and Hereford females prior to their first calving ($h^2 = 0.45$ to 0.57). The heritability for post-weaning P8 was lower than that for Hereford females prior to their first calving reported for that study ($h^2 = 0.64$), but heritability for BCS was comparable ($h^2 = 0.29$). The technicians employed to collect ultrasound data describing ovarian traits were not accredited BREEDPLAN carcass scanners, and this may explain the slightly lower than expected heiritability for the scanned fat depth trait.

Table 1. Number of records analysed (N), mean and standard deviation (SD), with additive variance (σ_a^2) and heritability (h^2) (and standard error (s.e.)) for post-weaning growth and body composition and ovarian scanned traits in Hereford heifers

Traits	Units	N	Mean	SD	σ_a^2	h^2	s.e.
Post-weaning growth and body composition							
AGE	Days	922	321.4	27.9	.	.	.
LWT	kg	922	262.9	35.0	460.4	0.55	0.11
HH	cm	921	116.7	4.6	6.8	0.49	0.11
P8	mm	837	3.6	1.8	0.6	0.29	0.10
BCS	Score (1 – 5)	922	2.8	0.6	0.03	0.20	0.08
Ovarian scanned traits							
AP	Days	481	365.8	38.3	363.0	0.26	0.13
APP	Days	902	396.2	44.3	588.7	0.38	0.10
PUB[A]	1/0	917	0.52	0.50	0.05	0.36	0.11
FC	Count	729	23.3	7.1	21.1	0.42	0.13

[A] Variance components for PUB estimated on the observed scale.

Ovarian scanned traits. Summary statistics, additive variances and heritabilities for ovarian scanned traits are also presented in Table 1. A key result from this work was the proportion of heifers which were pubertal into mating (PUB = 0.52). This reinforces the need to investigate the genetics of puberty traits in temperate breeds and for subsequent analyses, which will examine relationships of the trait with first mating outcomes. The phenotypic and additive variance for APP (1549.2 and 588.7 days respectively) were substantially lower than those reported by Johnston *et al.* (2009) for troppically adapted heifers, which was consistent with the much shorter scanning period in temperate breeds where maiden matings occur approximately 12 months earlier. The moderate heritability estimated for APP ($h^2 = 0.38$) suggested that opportunities exist to improve the trait by selction in the Hereford breed. Both AP and APP were under significantly greater genetic control than days to calving ($h^2 \sim 0.05$) which is currently the key descriptor of female reproductive performance in the BREEDPLAN genetic evaluation for the breed.

For sires with 10 or more progeny, EBVs for APP ranged from -42 to 28 days. The heifers available for this study were a reasonably small sample of the breed, but these results suggest that sire selection could impact age at puberty in the resulting progeny by at least 35 days. With only 52% of females pubertal into their first mating, and mating periods as low as 2 months in commercial beef breeding herds in southern Australia, this could have implications for reproductive outcomes for naturally mated maiden heifers.

Mean and standard deviation for post-weaning FC were consistent with those reported by Walsh *et al.* (2014) for dairy heifers in the US and Ireland, with heritabilities also comparable ($h^2 = 0.25$ and 0.31 respectively). FC was recorded in this project to investigate its genetic associations with economically important female reproduction traits and this will be the subject of future analyses.

CONCLUSIONS

This study presents an initial investigation of the genetics of age at puberty and associated traits in Australian Hereford seedstock heifers. Results showed that there are opportunities to improve (reduce) age at puberty by selection in the breed and, by including the trait in the breed's genetic evaluation, to monitor this aspect of female reproduction as selection is applied to improve other economically important traits. The proportion of heifers which were not pubertal as they entered their first mating was a key result of this study. The increasing prevalence of artificial insemination and the associated treatments to synchronise (and possibly induce) first oestrous, suggest that genetic and environmental factors which impact a heifer's capacity to conceive early in their first mating season may warrant monitoring and inclusion in the genetic evaluation for temperate beef breeds. It is acknowledged that serial ultrasound scanning to detect first oestrous is an expensive and labour intensive operation, making it a candidate for evaluation in intensively recorded reference populations, and for further research to economise the recording regime.

ACKNOWLEDGEMENTS

This research was supported by co-funding from Beef and Lamb New Zealand Genetics and Meat and Livestock Australia under project P.PSH.0869. The co-operation of the principals and employees of the Hereford seedstock herds which made animals available for this study, as well as their breed representative body (Herefords Australia Ltd.) is also gratefully acknowledged, as is the technical expertise provided by staff of NSW DPI, Holbrook Veterinary Centre and Terang and Mortlake Veterinary Clinic.

REFERENCES

Donoghue K.A., Lee S.J., Parnell P.F. and Pitchford W.S. (2018) *Anim. Prod. Sci.* **58:** 145.

Johnston D.J., Barwick S.A., Corbet N.J., Fordyce G., Holroyd, R.G., Williams, P.J. and Burrow, H.M. (2009) *Anim. Prod. Sci.* **49**: 399.

Johnston D.J., Corbet N.J., Barwick S.A., Wolcott M.L. and Holroyd R.G. (2014) *Anim. Prod. Sci.* **54**: 74.

Johnston D.J., Moore K., Cook J. and Grant T. (2019) MLA Project B.NBP.0759 final report.

Gilmour A.R., Gogel B.J., Cullis B.R. and Thompson R. (2009) 'ASReml user guide. Release 3.0.' VSN International: Hemel Hempstead, UK.

Graser H.U., Tier B., Johnston D.J. and Barwick S.A. (2005) *Anim. Prod. Sci.* **45**: 913.

Morris C.A., Wilson J.A., Bennett G.L., Cullen N.G., Hickey S.M. and Hunter J.C. (2000) *NZ J. Agric. Res.* **43.1**: 83.

Walsh S.W., Mossa F., Butler S.T., Berry D.P., Scheetz D. *et al.* (2014) *J. Dairy Sc.* **97**: 4503.

Proc. Assoc. Advmt. Anim. Breed. Genet. 23:95-98

HOW DOES MATERNAL WEANING WEIGHT (MILK) AFFECT BODY CONDITION SCORE AT WEANING IN ANGUS CATTLE

T. Granleese[1,2] and S.A. Clark[1,2]

[1]University of New England, Armidale, NSW 2351, Australia
[2]Angus Australia, Armidale, NSW 2351, Australia

SUMMARY

This paper reports the heritability estimates for mature body composition traits in Australian Angus cattle and the interaction with the maternal component of weaning weight. These traits include mature body weight (MCW), mature cow fat score (MBC), mature cow height (MCH), weaning weight (WWT) and the maternal component of weaning weight (MILK) with industry data obtained from breeders registered with Angus Australia. Heritability for MCW, MBC, MCH, WWT and MILK was estimated to be 0.44, 0.49, 0.15 and 0.13, respectively. MCW had a moderate genetic correlation with MBC of 0.61 and 0.46 for MCH. MBC and MCH had a genetic correlation of -0.06 but had a large standard error due to low cross-over of cows with phenotypes for the two traits. MILK had a negative genetic correlation with MBC of -0.48 and small positive genetic correlations of 0.12 and 0.23 with MCW and MCH, respectively. These results indicate that selection for mature body composition traits is possible but care should be taken when considering interactions with the maternal trait MILK.

INTRODUCTION

Mature body composition traits of mature beef cows have has not been as well measured in seedstock populations as compared to corresponding traits at younger ages (Donoghue *et al.* 2018). However, a high percentage of female cattle in a self-replacing herd will spend up to significantly more time as a mature animal compared to their first two years. Australian seedstock beef breeders have made significant gains in production traits where traits can be easily measured in large and complete cohorts in the first two years of age (Walmsley *et al.* 2018). Given the challenges of measuring traits on mature cows, the estimation of breeding values in routine analysis for mature body composition traits has been challenging. Mature cow body composition traits are an important aspect for self-replacing herds to focus on in their breeding objective. Traits such as mature body condition score (MBC), mature weight (MCW), mature hip height (MCH) are associated with feed maintenance costs and reproduction rates (Walmsley *et al.*, 2018). Previous studies in Angus research populations in Australia and United States of America have estimated heritability from MBCS, MCW and MCH to be 0.11-0.21, 0.40-0.71, and 0.62-0.83, respectively (Choy *et al.* 2002; Decker *et al.* 2012; Donoghue *et al.* 2018). Genetic correlations of mature body composition traits has not been well investigated in beef but dairy cattle analysis suggest a negative genetic correlation of 0.50 between milk yield and MBC (Berry *et al.* 2003).

The aim of this paper is to estimate genetic and residual variation and correlations of the three body composition traits above and the maternal component of weaning weight using industry data from the Angus Australia database.

METHODS

Animals. Industry data from animals in the Angus Australia database born in 2003 and after were included in the analysis. All males were excluded for MCW, MBC and MCH. For individuals to be included for these traits, there needed to be at least 10 females in a contemporary group. Records for WWT to be included were the corresponding progeny measured at the same point of MCW or MBC measurement. Contemporary grouping for all traits were created based on Graser *et al.* (2005) where the concatenation of herd, day of measurement and breeder-defined management group was used. Females were excluded if the first measurement for any trait was older than 5 years of age. Table 1 displays the number of records for traits and pairs of traits. The number of contemporary groups for each trait is found in Table 1.

Measurements. Weaning weight was measured between 60 and 300 days of age. Both MBC and MCW were measured within two weeks of weaning. The minimum age for MBC and MCW was 830 days of age while MCH was 730 days of age. All traits were measured following Angus Australia measuring protocols.

Statistical analysis. Single measurements were used for all traits. Records more than 4 standard deviations from the database mean were identified as outliers and removed from the analysis. Contemporary group was fitted as described above and age at measurement was fitted for all traits. Age of calf was fitted for MCW and MBC, while age of dam was fitted for WWT. The maternal component for weaning weight was fitted to estimate the maternal affect (MILK). Genetic parameters and predicted means were estimated using an animal model in WOMBAT (Meyer 2007). Genetic covariance was not estimated between WWT direct and maternal component of WWT (MILK). A numerator relationship matrix based on a four generation pedigree was used. Genetic and phenotypic parameters were estimated from a multivariate analyses.

Table 1. Means, standard deviations, minimums, maximums and number of contemporary groups for each trait

	Records	Mean	SD	Min.	Max.	Cont. Groups
WWT	56409	237.5	52.0	56	445	2892
MCW	31455	538.2	86.8	316	966	1044
MBC	4915	3.1	0.75	1	6	188
MCH	2952	133.1	5.16	117	155	116

RESULTS AND DISCUSSION

Table 2 contains summary statistics for mature body composition traits. All traits were heritable with low standard errors which suggests that selection for genetic progress can be made. Heritabilities from the multivariate analysis matched univariate analysis with small and insignificant differences between standard errors.

Heritability for MCW was 0.43. This is similar to Johnston *et al.* (1996) and Choy *et al.* (2002) but lower than Decker *et al.* (2012) and Donoghue *et al.* (2018). Heritability for MCH was 0.44 which was lower compared to most other Angus genetic parameter studies in Australia (Donoghue *et al.* 2018) and the USA (Choy *et al.* 2002; Decker *et al.* 2012) which estimated heritabilities ranging from (0.58-0.82). Heritability for MBC was 0.16 and was similar to other Angus genetic parameter studies (Choy *et al.* 2002; Donoghue *et al.* 2018). Results from this study are in agreement with past published studies that there is the potential to select for mature body composition traits. Weaning weight and its maternal component was 0.18 and 0.13, respectively. This is in agreement with Meyer's (1992) study where covariance between direct and maternal effect is not estimated.

Table 2. Phenotypic variance and heritabilites (with standard errors) for multivariate analysis

	σ_p	h^2■
WWT	90.5	0.18 (0.01)
MILK	72.8	0.13 (0.01)
MCW	901	0.43 (0.02)
MBC	0.043	0.16 (0.03)
MCH	6.15	0.44 (0.05)

Table 3. Genetic correlation above diagonal and phenotypic below diagonal from multivariate analysis with (standard errors)

	WWT	MILK	MCW	MBC	MCH
WWT	-	-	0.49 (0.03)	-0.09 (0.11)	0.46 (0.09)
MILK	-	-	0.12 (0.03)	-0.48 (0.09)	0.23 (0.08)
MCW	0.44 (0.01)	0.41 (0.01)	-	0.62 (0.07)	0.47 (0.08)
MBC	0.15 (0.02)	0.10 (0.02)	0.44 (0.01)	-	-0.01 (0.16)
MCH	0.31 (0.03)	0.31 (0.00)	0.45 (0.01)	0.07 (0.03)	-

Phenotypic and genetic correlations and their associated standard errors between mature body composition traits and weaning weight are reported in Table 3. MCW was moderately genetically correlated with MBC (0.62), MCH (0.47) and WWT (0.49). These genetic correlations are lower when compared to previous literature. Lower heritabilities and genetic correlations in this study could be caused by using industry data where culling takes place. The other cause may be due to using single measurement only where permanent environment effect is not accounted for (Kaps *et al.* 1999). Furthermore we did not fit sire-by-herd. Repeated measures to account for permanent environment effects as well as fitting sire-by-herd should give better genetic estimates. The genetic correlation between MBC and MCW was close to zero which corresponds to previous studies (Donoghue *et al.* 2018). Moving forward, Angus Australia members will need to make sure they are measuring both MBC and MCH on the same animals to provide more accurate genetic correlations between the two traits.

The genetic correlation between the maternal component of WWT (MILK) and MBC (which was measured at weaning) suggest that high milking cows will genetically lower body condition score animals with a genetic correlation of -0.48 (Table 3). However, this correlation could be broken with body condition score measured at weaning and selection for higher MILK and higher MBC possible. This genetic correlation is in agreement to Berry *et al.'s* (2003) study in milking cattle where they estimated a genetic correlation of -0.50 between milk yield and body condition score. The phenotypic correlation between these two traits is 0.10. This suggests Angus breeders are managing body condition score to make their performance recorded animals are in good body condition when raising a calf.

We did not estimate genetic covariance of WWT and MILK because our model did not use repeated records and we could not fit permanent environment effects as well as sire-by-herd effects. Bijma (2006) and Meyer (1992) explain the difficulties of estimating covariance of direct and maternal components of traits and is the next step for research with this study.

Mature body condition and MCW were measured at the weaning of their calves or within two weeks of weaning. The measuring of weight helps with the maternal 200 day weight EBVs of their calves. However, studies suggest that MBC are different traits at different stages of lactation. Donoghue *et al.* (2018) suggested that weaning MBC and pre-calving MBC in Australian Angus cattle while Wolcott *et al.* 2013 demonstrated that joining and weaning MBC were also two different traits in *Bos Indicus* cattle.

Investigation of mature body composition traits at joining and the genetic relationships with maternal productivity traits such as MILK and fertility would be a logical next step.

CONCLUSION

Genetic variation in mature composition traits is present in the Angus Australia database, confirming there is potential to select for more efficient females in a self-replacing herd. More cows in the database need phenotypes for both body condition (fat) score and mature hip height to be able to calculate genetic correlations with confidence. Including some early in life measurements will give these mature body composition breeding values some context.

ACKNOWLEDGEMENTS

The authors acknowledge Angus Australia for the funding of this research.

REFERENCES

Berry D. P., Buckley F., Dillon P., Evans R. D., Rath M., and Veerkamp R.F. (2003) *J. Dairy Sci.* **86**: 2193.

Bijma P. (2006) *J. Anim. Sci.* **84**: 800.

Choy Y. H., Brinks J.S. and Bourdon R.M. (2002) *J. Anim. Sci.* **80**: 2071.

Decker J.E., Vasco D.A., McKay S.D., McClure M.C., Rolf M.M., Kim J., Northcutt S.L., Bauck S., Woodward B.W., Schnabe R.D.I. and Taylor J. F. (2012) *BMC Genomics* **13**: 606.

Donoghue K.A., Lee S.J., Parnell P.F. and Pitchford W. S. (2018) *Anim. Prod. Sci.* **58**: 145.

Graser H.-U., Tier B., Johnston D. J. and Barwick S.A. (2005) *Aust. J. Exp. Agric.* **45**: 913.

Johnston D.J., Chandler H. and Graser H.-U. (1996) *Aust. J. Agric. Res.* **47**: 1251.

Kaps M., Herring W.O. and Lamberson W.R. (1999) *J. Anim. Sci.* **77**: 569.

Meyer K. (1992) *Livest. Prod. Sci.* **31**: 179.

Meyer K. (1995) *Livest. Prod. Sci.* **44**: 125.

Meyer K. (2007) *J. Zhejiang Uni. SCI. B* **8**: 815.

Walmsley B.J., Lee S.J., Parnell P.F. and Pitchford W.S. (2018) *Anim. Prod. Sci.* **58**: 1.

Wolcott M. L., Johnston D. J. and Barwick S. A. (2013) *Anim. Prod. Sci.* **54**: 60.

Proc. Assoc. Advmt. Anim. Breed. Genet. 23:99-102

CONTRIBUTIONS FROM GENETIC GROUPS AND OUTCROSSING TO COMPONENTS OF REPRODUCTION IN MATERNAL SHEEP BREEDS

K.L. Bunter[1], A.A. Swan[1] and D.J. Brown[1,2]

[1]Animal Genetics & Breeding Unit*, University of New England, Armidale, NSW, 2351 Australia
[2]Meat and Livestock Australia, Armidale, NSW, 2350 Australia

SUMMARY

Industry data for traits included in the new multi-trait genetic evaluation for reproductive traits provided by Sheep Genetics were used to investigate variation due to sub-populations (genetic groups) and due to outcross ewe genotypes in maternal sheep breeds. Substantial variation due to genetic groups (gg^2: typically 11-30% of the phenotypic variance) for traits reflecting development (*eg* weight, condition score, muscle depth) were not accompanied by comparable variation for reproductive traits (gg^2: 0-8%). Variation due to outcross ewe genotypes ranged from 0 to 8% across traits, being highest for adult ewe weight (8%) and yearling conception (6%) traits, which are expected to be affected by heterosis. Accommodating these sources of variation appropriately may be important for the genetic evaluation of data affected by admixture of populations.

INTRODUCTION

Two key issues for genetic evaluation of reproductive traits for maternal sheep breeds (referred to as the MATL evaluation) are the extent of variation between sub-populations described by genetic groups, as well as fair comparison of 'homebred' ewes with outcrossed contemporaries. The diversity of breeds and breed composition within the MATL evaluation is increasing. Breeds occur in sub-populations (*eg*. Australia vs New Zealand) and have also contributed to outcrossing and composite populations, increasing diversity of breed composition and expression of heterosis. Further, outside introductions can be accompanied by absence of pedigree and therefore creation of additional genetic groups. Preliminary investigation of breed composition demonstrated considerable variability in the genetic architecture of individual flocks (*eg*. composite vs pure-breeding) and the breed choice of outcross or introduced sires. Therefore, a general strategy to accommodate variation in the effects of heterosis is required. In this paper, we provide estimates of genetic parameters for traits included in the new single-step, multi-breed analyses used to produce breeding values for ewe reproductive performance traits (Bunter *et al.* 2019), including variances for genetic group effects and flock-outcross ewe genotypes.

MATERIALS AND METHODS

Data included in these analyses commenced in 2000, with pedigree and genetic groups extended back to 1998. Briefly, component traits were defined annually for conception of ewes joined (CON: 0=failed to conceive, 1= conceived) along with litter size (LS: 1 to n lambs born) and ewe rearing ability (ERA: lambs surviving/lambs born) for pregnant ewes. Pregnancy scan data was a secondary data source to define CON or LS when lambs were not recorded individually. Additional traits included maternal behaviour score of the ewe (MBS: scored from 1: good to 5: poor) as well as pre-joining weight (WT) and condition score (CS) recorded within the 30 days prior to joining. Data describing development of the young ewes and/or their male relatives was obtained for the subset of flocks included in reproductive analyses and included scanned post-weaning carcase fat (PFAT) and eye muscle depth (PEMD), along with post-weaning (PSC) or yearling (YSC) scrotal circumferences.

* A joint venture of NSW Department of Primary Industries and the University of New England

Based on previous analyses (Bunter and Brown 2013), yearling and adult performances of CON, LS and ERA were treated as separate traits. Models for reproductive traits accounted for the systematic effects of CG + age, where CG refers to joining (CON, LS) or lambing (ERA) contemporary groups (based on site-year-timegp-mgp details) and age refers to age at recording in years (adult ewes). Time group (timegp) was assigned based on lambing dates, to accommodate evidence of gaps between joining events, and management groups (mgp) were as specified by breeders. Contemporary groups for reproductive traits were further refined to include: 1) month of birth and dam age group (yearling, adult, unknown) in the CG for yearling traits, and 2) previous status of the ewe (no lamb, lambed and lost or weaned, or unknown) in the CG for 2-year-old traits, enabling flock specific differences with respect to these factors. Additional model terms included birth-rearing type group for yearling but not adult reproductive traits and litter size group (1, 2 and 3 or more) at birth for ERA, since litter size alters the rearing challenge for ewes (Bunter *et al.* 2018). For the remaining traits (PFAT, PEMD, PSC and YSC), contemporary groups were as previously defined for these traits (Brown *et al.* 2007), and additional model terms included regressions on age, but not weight, where $P<0.05$.

Specific model comparisons were made using univariate analyses. Trait dependent base models (model A) included animal genetic effects for all traits, permanent environmental effect of the dam (subset of traits), and permanent environmental effects to accommodate repeated records for adult ewes. Additional random effects subsequently added to base models included genetic groups (GG), defined as per Swan *et al.* (2016), and a flock×outcross term intended to represent a pure- or crossbred (PC) genotype for the individual ewe. Genetic groups were as assigned for the genetic evaluation of maternal breeds, which are currently kept constant across all relevant analyses and trait sets. Ewes were considered an outcross if their sire was identified by a different flock code; different types of outcrosses (*ie* sire breeds) were not distinguished. The full model (model GGPC) was only fitted for traits where each of these terms significantly ($P<0.05$) improved model fit.

RESULTS AND DISCUSSION

Estimates of heritabilities for early in life development traits (PFAT, PEMD, YWT, YCS) and scrotal measures (PSC, YSC) were generally consistent with expectation and are not discussed further. Model comparisons for pre-joining weight and condition score or maternal behaviour score are currently hindered by relatively low record numbers, but heritabilities were moderate.

Yearling vs adult expressions of reproductive traits. The order of magnitude for heritability estimates was YERA<YLS<YCON for yearling ewes (Table 1) and CON<ERA<LS for adult ewes (Table 2). Heritability for ERA was consistently lower than for litter size, reflecting an increase in environmental contributions to ERA. The relatively higher heritabilities for YCON vs CON and LS vs YLS support the strong influence of age at puberty, which is a moderately heritable trait, on YCON for yearling but not adult ewes, and an increased expression of genetic differences for litter size in adult compared to yearling ewes.

Genetic group effects. Pedigree is generally well known for current animals included in MATL analyses. Therefore, genetic groups predominantly represent within flock base populations and missing historical pedigree. Estimates of variances due to genetic group effects for early development traits ranged from negligible (PCF) to substantial (YWT) and the ratio of genetic group to additive (rgga) variance increased in magnitude from 0.20 (PCF)<YCS<PEMD< 3.96 (YWT). Considerable variance due to genetic groups was also evident for AWT and CS of adult ewes (Table 2), but rgga were lower (<1.5) than for corresponding yearling traits. With respect to reproductive traits, the range in rgga from GG models was much lower (0.06 to 1.83) across both yearling and adult ewes, and this ratio was largest when flock-outcross variances were present and not accounted for (Model GG vs GGPC). This result implies that the ratio of genetic group variance (gg^2) is potentially inflated due to the effects of multi-breed outcrossing. The ratio gg^2 was substantial for YCON but not CON, but generally negligible for all other reproductive traits. The ratios of genetic group to additive variances were somewhat similar to those reported by Swan *et al.* (2016) within trait groups.

Table 1. Parameter estimates for post-weaning fat (PCF), muscle depth (PEMD), post-weaning (PSC) and yearling (YSC) scrotal circumference, yearling conception (YCON), litter size (YLS) and ewe rearing ability (YERA), along with pre-joining weight (PWT) and condition score (PCS). The number of records is presented in brackets

		Variances						Ratios				
Trait	Model	σ^2_a	σ^2_{gg}	σ^2_{pc}	σ^2_{ped}	σ^2_e	σ^2_p	h^2	gg^2	pc^2	ped^2	rgga
PCF	A	0.135	-	-	0.020	0.460	0.614	0.22	-	-	0.03	-
(302747)	APC	0.135	-	0.005	0.020	0.466	0.625	0.22	-	0.01	0.03	-
	GG	0.134	0.027	-	0.020	0.467	0.647	0.22	0.04	-	-	0.20
PEMD	A	1.41	-	-	0.31	4.49	6.21	0.23	-	-	0.05	-
(301908)	APC	1.37	-	0.18	0.32	4.50	6.36	0.22	-	0.03	0.05	-
	GG	1.31	2.66	-	0.32	4.54	8.84	0.21	0.30	-	0.05	2.03
	GGPC	1.32	2.44	0.11	0.31	4.52	8.71	0.21	0.28	0.02	0.05	1.85
PSC	A	1.60	-	-	0.31	3.59	5.50	0.29	-	-	0.06	-
(69400)	APC	1.60	-	0.15	0.30	3.58	5.50	0.28	-	0.03	0.05	-
	GG	1.59	0.001	-	0.31	3.59	5.49	0.29	0.00	-	0.06	0
YSC	A	1.19	-	-	0.14	2.45	3.79	0.31	-	-	0.04	-
(42637)	APC	1.16	-	0.07	0.15	2.46	3.85	0.30	-	0.02	0.04	-
	GG	1.16	0.64	-	0.14	2.46	4.41	0.31	0.14	-	0.04	0.55
	GGPC	1.15	0.49	0.05	0.15	2.47	4.30	0.30	0.11	0.01	0.04	0.42
YCON	A	0.021	-	-	-	0.151	0.172	0.12		-	-	-
(24826)	APC	0.021	-	0.010	-	0.151	0.181	0.12		0.05	-	-
	GG	0.018	0.033	-	-	0.153	0.204	0.10	0.16	-	-	1.83
	GGPC	0.016	0.011	0.011	-	0.153	0.191	0.09	0.06	0.06	-	0.69
YLS	A	0.016	-		-	0.233	0.249	0.06	-	-	-	-
(58068)	APC	0.016	-	0.001	-	0.233	0.249	0.06	-	0.00	-	-
	GG	0.016	0.001		-	0.233	0.250	0.06	0.00		-	0.06
YERA	A	0.005	-		-	0.123	0.128	0.04			-	-
(41955)	APC	0.005	-	0.001	-	0.123	0.128	0.04	-	0.00	-	-
	GG	0.005	0.003		-	0.123	0.130	0.04	0.02		-	0.60
YWT	A	10.4	-	-	4.17	9.44	24.0	0.43	-	-	0.17	-
(4515)	APC	FTC	-	-	-	-	-	-	-	-	-	-
	GG	5.66	22.4	-	5.01	11.8	44.9	0.25	0.50	-	0.22	3.96
YCS	A	0.028	-	-	0.001	0.149	0.178	0.16	-	-	0.01	-
(2803)	APC	0.028	-	0.001	0.001	0.148	0.178	0.16	-	0.01	0.01	-
	GG	0.022	0.034	-	0.001	0.151	0.207	0.11	0.16	-	0.00	1.70

Variances due to additive genetic (σ^2_a), genetic group (σ^2_{gg}), flock-outcross (σ^2_{pc}), and maternal permanent environment (σ^2_{ped}) effects, along with the residual (σ^2_e) and phenotypic variances (σ^2_p). Variance ratios are heritabilities (h^2: σ^2_a / σ^2_p), variance due to genetic groups (gg^2: σ^2_{gg} / σ^2_p), flock-outcross (pc^2: σ^2_{pc} / σ^2_p) or permanent environmental effects of the dam (ped^2: σ^2_{ped} / σ^2_p), excluding σ^2_{gg} from σ^2_p for ratios not involving σ^2_{gg} in GG and GGPC models, and rgga= σ^2_{gg} / σ^2_a; FTC: failed to converge.

Flock-outcross effects. There is likely little advantage for accuracy of selection in correcting for differences in retained heterosis within stabilised composites. However, fair comparison of outcross with homebred ewes is warranted. Ratios of variances due to flock-outcross terms (pc2) were largest for fertility (YCON: 6%, CON: 3%) and ewe weight traits (AWT: 8%). For comparison, heterosis for fertility (17-21%), lamb survival (2-8%) but not litter size, was previously observed in structured data involving divergent maternal breeds by Fogarty *et al.* (1984). The absence of substantial ratios for pc2 for many traits implies that across the wide range of flocks and crosses, alternative ways to model heterosis may be required.

Table 2. Parameter estimates for adult conception (CON), litter size (LS) and ewe rearing ability (ERA), along with maternal behaviour score (MBS), pre-joining weight (WT) and condition score (CS). The number of records is presented in brackets

		Variances							Ratios			
Trait	Model	σ^2_a	σ^2_{pe}	σ^2_{gg}	σ^2_{pc}	σ^2_{ped}	σ^2_e	σ^2_p	h^2	gg^2	pc^2	rgga
CON	A	0.002	0.004	-	-	-	0.073	0.079	0.03	-		-
(144803)	APC	0.002	0.004	-	0.002	-	0.073	0.081	0.03	-	0.03	-
	GG	0.002	0.004	0.001	-	-	0.073	0.080	0.03	0.01	-	0.50
	GGPC	0.002	0.004	0.001	0.002	-	0.073	0.082	0.02	0.01	0.03	0.53
LS	A	0.019	0.012	-	-	-	0.304	0.335	0.06	-		-
(685962)	APC	0.019	0.012	-	0.002	-	0.303	0.336	0.06	-	0.01	-
	GG	0.018	0.013	0.015	-	-	0.304	0.350	0.06	0.04	-	0.83
	GGPC	0.018	0.013	0.018	0.002	-	0.304	0.354	0.05	0.05	0.01	1.00
ERA	A	0.001	0.003	-	-	-	0.081	0.085	0.02	-	-	-
(536320)	APC	0.001	0.003	-	0.001	-	0.081	0.086	0.01	-	0.01	-
	GG	0.001	0.003	0.001	-	-	0.081	0.086	0.01	0.01	-	1.00
MBS	A	0.101	0.074	-	-	-	0.501	0.676	0.15	-	-	-
(10293)	APC	0.101	0.074	-	0.001	-	0.501	0.677	0.15	-	0.00	-
	GG	0.100	0.075	0.021	-	-	0.501	0.696	0.15	0.03	-	0.21
AWT	A	18.2	3.04	-	-	2.53	25.7	49.5	0.37	-	-	-
(10709)	APC	18.1	2.99	-	11.2	2.41	25.6	60.3	0.30	-	0.19	-
	GG	16.7	3.72	21.0	-	2.57	25.6	69.6	0.34	0.30	-	1.25
	GGPC	16.6	3.68	18.8	11.1	2.48	25.6	78.3	0.28	0.24	0.08	1.13
CS	A	0.043	0.017	-	-	0.002	0.177	0.239	0.18	-	-	-
(14959)	APC	0.043	0.017	-	0.001	0.002	0.177	0.240	0.18	-	0.00	-
	GG	0.041	0.018	0.028	-	0.002	0.177	0.266	0.17	0.11	-	0.68

Variance due to repeated records (σ^2_{pe}); accompanying ratios ranged between 0.03 and 0.11. All other abbreviations as per Table 1. Range for ped^2: 0.01 to 0.05.

CONCLUSIONS

Admixture of populations within data used by Sheep genetics for MATL breed analyses requires strategies to accommodate variance due to genetic groups and outcrossing within flocks. For reproductive traits without a long and effective selection history within flocks, variances due to genetic groups were generally lower than or similar to estimates of additive variances. Variation in performance due to outcrossing explained relatively little variation for all traits except AWT and YCON. Alternative ways to model heterosis may be required.

REFERENCES

Brown D.J., Huisman A.E., Swan A.A., Graser H.U., Woolaston R.R., Ball A.J., Atkins K.D. and Banks R.G. (2007). *Proc. Aust. Assoc. Anim. Breed. Genet.* **17**: 268.

Bunter K.L., Swan A.A, Brown D.J., Brien F.D. and Smith J. (2018). *Anim. Prodn. Sci.* **58:** 791.

Bunter K.L., Swan A.A., Gurman P.M., Boerner V., McMillan A.J. and Brown D.J. (2019). *Proc. Aust. Assoc. Anim. Breed. Genet.* **23:** (in press).

Bunter K.L. and Brown D.J. (2013). *Proc. Aust. Assoc. Anim. Breed. Genet.* **20**: 82.

Fogarty N.M., Dickerson, G.E. and Young L.D. (1984). *J. Anim. Sci.* **58**: 301.

Swan A.A., Brown D.J. and van der Werf J.H.J. (2016). *Anim. Prodn. Sci.* **56**: 87.

Proc. Assoc. Advmt. Anim. Breed. Genet. 23:103-106

GRAPHICAL MODELLING OF THE RELATIONSHIP BETWEEN BODY RESERVES AND YEARLING REPRODUCTION IN MATERNAL SHEEP

S.F. Walkom[1], K.L. Bunter[1], A.A. Swan[1] and S.A. Clark[2]

[1]Animal Genetics & Breeding Unit*, University of New England, Armidale, NSW, 2351 Australia-
[2]School of Environmental and Rural Science, University of New England, Armidale, NSW, 2351

SUMMARY

The underlying causal relationship between traits associated with energy reserves and yearling reproduction is often a cause of conjecture within the sheep industry in part due to anecdotal evidence often mistaking phenotypic associations with genetic. The use of graphical models to disentangle the underlying causal relationships between traits associated with energy reserves and yearling reproduction showed that selection for sires with high post-weaning fat and muscle will have little impact on the reproductive performance of the resulting progeny as yearling dams.

INTRODUCTION

Body weight and traits associated with body composition and energy reserves (fat and muscle) are important components of breeding objectives in sheep. These traits influence the amount of saleable meat and therefore have a direct economic value to the production system. However, the economic value placed on fat to produce a leaner carcase is at odds with the desired direction of change in fat as an indirect selection criterion for other traits, primarily reproductive performance and maternal efficiency. Previous studies have illustrated that body weight and body composition traits are associated with reproductive performance. The relationship between these traits and reproduction can be moderate at the phenotypic level but is often lower at the genetic level (Walkom *et al.* 2014; Walkom and Brown 2016).

Mixed effects models (often solved using REML) have been commonly used to estimate the associations between the traits at both a phenotypic and genetic level. However, such estimates indicate a correlation between traits rather than discover or define underlying causality. An alternative way to model the association between multiple traits is using graphical models (Valente *et al.* 2011). Graphical models, such as structural equation models and Bayesian networks including Incremental Association Markov Blankets (IAMB) (Tsamardinos *et al.* 2003), attempt to model all possible pathways in which two traits are associated. Hence, they provide insight into possible causal relationships that may exist, rather than association indicated by correlation alone (Valente *et al.* 2011). In this study, we use a graphical model to explore the underlying causal relationship between traits associated with energy reserves and yearling reproduction at both the phenotypic and genetic levels.

MATERIALS AND METHODS

Data used for the study were provided by maternal sheep breeders to Sheep Genetics as part of the routine LAMBPLAN genetic evaluation (Brown *et al.* 2007). The analysis focussed on six core traits: post-weaning weight (PWT), post-weaning ultrasound fat (PCF*) and eye muscle depth (PEMD*), yearling conception (YCON), yearling number of lambs born (YNLB) and yearling number of lambs weaned (YNLW) (Table 1).

Statistical Analysis. For each trait phenotypic, residual and genetic variances were estimated from univariate animal models. A series of bivariate analyses where then used to estimate correlations between traits. The initial genetic analyses were conducted using ASReml (Gilmour *et al.* 2009)

* A joint venture of NSW Department of Primary Industries and the University of New England

using a pedigree 22 generations deep and containing 674,028 animals. The model fitted to the post-weaning traits had fixed effects of contemporary group (as per Brown and Swan 2016), birth and rearing type group (SS, MS, MM), dam age (years), and linear and quadratic age terms. Unlike the model for standard Sheep Genetics traits, weight was not fitted as a covariate for the scan traits (represented by *). The permanent environment of the dam was also fitted as a random effect. Yearling reproduction traits were adjusted for a contemporary group, which was formed based on site, flock, year grouping and developmental factors as discussed by Bunter *et al.* (these proceedings).

Table 1. Summary of records available and genetic parameters from a univariate animal model for maternal sheep breeds. Phenotypic variance (σ^2_p), direct additive variance (σ^2_a), maternal permanent environment variance (σ^2_c), residual variance (σ^2_e) and heritability (h^2)

Trait	Records	Mean	SD	σ^2_p	σ^2_a	σ^2_c	σ^2_e	h^2
PWT	279,872	45.70	8.98	25.1	5.18	2.27	17.63	0.29 ± 0.01
PCF*	282,251	3.20	1.31	0.61	0.13	0.02	0.46	0.27 ± 0.01
PEMD*	263,555	26.70	4.14	6.20	1.42	0.32	4.45	0.32 ± 0.01
YCON	68,669	0.90	0.35	0.06	0.01	-	0.06	0.08 ± 0.01
YNLB	68,085	1.20	0.70	0.33	0.02	-	0.31	0.07 ± 0.01
YNLW	51,496	0.90	0.72	0.37	0.02	-	0.35	0.05 ± 0.01

Graphical Modelling. A subset of the data were used in the graphical model analyses, restricted to animals with a phenotype for all 6 traits (20,093 animals). For the 'genetic' graphical model sires with single trait breeding values, calculated from the univariate analysis, for all 6 traits were used (2,261 sires). The graphic models in Figures 1 and 2 provide a graphical representation of Bayesian networks at the phenotypic and genetic (sire) levels, respectively, and were developed using the bnlearn package implemented in R (Scutari 2010). The networks were estimated using a constraint-based structure learning algorithm based on the Markov blanket detection algorithm, which is based on a two-phase selection scheme (a forward selection followed by an attempt to remove false positives) (IAMB, Tsamardinos *et al.* 2003). The need for every animal to have an observation for all traits resulted in the use of YNLB and YNLW instead of the component traits as per Bunter *et al.* (these proceedings). The probability of the connections (strength & direction) between the trait nodes was estimated using bootstrap sampling with the IAMB learning algorithm (Friedman *et al.* 1999).

RESULTS AND DISCUSSION

Phenotypic association. The phenotypic correlations from the bivariate analysis are shown in Table 2. Moderate to strong phenotypic correlations exist between the post weaning traits (PWT, PCF*, PEMD*) and between the reproduction traits (YCON, YNLB, YNLW). However, the correlations between the two trait groups were weak.

The graphical model based on the phenotypic associations, using raw phenotypes, an indication of the observed variation, is represented in Figure 1. PCF* has a causal effect on PWT. Thus, changes in PCF* will cause a change in the PWT, but changes in PWT can occur without a responding change in PCF*. The relationship for PCF* on PEMD* is also causative, with PCF* having both a direct and indirect association via PWT on PEMD*. The graphical model identifies no direct causative effect of PCF* or PEMD* on the yearling reproduction traits. The graphical model shows that once you condition on PWT (remove variation associated with PWT), changes in PCF* or PEMD* had no impact on yearling reproduction.

Table 2. Estimates of phenotypic (above diagonal) and genetic correlations (below diagonal) between body composition and yearling reproduction traits

	PWT	PCF*	PEMD*	YCON	YNLB	YNLW
PWT		0.48 ± 0.01	0.63 ± 0.01	0.07 ± 0.01	0.11 ± 0.01	0.09 ± 0.01
PCF*	0.42 ± 0.01		0.49 ± 0.01	0.05 ± 0.01	0.06 ± 0.01	0.04 ± 0.01
PEMD*	0.58 ± 0.01	0.51 ± 0.01		0.06 ± 0.01	0.07 ± 0.01	0.07 ± 0.01
YCON	0.04 ± 0.04	0.09 ± 0.04	0.06 ± 0.04		0.60 ± 0.01	0.43 ± 0.01
YNLB	0.06 ± 0.04	0.07 ± 0.04	-0.03 ± 0.04	0.69 ± 0.03		0.64 ± 0.01
YNLW	0.17 ± 0.06	0.09 ± 0.06	0.08 ± 0.05	0.70 ± 0.05	0.78 ± 0.04	

The ability to achieve conception (YCON) has a causal effect on values for NLB and NLW, as expected (Figure 1). The relationship from YCON and YNLW to PWT indicates that there is a phenotypic association between these traits. However, the direction of the relationship shows that the mechanisms behind increasing fertility and number of lambs weaned is associated with heavier PWT but increasing PWT will not necessarily cause a response in YCON or YNLW. The causative relationship of PWT on YNLB suggests increased weight, possibly as an indicator of maturity, is leading to increased litter sizes.

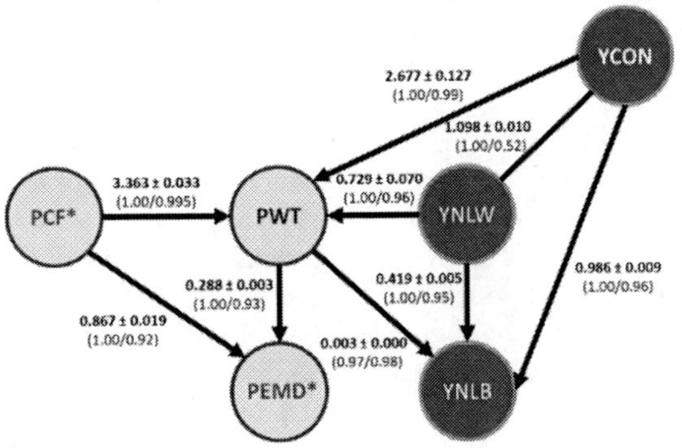

Figure 1. Graphical model of the phenotypic relationship between body composition and yearling reproduction. Size of the effect in bold with the probability of the relationship and then direction of causation using bootstrap techniques shown in parentheses

Genetic Association. The genetic correlations between the post weaning traits and the reproduction traits were weak in maternal breeds (Table 2), which relative to other breeds (eg. Merino) are heavier and fatter at young ages. This suggests that genetic selection for post-weaning body composition is likely to have a limited impact on yearling reproduction. The graphical model of the genetic association between the traits (Figure 2) is very different to Figure 1, indicating that the relationships are different at the genetic level. As observed in the phenotypic model (Figure 1) the association between the post-weaning traits remains strong but the causative direction between the traits could not be determined. A causative association between PWT and YNLW was detected and whilst the association was highly probable, the observed effect was very small, with a 1 kg increase in the sires's PWT breeding value (EBV) associated with an increase of only 0.002 in the sire's YNLW breeding value. An indirect association, via PWT, between PCF* and NLW would only see an extra 0.000128 lambs weaned per ewe joined for every extra (genetic) mm of PCF*. Variation in genetic merit for post-weaning traits is largely independent of genes which affect ovulation rate, litter size or lamb survival (Bunter *et al.*, these proceedings). The causative nature of YNLB and YCON on YNLW also means that the association between post-weaning traits and YNLW is due to the litter survival component of YNLW and not the variation associated with fertility or litter size. This may be related to known associations between lamb birth weight (which affects survival) and post-weaning development traits (Bunter *et al.*, these proceedings).

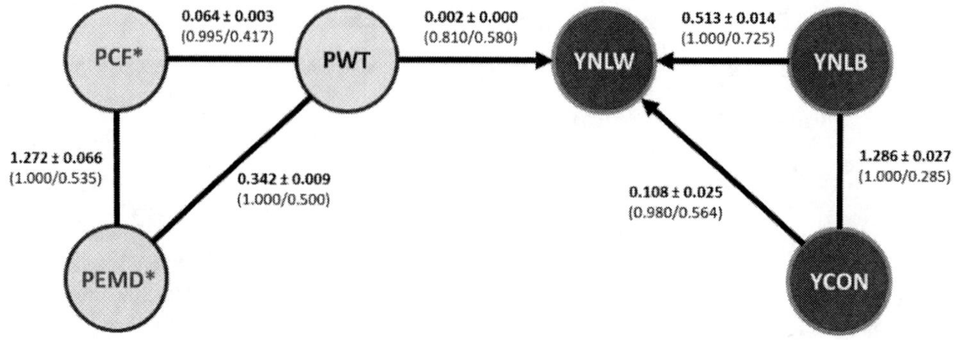

Figure 2. Graphical model of the genetic (sire breeding values) relationship between body composition and yearling reproduction. Size of the effect in bold with the probability of the relationship and then direction of causation using bootstrap techniques shown in parentheses

CONCLUSIONS

This study shows that modelling the relationship between body composition and yearling reproduction can be complex and not simple to interpret and the association between traits, and the causative associations between the traits, are strongly associated with the ability to disentangle the environmental and genetic components. In both phenotypic and genetic graphical models the effect of PCF* and PEMD* appears to be moderated through PWT. As has been shown from the genetic correlations and the graphical modelling, selection for higher PCF* and PEMD* sires will have little direct genetic impact on the reproductive performance of the resulting progeny as yearling dams, although it may influence the ease with which target weights are met pre-joining.

ACKNOWLEDGEMENTS

Sheep Genetics is funded by Meat and Livestock Australia. The authors acknowledge the data contribution from the Australian sheep breeders.

REFERENCES

Brown D.J., Huisman A.E., Swan A.A., Graser H.U., Woolaston R.R., Ball, A.J., Atkins, K.D. and Banks RG (2007) *Proc. Assoc. Adv. Anim. Breed. Genet.* **17**: 187.

Brown D.J. and Swan A.A. (2016) *Anim. Prod. Sci.* **56**: 690.

Bunter K.L., Swan A.A., Gurman P.M., Boerner V., McMillian A.J. and Brown D.J. (2019) *Proc. Assoc. Adv. Anim. Breed. Genet.* **23**: these proceedings.

Friedman N, Goldszmidt M. and Wyner A (1999) *Proc. 15th Annual Conf. Uncert. Art. Intl.*, pp. 196.

Gilmour A.R., Gogel B.J., Cullis B.R. and Thompson R. (2009) 'ASReml User Guide Release 3.0.' (VSN International Ltd: Hemel Hempstead).

Scutari M (2010) *J. of Stat. Softw.* **35**: 1.

Tsamardinos I, Aliferis CF and Statnikov A (2003). *Proc. of the 16th Int. FL. Art. Intl. Res. Soc. Conf.*, 376.

Walkom S.F., Brien F.D., Hebart M.L., Fogarty N.M., Hatcher S. and Pitchford W.S. (2014) *Anim. Prod. Sci.* **54**: 802.

Walkom S.F. and Brown D.J. (2016) *Anim. Prod. Sci.* **57**: 20.

Valente B.D., Rosa G.J.M., Silva M.A., Teixeira R.B. and Torres R.A. (2011) *Gene. Sel. Evol.* **43**: 37.

Proc. Assoc. Advmt. Anim. Breed. Genet. 23:107-110

GENETIC PARAMETERS FOR GROWTH TRAITS IN HAMPSHIRE SHEEP IN MEXICO

L. De la Cruz[1,2], S.F. Walkom[1], H.G. Torres[3] and A.A. Swan[1]

[1]Animal Genetics & Breeding Unit*, University of New England, Armidale, NSW, 2351 Australia
[2]National Institute of Forestry, Agriculture and Livestock Research in Mexico. Pachuca de Soto, Hidalgo, Mexico
[3]Colegio de Postgraduados. Carretera México-Texcoco Km. 36.5, Montecillo, Texcoco 56230, Estado de México

SUMMARY

Univariate and bivariate linear models via Restricted Maximum Likelihood (ASReml) were used to estimate heritability, phenotypic and genetic correlation for growth traits measured at birth (BW), weaning (WW) around 60 days, 90 days (W90), 120 days (W120) and 150 days (W150) in Hampshire sheep raised in Mexico. From 2005 to 2009 a total of 1,133 individual records of lambs born on 10 farms from 612 ewes and 63 sires were analysed. Direct heritability estimates for BW, WW, W90, W120 and W150 were 0.38±0.11, 0.15±0.08, 0.17±0.09, 0.18±0.07 and 0.14±0.06, respectively. All direct and maternal permanent environmental effect correlations were positive for BM, WW, W90, W120 and W150. The phenotypic correlations between all traits were positive and ranged from 0.29 to 0.96. The genetic correlations among growth traits were positive ranging from 0.35 to 0.94. The genetic parameter estimates presented here can be used to estimate breeding values to support genetic improvement programs for the Hampshire breed in Mexico.

INTRODUCTION

Sheep production in Mexico has increased over recent years, partly because of the demand created by a growing population with an increased desire for consumption of a traditional dish called Barbacoa. The Mexican sheep sector is mainly focused towards meat production (Partida *et al.* 2012) with growth in the use of specialized breeds such as the Hampshire (approximately 70% of commercial flocks in central Mexico) leading to recent increases in both productivity and profitability. The Mexican Hampshire breed has a database of 11,529 animal registrations (UNO 2016). However, knowledge of genetic parameters for key traits is very limited and thereby, limits the ability to implement any systematic breeding programs on farm to increase growth rates and meat production. The objective of this study was to estimate genetic parameters for growth traits at different ages, from birth until 150 days for Mexican Hampshire sheep.

MATERIALS AND METHODS

Weights records for 1,133 lambs were obtained from 10 Hampshire sheep breeding farms in the central part of Mexico (States of Hidalgo, Tlaxcala and Puebla), which participated in the regional reference sire program between 2005 and 2009 (UNO 2016). The 1,133 lambs were progeny of 63 sires and 612 ewes with a pedigree of 1,711 over 3 generations available for the Mexican Hampshire sheep population. Traits considered in this study were birth weight (BW), weaning weight (WW) around 60 days, weight at 90 days (W90), weight at 120 days (W120) and weight at 150 days (W150). Data editing and descriptive statistics were performed in R (R Core Team 2018) prior to using an animal model evaluation in ASReml (Gilmour *et al.* 2009) in a series of uni-variate and bi-variate analyses between the weight traits. Significant fixed effects fitted in the model included gender (male

* A joint venture of NSW Department of Primary Industries and the University of New England

and female), birth type (simple, twins and triplets), age of the animal and contemporary group. Contemporary group was defined by flock, year and season of birth (early, normal and later) for all the weight traits. For the weight traits from weaning onwards, the season at weaning was included in the contemporary group definition. For BW, WW, W90, W120 and W150 a total of 52, 69, 64, 68 and 66 contemporary groups were fitted, respectively. Variances and covariances were estimated to get the genetic parameters via Restricted Maximum Likelihood (ASReml) using uni-variate linear models with phenotypic and genetic correlations between traits estimated from a series of bivariate analysis. The general animal model fitted to the weight traits was:

$$Y_{ijklmn}=cg_i + g_j + bt_k + c_l + a_m + pe_n + e_{ijklmn}$$

where: Y_{ijklmn} is the observation for the growth traits (BW, WW, W90, W120, W150) measured on animal m, cg_i is the effect of the contemporary group i, g_j is the effect of the gender j, bt_k is the effect of the birth type k, c_l is the age of animal as a covariate (not fitted for BW), a_m is the random additive genetic effect of animal m, pe_n is the random permanent environmental effect of dam and e_{ijklmn} is the random error associated with each observation. Variance structures assumed for the random effects were: var(a) = $A\sigma^2_a$, var(m) = $A\sigma^2_m$, var(pe) = $I\sigma^2_{pe}$, and var(e) = $I\sigma^2_e$ where A is the matrix of pedigree relationships, and I refers to identity matrixes of appropriate order. Log likelihood ratio tests were used to test the significance of maternal genetic and permanent environment effects on each trait in univariate models.

RESULTS AND DISCUSSION

The mean weights at BW, WW, W90, W120 and W150 were 4.13, 24.0, 32.5, 41.4 and 50.1 kg, respectively (Table 1). Similar values to the means for BW, WW and W150 were reported in another study in Mexican Hampshire lambs (UNO 2016).

Table 1. Mean (kg), standard deviation, coefficient of variation (%), minimum, maximum weight and mean age (days) of growth traits in Mexican Hampshire sheep

Traits*	No animals	Mean	SD	CV (%)	Minimum	Maximum	Age (±SD)
BW	1133	4.1	1.1	27	1.0	8.0	-
WW	1133	24.0	5.5	23	11.0	43.0	63.7±5.5
W90	1133	32.5	7.0	22	14.0	55.0	91.1±6.5
W120	1133	41.4	8.8	21	19.0	73.0	122.1±5.7
W150	1133	50.1	10.4	21	21.0	84.0	154.4±10.1

*BW: Birth Weight; WW: Weaning Weight; W90: Weight at 90 days; W120: Weight at 120 days; W150: Weight at 150 days; CV: Coefficient of variation: SD: Standard deviation

Based on the log likelihood ratio test, maternal permanent environment effects were significant for BW, WW and W90. The shallow pedigree, low progeny per dam and a lack of weight records on the dams limited the ability to estimate a maternal genetic effect. Previous studies have shown that maternal genetic variation exists for weight traits (Brown and Swan 2016), reported low material heritabilities for growth traits (from 0.18±0.01 to 0.20±0.02).

Direct heritability estimates for BW, WW, W90, W120 and W150 were 0.38±0.11, 0.15±0.08, 0.17±0.08, 0.18±0.07 and 0.14±0.06, respectively (Table 2). The estimate for BW is inconsistent with previous studies, where authors generally found lower heritability estimates ranging from

0.14±0.03 to 0.21±0.03 (Fogarty 1995; Safari *et al.* 2005; Manzanilla Pech *et al.* 2012). The slightly higher heritability is likely to be in part due to inability to disentangle maternal and additive genetic variation. Brown and Swan (2016) reported a similar heritability estimate of 0.18±0.01 for weight at 120 days. However, for W150 estimated heritability was lower than the literature estimates ranging from 0.21±0.01 to 0.33±0.02 (Fogarty 1995; Safari *et al.* 2005). In general, the tendency for estimates of direct heritability to increase with age (Yazdi *et al.* 1997) was not observed in this study. The reason for this inconsistency may be due to the relatively shallow pedigree information (3 generations) and small size of the data set.

Table 2. Estimated additive variance (σ^2_d), maternal permanent environmental variance (σ^2_{pe}), phenotypic variance (σ^2_p), estimated heritability (h^2_d) for direct genetic effect and the variance ratio for permanent environment effects (c^2) for growth traits in Hampshire breed in Mexico

Traits*	σ^2_d	σ^2_{pe}	σ^2_p	h^2_d	c^2
BW	0.33±0.10	0.07±0.10	0.87±0.04	0.38±0.11	0.09±0.04
WW	2.67±1.52	1.43±0.80	17.91±0.83	0.15±0.08	0.08±0.05
W90	4.45±2.30	2.15±1.21	26.85±1.25	0.17±0.09	0.08±0.05
W120	6.81±2.81	-	37.41±1.72	0.18±0.07	-
W150	7.02±3.34	-	50.12±2.27	0.14±0.06	-

*For the trait abbreviation see Table 1.

Table 3. Direct genetic and permanent environmental of dam correlations (above diagonal) and phenotypic correlation (below diagonal) of growth traits in Hampshire sheep breed in Mexico

Trait*	Direct genetic and phenotypic					Permanent environmental of dam		
	BW	**WW**	**W90**	**W120**	**W150**	**BW**	**WW**	**W90**
BW		0.64±0.23	0.35±0.26	0.40±0.20	0.43±0.22		0.34±0.34	0.69±0.33
WW	0.38±0.03		0.85±0.10	0.83±0.08	0.79±0.11	-		0.99±0.05
W90	0.33±0.03	0.89±0.01		0.90±0.05	0.87±0.08	-	-	
W120	0.29±0.03	0.80±0.01	0.90±0.01		0.94±0.02	-	-	-
W150	0.29±0.03	0.75±0.01	0.85±0.01	0.96±0.00		-	-	-

*For the trait abbreviation see Table 1.

The phenotypic correlation between the weight traits were positive and moderate to strong ranging from 0.29 to 0.96 (Table 3). The weakest correlations were observed between birth weight and the other weight traits ranging from 0.29 to 0.38. Similar values were estimated in previous studies ranging between 0.21 and 0.90 (El Fadili *et al.* 2000; Brown and Swan 2016). The genetic correlations between the weight traits ranged from 0.35 to 0.94±0.02. These results are similar to other previous finding in other breeds, which were in a wide range from 0.29 to 0.92 (Kariuki *et al.* 2010). Low to high genetic correlations were estimated between BW and the other weight traits (range 0.35 to 0.64). High genetic correlation between BW and WW (0.64) indicates that selection for WW will result in a significant correlated response in BW. This will allow Hampshire sheep breeders in Mexico to improve growth rates and weights in the lambs without increasing the rate of dystocia, a common issue due to broad shoulders (UNO 2016). High genetic correlations between the later weights at W90, W120 and W150 suggest that Mexican sheep breeders looking to breed for higher growth rates and larger lambs can get away with a single later recording point to improve weights at all stages after birth. We recommend that this occur at W150 due to the proximity of the weight to the final sale age.

CONCLUSIONS

The heritabilities estimated in this study were reasonably consistent with estimates presented in a range of studies, albeit slightly lower. However, in order to develop genetic evaluation programs for Hampshire sheep, it is recommended that the Mexican sheep breeders continue to collect weight records on lambs across ages for future analyses. High correlations between the later weights at W90, W120 and W150 suggested that selection for W90 and W120 days will improve W150 days at sale age.

ACKNOWLEDGMENTS

The authors acknowledge the support and data provided from the Reference Sire Program established by the National Institute of Forestry, Agriculture and Livestock Research. The first author's PhD studies are supported by the Mexican National Council for Science and Technology (CONACYT).

REFERENCES

Brown D. and Swan A. (2016) *Anim. Prod. Sci.* **56:** 1449.

El Fadili M., Michaux C., Detilleux J. and Leroy P. (2000) *Small Ruminant Res.* **37:** 203.

Fogarty N. (1995) 'Genetic parameters for live weight, fat and muscle measurements, wool production and reproduction in sheep: a review, Animal Breeding Abstracts.'

Gilmour A.R., Gogel B., Cullis B., Thompson R. and Butler D. (2009) *VSN International Ltd, Hemel Hempstead, UK*

Kariuki C., Ilatsia E.D., Kosgey I.S. and Kahi A.K. (2010) *Trop. Anim. Health Prod.* **42:** 473.

Manzanilla Pech C.I.V., Torres Vazquez, J.A., Borrayo Zepeda A., Rios Utrera A., Rodriguez, B., Jose J., Martinez Vealzquez G., Vega Murillo V.E. and Montano Bermudez m. (2012) *Revista mexicana de ciencias pecuarias* **3**: 487.

Partida J.A., Vázquez E., Rubio M.S. and Méndez D. (2012) *J. Food Res.* **1:** 141.

R Core Team (2018) R: A language and environment for statistical computing. R Foundation for Statistical Computing. Vienna, Austria.

Safari E., Fogarty N. and Gilmour A.R. (2005) *Livest. Prod. Sci.* **92:** 271.

UNO (2016) http://www.uno.org.mx/catalogos/catalogoovinosjunio2014.pdf.

Yazdi M., Engström G., Näsholm A., Johansson K., Jorjani H. and Liljedah L.-E. (1997) *Anim. Scie.* **65:** 247.

Proc. Assoc. Advmt. Anim. Breed. Genet. 23:111-114

THE INHERITANCE OF FLIGHT DISTANCE AS A MATERNAL BEHAVIOUR SCORE OF THE DAM AND ITS IMPACT ON LAMB SURVIVAL

J.C. Greeff[1], A.C. Schlink[1], D. Blache[2] and G.B. Martin[2]

[1]Department of Primary Industries and Regional Development, South Perth, WA, 6151, Australia
[2]UWA Institute of Agriculture, University of Western Australia, Crawley, WA, 6009, Australia

SUMMARY

This study was carried out to test two hypotheses: 1) whether flight distance of the dam (scored at tagging of her lamb within 24 hours of birth) is an indicator of her maternal behaviour and is a dam trait correlated with the survival rate of her lambs; and 2) whether the genetic and permanent maternal environmental effects of survival rate differ between single and multiple born lambs. The results showed that flight distance was genetically correlated only with survival rate at marking. The direct genetic effects for survival rate at birth, marking or weaning did not differ significantly between single- and multiple-born lambs, but the permanent maternal environmental effects were more important in multiple- than in single-born lambs. These observations support the notion that ewes that rear multiple-born lambs should be retained as replacement ewes in breeding programs.

INTRODUCTION

Lamb survival is key determinant of a profitable lamb production system, yet it is estimated that lamb losses in Australia amount to $540 million annually (Lane *et al.* 2015). Lamb survival is affected by a variety of genetic and environmental factors (Brien *et al.* 2014). Several lines of evidence demonstrate the importance of the genetics of maternal behaviour: 1) Ewe temperament is a heritable trait and survival rate is higher for lambs from calm ewes than for lambs from nervous ewes (Murphy 1999); 2) Maternal rearing ability to weaning is also heritable and can be improved by selection (Cloete *et al.* 2009); 3) Maternal behaviour score is heritable (Brown *et al.* 2016) suggesting that it could be improved by selection.

On the other hand, Bunter *et al.* (2018) reported that litter size at lambing influences genetic evaluation of maternal rearing ability and suggested that rearing ability traits should be defined separately by litter-size class to improve the accuracy of genetic evaluation for rearing ability. This paper therefore investigates the inheritance of flight distance as a maternal behaviour trait and survival rate of the lambs, both as traits of the dam, and this aims to elucidate the effect of litter size on the genetic parameters of lamb survival and flight distance.

MATERIALS AND METHODS

Resources. The data were collected on the Breech Strike Resource flocks of the Department of Primary Industries and Regional Development (previously the Department of Agriculture and Food) in Western Australia. This flock consisted of approximately 1,000 ewes that were annually mated to 22 sires. The total dataset consisted of 16,788 repeated records that were collected over the lifetime of 4,767 dams that had been mated annually, from 2005 to 2018, to one of 243 sires.

Management and measurements. Ewes were mated in February/March and lambs were born in July/August. Body weights and body condition score (1 to 5) were recorded on all ewes pre- and post-mating. Two weeks prior to lambing, each sire's lambing ewe group was drafted off, weighed, condition scored and placed on a lambing plot to obtain accurate pedigrees of the lambs at lambing. This resulted in lambing plot and sire of the lambs being confounded within year. However, link sires across years were rotated between mating groups to ensure that repeat mating groups don't lamb in the same paddock again. Between 4 and 18 linked sires were used between years.

The lambs were tagged and weighed within 24 hours after birth. Date of birth, dam identification, gender of lamb, birth status (single, twin or triplet) and lambing difficulties were recorded at birth. During the process of tagging and recording birth information, flight distance, as an indicator of mothering ability, was scored from 1 (dam stays close to lamb) to 6 (>50 meters away from scorer) based on the average distance between the dam and the lamb. At marking, approximately 4 to 5 weeks after lambing, the lambs were tail docked, weighed and identification checked. They were weaned and weighed at an average age of 110 days. All deaths from birth to weaning were recorded.

Statistical analyses. Three survival rate categories were created as traits of the dam: survival at birth (within 24 hours after birth), survival from birth to marking, and survival from marking to weaning. The data were analysed with ASREML (Gilmour *et al.* 2015). A sire model for dam of the lamb, with repeat measurements of the dam (of the lamb) as an additional random factor, was fitted to estimate additive genetic variance and permanent maternal environmental effects for flight distance, survival at birth, survival from birth to marking, and survival from marking to weaning, as traits of the dam. The flight distance data were treated as normally distributed. By contrast, the survival data were binary (alive = 1; dead = 0) so were subjected to a binomial analysis with a logit link function. Year of birth, lamb gender, litter size, dam age, lambing paddock, and dam body weight and condition score (pre-mating, post-mating, pre-lambing) were fitted as covariates. All interactions between fixed effects were initially fitted. Statistically non-significant ($P < 0.05$) factors were dropped from the model until the final model only contained statistically significant factors.

The same analyses were carried out where the dataset was split into sets containing only singletons or only multiples. The phenotypic variance (Vp) was calculated as the sum of the sire variance, permanent maternal environmental variance and error variance. As this analysis was on a logistic scale, a variance of 3.289 was used for the error. The heritability of survival rate was calculated as 4 times the sire variance as a proportion of the phenotypic variation. The importance of the permanent maternal environmental effect was calculated as the proportion of the permanent maternal environmental variation relative to the phenotypic variation. A series of bivariate analyses were then carried out between the survival traits and flight distance, using the significant fixed factors from the univariate analyses of the different traits in the model to estimate the genetic covariance between flight distance and survival traits. The genetic correlation (r_g) was estimated as the covariance between flight distance and the survival traits divided by the square root of the product between the variance of flight distance and that of the survival traits.

RESULTS-AND DISCUSSION

Table 1 shows the number of records, means and variances for the three survival traits and genetic parameters for the combined dataset, separately for single- and multiple-born lambs. Survival rates were 0.95 at birth, 0.87 from birth to marking, and 0.98 from marking to weaning, resulting in 81 lambs surviving per 100 lambs born. Year of birth affected all traits ($P < 0.01$). Larger litters had lower survival rates at birth, marking and weaning ($P < 0.01$). Survival rates at marking and at weaning were lower ($P < 0.01$) for older ewes. However, older ewes stayed closer to their lambs at tagging than younger ewes ($P < 0.001$). For flight distance, interactions ($P < 0.001$) were observed between year of birth and litter size, and between year of birth and age of dam.

Heritability estimates (h^2_D). Where the dataset was split into single- and multiple-born lambs (Table 1), multiple lambs had higher phenotypic variances at birth, marking and weaning. Heritability estimates of survival rate in the total dataset were moderate (0.24 ± 0.09) at birth, low (0.09 ± 0.04) at marking, and not significantly different from zero at weaning. The heritability estimates of survival rate of multiple-born lambs at birth, marking and weaning were higher than those of single-born lambs, but all these estimates had large standard errors so were not significantly (2 x SE) different from zero.

The heritability of flight distance was low (0.07 ± 0.02). However, for single- and multiple-born lambs, the heritability estimates were considerably higher than those of the combined dataset. The heritability of flight distance of multiple-born lambs was higher ($P < 0.001$) than that of single-born lambs (0.33 ± 0.06 vs 0.17 ± 0.04), suggesting that survival rate as a dam trait may not be genetically the same trait for single and multiple born lambs.

Permanent environmental effects (m^2_{pe}). A moderate permanent maternal environmental effect of 0.25 (± 0.03) was found for survival rate at birth, which decreased to 0.10 (± 0.02) at marking. It had no effect on survival rate at weaning, showing the importance of maternal behaviour early in life. For flight distance a moderate permanent maternal environmental effect of 0.26 (± 0.01) was found in the combined dataset. The effect was more than five times that for multiple born lambs (0.47 ± 0.02) compared to single born lambs (0.09 ± 0.02).

Table 1. Number of records, means ± standard deviation (sd), variances and genetic parameters of survival rate as a trait of the dam at birth, marking and at weaning, for the combined dataset and for single- and multiple-born lambs

Parameter	Survival rate			Flight
	Birth	Marking	Weaning	distance
Total dataset				
No. of records	15,224	14,445	12,819	14,682
Mean ± sd	0.95 ± 0.22	0.87 ± 0.33	0.98 ± 0.12	3.69 ± 1.50
Vp	4.74	3.76	3.56	0.82
h^2_D ± SE	0.24 ± 0.09	0.09 ± 0.04	0.19 ± 0.16	0.07 ± 0.02
m^2_{pe} **± SE**	0.25 ± 0.03	0.10 ± 0.02	0.03 ± 0.07	0.26 ± 0.01
Single births				
No. of records	6,763	6,503	5,918	6,555
Mean ± sd	0.96 ± 0.19	0.91 ± 0.29	0.99 ± 0.09	3.70 ± 1.40
Vp	3.54	3.61	3.74	0.80
h^2_D ± SE	0.02 ± 0.18	0.05 ± 0.08	0.33 ± 0.21	0.17 ± 0.04
m^2_{pe} **± SE**	0.07 ± 0.09	0.07 ± 0.04	0.04 ± 0.07	0.09 ± 0.02
Multiple births				
No. of records	8,461	7,942	6,901	8,127
Mean ± sd	0.94 ± 0.24	0.84 ± 0.37	0.98 ± 0.15	3.69 ± 1.59
Vp	5.57	3.93	3.61	0.99
h^2_D ± SE	0.17 ± 0.12	0.09 ± 0.05	0.12 ± 0.19	0.33 ± 0.06
m^2_{pe} ± SE	0.37 ± 0.04	0.14 ± 0.02	0.06 ± 0.08	0.47 ± 0.02

Correlations. Table 2 shows the phenotypic, genetic and environmental correlations between flight distance and survival rate as a trait of the dam at birth, marking and at weaning. Correlations between flight distance and survival rate traits at birth and weaning were very low or not significantly different from zero, as were the genetic correlations at birth and weaning. The only significant genetic correlation was between flight distance and survival rate at marking (0.64 ± 0.20).

Table 2. Phenotypic (r_p), genetic (r_g) and environmental (r_e) correlations and standard errors (SE) between flight distance at tagging and survival rate at birth, marking and weaning as traits of the dam

Trait	Flight distance		
	$r_p \pm$ SE	$r_g \pm$ SE	$r_e \pm$ SE
Survival rate at birth	-0.03 ± 0.01	-0.12 ± 0.24	-0.03 ± 0.00
Survival rate at marking	-0.01 ± 0.01	-0.64 ± 0.20	0.01 ± 0.00
Survival rate at weaning	0.05 ± 0.02	-0.30 ± 0.45	0.06 ± 0.01

CONCLUSIONS

Survival rate as a trait of the dam at birth, was a heritable trait. This study did not show major differences in direct heritability estimates for survival rate at birth, marking and weaning, in the separate estimates for single- and multiple-born lambs. However, the permanent environmental effect was more important for survival rate in multiple-born than in single-born lambs at both birth and marking.

The direct heritability estimate, and the permanent environmental effect of flight distance were also significantly greater in multiple-born than in single-born lambs, suggesting that maternal behaviour as scored by flight distance is an important factor in the survival of multiple-born lambs. These observations support the conclusion of Hatcher *et al.* (2014) that ewes that consistently rear twins should be retained rather than ewes that consistently rear a single lamb. We conclude that, in breeding programs, permanent environmental effects should be accounted for more accurately to identify ewes that consistently rear multiple born lambs. More research on the inheritance and importance of permanent environmental factors is required on this issue as well as the underlying physiological causes of this phenomenon.

REFERENCES

Brown D.J., Fogarty N.M., Iker C.L., Ferguson D.M., Blache D. and Gaunt G.M. (2016) *Anim. Prod. Sci.* **56**: 767.

Bunter K.L., Swan A.A., Brown D.J., Brien F.D. and Smith J. (2018) *Anim. Prod. Sci.* **58**: 791.

Cloete S.W.P., Misztal I. and Olivier J.J. (2009) *Proc. Assoc. Advmt. Breed. Genet.* **18**: 104.

Forbes F.D., Cloete S.W.P, Fogarty N.M., Greeff J.C., Hebart M.L., Hiendleder S., Hocking Edwards J.E., Kelly J.M., Kind K.L. Kleeman D.O. Plush K.L. and Miller D.R. (2014) *Anim. Prod. Sci.* **56:** 656.

Gilmour, A. R., Gogel, B. J., Cullis, B. R., Welham, S. J. and Thompson, R. (2015). ASReml User Guide Release 4.1 Functional Speciation, VSN International Ltd, Hemel Hempstead, HP1 1ES, UK www.vsni.co.uk.

Hatcher S., Atkins K.D. and Mortimer S.I. (2014) Paper 884. *Proc. 10th Wrld. Congr. Genet. Appl. Livest. Prod.*

Lane J., Jubb T., Shephard R., Webb-Ware L. and Fordyce G. (2015) 'Priority list of endemic diseases for the red meat industries.' Meat and Livestock Australia: Sydney.

Murphy P.M. (1999) PhD thesis, University of Western Australia.

Proc. Assoc. Advmt. Anim. Breed. Genet. 23:115-118

GENETIC EVALUATION AND RELATIONSHIP ACROSS AGES FOR DAG SCORE IN MATERNAL SHEEP

A.J. McMillan, S.F. Walkom and D.J. Brown

Animal Genetics & Breeding Unit*, University of New England, Armidale, NSW, 2351 Australia

SUMMARY

Daginess in sheep is an undesirable trait with both economic and welfare implications. While the trait has been investigated in Merino sheep, this is not the case for maternal sheep. With sufficient records now available from industry and research flocks via Sheep Genetics the genetic parameters for dag score can now be estimated. The heritability of dag score ranged from 0.13 ($\pm$0.01) to 0.38 ($\pm$0.02) across age stages with the highest heritability occurring at the yearling stage. Given the heritability it should be possible for breeders to make genetic progress towards less daggy maternal sheep, which as an indirect selection tool will potentially assist to reduce labour costs, wool losses and flystrike incidence. Positive moderate genetic correlations between age classes (0.08 to 0.83) indicate that selection based on phenotypes recorded at any age will lead to reduced dag score across investigated stages. The results suggest that breeders should be focussed on recording dag score when the environmental conditions promote the greatest expression of genetic merit, than scoring at a specific age class. However, in maternal sheep the greatest phenotypic variation in dag score appears to occur in yearling sheep.

INTRODUCTION

Dag (measured by dag score) is the accumulation of faecal matter in wool around the breech of the animal, which is associated with increased flystrike incidence within the Australian sheep population. Previous studies have indicated that flystrike is costing $280m dollars annually (Sacket *et al.* 2006) as a result of sheep losses, cost of treatment and loss of wool as well as carcase production and value. In response to public concern and desire for management practices such as mulesing to be phased out sheep breeders are utilising indicator traits like dag score to reduce flystrike incidence (Brown *et al.* 2010). Dag score (scouring) has been shown to be related to flystrike in previous studies (James 2008; Greeff and Karlsson 2009; Smith *et al.* 2009) there are also costs associated with loss of production as well as crutching costs (Sacket *et al.* 2006).

The genetic evaluation of dag score has previously been reported in the Australian Merino population (Brown *et al.* 2010), however, the growth of maternal cross merino ewe flocks and self-replacing maternal flocks has influenced the interest within maternal stud breeders to utilise dag score in their breeding programs. The heritability of dag score has been shown to be moderate in Australian Merino sheep (0.20-0.26, Brown *et al.* 2010), (0.37-0.63, Greeff *et al.* 2013) and in another study where it ranged from 0.07 to 0.32 for animals recorded at 30 day intervals from weaning to hogget stage (Pollot *et al.* 2004).

The recent increase in dag score recording by maternal sheep breeders and records from the Information Nucleus and resource flocks (Fogarty *et al.* 2007) has led to an increase in dag score phenotype submission to Sheep Genetics, the paper investigates the genetic parameters for dag score within the maternal population and the relationship of dag score in the Weaning (Wdag), Post-weaning (Pdag), Yearling (Ydag) and Hogget (Hdag). The effect of modelling genetic groups and sire x flock effects were also explored.

---

* A joint venture of NSW Department of Primary Industries and the University of New England

MATERIALS AND METHODS

Data. Pedigree and performance were extracted from the Sheep Genetics LAMBPLAN (Maternal) database (Brown *et al.* 2007). This database stores pedigree and performance records submitted by ram breeders that are used for LAMBPLAN (Maternal) national genetic evaluations.

A summary of records from the maternal database are presented in Table 1. Currently there are 22 flocks who have submitted 39,035 dag phenotypes across the 4 age stages investigated. The most popular stage for dag score recording was at weaning with almost 18,519 records with yearling records and post weaning records combined made up the majority of the remaining records. Most flocks who recorded dag score had records from a number of stages across different years in this dataset. There were 6,301 animals which had records across multiple stages.

The dag score phenotype are visually scored from 1-5 with a score 1 having no dags in the breach area up to a score of 5 which has an accumulation of dags in breech area and down the legs of the animal (AWI. 2013). With increasing age the dag score phenotypes showed both an increase in score but also an increase in the variation for score.

Table 1. Summary of dag score phenotypes submitted to Sheep Genetics by maternal sheep breeders across age classes.ncg; number of contemporary groups, ngg; number of genetic groups

Trait	Records	Mean	sd	Pedigree	Sires	Dams	Flocks	Ncg	ngg
Wdag	18519	1.40	0.67	39321	4696	19506	12	213	35
Pdag	7762	1.64	0.85	28283	4843	16765	10	70	38
Ydag	11784	1.83	0.93	36222	5445	20448	11	159	42
Hdag	970	1.98	0.87	11243	3300	7058	9	21	21

Statistical Analysis. Parameters were estimated in bivariate analyses for each trait combination, ASReml (Gilmour *et al.* 2015) was used fitting an animal model. The model included direct genetic, dam permanent environment effects. Fixed effects of age of animal and age of dam were fitted as covariates with both linear and quadratic effects for dam age. Birth and rearing effects were treated as non-interacting fixed effects ranging from 1-4. Flock, year of birth, sex, the date of measurement and breeder management group were used to define contemporary group.

RESULTS AND DISCUSSION

Genetic parameter estimates and the genetic and phenotypic correlations across age classes are presented for the base model (Table 2) and the extended model which included genetic groups and the sire by flock interaction fitted as random (Table 3). These terms are used within Sheep Genetics Evaluations and improve the fit of models especially for analysing industry data structure and recording are not always balanced (Brown et al. 2007). The results showed that the inclusion of genetic groups and the sire x flock term within the model had no significant impact on the additive genetic variance nor the heritability.

The heritability of dag score at weaning, post-weaning, yearling and hogget stage was 0.13, 0.27, 0.38, 0.20, respectively (Table 2). These pattern are similar to that estimated by (Pollot *et al.* 2004) in Merino sheep which had a low heritability at weaning (0.07) before peaking at a moderate heritability (0.32) at 270 days and then declining to lower estimates of 0.08, 0.13 and 0.16 for 300, 330 and 360 days of age respectively. Although the heritability for these maternal animals may be higher due to industry recording likely only being undertaken when dag was more strongly expressed with the mean dag score ranging from 1.4 to 1.98 vs 0.36 to 1.50 (Pollot *et al.* 2004) although those merino animals were scored with a slightly different scoring method (Larsen *et al.* 1994).

The highest heritability estimate, 0.38, was observed for Ydag this is almost double the estimate for late dag score (ldag) (0.20) in Merinos although it should be noted that ldag includes dag score from across yearling, hogget and adult stages (Brown *et al.* 2010).

Phenotypic correlations were positive and range from low (0.19) to moderate (0.45) with stronger correlations for the later stages, environmental effects have a large impact for the trait to be expressed.

Genetic correlations were generally high and positive although there were some deviations between the base model and the extended model. The extended model showed slightly higher correlations between traits although not all of these were significant. However genetic correlations with Hdag were compromised by the small number of records at this stage leading to high standard errors. Given these correlations recording and selection for dag score at any of the stages would result in a positive impact on the other stages.

Table 2. Phenotypic variance ($\hat{\sigma}_p^2$) and direct heritability (h^2) with phenotypic correlations above the diagonal and genetic correlations below the diagonal for base model

Trait	$\hat{\sigma}_p^2$	h^2	Wdag	Pdag	Ydag	Hdag
Wdag	0.35±0.00	0.13±0.01		0.19±0.01	0.17±0.02	0.08±0.03
Pdag	0.66±0.01	0.27±0.02	0.66±0.07		0.45±0.02	0.19±0.04
Ydag	0.83±0.01	0.38±0.02	0.54±0.07	0.83±0.04		0.66±0.26
Hdag	0.55±0.02	0.20±0.04	0.75±0.16	0.08±0.18	0.60±0.15	

Table 3. Phenotypic variance ($\hat{\sigma}_p^2$) and direct heritability (h^2) with phenotypic correlations above the diagonal and genetic correlations below the diagonal for the model that included genetic groups and the sire by flock interaction

Trait	$\hat{\sigma}_p^2$	h^2	Wdag	Pdag	Ydag	Hdag
Wdag	0.35±0.00	0.13±0.01		0.19±0.01	0.18±0.02	0.08±0.03
Pdag	0.66±0.01	0.26±0.02	0.77±0.08		0.45±0.02	0.18±0.04
Ydag	0.84±0.01	0.38±0.02	0.62±0.07	0.87±0.05		0.81±0.16
Hdag	0.56±0.02	0.16±0.05	0.85±0.24	-0.12±0.23	0.63±0.21	

Given the results it appears that the extended model is appropriate for analysis of the data although the effects estimated were small especially considering the stronger genetic groups effects previously seen in Merinos. Measurements made at the weaning stage had the lowest heritability estimate and also the smallest phenotypic variances however still had strong genetic associations with dag score measured at the other stages. To improve dag score, recording could be at any of the stages with a preference for later recording. However recording would be best when trait expression is maximised regardless of stage.

The LAMBPLAN (Maternal) genetic evaluation has been following the MERINOSELECT analysis approach. Analysing early (edag) and late (ldag) traits with edag including marking and weaning stage records and late dag including records from yearling, hogget and adult records. The estimated genetic correlations are moderate-high across stages with the exception of Pdag-Hdag which particularly suffers from high standard errors due to small number of records. Given these correlations the traits could be analysed either following the MERINOSELECT model with combined late and early trait groups or as individual stage based traits.

CONCLUSIONS

It should be possible to genetically improve dag scores in maternal sheep with appropriate selection and recording as moderate heritability estimates were observed across stages with moderate- high genetic correlations between age stages. Given this visual scoring for dag score should be performed when the trait is showing its greatest expression.

ACKNOWLEDGEMENTS

Sheep Genetics is funded by Meat and Livestock Australia and was made possible through the support of the Australian sheep industry. The authors would also like to acknowledge the contribution of the data from the Sheep CRC Information Nucleus Flock and other industry funded research flocks.

REFERENCES

Australian Wool Innovation (2013) 'Visual sheep scores.' Australian Wool Innovation Ltd: Sydney.

Brown D.J., Huisman A.E., Swan A.A., Graser H-U, Woolaston R.R, Ball, A.J., Atkins K.D. and Banks R.G. (2007) *Proc. Assoc. Advmt. Anim. Breed. Genet.,* **17**: 187.

Brown D.J, Swan A.A and Graser H-U (2010) *Proc. 9th Wrld. Cong. Gen. Appld. Livst. Prod.*:**0924.**

Fogarty N.M., Banks R.G., van der Werf J.H.J., Ball A.J. and Gibson J.P. (2016) *Proc. Assoc. Advmt. Anim. Breed. Genet. **17**: 29.*

Gilmour A.R., Gogel B.J., Cullis B.R., Welham S.J. and Thompson R. (2015) 'ASReml User Guide Relase 4.1'(VSN International Ltd: Hemel Hempstead)

Greeff J.C. and Karlsson, L.J.E. (2009) *Proc. Assoc. Advmt. Anim. Breed. Genet.* **18**: 272.

Greeff J.C., Karlsson, L.J.E. and Schlink, A.C. (2013) *Anim. Prod. Sci.* **54**: 125.

James P. (2008) *Aust. J. Exp. Agric.* **46**: 1.

Larsen J.W.A., Anderson N, Vizard A.L. Andersen G.A. and Hoste H. (1994) *Aust. Vet. J.* **71**: 365.

Pollott G.E., Karlsson L.J.E, Eady S. and Greeff J.C., (2004) *J. Anim. Sci.* **82**: 2852.

Smith J., Brewer, H. and Dyall, T. (2009) *Proc. Assoc. Advmt. Anim. Breed. Genet.* **18:** 334.

Sackett, D., Holmes, P., Abbot, K, Jephcott, S. and Barber, M. (2006). 'AHW.087 Final Report', Sydney:MLA.

Proc. Assoc. Advmt. Anim. Breed. Genet. 23:119-122

GROWTH, CARCASS AND MEAT QUALITY TRAITS OF DORMER AND SOUTH AFRICAN MUTTON MERINO LAMBS

A. Muller[1], T.S. Brand[1,2], J.J.E. Cloete[2] and S.W.P. Cloete[1,2]

[1]Department of Animal Sciences, University of Stellenbosch, Private Bag X1, Matieland, 7602, South Africa
[2]Directorate: Animal Sciences: Elsenburg, Department of Agriculture, Western Cape Government, Private Bag X1, Elsenburg, 7607, South Africa

SUMMARY

This study compared animals from the Dormer and South African Mutton Merino (SAMM) breeds for birth weight, weaning weight, yearling weight, carcass and meat quality traits. Dormers were lighter at birth but heavier subsequently than their SAMM contemporaries. Dormer carcasses had greater fat depths than SAMM's. SAMM meat was lighter with a slightly higher cooking loss than Dormers. The observed breed differences reflect the roles the breeds play in the South African sheep industry.

INTRODUCTION

In South Africa, the Dormer is the most prominent terminal sire breed, while the South African Mutton Merino (SAMM) is the dominant dual-purpose breed (Cloete *et al.* 2014). The Dormer was developed at the Elsenburg Agricultural College in the 1940s when Dorset Horn rams were crossed with German Merino ewes to establish the composite breed (Van Wyk *et al.* 2003). The Dormer plays an important role as a terminal sire breed for crossbreeding with wool breeds. The SAMM originated from the German Merino, which was imported to South Africa in 1932 (Cloete *et al.* 2004c). The foundation flock was kept at Elsenburg, from where it spread throughout South Africa and to other countries such as Australia (Brown and Asadi Fozi 2005). The traits recorded in both breeds in the National Small Stock Evaluation Scheme include birth weight, weaning weight, postweaning weight and reproduction (Schoeman *et al.* 2010). No emphasis is thus directed to wool traits in either breed. Both breeds have a high growth rate and grow out to a high mature weight compared to other South African ovine genetic resources (Van der Merwe *et al.* 2019). Previous studies comparing these breeds for meat traits were based on small sample sizes and animals slaughtered at an age of 18 to 20 months (Cloete *et al.* 2004a; 2012). There is a need to update the earlier results on slaughter traits with information of animals slaughtered at a more reasonable age.

This study therefore aims to evaluate these breeds in terms of growth, as well as carcass and meat traits at an age aligned with industry practice. This aim excluded discussion of other fixed effects or genetic parameters.

MATERIALS AND METHODS

Data were collected from the Dormer and SAMM resource flocks at Elsenburg research farm, Western Cape, South Africa. The background of flocks was reported by respectively van Wyk *et al.* (2003) and Cloete *et al.* (2004c). Selection in both breeds was mostly based on early growth and conformation. Expressed relative to the overall means for weaning weight, mediocre annual genetic gains of 0.2% in Dormers (Van Wyk *et al.* 1993) and 0.1% in SAMM's (Zemuy 2002) were realised. No direct selection pressure was applied to any meat trait. Both breeds remained in the same flock during the study, except when mated within breeds in single-sire groups to rams of the same breed. Both breeds utilised either dryland lucerne or oat fodder crop paddocks during winter and spring, and irrigated pastures that mainly consisted of kikuyu for the rest of the year. Data collection for the weight traits took place from 2007 to 2018. The breed, sex, birth type, age of dam and year of birth of the lambs were recorded. Lamb birth weights of 3,043 lambs were recorded within 24 hours of birth, at weaning (at 103 ± 14 days; n = 2,765) and again as yearlings (at 356 ± 0.44 days; n = 2,155). A total of 201 Dormers and SAMM yearlings, born in 2015 and 2016, were slaughtered at an average age of 392 ± 51 days to assess meat traits. Lambs were weighed 24 hours prior to slaughter (slaughter weight). The sheep were slaughtered at a commercial abattoir, using the techniques previously described by Cloete *et al.* (2004a). The *ante mortem* treatment was similar for all the sheep within year-sex contemporary groups and sheep were slaughtered at random after electrical stunning at 200 V for 4 seconds. The sheep were exsanguinated, 0and carcasses allowed to bleed out before dressing. No electrical stimulation was applied. The dressed carcasses were hung in a chiller at 2^0C for 48 hours (McGeehin *et al.* 2011). The carcass weight, temperature and pH were determined after 48 hours and the dressing percentage was calculated as carcass weight divided by slaughter weight. At this stage, fat depth 25mm off the midline at the 13th rib and at the rump between the 3rd and 4th lumbar vertebrae was measured as described by Cloete *et al.* (2004a). Loin samples of 8 cm were excised from the left side of the *M. Longgissimus lumborum* between the 13th rib and 3rd and 4th lumbar vertebrae. Two 1.5cm thick slices were cut from these steaks and used to measure cooking loss and shear force on one and meat colour and drip loss on the other (Honikel 1998**)**. Individual 20 to 30g meat portions from the first slice were used to determine cooking loss. Samples were placed in thin-walled plastics bags and put in a water-bath at 80^0C for 1 hour. Cooked samples were removed from the water-bath, cooled in cold water, blotted dry and weighed again. Cooking loss was calculated as the difference in sample weight before and after cooking and expressed as a percentage of initial weight. Shear force was determined on these cooked samples using an Instron machine equipped with a Warner-Bratzler shear head (Honikel 1998). Three subsamples with a diameter of 1 cm were removed from the core of each cooled (4°C) sample. Maximum shear force values (N) were recorded for each sample and the mean was calculated. Shear force and tenderness is inversely correlated. The second slice was used to first measure colour by using a colour-guide $45^0/0^0$ colorimeter (BYK-Gardner, USA) to determine L* (lightness), a* (red-green range) and b* (blue-yellow range). Drip loss was then determined by attaching a 20 to 50g meat sample to a string and suspending it in an inflated plastic bag. These bags were left at 4^0C for 24 hours and weighed again to derive drip loss as explained for cooking loss (Honikel 1998).

Data were analysed using ASREML (Gilmour *et al.* 2015). Fixed effects included in the models for all traits were breed (SAMM or Dormer), year of birth (2007-2018 for body weights, 2015-2016 for carcass and meat quality traits), age of dam (2-5 years), sex (male or female) and birth type (single or multiple), two-factor interactions between birth year and sex as well as between birth year and breed as well as age at measurement as linear covariates. The random effects of sire and dam were included throughout for the variation it controlled.

RESULTS AND DISCUSSION

SAMM lambs were 7.3% heavier at birth than Dormers ($P < 0.05$; Table 1). A previous study by Brand *et al.* (1985) also reported that Dormers were significantly smaller than SAMM lambs at birth. In contrast, Dormers were heavier than SAMM contemporaries at weaning (6.8%) and yearling (13.9%) ages ($P<0.05$). Slaughter weight of Dormers tended ($P=0.054$) to be heavier than those of SAMM contemporaries, bearing in mind that this was based on much fewer records compared to the other weight traits. Carcass weight was increased by 10.1% in Dormers compared to SAMM contemporaries. Dressing percentage did not differ between the breeds. Previous studies by Cloete *et al.* (2004a; 2012) on these breeds suggested no significant difference between the two breeds for slaughter weight. However, carcass weight and dressing percentage differed significantly in favour of Dormers in the former study. The present results thus concur with those of Cloete *et al.* (2004a) for carcass weight.

Table 1. Predicted means (± SE) for the effect of breed (Dormer or SAMM) on growth and carcass traits

Trait	Breed		
	Dormer	SAMM	Significance
Birth weight (kg)	4.59 ± 0.06	4.95 ± 0.07	**
Weaning weight (kg)	29.7 ± 0.4	27.8 ± 0.4	**
Yearling weight (kg)	52.5 ± 0.4	46.1 ± 0.5	**
Slaughter weight (kg)	49.3 ± 1.6	44.9 ± 2.4	0.054
Carcass weight (kg)	22.8 ± 0.8	20.7 ± 1.1	*
Dressing percentage (%)	45.8 ± 0.7	45.5 ± 1.1	0.443

* $P < 0.05$; ** $P < 0.01$; actual significance for $P > 0.05$

The ultimate pH recorded 48 h post slaughter did not differ between the breeds (Table 2). An ultimate pH between 5.8-6.0 is considered as undesirable (Devine *et al.* 1993) and the ultimate pH of both breeds was below this range. The tenderness and texture deceases at an ultimate pH of 5.8-6.0. An ultimate pH above 5.8 also influences the flavour, juiciness and aroma of the meat. The proportion of high pH carcasses amounted to 0.075 in Dormers and 0.101 in SAMM's (Chi^2=0.98; degrees of freedom=1; P=0.45). Undesirable high pH carcasses were thus quite infrequent in both breeds. Ultimate pH was heritable in South African sheep (Naudé *et al.* 2018), allowing opportunities for selective breeding.

Table 2. Predicted means (± SE) for the effect of breed (Dormer or SAMM) on meat quality traits

Trait	Breed		Significance
	Dormer	SAMM	
pH48 hr	5.60 ± 0.01	5.58 ± 0.03	0.31
Fat 13th rib (mm)	2.04 ± 0.22	1.21 ± 0.34	*
Fat rump (mm)	5.31 ± 0.49	3.02 ± 0.66	**
Cooking loss (%)	29.1 ± 0.9	31.8 ± 1.4	*
Drip loss (%)	1.91 ± 0.21	1.82 ± 0.27	0.96
Colour L*	34.1 ± 0.5	35.8 ± 0.8	**
Colour a*	13.4 ± 0.3	13.9 ± 0.4	0.09
Colour b*	9.65 ± 0.21	9.87 ± 0.29	0.12
Shear force (N)	50.4 ± 3.2	56.2 ± 4.3	0.14

* $P < 0.05$; ** $P < 0.01$; actual significance for $P > 0.05$

Fat depth differed significantly between breeds at both sites, with Dormers being fatter than SAMM contemporaries. Fat depth at 20 months was independent of breed in a previous study on Dormer and SAMM sheep (Cloete *et al.* 2012). In contrast, Cloete *et al.* (2004a) also reported that Dormers were fatter (P<0.05) than SAMM contemporaries at 18 months. The present analyses use a substantially larger data set that any of the previous studies, while the animals were also slaughtered younger. Age and maturity type possibly combined to give the results that were obtained. Carcasses with subcutaneous fat depth of 1-4 mm fat measured 25mm from the midline at the 13th rib are considered as acceptable in South Africa (Government Gazette 14060 1992). The frequency of carcasses of each breed conforming to this desired fat distribution did not differ (Dormer=0.644 vs. SAMM=0.522, Chi^2=2.33; P=0.13). However, SAMM carcasses were more likely to be leaner (Dormer=0.197 vs. SAMM=0.478, Chi^2=16.0; P<0.01) and Dormer carcasses fatter (Dormer=0.159 vs. SAMM=0.000, Chi^2=10.6; P=0.01) than the desired range. The mean cooking loss of SAMM meat was higher than that of Dormer meat (P<0.05; Table 2). Drip loss was not affected by breed (P>0.05). Cloete *et al.* (2004a; 2012) found no differences for cooking loss between Dormers and SAMM's (P>0.05). This study involved younger sheep and a larger sample size, both of which could be causative in the result obtained. Further research is therefore needed. Although Dormer meat may be slightly darker than that of SAMM, the values differ by such a small margin that a consumer might not be able to visually perceive the differences (Cloete *et al.* 2012). The a* and L* values for Domers and SAMM are regarded as acceptable for the average consumer at respectively 9.5 and 34.0 or higher (Khliji *et al.* 2010). There was no significant difference between Dormer and SAMM for meat tenderness.

CONCLUSIONS

This study showed that, although SAMM lambs were heavier at birth, Dormers had higher subsequent weights. The observed breed differences reflect the different roles of the two breeds within the South African sheep industry. The thicker fat cover of Dormers compared to their SAMM contemporaries probably indicate that the focus of selection for growth in this breed was not for lean growth, as in many other sheep-producing countries. This result stems from the absence of meat quality as a selection trait in South Africa's formal recording scheme (Schoeman *et al.* 2010). Clearly this state of affairs is undesirable and requires further effort to align sheep recording in South Africa with international benchmarks.

REFERENCES

Brand A.A., Cloete S.W.P. and De Villiers T.T. (1985) *S. Afr. J. Anim. Sci.* **15**: 155.
Brown D.J. and Asadi Fozi M. (2005) *Proc. Assoc. Advmt. Anim. Breed. Genet.* **16**: 310.
Cloete J.J.E., Hoffman L.C., Cloete S.W.P. and Fourie J.E. (2004a) *S. Afr. J. Anim. Sci.* **34**: 44.
Cloete J.J.E., Hoffman L.C. and Cloete S.W.P. (2012) *Meat Sci.* **91**: 318.
Cloete S.W.P., Olivier J.J., Sandenbergh L. and Snyman M.A. (2014) *S. Afr. J. Anim. Sci.* **44**: 308.
Cloete S.W.P., Van Wyk J.B. and Neser F.W.C. (2004b) *S. Afr. J. Anim. Sci.* **34**: 37.
Devine C.E., Graafhuis P.H., Muir P.D. and Chrystall B.B. (1993) *Meat Sci.* **35**: 63.
Gilmour A.R., Gogel B.J., Cullis B.R., Welham S.J., and Thompson R. (2015) ASREML-user Guide Release 4.0 VSN International Ltd, Hemel Hempstead, UK.
Government Gazette (GG) (1992). No. R. 1748. Agricultural Products Standards Act, 1990 (Act No. 119 of 1990). In: Government Gazette, 324 (14060), 4.
Honikel K.O. (1998) *Meat Sci.* **49**: 447.
Khliji S., Van de Ven R., Lamb T.A., Lanza, M and Hopkins D.L. (2010) *Meat Sci.* **85**: 224.
McGeehin B., Sheridan J.J. and Butler F. (2001) *Meat Sci.* **58**: 79.
Naudé C.S., Visser C., Hoffman L.C., Cloete J.J.E. and Cloete S.W.P. (2018) *Proc. World Cong. Gen. Appl. Livest. Prod.* 11. 931. Auckland, New Zealand, 11-16 February 2018.
Schoeman S.J., Cloete S.W.P. and Olivier J.J. (2010) *Livest. Sci.* **130**: 70.
Van der Merwe D.A., Brand T.S. and Hoffman L.C. (2019) *Small Rumin. Res.*, in press.
Van Wyk J.B., Erasmus G.J. and Konstantinov K.V. (1993) *S. Afr. J. Anim. Sci.* **23**: 85.
Van Wyk J.B., Fair M.D. and Cloete S.W.P. (2003) *S. Afr. J. Anim. Sci.* **33**: 213.
Zemuy E.G. (2002) M.Sc. (Agric.) Thesis, Free State University, Bloemfontein, South Africa.

Proc. Assoc. Advmt. Anim. Breed. Genet. 23:123-126

BREEDING VALUES OF THE 1000-BULL-GENOME CATTLE ESTIMATED BY DAIRY PLEIOTROPIC VARIANTS

R. Xiang[1,2] and M.E. Goddard[1,2]

[1]Faculty of Veterinary & Agricultural Science, The University of Melbourne, Parkville 3052, Victoria, Australia
[2]Agriculture Victoria, AgriBio, Centre for AgriBiosciences, Bundoora, Victoria 3083, Australia

SUMMARY

The 1000-Bull-Genome (1KBull) project contains whole genome sequence data of thousands of cattle with different breeds from various countries. While most 1KBull cattle do not have phenotypic data, different breeds display distinct phenotype due to artificial and/or natural selections. For example, the milk production of Holstein cattle is expected to be higher than that of Angus cattle. Such expected phenotypic differences between breeds may be useful for validating the informativeness of a set of prioritised variants. Via meta-analysis of GWAS with 17.6 million imputed sequence variants with over 44,000 Australian dairy cattle, we prioritised a set of 92.5K pleiotropic variants associated with multiple traits including milk production, reproduction, management and linear assessment. With these pleiotropic variants, the genomic best linear unbiased prediction (gBLUP) was used to estimate dairy-trait breeding values (gEBV) for 2,334 1KBull cattle (Run 6). Based on principal components analysis, the dairy-trait gEBVs separated the dairy from beef breeds as well as the separation using whole genome sequence data. For individual trait gEBVs in the 1KBull cattle, while milk, protein and fat yield, somatic cell count, stature and angularity were significantly higher in dairy than in beef cattle, the milk protein and fat percentages, muzzle width and teat length were significant lower in the dairy than in the beef cattle. Compared to 1KBull Jersey cattle gEBVs, Holstein cattle had significantly higher milk, protein and fat yield and stature, but significantly lower fat and protein percentages and somatic cell count. Our study provides valuable insights into the genomic prediction of breed differences using within-breed trained equations. Our work also provides alternative validation strategies for prioritised markers.

INTRODUCTION

The 1000-Bull-Genome (1KBull) project collects whole genome sequence data worldwide via donations from consortium members. Since 2012 (Daetwyler *et al.* 2014), the dataset has grown to over 2,000 cattle from more than 100 breeds of *Bos taurus* and *Bos indicus*. Up to 44 million sequence variants have been identified in the 1KBull cattle and these variants are used as the basis for sequence variant imputation in large cattle populations. Large cattle populations with sequence variants have facilitated genome-wide association studies (GWAS) (Bouwman *et al.* 2018) and genomic prediction (VanRaden *et al.* 2017) of complex traits. Here we examine a new use for the 1KBull database; the prediction of trait differences between breeds.

Genomic prediction is usually used to predict differences in breeding value within a breed and it is unknown if it would correctly predict differences between breeds. One of the aims of this paper is to test the ability of within breed genomic prediction to predict differences between breeds. We develop prediction equations within breeds of dairy cattle and combine them with the genotypes of bulls in the 100KBull database to predict the differences between breeds. These predicted breed differences are compared to expectations such as higher milk yield in dairy breeds than in beef breeds.

MATERIALS AND METHODS

The 1KBull data used in this study was part of the Run 6 (http://www.1000bullgenomes.com/). In total the whole genome sequence data of 2,334 *Bos taurus* cattle were used. Dairy and beef cattle breeds and their sample sizes were defined as in Table 1. The defined dairy and beef cattle breeds were used for gEBV comparisons described later on.

Table 1. Sample size of defined dairy and beef cattle breeds

Dairy cattle		Beef cattle	
Holstein	567	Angus	266
Brown Swiss	148	Simmental	225
Jersey	66	Charolais	128
Montbeliarde	54	Limousin	82
Normandy	44	Hereford	75
Finnish Ayrshire	25	Guelph composite	30
Norwegian Red	24	Beef Booster	29
Guernsey	20	Blonde dAquitaine	26
Swedish Red	16	Belgian Blue	16
		Angus Red	6
		Maine Anjou	5
		BraunviehBeef	4

A set of pleiotropic sequence variants (92.5K) associated with 34 dairy traits were identified using Australian dairy bull (N>11,000) and cow populations (N>33,000) and 17.7 million imputed sequence variants with accuracy R2 > 0.4. The detail of the data and the GWAS model used can be found in (Xiang et al. 2019). Briefly, the traits were decorrelated by Cholesky transformation (Xiang et al. 2017). GWAS fitting breed as the fixed effects were conducted for each one of the 34 traits separately in bulls and cows. For the GWAS results of each trait from two sexes, a weighted t value was calculated to combine the variant effects with $t_w = \frac{\frac{\frac{B_{bull}}{se_{bull}^2}+\frac{B_{cow}}{se_{cow}^2}}{\frac{1}{se_{bull}^2}+\frac{1}{se_{cow}^2}}}{\sqrt{\frac{1}{\frac{1}{se_{bull}^2}+\frac{1}{se_{cow}^2}}}}$ (Xiang *et al.* 2018) where B_{bull} and se_{bull} were the beta and standard error (se) of the bull GWAS and B_{cow} and se_{cow} were the beta and se of the cow GWAS. The weighted t value across traits and variants were analysed by the multi-trait meta-analysis method (Bolormaa *et al.* 2014). Variants with the meta-analysis P-value < 1e-6 and MAF > 0.001 were retained as significant pleiotropic variants.

The genomic best linear unbiased prediction (gBLUP) implemented in MTG2 (Lee and Van der Werf 2016) was used to train prediction equations in the dairy dataset. A genomic relationship matrix (GRM) was calculated from the prioritized pleiotropic variants. Original traits (deregressed proofs) were used to perform gBLUP in Australian bulls and cows. The gBLUP model used was $y = mean + breed_i + a + error$, where y=vector of phenotypes for bulls or cows, $breed_i$=three breeds for bulls, Holstein, Jersey and Australian Red and four breeds for cows (Holstein, Jersey, Australian Red and MIX), a=polygenic random effects ~N(0, Gσg2) where G=GRM. This estimated the total genetic value of Australian bulls and cows and was followed by the back-solving for the variant solution in the Australian data. Then, the variant solutions were combined with the sequence genotypes to calculate dairy-trait gEBV of the 1KBull cattle.

RESULTS AND DISCUSSION

A principle component analysis (PCA) was carried out on the sequence genotypes of the 1KBull database and dairy-trait gEBVs (Figure 1). Overall, the first PC separated Holstein from other breeds and the 2nd PC separated Angus from other breeds. This may reflect that these two breeds were the most common in the database. The 1st PC of gEBVs (X-axis of the right panel of Figure 1) associated with milk production traits separated some dairy cattle breeds but did not separate beef cattle breeds. This also suggested that the 37 gEBV of dairy traits can be used to distinguish the phenotypic difference between dairy and beef cattle.

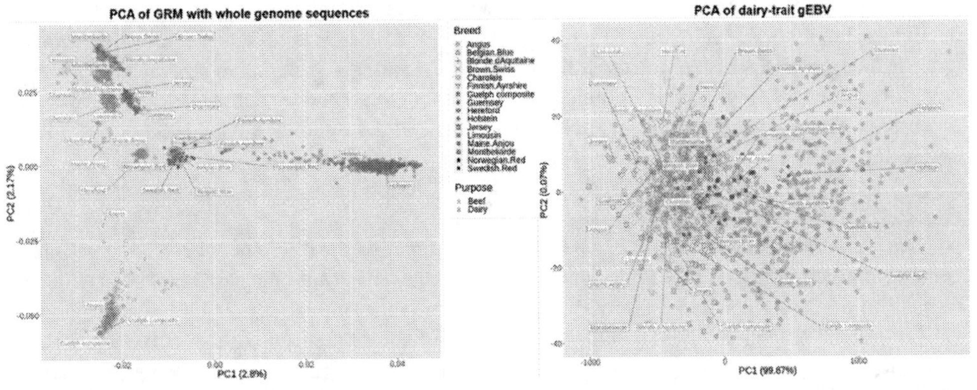

Figure 1. Principal components analysis results of the genomic relationship matrix and the dairy trait gEBVs of the 1000-bull-genome cattle

Individual dairy-trait gEBVs were compared between dairy and beef cattle breeds and were also compared between Holstein and Jersey breeds in the 1KBull individuals (Figure 1 and Table 2).

Table 2. gEBV difference. ns: not significant

gEBVs	Trait full name	Dairy VS Beef	Holstein VS Jersey
Prot	Protein yield	+	+
Fat	Fat yield	+	+
Milk	Milk yield	+	+
FatP	Fat percentage	-	-
ProtP	Protein percentage	-	-
SCC	Somatic cell count	+	-
Temp	Temperament	-	-
MSpeed	Milking speed	+	+
Stat	Stature	+	+
Like	Likeability	-	-(ns)
Angul	Angularity	+	+
MuzW	Muzzle width	-	+(ns)
TeatL	Teat length	-	+(ns)
UdTex	Udder texture	+	+
UdDep	Udder depth	+	+(ns)
RumpL	Rump length	+	+
OType	Overall type	+	+
Mamm	Mammary systems	+	+

Most dairy trait gEBVs were higher in the 1KBull dairy cattle than those in the 1KBull beef cattle. Thus, the within breed genomic predictions do predict qualitative differences between breeds. This result also supports the informativeness of the retained pleiotropic variants. The lower fat (FatP) and protein percentages (ProtP) in the diary breeds than in the beef breeds was due to that their higher milk yield. The somatic cell count (SCC) score and milk speed (MSpeed) was higher in the dairy cattle than in the beef cattle. The dairy cattle are predicted to have better overall type (OType) and mammary system (Mamm), to be more Angular and have shorter teat length (TeatL). These differences appeared to be consistent with the common expectations.

In the gEBV comparisons between Holstein Jersey breeds, Holstein cattle had higher milk productivities, but lower somatic cell count score, fat and protein percentages than Jersey cattle. Holstein cattle had better assessment of the overall type and the mammary system. No significant differences were found for likability, muzzle width (MuzW), teat length and udder depth (UdDep) between the two breeds. These observations appeared to be consistent with the common knowledge about Holstein and Jersey cattle.

CONCLUSIONS

Overall, our results show that it is possible to predict qualitative differences between breeds using genomic prediction based on a set of sequence variants chosen because they are associated with dairy traits. This study also provides alternative insights into efficient use of available data to conduct validation analysis. Our analysis included ~900 beef cattle from the Run6 of the 1KBull project. It is recommended to extend such genomic prediction analysis in a large beef cattle population where the allele frequency of the prioritised dairy pleiotropic variants can be properly examined and accounted for.

REFERENCES

Bolormaa S., Pryce J.E., Reverter A., Zhang Y., Barendse W., Kemper K., *et al.* (2014) PLOS *Genetics* **10**: e1004198.

Bouwman A.C., Daetwyler H.D., Chamberlain A.J., Ponce C.H., Sargolzaei M., Schenkel F.S., *et al.* (2018) *Nature Genetics* **50**: 362.

Daetwyler H.D., Capitan A., Pausch H., Stothard P., Van Binsbergen R., Brøndum R.F., *et al.* (2014) *Nature Genetics* **46**: 858.

Lee S.H. and Van der Werf J.H. (2016) *Bioinformatics* **32**: 1420.

VanRaden P.M., Tooker M.E., O'Connell J.R., Cole J.B. and Bickhart D.M. (2017) *Genet. Sel. Evol.* **49**: 32.

Xiang R., Berg I.v.d., MacLeod I.M., Hayes B.J., Prowse-Wilkins C.P., Wang M., *et al.* (2019) *Proc. Nat. Academy of Sciences* 201904159.

Xiang R., Hayes B.J., Vander Jagt C.J., MacLeod I.M., Khansefid M., Bowman P.J., *et al.* (2018) BMC *Genomics* **19**: 521.

Xiang R., MacLeod I.M., Bolormaa S. and Goddard M.E. (2017) *Scientific Reports* **7**: 9248.

Proc. Assoc. Advmt. Anim. Breed. Genet. 23:127-130

GENETIC PARAMETERS OF FIRST LACTATION MILK YIELD UNDER LOW, MEDIUM AND HIGH PRODUCTION SYSTEMS IN KENYA, USING TEST-DAY RANDOM REGRESSION MODEL

P.K. Wahinya[1,2], T.M. Magothe[3], A.A. Swan[1] and M.G. Jeyaruban[1]

[1]Animal Genetics & Breeding Unit*, University of New England, Armidale, NSW, 2351 Australia
[2]Karatina University, Department of Agricultural Sciences, Kenya
[3]Livestock Recording Centre, Ministry of Agriculture, Livestock and Fisheries, Kenya

SUMMARY

The aim of this study was to estimate genetic parameters for test-day milk yield in different production systems in Kenya. 10,923, 19,049 and 26,287 first lactation test-day records from multiple breeds under low, medium and high production systems, respectively, were analysed. On average cows under high production systems were younger and had a higher test-day milk yield than in low and medium production systems. A model fitting fourth order Legendre polynomials was found to be the most parsimonious and was therefore used to model the data. Additive genetic and permanent environmental variances were heterogeneous along different days in milk and between production systems. Heritability and repeatability were also different between days in milk and production systems. Heritability was on average 27%, 48% and 48% and repeatability 72%, 83% and 78% under low, medium and high production systems, respectively. Genetic correlations ranged from -32%, 34% and 45% to unity between daily milk yield in different days in milk under low, medium and high production systems, respectively. These parameters indicate that random regression using Legendre polynomial order four can be used to model test-day milk yield under the three production systems in Kenya. The observed heterogeneity of variance indicates that genetic parameters should be estimated within production systems for sustainable genetic improvement.

INTRODUCTION

Genetic evaluation using test-day milk yield allows better modelling of environmental factors affecting yield and variation in the lactation curve in addition to providing accurate genetic evaluation (Ptak and Schaeffer 1993). Random regression models using orthogonal Legendre polynomials are commonly used to model the covariance structure between test-day records. The models should include the general shape of the lactation curve, variation in test-day yields, effects specific to cows on the same test-day, and production levels if known (Ptak and Schaeffer 1993). In Kenya dairy production systems vary in terms of the level of inputs and outputs such that production systems can be classified into low, medium and high production systems (Wahinya *et al.* 2018). Genetic parameters of milk yield and persistency using test-day records under these production systems are not available. This paper, therefore, aims at estimating genetic parameters for milk yield using test-day Legendre polynomial random regression models.

MATERIALS AND METHODS

Data. 56,259 first lactation test-day records were received from the Livestock Recording Centre (LRC) in Kenya. The records were observed from 5,179 multi-breed cows in 142 herds from 1990 to 2014. The cows were managed under different production systems and in different geographical regions of the country. Records that were retained for this analysis ranged from 5 to 365 days post-partum with twelve records on average per cow and a range of three to twenty two records per

* A joint venture of NSW Department of Primary Industries and the University of New England

cow. Test-day yields were cleaned for outliers using a threshold of four standard deviations from the mean within the production systems.

Classification of herds into production systems was done in a separate analysis using average predicted herd 305-day milk yield. To obtain the predicted herd milk yield productivity level, individual cows 305 days milk yield were estimated using a linear mixed model with calving year, parity and breed group as fixed effects and herd as a random effect. K-means cluster analyses in R software (R Core Team 2017) using the Hartigan and Wong (1979) algorithm was then used to group predicted herd means into three groups here described as low, medium and high production systems, as summarised in Table 1.

Statistical analysis. Variance components were estimated using univariate animal test-day models using the ASReml software package (Gilmour *et al.* 2015). Contemporary group (CG) was defined based on herd-test month of milk sampling. Test-day milk yield was regressed on days in milk to account for the lactation curve. A random regression test-day model was fitted as:

$$y_{ijkl} = CG_i + Age + \beta t_j + g_l + \alpha_{kn}\varphi_n(t_j) + p_{kn}\varphi_n(t_j) + e_{ijkl}$$

where y_{ijk} is the test-day milk yield sampled on animal k, on t_j days in milk within the ith CG, with age at calving (Age) and in genetic group g_l; β, α_{kn} and p_{kn} are regression coefficients for days in milk, additive and permanent environmental random effects of each cow k, respectively; $\varphi_n(t_j)$ is the covariate of the regression function of nth Legendre polynomial order for the day in milk; and e_{ijkl} is the residual term. Seventy-four g_l were defined separately for sires and dams within six categories based on year of birth: before 1986, between 1986 to 1990, 1991 to 1995, 1996 to 2000, 2001 to 2005 and after 2005, and Friesian, Ayrshire, Guernsey, Jersey, Sahiwal, Brown Swiss, and Unknown breeds using Westell-Quaas method (Westell *et al.* 1988). Residual variance was assumed to be heterogeneous considering 11 classes of 5 – 15, 16 – 30, 31 – 60, 61 – 90, 91 – 120, 121 – 150, 151 – 180, 181 – 210, 211 – 240, 241 – 270, and 271 – 365 days in milk, however, genetic parameters were estimated up to 305 days in milk.

Based on log likelihood ratio test, AIC, BIC and variance estimates, a model fitting Legendre polynomial order 4 (LP4) was found to be the most parsimonious and therefore, was used to estimate genetic parameters.

Table 1. Test-day data structure and average age (days) and test-day milk yield (kg) (standard deviation in brackets) under low, medium and high production systems

System	Records	Cows	Herds	Sire	Dam	CG	Age	Milk yield
Low	10,923	1,034	50	385	916	587	1,112(277)	7.9(3.6)
Medium	19,049	1,659	55	450	1,283	638	990(228)	12.3(4.8)
High	26,287	2,486	37	626	1,580	434	910(140)	16.5(5.8)

RESULTS AND DISCUSSION

Low, medium and high production systems had different phenotypic means and variances for test-day milk yield (Table 1). Table 2 illustrates variance components, heritabilities and repeatabilities from model LP4 for selected test-days under the low, medium and high production systems. Variance components were heterogenous between and within production systems. Additive genetic variances ranged from 1.3 to 6.7, 6.9 to 14.6 and 7.9 to 17.2 under low, medium and high production systems, respectively. Within low and medium production systems, additive genetic variance was highest at the beginning of the lactation period which is consistent with other reports in the literature (Muasya *et al.* 2014). In the high production system additive variance increased from the beginning of lactation to a peak on day 100 then gradually decreased towards the end of the lactation. Andonov *et al.* (2013) reported higher additive variance in the mid-lactation while Berry *et al.* (2003) observed a similar trend to what was observed in this study. The trend for permanent environmental variance was different between production systems. In the low production system, it was highest around day 60 then it decreased to the end of the lactation. A similar trend was observed by Andonov *et al.* (2013) although in their study they estimated peaks at the beginning and end of the lactation period. Permanent environmental variance increased gradually from the beginning to the end of the lactation under the medium production system while under the high production system, it remained constant with peaks at the beginning and end of the lactation period as reported by Muasya *et al.* (2014). Residual variance was constant along the days in milk except for higher residual variances observed in the early stage of the lactation in high production systems.

Table 2. Additive, permanent environment (Pe) and residual variances, heritability (h^2) and repeatability (r) for daily milk yield in selected days in milk (DIM) under low (L), medium (M) and high (H) production systems (variances are rounded to the nearest whole number)

DIM	Additive			Pe			Residual			h^2 (%)			r (%)		
	L	M	H	L	M	H	L	M	H	L	M	H	L	M	H
5	7	15	8	4	2	14	2	3	4	53	74	30	84	85	84
60	2	7	15	5	6	7	2	3	13	22	46	43	79	81	63
100	1	7	17	4	6	7	2	2	7	18	47	55	71	86	76
180	1	8	14	3	6	7	2	3	5	23	46	53	70	81	80
260	2	7	11	2	6	8	2	2	5	32	44	47	70	82	80
305	2	8	11	3	7	8	2	3	5	28	45	47	66	85	81

Heritability estimates in this study ranged from 18% to 53%, 44% to 74% and 30% to 55% under low, medium and high production systems respectively. Similar results have been reported in literature (Costa *et al.* 2005). Higher estimates, especially for the medium production system, were reported here than were reported for Holstein-Friesian cattle in Kenya (Muasya *et al.* 2014). This can be attributed to the multi-breed data used in this study which is expected to have a higher genetic variance than in a single breed population (Gebreyohannes *et al.* 2016). Heritability estimates were highest at the beginning of the lactation under low and medium production systems (Bignardi *et al.* 2009). Under the high production systems heritability estimates were lowest at the beginning of the lactation, increased to a peak then gradually decreased to the end of the lactation period. Daily milk yield was highly repeatable (63% – 87%) and decreased from the start of the lactation to the end in low production systems but increased gradually from the beginning to the end of the lactation in the medium and high production systems.

Genetic correlations estimated using model LP4 for test-days up to 305 days are illustrated in Figure 1. The trend of correlation was different between production systems. In general, correlations were higher between adjacent days in milk but declined with increasing distance between days of lactation in all production systems. Most of the correlations were positive except for correlations between milk yield at the beginning and end of lactation in the low production system. Negative correlations have been reported in literature (Rekaya *et al.* 1999) indicating that improvement of yield at the beginning of the lactation would result in reduced yield at the end and therefore lower persistency. Positive correlations indicate that selection for high milk yield at the end of the lactation can be effective based on yield at the beginning of the lactation especially under medium and high production systems with moderate correlations up to 0.3 and 0.5 respectively.

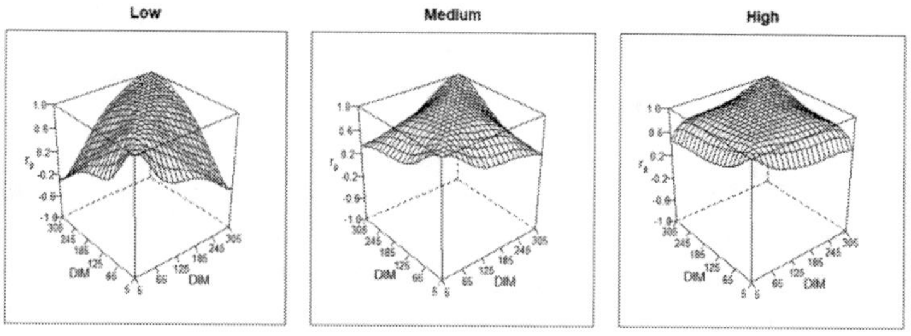

Figure 1. Genetic correlation (r_g) between milk yields in different days in milk (DIM) under low, medium and high production systems

CONCLUSIONS

A fourth order random regression model was most appropriate for modelling milk yield test-day records in this study. Genetic and permanent environmental variances were heterogenous along the trajectory of days in milk. Genetic and permanent environmental variances, heritabilities and repeatabilities were different in low, medium and high production systems. Genetic correlations between milk yields in different days of a lactation indicate that selection for improved milk yield at the beginning of the lactation period in the medium and high production systems would result in improved yield at the end of the lactation and therefore improved persistency, whereas under low production systems negative correlations were estimated between early and late lactation. Further analysis of the test day records is recommended using alternative models such as cubic splines. This study showed that genetic parameters should be estimated within production systems for sustainable genetic improvement and selection for milk yield can be effective based on yield at the beginning of the lactation.

ACKNOWLEDGEMENT

The authors thank the University of New England for providing PhD Research Award to the first author and Livestock Recording Centre (LRC) of Kenya for supplying the data.

REFERENCES

Andonov S., Ødegård J., Svendsen M., Ådnøy T., Vegara M. and Klemetsdal G. (2013) *J Dairy Sci* **96:** 1834.

Berry D., Buckley F., Dillon P., Evans R., Rath M. and Veerkamp R. (2003) *J Dairy Sci* **86:** 3704.

Bignardi A.B., El Faro L., Cardoso V.L., Machado P.F. and de Albuquerque L.G. (2009) *Livest Sci* **123:** 1.

Costa C.N., Melo C.M.R.d., Machado C.H.C., Freitas A.F.d., Packer I.U. and Cobuci J.A. (2005) *Braz. J. Vet. Res. Anim. Sci.* **34:** 1519.

Gebreyohannes G., Koonawootrittriron S., Elzo M.A. and Suwanasopee T. (2016) *ANRES* **50:** 64.

Gilmour A., Gogel B., Cullis B., Welham S. and Thompson R. (2015) VSN international ltd.

Hartigan J.A. and Wong M.A. (1979) *J R Stat Soc* **28:** 100.

Muasya T.K., Peters K.J., Magothe T.M. and Kahi A.K. (2014) *Livest Sci* **169:** 27.

Ptak E. and Schaeffer L. (1993) *Livest Prod Sci* **34:** 23.

R Core Team (2017) R Foundation for Statistical Computing.

Rekaya R., Carabano M. and Toro M. (1999) *Livest Prod Sci* **57:** 203.

Wahinya P., Magothe T., Swan A. and Jeyaruban G. (2018) *In Proc. 11th World Cong. Genet. Appl. Livest. Prod.* 2: 788.

Westell R., Quaas R. and Van Vleck L.D. (1988) *J Dairy Sci* **71:** 1310.

Proc. Assoc. Advmt. Anim. Breed. Genet. 23:131-134

GENETIC CORRELATION BETWEEN MILK UREA AND EFFICIENCY OF CRUDE PROTEIN UTILIZATION ESTIMATED FROM A RANDOM REGRESSION MODEL

H.B.P.C. Ariyarathne, M. Correa-Luna, H.T. Blair, D.J. Garrick and N. Lopez-Villalobos

School of Agriculture and Environment, Massey University, Private Bag 11-222, Palmerston North, 4442 New Zealand

SUMMARY

The concentration of urea in milk (MU) can be predicted from mid infrared spectra in routine herd testing of dairy cattle. A high positive phenotypic correlation between MU nitrogen and urinary nitrogen has been estimated in controlled indoor trials suggesting selection for low MU nitrogen might reduce urinary nitrogen excretion. Estimates of genetic correlations (r_G) of MU and other traits are required to evaluate the effects of selection for low MU. The aim of the current study was to estimate r_G between MU and efficiency of crude protein utilization (ECPU; ratio between crude protein yield in milk and crude protein intake) throughout the lactation using a random regression animal model (RRM). Results show that r_G between MU and ECPU was positive in early and late lactation but was mostly negative from day 40 to 180 of the lactation (mean=-0.09). The r_G of MU with crude protein (mean=-0.15) and fat (mean=-0.27) percentages were negative. Further research is required to confirm if MU can be used in selection to reduce urea nitrogen excretion and increase ECPU without reducing cow productivity and farm profitability.

INTRODUCTION

New Zealand cows graze almost exclusively on pasture all year round. Consequently, cows consume feed with more protein relative to energy than they require. The efficient conversion of feed protein to milk protein is sensitive to the ratio of protein to energy in the diet. Protein being eaten by the cows is degraded to amino acids and ammonia by rumen microbes. If the diet has an excess of protein and is deficient in energy, rumen microbes are less efficient in capturing available ammonia, therefore the surplus enters the bloodstream and is converted to urea in the liver. The majority of the urea produced in the liver is excreted as urine, however a proportion of urea is diffused to milk (milk urea, MU) through the bloodstream (Roseler *et al.* 1993). New Zealand cows fed at pasture produce greater levels of MU than levels produced by cows fed balanced mixed rations (Garcia-Muniz *et al.* 2013). Urea enters the environment as urine breaks down to ammonia and nitrous oxide at the site of the urine patch making it a major source of air and water pollution in New Zealand. Averaged across the year, 20% of the nitrogen (N) load is leached through the soil (Selbie *et al.* 2015).

Reducing N pollution is an urgent national need and one option may be genetic selection for less urea in urine (UU) thereby reducing the amount of urea reaching the environment. Measuring UU is not feasible in outdoor farming systems and even if practical it is very expensive to measure. However, MU could be useful as an indicator of UU if the strong positive correlation between MU nitrogen and UU nitrogen in controlled indoor experiments (Jonker *et al.* 1998) is confirmed in pastoral circumstances. Milk urea can be determined from mid infrared spectra generated from milk samples used for routine herd testing, but it is not reported to dairy farmers.

Milk urea has also been proposed as an indicator of efficiency of crude protein utilization (ECPU) (Baker *et al.* 1995), which can be defined as the proportion of crude protein produced in milk in relation to the intake of crude protein. Cows with high ECPU likely divert more absorbed protein for milk production rather than excreting and therefore wasting it as urea in urine.

Some authors have reported negative r_G between the concentration of MU and milk production traits (Miglior *et al.* 2007). However, few studies have reported estimates of r_G between MU and protein utilization efficiency traits, other than the negative but non-significant correlations found by Vallimont *et al.* (2011) between MU nitrogen and three feed efficiency traits.

There is evidence to suggest that, the additive genetic variance of longitudinal traits can change over time. Therefore, it is sensible to expect variability in r_G between traits over the different stages of lactation. Random regression models based on test-day records can capture variability in additive genetic and permanent environmental effects over stages of lactation. To our knowledge there is no literature on the variability of r_G between MU and ECPU traits at different stages of the lactation profile. The objective of this study was to estimate r_G between MU and each of ECPU, yields of milk (MY), fat (FY) and crude protein (CPY), percentages of fat (FP) and crude protein (CPP) for every day of lactation in grazing dairy cows in New Zealand using a test-day RRM.

MATERIALS AND METHODS

Data originated from 468 cows on two mixed-breed herds from Massey University in Palmerston North, New Zealand for the 2016 and 2017 production seasons were included in this study. Details of animal management and feeding can be found in Correa-Luna *et al.* (2018).

Daily MY, FY, CPY, FP and CPP were derived from monthly herd-test records. Three additional milk samples from each cow were collected in each production season representing early, mid and late lactation for determination of MU content. Daily ECPU was defined as the ratio of CPY to daily dietary intake of crude protein and expressed as a percentage. For both herds, daily live weight measurements were obtained from an automatic walk over scale in the exit race of the milking shed and body condition scores measurements on a 10-point scale were assigned in synchrony with each herd-test by a single research technician.

Apparent dry matter intake (kg DM consumed/cow/day) was obtained based on summing up the estimated metabolisable energy (ME) requirements for maintenance, pregnancy, production and daily weight variation and then dividing by ME content of the feed offered. Content of crude protein from feed quality analyses were used to calculate crude protein intake (CPI). Cows with a minimum of 3 herd test records and lactation lengths of not less than 150 days and up to 240 days in milk (DIM) were included in the analysis. After editing the data 380 cows remained in the data set.

Co(variance) components corresponding to additive genetic effect for MU, ECPU, MY, FY, CPY, FP and CPP was estimated using bivariate random regression test-day animal models. The model included the fixed effects of herd-test-date and parity, and as covariates, deviation from median calving date, proportion of Holstein-Friesian breed and coefficient of heterosis of Holstein-Friesian and Jersey breeds, and days in milk modelled as second-order orthogonal polynomial. Random effects included in the model were the animal additive genetic, cow-lactation permanent environment, cow permanent environment and residual effects. Animal additive genetic effect was modelled using second-order orthogonal polynomials for all the traits except for MY where a third order polynomial was used. Constant cow permanent environment, cow-lactation permanent environment variances and residual variances were also fitted in the model. Variance and covariance components were estimated using the ASReml package (Gilmour *et al.* 2009). The matrix of additive genetic (co)variances (**C**) for each day of lactation was estimated using the following covariance function, $\mathbf{C} = \mathbf{\Phi} \otimes \mathbf{G} \otimes \mathbf{\Phi}'$ where **G** is the matrix of variances of the random regression coefficients for additive genetic effects between two traits and $\mathbf{\Phi}$ is the matrix of orthogonal polynomial coefficients.

RESULTS AND DISCUSSION

The r_G between MU and ECPU, MY, FY, CPY, FP and CPP fluctuated over the lactation from 1 to 240 DIM (Figure 1). The overall genetic correlation between MU and ECPU was negative (-0.09) but point estimates at specific stages of lactation fluctuated from -0.46 to 0.81 (Table 1). Although the r_G was positive at the beginning, it turned moderately negative by mid-lactation (-0.46). The high positive correlation at the end of the lactation could be an artefact of the mathematical properties of polynomial random regression and reflect the lesser number of herd-test records towards the end of lactations. However, the strong negative r_G between MU and ECPU at the middle of the lactation when cows are producing more milk suggests that efficient cows convert more feed protein into milk protein and produce milk with low content of urea. These cows may divert absorbed proteins in a different manner compared to inefficient cows.

Table 1. Estimates of genetic correlation (r_G) between milk urea (MU) and efficiency of crude protein utilization (ECPU), yield of milk (MY), fat (FY), crude protein (CPY), and percentage of fat (FP) and crude protein (CPP) at different days in milk (DIM) in grazing dairy cows in New Zealand

DIM	$r_{G(MU-ECPU)}$	$r_{G\ (MU-MY)}$	$r_{G\ (MU-FY)}$	$r_{G\ (MU-CPY)}$	$r_{G\ (MU-FP)}$	$r_{G\ (MU-CPP)}$
1	0.25	0.29	0.06	0.27	-0.36	-0.15
60	-0.23	0.16	-0.16	-0.04	-0.23	-0.13
120	-0.46	0.18	-0.14	0.02	-0.29	-0.17
180	-0.11	0.21	0.17	0.16	-0.27	-0.18
240	0.81	0.29	0.55	0.64	-0.28	-0.09

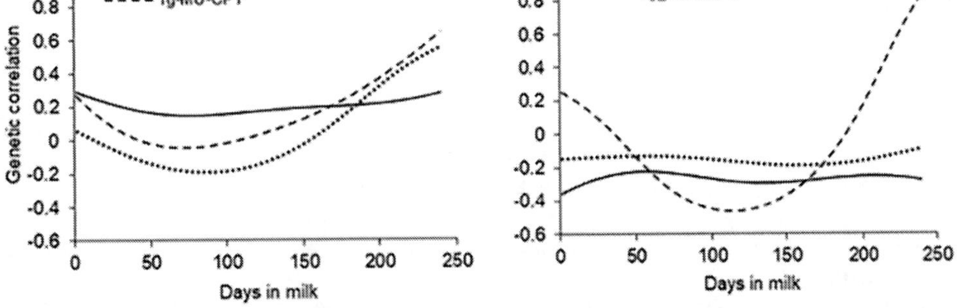

Figure 1. Daily genetic correlations (r_G) between milk urea (MU) and yield of milk (MY), fat (FY), crude protein (CPY), percentage of fat (FP), crude protein (CPP) and efficiency of crude protein and utilization (ECPU) at different days in milk in grazing dairy cows in New Zealand

The estimates of r_G of MU and FP and CPP were negative with some small fluctuations throughout the lactation. The r_G between MU and FP fluctuated from -0.23 to -0.36 and MU and CPP from -0.09 to -0.18 (Table 1). These negative correlations indicate that cows produce milk with high CPP and

FP and low content of MU. Together with the r_G between MU and ECPU these results suggest that cows that allocate more CP from the diet into milk protein divert less urea in milk. However, the r_G between MY and MU was positive throughout the lactation (0.19) with small fluctuations (0.16 to 0.29) (Table 1). A positive relationship between MU and MY has been previously reported by Godden *et al.* (2001) and this might be explained by the increased level of feeding that involuntarily increases the level of protein intake and this also increases the production of milk. Diets with a high CP:energy ratio reduce the efficiency of rumen microbes, with more ammonia converted into urea instead of proteins in milk (Baker *et al.* 1995).

The estimates of r_G of MU with FY and CPY ranged from -0.16 to 0.55 and -0.04 to 0.64 respectively (Table 1). Despite the small sample size used in this study, the average r_G of MU with FY and CPY estimated in this study were within the range reported by Beatson *et al.* (2019) using a much larger data set comprising several mixed-breed dairy herds in New Zealand. Studies by Wood *et al.* (2003) reported r_G of MU and FY and CPY not different from zero.

CONCLUSIONS

The genetic correlation between MU and ECPU was positive in early and late lactation but was mostly negative from day 40 to 180 of the lactation indicating that inclusion of MU in a selection index can cause correlated responses in ECPU. Further research is required to estimate the genetic correlation between MU and urine urea to fully evaluate if MU can be used as a trait to reduce nitrogen excretion in grazing dairy cows.

REFERENCES

Baker L., Ferguson J. and Chalupa W. (1995) *J. Dairy Sci.***78:** 2424.

Beatson P., Meier S., Cullen N. and Eding H. (2019) *Anim.* **13:**1.

Correa-Luna, M., Lopez-Villalobos, N., Almeida Jr, G.A., Donaghy, D.J. and Kemp, P.D. (2018) *N. Z. J. Anim Sci Prod.* **78:** 76.

Garcia-Muniz J.G., Lopez-Villalobos N., Burke J.L., Sandbrok T. and Vazquez-Pelaez C.G. (2013) *Proc. N.Z. Soc. Ani. Prod.* **73:** 108.

Gilmour A.R., Gogel B.J., Cullis B.R. and Thompson R. (2009) ASReml User Guide Release 3.0. VSN International Ltd, Hemel Hempstead, UK.

Godden S., Lissemore K., Kelton D., Leslie K., Walton J. and Lumsden J. (2001) *J. Dairy Sci.* **84:** 107.

Jonker J., Kohn R. and Erdman R. (1998) *J. Dairy Sci.* **81:** 2681.

Miglior F., Sewalem A., Jamrozik J., Bohmanova J., Lefebvre D. and Moore R. (2007) *J. Dairy Sci.* **90:** 2468.

Roseler D.K, Ferguson J.D, Sniffen C.J. and Herrema J. (1993) *J. Dairy Sci.* **76:** 525.

Selbie D.R., Buckthought L.E. and Shepherd M.A. (2015) *Adv. Agron.* **129:** 229.

Vallimont J., Dechow C., Daubert J., Dekleva M., Blum J., Barlieb C., Liu W., Varga G., Heinrichs A. and Baumrucker C. (2011) *J. Dairy Sci.* **94:** 2108.

Wood G., Boettcher P., Jamrozik J., Jansen G. and Kelton D. (2003) *J. Dairy Sci.* **86:** 2462.

Proc. Assoc. Advmt. Anim. Breed. Genet. 23:135-138

GENETIC PARAMETERS FOR MILK YIELD, MILK ELECTRICAL CONDUCTIVITY AND MILK FLOW RATE IN FIRST-LACTATION JERSEY COWS IN SRI LANKA

A.M. Samaraweera[1,3], V. Boerner[1], S. Disnaka[4], J.H.J. van der Werf[2] and S. Hermesch[1]

[1]Animal Genetics & Breeding Unit*, University of New England, Armidale, NSW, 2351 Australia
[2]School of Environmental & Rural Science, University of New England, Armidale, NSW, 2351
[3]Uva Wellassa University, Badulla, 90000 Sri Lanka
[4]National Livestock Development Board, Narahenpita, 00500 Sri Lanka

SUMMARY

Milk electrical conductivity is an indicator trait for mastitis, and for maintaining udder health, moderate milking speed is important. The heritabilities of sum of daily milk yields, mean milk electrical conductivity and mean milk flow rate for each 30-day period along the lactation trajectory in Jersey cows milking in their first lactation in Sri Lanka were estimated. The data included 248,854 daily records and 362,754 morning and evening records from 991 cows that calved from 2015 to 2018. Variance components and variance ratios were estimated from posterior means obtained from a Gibbs sampler. The heritability as estimated by univariate analyses for milk yield, milk electrical conductivity and milk flow rate ranged from 0.04 ± 0.01 to 0.13 ± 0.03, from 0.06 ± 0.02 to 0.09 ± 0.02, and from 0.06 ± 0.02 to 0.18 ± 0.05, respectively. Additive genetic correlations between milk yield and milk electrical conductivity or milk flow rate along the lactation ranged from -0.31 ± 0.49 to 0.77 ± 0.19 and from 0.46 ± 0.29 to 0.89 ± 0.12, respectively. Present heritability estimates were sufficiently high for milk electrical conductivity and flow rate to be used in a selection index. However, these estimates should be confirmed with more data.

INTRODUCTION

Mastitis is an important disease among dairy cows in the tropics which causes substantial economic losses (Bangar *et al.* 2015). Selective breeding against mastitis susceptibility is important to increase mastitis resistance in dairy cows. Milk electrical conductivity has been used as an indirect trait to reflect mastitis incidence (Norberg 2005). Fast milking is associated with a wider teat canal, which could lead to the entry of pathogens, and increased somatic cell score (Carlström *et al.* 2016). Therefore, moderate milk flow rate is important for udder health. The increasing use of modern milking systems in developing countries provides an opportunity to use automatically recorded data such as daily milk yield and milk electrical conductivity in genetic evaluation (Samaraweera *et al.* 2018). In Sri Lanka, milking systems with automatic recording are becoming popular, alongside recent importation of dairy cows to large-scale farms. The aim of this study was to estimate genetic parameters for milk yield, milk electrical conductivity and milk flow rate in first-lactation Jersey cows in an intensive dairy farm in Sri Lanka.

MATERIALS AND METHODS

Data. Milk yield records were obtained from a dairy farm located 37 meters above sea level of Sri Lanka, using Jersey cows imported from Australia as pregnant heifers. Milk yield, milk electrical conductivity and milking duration were recorded automatically in a DeLaval™ milking parlour. Daily milk yield and milking duration data were available from 248,854 daily records and milk electrical conductivity was available from 362,754 morning and evening (session) records from days five to 305

* A joint venture of NSW Department of Primary Industries and the University of New England

in first lactation. Data were available for 991 Jersey cows that calved from July 2015 to January 2018.

Data cleaning. Any negative milking durations for daily milk yields were removed and assumed to be an error in recording. Milking duration (measured in sec) was used to calculate milk flow rate (as kg/min). Any records with zero milk electrical conductivity were removed assuming a failure to record the milk electrical conductivity. Daily averages for the milk electrical conductivity (mS/cm) were calculated. The lactation length was divided into ten, 30-day periods starting from day five and going through to day 305. The total milk yield (MY), mean milk electrical conductivity (EC) and mean milk flow rate (FR) were calculated for each period. Outliers that differed by more than four standard deviations from the mean were excluded from the analyses.

Genetic parameter estimation. For each 30-day period, MY, EC and FR were considered as separate traits. The univariate animal model fitted was $\boldsymbol{y} = \boldsymbol{Xb} + \boldsymbol{Za} + \boldsymbol{e}$, where $\boldsymbol{y}$ is the vector of observations, $\boldsymbol{b}$ is the vector of estimates for fixed effects of year-season of calving (YS, for all traits) and lactation length as a covariate (for milk yield), $\boldsymbol{a}$ is the vector of random animal additive genetic effects estimates, $\boldsymbol{X}$ and $\boldsymbol{Z}$, the incidence matrices relating records to the fixed effects and random animal effects, and $\boldsymbol{e}$, the vector of random residual effects. The YS was used as the contemporary group and any contemporary groups with less than eight cows were discarded. There were two seasons as dry (from Dec to April next year) and wet (from May to Nov) and five YS combinations. The total number of animals in the pedigree was 1572 with information up to 3 generations. Cows with phenotypic records (991) descended from 39 sires and 521 cows out of total cows with phenotypes were related to one of 38 maternal grandsires. No maternal grandsires were used as sires. All sires for cows with data were known, but all dams were unknown. Therefore, maternal grandsires were fitted into the pedigree using dummy dams assuming a unique dam for each offspring.

Variance components for the three traits were estimated via the univariate model described above, using a Bayesian approach implemented in the BESSiE software (Boerner and Tier 2016). A blocked Gibbs sampler was run for 50,000 cycles, with scaled inverted Wishart distributions assigned as prior processes to the residual and additive genetic co-variance matrices with parameter "ν" set to "x" and "y", respectively (see Sorensen and Gianola (2002, pp. 576-588) for further details). The additive genetic and residual variances were calculated as posterior means by averaging the sum of every 100th iteration omitting the first 1000 iterations as burn-in. The additive genetic correlations between MY and EC and MY and FR for each period were estimated with bivariate animal models. The additive genetic correlations between periods within the same traits were estimated with ten-trait animal model.

RESULTS AND DISCUSSION

Milk yield (MY) was highest in the second and third 30-day periods, close to the peak milk production (around 60 days in milk) (Table 1). The coefficient of variation (CV) for MY was highest at the beginning (0.32) and at the end of lactation (0.35) whereas in the middle of the lactation the CV was around 0.22.

Mean EC across the whole lactation was 6.2 mS/cm and EC was highest at the beginning of lactation (6.4 mS/cm) and slightly decreased towards the end of lactation (6.2 mS/cm) (Table 1), with little variation in EC over the lactation. Similar ECs were observed in the literature for mastitis-infected cows, e.g. Norberg *et al.* (2004) found healthy, sub-clinically infected and cows with clinical mastitis had ECs (mS/cm) of 5.30 ± 0.03, 5.75 ± 0.04 and 6.73 ± 0.06, respectively ($P<0.001$). Therefore, the relatively high EC values in this study suggests that some cows had mastitis. However, there are a number of other factors that affect the milk EC such as milk temperature, bacterial strain, milk fat content etc. (Nielen *et al.* 1992; Woolford *et al.* 1998; Mabrook and Petty 2003). Therefore, changes in milk EC need to be validated with mastitis incidences.

Mean FR for 30-day milking periods throughout the lactation ranged from 1.08 to 1.41 kg/min (Table 1) and minimum and maximum values ranged from 0.40 to 2.30 kg/min, respectively. A study with higher mean MY (30 kg/d) than this study (14 kg/d) reported higher mean (2.2 ± 0.5 kg/min) and variation (from 0.3 to 8.2 kg/min) for FR (Firk *et al.* 2002).

Table 1. Descriptive statistics of milk yield (kg), milk electrical conductivity (mS/cm) & milk flow rate (kg/min) in each 30-day days in milk class

Days in	Milk yield			Milk electrical conductivity			Milk flow rate		
milk	# cows	Mean	SD	# cows	Mean	SD	# cows	Mean	SD
5-34	967	408	131	961	6.40	0.36	966	1.26	0.23
35-64	944	469	113	944	6.31	0.34	831	1.17	0.27
65-94	929	459	96	931	6.26	0.35	728	1.08	0.27
95-124	919	431	89	923	6.23	0.35	945	1.41	0.25
125-154	914	409	85	917	6.21	0.33	932	1.40	0.25
155-184	914	388	87	912	6.18	0.33	924	1.36	0.25
185-214	896	370	86	901	6.18	0.35	918	1.35	0.24
215-244	891	344	88	889	6.16	0.36	917	1.35	0.24
245-274	880	315	88	877	6.20	0.37	898	1.30	0.25
275-305	831	285	100	819	6.19	0.37	891	1.25	0.25

The heritability estimates for MY and EC were low compared to literature (Table 2). For example, moderate and high heritability for EC (ranged from 0.15 to 0.39) has been reported in Norberg (2005). FR was moderately heritable (0.10) and our estimates were consistent with Zwald *et al.* (2005) (milking duration, 0.17 ± 0.03). The phenotypic variance for EC and FR (Table 1) was close to the observed variance (Table 2) indicating that the YS did not explain much of the variance of EC and FR. Heritability estimates were slightly higher for all traits in the bivariate and multivariate analyses but differences from those from the univariate analysis were small.

Table 2. Heritability ± standard errors ($h^2 \pm seh^2 \pm se$) & phenotypic variance ($\sigma_p^2\sigma_p^2$) from univariate analyses for milk yield, milk electrical conductivity and milk flow rate in Jersey cows

Days in	Milk yield		Milk electrical conductivity		Milk flow rate	
milk	$h^2 \pm se$	σ_p^2	$h^2 \pm se$	σ_p^2	$h^2 \pm se$	σ_p^2
5-34	0.08 ± 0.03	5099	0.09 ± 0.02	0.13	0.06 ± 0.02	0.05
35-64	0.13 ± 0.03	5675	0.06 ± 0.02	0.12	0.09 ± 0.03	0.07
65-94	0.08 ± 0.02	5196	0.08 ± 0.03	0.12	0.07 ± 0.02	0.07
95-124	0.12 ± 0.03	5385	0.09 ± 0.02	0.12	0.09 ± 0.03	0.06
125-154	0.11 ± 0.03	4809	0.06 ± 0.02	0.11	0.13 ± 0.04	0.06
155-184	0.10 ± 0.03	5067	0.08 ± 0.02	0.11	0.18 ± 0.05	0.06
185-214	0.08 ± 0.02	5564	0.08 ± 0.03	0.12	0.15 ± 0.04	0.05
215-244	0.04 ± 0.01	5593	0.08 ± 0.02	0.13	0.10 ± 0.03	0.06
245-274	0.04 ± 0.02	5138	0.06 ± 0.02	0.14	0.11 ± 0.03	0.07
275-305	0.06 ± 0.02	5443	0.06 ± 0.02	0.14	0.07 ± 0.02	0.06

The low additive genetic correlations (<0.30, results not shown) within the same trait across 30-day periods of lactation show that they were independent traits. Additive genetic correlations between MY and EC ranged from -0.31 ± 0.49 to 0.77 ± 0.19 with high standard errors (Table 3). The additive genetic correlation between MY and EC was higher and positive around peak milk production (from 65 to 94 days). Significant positive additive genetic correlations between MY and FR were also observed, and additive genetic correlations ranged from 0.46 ± 0.29 to 0.89 ± 0.12 (Table 3). Therefore, selecting cows solely for high milk yield would lead to a correlated response of increased FR and EC. Therefore, selection emphasis would need to balance the value of increasing milk yield with electrical conductivity and an intermediate optimum for milk flow rate.

Table 3. Additive genetic ($r_a r_a$) and phenotypic ($r_p r_p$) correlations between milk yield (1), milk electrical conductivity (2) & milk flow rate (3) for each days in milk class in Jersey cows

	Days in milk									
	5-34	35-64	65-94	95-124	125-154	155-184	185-214	215-244	245-274	275-305
$r_{a,12}$	-.11±.40	.25±.46	.77±.19	.44±.32	-.03±.49	.26±.49	.03±.45	.02±.60	-.31±.49	-.31±.48
$r_{p,12}$	-.04±.03	.08±.04	.12±0.03	.09±.03	.09±.03	.08±.03	-.01±.04	-.05±.04	-.09±.05	-.19±.04
$r_{a,13}$	.89±.12	.46±.29	.54±.35	.54±.33	.78±.22	.53±.36	.79±.20	.66±.33	.53±.39	.81±.19
$r_{p,13}$	.69±.02	.50±.03	.48±.03	.51±.03	.56±.02	.64±.02	.63±.02	.65±.02	.67±.02	.67±.02

CONCLUSIONS

A significant positive additive genetic correlation between MY and EC was found around peak milk production and the same between MY and FR was positive. The heritabilities for MY and EC from this data were lower than anticipated. However, present heritability estimates were adequate to use EC and FR in a selection index. The genetic parameters for MY, EC and FR should be confirmed with more data.

ACKNOWLEDGMENTS

The authors thank the University of New England for providing the PhD Research Award (UNE IPRA) for the first author and National Livestock Development Board (NLDB), Sri Lanka for supplying the data.

REFERENCES

Bangar Y.C., Singh B., Dohare A.K., and Verma M.R. (2015) *Trop. Anim. Health Prod.* **47:** 291.

Boerner V. and Tier B. (2016) *Genet. Sel. Evol.* **48:** 63.

Carlström C., Strandberg E., Johansson K., Pettersson G., Stålhammar H. and Philipsson J. (2016) *Acta Agric. Scand. A.* **66:** 84.

Firk R., Stamer E., Junge W. and Krieter J. (2002) *Archiv fur Tierzucht* **45:** 213.

Mabrook M. and Petty M. (2003) *J. Food Eng.* **60:** 321.

Nielen M., Deluyker H., Schukken Y. and Brand A. (1992) *J. Dairy Sci.* **75:** 606.

Norberg E. (2005) *Livest. Prod. Sci.* **96:** 129.

Norberg E., Hogeveen H., Korsgaard I.R., Friggens N., Sloth K. and Løvendahl P. (2004) *J. Dairy Sci.* **87:** 1099.

Samaraweera M., Boerner V., Cyril H.W., van der Werf J. and Hermesch S. (2018) *Proc. World Congr. Genet. App. Livest. Breed.* **11**:741.

Sorensen D., and Gianola D. (2002) In 'Likelihood, Bayesian, and MCMC methods in quantitative genetics.' pp. 576-588, Springer Science & Business Media: New York.

Woolford M.W., Williamson J.H. and Henderson H.V. (1998) *J. of Dairy Res.* **65:** 187.

Zwald N., Weigel K., Chang Y., Welper R. and Clay J. (2005) *J. Dairy Sci.* **88:** 1192.

Proc. Assoc. Advmt. Anim. Breed. Genet. 23:139-142

EFFECTS OF SELECTION FOR FERTILITY ON MILK PRODUCTION TRAITS

E.M. Strucken[1], G.A. Brockmann[2] and Y.C.S.M. Laurenson[1]

[1]School of Environmental and Rural Science, University of New England, Armidale, NSW, Australia
[2]Breeding Biology and Molecular Genetics, Humboldt-Universität zu Berlin, Berlin, Germany

SUMMARY

A previous study investigated the impact of selection for fertility upon milk yield in the first lactation. The current study extends this analysis to include the yield and content of fat and protein. Daughter test-day records were used to estimate Wilmink curve parameters of each trait for 2,405 sires. The sires also had breeding values for the production traits and their fertility index. Correlations and linear regression between curve parameters and breeding values were carried out with and without correction for environmental effects. Selection for fertility was found to negatively affect milk, fat and protein yield. Improved fertility was found to result in an increased initial fat and protein content, but also increased the rate of decline during early lactation causing a reduced nadir. The persistency of protein content reduced with increased fertility; whilst, fat content rebounded to a greater extent in fertile cows than those with lower fertility. Fat-to-protein ratio reached its maximum about 5 weeks before peak milk production and was higher for less fertile cows, coinciding with time of strongest energy imbalance. Correction for environmental effects resulted in overall lower production curves for yield traits and fat content, but higher protein content. After correction, cows with higher fertility produced more milk compared to lower fertile cows purely on their genetic merit. Similar patterns were found for fat and protein yield. Fat-to-protein ratio was lower for higher fertile cows throughout the entire lactation.

INTRODUCTION

With the advent of modern cattle breeding in the mid to late 20th century, milk production has seen a dramatic increase (Brotherstone and Goddard 2005). With modern breeding, a whole array of factors such as nutrition, health and fertility came into focus, and it was observed that fertility declined with increasing milk production (De Kruif and Mijten 1992; Crowe *et al.* 2018). Consequently, such factors have been included in breeding schemes which have incorporated weighted indices with health and fertility traits (Osteras *et al.* 2007, Boichard and Brochard 2012).

Strucken *et al.* (2015) concluded that the observed impact of milk production on fertility had both a functional (to provide optimal birth spacing) and causal (energy deficit) explanation. Other studies have shown the impact of milk fat and protein on fertility traits, with the fat-to-protein ratio being an accepted measure for energy balance. The fat-to-protein ratio was shown to affect days-open (Buckley *et al.* 2003, Puangdee *et al.* 2017); higher fat and protein yields were genetically correlated with longer calving intervals (Albarran-Portillo and Pollott 2013), and lower protein content was associated with an increased risk of delayed ovulation (Opsomer *et al.* 2000).

This study follows on from Strucken *et al.* (2015) and investigates whether selection for fertility has resulted in observable effects on the lactation curves for milk, fat and protein yield, and fat and protein content; or whether the application of indices allowed breeders to break the genetic link between milk production and fertility.

MATERIALS AND METHODS

Data. Estimated breeding values (EBVs) and the fertility index (RZR) were available for 2,405

Holstein Friesian sires as provided by VIT, Verden (Germany). EBVs for five milk production traits represented actual deviations from the population mean at 305 days in milk (DIM). The RZR summarizes pre-corrected breeding values for six fertility traits and is standardized to a mean of 100 with a standard deviation of 12. Additionally, test-day records of five milk production traits were available for 1,797,852 daughters (Table 1). Each sire had an average of 747 daughters (min=50, max=84.387), with a minimum of 386 and a maximum of 731,431 test-day records per sire.

Table 1. Test-day records of 1.8m cows in the first lactation for five milk production traits and the fertility index (RZR) for 2405 sires

	Milk yield (kg)	Fat yield (kg)	Protein yield (kg)	Fat content (%)	Protein content (%)	RZR
mean	25.57	1.04	0.87	4.14	3.42	101
min	2.00	0.04	0.05	1.60	2.00	62
max	98.80	5.48	3.84	10.50	7.97	136
SD	6.54	0.25	0.20	0.74	0.35	9.9
# test-days	14,862,232					

Analyses. Test-day records for each trait were used to fit 38 lactation curve models with a mechanistic or biological interpretation of curve parameters, and goodness of fit was assessed using 7 criteria. All selection criteria provided the same ranking of models except the Durbin-Watson coefficient. The Wilmink curve (Wilmink 1987) was among the top 10 models for all traits and was selected to allow for comparison of selection effects between traits.

The Wilmink curve was adjusted to allow for better interpretation of parameters, such that:

$$y = a + (b-a) * (1-exp^{-k*DIM}) - c * DIM$$

where y is the test-day record of yield (kg); a is the y-intercept (kg), i.e. starting yield; b is the potential maximum daily yield (kg); c is the gradient of the linear decay in yield (kg d^{-1}); k is the increase in yield prior to peak production; and DIM are the days in milk.

Pearson's correlation coefficients between production EBVs and the RZR, and between the curve parameters and the EBVs and RZR were calculated. A linear regression of EBVs and RZR on the curve parameters was used to further assess the impact of selection on the shape of the curve. To separate environmental from genetic effects, we estimated curve parameters per sire within a linear mixed model which required the fixation of parameter k based on estimates retrieved from the non-linear curve previously used. Fixed effects included *age at calving*, *year season*, and *milk recording system* nested within *farm*. These calculations were carried out across the top and bottom 9% of sires (216 sires) for the fertility index which showed significant differences based on an unpaired two-sided t-test assuming unequal variances.

RESULTS

The pseudo-genetic correlations between yield EBVs and RZR were significantly negative (milk yield = -0.282, fat yield = -0.231, protein yield = -0.305), whilst the content EBVs were significantly positively correlated with RZR (fat content = 0.077, protein content = 0.049), confirming previous reports (Oltenacu & Broom 2010).

Correlations between uncorrected curve parameters and RZR described a similar relationship as the linear regression of RZR on curve parameters (Table 2). Parameter a, determining the y-intercept, was not significantly affected by fertility for any of the analysed traits. Parameter b, describing the potential maximum, was strongly influenced by the level of RZR showing that a better fertility resulted in lower production for the yield traits (except fat yield), and an increase for the content traits (Table 2). Associations of fertility with parameters *k*, describing the production slope before the nadir, showed that better fertility resulted in a stronger increase and earlier peak for milk yield, and a lesser decrease in early lactation for fat and protein content. Parameter *c*, describing the slope after the nadir, showed that better fertility resulted in a stronger decrease in fat yield, stronger increase in fat content and a lesser increase in protein content (Table 2). Fat-to-protein ratio spiked at lactation day 12, after which it dropped and almost stabilized around lactation day 65. Cows with better fertility showed a lower fat-to-protein ratio at peak, and higher and slightly increasing ratio after lactation day 65 (Figure 1).

Table 2. Correlation/Regression coefficient for RZR on uncorrected lactation curve parameters in the first lactation

	Milk yield	Fat yield	Protein yield	Fat content	Protein content
a	-0.031/ -0.01	0.002/ 0.017	0.017/ 0.029	-0.017/ -0.023	-0.008/ -0.002
b	**-0.181***/ -0.04*****	-0.048/ -0.0009	**-0.052†/ -0.002***	**0.077**/ 0.002*****	**0.167***/ 0.001*****
c	-0.11/ -0.000006	**0.068†/ 0.000002***	0.012/ 0.0000006	**-0.054*/ -0.000004****	**0.12***/ 0.000004*****
k	**0.064*/ 0.00017****	0.032/ 0.0008	-0.018/ -0.00006	0.036/ **0.0001†**	**0.05†/ 0.0003***

***P>0.0001, **P>0.001, *P>0.01, †P>0.05

Estimating curve parameters under the consideration of environmental effects showed that cows with a higher fertility also produced more milk (Figure 1), fat and protein yield, less fat content, and almost no difference for protein content. This being the inverse of the observed negative correlations between yield and fertility traits for uncorrected parameters. Correction for environmental effects showed that higher fertile cows have a strongly decreased peak and lower ratio throughout the entire lactation (Figure 1).

DISCUSSION

Reductions in fertility have been largely attributed to an increase in milk production and inadequate nutrition, which (especially at the beginning of the lactation) causes an energy deficit for the cow. This energy deficit forces the metabolism of the cow to shift energy partitioning in favour of milk production and results in the observed negative correlation with fertility traits. (Strucken *et al.* 2015). As such, it may be expected that breeding for better fertility slows milk production in early and peak lactation, unless the genetic link between these traits has been broken. We found that better fertility decreased milk production (especially around its peak), as seen by the significant effects on parameter *b* (parameter *a* in Strucken *et al.* (2015)); and moreover, similar effects were observed for fat and protein yield. Fat and protein content increased in early lactation with a better fertility, however, fat-to-protein ratio was lower for more fertile cows, all confirming the hypothesis of an energy deficit causing the negative trait correlation.

Correction for environmental effects revealed that highly fertile cows produced more milk, fat, and protein yield than less fertile cows, however, both high and low fertility cows profited from the environment. After correction for environmental effects, cows with a low fertility had a higher fat content, whilst protein content remained nearly unchanged. The fat-to-protein ratio strongly increased in early lactation around the time when the energy deficit can be expected to be most developed

(Negussie *et al.* 2013). After correction for environmental effects, cows with the highest fertility showed an overall decreased in fat-to-protein ratio, whilst the environment did not seem to affect cows with a poorer fertility (Figure 1).

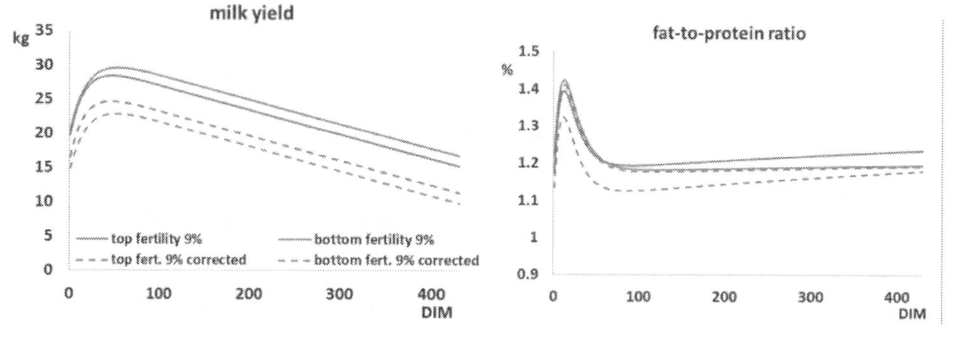

Figure 1. Lactation curves for milk yield and fat-to-protein ratio predicted with corrected and uncorrected Wilmink curve parameters for bulls ranking at the top and bottom of fertility

CONCLUSIONS

Highly fertile cows seem to be capable of producing more milk compared to low fertile cows purely based on the genetic merit. This suggests that the negative genetic link between high milk production and low fertility can be broken. The environment, i.e. favourable management, is not as optimal for high fertile cows and a limiting factor that can be overcome with better management, but sufficient for less fertile cows. This is also reflected in the fat-to-protein ratio as a measure of energy balance, which shows that especially highly fertile cows experience a strong energy deficit in early lactation.

ACKNOWLEDGEMENTS

We thank A. Bünger and F. Reinhardt (VIT, Verden, Germany) for providing the EBVs and daughter records.

REFERENCES

Albarran-Portillo B. and Pollott G.E. (2013) *J. Dairy Sci.* **96**: 635.
Boichard D. and Brochard M. (2012) *Animal* **6**: 544.
Brotherstone S. and Goddard M. (2005) *Philos. Trans. R. Soc. Lond. B. Biol. Sci.* **360**: 1479.
Buckley F., O'Sullivan K., Mee J.F., Evans R.D. & Dillon P. (2003) *J. Dairy Sci.* **86**: 2308
Crowe M.A., Hostens M. and Opsomer G. (2018) *Irish Vet. J.* **71**: 1.
De Kruif A. and Mijten P. (1992) *Berl. Munch. Tierarztl. Wochenschr.* **105**: 271.
Negussie E., Stranden I. and Mantysaari E.A. (2013) *J. Dairy Sci.* **96**: 1237.
Oltenacu P.A. and Broom D.M. (2010) *Anim. Welfare* **19(S)**: 39.
Opsomer G., Grohn Y.T., Hertl J., Coryn M., Deluyker H. and de Kruif A. (2000) *Theriogen.* **53**: 841.
Osteras O., Solbu H., Refsdal A.O., Roalkvam T., Filseth O. and Minsaas A. (2007) *J. Dairy Sci.* **90**: 4483
Puangdee S., Duangjinda M., Boonkum W., Katawatin S., Buaban S. and Theppart M. (2017) *Anim. Sci. J.* **88**: 723.
Strucken E.M., Laurenson Y.C.S.M. and Brockmann G.A. (2015) *Proc. Advmt. Anim. Breed. Genet.* **21**: 362.
Wilmink J.B.M. (1987) *Livest. Prod. Sci.* **16**: 335.

Proc. Assoc. Advmt. Anim. Breed. Genet. 23:143-146

AGE AT CULLING AND REASONS OF CULLING IN AUSTRALIAN DAIRY COWS

Z.W. Workie, J.P. Gibson and J.H.J. van der Werf

School of Environmental and Rural Science, University of New England, Armidale, NSW, Australia

SUMMARY

Culling reasons in Australian dairy cattle were examined using culling records from 1995 through 2016. A total of 2,452,124 individual cow culling observations were obtained of which 2,140,337 were Holstein and 311,787 were from Jersey cows. The most important culling reasons identified were infertility (17.2%), mastitis (13.3%), low production (9.7%), sold for dairy purpose (6.4%) and old age (6.5%) while 38.7% were "other reasons not reported". The average age at culling was nearly the same for Holsteins (6.75 years) and Jerseys (6.73 years). The trend in age at culling over the last twenty years showed a slight increase for Holstein cows (by 0.01 years) and a decrease for Jersey cows (by 0.03 years). Over the last two decades, culling age has changed little in both breeds, whereas culling reasons have changed with low production becoming a less important reason for culling (decreasing by 29% and 37% in Holsteins and Jerseys, respectively) and infertility increasing in both breeds by 13% and 19% in Holsteins and Jerseys, respectively).

INTRODUCTION

A key objective of dairy farmers is to reduce replacement costs, by keeping productive and fertile cows in their herds. However, a number of reasons may trigger farmers to cull cows from their herd; such reasons for culling can be classified as voluntary, or involuntary culling (Weigel *et al.* 2003; Fetrow *et al.* 2006). Involuntary culling happens when the farmer is coerced to cull a productive, profitable cow due to illness, injury, infertility, or death. Voluntary culling, on the other hand, occurs when a farmer chooses to remove a cow due to poor milk production, old age and replacement. Longevity of a cow is also an important trait affecting dairy farm profitability. Increased longevity of dairy cattle helps dairy farmers to get more economic return and reduce replacement cost (Allaire and Gibson 1992; Pritchard *et al.* 2013). Protein yield and fertility are important traits in the breeding objective, in addition to being possible reasons for culling. A previous study (Haile-Mariam and Pryce 2015) estimated genetic parameters for survival traits over time, however, information on reasons for culling and their trend over time is limited in the Australian dairy herd. The aim of this study was therefore to investigate the main causes of culling in Australian dairy herds and thereby to evaluate trends in age of culling and culling reasons.

MATERIAL AND METHODS

Data source. For this study, data on culling reasons were provided by DataGene (previously ADHIS). The data used for this study were extracted from milk recorded herds in Australia. The data were collected based on farmers' recording about each culling reason. A total of 2,502,258 records were received with each record including data on cow identification number, national herd identification number, breed, date of birth of a cow, disposal date of a cow and a code for individual culling reasons. Only a single reason of culling was recorded for each cow removed from the herd. Records of all culled cows were examined across year of culling. Analysis based on year of birth was not considered due to the effect of censored data in recent birth years and relatively older cows in the data for herds in those earlier birth years.

For an evaluation of trend in age at culling over time, we undertook an analysis based on ordering cows by culling year and counting the total number of cows culled for all years from 1970 through to 2016. However, for the years from 1970 to 1994, there were very few recorded reasons for culling. As a result, records that had a year of culling before 1995 were disregarded. In the final data set, a total of 2,452,124 records (2,140,337 Holstein and 311,787 Jersey cows) were retained from cows in 11,145 herds culled between 1995 and 2016. Birth season of a cow was classified into two categories; season 1 contained the records of cows that were born from January to June while season 2 covered the period from July to December, as in (Visscher and Goddard 1995b). For evaluation of trend of culling reasons, data were also split into two year groups (1995-2005 and 2006-2015) based on the differences observed in proportion of culling reasons on these periods.

Statistical analysis. Descriptive statistics were carried out to identify and describe the main disposal reasons stated by farmers. Trends in age at culling were analysed based on year of culling to evaluate how herdlife has changed over time according to the animal's culling year and how it differed between breeds. Age at culling was analysed using a univariate analysis with the following linear model,

$$y_{ijkm} = \mu + B_i + Y_j + HS_k + e_{ijkm}$$

where, y_{ijkm} = is an observed age (in years) on animal ijkm in breed i, year j and in herd-season k, μ = the overall mean, B_i = effect of breed, Y_j = effect of culling year and HS_k = the fixed effect of herd-season, e_{ijkm} = error term.

RESULTS

Reasons for culling. About 38.7% of the cows left the herd for 'other reasons' (Table 1). Aside from 'other reasons', the main reasons for culling across breeds were infertility, mastitis, low production, sale for dairy purpose and old age. The proportion of cows culled due to infertility was slightly higher for Holstein than Jersey cows. The proportion of cows culled for infertility in both breeds increased in the culling year group (2006-2015) compared with the culling year group (1995-2005). Next to infertility, the second and third most common reasons of culling in Holstein cows were mastitis and low production. By contrast, the second and third causes for culling were reversed in Jersey cows. Culling due to low production decreased from 10.4 to 7.4% in Holstein and 18.2 to 11.4% in Jersey cows, between the decades 1995-2005 and 2006-2015. In contrast, the proportion of cows culled due to sale reasons increased from 1995-2005 to 2006-2015. Culling of cows for involuntary culling (IC) reasons included infertility, mastitis and accident, which together accounted for 33.0% of culling reasons for the 2 breeds. Voluntary culling (VC) accounted for 22.5% of reasons, with about 9.7%, 6.4% and 6.4% of cows removed because of low production, sale for dairy purpose and old age, respectively.

Table 1. Proportion (%) of culling reason types by breed and year of culling

Culling reasons	Year of culling (1995-2005)		Year of culling (2006-2015)		Overall
	Holstein	Jersey	Holstein	Jersey	
Other reasons	39.8	37.5	38.5	33.8	38.7
Infertility	16.4	14.5	18.5	17.3	17.3
Mastitis	14.0	11.9	12.8	13.6	13.3
Low production	10.4	18.2	7.4	11.4	9.7
Old age	7.0	7.4	5.7	6.4	6.6
Type defect	3.3	2.9	3.2	4.2	3.3
Sold for dairy purpose	3.3	3.5	9.7	9.6	6.4
Accident	2.7	1.6	1.8	1.4	2.2
Poor temperament	1.9	2.1	1.2	1.8	1.6
Calving difficulties	1.2	1.2	1.2	0.6	1.1

Trend in culling age. The least squares mean of age at culling across year of culling for Holstein and Jersey cow breeds is shown in Figure 1. The difference in estimated age at culling was small between breeds but significant ($P<0.05$). Holstein dairy cows had slightly higher estimated mean age at culling compared with their Jersey counterparts (6.75 years for Holstein and 6.73 years for Jersey cows). The minimum and maximum mean estimated ages at culling for Holstein cows were observed in the year 2003 (6.55 years) and in 1997 (7.0 years), whereas for Jersey cows the corresponding average values were 6.45 years in 2005 and 7.10 years in the year 1997, respectively. Holstein cows had a slightly increasing trend in age at culling (0.01 years) over the last 20 years whereas the estimated age at culling had declined little in the same period for Jersey cows. Overall, Holstein cows were culled at a slightly older age, especially in the last decade, and that the age of voluntary culling (VC) was overall slightly lower than involuntary culling (IC) with more difference for the Jersey cows. There was also an association between culling reasons, whereby younger cows were culled for low production and infertility, while older cows were culled for mastitis.

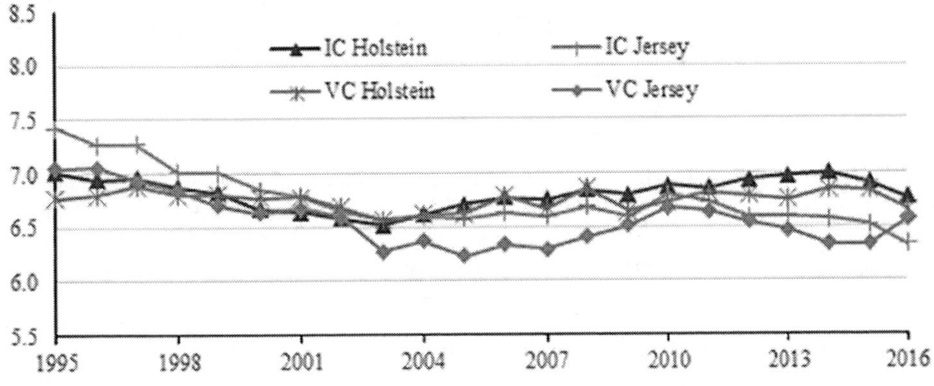

Figure 1. Least squares means of age at culling in each year for Holstein and Jersey dairy cows by year of culling

DISCUSSION

Descriptive statistics were used to calculate the proportion of culling reasons recorded for the two dairy breeds. Identifying reasons for culling cows could also be useful in determining the main problems in dairy herds and in identifying breeding objectives and evaluating results of selection. Excluding other reasons not reported, the most prevalent reason for culling dairy cows was infertility followed by mastitis and low production. In agreement to the current study, previous research findings identified infertility as the main reason of culling dairy cows in Sweden (Ahlman *et al.* 2011) and USA (Bascom and Young 1998; Smith *et al.* 2000). In this study, the phenotypic trend of culling cows due to infertility has increased for both dairy cow breeds from 1995-2005 to 2006-2015, whereas low production has shown a sharp decline. Culling due to low production could be part of the economic and management decisions to maintain a required number of dairy cows in a particular farm where good producing cows might have low chance to be culled (Roxström and Strandberg 2002; Pinedo *et al.* 2014). The proportion of Jersey cows culled for mastitis increased over the year groups. Previous studies regarding the proportion of cows leaving for mastitis of 12.1% (Hadley *et al.* 2006) and 12.0% (Smith *et al.* 2000) in the US dairy cows are closer to the levels in this study.

In the current study, culling related to other reasons not reported had the highest proportion for both dairy breeds. In terms of making management decisions, this category yields no information. A more descriptive category needs to be developed that can account for the list of reasons that these cows were removed from the herd.

The pattern of age at culling over time for both dairy breeds was evaluated with the year of birth and year of culling. When age at culling was evaluated against year of birth, the estimated trend of age at culling sharply declined (results not presented) but this estimate was deemed to be biased because of censoring. A censored record can be seen as the minimum survival the cow reaches and this could be a problem in prediction of breeding values for survival because estimated breeding values are required for live animals. By fitting year of culling in the model, all age groups of culled cows were included in the analysis. In the same way, the trend of estimated age at culling for the two dairy breeds for the last 20 years was less varied (Figure1). The overall estimated least squares mean for age at culling was about 6.65 years. By assuming the average age of 2 years at first calving for most of the heifers, the productive life of cows in the present study estimated to be 4.6 lactations, which is comparable with earlier reports of average productive life of 4.6 and 4.3 lactations for Holstein and Jersey cows, respectively in Australian dairy cattle (Visscher and Goddard 1995a). The average herdlife observed in the current study is higher than the average herd life observed in Dutch dairy cattle (Van Pelt *et al.* 2015), which was found to be 3.2 lactations.

CONCLUSION

Phenotypic analysis of culling data showed that the estimated average age at culling has changed little between 1995 and 2016. The proportion of major culling reasons such as infertility, low production and mastitis in both dairy breeds have changed over the past two decades; which might indicate a change in survival traits over time and a likely change in correlation of survival with other objective traits such as yield and fertility.

ACKNOWLEDGMENTS

The authors thank Michelle Axford and her team at DataGene organization for providing culling data on dairy cows. The first author duly acknowledges the support of the Australian commonwealth Government, RTP scholarship.

REFERENCES

Ahlman T., Berglund B., Rydhmer L. and Strandberg E. (2011) *J Dairy Sci.* **94**: 1568.
Allaire F. and Gibson J. (1992) *J Dairy Sci.* **75**: 1349.
Bascom S. and Young A. (1998) *J Dairy Sci.* **81**: 2299.
Fetrow J., Nordlund K. and Norman H. (2006) *J Dairy Sci.* **89**: 1896.
Hadley G., Wolf C. and Harsh S. (2006) *J Dairy Sci.* **89**: 2286.
Haile-Mariam M. and Pryce J.E. (2015) *J Dairy Sci.* **98**: 7364.
Pinedo P., Daniels A., Shumaker J. and De Vries A. (2014) *J Dairy Sci.* **97**: 2886.
Pritchard T., Coffey M., Mrode R. and Wall, E. (2013) *J Dairy Sci.* **96**: 3296.
Roxström A. and Strandberg E. (2002) *Livest. Prod. Sci.* **74**: 125.
Smith J., Ely L. and Chapa A. (2000) *J Dairy Sci.* **83**: 2980.
Van Pelt M., Meuwissen T., de Jong G. and Veerkamp R. (2015) *J Dairy Sci.* **98**: 4117.
Visscher P. and Goddard M. (1995a) *Animal Science* **61**: 9.
Visscher P. and Goddard M. (1995b) *J Dairy Sci.* **78**: 205.
Weigel K., Palmer R. and Caraviello D. (2003) *J Dairy Sci.* **86**: 1482.

Proc. Assoc. Advmt. Anim. Breed. Genet. 23:147-150

LATE GESTATION HEALTH STATUS IS CORRELATED WITH LACTATION OUTCOMES FOR SOWS

L. Vargovic[1], K.L. Bunter[1], S. Hermesch[1], J. Harper[2] and R. Sokolinski[3]

[1]Animal Genetics & Breeding Unit*, University of New England, Armidale, NSW, 2351 Australia
[2]Rivalea Australia Pty Ltd, Corowa, NSW, 2646, Australia
[3]PIC Australia, Grong Grong, NSW, 2652 Australia

SUMMARY

Gilts and sows from two nucleus farms (N=1103) were recorded after transfer to the farrowing shed for a range of health-related traits and subsequent lactation outcomes. Traits recorded pre-farrowing included fight lesions (FIGHT), caliper score (CAL), udder condition (MAST), haemoglobin level (HB), respiration rate (RESP), rectal temperature (RECT) and feed refusal before farrowing (FRBF). Lactation outcomes included the number of weaned piglets (NWEAN) and lactation failure (LFAIL). The highest heritabilities (h^2) were estimated for CAL (0.34±0.08), FRBF (0.21±0.08) and RESP (0.20±0.09), while the remaining traits were lowly heritable. Antagonistic genetic (rg) and/or phenotypic (rp) correlations were estimated for NWEAN with FRBF (rg: -0.36±0.30; rp: -0.10±0.03) and for CAL with HB (rg: 0.33±0.41; rp: 0.15±0.03). The absence of pre-farrowing mastitis was associated with higher NWEAN both genetically (-0.74±0.30) and phenotypically (-0.05±0.03), indicating that selection for healthy udder led to increase in NWEAN. Sows with higher levels of HB and fewer feed refusals had increased NWEAN. Non-zero heritabilities demonstrate that health-related traits have a genetic component, but evaluation of their potential use as selection criteria to improve lactation outcomes for sows requires additional data to obtain more accurate estimates of genetic correlations.

INTRODUCTION

Lactation outcomes can be defined by the number of weaned piglets, lactation length or removal reasons related to poor mothering ability. Selection for litter size in pigs is aimed at increasing the number of weaned piglets, which can have detrimental effects for health of both sows and piglets and lead to a poor lactation outcome. Previous studies reported genetic associations between piglet survival and traits such as body condition, fight lesions, appetite or rectal temperatures of sows (Tabuaciri *et al.* 2010). In a phenotypic study, Anil *et al.* (2008) reported negative correlations between lactation outcomes and lactation feed intake, elevated rectal temperature or health issues.

The objective of this study was to test whether health traits (haemoglobin, fight lesions, respiration rate, mastitis, rectal temperature, appetite or body condition) were heritable and accompanied by negative genetic correlations with lactation outcomes. The hypothesis was that those traits are heritable and can be considered for developing breeding goals that balance high production performance with improved health and welfare of sows and piglets.

MATERIALS AND METHODS

Data. The data used in this study were recorded at two nucleus farms operated by independent companies, collected during the period October-December 2017 (Farm A, N=558 sows) and March-June 2018 (Farm B, N=545 sows). The sows recorded included both primi- and multi-parous sows and represented a total of 10 (maternal or terminal) lines across both farms. Farms differed generally in their production environment, management, housing, feeding regimes and health status, which are not described further here. Sows were transferred from gestation housing to the farrowing shed at an

* A joint venture of NSW Department of Primary Industries and the University of New England

average gestation length of 110 days and recorded for a range of health and welfare characteristics by a single operator. Subsequently, sows farrowed naturally and were managed according to each farm's commercial protocols. The targeted lactation lengths were four (Farm A) and three weeks (Farm B).

Late gestation characteristics. The extent of fight lesions (FIGHT) was scored as 0: no lesions; 1: 1-5 lesions; 2: 6-10 lesions; and 3: 10+ lesions (Bunter 2017). Body condition (CAL) was measured as caliper increments, using procedures described by Knauer *et al.* (2015), with increasing value corresponding to increasing body condition. Udder health was assessed by recording pre-farrowing mastitis (MAST, 0/1), considered to be present (score=1) for sows with a hard and swollen udder, irrespective of whether this was accompanied by an elevated rectal temperature. Resting respiration rate (RESP) was recorded as the number of expirations per 30 seconds, expressed per minute. Rectal temperatures (RECT) were obtained when sows were at rest ensuring the thermometer was in contact with the bowel wall. Haemoglobin (HB) level was measured using the Hemocue H201+ (HemoVue AB, Angeloholm, Sweden) using a single drop of blood obtained from a skin prick on the sow's ear (Hermesch and Tickle 2012). Sows which farrowed prior to the measurement date or which appeared distressed at the time of procedure were not sampled for HB. Feed refusal before farrowing (FRBF) was recorded as the proportion of days observed where less than half the meal was eaten, assessed 3-4 hours after the first feed delivery in the morning. Sows were observed for FRBF for 5.62±2.14 days, on average.

Lactation outcomes. Lactation failure (LFAIL, 0/1) was defined to occur (score=1) for any combination of: weaned piglets <7; lactation length <15 days; or if removal reasons included poor mothering ability, bad udder or no milk. A trait frequently used to describe lactation performance is the number of weaned piglets (NWEAN). Sows which weaned no piglets (due to piglet deaths) or had all piglets removed prematurely were assigned NWEAN=0. For sows which were used to foster a second litter (N=4), NWEAN was based on the first litter only. If the sow did not lactate at all (culled or died), LFAIL and NWEAN were considered missing (N=3). Records clearly identified with recording errors were excluded from analyses (N=12).

Analyses. Data preparation and summary statistics were obtained using R (R Core Team 2018). Raw data were firstly examined for errors and outliers, which were excluded from analyses (HB: N=4) if trait values were more than four standard deviations from the mean, within farm. The combined farm dataset was then used for analyses. Estimates of variance components were obtained by fitting a linear mixed animal model using residual maximum likelihood procedures in ASReml (Gilmour *et al.* 2014). Systematic effects fitted for all traits included parity group (4 levels: parities 1, 2, 3-4 and >4) and the interaction between breed and farm (10 levels). Estimates for heritabilities were obtained from univariate analyses. Correlations between traits were estimated using a series of bivariate analyses. Sows were progeny of 352 sires and 852 dams, and the pedigree was extended over 5 generations to contain 1261 sires and 3274 dams in total. There were 104 commercial sows without pedigree retained in the data.

RESULTS AND DISCUSSION

Characteristics of the data. The incidence of sows which experienced undesirable lactation outcomes (LFAIL) was <10% (Table 1), consistent with results from a different population (Bunter *et al.* 2018). Fight lesions were observed on a relatively high percentage of sows, demonstrating aggression exists amongst group-housed sows in late gestation. The average value for HB was 106 g/l, with 2.71% of sows considered borderline anaemic (< 80 g/l). The average values for HB align with previous study by Hermesch and Tickle (2012). The extent of feed refusal was variable (CV=141%), with an average of 20% of meals observed pre-farrowing with feed refused.

Table 1. Raw data characteristics, distribution (%×100) of scores, estimates of heritability (h^2) and phenotypic variance (σ^2_p) from univariate model, with model R^2

Trait	N	Mean (SD)	CV%	Distribution of scores				h2 (SE)	σ2p	R2 (%)
				0	1	2	3			
NWEAN	1088	9.38 (2.62)	28	na	na	na	na	0.16 (0.08)	6.65	4.50
LFAIL	1100	na	na	90.2	9.8	na	na	0.09 (0.08)	0.09	2.51
CAL	1098	14.4 (2.66)	19	na	na	na	na	0.34 (0.08)	5.90	16.5
FIGHT	1103	na	na	26.5	36.3	26.2	11.0	0.14 (0.07)	0.65	29.3
MAST	1103	na	na	93.7	6.3	na	na	0.15 (0.08)	0.52	10.8
RESP	1067	25.4 (16.7)	68	na	na	na	na	0.20 (0.09)	225	19.6
RECT	1067	37.8 (0.47)	1	na	na	na	na	0.12 (0.08)	0.19	13.6
HB	960	106 (14.0)	13	na	na	na	na	0.06 (0.07)	171	12.9
FRBF	1076	0.20 (0.28)	141	na	na	na	na	0.21 (0.08)	0.80	0.56

Abbreviations: NWEAN: count of weaned piglets, LFAIL: lactation failure (0/1), CAL: caliper increments (count), FIGHT: fight lesion scores (0-3), MAST: pre-farrowing mastitis (0/1), RESP: count of expirations/ minute, RECT: rectal temperature (°C), HB: haemoglobin level (g/l), FRBF: proportion of days observed where less than half the meal was eaten, na: not applicable

Heritability estimates. Overall, results presented in Table 1 demonstrate genetic contributions to performance (LFAIL, NWEAN), as well as feeding or interactive behaviours (FRBF, FIGHT), health or condition (MAST, CAL), and physiological traits (RESP, RECT, HB) recorded prior to farrowing. LFAIL and NWEAN were two traits for assessing sow performance as nursing sow. Heritability estimate for LFAIL was 0.09±0.08, which was higher than previously reported ($h^2 = 0.00$) for crossbred sows (Bunter *et al.* 2018). The heritability estimate for NWEAN was higher (0.16±0.08) than the average of 0.07 reported in the review of Rydhmer (2000), and is potentially influenced by the minimum cross-fostering, diversity of lines, combined with phenotypes which included zero values for sows which weaned no piglets. Moderate h^2 (0.21±0.08) for FRBF suggests that when sows are observed pre-farrowing for feed refusals following fixed delivery, phenotypic differences between animals may be accurately observed, revealing differences in appetite before farrowing. Estimate of heritability for CAL was high (0.34±0.08), consistent with similar traits like sow weight or back fat (Tabuaciri *et al.* 2010). Heritability for FIGHT was moderate (0.14±0.07) and align with previously reported by Bunter (2017).

Correlations. Large genetic (-0.97±0.18) and residual (-0.73±0.03) correlations between NWEAN and LFAIL are consistent with the use of NWEAN to define LFAIL phenotypes (Table 2). All other correlations were of lesser magnitude. Genetic correlations were only consistent in direction or magnitude with phenotypic correlations for some trait combinations, which probably reflects relatively small sample size. The genetic correlation between NWEAN and MAST was strong (-0.74±0.30), indicating selection for udder health could contribute to increased NWEAN. Genetic and phenotypic correlations were positive between CAL and HB, and between FRBF and RECT. Sows with lower FRBF (rg: -0.36±0.30; rp: -0.10±0.03) or higher HB (rp: 0.08±0.03) weaned more piglets. Iron status influences appetite and vitality of piglets at birth (cited in Hermesch and Tickle (2012)). Rectal temperature, RESP and FRBF were positively correlated phenotypically, consistent with the expectations that animals with elevated body temperature will breathe faster and reduce feed intake.

CONCLUSIONS

Traits related to health of sows (MAST, CAL, FRBF, RESP, RECT, HB) were heritable. Genetic correlations in this study were preliminary estimates, had high standard errors, and were frequently inconsistent in magnitude or direction with phenotypic correlations. More data are required to obtain more accurate estimates of genetic correlations, particularly for trait combinations where phenotypic correlations between traits were substantial. However, negative genetic and phenotypic correlations between NWEAN and FRBF or NWEAN and MAST were implying that feed refusals and udder health have implications for current performance and for breeding programs.

Table 2. Estimates of genetic (above diagonal), residual (1st row) and (2nd row) phenotypic (below diagonal) correlations (SE in subscript) between traits

	NWEAN	LFAIL	CAL	FIGHT	MAST	RESP	RECT	HB	FRBF
NWEAN		$-0.97_{(0.18)}$	$-0.56_{(0.29)}$	$-0.03_{(0.37)}$	$-0.74_{(0.30)}$	$0.89_{(0.37)}$	$-0.16_{(0.40)}$	$-0.69_{(0.69)}$	$-0.36_{(0.30)}$
LFAIL	$-0.73_{(0.03)}$		$0.65_{(0.41)}$	$-0.04_{(0.47)}$	$0.53_{(0.42)}$	$-0.48_{(0.52)}$	$0.20_{(0.50)}$	$0.62_{(0.76)}$	$0.47_{(0.41)}$
	$-0.75_{(0.01)}$								
CAL	$0.17_{(0.08)}$	$-0.18_{(0.07)}$		$-0.43_{(0.26)}$	$0.07_{(0.46)}$	$-0.42_{(0.26)}$	$-0.04_{(0.29)}$	$0.33_{(0.41)}$	$-0.13_{(0.23)}$
	$0.001_{(0.03)}$	$-0.02_{(0.03)}$							
FIGHT	$0.02_{(0.07)}$	$-0.01_{(0.06)}$	$-0.004_{(0.07)}$		$-0.62_{(0.34)}$	$-0.21_{(0.33)}$	$-0.21_{(0.41)}$	$-0.33_{(0.54)}$	$-0.47_{(0.34)}$
	$0.02_{(0.03)}$	$-0.02_{(0.03)}$	$-0.09_{(0.03)}$						
MAST	$0.11_{(0.07)}$	$-0.02_{(0.07)}$	$0.13_{(0.06)}$	$0.09_{(0.07)}$		$-0.03_{(0.35)}$	$0.16_{(0.43)}$	$-0.25_{(0.58)}$	$0.21_{(0.32)}$
	$-0.05_{(0.03)}$	$0.06_{(0.03)}$	$0.12_{(0.03)}$	$-0.03_{(0.03)}$					
RESP	$-0.16_{(0.07)}$	$0.05_{(0.07)}$	$0.20_{(0.08)}$	$-0.07_{(0.07)}$	$-0.07_{(0.07)}$		$-0.12_{(0.43)}$	$-0.69_{(0.76)}$	$-0.49_{(0.35)}$
	$0.02_{(0.03)}$	$-0.01_{(0.03)}$	$0.04_{(0.03)}$	$-0.09_{(0.03)}$	$-0.06_{(0.03)}$				
RECT	$-0.01_{(0.07)}$	$0.02_{(0.06)}$	$0.14_{(0.07)}$	$-0.02_{(0.06)}$	$-0.05_{(0.07)}$	$0.30_{(0.06)}$		$0.98_{(0.64)}$	$0.20_{(0.34)}$
	$-0.03_{(0.03)}$	$0.04_{(0.03)}$	$0.10_{(0.03)}$	$-0.05_{(0.03)}$	$-0.02_{(0.03)}$	$0.24_{(0.03)}$			
HB	$0.16_{(0.07)}$	$-0.13_{(0.06)}$	$0.13_{(0.07)}$	$-0.06_{(0.06)}$	$-0.02_{(0.07)}$	$0.19_{(0.07)}$	$-0.05_{(0.06)}$		$-0.08_{(0.50)}$
	$0.08_{(0.03)}$	$-0.06_{(0.03)}$	$0.15_{(0.03)}$	$-0.08_{(0.03)}$	$-0.04_{(0.03)}$	$0.01_{(0.03)}$	$0.04_{(0.03)}$		
FRBF	$-0.04_{(0.07)}$	$0.04_{(0.07)}$	$0.06_{(0.08)}$	$0.04_{(0.07)}$	$-0.12_{(0.07)}$	$0.24_{(0.07)}$	$0.12_{(0.07)}$	$0.15_{(0.06)}$	
	$-0.10_{(0.03)}$	$0.10_{(0.03)}$	$0.01_{(0.03)}$	$-0.04_{(0.03)}$	$-0.06_{(0.03)}$	$0.10_{(0.03)}$	$0.13_{(0.03)}$	$0.12_{(0.06)}$	

For trait name abbreviations see Table 1.

ACKNOWLEDGEMENT

This research was funded by the Australian Pork CRC under the project 2A-116. The first author is supported by UNE via International Postgraduate Research Award (UNE IPRA).

REFERENCES

Anil S.S., Anil L. and Deen J. (2008) *Can J Anim Sci* **88:** 381.

Bunter K.L. (2017) *Proc. Assoc. Advmt. Anim. Breed. Genet.* **22**: 229.

Bunter K.L., Vargovic L., Athorn R.Z., Henman D. and Luxford B.G. (2018) In *Proc. World Congr. Genet. App. Livest. Breed* **11**: 992.

Gilmour A., Gogel B., Cullis B., Welham S. and Thompson R. (2014) ASReml User Guide Release 4.1 VSN International Ltd, Hemel Hempstead, HP1 1ES, UK.

Hermesch S. and Tickle K.M. (2012) In *Pig Genetic Workshop Notes*. Armidale, Australia. pp. 31:38.

Knauer M.T. and Baitinger D.J. (2015) *Appl. Eng. Agric.* **31:** 175.

R Core Team (2018) *R Foundation for Statistical Computing*, Vienna, Austria.

Rydhmer L. (2000) *Livest. Prod. Sci.* **66:** 1:12.

Tabuaciri P.,Bunter K.L. and Graser H.U.(2010) *Proc. World Congr. Genet. App. Livest. Breed* **9**: 684.

Proc. Assoc. Advmt. Anim. Breed. Genet. 23:151-154

POST-FARROWING HEALTH STATUS OF SOWS AND PIGLETS IS CORRELATED WITH LACTATION OUTCOMES OF SOWS

L. Vargovic[1], K.L. Bunter[1], S. Hermesch[1], R. Sokolinski[2] and J. Harper[3]

[1]Animal Genetics & Breeding Unit*, University of New England, Armidale, NSW, 2351 Australia
[2]PIC Australia, Grong Grong, NSW, 2652 Australia
[3]Rivalea Australia Pty Ltd, Corowa, NSW, 2646

SUMMARY

The genetic (rg) and phenotypic (rp) correlations between piglet vitality at birth, traits measured on sows 2 and 5 days post-farrowing and lactation outcomes were estimated using the data from 2 nucleus farms (N=1103). All observations were analysed as traits of the sow. The highest heritabilities (h2) were estimated for functional and un-suckled teats (0.36±0.09 and 0.24±0.09) and for the number of vital piglets (0.09±0.07). Detrimental piglet attributes were genetically and phenotypically associated with each other and with a lower number of weaned piglets. High respiration rate and rectal temperature were genetically (0.81±0.31 and 0.73±0.30), but not phenotypically, associated with the number of weaned piglets. Correlations between other traits were not significantly different from zero, or had high standard errors and therefore required more data for more accurate estimation of variance components.

INTRODUCTION

Examination of sows and piglets shortly after farrowing can be used to identify risk-factors, which might have an impact on lactation outcomes (Madec *et al.* 1992). Lactation outcome can be defined by the number of weaned piglets, lactation length or removal reasons related to poor mothering ability. While numerous studies reported the association between birth weight and the number of weaned piglets, relatively fewer studies have considered the implications of other piglet vitality traits at birth and post-farrowing health indicators of sows for the lactation outcomes. The objective of this study was to estimate the genetic parameters for the health-related post-farrowing predictors and to obtain preliminary estimates of the genetic associations with lactation outcomes.

MATERIALS AND METHODS

The data used in this study were recorded at 2 nucleus farms operated by independent companies, collected between October-December 2017 for Farm A (N=558 sows) and March-June 2018 for Farm B (N=545 sows). Further details were provided in Vargovic *et al.* (2019). After farrowing, but before cross-fostering, sows and their piglets were recorded for a range of characteristics. All observations were treated as traits of the sow. Sows were progeny of 352 sires and 852 dams and the pedigree was extended over 5 generations containing 1,261 sires and 3,274 dams in total. There were 104 commercial sows without pedigree retained in the data.

Characteristics of piglets. The vitality of piglets within the birth litter was assessed within 12 hours of the completion of farrowing. Negative indicators for piglet vitality included the number of pale (NPALE) or thin (NTHIN) piglets, whereas the number of vital piglets (NVITAL) was recorded as the total number of piglets without any detrimental attributes.

Characteristics of sows. Sows were recorded for a range of attributes, on days 2 and 5 post-farrowing. Resting respiration rates (RESP2, RESP5) were recorded as the number of expirations per 30 seconds, expressed per minute. Rectal temperatures (RECT2, RECT5) were recorded ensuring

* A joint venture of NSW Department of Primary Industries and the University of New England

the thermometer was in contact with the bowel wall. Mastitis (MAST2, 0/1) was considered to be present (score=1) for sows with a hard and swollen udder. Indicators of suckling load included the count of un-suckled (TEATU2) and functional teats (TEATF2). Feed refusal after farrowing (FRAF) was recorded as the proportion of days observed where less than half the meal was eaten, assessed 3-4 hours after the fixed feed delivery. Sows were observed for FRAF over 2.95±2.80 days on average. Lactation failure (LFAIL) and the number of weaned piglets (NWEAN) were defined as described by Vargovic *et al.* (2019).

Analyses. Data preparation and summary statistics were obtained using R (R Core Team 2018). Estimates of variance components were obtained by fitting a linear mixed animal model using residual maximum likelihood procedures in ASReml (Gilmour *et al.* 2014). Systematic effects fitted for sow traits included parity group (4 levels: parities 1, 2, 3-5 and >5) and the interaction between breed and farm (10 levels). For piglet vitality traits, models included total piglets born fitted as a linear covariate. Estimates for heritabilities were obtained from univariate analyses. Correlations were estimated using a series of bivariate analyses.

RESULTS AND DISCUSSION

Characteristics of the data. Traits that represent piglet vitality (NPALE, NTHIN) and the resulting un-suckled teats (TEATU2) were highly variable between litters (Table 1). However, no detrimental attributes were observed on 77.5% of born alive piglets. This study showed that un-suckled teats can be observed early post-farrowing, which could result in rapid regression (Kim *et al.* 2001). Mastitis was recorded in 15.5% of sows, and 5.49/15.5=35% of these sows also had elevated rectal temperatures. However, farrowing followed by physiological hyperthermia can cause misinterpretation as to whether mastitis is present or not (Friendship *et al.* 2015).

Table 1. Raw data characteristics, estimates of heritability (h^2) with standard errors (SE) and phenotypic variance (σ^2_p) from univariate model, with model R^2

Trait	N	Model effects	Mean (SD)	CV%	$h^2_{(SE)}$	σ^2_p	R^2(%)
NWEAN	1088	P, BF	9.38 (2.62)	28	0.16 $_{(0.08)}$	6.65	4.50
LFAIL	1100	P, BF	0.098 (0.30)	303	0.09 $_{(0.08)}$	0.09	2.31
NVITAL	1072	P, BF, TB	8.83 (2.82)	32	0.09 $_{(0.07)}$	5.31	33.2
NPALE	1072	P, BF, TB	0.93 (1.59)	171	0.04 $_{(0.06)}$	2.21	12.0
NTHIN	1072	P, BF, TB	2.70 (2.63)	97	0.08 $_{(0.07)}$	4.96	28.6
RESP2	1025	P, BF	23.7 (12.3)	52	0.17 $_{(0.09)}$	145	3.03
RESP5	973	P, BF	28.1 (15.4)	55	0.10 $_{(0.08)}$	236	11.3
RECT2	1064	P, BF	38.9 (0.51)	1	0.21 $_{(0.09)}$	0.23	0.62
RECT5	1060	P, BF	38.9 (0.57)	2	0.12 $_{(0.08)}$	0.24	24.5
MAST2	1059	P, BF	0.155 (0.36)	234	0.05 $_{(0.06)}$	0.13	3.41
TEATU2	1059	P, BF	1.26 (1.33)	105	0.24 $_{(0.09)}$	1.73	1.39
TEATF2	1059	P, BF	13.8 (1.17)	9	0.36 $_{(0.09)}$	1.26	8.18
FRAF	1065	P, BF	0.35 (0.39)	114	0.01 $_{(0.07)}$	0.14	10.3

Abbreviations: NWEAN: count of weaned piglets; LFAIL: lactation failure (0/1); NVITAL, NPALE, NTHIN: count of vital, pale and thin piglets; RESP2 and RESP5: count of expirations/minute; RECT2 and RECT5: rectal temperature (oC); MAST2: mastitis (0/1); TEATU2 and TEATF2: count of un-suckled and functional teats; FRAF: feed refusal after farrowing; P: parity group; BF: breed:farm; TB: total born piglets

Heritability estimates. After accounting for systematic effects, heritability estimates (h^2) were low (<0.07) for NPALE, MAST2 and FRAF (Table 1). The h^2 for NPALE was similar to that reported by

Tabuaciri *et al.* (2011). The highest h^2 was for TEATF2 (0.36±0.09), consistent with Lundeheim *et al.* (2013). With respect to sow attributes, RECT2 and RESP2 were moderately heritable (0.21±0.09 and 0.17±0.09), and lower than reported by Gourdine *et al.* (2017), averaged across lactation (0.35±0.09 and 0.39±0.13). The h^2 for NWEAN was higher (0.16±0.08) than the mean (h^2=0.07) previously reported by Rothschild *et al.* (1998).

Correlations for piglet attributes. NTHIN and NPALE were positively correlated with each other and negatively with NVITAL (Table 2). Both phenotypic (rp) and genetic (rg) correlations indicated that NVITAL was positively correlated with NWEAN and negatively correlated with LFAIL and TEATU2. Piglet vitality at birth is an important contributor to successful lactation outcomes assessed for sows. Lower rg and rp were estimated between piglet traits (NTHIN, NVITAL, NPALE) and sow health-related traits (RESP2, RESP5, MAST2), suggesting independence of these traits genetically.

Correlations for sow attributes. Rectal temperature and respiration rate were strongly correlated with each other (Table 2), and favourably associated with NWEAN, while attributes measured day 5 were less informative, due to lower h^2 and higher standard errors. Sows with high genetic potential for NWEAN had genetically higher RESP and RECT, suggesting better environmental management may be required for genetically superior sows. Moderate to high rg between MAST2 and NWEAN/LFAIL were favourable, indicating that visual observation of udder for mastitis (even without confirmation by taking rectal temperature) was correlated with the number of weaned piglets. Moderate rg between NWEAN and TEATF2 demonstrated that the number of functional teats post-farrowing was favourably associated with the number of weaned piglets. Large rg (-0.97±0.18) and re (-0.73±0.03) between NWEAN and LFAIL are consistent with the use of NWEAN to define LFAIL phenotypes.

CONCLUSIONS

Results presented in this study demonstrated that piglet vitality contributes to sow lactation performances. Sows which wean more piglets were genetically predisposed to higher rectal temperature and respiration rate. Visually assessed presence of mastitis was genetically associated with the lactation outcomes. Large standard errors in genetic parameters were observed, with further data required to reduce this error.

ACKNOWLEDGEMENT

This research was funded by the Australian Pork CRC (project 2A-116). The first author is supported by UNE via International Postgraduate Research Award (UNE IPRA).

REFERENCES

Friendship R.M., O'Sullivan T.L. (2015) In 'The gestating and lactating sow.' pp. 409:421, editor C Farmer, Wageningen Academic Publisher.

Gilmour A., Gogel B., Cullis B., Welham S. and Thompson R., (2014). ASReml User Guide Release 4.1 Functional Specification. VSN International Ltd, Hemel Hempstead, HP1 1ES, UK.

Gourdine J., Mandonnet N., Giorgi M. and Renaudeau D. (2017) *Animal* **11:** 365.

Kim S.W., Easter R.A. and Hurley W.L. (2001) *J. Anim. Sci.* **79:** 2659.

Lundeheim N., Chalkias H. and Rydhmer L. (2013) *Acta Agric Scand A-Anim Sci* **63:** 121.

Madec F. and Leon E. (1992) *J Vet Med Ser A* **39:** 433.

R Core Team (2018) R Foundation for Statistical Computing, Vienna, Austria

Rothschild M. and Bidanel J. (1998) In 'Genetics of the Pig.' pp. 313:343

Tabuaciri P., Bunter K.L. and Graser H.U. (2011). *Proc. Assoc .Advmt. Anim. Breed .Genet.* **19:**. 115.

Vargovic L., Bunter K.L., Hermesch S., Harper J., Sokolinski R. (2019). *Proc. Assoc. Advmt. Anim. Breed. Genet.* **23**: (in press)

Table 2. Estimates of genetic (above diagonal) and below diagonal (1st row) residual and (2nd row) phenotypic correlations with SE in subscript for the traits

	NWEAN	LFAIL	NVITAL	NPALE	NTHIN	RESP2	RESP5	RECT2	RECT5	MAST2	TEATU2	TEATF2	FRAF*
NWEAN		-0.97 (0.18)	0.57 (0.45)	-0.16 (0.61)	-0.35 (0.50)	0.81 (0.31)	0.03 (0.46)	0.73 (0.30)	0.55 (0.41)	-0.37 (0.62)	-0.02 (0.33)	0.40 (0.28)	
LFAIL	-0.73 (0.03)		-0.62 (0.57)	0.40 (0.78)	0.28 (0.63)	-0.56 (0.49)	0.74 (0.57)	-0.94 (0.48)	-0.49 (0.53)	0.91 (0.74)	0.07 (0.41)	-0.36 (0.37)	
	-0.75 (0.01)												
NVITAL	0.13 (0.06)	-0.11 (0.06)		0.09 (0.79)	-0.86 (0.38)	0.25 (0.46)	0.19 (0.56)	0.38 (0.40)	-0.87 (0.81)	-0.50 (0.86)	-0.83 (0.27)	-0.14 (0.33)	
	0.18 (0.03)	-0.16 (0.03)											
NPALE	-0.07 (0.06)	0.02 (0.06)	-0.23 (0.05)		-0.43 (1.15)	-0.14 (0.67)	-0.90 (1.02)	-0.28 (0.58)	-0.87 (0.81)	-0.43 (0.96)	0.66 (0.75)	0.25 (0.49)	
	-0.07 (0.03)	0.04 (0.03)	-0.21 (0.03)										
NTHIN	-0.13 (0.06)	0.12 (0.06)	-0.45 (0.05)	0.32 (0.05)		-0.61 (0.57)	0.05 (0.60)	-0.22 (0.42)	-0.76 (0.67)	-0.06 (0.75)	0.68 (0.45)	0.10 (0.38)	
	-0.15 (0.03)	0.14 (0.03)	-0.48 (0.02)	0.28 (0.03)									
RESP2	-0.07 (0.07)	0.03 (0.07)	0.08 (0.07)	0.01 (0.06)	0.04 (0.07)		1.0 (0.49)	0.09 (0.33)	0.39 (0.41)	0.32 (0.59)	0.37 (0.31)	0.03 (0.29)	
	0.09 (0.03)	-0.04 (0.03)	0.10 (0.03)	-0.01 (0.03)	-0.03 (0.03)								
RESP5	-0.01 (0.07)	-0.04 (0.06)	0.08 (0.06)	0.08 (0.06)	-0.04 (0.06)	0.03 (0.07)		-0.13 (0.41)	-0.27 (0.56)	0.36 (0.63)	0.72 (0.39)	0.64 (0.34)	
	-0.01 (0.03)	0.04 (0.03)	0.09 (0.03)	0.02 (0.03)	0.03 (0.03)	0.14 (0.03)							
RECT2	-0.16 (0.08)	0.16 (0.07)	-0.07 (0.07)	0.03 (0.06)	0.02 (0.07)	0.17 (0.07)	0.10 (0.07)		0.76 (0.29)	0.39 (0.52)	0.42 (0.29)	0.30 (0.23)	
	0.02 (0.03)	0.00 (0.03)	0.00 (0.03)	0.00 (0.03)	-0.01 (0.03)	0.15 (0.03)	0.07 (0.03)						
RECT5	-0.06 (0.07)	0.06 (0.06)	-0.06 (0.06)	0.06 (0.06)	0.05 (0.06)	-0.02 (0.07)	0.38 (0.06)	0.24 (0.06)		0.73 (0.77)	-0.26 (0.34)	0.31 (0.30)	
	0.03 (0.03)	0.01 (0.03)	0.03 (0.03)	-0.02 (0.03)	-0.02 (0.03)	0.04 (0.03)	0.31 (0.03)	0.32 (0.03)					
MAST2	-0.06 (0.06)	0.06 (0.06)	-0.01 (0.06)	-0.03 (0.05)	-0.03 (0.06)	0.01 (0.06)	0.08 (0.06)	0.04 (0.06)	-0.08 (0.06)		-0.08 (0.52)	-0.07 (0.45)	
	-0.09 (0.03)	0.12 (0.03)	-0.04 (0.03)	-0.04 (0.03)	-0.03 (0.03)	0.04 (0.03)	0.10 (0.03)	0.08 (0.03)	-0.02 (0.03)				
TEATU2	-0.08 (0.08)	0.07 (0.07)	0.07 (0.08)	-0.16 (0.06)	-0.14 (0.07)	-0.16 (0.08)	-0.04 (0.07)	-0.13 (0.08)	0.02 (0.07)	0.06 (0.07)		0.55 (0.19)	
	-0.07 (0.03)	0.07 (0.03)	-0.11 (0.03)	-0.08 (0.03)	-0.01 (0.03)	-0.05 (0.03)	0.08 (0.03)	0.00 (0.03)	-0.03 (0.03)	0.04 (0.03)			
TEATF2	-0.02 (0.08)	0.02 (0.08)	-0.01 (0.08)	-0.07 (0.07)	-0.18 (0.08)	-0.17 (0.08)	-0.13 (0.08)	-0.07 (0.09)	-0.03 (0.08)	-0.03 (0.07)	0.20 (0.08)		
	0.08 (0.03)	-0.05 (0.03)	-0.03 (0.03)	-0.02 (0.03)	-0.04 (0.03)	-0.12 (0.03)	0.03 (0.03)	0.04 (0.03)	0.05 (0.03)	-0.03 (0.03)	0.30 (0.03)		
FRAF	-0.09 (0.06)	0.10 (0.06)	0.09 (0.06)	0.03 (0.05)	0.00 (0.06)	-0.06 (0.06)	-0.07 (0.06)	0.08 (0.06)	0.02 (0.06)	-0.04 (0.06)	-0.08 (0.07)	-0.07 (0.08)	
	-0.07 (0.03)	0.06 (0.03)	0.10 (0.03)	0.04 (0.03)	0.01 (0.03)	-0.05 (0.03)	-0.05 (0.03)	0.09 (0.03)	0.00 (0.03)	0.02 (0.03)	-0.04 (0.03)	0.07 (0.03)	

For trait name abbreviations see Table 1; * standard errors >1.00

Proc. Assoc. Advmt. Anim. Breed. Genet. 23:155-158

GENOTYPE BY TEMPERATURE GROUPING INTERACTION FOR FARROWING RATE AT FIRST INSEMINATION

A.M.G Bunz[1,2], K.L. Bunter[2], R. Morrison[1], B.G. Luxford[1] and S. Hermesch[2]

[1]Rivalea Australia (Pty Ltd), Corowa, NSW, 2646
[2]Animal Genetics & Breeding Unit*, University of New England, Armidale, NSW, 2351 Australia

SUMMARY

This study examined the effect of temperature grouping (T-group) on genetic parameters for farrowing rate from first insemination (FR). Further, this study investigated if genotype by T-group interaction for FR exists. The lowest FR was observed in T-group 1 and 3, which were both characterised by high mean maximum temperature (>29°C) prior to mating. The heritability of FR across all T-groups differed only marginally from each other and were low (0.03, 0.00, 0.03, 0.02, and 0.03 for T-group 1, 2, 3, 4 and 5, respectively). Genetic correlations between FR recorded in different T-groups were generally positive and high (>0.70), with the exception of the genetic correlation for FR between T-group 1 and 5 which was lowly negative and close to zero (-0.10±0.27). This is an indication that FR in T-group 1 and T-group 5 were two genetically different traits and should be treated as separate traits in pig breeding programs.

INTRODUCTION

Seasonal infertility in pigs has been described as a reduction in reproductive performance during late summer and early autumn (Love 1978). In domestic pigs, seasonal infertility seems to be mainly explained by changes in photoperiod, but can be elevated or alleviated by multiple factors, such as heat stress or management strategies (for example shed cooling systems) (Auvigne *et al.* 2010). The heat stress component of seasonal infertility is becoming more important in Australia, as severity and frequency of extreme heat events have increased across large parts of the country (Whetton *et al.* 2011). Since environmental modification and management seem unlikely to eliminate all heat stress effects or their consequences for seasonal infertility, selection for reduced seasonal infertility in pigs should be explored.

Seasons are classically defined by grouping calendar months according to specific climate characteristics. Most studies have used the classic definition of season to analyse seasonal differences in reproduction performance (Lewis and Bunter 2011). However, seasonal variation may not be well described by this classic definition of season and a more flexible approach is required. A methodology has been developed using cluster analysis to define temperature groupings (T-group) influencing farrowing rate (Bunz *et al.* 2019). These T-groups accounted for different maximum temperature histories that sows were exposed to around mating events. Farrowing rate is an indicator trait for seasonal infertility.

The objective of this study was to investigate the effect T-groups had on genetic parameters for farrowing rate (FR) at first insemination and if genotype by T-group interactions exist for FR.

MATERIALS AND METHODS

Mating data and outcomes from two maternal lines (Large White and Landrace origin) and one terminal line (Duroc origin) were collected from a single farm in southern New South Wales, Australia. The climate is characterised by very hot summers, cool winters and low humidity. The full pedigree information was used, extending over 18 generations. Data included 36,767 FR records of 17,090

* A joint venture of NSW Department of Primary Industries and the University of New England

sows (daughters of 977 sires) from the first insemination event within each mating cycle (FR: 0=fail, 1=pregnant) collected from 2012 to 2017. The data set was limited to records from the first three parities of sows. All mating events were performed using artificial insemination, with each sow receiving 2 inseminations of the same boar's semen, supplied from a single boar stud. Boars were housed in sheds with an evaporative cooling system and their semen had to pass quality-control checks before use. Sows were housed in naturally ventilated sheds and had drip cooling provided during their lactation period when shed temperature exceeded 30°C. The following steps outlined further in Bunz *et al.* (2019) were applied for defining 5 T-groups (n = 5): a) a generalized linear model with a logit link was used to identify the most informative days (p-value<0.05) for FR at first insemination regarding maximum ambient temperature (Tmax) in the time period 35 days prior to and 35 days post mating date; b) for every mating date the Tmax of significant days were extracted; and, c) a cluster approach based on partitioning around medoids (PAM; Kaufmann and Rousseeuw 1990) methods was applied to group temperature patterns for every mating date according to their similarity. Parameter estimates for each trait were obtained using an animal model applying ASREML (Gilmour *et al.* 2014). Using a general formulation, the model for FR at first insemination was:

$$y_{ik} = X\beta + Z_1 a_i + Z_1 p_i + Z_3 s_k + e_{ik} \quad (1)$$

where $y_{ik}$$y_{ik}$ ae observations for the *i*th animal inseminated by the *k*th service sire, X is an incidence matrix of factors (β). Z_1 is the incidence matrices relating records to additive genetic and permanent environment effects and, Z_3 is the incidence matrices relating records to service sire effect, and *a, p* and *s* are vectors of additive genetic, permanent environment and service sire effects, respectively. Significant systematic effects included first insemination year-quarter (24 levels, contemporary groups), breed (3 levels) and sow parity (3 levels). Effects were distributed as $Var(a) = A\sigma_a$, where A is the numerator relationship matrix, $Var(p) = I\sigma_p$, $Var(s) = I\sigma_{service\ sire}$ and $Var(e) = I\sigma_e$ where I is an identity matrix.

To investigate the genotype by T-group interaction data on FR was subsequently split into five traits based on T-group at first insemination, as outlined by Bunz *et al.* (2019). Estimates of genetic correlations between FR in each T-group were then obtained from a series of bivariate analyses. For the bivariate analysis only one record per season per sow was kept avoiding multiple records per sow in one season, leading to 34,838 records. The permanent environment effect of the sow was therefore not fitted in bivariate analyses.

RESULTS AND DISCUSSION

The lowest mean FR was observed in temperature group 1 and 3, which were both characterised by high maximum temperature prior to mating (Table 1). Observation in T-groups were independently distributed from season of the year (Table 2). This study found low heritabilities for FR (Table 3), similar to those reported by Sevillano *et al.* (2016). Farrowing rate was not heritable in T-group 2. However, heritability estimates for FR differed only marginally between T-groups.

Further, the phenotypic variance and the ratio between service sire variance and phenotypic variance was larger in more stressful environments (T-groups 1 and 3) than in less stressful environments (T-groups 2,4,5), which is consistent with results from Sevillano *et al.* (2016).

Estimates of genetic correlations between the same trait recorded in different T-groups are shown in Table 5. The standard errors for genetic correlations were high due to the low heritability in T-group and partially low representation of sows and sire of sows across T-groups (Table 4). Further, it was not possible to estimate genetic correlations between T-group 2 and other T-groups due to non-existence of additive genetic variation in T-group 2. The genetic correlations between FR recorded in different T-groups were high with one exception; the genetic correlation between T-group 1 and 5. This genetic correlation was negative and close to zero, suggesting the existence of a genotype by temperature grouping interaction. T-group 1 and T-group 5 were the opposite in the maximum temperature characteristics, which is a possible explanation for the low genetic correlation for farrowing rate at first insemination between these two T-groups. Sevillano *et al.* (2016) found a higher genetic correlation (0.76±0.19) of FR between opposite environments (stressful and non-stressful) using a bivariate model.

Table 1. Data characteristics for farrowing rate according to temperature group (T-group)

T-group	Temperature characteristics of T-group	n records	n sows	Mean (sd)	CV
1	high prior and post mating	8686	8080	0.77 (0.42)	54.6
2	low prior and medium post mating	6989	6648	0.85 (0.35)	41.5
3	high prior and medium post mating	5471	5399	0.75 (0.43)	57.5
4	medium prior and low post mating	5204	5093	0.84 (0.37)	44.2
5	low prior and post mating	10411	9618	0.86 (0.34)	40.0

Abbreviations: Mean Maximum temperature characteristics: high >29ºC; medium 21-29°C; low <21ºC

Table 2. Distribution of records across T-groups and season

T-group	n records	Summer (Jan-Mar)	Autumn (April-Jun)	Winter (July-Sept)	Spring (Oct-Dec)
1	8686	4859			3827
2	6989	31		1377	5581
3	5471	4190	1152		129
4	5204	48	5081	39	36
5	10411		2766	7434	211

Table 3. Estimates of variances due to additive genetic (σ^2_a) and service sire effects on farrowing rate, along with the residual (σ^2_e) and phenotypic (σ^2_p) variances and ratios of heritability (h^2: se in brackets) or service sire effects by temperature grouping (T-group)

T-group	σ^2_a	σ^2_{pe}	σ^2_{ss}	σ^2_e	σ^2_p	h^2	pe^2	ss^2
1	0.0051	0.0101	0.0041	0.1550	0.1642	0.0311 (0.012)	0.0615 (0.053)	0.0252 (0.006)
2	0.0000	0.0080	0.0026	0.1128	0.1154	0.000 (0.000)	0.0693 (0.069)	0.0223 (0.006)
3	0.0050	0.0071	0.0037	0.1677	0.1763	0.0284 (0.017)	0.0407 (0.112)	0.0208 (0.007)
4	0.0023	0.0000	0.0012	0.1321	0.1356	0.0172 (0.017)	0.0000 (0.000)	0.0086 (0.006)
5	0.0031	0.0000	0.0010	0.1139	0.1179	0.0261 (0.099)	0.0000 (0.000)	0.0085 (0.004)

Abbreviations: $h^2 = \sigma^2_a/\sigma^2_p$; $pe^2 = \sigma^2_{pe}/\sigma^2_p$; $ss^2 = \sigma^2_{ss}/\sigma^2_p$

The current study focused only on the temperature component of seasonal infertility. However, the methodology can be further developed accounting also for the photoperiod component. The presence of genotype by T-group interaction can be explore for other traits.

Table 4. Number of sows by temperature grouping (T-group) on the diagonal; the number of sows in common between T-group above the diagonal, the number of sire of sows in common between T-group below the diagonal

T-group	1	2	3	4	5
1	8080	1827	1310	2153	5415
2	437	6648	2226	2669	2392
3	332	494	5399	695	3279
4	495	545	254	5093	1656
5	672	541	612	436	9618

Table 5. Genetic correlations (above diagonal), residual correlations (below diagonal 1st row) (SE) and phenotypic correlations (below diagonal 2nd row) and for farrowing rate at first insemination between temperature groupings (T-group)

T -group	1	3	4	5
1		0.82(0.23)	0.85(0.48)	-0.10(0.27)
3	0.00(0.04) 0.04(0.03)		0.98(0.55)	0.79(0.34)
4	0.09(0.03) 0.11(0.03)	0.04(0.05) 0.07(0.04)		0.89(0.39)
5	0.03(0.02) 0.03(0.02)	-0.03(0.02) 0.00(0.02)	0.04(0.03) 0.00(0.03)	

CONCLUSIONS

This study was able to show that genotype by T-group interactions exist for FR, which is a trait used to indicate seasonal infertility. Farrowing rates observed in T-group 1 and 5, which were characterised by opposite mean temperature patterns around mating events, were genetically two different traits. The results of this study show that using trait-specific T-groups can provide an opportunity to improve the heat stress component of seasonal infertility in pigs genetically. Additionally, this methodology can be extended to include photoperiod information and applied to other reproduction traits.

ACKNOWLEDGEMENT

Funded by Australian Pork Limited (Project 2017/2208) and student scholarship (2018-007). Support from Joel Lisonbee from the Bureau of Meteorology is appreciated.

REFERENCES

Auvigne V., Leneveu P., Jehannin, C., Peltoniemi O. and Salle E. (2010) *Theriogenology* **74:** 60

Bunz A.M.G., Bunter K.L., Morrison R., Luxford B.G. and Hermesch S. (2019) *Adv. Anim. Bio.*. Submitted.

Gilmour A., Gogel B., Cullis B., Welham S., Thompson R. (2014). ASReml User Guide Release 4.1 Functional Specification. VSN International Ltd, Hemel Hempstead, HP1 1ES, UK

Kaufman L. and Rousseeuw P.J. (1990) 'Finding groups in data: an introduction to cluster analysis' 1st ed, John Wiley & Sons.

Lewis C.R. and Bunter K.L. (2011) *Anim. Prod. Sci.* **51:** 615

Love R.J. (1978) *The Veterinary record* **103:** 443

Sevillano C.A., Mulder H.A., Rashidi H., Mathur P.K. and Knol E.F. (2016) *J. Anim. Sci.* **94:** 3185

Whetton P., Moise A., Abbs D. (2011) In 'Climate Science Update: A report to the 2011 Garnaut Review. CAWCR Technical Report No. 036', pp. 73-80, editor T.D. Keenan and H.A. Cleugh, The Centre for Australian Weather and Climate Research: a partnership between CSIRO and the Bureau of Meteorology, Australian Government.

Proc. Assoc. Advmt. Anim. Breed. Genet. 23:159-162

GENETIC PARAMETER ESTIMATES FOR PRE- AND POST-WEANING PIGLET MORTALITY

J. Harper[1], K.L. Bunter[2] and S. Hermesch[2]

[1]Rivalea Australia, Corowa, NSW, 2646 Australia
[2]Animal Genetics & Breeding Unit*, University of New England, Armidale, NSW, 2351 Australia

SUMMARY

Alternative models for genetic evaluation of pre- and post-weaning mortality traits were investigated. For pre-weaning mortality, the best model accounted for direct piglet effects, common litter effects of both the nurse sow and biological dam, repeated records of the nurse sow and the maternal nurse sow genetic effects. For post-weaning mortality, the most parsimonious model included only direct piglet effects and the common litter effects of both the nurse sow and biological dam. After accounting for systematic effects, genes of the piglet contribute to both pre- and post-weaning mortality (direct $h^2 = 0.02 \pm 0.002$ for pre- and post- weaning), whereas the nurse maternal genes only contribute to pre-weaning (maternal $m^2 = 0.01 \pm 0.002$). While heritabilities were low, there is potential for genetic improvement of both pre- and post-weaning mortality traits.

INTRODUCTION

Selection for efficient, lean growth and increased litter size can increase piglet pre-weaning mortality (Bunter 2009), with recent pre-weaning mortality rates reported as high as 18% in Australian herds (Australian Pig Industry Benchmarking Report, 2018). Therefore, breeding values for survival have become an important component of breeding programs. It is possible to make improvements by genetically enhancing a piglet's ability to survive (Mesa *et al.* 2006), while also improving litter size, although an antagonistic relationship occurs between the two traits (Bunter 2009). Piglet survival involves different phenotypes and genes, including that of the piglet's biological dam, the sow nursing the piglet, and the genotype of the piglet itself (Knol *et al.* 2002). In addition, piglets born and/or nursed within a common litter have common environmental effects contributing to their mortality (Bunter 2009). In the review of Bunter (2009), heritability estimates were on average 0.05 at the piglet level, and 0.11 at the sow level, indicating that both direct and maternal components should be considered. The purpose of this study was to investigate alternative models for genetic evaluation of piglet pre- and post-weaning mortality, treated as a trait of the piglet.

MATERIALS AND METHODS

Data. Data on individual piglet mortality (alive = 0, dead = 1) before weaning or post-weaning and other related traits, were recorded at a commercial piggery located in southern New South Wales, Australia. Data included 466,012 individually identified pedigreed piglets born between 2009 and 2018, from two Maternal (Large White and Landrace) and one Terminal (Duroc) selection line. This data set represented progeny of 1,535 sires, 19,867 dams and 28,228 nurse sows, which were included in the pedigree, extending over 10 generations, born in 43,462 litters. Piglets were individually identified within 24 hours from birth, with individual birth weights and sex recorded. Cross fostering occurred after identification, and all movements and deaths of individual piglets were recorded, along with corresponding dates. A piglet was recorded as a pre-weaning death if it was born alive and died before weaning (average of 26 days). A post weaning death was recorded if the piglet had been weaned and was less than 70 days of age at death. Piglets with a pre-weaning record equal to 1 do not have a post-weaning mortality record.

---

* A joint venture of NSW Department of Primary Industries and the University of New England

Statistical Analysis. Preliminary analyses confirmed that the following fixed effects significantly ($P<0.0001$) contributed to mortality outcomes: piglet breed; gender (2 levels: male and female); piglet fostering status (un-fostered = 0, fostered by day 2 = 1, fostered after day 2 = 2), which was concatenated with sow (birth-nurse) parities and farrowing farm (totalling 48 levels); and birthweight class (6 levels: 0.60-1.21, 1.22-1.39, 1.40-1.54, 1.55-1.69, 1.70-1.90, 1.91-3.00 kilograms). Additionally, gestation length (6 levels: 105-114, 115, 116, 117, 118-125 days); total born group (5 levels: 1-5, 6-10, 11-15, 16-20, 20-25) and birth year quarter (40 levels: accounting for the managerial and seasonal differences) were accounted for in models for analysis. Additional factors for post-weaning mortality included age group when individual piglets were weaned (5 levels: 0-14, 15-21, 22-28, 29-35 and 36-60 days), and the farm that piglets were weaned into (7 levels).

To estimate genetic parameters for pre- and post-weaning mortality, an univariate analysis was performed using linear models in ASReml (Gilmour *et al.* 2015), where either nurse sow, biological dam models or elements of both were investigated. Random effects tested included terms for the animal (piglet), the common litter effect of either nurse litter, birth litter or a combination of both, the permanent environment of the nurse sow or the biological dam and the maternal genetic effect of either nurse sow or biological dam. The mixed model is represented by:

$$\mathbf{y} = \mathbf{X\beta} + \mathbf{Z_1a} + \mathbf{Z_2c} + \mathbf{Z_3mpe} + \mathbf{Z_4m} + \mathbf{e} \quad (1)$$

where **y** is the vector of the observations; **X** is the incidence matrix for the vector of fixed effects in **β**; **Z** are incidence matrices associated with vectors of random effects including additive genetic effect (piglet) in **a**, common nurse and/or biological litter effect in **c**, nurse sow or biological dam permanent environment effect in **mpe**, nurse sow or biological dam maternal genetic effect in **m** and **e** is a vector residuals. Effects were distributed as $Var(\mathbf{a}) = \mathbf{A}\sigma_a^2$, where **A** is a matrix describing the relationships between animals, and the remaining effects: $Var(\mathbf{c})=\mathbf{I}\sigma_c^2$, $Var(\mathbf{mpe})=\mathbf{I}\sigma_{mpe}^2$ and $Var(\mathbf{m})=\mathbf{I}\sigma_m^2$, where **I** is an identity matrix. The Akaike Information Criteria (AIC test) was used to test if the inclusion of additional random effects was significant, with the preferred model having the lowest AIC value (Mendenhall *et al.* 1996).

RESULTS AND DISCUSSION

To avoid double counting, models for piglet mortality traits accounted for sow traits already used as selection criteria, such as birth weight and litter size, and also accounted for fostering status, which is typically ignored in other studies. Accounting for these systematic effects might have reduced estimates of heritabilities in this relative to other studies (Table 1).

Pre-weaning mortality. On average, 87.4% of piglets were nursed by their biological dam, based on fostering events in the first two days of birth, but 19.3% of litters contained piglets with different parentage. Based on the AIC values, the nurse sow model (N3b) fit the data significantly better than the corresponding biological sow model (D3). Model N3b included the common litter effects of both the nurse sow and the biological dam, accommodating the effects of both the gestation and lactation environment experienced by the piglets. The model was improved substantially by the addition of maternal effects (mpe or m), accommodating repeated records and maternal genetic effects of the nurse sow. Maternal genetic effects reflect heritable traits like uterine nutrient supply and capacity, milk quality and quantity and general maternal care (Kaufmann *et al.* 2000). Removal of the model terms related to gestation length, birth weight and litter size (model N3c), which are typically considered as sow traits, created an increase in the phenotypic variance for pre-weaning mortality along with increases in maternally mediated variance ratios (cnl^2 and m^2), demonstrating that these maternal factors influence pre-weaning mortality. Direct heritability of piglets was improved with a nurse sow model (N1 vs D1), as the survival of piglets from different litters were compared within common nurse litters. Direct ($h^2 = 0.002 \pm 0.002$) and maternal ($m^2 = 0.01 \pm 0.002$) heritabilities for piglet mortality estimated from model N3b were low in comparison to average literature estimates (Bunter 2009). Knol *et al.* (2002) reported that direct heritabilities were 0.03 for dam lines and 0.22 for maternal heritabilities, where the genetic effects of the biological dam, maternal permanent environmental and common birth litter effects were accounted for. Roehe *et al.* (2010) reported a direct heritability of 0.24 and maternal heritability 0.14 using a biological dam model, without accounting for nurse sow effects, making direct comparisons with the literature difficult.

Post-weaning mortality. The best model for post-weaning mortality was D6b based on AIC values, which was a biological sow model including common nurse litter effects. However, in contrast to pre-weaning mortality, variances for permanent environmental and maternal genetic effects were much lower and contributed no substantial improvement to the model fit (for example model N4b vs N6b; D4 vs D6a). Furthermore, the ratios for maternal permanent environment and direct maternal effects were very small, indicating sows only contribute to post-weaning outcomes through additive genetic effects, and any carry over from prior gestation and lactation periods was accounted for in the common litter effects. This was also inferred using model N4c, where removing gestation length, total born and piglet birth weight terms from the model resulted in no changes to variance estimates. Since sampling correlations between direct and biological dam effects are high (not shown) and taking these results into account, a more parsimonious model should be used, leading to model N4b being the best model for the data. This model is a nurse sow model accounting for common nurse and biological litter effects, which resulted in direct heritability estimates to be 0.02 ± 0.002, with no literature available for direct comparison of post-weaning mortality estimates in pigs.

CONCLUSIONS

This large data set enabled separation of direct from maternal effects, along with common litter effects and permanent environmental effects of the sow and fostering status. Failing to separate all of these effects could lead to overestimates of direct or maternal heritability but is complicated due to high sampling correlations for mortality traits. Nevertheless, the genetic parameters presented in this study suggest that there is potential for genetic improvement of pre- and post-weaning mortality traits in commercial breeding programs, independent of other important traits such as birth weight, gestation length and litter size.

ACKNOWLEDMENTS

Jo-Anne Harper is supported by Australian Pork Limited (APL) through a study award, project number 2015/037 and Rivalea Australia.

REFERENCES

Bunter, K.L. (2009) *Manipulating pig production XII'.(Ed. RJ van Barneveld) pp* 149

Gilmour, A.R., Gogel, B.J., Cullis, B.R., Welham, S.J. and Thompson, R. (2015) *Hemel hempstead: VSN international ltd*

Kaufmann, D., Hofer, A., Bidanel, J.P. and Künzi, N. (2000) *J. Anim. Breed. Genet.* **117:** 121

Knol, E.F., Ducro, B.J., Van Arendonk, J.a.M. and Van Der Lende, T. (2002) *Livest. Prod. Sci.* **73:** 153

Knol, E.F., Leenhouwers, J.I. and Van Der Lende, T. (2002) *Livest. Prod. Sci.* **78:** 47

Mendenhall, W., Sincich, T. and Boudreau, N.S. (1996) A second course in statistics: regression analysis Prentice Hall Upper Saddle River, NJ

Mesa, H., Safranski, T.J., Cammack, K.M., Weaber, R.L. and Lamberson, W.R. (2006) *J. Anim. Sci.* **84:** 32

Roehe, R., Shrestha, N., Mekkawy, W., Baxter, E., Knap, P., Smurthwaite, K., Jarvis, S., Lawrence, A. and Edwards, S. (2010) *J. Anim. Sci.* **88:** 1276

Table 1. Variance component estimates (×100) and resulting ratios of variance components under different models for pre- and post-weaning mortality

Level	Model	σ^2_a	σ^2_{cnl}	σ^2_{cbl}	σ^2_{mpe}	σ^2_m	σ^2_p	$h^2 \pm SE$	$cnl^2 \pm SE$	$cbl^2 \pm SE$	$mpe^2 \pm SE$	$m^2 \pm SE$	AIC
Pre-wean	N1	0.93	1.31	-	-	-	14.1	0.07±0.003	0.09±0.001	-	-	-	-471314.3
Nurse	N2	0.58	1.14	-	0.26	-	14.0	0.04±0.003	0.08±0.002	-	0.02±0.001		-471547.46
	N3a	0.50	1.14	-	0.13	0.17	14.0	0.04±0.003	0.08±0.002	-	0.01±0.002	0.01±0.001	-471669.45
	N3b	0.29	0.70	0.53	0.13	0.20	14.0	0.02±0.002	0.05±0.002	0.04±0.002	0.01±0.002	0.01±0.002	-472115.44
	N3c	0.27	0.82	0.40	0.11	0.23	14.8	0.02±0.002	0.06±0.002	0.03±0.002	0.007±0.002	0.02±0.002	
Pre-wean	D1	0.67	-	1.25	-	-	14.0	0.05±0.003	-	0.09±0.001	-	-	-469850.36
Dam	D2	0.32	-	1.09	0.27	-	14.0	0.02±0.002	-	0.08±0.001	0.02±0.001	-	-470075.54
	D3	0.26	-	1.09	0.14	0.17	14.0	0.02±0.002	-	0.08±0.001	0.01±0.002	0.01±0.001	-470193.32
Post-wean	N4a	0.14	0.24	-	-	-	5.09	0.03±0.002	0.05±0.001	-	-	-	-756059.93
Nurse	N4b	0.11	0.10	0.16	-	-	5.09	0.02±0.002	0.02±0.002	0.03±0.002	-	-	-756306.33
	N4c	0.11	0.12	0.16	-	-	5.11	0.02±0.002	0.02±0.002	0.03±0.002	-	-	
	N5	0.13	0.24	-	0.001	-	5.09	0.03±0.002	0.05±0.001	-	0.002±0.001	-	-756061.24
	N6a	0.13	0.24	-	0.0004	0.0007	5.09	0.02±0.002	0.05±0.001	-	0.000±0.001	0.001±0.001	-756064.37
	N6b	0.10	0.09	0.17	0.0006	0.0009	5.09	0.02±0.002	0.02±0.002	0.03±0.002	0.002±0.001	0.002±0.001	-756317.81
Post-wean	D4	0.11	-	0.25	-	-	5.09	0.02±0.002	-	0.05±0.001	-	-	-756199.49
Dam	D5	0.09	-	0.24	0.02	-	5.09	0.02±0.002	-	0.05±0.001	0.004±0.001	-	-756213.85
	D6a	0.09	-	0.24	0.01	0.01	5.09	0.02±0.002	-	0.05±0.001	0.002±0.001	0.003±0.001	-756225.55
	D6b	0.09	0.11	0.15	0.01	0.01	5.09	0.02±0.002	0.02±0.002	0.03±0.002	0.002±0.001	0.003±0.001	-756332.69

Abbreviations: N = nurse sow model; D = biological dam model; σ^2_a = additive genetic variance; σ^2_{cnl} = common nurse litter variance; σ^2_{cbl} = common biological litter variance; σ^2_{mpe} = maternal permanent environmental variance either attributed to nurse or biological dam; σ^2_m = maternal genetic variance either attributed to nurse or biological dam; σ^2_p = phenotypic variance; h^2 = heritability estimate; cnl^2 = proportion of phenotypic variance attributed to nurse litter effect; cbl^2 = proportion of phenotypic variance attributed to biological litter effect; mpe^2 = proportion of phenotypic variance attributed to permanent environmental effect (either nurse or biological dam); m^2 = proportion of phenotypic variance attributed to maternal genetic variance (either nurse or biological dam); AIC = Akaike Information Criterion

Proc. Assoc. Advmt. Anim. Breed. Genet. 23:163-166

GENETIC ASSOCIATIONS BETWEEN EARLY AND LATE GROWTH WITH SEXUAL MATURITY IN THAI NATIVE CHICKENS

S. Tongsiri[1,2], M.G. Jeyaruban[1], J.H.J. van der Werf[2], L. Li[1], S. Hermesch[1] and T. Chormai[3]

[1]Animal Genetics & Breeding Unit*, University of New England, Armidale, NSW, 2351 Australia
[2]School of Environmental and Rural Science, University of New England, Armidale, NSW, 2351
[3]Department of Livestock Development, Thailand

SUMMARY

The associations between early and late growth rates with sexual maturity of Lueng Hang Kao Kabinburi (LHKK) native chicken in Thailand were explored. Five generations of data from 2003 to 2007, involving 11,588 chickens, were collected at Kabinburi Livestock Research and Breeding Centre (KLRBC). Body weight measured from day-old (BW1D) to 24 weeks of age at 4 weekly intervals of 4 (BW4), 8 (BW8), 12 (BW12), 16 (BW16), 20 (BW20), 24 (BW24) weeks, and sexual maturity traits, age at first egg (AFE) and egg weight at first egg (EWFE), were recorded. The growth rates were grouped into 5 categories: BW1D to BW8 (Growth_1), BW4 to BW12 (Growth_2), BW8 to BW16 (Growth_3), BW12 to BW20 (Growth_4), and BW16 to BW24 (Growth_5). Growth_1 to 3 represented early growth and Growth_4 and 5 represented late growth. Genetic correlations were estimated between early and late growth rates against AFE and EWFE using Restricted Maximum Likelihood. Growth_1 had a favourable genetic correlation of -0.15 with AFE and a high positive (favourable) genetic correlation of 0.42 with EWFE. Growth rate between Growth_4 and Growth_5 had unfavourable genetic correlations of 0.08 and 0.30, respectively, with AFE, and favourable genetic correlations of 0.28 and 0.31, respectively, with EWFE. This study indicated that selecting for higher growth rate between day-old to 8 weeks of age would also improve sexual maturity by reducing the AFE and increasing the EWFE of LHKK chicken in Thailand.

INTRODUCTION

The Thai native chicken, Lueng Hang Kao Kabinburi (LHKK), is a dual-purpose chicken breed used to produce meat and eggs. However, the growth and egg production rate of LHKK are low under the backyard production systems in Thailand. Therefore, there is a need to improve both growth and egg production to improve productivity and profitability of the LHKK chickens. Tongsiri *et al.* (2019) showed that body weights at various ages and sexual maturity traits are moderately heritable (0.10 to 0.37) and, therefore, selecting for higher body weight and early sexual maturity are expected to improve both traits of LHKK chickens. A number of studies have examined the genetic association between growth and egg production of native chickens at various ages. Bahmanimehr (2012) and Niknafs *et al.* (2012) found positive genetic correlations between early growth and egg weight first egg (EWFE) by 0.30 to 0.51 in Iranian native chickens. However, Niknafs *et al.* (2012) found an unfavourable genetic relationship between age at first egg (AFE) and body weight in the early growth period (day-old). By contrast, Bahmanimehr (2012) and Niknafs *et al.* (2012) concluded that both body weight and egg production could be improved simultaneously by selecting based on body weights measured during the early growth period.

Selecting replacement chickens based on growth information at an early age would decrease the number of chickens required to be kept as replacement stock and, thereby, reduce management cost and increase the profitability of the LHKK nucleus flock. Thus, a better understanding of the genetic relationship between growth rates during various growth periods and their relationship with sexual

* A joint venture of NSW Department of Primary Industries and the University of New England

maturity will help to develop a suitable breeding program for the LHKK chickens. Therefore, the aim of this study is to identify an appropriate age range to select for higher growth and early sexual maturity to improve both growth and egg production in LHKK chicken.

MATERIALS AND METHODS

Animal and Data. Five generations of growth and egg production data from 2003 to 2007 on 11,588 individual chickens were collected. The chickens were descendants of 486 cocks and 1,461 hens. The pure-bred dual-purpose LHKK chickens were housed on a Thai government farm at the Kabinburi Livestock Research and Breeding Centre (KLRBC) in the Eastern region of Thailand. The chickens were subjected to the management and vaccination program (Marek's disease, New Castle disease (ND), Infectious Bronchitis disease (IB) and Fowl Pox disease) recommended by the Department of Livestock Development, Thailand. All chicks were wing-banded on day-old and body weights were measured from day-old to 24 weeks of age at 4 weekly intervals (day-old (BW1D), 4 (BW4), 8 (BW8), 12 (BW12), 16 (BW16), 20 (BW20) and 24 (BW24)). The sexual maturity was measured as AFE and EWFE. The number of chickens measured for body weight at different ages reduced gradually as age increased (Table 1). This is mainly due to selective culling of chickens based on phenotypic characteristics such as colour of face, eyes, beak, skin, shank, body plumage, neck, tail, saddle and type of comb.

Statistical analyses. Genetic parameters and variance components were estimated using a mixed linear model using Restricted Maximum Likelihood (REML) in the WOMBAT software (Meyer 2007). For growth rate, model included year and hatch within year and sex as fixed effects, and direct additive genetic, maternal genetic, maternal permanent environmental and residual as random effects. For AFE and EWFE, sex, maternal genetic and maternal permanent environmental were not fitted (Tongsiri *et al.* 2019). A series of bivariate analyses between growth rate at different ages and AFE and EWFE were conducted using REML in WOMBAT. Five growth periods were identified: Growth_1 was between BW1D and BW8, Growth_2 was between BW4 and BW12, Growth_3 was between BW8 and BW16, Growth_4 was between BW12 and BW20, and Growth_5 was between BW16 and BW24. Growth_1, Growth_2 and Growth_3 were grouped as 'Early growth' and Growth_4 and Growth_5 were grouped as 'Late growth'. Body weight measured at various ages was used to calculate growth rates for early and late growth periods. Genetic correlations (r_g) between early and late growth rates with AFE and EWFE were estimated.

RESULTS AND DISCUSSION

Estimated heritabilities for body weight were constant across the different ages, except for BW1D and BW4 (Table 1).

Early growth rate and AFE. Genetic correlations between early growth rates and AFE ranged from -0.15 to 0.07 (Table 2). Growth rates in Growth_1 and Growth_2 were negatively correlated with AFE, indicating that chickens with higher growth rates during these periods will grow faster and reached sexual maturity earlier than their contemporaries. The favourable associations of early growth rate in Growth_1 and Growth_2 with AFE agreed with the correlations reported by El-Dlebshany (2008) for native chickens in Egypt.

Late growth rate and AFE. Estimated genetic correlations between late growth and AFE ranged from 0.08 to 0.30, indicating that chickens with higher growth rates after 16 weeks of age will have older age at sexual maturity. Sang *et al.* (2006) reported unfavourable correlations ranging from 0.14 to 0.72 between late growth and AFE on native chickens in Korea. Kinney (1969) observed a favourable association of growth rate with AFE and the magnitude of the association reduced with increasing age. In contrast, Lwelamira *et al.* (2009) estimated genetic correlations between growth and AFE on two native chicken breeds in Tanzania, and they reported the associations were more favourable with increasing age (-0.03 to -0.23). Overall, the magnitude of the genetic association of growth rate with AFE increased when growth reached its final stage.

The results of this study indicate that growth rate and AFE are genetically related. Therefore, selecting chickens with higher growth rates during the early growth periods (Growth_1 and Growth_2) will also reduce AFE. Selecting for higher growth rates in the late growth periods will delay sexual maturity in LHKK chickens.

Table 1. Descriptive statistics with number of records, mean, standard deviation (SD), minimum (Min), maximum (Max), direct additive genetic variance (σ^2) and heritability (h^2) estimated for body weight and sexual maturity traits of LHKK chickens

Traits	Number of records	Mean	SD	Min	Max	σ^2	h^2
BW1D (g)	11,588	30.93	3.38	22	40	1.20	0.10±0.02
BW4 (g)	11,201	218.9	56.7	46	379	232.7	0.20±0.03
BW8 (g)	10,807	642.1	139	260	1,034	2,646	0.34±0.03
BW12 (g)	9,777	1,098	210	520	1,673	10,503	0.37±0.03
BW16 (g)	8,948	1,486	306	586	2,376	12,076	0.30±0.03
BW20 (g)	7,643	1,809	406	640	3,000	18,814	0.30±0.03
BW24 (g)	6,157	2,123	470	893	3,520	28,242	0.30±0.04
AFE (day)	1,395	199.1	21.0	138	260	40.49	0.16±0.06
EWFE (g)	1,393	36.94	4.85	26	48	3.03	0.16±0.05

Early growth rate and EWFE. Estimated genetic correlations between early growth rate and EWFE ranged from 0.19 to 0.42 (Table 2). Moderate to high genetic correlations between early growth rate and EWFE indicated that chickens with higher growth rate at Growth_1 are expected to lay heavier eggs at the onset of lay. Positive genetic correlations between early growth rate and EWFE is in agreement with the values of 0.30 to 0.39 reported by Niknafs *et al.* (2012) for native chickens in Iran. The lowest correlation was observed between growth rate measured at Growth_3 and EWFE indicating that selecting on growth rate between 8 to 16 weeks will have less favourable response on EWFE compared with selecting on Growth_1. This finding agrees with those published by Hosseini and Tahmoorespur (2013).

Late growth rate and EWFE. Positive genetic correlations were estimated between late growth rate and EWFE which ranged from 0.28 to 0.31 (Table 2). Kinney (1969) also reported moderate positive correlations (0.13 to 0.29) between late growth and EWFE. This finding indicates that the genetic correlation growth rate has with EWFE was moderate when chickens reached maturity. Thus, selecting heavy chickens in the late growth period, especially between 16 and 24 weeks of age, would lead to chickens laying heavier eggs at the onset of lay. Both early and late growth rates had positive genetic relationships with EWFE. However, selection based on Growth_1 to improve EWFE would be expected to reduce production costs associated with keeping extra-replacement chickens.

Growth rate and sexual maturity traits. Selection for earlier AFE would be achieved more effectively by selecting chickens with higher growth rates at earlier ages (between day-old and 8 weeks). Growth rate in this early growth period is also favourably related to EWFE. Estimated heritabilities listed in Table 1 suggest that both body weight and sexual maturity are heritable. Therefore, both growth and sexual maturity traits could be improved by selection. Thus, selection based on early growth rate will increase AFE and EWFE simultaneously. Bahmanimehr (2012), El-Dlebshany (2008) and Magothe *et al.* (2006) also suggested that selecting heavy weight juvenile chicks may help the breeder to increase the final weight and egg weight while the chickens reach early sexual maturity.

Table 2. The genetic correlations between growth rates in the early and late growth periods and sexual maturity traits

Body weight	Genetic correlations (r_g)	
	AFE	EWFE
	Early growth period	
BW1D to BW8 (Growth_1)	-0.15±0.15	0.42±0.13
BW4 to BW12 (Growth_2)	-0.12±0.15	0.27±0.14
BW8 to BW16 (Growth_3)	0.07±0.18	0.19±0.16
	Late growth period	
BW12 to BW20 (Growth_4)	0.08±0.19	0.28±0.17
BW16 to BW24 (Growth_5)	0.30±0.25	0.31±0.20

CONCLUSIONS

This study showed that growth rates in the early and late growth periods had different associations with sexual maturity of LHKK chickens. Selecting chickens for early growth is the most economically attractive option because the trait is favourably correlated with sexual maturity, which provides greater potential for improvement of meat and egg productivity compared to selecting for late growth (between 16 and 24 weeks of chicken age). Late growth expressed an unfavourable relationship with AFE, as it delayed the onset of AFE but it increased EWFE. Moreover, selection in early growth between day-old and 8 weeks would reduce the number of chickens required as replacements, therefore, there would be less need to measure body weight in the later growth periods, which would reduce management cost in LHKK nucleus flock.

ACKNOWLEDGEMENTS

We would like to thank KLRBC for data support, and the Agricultural Research Development Agency (Public Organization), ARDA, for funding the PhD study of Ms Siriporn Tongsiri.

REFERENCES

Bahmanimehr A. (2012) *World Appl. Sci. J.* **20:** 1332.

El-Dlebshany A.E. (2008) *Egypt Poult. Sci.* **28:** 1253.

Hosseini S.A. and Tahmoorespur M. (2013) *Indian J. Anim. Res.* **47:** 205.

Kinney T.B. (1969) 'A summary of reported estimateds of heritabiltity and of genetic and phenotypic correlation for traits of chickens.' (Ed.) Agricultural Reserch Service, U.S.A.

Lwelamira J., Kifaro G. and Gwakisa P. (2009) *Trop. Anim. Health and Prod.* **41:** 51.

Magothe M., Muhuyi W.B. and Kahi A.K. (2006) 8th WCGALP. **2006:** 32.

Meyer K. (2007) *Zhejiang Univ. Sci. B* **8:** 815.

Niknafs S., Nejati-Javaremi A., Mehrabani-Yeganeh H. and Fatemi S.A. (2012) *Trop. Anim. Health and Prod.* **44:** 1437.

Sang B.D., Kong H.S., Kim H.K., Choi C.H., Kim S.D., Cho Y.M., Sang B.C., Lee J.H., Jeon G.J. and Lee H.K. (2006) *Asian-Australas. J. Anim. Sci.* **19:** 319.

Tongsiri S., Jeyaruban G., Hermesch S., van der Werf J.H.J., Li L. and Chormai T. (2019) *Asian-Australas. J. Anim. Sci.* **32:** 930.

Proc. Assoc. Advmt. Anim. Breed. Genet. 23:167-170

GENETIC ANALYSES OF SOW LONGEVITY TRAITS, AGE AT FIRST FARROWING AND FIRST-LITTER CHARACTERISTICS

N. Kerssen[1,2], B.J. Ducro[2] and S. Hermesch[1]

[1]Animal Genetics & Breeding Unit*, University of New England, Armidale, NSW, 2351 Australia
[2]Animal Breeding and Genomics, Wageningen University & Research, 6708 PB Wageningen, the Netherlands

SUMMARY

Sow longevity is a vital trait in the pig production sector because of its economic and welfare importance. However, this trait is recorded late in a sow's life and early selection criteria associated with sow longevity are beneficial for genetic improvement of sow longevity. The aim of this study was to estimate genetic parameters of sow longevity and other sow reproduction traits. Data included 14,284 purebred sows recorded from 1996 to 2016 in 7 commercial herds across Australia. Traits describing sow longevity included the number of maximum parities reached, length of productive lifetime in days, total number of piglets born alive per sow over her lifetime, and stayability from parity 1 to parity 4. Further traits considered were number of piglets born alive (litter size) and average piglet birth weight (both recorded in the first litter), and age at first farrowing. Sow longevity traits were genetically the same traits and had low heritabilities (0.07 to 0.13). Genetic correlations were lowly negative between sow longevity and age at first farrowing (-0.13 to -0.22), and between sow longevity and average piglet birth weight (-0.19 to -0.26). First litter size had positive genetic correlations with sow longevity traits (0.49 to 0.65). This study showed favourable genetic correlations of the traits first litter size and age at first farrowing with sow longevity, suggesting that these two traits could be suitable genetic indicators for sow longevity.

INTRODUCTION

Improving sow longevity is important not only for welfare but also for economic reasons, as it will enhance the proportion of sows in higher production phases and reduce the number of replacement gilts (Hoge and Bates 2011). Sow longevity has been defined in different ways over time, but a broad definition is the ability of a sow to live a long and healthy life while producing good quality litters. Traits that define sow longevity include the length of productive lifetime, the maximum number of parities a sow reached, total number of piglets born alive per sow, and stayability up to parity 4. However, genetic improvement of these sow longevity traits is a slow process, since they can only be assessed when a sow has had the chance to reach a certain number of parities. To enable earlier genetic selection for sow longevity, there is a need to select other traits that are good indicators of sow longevity and that can be measured earlier in a sow's life. Previous studies have shown significant effects of the age at first farrowing on sow longevity traits (Serenius *et al.* 2008; Engblom *et al.* 2016), but genetic correlations were low and somewhat variable ranging from -0.21 to -0.03. First-litter characteristics, including litter size and average piglet birth weight, could be indicators for sow longevity as well, but inconclusive results have been found so far (e.g. Tholen *et al.* 1996; Hoge and Bates 2011; Engblom *et al.* 2016). Therefore, genetic relationships between sow longevity and age at first farrowing and first-litter traits need further quantification. The aim of this study was to estimate genetic parameters of sow longevity, age at first farrowing, and first-litter characteristics.

---

* A joint venture of NSW Department of Primary Industries and the University of New England

MATERIALS AND METHODS

Animals and data. Pedigree and phenotypic data were available from 7 commercial sow herds in Australia. Only sows that had at least 1 parity record were included in the dataset. Purebred sows born between 1996 and 2016 were included in the analysis, so all sows had the chance to reach at least parity 4. This resulted in 14,284 records of Large White sows.

Traits. Four traits were defined for sow longevity including: maximum number of parities, length of productive lifetime in days (calculated from first farrowing until last weaning), total number of piglets born alive per sow during her lifetime, and stayability from parity 1 to parity 4, as a binary trait. Data were not censored for the trait stayability 1-4 because only records of sows that had the opportunity to reach 4 parities were included. Three early reproduction traits were identified: age of the sow at first farrowing in days, the number of piglets born alive at first farrowing (first litter size), and average piglet birth weight at first farrowing in grams based on piglets born alive recorded within 24 hours after farrowing.

Statistical Analyses. For all traits, fixed effects were defined as the herd-year-season interaction, and the only random effect fitted was animal. There were no repeated records for traits, and year and season were based on the first farrowing of each sow. Parameter estimates were obtained using linear mixed models under an animal model with ASReml (Gilmour *et al.* 2014), for all traits except stayability 1-4. This trait was analysed as a binary trait using a generalized mixed linear animal model with the logistic link function. Further, stayability 1-4 was also analysed as a continuous trait on the linear scale. Genetic and phenotypic correlations between the traits were estimated with use of a series of bivariate animal models, using the same fixed effects as for the univariate models. Bivariate analyses involving the trait stayability 1-4 were carried out on the linear scale.

RESULTS AND DISCUSSION

Descriptive statistics for all traits are displayed in Table 1. Sows had 4.09 parities on average, corresponding with a productive lifetime of 480.5 days during which they farrowed 46.9 live piglets. The relatively large coefficients of variation (CV) found for the sow longevity traits can be explained by the fact that there were several extreme values in the data. Skewness of the data was reflected in the minimum and maximum values of the different traits which depended mainly on management decisions. High variation in sow longevity traits has been found in previous studies (e.g. Serenius *et al.* 2008; Hoge and Bates 2011). Nonetheless, the SD and CV for age at first farrowing were small, especially compared to the other traits. This probably reflects management decisions. Age at first farrowing is strictly managed by selecting gilts in a short age range at puberty or mating, which therefore decreases the variability of age at first farrowing in sows.

Table 1. Descriptive statistics of the traits with the number of observations (N), the standard deviations (SD) of the means, coefficients of variation in % (CV), and the minimum and maximum values of each trait

Trait*	N	SD	CV (%)	Min	Max
MNP	14284	2.47	60.29	1	13
LPL (days)	14284	363.65	75.69	2	1744
TNBA	14284	31.90	68.04	1	165
STAY14	13920	0.50	93.98	0	1
AFF (days)	14284	21.31	6.2	264	417
LS	14284	3.02	29.29	1	22
PBWT (g)	7782	242.30	18.68	600	3000

**MNP=Maximum number of parities; LPL=Length of productive lifetime; TNBA=Total number of piglets born alive per sow per lifetime; STAY14=Stayability 1-4; AFF=Age at first farrowing in days; LS=first litter size including only live-born piglets; PBWT=Average piglet birth weight in grams in the first litter.*

Heritability estimates for the sow longevity traits ranged from 0.07 to 0.13 (Table 2). Genetic gain in these traits is expected to be slow, not only due to low genetic variance but also due to late expression of the traits. These estimates for the sow longevity traits were in agreement with previous estimates presented in literature which ranged from 0.02 to 0.22 (Serenius *et al.* 2008; Engblom *et al.* 2016). Heritability for stayability 1-4 estimated on the logistic and the linear scale were 0.09 and 0.07, respectively, which is in agreement with previous findings of Tholen *et al.* (1996) who found heritabilities of 0.08 and 0.09 for this stayability trait using a linear model. Estimates from both a logistic and linear scale are presented. The logistic scale takes the non-normal distribution of this binary trait into account. However, a linear model may be used in genetic evaluation systems.

Age at first farrowing had a heritability estimate of 0.10 ± 0.01, which was within the range of heritabilities (0.04 to 0.10) estimated in previous studies (e.g. Serenius *et al.* 2008; Engblom *et al.* 2016). However, the result in this study may be partly affected by the low variation in age at first farrowing resulting from management strategies. Further, the heritabilities for first litter size of 0.07 ± 0.01 and for average piglet birth weight of 0.19 ± 0.02 corresponded with previous findings in literature as well (Tholen *et al.* 1996; Engblom *et al.* 2016).

Table 2. Heritability estimates (h^2) with standard errors (SE), phenotypic variance (V_p), additive genetic variance (V_a) and residual variance components (V_e) per trait

Trait*	h^2 ± SE	V_p	V_a	V_e
MNP	0.10 ± 0.01	5.46	0.57	4.89
LPL (d)	0.10 ± 0.01	118830	12205.60	106627
TNBA	0.13 ± 0.01	937.85	124.00	813.85
STAY14[1] (logistic)	0.09 ±0.001	3.60	0.31	3.29
STAY14 (linear)	0.07 ± 0.01	0.23	0.016	0.22
AFF (d)	0.10 ± 0.01	293.88	29.50	264.38
LS	0.07 ± 0.01	8.74	0.57	8.17
PBWT (g)	0.19 ± 0.02	55500	10743	7782

**For abbreviations see Table 1; [1]Genetic estimates for stayability 1-4 were derived on both a logistic and linear scale.*

Sow longevity traits were genetically the same traits with high genetic correlations between them (range: 0.95 to 0.99). This is reflected in similar genetic and phenotypic correlations of the sow longevity traits with the other sow reproduction traits investigated (Table 3). For the first-litter traits, the correlations were all found to be positive between litter size and the sow longevity traits, with genetic correlations ranging from 0.49 to 0.65. For the trait average piglet birth weight, the genetic correlations with the sow longevity traits were all negative, ranging from -0.19 to -0.26. The genetic correlations between age at first farrowing and the sow longevity traits were negative as well, ranging from -0.13 to -0.22.

The relatively high positive genetic correlations between the sow longevity traits and first litter size were not expected compared to previous studies (e.g. Tholen *et al.* 1996; Engblom *et al.* 2016), with non-significant or even negative associations between first litter size and sow longevity traits. The results from this study showed that the first litter size could be a good selection criterion for sow longevity. Moreover, litter size is an important economic trait in commercial pig production systems, and sows with low first litter sizes are unlikely to be kept in the herd for subsequent parities. Even though genetic correlations were only found to be low to moderate between age at first farrowing and the sow longevity traits, all correlations did show negative associations. This implies that sows that farrow their first litter early in life have increased longevity. These results suggest that the trait, age at first farrowing may be incorporated as an indicator for sow longevity as well.

The average piglet birth weight was negatively correlated with sow longevity, suggesting that high piglet birth weights in the first litter are undesirable concerning the longevity of the sows. These correlations between average piglet birth weight and the sow longevity traits were slightly stronger than previous results (-0.02 and -0.15) found by Tholen *et al.* (1996). Further, negative correlations, both genetic and phenotypic, were estimated between the average piglet birth weight and the first litter size of -0.36 ± 0.09 and -0.45 ± 0.01, respectively. This indicates that there is a trade-off between litter size and average piglet birth weight in parity 1. As expected, since the uterine capacity of sows is limited and selecting sows only on litter size may reduce the average piglet birth weight, found in previous studies (e.g. Tholen *et al.* 1996). Even so, the strong positive genetic correlations found in this study between first litter size and sow longevity suggest that these sows were primarily selected on the number of piglets born alive, since this is an important production trait. Furthermore, autocorrelation between first litter size and subsequent litter sizes is expected, which influences retention to the next parity.

Table 3. Genetic and phenotypic (1st and 2nd row for each trait) correlations between the reproduction and the sow longevity traits, along with their standard errors (±SE)

Trait*	MNP	LPL	TNBA	STAY14[1]	AFF	LS
AFF	-0.16 ± 0.09	-0.14 ± 0.10	-0.13 ± 0.09	-0.22 ± 0.11		
	-0.07 ± 0.01	-0.07 ± 0.01	-0.06 ± 0.01	-0.06 ± 0.01		
LS	0.51 ± 0.10	0.53 ± 0.10	0.65 ± 0.08	0.49 ± 0.12	0.21 ± 0.11	
	0.16 ± 0.01	0.16 ± 0.01	0.27 ± 0.01	0.15 ± 0.01	0.05 ± 0.01	
PBWT	-0.20 ± 0.09	-0.19 ± 0.09	-0.26 ± 0.08	-0.26 ± 0.11	-0.15 ± 0.10	-0.36 ± 0.09
	-0.08 ± 0.01	-0.08 ± 0.01	-0.14 ± 0.01	-0.07 ± 0.01	0.07 ± 0.01	-0.45 ± 0.01

**For abbreviations see Table 1; [1]Correlations with stayability 1-4 were derived on the linear scale.*

When including age at first farrowing and first-litter characteristics as selection criteria for sow longevity, it should be taken into account that age at first farrowing is a highly managed period in pig production systems, and genetic variation in this trait is low. On a phenotypic level, gilts that are mated too early, are more prone to anoestrus after the first litter (e.g. Hoge and Bates 2011). In addition, sows with high litter size or high litter birth weights in parity 1 are exposed to high stress levels during gestation, farrowing, and subsequent lactation, which may lead to prolonged weaning to conception intervals (Tholen *et al.* 1996). It is likely that there is an optimum not only for first mating and farrowing age but also for first litter size and average piglet birth weight, and this should be taken into consideration for future research.

CONCLUSIONS

This study showed that first-litter characteristics had significant genetic associations with sow longevity. Both age of the sow at first farrowing and litter size are important selection criteria for sow longevity because these two traits can be measured earlier.

REFERENCES

Engblom, L., Calderón Díaz, J.A., Nikkilä, M., Gray, K., Harms, P., Fix, J., Tsuruta, S., Mabry, J. and Stalder, K.J. (2016) *J. Anim. Breed. Genet.* **133**: 138.

Gilmour, A.R., Gogel, B.J., Cullis, B.R., Welham, S.J. and Thompson, R. (2014) 'ASReml User Guide Release 4.1 Functional Specification' VSN International Ltd, Hemel Hempstead, HP1 1ES, UK.

Hoge, M.D. and Bates, R.O. (2011) *J. Anim. Sci.* **89**: 1238.

Serenius, T., Stalder, K.J. and Fernando, R.L. (2008) *J. Anim. Sci.* **86**: 3324.

Tholen, E., Bunter, K.L., Hermesch, S. and Graser, H.U. (1996) *Aust. J. Argic. Res.* **47**: 1275.

Proc. Assoc. Advmt. Anim. Breed. Genet. 23:171-174

INTEGRATING GENDER CONSIDERATIONS INTO LIVESTOCK GENETIC IMPROVEMENT PROGRAMS IN LOW TO MIDDLE INCOME COUNTRIES

K. Marshall, N. de Haan and A. Galiè

International Livestock Research Institute, Nairobi, 00100 Kenya

SUMMARY

Adoption of new or improved animal breeds by resource-poor farms is likely to increase when these breeds provide tangible benefits for the women and men involved in their production, consumption, and marketing. Yet, little consideration is often given by genetic programs to how gender dynamics and norms affect women's and men's preferences for species breeds, and traits, as well as women's and men's ability to participate in, and benefit from, livestock genetic improvement programs. Gender dynamics and norms refer to the ways in which women, men, boys and girls interact based on socio-cultural perceptions of what is considered appropriate behaviour for each group (e.g. roles, jobs, control over resources, decision-making etc.). Here we begin to fill the gap on where and how gender matters in the implementation of livestock genetic improvement programs in low to middle income countries by providing a conceptual framework. This framework stresses that gender considerations are relevant at all stages of implementation of a genetic improvement strategy, from targeting to ensuring equitable benefits.

INTRODUCTION

Adoption of new or improved animal breeds by resource-poor farms is likely to increase when these breeds provide tangible benefits for the women and men involved in their production, consumption, and marketing. Understanding how new breeds can benefit both women and men all livestock keepers is therefore important for two mutually supporting goals: to increase adoption, and to ensure equitable benefits from this. Gender dynamics affect both. Gender dynamics are the ways in which boys, girls, women, and men relate and interact. Gender dynamics are informed by socio-cultural ideas about what it means to be a 'man' or a 'woman', a 'boy' or a 'girl'; what are considered to be appropriate behaviours for each group; and the power relationships that define these groups. Gender dynamics affect all aspects of life, for example, who can own or claim property; who is considered to be in charge of providing for the family or child care; who is more likely to access opportunities for work or migration; and who is supposed to take decisions in a household. These gender dynamics also depend on other social factors such as ethnicity, wealth, marital status, or age. Such gender interactions, when persistent over time, become informal, unspoken and 'normative' rules of behaviour (or gender norms) that regulate what men and women, boys and girls can do at different stages of their life.

Gender norms and dynamics matter in livestock research for development because they affect, for example, who is considered to be a livestock keeper by communities and development programs, and is therefore involved as a participant; the financial resources women and men may access and control when investing in livestock improvement or the information they may access; and who can sell livestock or livestock products and decide how to utilize the money earned. These gender dynamics have strong implications for development goals such as improved nutrition. Evidence shows, for example, that rural women tend to invest agricultural revenues on child nutrition more than men, yet men need to be involved in supporting nutrition goals, in their traditional roles of decision-makers and family providers, for women to be able to allocate agricultural revenues to household nutrition (Galiè *et al.* 2019).

To-date there has been little attention to where and how gender matters in the implementation of livestock genetic improvement programs in low to middle income countries (LMICs). Here we begin to address this by providing a conceptual framework to this end. The focus of the framework is on women and men livestock keepers, as key beneficiaries of improved livestock genetics. This conceptual framework is intended to be used by researchers and practitioners involved in the implementation of livestock genetic improvement strategies in LMICs, to ensure that gender considerations are appropriately considered and acted on for maximal and equitable benefit from the livestock genetic technologies.

THE CONCEPTUAL FRAMEWORK (see also Figure 1).

Targeting of genetic improvement strategies. Targeting of genetic improvement strategies is an important issue to consider in LMIC contexts due to the large number of livestock systems that would benefit from genetic improvement and the limited resources available to support them. Strategic choices need to be made on where, with and for whom, and the species and breed focus. Here it is important to consider that women and men livestock-keepers can differ in their reasons for keeping, preferences around, aspirations for, and benefits from, different livestock species, breeds, and traits. For example, a study on gendered preferences for chicken in Ethiopia shows that women valued traits such as behaviour and feathers and that their preferences for these traits affected whether a breed was adopted by a household or not (Ramasawmy *et al.* 2018). Traits valued by males focused on productivity, health, and marketing of chickens with a view to scaling up their poultry keeping to an intensive system of production for business. In contrast, women responders aspired to increase the scale of their poultry keeping within their household level only, and thus valued traits that allowed chickens to be kept in an extensive system while increasing productivity. Women were not interested in making poultry into a business because of: the related high labour requirements (mostly their responsibility); their lack of land to keep chickens intensively or assets to make financial investments needed for intensification; and their loss of control over the benefits provided by chickens when, with intensification, men took on the marketing of the birds. Another example is from a study in Somaliland on the livestock keeping objectives of male and female pastoralists for goat, sheep, cattle and camel (Marshall *et al.* 2014). This study showed that each of these species were kept for multiple (up to 14) and gender differentiated reasons. For example, the livestock keeping objectives of 'savings and insurance' and 'sale of breeding animals' were more important to female and male pastoralists, respectively. The same study also showed trait preferences to be gender differentiated, for example the camel trait of 'good quality and tasty meat' was more important to men, reflecting men being the main consumers of camel meat (Marshall *et al.* 2016). This makes it vital to include both women and men in determining the genetic improvement priorities, both in terms of the livestock system targeted as well as type of genetically improved animal produced (influenced by the choice of breeding objective). Livestock breeds that better respond to local and gendered needs are more likely to be adopted and contribute to gender-equitable livestock development.

Choice of the type of genetic improvement strategy. The choice of genetic improvement strategy (whether breed substitution, within-breed improvement, cross-breeding etc.) to be implemented should be taken jointly by all stakeholders. Gender considerations here include who can participate in the genetic improvement program, the required investment level by household and individuals including on labour and financial resources, and the expected benefit. All of these are likely to be affected by gender dynamics and norms that influence intra-household sharing of resources, decision-making and opportunities. For example, the choice of focusing a breeding program on local versus exotic breeds may enhance the participation of women versus men, respectively, if women are the main controllers of local breeds and men the exotic (Njuki and Sanginga 2013). A study in Tanzania found that the introduction of new exotic breeds of goats shifted livestock labour from men to women because the goats were to be kept in the courtyard, a space assigned to women (Galiè and Kantor 2016). While women enjoyed increased access to goat milk, the overall decision-making on the new breed stayed with the men. More productive breeds also often entail higher financial investments to procure livestock inputs, in comparison to local breeds. This may exclude poorer farmers, of which rural women are the majority, from participating. These gender considerations need to be included when selecting a genetic improvement strategy to help ensure its effectiveness, sustainability and equity.

Implementation of the genetic improvement strategy. During implementation of participatory breeding programs, data is captured from, and shared back with, farmers. Incentives for continued farmer participation in the program are key to sustainability, and these may differ depending on whether women, men or both are involved. For example, incentives around the provision of feedback information for improved farm management, via mobile devices, are currently being tested, and the more relevant information to women or men livestock keepers may depend on how they are involved in the household livestock enterprise in terms of decision making, provision of labour, payment of costs and control of benefits. As an example of this, for some households keeping dairy cattle in Senegal, men are the main decision makers and providers of labour for cattle husbandry aspects (feeding, watering etc.), as well as well as control the income from animal sale, whereas women are the main decision makers, labourers and income controllers for sale of milk (Marshall *et al.* 2017). Here information that helps improve animal husbandry practices is likely to be more appreciated by the men, whilst information on milk quality and milk sale price is likely to be more appreciated by the women. Another related concern is who within the household can access the mobile devices and information, as this will not necessarily be shared within a household (FAO 2018). A further example that may affect who meaningfully participates in a genetic improvement program is simple logistical choices around program meetings (such as location, timing, and group composition). This can be affected by, for example, the spaces women and men can frequent, their availability for a meeting vis-a-vis other commitments, and their ability to speak out in groups where social hierarchies often establish who this is acceptable for, etc. Participation in meetings affects participation in decision-making.

Adoption and use of the improved genetics. In LMICs livestock keepers often cite the inability to access improved genetics (whether via artificial insemination, sire service, or live animal purchase), either because it is unavailable or unaffordable, as a key constraint to their livestock enterprise. In addition to being able to access the improved genetics, the livestock keepers also need to manage them appropriately (feeding, health care etc.) and market the animal or its products, to maximise benefit. To ensure that those who wish to adopt the improved livestock genetics can do so and enjoy benefits, access to the improved livestock genetics, as well as the resources need to maximize benefit, should be gender equitable. Here issues of gendered control over the household livestock enterprises may need to be addressed. These include on gendered constraints in decision-making over household investments and engagement with genetic technologies, access to information, access to credit, mobility, interaction with service providers, and market access for the products. Women's reduced mobility as compared to men is well documented in the literature, and affects, for example, access to genetic material, animal services and markets (Galiè *et al.* 2017). Further, many studies have shown that women in LMICs commonly do not have the same access to technologies, information, and service providers, including on credit, as men (Fletschner and Mesbah 2011).

Ensuring equitable benefit from the improved genetics. Finally, it should be ensured that the intrahousehold benefits from use of the improved genetics are equitable. Here a key concern is the shift in benefits between intrahousehold members associated with adoption of the genetic technology. Many studies have shown that as household enterprises that benefit women become increasingly commercially oriented, there is a shift in the control of benefits from women to men (Galiè and de Haan 2019). This was demonstrated to be the case for smallhold dairy cattle enterprises in Senegal, where higher levels of market orientation were associated with the keeping of cross-breed indigenous by exotic dairy cattle (as opposed to the traditional indigenous breeds) and a shift in the control of the income from milk sales from women to men (Marshall *et al.* 2017).

Figure 1. Key stages for integration of gender considerations into livestock genetic improvement strategies

CONCLUDING COMMENTS

The framework presented serves to highlight key gender issues around livestock genetic improvement strategies that need to be acted upon to ensure maximal and equitable benefit from genetic technologies and therefore increased adoption. Approaches that can be applied include accommodative or transformative approaches. Accommodative approaches recognize and respond to the specific needs and realities of men and women, based on their existing roles and responsibilities shaped by existing social and economic structures. Accommodative approaches do not question the barriers put up by the context they live in (Cornwall & Edwards, 2010). Transformative approaches aim to deeper social change by addressing some of the norms that constrict a particular group (Galiè and Kantor, 2016).

Whilst the focus of our analysis is gender, we recognise that men and women are not homogenous groups and that other social markers (such as age, social status, ethnicity, caste etc.) intersect with gender and affect interaction with, and benefit from, livestock genetic technologies. Future versions of this framework will be extended to include consideration of these.

REFERENCES

Cornwall A. and Edwards J. (2010) *IDS Bulletin*. **41**:1.

Fletschner D. and Mesbah D. (2011) *World Dev.* **39**:1422.

FAO (2018) Gender and ICTs, by Treinen, S and Van der Elstraeten, A. Rome, Italy

Galiè A., Teufel N., Webb Girard A., Baltenweck I., Price M.J., Dominguez-Salas P., Jones R., Lukuyu B., Korir L., Raskind I., Smith K. and Yount, K. (2019) *Glob. Food Sec., Gender and Food Security special issue.* In print.

Galiè A., Distefano F., Kangogo D., Mattioli, R.C., Wieland B. and Baltenweck I. (2017) *Agri-Gender* **2**: 43.

Galiè A. and Kantor P. (2016) In: Njuki, J. J. R. Parkins, A. Kaler (eds), Transforming Gender and Food Security in the Global South, International Development Research Centre and Routledge.

Galiè A. and de Haan N. (2019) In: Ferranti P., Berry E.M. and Anderson, J.R. (Eds.), Encyclopedia of Food Security and Sustainability **3**: 426.

Marshall K., Mtimet N., Wanyoike F. and Ndiwa, N. (2014) 10th WCGALP, paper 299

Marshall K., Mtimet N., Wanyoike F., Ndiwa N., Ghebremariam H., Mugunieri L. and Costagli, R., (2016) *JABG* **133:** 534.

Marshall, K., Tebug, S., Salmon, G.R., Tapio, M., Juga, J. and Missohou, A. (2017) *ILRI Policy Brief 22. Nairobi, Kenya: ILRI.*

Njuki, J. and Sanginga, P.C. 2013. 'Women, Livestock Ownership and Markets'. Earthscan USA/ International Development Research Centre, Canada, New York, USA

Ramasawmy M., Galiè A., and Dessie, T. (2018) In: Tufan H.A., Grando S. and Meola C. (Eds.), State of the Knowledge for Gender in Breeding: Case Studies for Practitioners. Lima (Peru).

CGIAR Gender and Breeding Initiative. Working Paper. No. 3.

Proc. Assoc. Advmt. Anim. Breed. Genet. 23:175-178

PERCEPTIONS OF WOMEN AND MEN SMALLHOLDER PIG KEEPERS IN UGANDA ON PIG KEEPING OBJECTIVES, AND BREED AND TRAIT PREFERENCES

B.M. Babigumira[2,3], E. Ouma[2], J. Sölkner[3] and K. Marshall[1]

[1]International Livestock Research Institute, P.O. Box 30709, Nairobi 00100, Kenya
[2]International Livestock Research Institute, P.O. Box 24384, Kampala, Uganda
[3]Division of Livestock Sciences, Department of Sustainable Agricultural Systems, BOKU - University of Natural Resources and Life Sciences, Vienna, Austria

SUMMARY

The objective of this study was to compare the perceptions on livestock keeping objectives, breed and trait preferences of smallholder women and men pig farmers in Uganda. To this end, the study interviewed adult males and females from 200 pig keeping households, within two study sites. The main pig keeping objectives of both women and men were savings and insurance, income from the sale of pigs for fattening or slaughter, or income from the sale of pigs for use as breeding animals. The most preferred breed-types for both women and men were the same with exotic breeds the most preferred, followed by the cross of the exotic and local breeds. Many key traits, such as those for reproduction, growth and disease resistance, were of similar importance to men and women. Overall, the results suggest that gender differentiated breeding objectives, and breed and trait focus, are not required as part of a pig breeding program in Uganda.

INTRODUCTION

Uganda, located in East Africa, is one of the world's poorest countries. Within Uganda, pig farming is an important livelihood to about 1.1 million poor smallholder farmers. Uganda's current pig population is about 3.2 million and rapidly increasing. Currently there are no structured pig breeding programs within Uganda, however efforts to establish one at a national level are underway. In low-and-middle income countries, there is strong recognition that the gender of stakeholders is important to consider in the design of rural development interventions. This ensures, for example, that both adoption and benefits are maximised and equitable. Gender has also been shown to matter in the design and implementation of livestock breeding programs in terms of ownership and control of animals (Marshall *et al.* 2019).

As background information feeding into the design of such a program, this paper compares Ugandan women and men smallholder pig farmers, in terms of their reasons for keeping pigs, and preferences for breeds and traits. The implication of these results on design of the potential Ugandan pig breeding program is also discussed.

MATERIALS AND METHODS

Study site and pig breeds present. The project study sites were within Kamuli and Hoima districts of Uganda, selected due to having a relatively high number of pig keeping households. Two hundred pig-keeping households, with 100 households in each site, participated in the study. The main pig types kept comprise local, exotic, and crosses between the two. The local breed is a small black pig, well adapted to the local environmental conditions. The exotic breeds comprise of Large White, Landrace, and the Camborough line from PIC. The exotic breeds are recent introductions to increase productivity. Various crossbreed types exist because of the unstructured crossbreeding between the local and exotic breeds.

Baseline survey. A baseline survey was administered separately to the female and male adult within each study household between April and May 2018. There were 200 female respondents (of which 19.5% were household heads and 80.5% were spouses) and 161 male respondents (all household heads). The survey comprised a questionnaire that collected data on household characteristics (such as structure, membership, education, livelihoods, asset base, food security etc.), as well as pig production practices.

Rating scales and statistical analysis. All ratings were based on Likert scales. For ratings of the importance of reasons for pig keeping and traits, the scale was 0 to 5, where 0 was no importance, 1 was the lowest importance, and 5 was the highest importance. For breed preference, the scale was 1 to 5, where 1 was strongly dislike and 5 was strongly like. To test differences between the average ratings by men and women, an independent t-test was applied under assumptions of normality and equal variances. The level of significance used was 0.05.

RESULTS AND DISCUSSION

Types of household pig enterprises, and their importance to livelihoods. Most households (92%) practiced a combination of farrow to wean and farrow to finish pig production systems. Most commonly, 1 to 3 sows were kept, with 2 to 16 piglets. Pig farming was the primary livelihood for 32% of women and 15% of men and secondary for 45% of women and 63% of men.

Reasons for keeping pigs. Women and men farmers rated reasons for keeping pigs, using a pre-defined list of reasons from literature (Ouma *et al.* 2015) with the option of including additional reasons (Table 1). The most important reasons for both genders were for savings and insurance purposes (keeping of pigs to sell in times of need) and income from sale of animals (both for fattening or slaughter, and as breeding animals). Women rated the keeping of pigs for savings and insurance purposes significantly more important than men, though the difference was small (Table 1). A similar result has been previously reported (Marshall *et al.* 2014). Both genders assigned lower importance to keeping pigs for income from boar sire service (likely because not all households keep boars) and manure for cropping. Keeping of pigs for home consumption of pig meat and income from manure sale was of almost no importance. This information was asked to help inform development of breeding objectives for the Ugandan pig breeding program. Results suggest that a common breeding objective, i.e. for both women and men, is appropriate. This objective would centre around ensuring pig keeping translates into household income from both planned and emergency pig sales. Further development of this objective will be performed in collaboration with stakeholders.

Table 1. Average ratings for reasons for keeping pigs, by women and men farmers. The P-value indicates the significance of the difference between women's and men's ratings

Reason for keeping pigs	Women	Men	P-value
Savings / insurance (keeping of pigs to sell in times of need)	4.2	3.9	0.04
Income from the sale of pigs for fattening or slaughter	3.7	3.8	0.71
Income from the sale of pigs for use as breeding animals	3.7	3.8	0.40
Income from boar sire service	1.2	1.1	0.66
Manure for cropping	0.9	0.7	0.40
Home consumption of pig meat	0.3	0.3	0.65
Income from the sale of manure	0.1	0.1	0.09

Breed preferences. Women and men respondents were asked to rate their preferences for the breed-types they were familiar with. The proportion of women familiar with local, crossbred and exotic pigs were 69%, 41% and 12%, respectively, whilst for men it was similar at 70%, 41% and 17%, respectively. Results (Table 2) showed breed preferences not to be significantly different between the genders, with the most preferred breed (for both sows and fattening pigs) to be the exotic, followed by the crossbreed. Combined results across both genders showed the average rating for the exotic breed was significantly (a difference of 0.5 and P=0.019) higher to that for the crossbreed. Also, the ratings for the local were significantly lower than for cross (-0.73, P<0.001) and exotic (-1.24, P<0.001).

The same respondents named the advantages and disadvantage of the different breeds. The main advantages of local pigs included being adapted to the local environmental conditions (disease resistance, general adaptation, eating local feedstuff) and not requiring special housing, whilst the main disadvantages included low performance (growth, weight, litter size) and low market prices. The main advantages for exotic breeds were high performance, high market price and demand, whilst the main disadvantages were poor adaptation to local environmental conditions, high feed intake, feed cost, and the requirement for housing. For the crossbreed, the named advantages and disadvantages were as for the exotic breed. Whilst both genders generally named similar breed advantages or disadvantages, the proportion of women versus men naming a particular advantage or disadvantage differed. Most notably more women than men named 'high litter size' and 'high market price and demand' as an advantage for the crossbred. On the other hand, more men than women named 'faster growth' and 'high market demand' as advantages for the exotic breed.

In terms of breeding program design, the breed advantages and disadvantages gives some weight to focusing the program on exotic rather than cross-bred or local breeds. However, this result will later be combined with other results from the same study (such as the profit from keeping different breed-types) before a final decision on this choice is made. Continual feedback from the pig-keepers on preferred breed is also recommended, as breed preferences may change as people become increasingly familiar with the breed options.

Table 2. Average ratings for breed preference, by women and men farmers. The P-value indicates the significance of the difference between women's and men's ratings

	Sows			Fattening pigs		
Breed	Women	Men	p-value	Women	Men	p-value
Local	3.6	3.6	0.68	3.6	3.5	0.50
Crossbred	4.4	4.3	0.42	4.3	4.3	0.65
Exotic	4.8	4.9	0.62	4.8	4.9	0.47

Trait importances. Women and men respondents rated the importance of traits of sows and fattening pigs, using a pre-set trait list based on Ouma *et al.* (2015), with the option of including additional traits. The traits comprised of reproduction, growth, size, adaptation, body features (which farmers use to help indicate the breed-type), and other. Trait ratings were not statistically different between the genders, with two exceptions (Table 3). Sow traits that were considered moderately or more important (average ratings of ≥ 3) by both genders were reproduction, and growth / size, as well as disease resistance, ear-shape and feed intake. Traits that were low to moderately important for both genders (average ratings of ≥ 1 and <3) were heat-resistance, other body feature traits, and temperament. Traits of importance for fattening pigs were similar to those for sows (barring the reproductive traits that are not relevant to fattening pigs). It is of note that temperament was significantly more important to women than men for fattening pigs (and almost for sows), though the difference in ratings was small. This may stem from women being the main labour providers in cooling pigs, which is commonly done via dousing the pigs with water, with the water sometimes fetched from far away. Further, women rated feed intake significantly higher than men did. However, both genders desired the same direction of change in the trait (see Table 3). In considering breeding program design, these results indicate no concerns in having a common trait focus for both women and men pig keepers.

Table 3. Average ratings for trait importance, by women and men farmers. The P-value indicates the significance of the difference between women's and men's ratings

Trait group	Trait	Direction[1]	Sows			Fattening pigs		
			Women	Men	P-value	Women	Men	P-value
Reproductive	Return to heat	Faster	4.0	4.0	0.80			
	Litter size	12,10	4.7	4.5	0.20			
	Teat number	14,12	3.6	3.7	0.51			
Growth, size	Growth rate	Faster	4.4	4.3	0.51	4.6	4.6	0.90
	Body length	Longer	4.3	4.5	0.06	4.4	4.5	0.48
	Wither height	Taller	3.2	3.0	0.48	3.0	2.7	0.26
Adaptation	Disease resistance	Higher	4.2	4.0	0.15	4.2	4.0	0.26
	Heat resistance	Higher	2.3	2.2	0.89	2.2	2.2	0.71
Body features	Ear-shape	Floppy	3.5	3.3	0.35	3.0	2.8	0.34
	Back-shape	-[2]	2.5	2.5	0.61	2.4	2.2	0.27
	Mouth-shape	Short	3.1	2.8	0.12	2.8	2.6	0.43
	Colour	White	2.8	2.7	0.51	2.4	2.2	0.26
Other	Temperament	Docile	1.9	1.6	0.06	2.0	1.6	0.03
	Feed intake	High	3.8	3.5	0.02	4.0	3.9	0.17

[1]Direction of desired trait change or optimal value. The most common answer, giving singularly if the same for women and men, else for women and men, respectively.
[2]Both a curved and straight back shape was almost equally cited by both women and men

CONCLUSION

This work adds to a small, but growing, body of work on whether / how gender matters in the implementation of livestock breeding programs within low-and-middle income countries. In this case gender differentiated breeding objectives, and breed and trait focus, do not appear necessary. However, other studies have found significant differences between women and men for livestock keeping objectives and trait preferences (Marshall *et al.* 2014; Ramasawmy *et al.* 2018), which could impact on breeding program design. Despite the similarities between women and men on the issues reported here, a gender-lens should still be applied when considering other aspects of the potential pig breeding program for Uganda (see Marshall *et al.*, 2019 for more details).

ACKNOWLEDGEMENTS

We gratefully acknowledge the women and men farmers who participated in this study, as well as the local district and project staff. This study is part of the Uganda Pig Genetics project, funded by the Austrian Development Agency and the CGIAR research program on livestock.

REFERENCES

Ouma E.A., Dione M.M., Lule P., Pezo D., Marshall K., Roesel K., Mayega L., Kiryabwire D., Nadiope G. and Jagwe J. (2015) ILRI Project Report. Nairobi, Kenya: ILRI.

Marshall K., de HaanN. and Galiè A. (2019) *Proc. 23rd Assoc. Advmt. Anim. Breed. Genet.* (submitted).

Marshall, K., Mtimet, N., Wanyoike, F. and Ndiwa, N. (2014) *Proc 10th WCGALP*. Ramasawmy M., Galiè A., and Dessie, T. (2018) In: Tufan H.A., Grando S. and Meola C. (Eds.), 'State of the Knowledge for Gender in Breeding: Case Studies for Practitioners. Lima (Peru)'. CGIAR Gender and Breeding Initiative. Working Paper. No. 3.

Proc. Assoc. Advmt. Anim. Breed. Genet. 23:179-182

GENETIC STRUCTURE AND DIFFERENTIATION AMONG AFRICAN *BOS TAURUS* CATTLE BREEDS

N.Z. Gebrehiwot[1], E.M. Strucken[1], H. Aliloo[1], K. Marshall[2] and J.P. Gibson[1]

[1]Centre for Genetic Analysis and Applications, School of Environmental and Rural Science, University of New England, Armidale, NSW, 2351 Australia
[2]International Livestock Research Institute and Centre for Tropical Livestock Genetics and Health, Nairobi, Kenya

SUMMARY

African taurine cattle populations are widely distributed in humid and sub-humid zones of West and Central Africa. We assessed the genetic structure and differentiation within and across 8 African *Bos taurus* populations: 4 N'Dama populations (N'Dama, N'Dama1, N'Dama2, N'Dama3), Lagunaire, Lagune, Somba, and Baoule. A total of 38k autosomal SNPs were used for principal component analyses (PCA), estimation of pairwise F_{ST} values and within population heterozygosity (F_{IS}), and neighbour-joining (NJ) tree construction. The first PC clearly differentiated Lagune and Lagunaire from N'Dama; PC2 separated Lagunaire, Lagune and one N'Dama population from the rest of taurine breeds; and PC3 separated N'Dama3 from Somba and Baoule. Estimates of pairwise F_{ST} values among the majority of populations ranged from 0.03 to 0.149, indicating low to moderate genetic differentiations, while a high genetic divergence between N'Dama3 and Lagune (F_{ST}=0.178), and N'Dama3 and Lagunaire (F_{ST}=0.168) was observed. No genetic subdivision was found between N'Dama and N'Dama1, and Lagune and Lagunaire. A higher heterozygosity (F_{IS} value from -0.011 to 0.025) was found in N'Dama, N'Dama1, N'Dama2, Lagune, Lagunaire, and Baoule breeds. The NJ tree clearly separated Lagune and Lagunaire as well as Somba and Baoule with a 100% and around 31% bootstrap value, respectively, from the other taurine populations. We highlighted that African taurine populations are diverse and genetic differences between sampling locations exists even within a breed. Therefore, choice of an African taurine breed to anchor African indigenous breeds should be carefully considered.

INTRODUCTION

Taurine cattle are known to have been first domesticated in the Near East and are believed to have been introduced to Africa through present day Egypt (Gifford-Gonzalez and Hanotte 2011). It is thought that the humpless taurine Hamitic Shorthorns and Longhorns arrived in sub-Saharan Africa around 4500 to 4000BP (Payne and Hodges 1995). Nowadays, African taurine cattle breeds have been categorized as: longhorn N'Dama of the far West forest savannah, and 14 humpless shorthorns such as Baoule, Somba and Lagune breeds which are widely distributed in the humid and sub-humid zones of West and Central Africa (Rege 1999). These zones are also known for their endemic trypanosomes, which may affect cattle distributions due to some breeds being tolerant of infection, whilst others are strongly affected by trypanosomes (Berthier *et al.* 2015). There is, however, no comprehensive study showing how genetically diverse African taurine breeds actually are. Up to now, only one African taurine breed, the longhorn N'Dama, has been largely studied and is often considered as a reference breed when other African indigenous breeds are analysed. The present study aimed at assessing genetic structure and differentiation within and among 8 African *Bos taurus* populations.

MATERIALS AND METHODS

SNP data for a total of 130 African *Bos taurus* individuals originating from different locations of West Africa. These included N'Dama (separated into 4 populations: N'Dama (n=20), N'Dama1 (n=14), N'Dama2 (n=14), and N'Dama3 (n=17), sampled in Guinea, Cote d'Ivoire, South-Eastern Burkina Faso and South-Western Burkina Faso, respectively), Lagune (n=20) from Benin, Baoule (n=20) from Burkina Faso, Somba (n=20) from Togo, and Lagunaire (n=5) from "West Africa". Pooled *Bos indicus* (n=105) and European *Bos taurus* (n=100) were used as reference populations.

Animals and genotypes were sourced from the Bovine HapMap consortium (777k) and from Decker *et al.* (2014, 50k). After quality control, the genotypes of 777k SNPs and 50k SNPs datasets reduced to 735k SNPs and 45k SNPs, respectively, of which 38,556 SNPs were in common and used in this study.

Principal component analyses (PCA) were performed using a genomic relationship matrix (GRM) according to the first method of VanRaden (2008). The PCA were carried out with and without the reference breeds. To estimate population differentiation, pairwise F_{ST} was calculated according to Weir and Cockerham (1984). The degree of inbreeding was inferred from the F_{IS} coefficient calculated according to Nei (1977). Allele frequencies in the African taurine populations were used to construct a neighbour-joining tree. Bootstrapping (1000 replicates) was carried out to assess the strength of support for the internal nodes. All data analyses were performed using the R software (R Core Team, 2018).

RESULTS AND DISCUSSION

The first two PCs obtained from the analysis of African taurine and the reference populations explained 79% and 16.7% of the total variation in the GRM, respectively, and differentiated the African taurine, *Bos indicus*, European taurine from each other (Figure 1a). In comparison, the first 5 PCs obtained from the analysis of African taurine cattle populations explained a total of 65.22% of the variation in the GRM (Figure 1b and 1c).

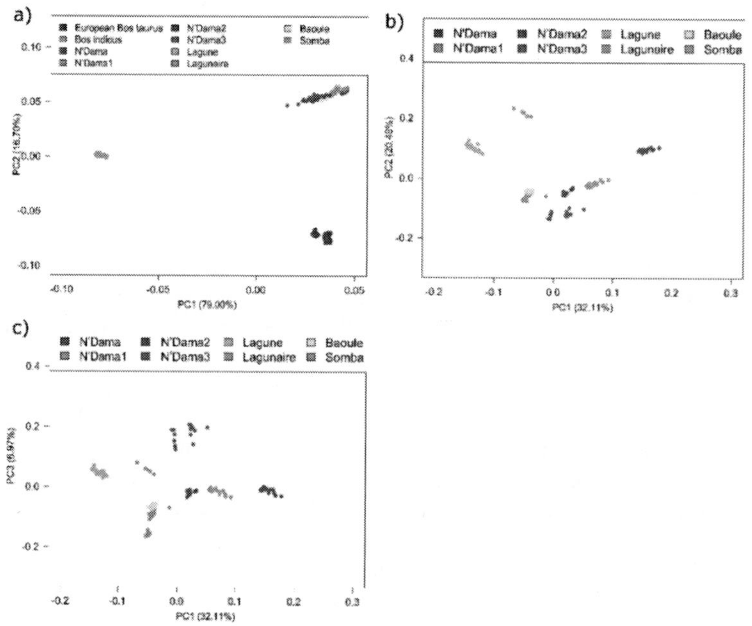

Figure 1. Plot PC1 vs PC2 for reference and African *Bos taurus* populations (a) Plot of PC1 vs PC2 (b) and PC1 vs PC3 (c) for only African *Bos taurus* populations

The first PC accounted for 32.11% of the total genetic variation and clearly differentiated Lagune and Lagunaire from N'Dama; whereas PC2 explained 20.48% of the total variation and separated Lagunaire, Lagune and N'Dama from the rest of the African taurine populations (Figure 1b). The third PC contributed 6.97% of the total variation and separated N'Dama3 from Somba and Baoule (Figure 1c). The majority of Somba and Baoule animals grouped tightly together, however, one Somba outlier was detected (Figure 1b and 1c). The N'Dama3 clusterd in two separate groups (Figure 1b and 1c), which might indicate that they were sampled from two different villages or sub-locations. Both Figure 1b and 1c showed that N'Dama1 and N'Dama2 clustered near each other, with N'Dama1 closer towards N'Dama.

Estimated F_{IS} values showed less inbreeding than expected under Hardy-Weinberg equilibrium for N'Dama1, N'Dama2, N'Dama3, and Baoule (Table 1). N'Dama3 showed the highest F_{IS}=-0.109 which confirms the separation of this breed into two clusters in the PC plots. The relatively high amount of within breed genetic variation indicated by F_{IS} values for all breed samples indicates a valuable reservoir of genetic diversity for future breeding endeavours and a viable target for conservation.

The pairwise F_{ST} estimates showed a high genetic divergence between N'Dama3 and Lagune (F_{ST} =0.178), and N'Dama3 and Lagunaire (F_{ST}=0.168), indicating a clear genetic difference among these breeds. Moderate pairwise F_{ST} values of 0.061 to 0.149 were observed between the majority of the African taurine populations (Table 1). The minimum possible genetic distance between populations is zero. Negative values of the estimate of genetic distance, F_{ST}, can arise by random sampling and can be interpreted as zero genetic distance between populations. Lagune and Lagunaire as well as N'Dama and N'Dama1 are inferred to have zero genetic distance.

Table 1. Estimated F_{IS} values (diagonal) within and pairwise F_{ST} (off-diagonals) between the African taurine cattle populations

	N'Dama	N'Dama1	N'Dama2	N'Dama3	Lagune	Lagunaire	Baoule	Somba
N'Dama	0.011	-0.017	0.030	0.099	0.145	0.133	0.085	0.067
N'Dama1		-0.005	0.030	0.100	0.149	0.135	0.086	0.067
N'Dama2			-0.011	0.083	0.127	0.109	0.061	0.044
N'Dama3				-0.109	0.178	0.168	0.112	0.095
Lagune					0.004	-0.006	0.123	0.100
Lagunaire						0.025	0.112	0.083
Baoule							-0.021	0.039
Somba								0.049

The neighbor-joining tree is consistent with the PCA and F_{ST} results (Figure 2). The African taurine populations were separated into two clades: Lagune and Lagunaire, which F_{ST} indicated can be regarded as one population, separated from the other breeds. Within the second clade, N'Dama3 separated on 100% of bootstraps from the other breeds, again confirming our previous findings. Somba and Baoule as well as the remaining N'Dama populations clustered together in 100% of bootstraps, with N'Dama and N'Dama1 clusters being consistent with the F_{ST}, which showed that they are effectively a single population. The fact that Somba and Baoule formed overlapping clusters in the PCA plots (Figure 1b and 1c) and had relatively low F_{ST} coupled with the geographical proximity of their sampling areas suggests that they might be considered sub-populations of the same breed.

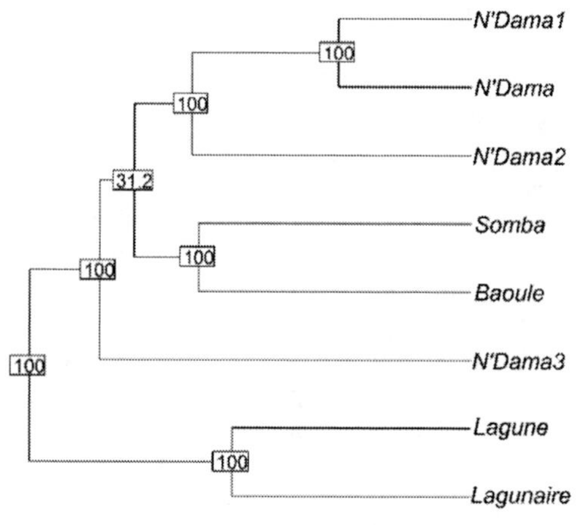

Figure 2. Phylogenetic tree constructed by the neighbour-joining method based on allele frequencies among the African *Bos taurus* cattle populations

The PCA (Figure 1a) showed that N'Dama2 and N'Dama3 spread towards the pooled *Bos indicus* reference breeds, showing that they are not pure African taurine breeds. In admixture analyses, results not shown here, where the African *Bos taurus* breeds were analysed along with many other African indigenous breeds, *Bos indicus* and European *Bos taurus* controls, N'Dama2 and particularly N'Dama3 appeared to have a small proportion of admixture with *Bos indicus* most likely coming from local zebu breeds (Gebrehiwot *et al.* in preparation). This likely explains N'Dama3 being clustered quite separately from other N'Dama samples.

CONCLUSIONS

Our study provides an insight into the genetics of the African *Bos taurus* breeds. The current research indicates that Lagune and Lagunaire, as well as N'Dama and N'Dama1 can be considered as single populations, respectively. The results presented are important for the design of conservation, improvement and breed management strategies of West African *Bos taurus* breeds.

REFERENCES

Berthier D., Peylhard M., Dayo G.-K., Flori L., Sylla S., Bolly S., Sakande H., Chantal I. and Thevenon S. (2015) *PloS One* **10**: e0126498.

Decker J.E., McKay S.D., Rolf M.M., Kim J., Alcalá A.M., Sonstegard T.S., Hanotte O., Götherström A., Seabury C.M. and Praharani L. (2014) *PLoS Genet.* **10**: e1004254.

Gebrehiwot N.Z., Strucken E.M., Aliloo H., Marshall K. and Gibson J.P. In preparation.

Gifford-Gonzalez D. and Hanotte O. (2011) *J. World Prehist.* **24**: 1.

Nei M. (1977) *Ann. Hum. Genet.* **41**: 225.

Payne W. and Hodges. J. (1995) ' Tropical cattle: origins, breeds and breeding policy ' Blackwell Science, Oxford.

Rege J. (1999) *Anim. Genet. Res.* **25**: 1.

VanRaden P.M. (2008) *J. Dairy Sci.* **91**: 4414.

Weir B.S. and Cockerham C.C. (1984) *Evol.* **38**: 1358

Proc. Assoc. Advmt. Anim. Breed. Genet. 23:183-186

ANALYSIS OF GROWTH OF TWO MAJOR BREEDS OF DOMESTIC CAMEL IN PAKISTAN: IMPLICATIONS FOR BREED IMPROVEMENT

S. Sabahat[1,2], M.S. Khatkar[3], A. Nadeem[1] and P.C. Thomson[1,2]

[1]Institute of Biochemistry and Biotechnology, University of Veterinary and Animal Sciences, Lahore, 54000 Pakistan
[2]School of Life and Environmental Sciences, University of Sydney, Camden, NSW, 2570 Australia
[3]Sydney School of Veterinary Science, University of Sydney, Camden, NSW, 2570 Australia

SUMMARY

The camel is an important domestic animal for producing valuable food and is well adapted to extremely harsh environments. The camel is gaining importance as a source of meat in Pakistan. However, the information on camel is very limited. We obtained blood samples and growth records of 136 animals of native Pakistani camel breeds, viz. Marecha camels from Punjab and Lassi camels from Baluchistan. In this study, we will present results on weight traits and growth modelling for these animals. This is part of a larger study investigating genetic diversity and genome-wide associations in these two important breeds of camel in Pakistan. We discuss the potential of the camel as a meat producer in Pakistan.

INTRODUCTION

Traditionally, the livestock sector is an important component of social structure in rural areas of Pakistan, in addition to its role in farming and commercial operations. In Pakistan, agriculture is the biggest sector of the economy, and in particular, the livestock sector has a great impact in Pakistan because 35-40 million people in rural populations depend on livestock and obtain about 30-40% of their earnings from livestock (Government of Pakistan 2017).

The camel is an important species well adapted to hot and dry environments and contributes appreciably to the food security of the nomadic rural households in Pakistan. Due to its unique adaptability, this species is well suited for management in arid and semi-arid environment. The camel remained a neglected species among livestock for scientific research. One of the main reasons for its neglect is that the camel is mainly found in areas of poor nutrition and dry environments of Asia and Africa rather than developed countries with stronger agricultural research programs (Sohail 1983).

The camel is also known as the "ship of the desert" because of its adaptability and suitability to thrive in the hot, dry and semi-arid region of the world. This animal has a distinctive ability to change the scarce vegetation of the desert into milk, fibre and meat. The camel has limited competition with other animals for feed, eats relatively less comparative to it body size (Khan *et al.* 2003).

The camel offers an opportunity to address food insecurity in Pakistan and other arid and semi-arid developing countries because of its biological and production characteristics. With climate change, this is becoming an imperative. To date, few systematic breed improvements have been made to breed superior camels for improved meat, milk, or other production characteristics. However, developments in genetic technologies in recent years have made it worthwhile to investigate the feasibility of applying these methods to breed camels for improved meat production, or improved milk production. Applications of these methods have been well developed in other livestock species to optimally select animals for breeding, using marker-assisted selection, and more recently with the availability of high-density genotyping arrays, using genomic selection. Much could be learnt from experiences in those other livestock sectors and applying them, to camel breeding.

This paper investigates the growth of the camel, *Camelus dromedarius*, as the first stage in providing phenotypic information for a quantitative genetic analysis of growth. The output from the growth analysis will allow a genetic evaluation at various ages of the camel, towards a genetic improvement program of the camel in Pakistan.

MATERIALS AND METHODS

Weight records at birth, weaning and monthly records for the most recent four years were obtained from farm records of Marecha camels (n = 83 female, n = 26 male) from the Camel Breeding and Research Station (CBRS) at Rakhmahni, Bhakkar Pakistan. Up to 48 monthly records were available. These were obtained using walk-over scales. Concurrently, blood samples were also collected from these animals for subsequent genomic analysis (not reported here). In addition, weight records and blood samples were obtained from Lassi camels (n = 27 females) on privately-owned farms in Lasbela, Baluchistan Pakistan.

Comparisons of birth weights and of weaning weights between the three groups of camels (Lassi females, Marecha females, and Marecha males) were made using linear models. For the growth records, linear mixed models were fitted to the data, with fixed effects for breed-sex group and age (as a covariate). To allow for the nonlinear growth curve, a spline function of age was included in the random effects model. Individual variability of growth curves was accommodated by inclusion of random intercepts, random slopes, as well as random splines for each camel. The analysis was conducted using the ASReml-R package (Butler *et al.* 2009) in R.

RESULTS AND DISCUSSION

Figure 1 shows the birth weights and weaning weights for female Lassi, female Marecha and male Marecha camels. There are significant differences in birth weight across the three groups ($P <$ 0.0001), with Marecha males being significantly heavier (45.42 ± 0.92 kg, mean ± SE) than females of either breed at birth (Lassi: 40.96 ± 0.91 kg; Marecha: 38.95 ± 0.52 kg). The difference in mean birth weights for females of the two breeds was marginally non-significant (P = 0.057). Similarly, for weaning weight, there were significant differences amongst the three groups of camels (P = 0.006), again with male Marecha camels (112.38 ± 3.01 kg) having significantly heavier weaning weights than females of either breed (Lassi: 99.70 ± 2.95 kg; Marecha: 102.66 ± 1.69 kg). There was no significant difference between the mean weaning weights of females of the two breeds (P = 0.39). However, caution need to be applied in making between-breed comparison of both birth weights and weaning weights, due to the different management and environment: the Marecha camels were managed in a research farm, while Lassi camels were managed by private farmers with close cultural ties to their animals. The relatively reduced significance of differences in mean weaning weights compared to between mean birth weights could be due to variation in ages of weaning of individual animals.

Figure 2 shows a sample of fitted growth curves for individual camels, representing each of the three groups. Table 1 shows selected model-based means weights at different ages, again for each of the three groups. What is immediately apparent is that many camels go through periods of losing weight. This is seen in the individual growth curves as well as the overall average of male Marecha camels. This decline may reflect decline in food availability, particularly over winter, but for Marecha males, another possibility of weight loss is when they are provided for breeder services. Also evident in Table 1 is the lack of substantial difference in mean growth profiles of the three groups of camels: divergence starts after two years, with male Marecha camels having a substantially higher growth rate than female Marecha camels. It is also seen that Lassi female had a higher growth rate than female Marecha camels. Again, the different management and environments of the two breeds need to be considered when interpreting between-breed differences.

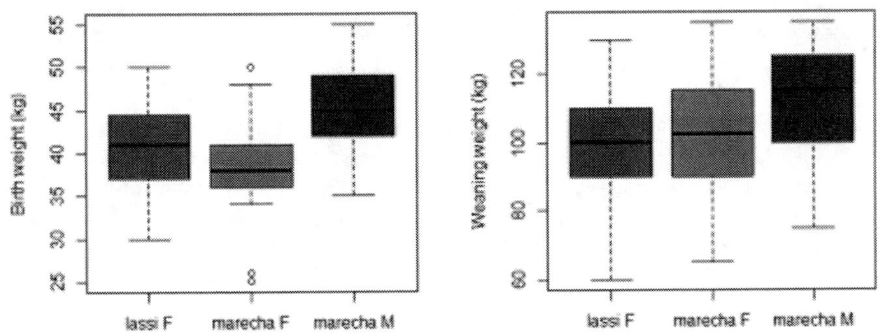

Figure 1. Distributiosn of birth weights (LHS) and weaning weights (RHS) for female Lassi, female Marecha and male Marecha camels

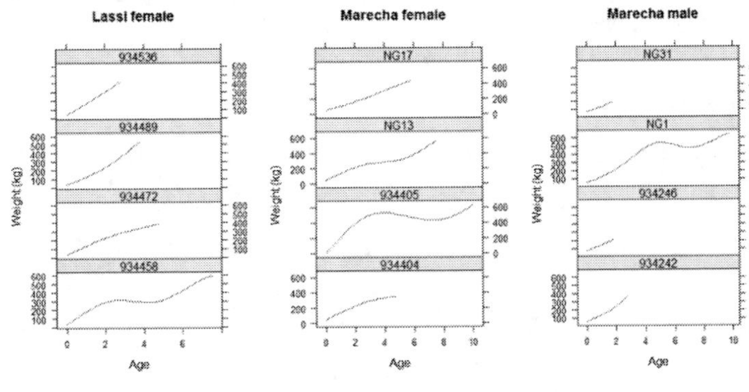

Figure 2. Fitted growth curves of a selection of female Lassi, female Marecha and male Marecha camels

As the body weight data were not collected at consistent ages across all animals, it would be difficult to use these raw data as phenotypic input into a GWAS for example. However, using model-based predictions for individual animals, it is possible to obtain values for all animals at specific ages. This allows an age specific GWAS to be conducted, and to track effects of genes over different times. These results will be reported in a subsequent study.

Table 1. Model-base mean body weight at selected ages of female Lassi, female Marecha and male Marecha camel

	Mean body weight ± SE (kg)		
Age (yr)	Lassi Female	Marecha Female	Marecha Male
0	40.66 ± 26.10	49.00 ± 17.03	48.61 ± 29.89
2	233.78 ± 16.61	227.32 ± 7.88	236.85 ± 13.68
4	429.56 ± 33.54	346.93 ± 16.25	508.58 ± 32.21
6	643.49 ± 44.06	434.87 ± 19.64	669.61 ± 45.81
8	791.27 ± 60.60	536.20 ± 17.35	592.11 ± 56.38

The camel in Pakistan represents an opportunity to 'future proof' the country against the risk of climate change and to add to increasing food security for the nation. The tools to develop a livestock industry have been well developed in other industries, particularly in developed counties. But with reductions in costs of genetic technologies and considering that not much in the way of breed improvement in the camel has been conducted, it would be expected a well-developed breeding program would have great financial and social benefits. It is also important to conserve the range of camel breeds that exist in Pakistan, as this genetic diversity represents another form of 'future proofing' production. This diversity should be considered in relation to the range of functions of the camel, i.e. meat, milk, skin, and even tourism. Genomic tools provide an efficient way to assess this genetic diversity and to plan how this can be managed in an optimal way.

CONCLUSIONS

This study provides information on growth traits in two major breeds of camel in Pakistan. A spline-based method is used to model growth and develop predictive models for individual animals and these predictions can be used as phenotypic input for a GWAS of camel growth. Little systematic breed improvement had been conducted on dromedary camels in Pakistan, but improvements in productivity have the potential to improve the economic situation for farmers in (semi-)arid regions but also to 'future proof' the nation in terms of its food security.

ACKNOWLEDGEMENT

The authors are grateful to Dr Ghulam Yasin and Dr Shahid Nabeel, Camel Breeding & Research Station Rakhmani, Bhakkar, as well as farmers in District Lasbela, Baluchistan, for helping in sample and data collection. SS was supported by the Higher Education Commission of Pakistan, for an IRSIP Research Fellowship to undertake part of this research at The University of Sydney.

REFERENCES

Butler, D, Cullis, BR, Gilmour, A, Gogel, B (2009) 'ASReml-R Reference Manual.' (Queensland Department of Primary Industries and Fisheries: Brisbane).

Government of Pakistan (2017). Economic Survey of Pakistan, 2016–17. Government of Pakistan, Economic Affairs Division, Ministry of Finance Islamabad.

Khan, B., Iqbal, A. & Riaz, M. (2003) Production and Management of Camels, Pakistan. TM Printers, Al-Rahman Plaza, Aminpur Bazar, Faisalabad.

Sohail, M. (1983) *World Rev. Animal Prod.* **19**: 38.

Proc. Assoc. Advmt. Anim. Breed. Genet. 23:187-190

GENETIC CHARACTERIZATION OF INDIAN INDIGENOUS CATTLE BREEDS

E.M. Strucken[1], M. Swaminathan[2], S. Joshi[2], and J.P. Gibson[1]

[1]School of Environmental and Rural Science, University of New England, Armidale, NSW, 2351 Australia
[2]BAIF Development Research Foundation, Pune, India

SUMMARY

Using the Illumina 770k bovine SNP assay, we genetically characterized 15 Indian indigenous cattle breeds and 1 non-descript indigenous population, which is the largest sample of Indian breeds yet studied. 15.6% of the animals were found to have more than 1% recent *Bos taurus* admixture and were removed or separately analysed. Inbreeding levels for the Indian indigenous breeds, based on F_{IS} and diagonal elements of the GRM, were similar compared to European taurine breeds. We did not find evidence for historical admixture with *Bos taurus* during breed formation. Only 1.4% of the genetic variance in allele frequencies was between breeds, compared to about 42.4% for European taurine breeds. Consequently, Indian breeds can be treated as a single population for some purposes, such as SNP assay design.

INTRODUCTION

Present day India is accepted to be one of the centres of cattle domestication, in particular where *Bos indicus* cattle have developed from its supposed ancestor *Bos primigenius nomadicus* some 100,000-850,000 years ago (e.g. MacHugh *et al.* 1997; Verkaar *et al.* 2004). Furthermore, archaeological evidence suggest that there might have been several centres of domestication within India, as phenotypic differences between cattle from the North and South were already described as early as the Neolithic time period (Naik 1978). Today, the National Bureau of Animal Genetic Resources in India (http://www.nbagr.res.in/nbagr.html) lists 43 registered Indian cattle breeds, however, the large majority of cattle used as milk, draught or dual purpose cattle are raised by smallholders and are of no descriptive breed (e.g. Sharma *et al.* (2015)).

Bos indicus cattle are well adapted to high temperatures and resistance to some prevailing parasites of tropical regions, and have therefore been exported, bred, and adapted in other parts of the world. Zebu cattle are believed to have entered Africa between 3,500 and 700 BCE through present day Egypt (Marshall 1989), and contributed to the formation of African indigenous Sanga and Zenga type breeds (Rege & Tawan 1999). Others, such as the Brazilian Nellore and Guzerat or Australian Brahman and Droughtmaster have been imported to these countries and crossed with other breeds during the last 200 years (Porter *et al.* 2016).

Despite the importance of indicine cattle breeds globally and their wide use especially for crossbreeding with taurine breeds, knowledge of the genetic diversity of the pure *Bos indicus* breeds in India itself is scarce. Many studies focussed on limited numbers of microsatellite or single nucleotide markers, on breeds outside India, or limited sample sizes (e.g. Dash *et al.* 2018; Nayee *et al.* 2018). Here, we have assembled and analysed the largest dataset on Indian indigenous breeds for genetic diversity and relationship, and compared these breeds with taurine and other indicine reference breeds. Lastly, we draw conclusions with regards to the requirement of genomic tools designed specifically for indicine cattle populations.

MATERIALS AND METHODS

Data. A total of 702 Indian indigenous cattle from 15 registered breeds and one non-descript (ND) population were sampled by the BAIF Development Research Foundation (Table 1). All animals were genotyped with the 777k-SNP BovineHD Beadchip (Illumina Inc., San Diego). Genotypes and animals were quality controlled (QC) based on a median GC score >0.6 and a call rate >0.9. Further, animals with more than 1% *Bos taurus* content (based on a preliminary Admixture analysis) were excluded, leaving 588 animals and 716,599 SNPs for subsequent analyses. Reference breeds included 6 *Bos taurus* breeds (each N=20), and 16 *Bos indicus* breeds (N=10 to 20), sourced from the bovine HapMap consortium, Canadian Dairy Network, SRUC, and Decker *et al.* (2014). Reference data were either previously quality controlled or subjected to the same QC criteria as the Indian indigenous breeds. The 770k Illumina assay has close to 300,000 SNPs that are at high minor allele frequency in *Bos indicus* breeds, so that it has much lower ascertainment bias than earlier versions of the 50k assay.

Analyses. Analyses included calculation of Pearson's correlation coefficient between observed (R_{obs}) and expected (R_{exp}) allele frequencies for each breed-pair. These calculations only included SNPs with frequencies 0.05<p<0.95 to reduce bias due to limited numbers of SNPs with small frequencies. R_{exp} was calculated as follows:

$$R_{exp} = V_p/[V_p+V_{e1}+V_{e2}],$$

where V_p is the variance of p in the meta-population (i.e. all Indian indigenous animals or all *Bos taurus* animals), V_{e1} and V_{e2} are the error variances of the estimates of p in the two breeds. V_{e1} and V_{e2} were estimated as the average across all loci of p(1-p)/2n, where n is the number of animals in the given breed and p is the meta-population value of p for each SNP. V_p was not corrected for the sampling error of p, which in all cases was less than 1% of the estimate of V_p. The variance of true SNP allele frequencies in one breed that was explained by the true SNP allele frequencies in another breed was estimated as the ratio of R_{obs}^2/R_{exp}^2.

Principal components were estimated using a GRM based on Van Raden (2008). Further analyses included supervised Admixture models including reference breeds as potential ancestors (Alexander *et al.* 2009). Genetic differentiation between and within breeds were estimated using F_{ST} (Weir & Cockerham 1984), and F_{IS} (Nei 1972), respectively (Table 1).

Table 1. Data information on Indian indigenous breeds and inbreeding levels (F_{IS})

Breed	N	Sampling location	# excluded / reason	F_{IS}
Dangi	68	Maharashtra	3 / taurine>0.01	-0.015 (±0.124)
Gaolao	20	Maharashtra	1 / taurine>0.01	0.022 (±0.226)
Gir	121	Gujarat	3 / taurine>0.01	0.012 (±0.101)
Hallikar	28	Karnataka	1 / taurine>0.01	0.002 (±0.179)
Hariana	17	Haryana	4 / taurine>0.01	0.0002 (±0.262)
Khillar	25	Maharashtra	1 / taurine>0.01	0.015 (±0.2)
Krishna Valley	22	Karnataka	5 / taurine>0.01	0.004 (±0.218)
Red Kandhari	35	Maharashtra		0.008 (±0.168)
Malnad Gidda	19	Karnataka	5 / taurine>0.01	0.001 (±0.274)
Ongole	50	Andhra Pradesh	4 / low call rates	0.028 (±0.153)
Rathi	1	Rajasthan		NA
Red Sindhi	63	Odisha	1 / low call rates 20 / taurine>0.01	-0.034 (±0.154)
Sahiwal	140	Punjab	36 / taurine>0.01	0.015 (±0.108)
Tharparkar	48	Rajasthan	3 / taurine>0.01	-0.024 (±0.144)
Vechur	1	Kerala		NA
Non-descript	43	Maharashtra, Odisha, Uttar Pradesh	27 / taurine>0.01	0.029 (±0.236)

RESULTS AND DISCUSSION

Principal components analysis and Admixture showed a clear separation between *Bos indicus* and *Bos taurus* breeds. The majority of the 109 Indian indigenous animals with more than 1% *Bos taurus* content belonged to Sahiwal, Red Sindhi and ND. Nayee *et al.* (2018) also found some of their Red Sindhi sample to have some taurine admixture. The otherwise tight clustering of the Indian indigenous breeds indicates that the taurine admixture is recent and not, as some sources speculate, a result of crossing *Bos indicus* with *Bos taurus* animals during the early history of breed formation.

Observed allele frequency correlations between Indian indigenous breeds were on average 0.92 (±0.02). In comparison, R_{obs} between the exotic taurine breeds was 0.65 (±0.04). The estimated proportion of variance of true allele frequency explained by the true frequency in another breed was on average 0.986 in the Indian indigenous breeds; i.e. most or all of the genetic variance at the SNP level is within breeds. The estimated proportion of variance that is within-breeds for the *Bos taurus* breeds was 0.576. These results suggest that, in contrast to *Bos taurus* breeds, Indian indigenous breeds can be treated as a single population for some purposes, such as SNP assay design.

Figure 1a) shows the estimated breed proportions of the Indian indigenous breeds based on the indicine reference breeds as a heatmap. Red Sindhi and Gir were both best represented by the Red Sindhi and Gir reference breeds. Hallikar and Khillar showed a strong Ongole signal, whilst Ongole were best represented by the Nelore reference, which confirms the connection that Brazilian Nelore were bred from imported Indian Ongole (Porter *et al.* 2016). Tharparkar were, however, not best represented by the Tharparkar reference but by Kankraj and Dhanni; and Sahiwal were represented as an admixture of Tharparkar, Sahiwal and Hissar, which stands in contrast to Nayee *et al.* (2018) and Gajjar *et al.* (2018) who reported their Sahiwal sample to have the least evidence for admixture. These and other analyses indicate that it is difficult to trace history and relationships among Indian indigenous breeds which is not unexpected given the low level of between-breed variation estimated for these populations.

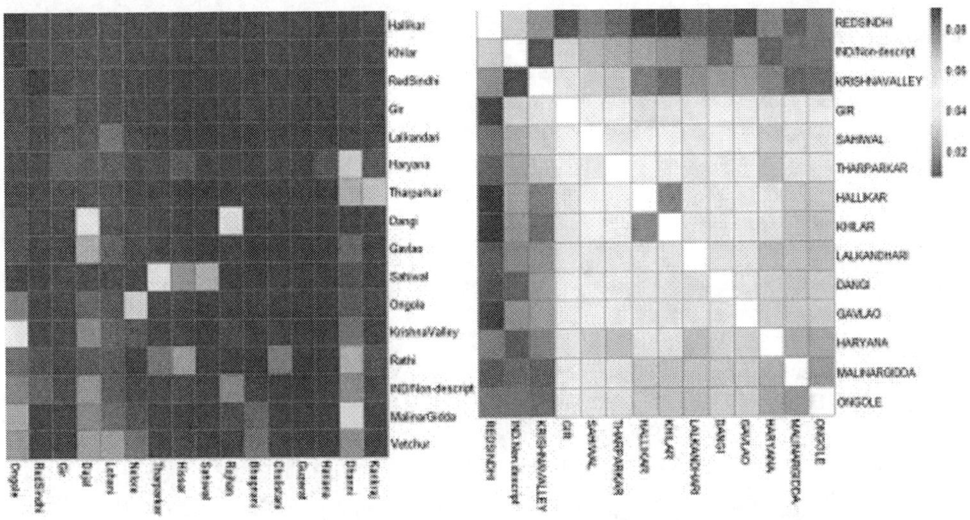

Figure 1. a) Heatmap of estimated breed proportions for 16 Indian indigenous populations (vertical) from a supervised Admixture analyses with 16 indicine reference populations (horizontal); b) heatmap of pair-wise F_{ST} values between 14 Indian indigenous populations

Figure 1b) shows pairwise F_{ST} values as a heat map with a phylogenetic tree based on hierarchical clustering, when only the Indian indigenous breeds were considered. This clearly shows Red Sindhi as an outgroup to the other indigenous breeds, whilst the ND followed by Krishna Valley are the least genetically distinct groups. The genetic distinction of Red Sindhi might reflect their sampling from a single central breeding farm. However, Nayee *et al.* (2018) also found Red Sindhi to be genetically different from other Indian indigenous breeds.

Levels of inbreeding as measured by F_{IS} are similar (-0.034 to 0.029) compared to the taurine reference breeds (-0.026 to 0.023). Studies based on microsatellite data found increased F_{IS} values (e.g. Sharma et al. 2015). Whilst exact F_{IS} values cannot be directly compared between these studies, we can confirm that higher inbreeding levels were found for Gaolao and Ongole and comparatively lower values for Hariana (Table 1).

CONCLUSION

Indian indigenous breeds show remarkably little between-breed variation, and therefore can be treated as a single population when developing genomic tools such as SNP assays.

ACKNOWLEDGEMENT

We acknowledge the Bill and Melinda Gates Foundation grant OPP1112185.

REFERENCES

Alexander D.H., Novembre J. and Lange K. (2009) *Genome Res.* **19**: 1655.

Danecek P., Auton A., Abecasis G., Albers C.A., Banks E., DePristo M.A., Handsaker R.E., Lunter G., Marth G.T., Sherry S.T., McVean G. and Durbin R. (2011) *Bioinfor.* **27**: 2156.

Dash S., Singh A., Bhatia A.K., Jayakumar S., Sharma A., Singh S., Ganguly I. and Dixit S.P. (2018) *Anim. Biotechnol.* **29**: 129.

Decker J.E., McKay S.D., Rolf M.M., Kim J., Molina Alcala A. and Sonstegard T.S. (2014) *PLoS Genet.* **10**.

Gajjar S.G., Guldbrandtsen B., Nayee N., Sudhakar A., Trivedi K., Lund M.S. and Sahana G. (2018) *Proc. World Cong. Genet. App.Livest.Prod. (WCGALP)*, Auckland.

Hill W.G. and Robertson A. (1968) *Theor. Appl. Genet.* **38**: 226.

MacHugh D.E., Shriver M.D., Loftus R.T., Cunningham P. and Bradley D.G. (1997) *Genet.* **146**: 1071.

Marshall F. (1989) *Current Anthropology* **30**: 235.

Naik S.N. (1978) *J. Hum. Evol.* **7**: 23.

Nayee N., Sahana G., Gajjar S., Sudhakar A., Trivedi K., Lund M.S. and Guldbrandtsen B. (2018) *J. Anim. Breed. Genet.* **135**: 432.

Nei M. (1972) *The Am. Naturalist* **106**: 283.

Porter V., Alderson L., Hall S.J.G. and Sponenberg D.P. (2016) 'Mason's World Encyclopedia of Livestock Breeds and Breeding', 2 Volume Pack. CABI.

Rege J.E.O. and Tawan C.L. (1999) *Anim. Genet. Res. Information Bulletin* **26**: 1.

Sharma R., Kishore A., Mukesh M., Ahlawat S., Maitra A., Pandey A.K. and Tantia M.S. (2015) *BMC Genet.* **16**: 73.

Sved J.A. (1971) *Theor. Popul. Biol.* **2**: 125.

Van Raden P.M. (2008) *J. Dairy Sci.* **91**: 4414.

Verkaar E.L., Nijman I.J., Beeke M., Hanekamp E. and Lenstra J.A. (2004) *Mol. Biol. Evol.* **21**: 1165.

Weir B.S. and Cockerham C.C. (1984) *Evol.* **38**: 1358.

Proc. Assoc. Advmt. Anim. Breed. Genet. 23:191-194

MORPHOMETRIC DIFFERENTIATION OF SELECTED INDIGENOUS CATTLE BREEDS IN NIGERIA

A.D. Oladepo[1], A.E. Salako[2], A.A. Adeoye[3] and O.A. Adeniyi[4]

[1]Department of Agricultural Science, Adeyemi College of Education, Ondo, Nigeria
[2]Department of Animal science, University of Ibadan, Ibadan, Nigeria
[3]Department of Animal Production and Health, Ondo State University of Science and Technology, Okitipupa, Ondo State, Nigeria
[4]Department of Animal Health and Production, Technology, Oyo State College of Agriculture and Technology, Igboora, Oyo State, Nigeria

SUMMARY

In Nigeria, population characteristics of selected indigenous cattle breeds have not been fully documented. Therefore, morphometric attributes of selected indigenous cattle breeds in Nigeria were assessed. Three hundred and one White Fulani (WF), 242 Red Bororo (RB), 247 Sokoto Gudali (SG), 233 Bornu Kuri (BK) and 184 Muturu cattle in the age group of 3 to $3^1/_2$ years were purposively sampled. Eighteen morphometric parameters were measured using standard procedures. Data were analysed using descriptive statistics, ANOVA at $\alpha_{0.05,}$ cluster analysis and Euclidean genetic distance. Differences existed in face length, rear leg length; wither height and rump height among the breeds.

INTRODUCTION

One of the ways of characterizing livestock breeds is to evaluate their morphostructural characteristics and determine genetic distance among contemporary populations (Metta *et al.* 2002). The initial step in characterization is identification of distinct populations using information on their geographic and ecological isolation, traditional nomenclature (traditionally, recognized populations), phenotypic distinctness and level of genetic differentiation among the population (Gizaw *et al.* 2011). Indigenous cattle breeds are important to preserve as they are well adapted to local climates, food supply and other local environmental factors, which often shows in their robustness and hardiness. Indigenous livestock resources are also strategic in the socio-economics of rural agricultural systems to ensure food security in developing countries. The selected cattle breeds are found throughout Nigeria but are most common in the northern part the country.

The wither height and rump height among these breeds differentiated them. This study is designed to unveil the phenotypic and genetic diversity among selected Nigeria indigenous cattle breeds using primary data obtained from field morphological survey to assess diversity of the selected indigenous populations in order to update published variations as well as document genetic distances between the populations.

MATERIALS AND METHODS

Data Collection. Animals were sampled from four different isolated areas where they were abundant in Nigeria. Each location was divided into clusters of ten units for easy measurements and adult animals that were within the age bracket of 3 to $3^1/_2$ years were sampled. A total of 1,207 indigenous cattle comprising 301 White Fulani 247 Sokoto Gudali, 242 Red Bororo, 233 Bornu Kuri and 184 Muturu were sampled.

Eighteen linear body measurements (cm) were taken on each sampled animal with the use of a measuring tape (Table 1). Quantitative variables measured in this study were adapted from the standard cattle descriptor list (FAO 2002) and a final list of variables were developed and used. Documented morphological features described by (Hall 1991) were used as base line markers to ascribe sampled animals to a breed. Individuals that did not strictly conform to primary breed characters; visibly pregnant, sick and castrated animals were excluded.

Statistical Analysis. Data collected were subjected to Generalized Linear Model Analysis of Variance (ANOVA) procedure of the Statistical Analysis System (SAS 2002) and cluster analysis of Palenotological Statistics (PAST). East Square Means (x) and Standard Error (SE) associated with each linear body measurements were estimated. Genetic distances among the five breeds based on their actual morphometric variables measured were calculated using Euclidean genetic distance measure. The statistical model used for analyzing quantitative phenotypic variations among the breed populations was as follows:

$$Y_i = \mu + B_i + e_i$$

where Y_i are the observed body measurements; μ is the overall mean, B_i is the fixed effect of breed ($i = 1,2,..5$) and e_i is the standard error.

RESULTS AND DISCUSSION

Mean values of morphological variables and their SE are depicted in Table 1 for aggregated gender. Pairwise mean comparison showed significant differences for most of the morphological variables among the breeds. Morphological variables such as wither height, rump height, body length, and tail length were significantly higher ($P<0.05$) for White Fulani and Sokoto Gudali as compared with any other breeds considered in this study. Muturu breeds were significantly ($P<0.05$) larger for face length and head width than other four breeds.

The genetic distance among the cattle population ranged from 43.77 to 145.52 (Table 2). The longest genetic distance was observed between Bornu kuri and Muturu while the shortest distance was observed between Red Bororo and Bornu Kuri. The cluster analysis generated showed two main clusters having Muturu in a cluster and White Fulani, Sokoto Gudali, Bornu Kuri and Red Bororo in the other.

The significantly higher values for most of the morphometric measurements of White Fulani, Sokoto Gudali, Red Bororo and Bornu Kuri that were predominantly in the northern part of the country than Muturu that is predominantly in southern part of the country seem to be an adaptation where tallness (wither height) and large body size (heart girth and body length) are suitable for trekking long distances to water and grazing points and this is in agreement with what Nwacharo (2006) and Hall (1991) reported. These observations could be as a result of genetic and species differences. The differences in body measurements of the five cattle breeds with respect to some morphological variables indicates that the five cattle breeds were sub-divided into distinct populations perhaps due to differences in availability of feed resources, breeding practices used and inherent genetic differences (Nwambene *et al.* 2012).

Table 1. Means of body measurements (±SEM) amongst the five cattle breeds

	WF±SEM	SG±SEM	RB±SEM	BK±SEM	MU±SEM
FL	49.37±0.44[b]	47.51±0.36[c]	40.35±0.33[c]	45.44±0.57[d]	54.03±0.39[a]
HW	22.62±0.16[c]	22.57±0.15[c]	21.21±0.14[d]	24.11±0.18[b]	28.53±0.68[a]
HL	45.16±0.84[a]	4.56±0.44[d]	38.05±0.32[b]	38.39 ±0.57[b]	8.30±0.35[c]
EL	23.86±0.52[a]	24.17±0.65[a]	24.07±0.52[a]	24.17±0.56[a]	14.61±0.24[b]
NL	54.56±0.58[a]	51.34±0.41[b]	38.97±0.40[c]	38.79±0.51[c]	25.40 ±1.09[d]
NC	54.71±0.51[c]	59.08±0.51[b]	76.85±0.32[a]	77.75 ±0.81[a]	60.93±0.18[b]
WH	130.39±0.21[a]	129.21±0.23[a]	125.65±0.11[b]	125.18±0.15[b]	82.41±0.07[c]
FLL	78.39±0.45[b]	80.39 ±0.45[b]	85.96±0.41[a]	87.59±0.43[a]	56.71±0.25[c]
HG	83.20[c] ±0.18[c]	95.92±0.28[b]	144.41±0.12[a]	144.40±1.12[a]	57.03±0.43[d]
BL	116.45±0.11[a]	116.03±0.08[a]	111.17 ±0.06[b]	112.64±0.41[b]	73.55±0.21[c]
RH	135.40±0.17[a]	134.22±0.21[a]	130.59 ±0.09[b]	131.53±0.50[b]	88.46±0.54[c]
RL	30.56±1.19[a]	30.58±0.26[a]	31.03±0.75[a]	30.97±0.21[a]	20.89±0.41[b]
RLL	105.45±0.21[a]	105.41±1.01[a]	104.64 ±0.19[a]	105.06±1.17[a]	73.93±0.63[b]
TL	96.10 ±0.49[a]	94.98±0.36[a]	91.89 ±0.17[b]	91.30±0.14[b]	55.98±0.41[c]
SC	25.06±0.43[a]	24.75 ±0.39[ab]	24.66 ±0.22[ab]	24.40±0.08[b]	21.86±0.14[c]
HoL	50.90±1.11[a]	50.05±0.71[b]	50.07 ±0.47[b]	49.41 ±0.43[b]	32.74±0.35[c]
RW	35.19±0.56[a]	35.38±0.42[a]	34.45 ±0.41[b]	35.18±0.20[a]	28.32±0.24[c]
SW	68.02±0.52[b]	69.59±0.34[b]	80.89 ±0.46[a]	81.85±0.11[a]	63.58±0.22[c]

Means with same superscript are not significantly different (P>0.05). SEM = Standard Error Mean
WF =White Fulani, SG = Sokoto Gudali, RB = Red Bororo, BK = Bornu Kuri, MU = Muturu
FL = Face Length, Head Width = HW, Horn Length = HL, Ear Length = ER, Neck Length =NL, Neck Circumference = NC, Wither Height = WH, Foreleg Length = FLL, Hearth Girth = HG, Body Length = BL, Rump Height = RH, Rump Length = RL, Rearleg Length = RLL, Tail Length =TL, Shin Circumference = SC, Hock Length = HoL, Rump Width = RW, Shoulder Width = SW

Table 2. Euclidean genetic distance estimate based on actual measured morphometric variables

	White Fulani	Sokoto Gudali	Red Bororo	Bornu Kuri	Muturu
Sokoto Gudali	54.36	-			
Red Bororo	70.44	65.63	-		
Bornu Kuri	54.54	54.54	43.77	-	
Muturu	114.87	83.86	136	145.52	-

A shorter genetic distance obtained between White Fulani and Sokoto Gudali suggests a close genetic relationship between the breeds while the longer genetic distance was observed between White Fulani and Muturu is an indication that an appreciable heterosis especially with regard to most body measurements which are of economic importance can be obtained by crossing any of the two breeds. The phylogenetic tree separated the five cattle breeds into two main clusters. In addition, the close genetic relationship between the breeds may also be attributed to similarity in ecological zones and production systems as well as the incidents of cross border livestock rustling contributing to the migration and movement of livestock and subsequent interbreeding between such livestock, this in agreement with had been reported by (Nwacharo *et al.* 2006). There were clear disparities in the wither height and body length of the five breeds. White Fulani was found superior to any of the other four breeds studied. Genetic distance based on morphological indices among the breeds as revealed by the cluster analysis showed that the breeds were genetically distinct.

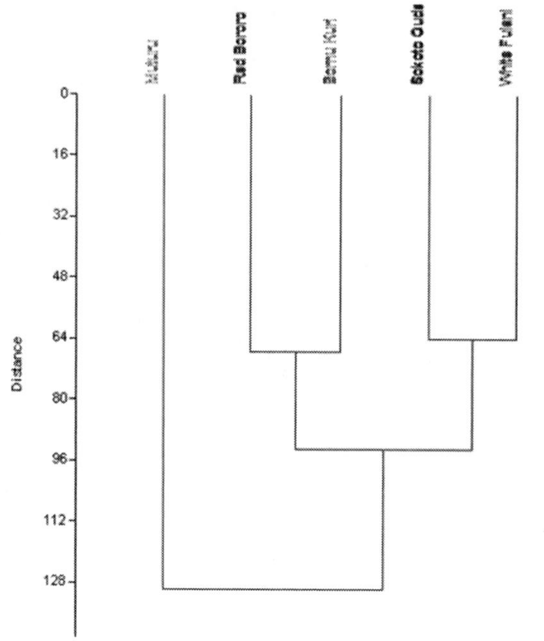

Figure 1. Dendogram showing genetic relationship among five indigenous cattle from Nigeria

CONCLUSIONS

Morphometric attributes are very good tools in differentiating cattle on the basis breeds. The closer genetic relationship among White Fulani, Sokoto Gudali, Red Bororo and Bornu Kuri five cattle breeds may be attributed to possible interbreeding among these populations that were predominantly abundant in the northern part of the country forming homogeneous population separated by no physical geographic boundaries.

REFERENCES

FAO. (2012). FAO Animal production and Health Guidelines No. 11, Rome.
Gizaw S., Komen H., Hanotte O., Van Arendonk J.A.P: Kemp S., Aynalem H. Mwai Ol and Dessie, T. (2011) ILRI Research Report No. 27. Nairobi, Kenya. ILRI.
Hall S.J.G. (1991) *Anim. Prod. (UK)* **53**: 61.
Metta M., Kanginakudru S., Gudiseva N. and Nagaraju, J. (2004) *BMC Genetics* **5**:16.
Mwacharo J.M., Okeyo A.M., Kamande G.K. and Rege J.E.O. (2006) *Trop. Anim. Health Prod.* **38**: 65.
Mwambene P.L., Katule A.M. Chenyambuga S.W. and Mwakilembe P.A.A. (2012) *AGRI.* **51**: 51.
SAS. (2002). User's guide: Statistic. SAS institute Inc. Cary, North Carolina, USA.

Proc. Assoc. Advmt. Anim. Breed. Genet. 23:195-205

DEVELOPING GENOMIC STRATEGIES FOR THE LIVESTOCK INDUSTRIES: ALL IMPLEMENTATIONS ARE CHALLENGING

D.A.L. Lourenco, S. Tsuruta, Y. Masuda and I. Misztal

Department of Animal and Dairy Science, University of Georgia, Athens, GA, 30602 USA

SUMMARY

Genomic selection (GS) using dense SNP panels was first implemented in 2009 for dairy cattle. Since then, GS has been extended to other livestock. However, different problems and challenges are always encountered during the implementation of GS in each population. In this paper, we show the issues in the implementation of GS and how they have been successfully solved in beef and dairy cattle, pigs, chicken, and fish. We also discuss changes in current methods and development of new algorithms to deal with large genomic data. Overall, complications for GS include, but are not limited to, selective genotyping, computing limitations, convergence problems especially for complex models, compatibility between pedigree and genomic information, among others.

BACKGROUND

Soller and Beckmann (1983) hypothesized, in the early 1980's, that DNA markers could be useful in constructing more precise genetic relationships, detecting causative variants, and determining parentage. After the first draft of the human genomic project was published in 2001 (The International SNP Map Working Group 2001), single nucleotide polymorphism (SNP) became the most important source of genome sequence variation, and therefore, the most important DNA marker. Concurrently, Meuwissen *et al.* (2001) anticipated that genomic information could help animal breeders to generate more accurate breeding values if a dense SNP assay that covered the entire genome could be constructed. It took almost 8 years for the first dense SNP assay to become available, and this was for dairy cattle (Matukumalli *et al.* 2009).

In January of 2009, researchers from AGIL-USDA released the first official genomic evaluation for Holstein and Jersey in the USA. This implementation brought a lot of excitement, especially because the top bull in the evaluation had no daughters with milking records, meaning his genetic merit was computed based on pedigree and genomic information. The superiority of this very bull was later confirmed when his progeny records became available. This endorsed the hypothesis that Meuwissen *et al.* (2001) had tested based on simulated data: the genetic merit of young animals can be computed with high accuracy if they are genotyped and SNP effects are available from a reference population.

With the release of genomic predictions based on dense SNP assays for Holsteins in the USA, the race for the implementation of genomic selection (GS) in livestock became official. Essentially, two main methods for genomic evaluation were developed: multi-step and single-step. The multi-step method was the first to be implemented (VanRaden 2008). The main advantage of the multi-step approach is that the traditional BLUP evaluation is kept unchanged and GS can be carried out by using additional analyses; however, only genotyped animals have genomic EBV (GEBV). As a result, several adjustments were proposed, especially in dairy cattle, to make EBV for non-genotyped animals comparable to GEBV under multi-step evaluations (Wiggans *et al.* 2011; Wiggans *et al.* 2012).

Intending to solve incompatibility problems and to reduce the burden in obtaining genomic predictions when only a fraction of animals is genotyped, Misztal *et al.* (2009) and Legarra *et al.* (2009) proposed a method that combines phenotypes, pedigree, and genotypes into a single evaluation. This method is called single-step genomic BLUP (ssGBLUP) and replaces the pedigree relationship matrix in the traditional BLUP by a realized relationship matrix (**H**), which combines pedigree and genomic relationships. Another class of single-step was also proposed by Fernando *et al.* (2014), which is based on a marker effect model and is called single-step Bayesian regression (ssBR). Under the same assumptions (e.g., all SNP have non-zero effect and constant variance), ssGBLUP and ssBR are equivalent models (Gao *et al.* 2018).

Over the past 4 or 5 years, ssGBLUP has become the preferred method for genomic evaluation in several species, namely beef cattle (Lourenco *et al.* 2015a), dairy cattle (Vukasinovic *et al.* 2017), pigs (Forni *et al.* 2011; Lourenco *et al.* 2016), broilers (Chen *et al.* 2011; Lourenco *et al.* 2015b), layers (Yan *et al.* 2018), dairy sheep and goat (Rupp *et al.* 2016), Australian sheep (Brown *et al.* 2018), and fish (Garcia *et al.* 2018). Possibly, in the near future, the great majority of genomic evaluations will all be based on single-step methods.

Although the idea and theory behind ssGBLUP are easily understandable, and the method seems to be simple because it just requires the change in the relationship matrix, its implementation for official genomic evaluations is quite challenging and demands several data-dependent adjustments. It is worthwhile to remember that even a small change to the genetic evaluation system can create issues. For example, a simple change in variance components can cause convergence problems and changes in scaling. Usually, the issues and challenges encountered during the change from traditional or multi-step evaluations to single-step are not disclosed. However, showing issues and strategies to solve them can help troubleshooting future implementations. In this paper, we show the problems in the implementation of GS and how they have been successfully solved in beef and dairy cattle, pigs, chicken, and fish.

GENOMIC STRATEGIES

Beef cattle. In 2009, Angus Genetics Inc. started to run multi-step genomic evaluations for American Angus Association (AAA) using a correlated approach described by Kachman (2008). In this approach, the trait phenotype and the direct genomic values (DGV) calculated based on SNP effects are used as phenotypic information in a 2-trait model, where heritability for DGV is assumed to be 0.99. Restricted maximum likelihood (REML) estimates are then obtained. Genetic correlations between each trait and DGV reflect accuracy of DGV, and solutions for the first trait are genomic-enhanced EBV. The drawback of this method is that it doubles the number of traits in the model. Additionally, genetic correlations between each trait and DGV can be overestimated, indicating the genomic information is explaining more of the genetic variance than expected, which can inflate predictions. Figure 1 shows genetic trends for marbling using traditional BLUP and the multi-step correlated approach. Inflated genetic trends for multistep predictions, together with big fluctuations for predictions every time SNP effects were recalculated (i.e., during calibration) and re-ranking of high accuracy bulls in subsequent evaluations urged Angus Genetics Inc. to find another method for their genomic evaluation.

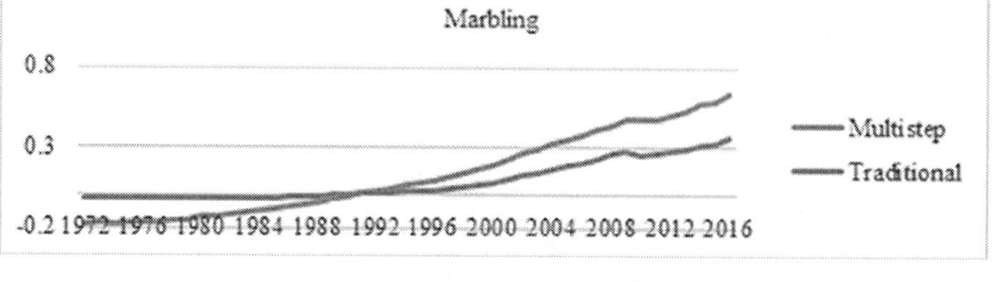

Figure 1. Genetic trends for marbling using traditional BLUP and the multi-step correlated approach in American Angus

Scaling factors and inbreeding. In 2014, we started testing ssGBLUP for growth and calving ease (CE) models in the American Angus population. Several datasets were used, but the first one was for growth traits and calving ease and comprised 8 million animals in the pedigree, along with 6 million birth (BW) and weaning weight (WW) records, 3.4 million records for post-weaning gain (PWG), 1.3M CE records, and genotypes for 52k animals. The first issue observed was the inflation of GEBV in a validation test. When adjusted phenotypes for animals considered young in 2013, but with phenotypes in 2014, were regressed on GEBV, regression coefficients were lower than 1. To solve this problem scaling parameters can be used in $\mathbf{H}^{-1}$ (Aguilar *et al.* 2010; Christensen and Lund 2010; Tsuruta *et al.* 2011):

$$\mathbf{H}^{-1} = \mathbf{A}^{-1} + \begin{bmatrix} 0 & 0 \\ 0 & \tau\mathbf{G}^{-1} - \omega\mathbf{A}_{22}^{-1} \end{bmatrix}$$

where $\mathbf{A}^{-1}$ is the inverse of the pedigree relationship matrix, $\mathbf{G}^{-1}$ is the inverse of the genomic relationship matrix, and $\mathbf{A}_{22}^{-1}$ is the inverse of the pedigree relationship matrix among genotyped animals; τ and ω were used to rescale the amount of information in $\mathbf{G}^{-1}$ and $\mathbf{A}_{22}^{-1}$, respectively. Primarily, ω controls inflation due to incompleteness of pedigree while τ controls additive genetic variance. Based on validation, we found the best combination of τ and ω for this data was 1.0 and 0.7. However, scaling parameters are completely *ad-hoc* and should be avoided generally. In the original implementation of ssGBLUP (Aguilar *et al.* 2010) inbreeding for $\mathbf{G}^{-1}$ and $\mathbf{A}_{22}^{-1}$ was considered, but not for $\mathbf{A}^{-1}$. After inbreeding was included in the computation of $\mathbf{A}^{-1}$, ***ω lower than 1*** was not needed anymore for this beef cattle data. Figure 2 shows coefficients for the regression of adjusted phenotypes on GEBV for BW, WW, PWG when inbreeding for $\mathbf{A}^{-1}$ is considered and ω is lower than 1. It is clear that GEBV are deflated if ω is lower than 1, proving that inbreeding in $\mathbf{A}^{-1}$ is enough to avoid inflation in this AAA dataset. Therefore, current official AAA evaluations use $\tau=\omega=1$.

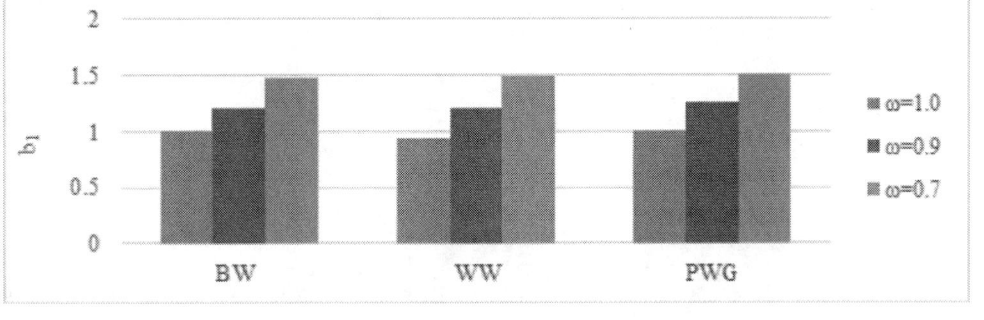

Figure 2. Regression coefficients (b_1) for birth weight (BW), weaning weight (WW) and post-weaning gain (PWG) with varying ω

Selective genotyping. Regarding the advantages of using genomic information, the average gain in predictive ability for growth traits, when moving from traditional BLUP to ssGBLUP, was 24%. In contrast, the gain in prediction accuracy for CE was only 8%, going from 0.12 to 0.13. This small increase in predictive ability is possibly because animals with difficult calving are unlikely to be retained for breeding and therefore would not be genotyped on a regular basis. In fact, only 0.35% of the animals with difficult calving were genotyped. Therefore, selective genotyping can compromise the gains that can be obtained with genomics and can also introduce some pre-selection bias into the evaluations. Similar problems occur for traits related to survival or other specific phenotypes (e.g., animals with undesirable phenotypes or dead). This means some traits might need a controlled reference set of animals, rather than relying on Industry genotyping strategies which are influenced by the perceived value of animals.

External information. For the growth model, another issue was related to the inclusion of external information in ssGBLUP. For traditional evaluations, the external EBV from Red Angus is used as prior information in the right hand side of the mixed model equations (MME), and the reliability is added to the pedigree relationships among external animals in the left hand side of the MME (Legarra *et al.* 2007). We changed the computing algorithm to support both genomic and external information, and the implementation of a genomic multi-breed model increased the computing time only by 2.5 hours, compared to a genomic single-breed model.

Many more genotyped animals. On a weekly basis, more than 2k genotyped animals are added to the AAA database. From July of 2014 to March of 2019, the number of genotyped animals increased from 82k to 627k. The most computationally expensive operation in ssGBLUP is the inversion of **G** and $\mathbf{A}_{22}$. This operation has an approximately cubic cost with the number of genotyped animals. With efficient computing algorithms, matrix inversions are feasible for up to 150,000 genotyped animals.

To overcome the limitation set by the number of genotyped animals in ssGBLUP, Misztal *et al.* (2014) proposed the algorithm for proven and young (APY) animals to construct $\mathbf{G}^{-1}$ without having to explicitly invert **G**. The logic behind the construction of $\mathbf{G}^{-1}_{APY}$ is that the genotyped animals are split into core (c) and noncore (n), and breeding values for noncore animals ($\mathbf{u}_n$) are functions of breeding values of core animals ($\mathbf{u}_c$):

$$\mathbf{u}_n = \mathbf{P}_{nc}\mathbf{u}_c + \mathbf{\Psi}_n$$

where $\mathbf{P}_{nc}$ is a matrix that relates breeding values for noncore to core animals, and $\mathbf{\Psi}_n$ is a diagonal matrix with estimation errors. The $\mathbf{G}^{-1}_{APY}$ can be constructed as:

$$\mathbf{G}^{-1}_{APY} = \begin{bmatrix} \mathbf{G}^{-1}_{cc} & 0 \\ 0 & 0 \end{bmatrix} + \begin{bmatrix} -\mathbf{G}^{-1}_{cc}\mathbf{G}_{cn} \\ \mathrm{I} \end{bmatrix} \mathbf{M}^{-1}_{nn} [-\mathbf{G}_{nc}\mathbf{G}^{-1}_{cc} \quad \mathbf{I}]\Psi$$

with $m_{nn_{ii}} = g_{ii} - g_{ic}G^{-1}_{cc}g_{ci}$. The APY algorithm creates a generalized sparse inverse of **G** at approximately a linear cost in computing and storage. However, if $\mathbf{G}^{-1}_{APY}$ is efficiently computed but A^{-1}_{22} is not, ssGBLUP still cannot be used for over 150,000 genotyped animals. To avoid explicit inversion of A_{22}, Masuda *et al.* (2017) proposed to compute an efficient inverse indirectly as a product of sparse matrices:

$$\mathbf{A}^{-1}_{22} = \mathbf{A}^{22} - \mathbf{A}^{21}(\mathbf{A}^{11})^{-1}\mathbf{A}^{12}$$

where A^{11}, A^{21}, and A^{22} are portions of A^{-1} for non-genotyped, between genotyped and non-genotyped, and for genotyped animals, respectively.

Without APY, ssGBLUP would not be feasible for the AAA evaluations. However, identifying core animals was not an easy task at the beginning of the implementation of ssGBLUP for AAA. Choosing core animals randomly or based on EBV accuracy resulted in correlations >0.99 between GEBV from regular and APY ssGBLUP, providing the core group had a minimum of 10k animals. Less optimal core definitions caused convergence issues. We ultimately chose to select core animals based on EBV accuracy for the official evaluations because those animals would have more progeny recorded. No differences were found in convergence and computing time for the growth model However, for the carcass model, which combines 9 different traits, some of them are sparsely recorded (e.g., fat and ribeye area), using a random core instead almost halved the number of rounds to convergence. Since 2017, Angus Genetics Inc. has been using APY ssGBLUP for weekly evaluations of around 18 traits, using a single set of core animals assigned based on EBV accuracy.

Computing SNP effects in ssGBLUP. Besides the official national Angus evaluation, genomic predictions or direct genomic values (DGV) based on SNP effects are provided for non-registered animals, usually females, for herd management. Although ssGBLUP provides GEBV as final output, estimates of SNP effects (**a**) can be obtained by back-solving GEBV (Wang *et al.* 2012):

$$\hat{\mathbf{a}} = \mathrm{k}\mathbf{Z}'\mathbf{G}^{-1}\hat{\mathbf{u}}$$

where: k is the ratio of SNP to additive genetic variance, **Z** is a centred matrix of SNP effects, and **u** is a vector of GEBV. As the calculation of SNP effects is done by standalone software from the BLUPF90 family (postGSf90), there is a need to save $\mathbf{G}^{-1}_{\mathrm{APY}}$ to disk. However, even half stored requirements were large. To overcome this problem, we investigated the use of the subset of $\mathbf{G}^{-1}_{\mathrm{APY}}$ only for core animals ($\mathbf{G}^{-1}_{\mathrm{CC}}$). Correlations between GEBV and DGV obtained with $\mathbf{G}^{-1}_{\mathrm{APY}}$ or $\mathbf{G}^{-1}_{\mathrm{CC}}$ were greater than 0.98 using the 2014 dataset. However, as the number of genotyped animals increased, we observed a decrease in correlation when $\mathbf{G}^{-1}_{\mathrm{CC}}$ was used. Based on that, we changed the algorithm in postGSf90 to work with blocks of $\mathbf{G}^{-1}_{\mathrm{APY}}$ instead of having to allocate the full matrix in memory. The new software requires less memory and is extremely fast.

Accuracy as a measure of GEBV risk. One of the benefits of using genomic information is to increase breeding value accuracy. Accuracies are calculated based on prediction error variance (PEV) and can be obtained from the inverse of the LHS of MME. If the number of animals in the pedigree is large, the inverse is not computationally feasible and an approximation has to be used. Approximating accuracy of GEBV requires the calculation of the combined contributions due to phenotypes, pedigree, and genomic information. An algorithm to approximate genomic contributions was developed based on diagonals of **G** and the average traditional accuracy for genotyped animals. Compared to the approximation based on pedigree and phenotypes only, the increase in computing time was irrelevant. Another advantage of this algorithm is that the diagonal of **G** or $\mathbf{G}_{\mathrm{APY}}$ can be easily saved and requires a small disk space. Correlations between accuracy from the new algorithm and true accuracy from PEV were higher than 0.85 for growth traits, using a sample dataset.

Dairy cattle. Single-step GBLUP is currently used by Zoetis for genomic evaluations of wellness traits in dairy cattle. Those traits have a binary response and each one is currently analysed separately using a univariate threshold model. Heritabilities are low, ranging from 0.06 to 0.08, and trait incidences vary from 2% to 25% (Vukasinovic *et al.* 2017). As for beef cattle, changes had to be made to accommodate an increasing number of genotyped animals. Although all models were single-trait, the time to convergence with APY core animals selected based on their accuracy of EBVs, was between 24 and 50 hours, which is un-acceptable. When the core animals were randomly selected, the computing time was between 4 and 10 hours. As was previously observed for carcase traits in beef cattle, the choice of core animals becomes an issue when few genotyped animals have phenotypes. Probably, $\mathbf{G}^{-1}_{\mathrm{APY}}$ with random core is better conditioned than with high EBV accuracy core.

Although official genomic evaluations in the US are still done with multi-step methods, several tests have been done by our group to investigate the feasibility of ssGBLUP for dairy cattle using data provided by the US Holstein Association and the Council on Dairy Cattle Breeding (CDCB). A common problem is the inflation of GEBV. Although using inbreeding for the calculation of $\mathbf{A}^{-1}$ eliminated inflation in the beef cattle evaluation, the same was not true for dairy evaluations. This is because the missingness of pedigree is greater in dairy cattle data. In the initial tests before implementing inbreeding, convergence could only be reached with $\omega<1$.

In BLUP-based methods, missing parents can be modelled by unknown parent groups (UPG). Such groups are also known as phantom parents or genetic groups, and are used to represent the average level of breeding value in a group where parents were missing. In ssGBLUP, when UPG are applied only to **A**, convergence may fail or the convergence rate can be slow. Alternatively, UPG can be assigned to both **A** and $\mathbf{A}_{22}$. For US Holstein with 18 type traits, using 10M animals in the pedigree and 570k genotyped, we observed that adding UPG for $\mathbf{A}_{22}$ helped to reduce inflation. However, the least GEBV inflation for young genotyped bulls was observed when inbreeding for UPG was also considered and the genetic variance was halved (Figure 3). The problem of reducing additive genetic variance is the shrinkage of GEBV for all animals, not only for young genotyped bulls. According to VanRaden *et al.* (2014), a reduction in genetic variance for yield traits reduced inflation caused by the inclusion of female genotypes.

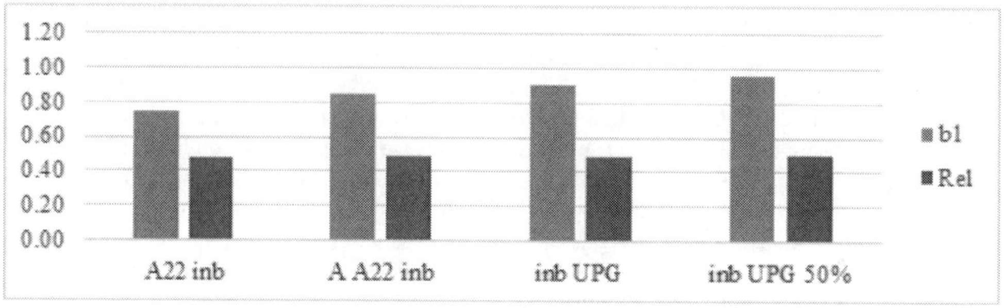

Figure 3. Average regression coefficient (b1) and reliability (Rel) for 18 type traits in US Holsteins A22 inb = inbreeding in $\mathbf{A}_{22}^{-1}$ + UPG; A A22 inb = inbreeding in $\mathbf{A}^{-1}$ and $\mathbf{A}_{22}^{-1}$ + UPG; inb UPG = inbreeding for UPG; Inb UPG 50% = inb UPG with 50% reduction of additive genetic variance

The US dairy industry has collected almost 3M Holstein genotypes by April 2019 (https://queries.uscdcb.com/Genotype/counts.html). Only 11% of those are for males and over 75% of the females will never have phenotypic records. Initial ssGBLUP tests using 2.3M genotyped animals, 13.5M animals in the pedigree, and 11M records on 18 type traits took 3 days to converge and required over 300 GB of memory. This was using APY with 15k randomly chosen core animals. Currently, the ssGBLUP software used to solve large systems of equations is undergoing changes for the implementation of a message-passing interface (MPI), which uses multi-processor architecture, allowing a higher level of parallelization. After these changes are completed, the convergence for the 18 type trait model is expected to be reached within 36 hours using about 30 GB of memory.

Pigs. The biggest challenge for genomic evaluations in pigs is to have accurate predictions for multi-breed or crossbred populations. In within breed ssGBLUP, **G** is constructed based on the average allele frequency. However, different breeds may have different allele frequencies, and construction of **G** must be modified. Using 2 breeds and their F1, we observed negative genomic relationships between breeds, which is an indicator of distinct allele frequencies. Breed-specific allele frequencies were subsequently used to centre and scale **G** in simulated and real pig datasets (Lourenco *et al.* 2016). Although the average relationship between the 2 breeds was zero when using breed-specific allele frequencies, accuracy of GEBV was similar to the default scenario that used across-breed allele frequencies to construct **G**. If there is a dominant breed, meaning one breed has many more genotyped animals than the other, the largest breed will likely have more accurate predictions. To avoid this issue, **G** can be constructed assuming SNP are not shared among breeds, which would create a block-diagonal **G**; however, this is less straightforward when genotypes for crossbreds are included in the evaluation (Steyn *et al.* 2019).

When APY ssGBLUP is used in multi-breed or crossbred evaluations, the choice of core animals becomes even more complex. This is because the appropriate number of core animals depends on the theory of limited dimensionality of genomic information and chromosome segments, which relies on effective population size. The suitable number of core animals in multi-breed populations can be accessed by the number of eigenvalues explaining 98% of the variance in **G**, considering all breeds together. If breeds are completely independent, the expectation for the number of eigenvalues across 2 breeds is the sum of eigenvalues within each breed (Figure 4). We observed that comparing the number of eigenvalues across and within breeds can indicate the ability to perform across-breed predictions because chromosome segments are shared (Pocrnic *et al.* 2019).

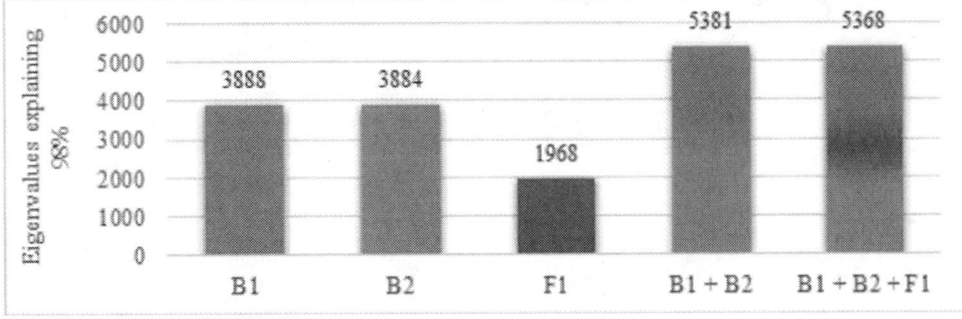

Figure 4. Number of eigenvalues explaining 98% of the variance of G across and within breeds (B1 and B2) and crossbred (F1)

Large pig breeding companies usually buy small farms/companies and combine the populations into a single evaluation. Assigning UPG for each population can help to account for the difference in base population. However, UPG are usually considered as fixed effects and a reasonable number of observations is needed for their accurate estimation. Datasets coming from small farms may have insufficient amount of information linked to UPG, leading to estimation errors and inflation of GEBV. In such a case, we observed that using random instead of fixed UPG solved estimation problems related to poor UPG connections (Pocrnic *et al.* 2018).

Poultry. Inflation of GEBV is not so evident in chicken datasets because old generations are removed and genotyped animals have complete pedigree. In fact, only 2 to 4 years of data are retained for genetic and genomic evaluation in chickens. However, in the first tests of ssGBLUP in chicken data (from Cobb-Vantress), back in 2013, we observed inflated genetic trends for GEBV compared to EBV, especially for young animals. The sources of this inflation were identified to result from the inclusion of unmapped SNP (i.e., mapped to chromosome 0) in the evaluation, the presence of imputation errors, and incorrectly labelled samples. If the imputation uses a family-based method but the pedigree has errors, the imputation can be compromised, resulting in low correlation between **G** and $\mathbf{A}_{22}$. This outcome illustrates more generally that quality control of SNP, samples, and genomic relationships is an important step before genomic evaluation, as small errors in the SNP data can be propagated, generating biased estimates.

Another issue observed in chicken data was the lower predictive ability (i.e., correlation between adjusted phenotypes and EBV or GEBV) for females compared to males for a growth trait, even though females had almost twice the number of genotypes. For an efficiency trait with the same amount of information and similar heritability, predictive ability was comparable between males and females. Separate genetic trends for males and females showed stronger selection for females than males in the growth trait but a similar trend in the efficiency trait (Figure 5). This shows that predictive ability takes the selection intensity into account and different predictive ability is expected if males and females have distinct selection differentials.

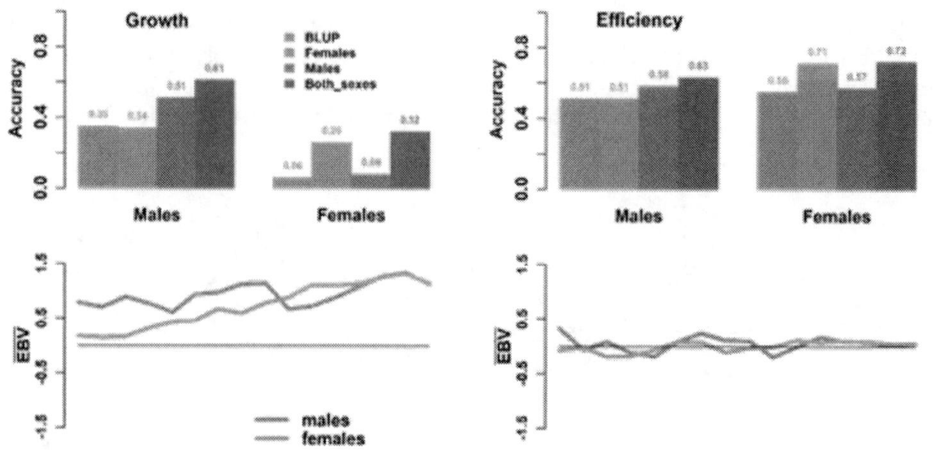

Figure 5. Accuracy and genetic trends for growth and efficiency in chickens

Fish. In fish production, several families may be raised in common ponds, making the identification of individuals difficult. Even though low density microsatellite panels are used for parentage determination with relatively high accuracy (Waldbieser and Bosworth 2012), pedigree errors still exist. As fish families are large, a single mistake in parentage assignment can be multiplied to thousands of individuals. In the implementation of genomic selection for catfish in the USA (ARS-USDA Warmwater Aquaculture Research Unit), we were able to identify mis-assigned parentage based on SNP markers. After pedigree corrections, heritability for carcass weight was adjusted from 0.27 to 0.21. Correcting variance components avoided the overestimation of genetic gains.

Another issue present in fish populations is the choice of individuals to be genotyped, given that genotyping is still expensive and full or half-sib families are large. We decided to genotype 40 fish per family, in a total of 75 families. As carcass weight is one of the most important traits in fish, genotyped individuals were also slaughtered to evaluate whether half or full-sib phenotypes would be enough to produce a high predictive ability for selection candidates. Using own genomic information combined with phenotypes and genotypes on siblings provided a 22% increase in predictive ability for carcass weight, compared to traditional BLUP.

Assessing predictive ability for disease traits either in fish or other species is quite challenging because phenotypes have a binary nature and breeding values have a normal distribution. Therefore, correlations between adjusted phenotypes and GEBV are usually very small or negative. Additionally, the regression coefficients are much lower than 1, which may not support the use of genomic selection. For binary and categorical traits, other validation methods may be more appropriate than predictive ability. For the initial tests on the feasibility of genomic selection for columnaris disease resistance in rainbow trout in the USA (ARS-USDA Cool and Cold Water Aquaculture), we adopted the LR validation (Legarra and Reverter 2018). This method is based on comparisons between complete and partial predictions. The relative increase in accuracy of GEBV compared to EBV was 40%, which encouraged the adoption of genomic selection to predict disease resistance in rainbow trout (Silva *et al.* 2019).

Common problems in large-scale genomic evaluations. Although different species usually require different strategies, common issues emerge when using large datasets. For ssGBLUP evaluations using a large number of genotyped animals, APY is one of the options. Another option is the ssGTBLUP (Mantysaari *et al.* 2017) that uses Woodbury formulas and requires only the inverse of a matrix with the size of the number of SNPs. An equivalent model that can also be used for large data is ssBR or hybrid model (Fernando *et al.* 2014), where SNP effects are estimated regardless of the number of genotyped animals.

When APY is used, even though the correlation between GEBV from regular ssGBLUP and APY ssGBLUP is greater than 0.99 when the appropriate number of core animals is used, re-ranking is still observed when different core groups are used. We investigated in beef and dairy cattle, and pig datasets different definitions of core and random core groups to identify which animals have the biggest changes in GEBV and how those changes can be minimized. In all datasets, larger changes in GEBV by using different core groups were observed for animals with lower accuracy. The observed changes relative to standard deviations of GEBV were, on average, 5%, but ranged from 0 to 100%. Increasing the number of core animals beyond the optimal value helped to asymptotically reduce changes in GEBV. Although core-dependent changes in GEBV exist, they are small and can be reduced with larger core groups.

Accounting for selected sequence variants in GBLUP-based methods. As sequence data is slowly becoming available for livestock, there is a question whether GBLUP-based methods can account for selected sequence variants and what is the possible gain in accuracy. Although the default assumption of GBLUP methods is that all SNP explain the same proportion of variance, it is possible to weight SNP differently. Recently, we observed that the increase in accuracy by SNP weighting is smaller in large populations, compared to small populations. This is because large genotyped populations allow more accurate estimation of chromosome segment effects; therefore, there is no advantage in selecting SNP and tagging segments with larger value (Lourenco *et al.* 2017).

Using a US Holstein dataset, Fragomeni *et al.* (2019) tested the performance of GBLUP and ssGBLUP when using nearly 54,000 SNP and when adding 17,000 significant variants discovered from a GWAS using sequence data involving 33 traits (VanRaden *et al.* 2017). Although VanRaden *et al.* (2017) reported an increase in reliability of GEBV of 4.3 points for stature by using non-linearA weights (i.e., a fast version of BayesA) in a multistep scenario, no gain was observed by Fragomeni *et al.* (2019) using either quadratic or non-linearA weight in GBLUP with heterogeneous residual variance and ssGBLUP. This is possibly because the amount of data used in ssGBLUP overwhelms any a priori assumption made about SNP effects, making this method less sensitive to SNP weighting in the presence of large data. Another hypothesis to explain the steady reliability is that not all causative variants were present among the 17,000 significant SNP. In a simulation study done by Fragomeni *et al.* (2017), including all simulated causative variants with respective true weights among 60k SNP, increased accuracy of ssGBLUP GEBV from 0.49 to 0.94. Although causative variants can be included in ssGBLUP assuming different weights for SNP, maximizing the accuracy of GEBV would require the true identification of all causative variants and their substitution effect.

CONCLUSIONS

Although the implementation of genomic selection seems to be straightforward, given genotypes are added to phenotypes and pedigree that are already in the evaluation system, several issues and challenges were raised during the initial application of this methodology to breeding programs of several species. Fortunately, solutions to most of the problems have come in a fast pace, enabling the widespread use of this methodology. Overall, the sources of problems include missingness of pedigree, selective genotyping, increasing number of genotyped animals, incompatibility between pedigree and genomic information, and difficulty in assessing predictive ability of genomic models for specific traits. It is expected that more issues will rise, and most of them may be related to the amount, type, and way the genomic data are being generated.

ACKNOWLEDGEMENTS

This paper is based on numerous studies done by several people from our animal breeding and genetics group at the University of Georgia, including current graduate students (Andre Garcia and Yvette Steyn), former postdocs (Ivan Pocrnic, Breno Fragomeni, and Rafael Silva), and the extended research group (Ignacio Aguilar and Andres Legarra). We thank the continued support from American Angus Association, US Holstein Association, Zoetis, PIC, Smithfield Premium Genetics, Cobb-Vantress, and USDA.

REFERENCES

Aguilar I., Misztal I., Johnson D.L., Legarra A., Tsuruta S. and Lawlor T.J. (2010) *J. Dairy Sci.* **93**: 743.

Brown D.J., Swan A.A., Li L., Gurman P.M., McMillan A.J., van der Werf J.H.J., Chandler H.R., Tier B. and Banks R.G. (2018) In Proc 11th WCGALP Auckland, New Zealand.

Chen C.Y., Misztal I., Aguilar I., Tsuruta S., Meuwissen T.H.E., Aggrey S.E., Wing T. and Muir W.M. (2011) *J. Anim. Sci.* **89**: 23.

Christensen O.F. and Lund M.S. (2010) *Genet. Sel. Evol.* **42**: 2.

Fernando R.L., Dekkers J.C.M., and Garrick D. J. (2014) *Genet. Sel. Evol.* **46**: 50.

Fragomeni B.O., Lourenco D.A.L., Masuda Y. and Misztal I. (2017) *Genet. Sel. Evol.* **49**: 59.

Fragomeni B.O., Lourenco D.A.L., Legarra A., VanRaden P.M., and Misztal I. *J. Dairy Sci.* (*Under review*).

Forni S., Aguilar I. and Misztal I. (2011) *Genet. Sel. Evol.* **43**: 1.

Gao H., Koivula M., Jensen J., Stranden I., Madsen P., Pitkanen T., Aamand G.P. and Mantysaari E.A. (2018) *J. Dairy Sci.* **101**: 10082.

Garcia A.L.S., Bosworth B., Waldbieser G., Misztal I., Tsuruta S. and Lourenco D.A.L. (2018) *Genet. Sel. Evol.* **50**: 66.

Kachman S.D., Spangler M.L., Bennett G.L., Hanford K.J., Kuehn L.A., Snelling W.M., Thallman R.M., Saatchi M., Garrick D.J., Schnabel R.D., Taylor J.F. and Pollack J. (2013) *Genet. Sel. Evol.* **45**: 30.

Legarra A., Bertrand J.K., Strabel T., Sapp R.L., Sanchez J.P. and Misztal I. (2007) *J. Anim. Breed. Genet.* **124**: 286.

Legarra A., Aguilar I. and Misztal I. (2009) *J. Dairy Sci.* **92**: 4656.

Legarra A. and Reverter A. (2018) *Genet. Sel. Evol.* **50**: 53.

Lourenco D.A.L., Tsuruta S., Fragomeni B.O., Masuda Y., Aguilar I., Legarra A., Bertrand J.K., Amen T., Wang L., Moser D.W. and Misztal .I (2015a) *J. Anim. Sci.* **93**: 2653.

Lourenco D.A.L., Fragomeni B.O., Tsuruta S., Aguilar I., Zumbach B., Hawken R.J., Legarra A. and Misztal I. (2015b) *Genet. Sel. Evol.* **47**: 56.

Lourenco D.A.L., Tsuruta S., Fragomeni B.O., Chen C.Y., Herring W.O. and Misztal I. (2016) *J. Anim. Sci.* **94**: 909.

Lourenco D.A.L., Fragomeni B.O., Bradford H.L., Menezes I.R., Ferraz J.B.S., Aguilar I., Tsuruta S. and Misztal I. (2017) *J. Anim. Breed. Genet.* **134**: 463.

Mantysaari, E. A., Evans R. D. and Stranden I. (2017) *J. Anim. Sci.* **95**: 4728.

Masuda Y., Misztal I., Legarra A., Tsuruta S., Lourenco D.A.L., Fragomeni B.O. and Aguilar I. (2017) *J. Anim. Sci.* **95**: 49.

Matukumalli L.K., Lawley C.T., Schnabel R.D., Taylor J.F., Allan M.F., Heaton M.P., O'Connell J., Moore S.S., Smith T.P L., Sonstegard T.S. and VanTassell C.P. (2009) *PLoS One* **4**: e5350.
Meuwissen T.H.E., Hayes B.J. and Goddard M.E. (2001) *Genetics* **157**:1819.
Misztal I., Legarra A. and Aguilar I. (2009) *J. Dairy Sci.* **92**: 4648.
Misztal I., Legarra A. and Aguilar I. (2014) *J. Dairy Sci.* **97**: 3943.
Pocrnic I., Lourenco D.A.L., Bradford H.L., Chen C.Y. and Misztal I. (2017) *J. Anim. Sci.* **95**: 3391.
Pocrnic I., Lourenco D.A.L., Chen C.Y., Herring W.O. and Misztal I. (2019) *J. Anim. Sci.* **97**: 1513.
Rupp R., Mucha S., Larroque H., McEwan J. and Conington J. (2016) *Anim. Frontiers* **6**: 39.
Saatchi M., McClure M.C., Mckay S.D., Rolf M.M., Kim J., Decker J.E., Taxis T.M., Chapple R.H., Ramey H.R., Northcutt S.L., Bauck S., Woodward B., Dekkers J.C.M., Fernando R.L., Schnabel R.D., Garrick D.J. and Taylor J.F. (2011) *Gen. Sel. Evol.* **43**: 40.
Silva R.M.O., Evenhuis J.P., Vallejo R.L., Gao G., Martin K.E., Leeds T.D., Palti Y. and Lourenco D.A.L. (2019) *Genet. Sel. Evol.* (*Under review*).
Soller M. and Beckman J.S. (1983) *Theor. Appl. Genet.* **67**: 25.
Steyn Y., Lourenco D.A.L. and Misztal I. (2019) *J. Anim. Sci.* (*Under review*).
The International SNP Map Working Group (2001) *Nature* **409**: 928.
Tsuruta S., Misztal I., Aguilar I. and Lawlor T.J. (2011) *J Dairy Sci.* **94**: 4198.
VanRaden P.M. (2008) *J. Dairy Sci.* **91**: 4414.
VanRaden P., Tooker M.E., Wright J.R., Sun C. and Hutchison J.L. (2014) *J. Dairy Sci.* **97**: 7952.
VanRaden P.M., Tooker M.E., O'Connell J.R., Cole J.B. and Bickhart D.M. (2017) *Gen. Sel. Evol.* **49**: 32.
Vukasinovic N., Bacciu N., Przybyla C.A., Boddhireddy P. and DeNise S.K. (2017) *J. Dairy Sci.* **100**: 428.
Waldbieser G.C. and Bosworth B.G. (2012) *Animal* **44**: 476.
Wang H., Misztal I., Aguilar I., Legarra A. and Muir W.M. (2012) *Genetics Research* **94**: 73.
Wiggans G.R,, Cooper T.A., VanRaden P.M. and Cole J.B. (2011) *J. Dairy Sci.* **94**: 6188.
Wiggans G.R., VanRaden P.M. and Cooper T.A. (2012) *J. Dairy Sci.* **95**: 3444.
Yan Y., Wu G., Liu A., Sun C., Han W., Li G. and Yang N. (2018) *Poult. Sci.* **97**: 397.

THE ACCURACY OBTAINED FROM REFERENCE POPULATIONS FOR GENOMIC SELECTION

J.H.J. van der Werf[1], S.A. Clark[1], S.H. Lee[1,2] and N. Moghaddar[1]

[1]School of Rural & Environmental Science, University of New England, Armidale, NSW, 2351 Australia
[2]Australian Centre for Precision Health, University of South Australia Cancer Research Institute, University of South Australia, Adelaide, SA, 5000 Australia

SUMMARY

For the design of breeding programs it is important to understand how trait measurement translates into selection accuracy. The introduction of genomic selection has created new challenges, in particular in relation to designing reference populations and valuing information sources for their contribution to genetic gain. The accuracy of genomic prediction depends on trait heritability, the number of phenotypes used (on genotyped animals) and the 'effective number of chromosome segments' that need to be estimated. The latter parameter is challenging to estimate but can in principle be derived from the variation in relationships between the reference set and the target animal. This paper attempts to validate that theory based on real data, with the aim to develop further insight into the value of a certain reference set for the genomic prediction of a certain target animal.

INTRODUCTION

Genomic selection has become an integral part of breeding programs. The information about genetic merit obtained from genomically tested animals depends on the accuracy of the genomic test itself, and that from various other sources of information such as performance data on an animal itself and (or) its relatives. There is good selection index theory about the value of various information sources, and the accuracy of estimated merit we can expect if we combine them in a prediction framework such as Best Linear Unbiased Prediction (BLUP). However, we are still struggling to get a good handle on the information that we can expect from a genomic test. A better understanding of the components that drive the accuracy of a genomic test is important, not only for the breeder who needs to decide whether to invest in it, but also for those setting up reference populations to facilitate a higher accuracy of genomic testing. Investment in reference population occurs through individual breeders or breeder groups, breed societies, and funding bodies. It is important to be able to value the contributions of different information sources, the possible advantages of further increasing the size of the reference population and the usefulness of a certain reference set for animals with varying degrees of relationship to that reference.

The purpose of this paper is to review the theory that has been proposed to predict the accuracy of genomic prediction and to validate this theory with some examples involving real data. This might lead to a way forward on how to decide about the size and structure of reference populations and how to value them in prediction of genetic merit in the context of breeding programs.

THEORY ON THE ACCURACY OF GENOMIC TESTING

The most frequently cited formula to predict the accuracy of genomic testing comes from Daetwyler *et al.* (2008), who proposed:

$$r_{\hat{g},g} = \sqrt{\frac{h^2}{h^2 + M_e/N}} \qquad [1]$$

where h^2 is the trait heritability, N is the number of individuals with an observed phenotype as well as genotype, and M_e is the 'effective number of chromosome segments'. The formula is remarkably simple. It is based on the accuracy of estimating a random effect, which is N/(N+λ), where λ is the ratio of the residual variance (V_e) and the variance of the effect to be estimated. Under a polygenic model quantitative trait loci (QTL) are spread across the whole genome, each with a small effect. The variance of each independent chromosome segment is the V_A/M_e, where V_A is the additive genetic variance. When estimating one segment at a time then V_e is approximately equal to the phenotypic variance and $\lambda \cong M_e/h^2$, such that (1) is equivalent to N/(N+λ). This will give a slight underestimation of accuracy if all segments are estimated jointly and $V_e < V_p$.

Further papers by Goddard (2009) and Goddard *et al.* (2011) have refined the theory, e.g. by accounting for lower density marker panels, where the LD between markers and QTL is insufficient such that the proportion of the genetic variance 'captured by markers' is $b = M/(M_e + M)$, where M is the number of genetic markers, and $r_{\hat{g},g} = \sqrt{bh^2/(h^2+M_e/N)}$. Note that with very many markers *b* approaches 1. For a given M_e and high values of b, there is limited dispute about predicting genomic accuracy. However, approximations for M_e vary widely, and various formulae have been presented all leading to quite different results (Table 1). In fact, variation between predictions of genomic accuracy almost entirely depend on the approximation of M_e.

Table 1. Predicted accuracy of genomic test ($r_{\hat{g},g}$), assuming 2500 observations (N), heritability h^2=0.30, Effective population size N_e = 250; average chromosome length L=1; number of chromosomes k=30, and number of markers M=50,000

	Reference and approximation for M_e				
	Daetwyler *et al.* 2008	Goddard 2009	Goddard *et al.* 2011	Meuwissen *et al.* 2013	Lee *et al.* 2017
Parameter[1]	$2N_eLk$	$2N_eLk/\ln(4N_eLk)$	$2N_eLk/\ln(N_eL)$	$2NeLk/\ln(2N_e)$	Eq(11)
Me	15000	1455	2717	2414	611
$b= M/(M_e + M)$	1.00	0.97	0.95	0.95	0.99
$\lambda = M_e/h^2$	50000	4991	9548	8434	2060
$\sqrt{(N/(N+\lambda))}$	0.22	0.58	0.46	0.48	0.74
$r_{\hat{g},g}$	0.22	0.57	0.44	0.47	0.74

[1] M_e = Effective number of chromosome segments; b= Proportion of genetic variance captured by markers; λ = variance ratio of residual and that of one chromosome segment; $\sqrt{(N/(N+\lambda))}$ is accuracy for b=1.

In the theory described so far the approximations of M_e assume the reference as a homogenous population where all individuals are more or less equally related to each other. However, genomic predictions are more accurate if the genomic relationship between the target animal and the reference population is higher (Habier *et al.* 2007; Clark *et al.* 2012). Van der Werf *et al.* (2015) noted that most reference populations are heterogeneous in their relationship towards the target animals they predict, i.e. some individuals in the reference are much more related to the target individual than others. They demonstrated in a simple model how a small group of more related individuals can contribute more information than a very large group of distantly related individuals. Heterogeneity also exists if the reference population consists of different breeds or crossbreds. Wientjes *et al.* (2015) have proposed deterministic prediction methods to accommodate information from different populations, where they also account for genetic correlations between populations being less than one.

The variation in relatedness is often hard to predict in advance in real world examples, and a pragmatic approach can be taken by looking at the variation in realised genomic relationships between the members of the reference population and the target individual to be predicted (Goddard *et al.* 2011). This 'empirical' M_e value derived from variation in genomic relationships implies that the M_e parameter is related to the data set used for genomic prediction rather than being a population parameter, e.g. related to a certain breed. Lee *et al.* (2017) showed via simulation of a full sib population structure that the variation in genomic relationship ($var(g_{ij})$) gives a reliable estimate of M_e as $M_e = 1/var(g_{ij})$. Using this M_e value in the Daetwyler formula gave satisfactory approximations of accuracy. However, calculating M_e from variation in relationships seemed to over predict the accuracy of a genomic test when simulating a typical nucleus breeding program with a nested full-sib/half sib design across multiple generations (Jack Dekkers, pers. comm). Van den Berg *et al.* (2019) also found over prediction when applying it to simulated and real data from mixed breeds of dairy cattle.

VALIDATING THEORY WITH EMPERICAL RESULTS

It is difficult to validate the genomic prediction theory in real data based on outcomes of industry genetic evaluations such as BREEDPLAN or LAMBPLAN because these are based on so-called single-step models where information via genomic relationships is combined with information through pedigree relationships. Moreover, these evaluations are based on multiple trait models where information from correlated traits is included in the estimated breeding value (EBV). To quantify the accuracy of the genomic test in a more designed way we compared the prediction of genomic breeding value accuracy for three different traits, with varying heritability, and using the same reference population and two different validation sets. We derived M_e from the variance in relationships (Lee *et al.* 2017) of the off-diagonal block of the genomic relationship matrix, i.e. between animals in the reference and animals in the validation set, and derived the predicted accuracy using [1]. The reference population consists of 5000 animals from multiple breeds from the CRC information Nucleus and MLA reference flocks. The validation population refers to 300 purebred merinos and 300 crossbred Border Leicester x Merino crosses. Predicted accuracies were compared with empirical accuracies derived from the correlation between predicted genomic breeding values and adjusted phenotypes of animals in the validation set, divided by h. Results are shown in Table 2.

The results show an obvious overestimation of the accuracy when using the variation in relationships to estimate the M_e value. A likely reason is that the reference population consists of multiple breeds, giving a much larger variation in relationship relative to using a purebred reference. Note that the accuracy is evaluated after correction for breed effects, i.e. it is a within breed accuracy. An accuracy 'across breeds' is much larger as from genotype data 'it is relatively easy to predict differences between breeds, or genetic groups within breeds. A next step is therefore to correct the G-matrix for effects of population structure by taking out a number of principal components, i.e. using $G^* = G - \Sigma E_i E_i' d_i$ where d_i is an eigenvalue of G and E_i is the associated eigenvector. Further testing can also occur using purebred reference populations, although such populations can still have an underlying group structure that needs to be taken into account. Van den Berg *et al.* (2019) also concluded that the variance in genomic relationships overestimated the accuracy, when they compared reference populations with various numbers of individuals from different breeds. They proposed an alternative method that seemed to be useful to predict accuracy from reference populations from combining breeds. However, there is also a need to evaluate the value of adding within breed cohorts to the reference, where these cohorts may vary in their relationship to the animals that are targeted in prediction.

Table 2. Realized genomic prediction accuracy and theoretical accuracy predicted from variation in relationships and effective number of chromosomes (M_e) for two validation sets and using a multi-breed reference population[1]

Test Set	Var(g_{ij})	Me=1/Var(g_{ij})	Predicted accuracy[2]	Realized accuracy[3]
BL x Merino	0.001989	502.7	0.86	0.21
Merino	0.001840	543.6	0.85	0.29

[1] *Using a multi-breed reference set of N = 5000 animals, trait is post weaning weight; $h^2 = 0.28$*

[2] *Accuracy predicted using the Daetwyler formula [1] and the estimated value for M¬e¬.*

[3] *Realized accuracy is correlation between predicted genomic breeding value and observed phenotype (corrected for fixed effect), divided by the square root of heritability.*

CONCLUSIONS

Further work is needed to validate the theory of deriving genomic prediction accuracy from the variation in genomic relationships, and to put a value on adding particular information sources to the reference population for genomic prediction. Although this approach requires a matrix with realised genomic relationships, it provides information about the contribution of various information sources, and this may be used to predict contributions of future cohorts. Moreover, this approach is flexible and can allow animals from multiple breeds or crossbreds.

REFERENCES

Clark S.A., Hickey J.M., Daetwyler H.D. and van der Werf J.H.J. (2012) *Genet. Sel. Evol.* **44**: 4.

Daetwyler H.D., Villanueva B., and Woolliams J.A.. (2008) *PLoS One* **3**: e3395.

Goddard M.E. (2009) *Genetica*. **136**: 245.

Goddard M.E., Hayes B.J. and Meuwissen T.H.E. (2011) *J. Anim. Breed. Genet.* **128**: 409.

Habier D., Fernando R.L. and Dekkers J.C.M. (2007) *Genetics* **177**: 2389.

Meuwissen T., B. Hayes and M. Goddard (2013) *Annu. Rev. Anim. Biosci.* **1**: 221.

Lee S. H., Clark S.A. and van der Werf J. H. J. (2017) *PLoS One* **12**: e0189775.

Van der Werf J.H.J, Clark S.A., Lee S.H. (2015) *Proc. Assoc. Advmt. Anim. Breed. Genet.* **21**:

Van den Berg I., Meuwissen T.H.E., MacLeod I.M. and Goddard M.E. (2019). *J. Dairy Sci.* **102**:1.

Wientjes Y.C.J., Bijma P., Veerkamp R.F. and Calus M.P.L. (2016) *Genetics* **202**: 799.

INCREASE OF POWER AND EFFICIENCY TO FINE-MAP GENETIC DEFECTS USING GENOTYPE PROBABILITIES THROUGH SEGREGATION ANALYSES

N. Duijvesteijn[1,2], S.A. Clark[2], B.P. Kinghorn[2] and J.H.J. van der Werf[2]

[1]Hendrix Genetics Research, Technology & Services B.V., Boxmeer, the Netherlands
[2]School of Environmental and Rural Science, University of New England, Armidale, NSW, 2351 Australia

SUMMARY

This simulation study shows a method which makes more efficient use of pedigree and genomic information to increase the chance to detect genetic disorders. We make use of Geneprob, a program which uses segregation analysis to calculate the genotype probabilities of pedigreed animals. The results show that our method, for a trait with a recessive inheritance pattern, is better in the detection of the region of the causative mutation compared to a method which used allele frequencies of cases and controls only. This method can be used across all pedigreed species.

INTRODUCTION

In recent years, the detecting of genetic disorders and lethal recessive conditions in livestock populations through the use of genomic tools, has increased (f.e. VanRaden *et al.* 2011 and Derks *et al.* 2017). The defects are mostly spread by intensive use of elite sires which are unknowingly carrier of an autosomal recessive defect. In most populations of cattle where artificial insemination has resulted in a very efficient distribution of the genetic material of superior sires, genetic disorders and lethal recessive conditions have been detected. The success of fine mapping an observed Mendelian genetic disorder requires another approach than that classically used to detect lethal recessive conditions.

A genetic disorder often gives the animal an abnormal phenotype and deprived performance. Accurate recoding of the phenotype by the farmer is essential and often targeted genotyping or sequencing of affected animals and related family members has resulted in successful fine mapping of genetic disorders (e.g. Daetwyler *et al.* 2014). In populations with less routinely genotyping and / or large populations which are extensively managed, success of detection has been compromised. For example, in sheep very few genetic disorders or lethal recessives based on genomic information have been identified. More efficient use of pedigree information and genomic information could increase the chance of detection of genetic disorders.

In this study we show a simple but effective application with the use of Geneprob, a program which uses segregation analysis, to calculate the genotype probabilities of animals within the pedigree, to facilitate the detection of genetic disorders. All animals genotyped within the pedigree are for a GWAS where the phenotype is a linear score derived from genotype probabilities (viz. the probable number of alleles carried). A simulation is done using sheep data to illustrate the application, but the method can be used on any pedigreed population.

MATERIALS AND METHODS

Genotypic data. Genotypes originated from various research flocks (Sheep Genomics, the CRC Information Nucleus Flock, and the MLA Resource Flocks) as well as from industry data collected by sheep breeders. For this study only genotypes of animals from the Merino breed were selected. In total 21,000 Merino sheep were genotyped and imputed up to sequence (Bolormaa *et al.* 2019). For the purpose of this study one chromosome was selected (OAR5) to demonstrate the detection of a recessive causative mutation.

Detection of the recessive causative mutation. A single simulation was run on sheep data to illustrate the concept. One SNP on OAR5 was selected to be the causative mutation for an unknown fictive genetic disorder. The minor allele frequency (MAF) of the SNP needed to be between 0.04 and 0.05 to reflect a mutation that is present within a population at low frequency. The SNP was located between 20 and 30 Mb. The randomly selected SNP was Chr5:29170109. The highest linkage disequilibrium (LD) between the causative SNP and a SNP located at the 50K SNP array was 0.333 and the SNP from the 50K SNP array was Chr5:29178193. For an unidentified recessive disease, the genotyping strategy will depend on the available budget and availability of identified cases, but in this study we assume that both cases and controls will be genotyped with the commercially available Illumina ovineSNP50 BeadChip.

From the Merino dataset, 54 cases (homozygous for the recessive allele of SNP Chr5:29170109) were identified. From those 54 cases, we selected 20 cases for our study based on criteria: 1) sires (father of the cases) needed to have more than 1 offspring genotyped, 2) the dam needed to be known and, 3) no full sibs were selected. Besides the 20 cases we selected 10 offspring from sires of cases and 10 random controls which came from the same flock and year as the cases and weren't sires or dams from cases. For this group of 40 sheep, the pedigree was pruned and phenotypes for the disease status was appointed to them. The phenotype code 0 was given to all controls and parents of cases (as they don't have the recessive disease), and all cases were appointed phenotype code 1. In total 31 animals had phenotype 0 and 20 had phenotype 1, all remaining animals from the pedigree got phenotype 8, which means they can be carrier but they are not homozygous for the recessive allele.

Method using genotype probabilities. For the scenarios of which we wanted to improve the power by using pedigree information and genotype data, we used the software program Geneprob (Kerr and Kinghorn 1996). It uses segregation analysis to calculate the genotype probabilities of animals within the pedigree. Every animal will get assigned a probability for each genotype class (*aa*, A*a* or AA). Following convergence of Geneprob, the estimated genotype probabilities were expressed as the Most Probable Allele Count (MPAC) using the following equation:

$$MPAC = 0 * p(aa) + 1 * p(Aa) + 1 * p(aA) + 2 * p(AA),$$

where p(aa) is the genotype probability for the genotype class aa, p(Aa) is the genotype probability of the genotype class Aa, p(aA) is the genotype probability of the genotype class aA and p(AA) is the genotype probability of the genotype class AA. The value of MPAC lies between 0 and 2, similar to a SNP genotype. The MPAC was regressed on the SNP genotypes. Similar to a traditional GWAS, $-\log_{10}$(Pvalues) can be plotted to indicate a possible QTL region.

Scenarios. Four different scenarios compared in their success to detect the region of the causative recessive mutation. Additionally, 2 scenarios were evaluated to compare the results when very few cases were genotyped (N= 2).

The *first* scenario reflects the traditional approach. In the field 20 cases and 20 controls have been collected and the difference in MAF between cases and controls is compared. Software program PLINK (Chang *et al.* 2015) was used with the Fisher's exact test to generate p-values and $-\log_{10}$(Pvalues) are plotted to indicate a possible QTL region.

The *second* scenario uses the 40 animals with an appointed phenotype status, but pedigree information is used to increase the power of the analysis. Software program Geneprob is used to calculate the MPAC and were regressed on the SNP genotypes to indicate a possible QTL region.

For the *third* scenario, no money was available to genotype cases, but phenotype status was available on the 43 animals from the pedigree as well as 20 controls which were routinely genotyped for the breeding program. Similar to scenario 2, Geneprob was used to calculate the MPAC and regressed on the SNP genotypes.

For the *fourth* scenario, the 20 cases, 20 controls and 43 animals from the pedigree were genotyped and similar to scenario 2, Geneprob was used to calculate the MPAC and regressed on the SNP genotypes.

Two scenarios were added where the traditional method (compare MAF between cases and controls) had 2 cases genotyped and 20 controls which was compared to the method where we used Geneprob to include all available data (2 cases, 20 controls and 43 animals from the pedigree were genotyped) and the MPAC was calculated and regressed on the SNP genotypes.

The 50K data of OAR5 was used for the animals in the different scenarios (1,900 SNPs).

RESULTS AND DISCUSSION

The SNP in highest LD with the causative mutation shows an incomplete inheritance pattern of the disease (Table 1). If selection was to exclude all animals with genotype 2 (homozygous for recessive allele), four animals would be excluded, while they don't have the recessive disorder.

Table 1. Count of animals per phenotype class and genotype class

Phenotype status	Genotype SNP 50K		
	0	1	2
0	8	18	4
1	0	0	20
8	27	6	0

The results of the chromosome-wide association study for each of the four scenario's is shown in Figure 1. The scenario which the largest -$\log_{10}$(Pvalue) was scenario 4 (-$\log_{10}$(Pvalue) =22.6), followed by scenario 2 (-$\log_{10}$(Pvalue)=8.9), then scenario 1 (-$\log_{10}$(Pvalue)=6.7), and scenario 3 has the lowest-$\log_{10}$(Pvalue) with 5.5. In all scenario's the SNP with the highest LD to the causative mutation was indicated. Although in scenario 3, another SNP along the chromosome showed a very similar -$\log_{10}$(Pvalue) and a misidentification of the region could easily have occurred.

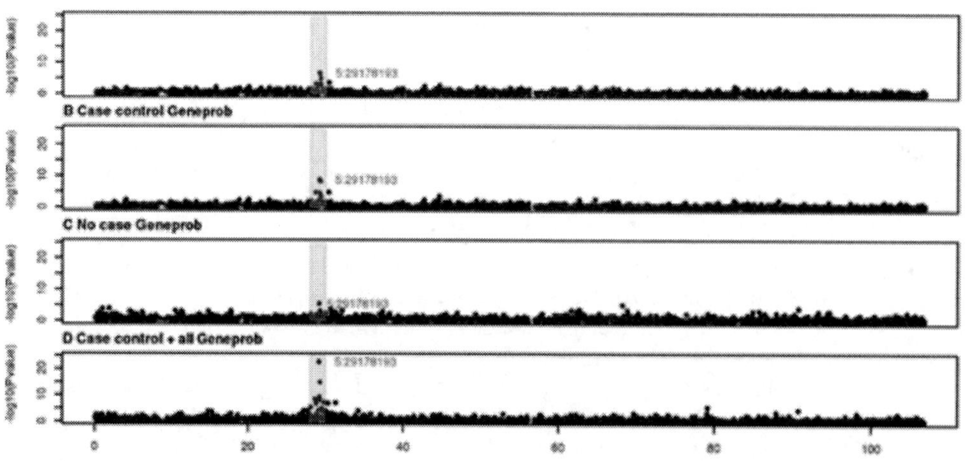

Figure 1. Chromosome-wide association of OAR 5. A) Association analyses of 20 cases and controls using PLINK. B) Association analyses using Geneprob on all 20 cases and 20 controls. C) Association analyses using Geneprob on only the controls and genotyped animals from the pedigree. D). Association analyses using Geneprob on all cases, controls and genotyped animals from the pedigree

Also, the 'value' of only genotyping 2 cases has been investigated (Figure 2) and the method using genotype probabilities had increased power compared to the traditional method using Fischer's exact test. The traditional method did not detect the region with the causative mutation (Figure 2A), while the method using Geneprob did detect the region with a clear signal (Figure 2B).

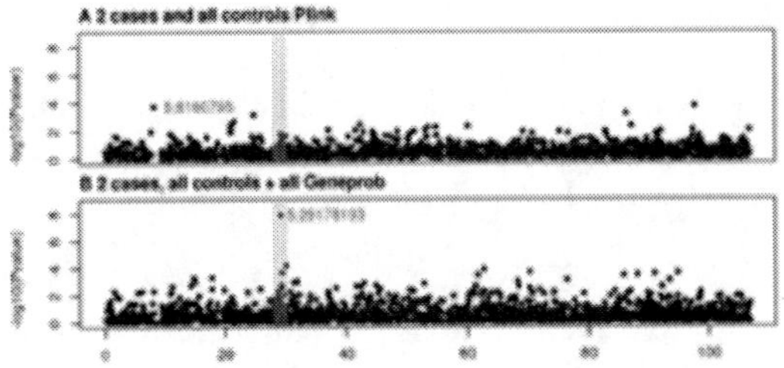

Figure 2. Chromosome-wide association of OAR 5. A) Association analyses of 2 cases and 20 controls using PLINK. B) Association analyses using Geneprob on 2 cases, 20 controls and genotyped animals from the pedigree

For the scenarios tested, we have shown the added power to detect a recessive mutation through the use of a segregation analyses and use all available data (pedigree, genotype data and phenotypic information; scenario 4). The results are especially valuable to use for pedigreed species where genotyping is still costly and additional genotyping of affected animals is not covered by available budgets. This study is relatively small and further testing is needed to determine to which extend this method is more beneficial compared to more traditional methods.

CONCLUSIONS

To conclude, we have demonstrated in this small simulation study that segregation analysis of a trait with a recessive inheritance pattern can lead to considerably power in a GWAS and therefore is better in the detection of the region of the causative mutation compared to a method which used allele frequencies of cases and controls only. We advise at least some cases need to be genotyped to be able to accurately determine the region of the recessive genetic disorder. This method can be used across all pedigreed species and is especially valuable for species where genotyping is still relatively expensive.

ACKNOWLEGEMENTS

The authors acknowledge the contributions of people from breeders and many CRC participants that contributed to the Sheep CRC Information Nucleus flocks. Also Nasir Moghaddar is acknowledged for his help retrieving data and helpful discussions.

REFERENCES

Bolormaa S., Chamberlain A. J., Khansefid M., Stothard P., Swan A. A., Mason B., ... and Daetwyler H. D. (2019) *Genet. Sel. Evol.* **51:** 1.

Chang C.C., Chow C.C., Tellier L.C., Vattikuti S., Purcell, S.M. and Lee, J.J. (2015) Second-generation PLINK: rising to the challenge of larger and richer datasets. *Gigascience.* **4**: 7.

Daetwyler H. D., Capitan A., Pausch H., Stothard P., Van Binsbergen R., Brøndum R. F., ... and Esquerré D. (2014) *Nat. Genet.* **46:** 858.

Derks M. F., Megens H. J., Bosse M., Lopes M. S., Harlizius B. and Groenen M. A. (2017) *BMC Genomics* **18:** 858.

Kerr R. J. and Kinghorn B. P. (1996) *J. Anim. Breed. Genet.* **113**: 457.

VanRaden P. M. , Olson K.M., Null D.J. and Hutchison J.L. (2011) *J. Dairy Sci.* **94**: 6153.

PROGENY OF ANDERSON RAMS SELECTED FOR RESISTANCE TO INTERNAL PARASITES IN AUSTRALIA ARE COMPARABLE IN OTHER TRAITS TO THAT FROM TALITAS RAMS SELECTED IN URUGUAY

A. Sánchez[1], F. De Brum[2], E. van Lier[1], A. Burton[1], J. Paperán[3], W. Bell[1] and R. Ponzoni[1]

[1]Dept. Prod. Anim., Facultad de Agronomía, Montevideo, Uruguay
[2]Principal, Talitas Rams, Artigas, Uruguay
[3]Central Lanera Uruguaya, Salto, Uruguay

SUMMARY

Gastrointestinal parasites constitute a serious problem in many sheep production systems. Two studs, Anderson Rams in Australia and Talitas Rams in Uruguay, have been selecting for resistance for about two decades with considerable success. We used semen from Anderson Rams in Uruguay and compared their progeny with that of Talitas Rams. The genetic merit of Anderson Rams for worm egg count per gram of faeces is comparable to that of the best in Talitas Rams. The same may be said about production traits and visually appraised characters. In particular, fleece rot and wool quality were feared to be a problem among the progeny of Anderson rams, but contrary to expectation, their performance was very good and comparable to that of the best Talitas rams. Because Anderson Rams and Talitas Rams have worked independently, their progeny are unrelated, thus mutually providing an opportunity to increase the effective population size without compromising genetic merit in resistance to gastrointestinal parasites, in production traits, or in visually assessed characters.

INTRODUCTION

Gastrointestinal parasites constitute a serious problem in many sheep production systems (some in Australia and most in Uruguay). Talitas Rams stud in Uruguay, has been successfully selecting for resistance to internal parasites for about two decades. Semen from Anderson Rams stud in Western Australia, which has been selecting in the same direction, has been imported to Uruguay and used in a number of flocks. Worm egg count per gram of faeces (WEC) is used as a selection criterion for resistance in sheep genetic evaluations. In the latest Uruguayan Genetic Evaluation (INIA and SUL 2018) the top ram for WEC was from Talitas. Three Anderson Rams were among the 10% best for WEC, and one of them ranked third (together with a Talitas ram). It is a remarkable performance given that the three Anderson rams have no ancestors or other relatives, except for the progeny they have produced in Uruguay. Their breeding values may be negatively biased since the model fitted in the Uruguayan evaluation does not include genetic groups (Westell *et al.* 1988). Despite the demonstrated genetic merit for resistance to internal parasites of Anderson rams, some breeders have reservations. The Australian and Uruguayan environments are different, and they are wary about the performance regarding production traits and visually assessed characters. In this paper we report the progeny performance of three Anderson rams and nine Talitas rams for wool and body traits.

MATERIALS AND METHODS

Sheep and the environment. Records were available from 326 progeny of 12 rams born in the Spring of 2017. All rams had at least 20 progeny, that were reared in two locations in northern Uruguay, a University Farm in Salto (Estación Experimental Facultad de Agronomía Salto), and at Talitas Rams stud in Artigas. Two rams had progeny at the University Farm, whereas eleven rams had progeny in Talitas. Anderson rams were coded A1 to A3 and had expected progeny deviations (EPDs) for WEC ranging -0.44 to -0.31 in the Uruguayan genetic evaluation (scale -0.5 most resistant, 0.5 most susceptible). Talitas rams were coded T1 to T9 and had WEC EPDs ranging -0.22 to -0.04. The national average is -0.13 and the best record is for a Talitas ram born in 2009 is -0.5. The ram coded as A1 (Table 4) had progeny in both locations (33 at the University Farm and 34 at Talitas), thus providing a genetic link between both locations.

The University Farm and Talitas Ram stud are at a latitude of 31 degrees south. Average rainfall is 1320 mm. Mean maximum and minimum temperatures are 24 and 12 degrees C, respectively. During the wool growth period rainfall was greater than the average. The spring of 2017 was very rainy (500 mm), followed by a relatively dry summer (370 mm). Later, in May alone, rainfall was 360 mm, accompanied by warm temperatures. Overall, wool growth took place in conditions that were conducive to wool discoloration and fleece rot.

Traits recorded. The objectively measured (yearling) traits recorded were: greasy fleece weight (GFW), yield (YLD), clean fleece weight (CFW), fibre diameter (FD) and post shearing live weight (LWT). Prior to shearing, the subjectively assessed traits were: overall visual appraisal (VISAP, 1= top, ..., 3=cull), face cover (FC, 1=open face, ..., 6=muffled face), pigmentation in non-wool areas (PGM, 1=free of pigmentation, ..., 5=highly pigmented), wool quality (WQUAL, 1=harsh poor quality, ..., 5= the best in terms of colour, handle and wool character), fleece rot (FR, 0=complete absence of fleece rot, ..., 5=high incidence of yellow or green bands). At the University Farm lambs were not shorn, visual appraisal was conducted in August 15, 2018, whereas shearing took place on September 10. At Talitas Ram stud lambs were shorn in December 2017, and visual appraisal and shearing took place in September and October 2018, respectively.

Data analyses. PROC MIXED in SAS (SAS Institute Inc., 2011) was used to fit a linear model to the data. Location, sire, sex, type of birth, age of dam, and management group within location were fitted as fixed effects, whereas date of birth was fitted as a linear covariate within location. This enabled the calculation of 'adjusted means' (least squares means) for sires, as is usually done in sire evaluation in Australia. We also analysed the visually appraised traits using PROC GLIMMIX in SAS, assuming a multinomial distribution. The results were almost identical to those obtained using PROC MIXED, except for small differences in a few and unimportant cases. Here we present the results obtained with PROC MIXED.

RESULTS AND DISCUSSION

Table 1 shows descriptive statistics for the traits studied. Fleece rot are not presented, only a very small proportion of animals were affected, and none with scores 3 to 5.

Table 1. Number of observations (N), mean (μ), minimum, maximum and standard deviation (σ) of GFW, YLD, CFW, FD, LWT, VISAP, FC, PGM), WQUAL, FR

Variable	N	μ	Min	Max	σ
GFW (kg)	318	2.69	1.30	4.20	0.47
YLD (%)	326	74.78	59.80	86.70	5.20
CFW (kg)	318	2.01	1.05	3.21	0.34
FD (μm)	326	16.81	13.30	21.30	1.57
LWT (kg)	316	34.93	16.00	53.00	6.43
VISAP (1-3)	319	1.83	1	3	0.60
FC (1-6)	319	1.93	1	4	0.74
PGM (1-5)	319	2.27	1	5	0.82
WQUAL (1-5)	319	4.41	1	5	0.70
FR (0-5)	326	0.04	0	2	0.23

Table 2 shows the analysis of variance for objectively measured traits. We mainly focus on the sire effect, which was statistically significant in all cases, except for YLD.

Table 2. Degrees of freedom (DF) and P values from the analysis of variance of GFW, YLD, CFW, FD and LWT

Effect	DF	GFW	YLD	CFW	FD	LWT
Location	1	0.8456	0.5076	0.6263	0.6951	0.1180
Sire	11	0.0002	0.1267	<.0001	0.0218	0.0206
Sex	1	<.0001	<.0001	<.0001	0.8223	<.0001
Birth type	2	<.0001	0.4725	<.0001	0.0296	0.0906
Age of dam	8	0.3734	0.6634	0.2365	0.8513	0.4220
Management group (location)	1	0.5458	0.8685	0.5936	0.0171	0.3513
Birth date (location)	2	0.0623	0.7214	0.1162	0.1164	0.0182

Table 3 shows the analysis of variance for subjectively assessed characteristics. The effect of sire was statistically significant for FC and PGM, whereas it bordered significance for WQUAL.

Table 3. Degrees of freedom (DF) and P values from the analysis of variance of VISAP, FC, PGM and WQUAL

Effect	DF	VISAP	FC	PGM	WQUAL
Location	1	0.8130	0.9576	0.6464	0.0419
Sire	11	0.2393	0.0220	0.0012	0.0955
Sex	1	0.3685	0.3528	0.1557	0.5220
Birth type	2	0.0016	0.6771	0.9721	0.0132
Age of dam	8	0.0401	0.5613	0.2983	0.9723
Management group (location)	1	0.2174	0.9151	0.6396	0.5161
Birth date (location)	2	0.3091	0.7025	0.7724	0.1225

Table 4 shows the least squares means for sires.

Table 4. Least squares means for GFW, YLD, CFW, FD, LWT, VISAP, FC, PGM and WQUAL. The three 'best' sires for each trait are indicated in bold type

Sire[1]	GFW	YLD	CFW	FD	LWT	VISAP	FC	PGM	WQUAL
A1	**2.51**	74.72	**1.88**	17.74	**35.84**	1.95	1.94	2.32	4.25
A2	2.08	74.51	1.55	**17.04**	34.18	2.16	2.05	2.22	**4.44**
A3	**2.47**	75.84	**1.86**	17.98	**37.73**	**1.68**	**1.57**	1.72	**4.45**
T1	2.31	72.28	1.65	**16.78**	33.90	2.18	2.19	1.99	4.07
T2	2.31	**76.01**	1.73	17.69	35.83	1.87	**1.61**	2.53	**4.32**
T3	2.32	73.86	1.70	17.37	**37.12**	**1.75**	**1.71**	2.02	4.22
T4	2.31	73.18	1.67	**17.10**	35.57	1.87	2.12	**1.43**	4.19
T5	2.36	**75.87**	1.78	17.87	33.22	1.96	1.90	**1.51**	3.99
T6	2.21	**75.98**	1.65	17.34	33.39	1.90	2.29	1.83	4.11
T7	**2.40**	75.32	**1.81**	17.61	33.50	**1.85**	1.73	1.61	4.21
T8	2.29	73.67	1.67	17.49	32.75	2.04	2.04	**1.34**	3.63
T9	2.09	74.90	1.55	17.42	34.21	2.07	2.11	1.62	3.81
SE	0.10-0.15	1.23-1.84	0.07-0.12	0.33-0.49	1.14-1.72	0.16-0.23	0.20-0.30	0.21-0.32	0.18-0.28

1- A: Anderson sires; T: Talitas sires; SE is the range in standard errors of least squares means

Talitas Rams is an Australian Merino stud of excellent reputation in Uruguay, selling 180 to 220 rams per year to a well-established clientele. It has been using objective measurement for decades and its sires always rank well in the Uruguayan genetic evaluation. It provides a valuable reference for the Anderson sires being introduced.

Anderson sires have expressed high genetic merit for resistance to internal parasites in the Uruguayan environment. The results presented in this paper should help allay concerns about the performance of their progeny with regards to production traits and visually assessed characters. Table 4 shows that for all traits considered, the progeny of Anderson rams compared well with that of Talitas. In the case of GFW and CFW, two of the heaviest cutting progeny were by Anderson rams. YLD was generally greater for the progeny of Talitas rams, but the difference was not large, and the lower yield could be advantageous if it conferred greater fibre protection. Fibre diameter among all progeny ranged between 17 and 18 microns. Progeny from one of the Anderson rams was the second finest, whereas for another one it was the strongest. However, all were within the range of the progeny of Talitas rams. Concerns about Anderson rams undoing the results of many years of selection for reduced fibre diameter seem unjustified. Anderson rams produced two of the heaviest progeny groups, one of them having the greatest LWT. Regarding VISAP, Anderson rams had the best scoring progeny, as well as one of the worst. However, the values were comparable to those of Talitas, and indicated that a high proportion of all progeny were deemed visually acceptable. FC scores of all progeny were low; the greatest value was 2.3, which still corresponds to an open face sheep. Pigmentation scores were low (greatest value 2.5 out of a maximum possible individual score of 5.0). Initial apprehension about the adequacy of fleeces bred in Western Australia for the Uruguayan environment is in principle justified. The environments notably differ in rainfall. We did not analyse the FR data because of its extremely low incidence. Coupled with the WQUAL results, this should put at rest fears about wool colour and quality generally. The two best scoring progeny groups were from Anderson rams. The results for FR and WQUAL suggest that in this regard, the Anderson rams performed as well as, if not better, than Talitas rams.

CONCLUSIONS

Although the progeny number produced to date is limited, the results from the Uruguayan genetic evaluation suggest that the genetic merit for resistance to internal parasites expressed in Australia by Anderson rams, is also expressed in Uruguay. Furthermore, the genetic merit of Anderson Rams for WEC is comparable to that of the best in Talitas Rams. The number of studs that have selected for resistance to gastrointestinal parasites is limited, so both, Anderson Rams and Talitas Rams face the problem of few or no alternative sources of stock to ensure long term continuity to their breeding programs. Because Anderson Rams and Talitas Rams have worked independently, their animals are unrelated, thus mutually providing an opportunity to increase the effective population size without compromising genetic merit in resistance to gastrointestinal parasites, in production traits, or in visually assessed characters. In the immediate future, the flow of genes from Anderson Rams to Uruguay will most likely continue, whereas in the more distant future, we should not rule out the possibility of a flow in both directions.

ACKNOWLEDGEMENTS

Lynley Anderson, Anderson Rams, for permission to publish these results.

REFERENCES

INIA, SUL (2018) Evaluaciones Genéticas Ovinas. Uruguay. Accessed 3 September 2019. www.geneticaovina.com.uy.

SAS Institute Inc. (2011) SAS/STAT® 9.3 User's Guide, SAS Inst. Inc. Cary NC, USA.

Westell R.A., Quaas R.L. and Van Vleck L.D. (1988) *J. Dairy Sci.* **71**: 1310.

INVESTIGATING NOVEL TRAITS IN SINGLE TRAIT SELECTION FOR THEIR POTENTIAL IN SELECTION INDEXES FOR FEED EFFICIENCY OF CROSSBRED PIGS

M.N. Aldridge[1], R. Bergsma[2] and M.P.L. Calus[1]

[1]Animal Breeding and Genomics, Wageningen University and Research, Wageningen, Gelderland, 6708 PB, the Netherlands
[2]Topigs Norsvin, PO Box 43, 6640 AA Beuningen, the Netherlands

SUMMARY

Here we considered selection on a single trait only, assuming a selection differential of one genetic standard deviation, to determine which novel traits to include in a multi-trait selection index for feed efficiency in crossbred pigs. The mean feed conversion ratio (FCR) is 2.52 (kg/kg), and a decrease of 5.56%, was observed if selection was based on FCR itself. Selection on other traits also reduced FCR but with a lower response: average daily gain (-2.9%), dry matter digestibility (-1.2%), nitrogen excreted (-0.40%), daily feed intake (-0.37%), group daily feed intake (-0.35%), eating time per day (-0.04%), and growth rate with social effect (-0.01%). Selection for the welfare traits increased FCR: joint lesions (0.2%), and total lesion count (0.06%). Further analysis will include additional traits and use selection index theory with multi-trait selection to determine an optimal selection index for feed efficiency.

INTRODUCTION

Selection for feed efficiency is of high importance for livestock species as it has direct effects on economic factors, reduces water and land requirements, and decreases greenhouse gas footprints (Hayes *et al.* 2013). There is a desire within the pig industry to improve the rate of genetic gain for feed efficiency of crossbred commercial animals, which requires an optimised selection index (Feed-a-Gene 2015). The main objective of this study was to identify indicator traits that make a promising contribution to such an index.

MATERIALS AND METHODS

Feed efficiency is defined as average daily gain (ADG) / daily feed intake (DFI). However, it is standard practice in pig breeding programs to select for a lower feed conversion ratio (FCR), where FCR = DFI / ADG. Because one is the reciprocal of the other, we assumed a genetic correlation of one between feed efficiency and FCR. Reported responses to selection are in terms of relative change to the across breed literature mean of 2.52 (kg/kg) (Mrode and Kennedy 1993; Cameron and Curran 1994; Labroue *et al.* 1997; Hoque and Suzuki 2008; Bates and Maechler 2010; Do *et al.* 2013; Saintilan *et al.* 2013; Gilbert *et al.* 2017).

Based on discussions with industry stakeholders we compiled a list of key traits. Production traits currently used in crossbred breeding programs included: FCR, ADG, and DFI. As a preliminary analysis a small number of traits from the list were selected as a representative of broader trait categories. These novel traits included: digestibility of dry matter (DIG), time spent eating per day (BEH), group daily feed intake (GFI), average daily gain with a social effect (GADG) which was selected as an indirect genetic effect (IGE), and nitrogen excreted (BIO). Two traits were selected as indicators of animal welfare including: joint lesions (JOINT) and total skin lesion count (WELF), where the latter represents impaired welfare inflicted by pen mates.

We reviewed the literature for parameter estimates of the indicator traits. Preference was given to estimates from sources that used crossbreds, had large numbers of progeny with phenotypic information and estimates with smaller standard errors. To determine which traits are likely to benefit the selection of improved feed efficiency, the correlated response to selection in FCR was evaluated for one indicator trait at a time.

For the analysis, genetic standard deviations and genetic correlations with FCR were required (Table 1). For most sources, the genetic variances or the heritability and the phenotypic variance were published, which were then converted to genetic standard deviations ($\sigma_G = \sigma_P * h$). There are no published estimates of genetic standard deviation for digestibility of dry matter, however, the authors of Ouweltjes *et al.* (2018) provided us with unpublished estimates of heritability which we used to estimate a genetic standard deviation.

Table 1. Genetic standard deviations (σ_G), genetic correlation with FCR, and summary of references used in the analysis

Trait	σ_G	Genetic correlation with FCR	σ_G references	Genetic correlation references
FCR	0.14	1.00	(Do *et al.* 2013)	
ADG	0.07	-0.44	(Do *et al.* 2013)	(Saintilan *et al.* 2013)
DFI	0.63	0.36	(Do *et al.* 2013)	(Saintilan *et al.* 2013)
DIG	0.41	-0.65	(Ouweltjes *et al.* 2018)	From broilers (Mignon-Grasteau *et al.* 2004)
BEH	3.35	0.17	(Do *et al.* 2013)	(Do *et al.* 2013)
GFI	0.17	0.12	(Canario *et al.* 2017; Sánchez *et al.* 2018)	(Peeters *et al.* 2013)
GADG	27.94	0.10	(Bergsma *et al.* 2008; Canario *et al.*, 2017)	(Canario *et al.* 2017)
BIO	0.23	0.16	(Saintilan *et al.* 2013)	(Saintilan *et al.* 2013)
JOINT	0.16	-0.09	(Luther *et al.* 2007)	(Luther *et al.* 2007)
WELF	0.34	-0.08	(Turner *et al.* 2006)	(Turner *et al.* 2006)

The response to selection for FCR was calculated as $\mathbf{R}=\mathbf{b}'\mathbf{G}/\sigma_I$, where **b** is a vector of weights for each trait, **G** is a covariance matrix calculated as a function of the genetic correlations and genetic standard deviations, and σ_I is the standard deviation of the index. As we were only interested in the change of a single trait this could be reduced to $\sigma_{I,R}/\sigma_I = r_{I,R} * \sigma_R$, where I is the indicator trait and R the response trait. The analysis was repeated for each of the traits, with the full weight placed on a single trait each time.

RESULTS AND DISCUSSION

The results showed that selecting for production traits had the largest impact on feed efficiency when included in a selection index (Table 2). Feed conversion ratio decreased (-5.6% relative to the literature mean), from 2.52 (kg/kg) to 2.38 (kg/kg), when 100% of selection was placed on FCR. Other traits reduced FCR in the following descending order: average daily gain (-2.9%), dry matter digestibility (-1.2%), nitrogen excreted (-0.4%), daily feed intake (-0.37%), group daily feed intake (-0.35%), eating time per day (-0.1%), and growth rate with social effect (-0.01%). Selection for joint lesions or total lesion count had the undesirable effect of increasing FCR (0.3% and 0.1%, response to FCR respectively).

Table 2. Response in feed conversion ratio due to a change of one genetic standard deviation (σ_G) in the selected trait

Trait	FCR (kg/kg) with one σ_G change in selected trait	Relative phenotypic change in FCR with one σ_G change in selected trait
Feed conversion ratio (FCR)	2.380 (Originally 2.520)	-5.56%
Average daily gain (ADG)	2.447	-2.90%
Daily feed intake (DFI)	2.511	-0.37%
Dry matter digestibility (DIG)	2.490	-1.20%
Eating time per day (BEH)	2.519	-0.04%
Group daily feed intake (GFI)	2.511	-0.35%
Growth rate with social effect (GADG)	2.520	-0.01%
Nitrogen excreted (BIO)	2.510	-0.40%
Joint lesions (JOINT)	2.525	0.20%
Total lesion count (WELF)	2.522	0.06%

We were interested in the traits that have the largest reduction in FCR and would therefore significantly contribute to a selection index. It is not surprising that the largest improvement to FCR occurred with direct selection, or that selecting for the component traits (ADG and DFI) also resulted in a significant response in FCR. As selection for dry matter digestibility had a reasonable impact on the response to selection for FCR, other digestibility traits such as energy or organic matter digestibility should be investigated further. If faeces are collected to include digestibility, it would be beneficial to also include nitrogen excreted. Unfortunately, there was limited research available on blood biomarkers but these could be worth exploring if they have similar genetic correlations as faecal biomarkers.

The traits that had limited impact on the response to selection of FCR, could still be beneficial. Selection for eating time per day had a limited impact on FCR, but feeding behaviour traits such as time per meal, and number of meals per day, have higher genetic correlations with FCR, have higher heritabilities, but have less accurate parameter estimates (Do *et al.* 2013). Group daily feed intake appears to be a good indicator of individual daily feed intake and had a similar benefit to the selection response of FCR. It is not logistically or economically possible to record DFI on crossbred pigs, but GFI would be much easier and cheaper to record, this would benefit a selection index for crossbred feed efficiency. Including an IGE with GADG appears to have limited benefit to selection for FCR but could be important for defining the ADG model used in animal evaluations. The low negative genetic correlation between the welfare traits is unfavourable. However, to address consumer concerns it is important they are added to future selection indexes to limit any negative trends.

For future analysis a genetic covariance matrix will be required, which is to be built with estimates available in the literature. Currently a data set is being analysed which will complete the missing variance components, genetic correlations between traits, and genetic correlations between purebred and crossbred pigs, which Wientjes and Calus (2017) showed to not be equal to one. When the parameter estimation is complete, an optimised multi-trait selection index for feed efficiency in crossbred pigs will be built, and will be based on selection index theory (Hazel 1943). This study used a limited number of traits, future work will include additional traits related to digestibility, i.e. eating behaviour, group records, welfare, biomarkers, perturbations (Putz *et al.* 2018), and microbiota

(Camarinha-Silva *et al.* 2017). Finally, the potential for selection based on variation, heritability, and ease of phenotyping will also be considered.

CONCLUSIONS

The objective of this study was to determine which indicator traits are likely to have a significant contribution to an optimised selection index for feed efficiency in crossbred pigs. From these results production traits are the most promising, but novel traits such as digestibility, group records, and biomarkers could also increase the rate of genetic gain. Before such an index is built genetic correlations between novel traits and FCR need to be estimated.

ACKNOWLEDGMENTS

This article is part of the project Feed-a-Gene. The project has received funding from the European Union's Horizon 2020 research and innovation program under grant agreement No 633531.

REFERENCES

Bates D. and Maechler, M. (2010) *R*

Bergsma R., Kanis E., Knol E.F. and Bijma P. (2008) *Genetics* **178**: 1559.

Camarinha-Silva A., Maushammer, M., Wellmann, R., Vital, M., Preuss, S. and Bennewitz, J. (2017) *Genetics* **206**: 1637.

Cameron N.D. and Curran M.K. (1994) *Anim. Sci.* **59**: 281.

Canario L., Lundeheim N. and Bijma P. (2017) *Heredity* **118**: 534.

Do D.N., Strathe A.B., Jensen J., Mark T. and Kadarmideen H.N. (2013) *J. Anim. Sci.* **91**: 4069.

Feed-a-Gene. (2015) 'WP5 Use of traits in animal selection (Genetic parameter estimations, genetic model developments & evaluation of breeding schemes).' Available at https://www.feed-a-gene.eu [Accessed 10/04/2019].

Gilbert H., Billon Y., Brossard L., Faure J., Gatellier P., Gondret F., Labussière E., Lebret B., Lefaucheur L. and Le Floch N. (2017) *Animal* **11**: 1427.

Hayes B.J., Lewin H.A. and Goddard M.E. (2013) *Trends. Genet.* **29**: 206.

Hazel L.N. (1943) *Genetics* **28**: 476.

Hoque M.A. and Suzuki K. (2008) *Anim. Sci. J.* **79**: 543.

Labroue F., Guéblez R. and Sellier P. (1997) *Genet. Sel. Evol.* **29**: 4510.

Luther H., Schwörer D. and Hofer A. (2007) *Animal* **1**: 1105.

Mignon-Grasteau S., Muley N., Bastianelli D., Gomez J., Péron A., Sellier N, Millet N., Besnard J., Hallouis J.-M. and Carre B. (2004) *Poultry Sci.* **83**: 860.

Mrode R. and Kennedy B. (1993) *Anim. Sci.* **56**: 225.

Ouweltjes W., Verschuren L., Pijlman J., Bergsma R., Schokker D., Knol E., van der Aar P., Molist F. and Calus M. (2018) *Livest. Sci.* **207**: 63.

Peeters K., Ellen E.D. and Bijma, P. (2013) *Genet. Sel. Evol.* **45**: 27.

Putz A.M., Harding J.C., Dyck M.K., Fortin F., Plastow G.S., Dekkers J.C. and Canada, P. (2018) *Front. Gent.* **9**

Saintilan R., Merour I., Brossard L., Tribout T., Dourmad J.-Y., Sellier P., Bidanel J., Van Milgen J. and Gilbert H. (2013) *J. Anim. Sci.* **91**: 2542.

Sánchez J.P., Piles M., Tulsá M., Reixach J. and Quintanilla R. (2018) *WCGALP* **10**: 574.

Turner S., White I., Brotherstone S., Farnworth M., Knap P., Penny P., Mendl M. and Lawrence A. (2006) *Anim. Sci.* **82**: 615.

Wientjes Y. and Calus M. (2017) *J. Anim. Sci.* **95**: 3467.

GENOMIC RELATIONSHIPS TO CONTROL INBREEDING IN OPTIMUM-CONTRIBUTION SELECTION REALISE MORE GENETIC GAIN THAN PEDIGREE RELATIONSHIPS WHEN INBREEDING CONTROL IS RELAXED AROUND QUANTITATIVE TRAIT LOCI

M. Henryon[1,2], P. Berg[3,4], H. Liu[4], G. Su[4], T. Ostersen[1] and A.C. Sørensen[4]

[1]Danish Pig Research Centre, SEGES, Denmark
[2]School of Agriculture and Environment, University of Western Australia, Australia
[3]Department of Animal and Aquaculture Sciences, Norwegian University of Life Sciences, Norway
[4]Institute for Molecular Biology and Genetics, Aarhus University, Denmark

SUMMARY

We tested the premise that optimum-contribution selection with genomic relationships to control inbreeding (GOCS) realises more genetic gain (ΔG) than optimum-contribution selection with pedigree relationships (POCS) at the same rate of true inbreeding (ΔF) when we relax inbreeding control in regions of the genome harbouring QTL. We used stochastic simulation to compare ΔG realised by GOCS with POCS at 0.01 ΔF when we relaxed inbreeding control around 18 major QTL. These QTL were unlinked and explained either 100 or 50% of the total additive-genetic variation (V_a) for a trait under selection. We found that GOCS with relaxed inbreeding realised up to 4.7% more ΔG than POCS at 0.01 ΔF when the 18 major QTL explained 100% V_a. When these QTL explained 50% V_a, GOCS with relaxed inbreeding control realised up to 1.1% more ΔG. Even though GOCS with relaxed inbreeding control realised more ΔG than POCS, we were surprised that the amount of extra ΔG was small, given that we simulated simple genetic models. This does not bode well for practical breeding schemes, where most traits under selection are controlled by many linked QTL and we don't know where most of these QTL are located. So, GOCS with relaxed inbreeding control is a concept that realises more ΔG than POCS at the same ΔF, but we have more to learn before it becomes applicable to practical breeding schemes. For these schemes, POCS remains a worthy method of optimum-contribution selection.

INTRODUCTION

Pedigree relationships to control inbreeding in optimum-contribution selection (OCS) realise more genetic gain (ΔG) than genomic relationships at the same rate of true inbreeding (ΔF), where the true inbreeding coefficient of an individual is the observed proportion of loci in its genome with alleles that are identical-by-descent (IBD) (Henryon *et al.* 2019). Using pedigree relationships to control inbreeding in OCS – hereafter referred to as POCS – realises more ΔG because it manages expected genetic drift without restricting selection at QTL. By contrast, genomic relationships – referred to as GOCS – penalises changes in allele frequencies at marker loci generated by genetic drift and selection. Because these marker alleles are in linkage disequilibrium with QTL alleles, GOCS restricts changes in allele frequencies at QTL. This implies that if GOCS is to realise more ΔG than POCS, we should allow changes in allele frequencies at some markers by varying the level of inbreeding control across the genome while controlling ΔF at acceptable levels. This will involve relaxing inbreeding control in regions of the genome that harbour QTL – allowing selection to increase the frequencies of favourable alleles at QTL – while increasing inbreeding control to reduce genetic drift in other regions. This reasoning led us to believe that GOCS realises more ΔG than POCS at the same ΔF when we relax inbreeding control in regions of the genome harbouring QTL. We tested this premise by stochastic simulation.

MATERIALS AND METHODS

Procedure. We used stochastic simulation of animal-breeding schemes to compare ΔG realised by GOCS with POCS at ΔF = 0.01 (0.01 ΔF) when we relaxed inbreeding control around 18 major QTL. These QTL were unlinked and explained either 100 or 50% of the total additive-genetic variation (V_a) for a single trait under selection. GOCS with relaxed inbreeding control was carried out by excluding markers located within 0, 1, 2, 5, 10, 20, 30, 40, and 50 cM of the 18 major QTL from genomic-relationship matrices used to control inbreeding (i.e., excluding markers in genome regions of 0, 2, 4, 10, 20, 40, 60, 80, and 100 cM centred around the 18 QTL). These GOCS are referred to as $GOCS_0$, $GOCS_1$ … $GOCS_{50}$, where $GOCS_0$ includes all markers and is the GOCS used in Henryon *et al.* (2019). ΔF was calculated as the increase in the observed proportion of IBD loci across the genome that were IBD. The trait under selection had a heritability of 0.2. Breeding values for the trait were predicted by GBLUP. Breeding schemes were run for 10 discrete generations ($t = 1 \ldots 10$) and replicated 500 times. Each replicate was initiated by sampling a unique base population from a founder population. Selection candidates were genotyped and phenotyped before selection.

Breeding scheme. A total of 25 matings were allocated to 125 selection candidates by OCS in each generation. There was no upper limit for the number of matings that were allocated to each male; males were allocated 0, 1, 2 … or 25 matings. Twenty-five females were allocated a single mating. The 25 sire and dam matings were paired randomly. Each pair (dam) produced five offspring, resulting in 25 full-sib families and 125 offspring. Offspring were assigned as males or females with a probability of 0.5.

Genetic models. The founder population was established using a Fisher-Wright inheritance model to generate linkage disequilibrium between QTL and markers. The genome was 30 M and consisted of 18 pairs of autosomal chromosomes; each chromosome was 167 cM long. The 18 major QTL were located on separate chromosomes. Each of these QTL had a minor-allele frequency of 0.25 (approx.) and explained equal proportions of V_a in the founder population. They each explained $\frac{1}{18}V_a$ when the 18 major QTL explained 100% V_a. When the major QTL explained 50% V_a, each QTL explained $\frac{1}{36}$ V_a; the remaining 50% V_a was explained by an additional 7684 minor QTL that were randomly distributed across the genome. The genome also contained 54218 biallelic markers that were randomly distributed across the genome. These markers were distinct from QTL and used in GOCS and GBLUP. A total of 6012 IBD loci were placed evenly across the genome in base populations. Unique alleles at these loci were used to calculate ΔF.

Optimum-contribution selection. POCS was carried out by maximising $\mathbf{U}_t(\mathbf{c}) = \mathbf{c}'\hat{\mathbf{a}} - \omega\mathbf{c}'\mathbf{A}\mathbf{c}$, where **c** is a vector of genetic contributions to the next generation, â is a vector of GBLUP-EBV, ω is a penalty applied to the average-estimated relationship of the next generation, and **A** is a pedigree-relationship matrix (after Henryon *et al.* 2019). The penalty, ω, was constant across generations. It was calibrated to realise 0.01 ΔF. GOCS was carried out by replacing **A** with a genomic-relationship matrix, **G**. **G** was constructed as described by VanRaden (2008) using marker-alleles frequencies in the base populations.

Data analyses. ΔG was calculated as the linear regression of G_t on t, where G_t is the average breeding value of animals born at times $t = 4 \ldots 10$. ΔG realised by POCS and GOCS differed when the 18 major QTL explained 100 and 50% V_a. We scaled ΔG by setting ΔG realised by POCS to 100 in the two genetic models. ΔF was calculated as 1-exp(β), where β is the linear-regression coefficient of $\ln(1-F_t)$ on t, and F_t is the average coefficient of true inbreeding for animals born at times $t = 4 \ldots 10$ (after Sonesson *et al.* 2004). We also present IBD profiles for POCS, $GOCS_0$, and $GOCS_{10}$ on chromosome 3 when the 18 major QTL explained 100% V_a. IBD profiles are presented as the change in realised IBD from generations $t = 4$ to 10 at the 6012 IBD loci. Scaled ΔG and IBD profiles are presented as means of the 500 replicates.

RESULTS AND DISCUSSION

Our findings supported our premise that GOCS realises more ΔG than POCS at the same ΔF when we relax inbreeding control in regions of the genome harbouring QTL. We found that $GOCS_5$ … $GOCS_{40}$ realised 2.7-4.7% more ΔG than POCS at 0.01 ΔF when 18 major QTL explained 100% V_a (Figure 1). When these QTL explained 50% V_a, $GOCS_{10}$ and $GOCS_{20}$ realised 0.3 and 1.1% more ΔG than POCS. Clearly, GOCS with relaxed inbreeding control – where we removed the penalty applied to changes in allele frequencies at markers located around major QTL – is a concept that worked. It worked for two reasons. First, selection increased the frequency of the favourable allele at each of the 18 major QTL with POCS and GOCS with relaxed inbreeding control, but GOCS with relaxed inbreeding control allowed selection to increase the frequencies of favourable alleles more than POCS. Second, GOCS with relaxed inbreeding control allowed selection to generate more IBD in genome regions around the major QTL than POCS. This was illustrated by our IBD profiles on chromosome 3 when the 18 major QTL explained 100% V_a (Figure 2). $GOCS_{10}$ generated a higher IBD peak around the major QTL on chromosome 3 than POCS and $GOCS_0$. At the same time, $GOCS_{10}$ generated, on average, less IBD than POCS and $GOCS_0$ in regions of the genome that lacked major QTL. It must have generated less IBD in these regions because the area under an IBD profile increases at the same rate at the same ΔF. These two reasons tell us that GOCS with relaxed inbreeding control allows more IBD in regions of the genome where we want to increase the frequency of favourable alleles at QTL, while controlling IBD and genetic drift in other regions. It is exactly how we want to control inbreeding in animal breeding when the aim is to maximise ΔG at acceptable ΔF. So, GOCS with relaxed inbreeding control realises more ΔG than POCS at the same ΔF because it allows inbreeding in regions of the genome that realise ΔG and controls it in other regions.

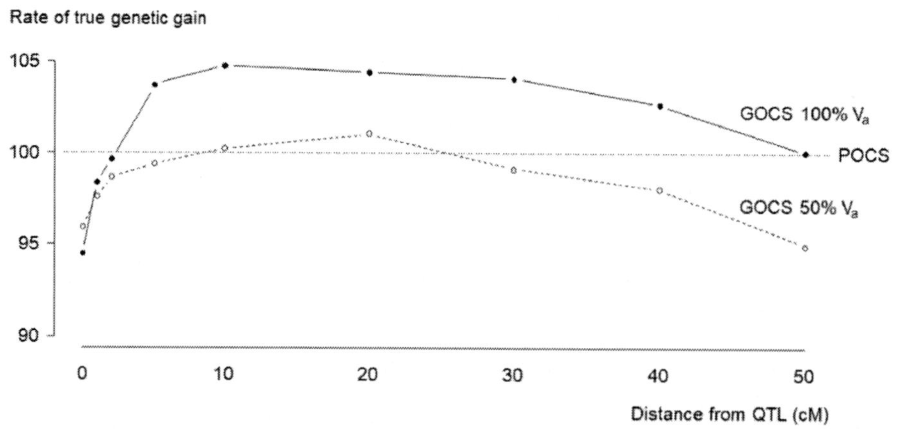

Figure 1. Rates of genetic gain realised by POCS and GOCS with relaxed inbreeding control at 0.01 rate of true inbreeding plotted against distance from 18 major QTL excluded from inbreeding control. The 18 QTL explained 100 and 50% of the additive-genetic variation (100% V_a, 50% V_a) for a single trait under selection. Rates of genetic gain were scaled by setting the rates of genetic gain realised by POCS with 100 and 50% V_a to 100. The rates are means of 500 simulation replicates. SD between the replicates ranged from 12.0-13.7

Even though GOCS with relaxed inbreeding control realised more ΔG than POCS, we were surprised that the amount of extra ΔG was small when we simulated a simple genetic model where 100% V_a was explained by only 18 unlinked QTL with known genome locations. This extra ΔG all but disappeared when the 18 major QTL explained 50% V_a. These findings are important because they imply that GOCS with relaxed inbreeding control only realises more ΔG than POCS at the same ΔF when traits are controlled by few unlinked QTL and we know where these QTL are located on the genome. It does not bode well for practical breeding schemes, where most, if not all, traits under selection are controlled by many linked QTL – each with small effects – and we don't know where most of these QTL are located. So, GOCS with relaxed inbreeding control is a concept that realises more ΔG than POCS, but we have more to learn before it becomes applicable to practical breeding schemes. For these schemes, POCS remains a worthy method of OCS.

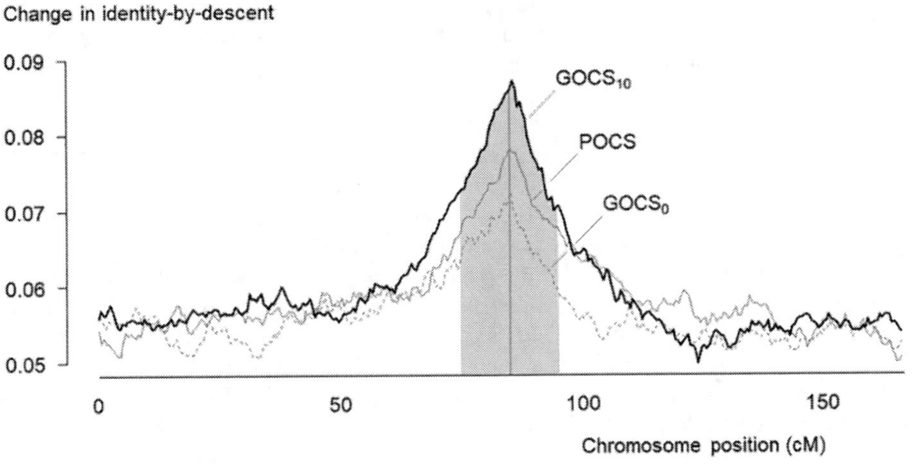

Figure 2. Identity-by-descent profiles for POCS, $GOCS_0$, and $GOCS_{10}$ on chromosome 3 at 0.01 rate of true inbreeding when 18 major QTL explained 100% of the additive-genetic variation for a single trait under selection. The profiles present the change in IBD realised at IBD loci located across the chromosome. The vertical line at 84.8 cM is the position of a single major QTL on chromosome 3; the shaded area represents the region of the genome that is within 10 cM of the major QTL. The profiles are means of 500 simulation replicates

ACKNOWLEDGEMENTS

This study was financed by the Center for Genomic Selection in Animals and Plants (GenSAP), which was partially funded by Innovation Fund Denmark (grant 0603-00519B); Danish Ministry of Food, Agriculture and Fisheries (grant 34009-12-0540); and SEGES, Danish Pig Research Centre.

REFERENCES

Henryon M., Liu H., Berg P., Su G., Nielsen H.M., Gebregiwergis, G.T. and Sørensen A.C. (2019) *Genet. Sel. Evol.* **51**, 39.

Sonesson A.K., Woolliams J., Meuwissen T.H.E. (2004) In 'Selection and breeding programs in aquaculture', pp. 73-87, editors T. Gjedrem and K. Doordrecht, Springer, Dordrecht.

VanRaden P.M. (2008) *J. Dairy Sci.* **91**: 4414.

FINE CONTROL OF BULL ALLOCATION TO HELP AVOID DYSTOCIA

B.P. Kinghorn[1] and A.L. Van Eenennaam[2]

[1]University of New England, Armidale, NSW, 2351 Australia
[2]Department of Animal Science, University of California, Davis, California, USA

SUMMARY

A method is presented for influencing mate selections according to phenotype. The example uses female body weight at or close to joining as an indicator of liability to dystocia in cattle. However, the method may also be applicable to sheep and other species. A common way to manage this issue is to allocate only good calving ease EBV bulls to heifers. However, the method presented here is more powerful, as it customises allocations according to the bodyweight of each female, with smaller heifers getting better calving ease bulls. In addition, the overall emphasis on calving ease can be controlled during a mate selection analysis, trading it off against all other issues.

INTRODUCTION

Breeders who want to manage calving ease in their herd (or flock) can choose bulls that have favourable EBVs for 'calving ease direct' (CED) – meaning that the genotype of the resulting calf is more favourable for calving ease at its own birth. A more refined solution is to use grouping to allocate only high CED bulls to heifers, as described in Figure 1.

However, the approach presented in this paper is more powerful than simple grouping, as it customises allocations according to the ***phenotype*** of each heifer for a liability indicator trait such as pre-joining bodyweight, with smaller heifers getting better CED bulls (Figure 1). In addition, when using a Mate Selection implementation (Kinghorn and Kinghorn 2019), the breeder can alter the overall emphasis on calving ease, trading it off against all other issues. This approach requires some upfront effort to define the parameters that reflect the breeder's desires in relation to calving ease for heifers of different weights, but it does allow for a more strategic use of sires.

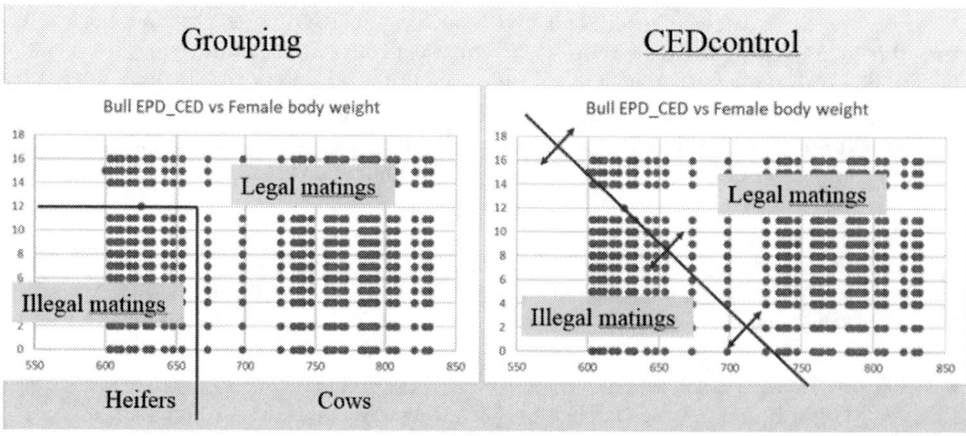

Figure 1. A contrast of simple grouping to allocate only high CED bulls to heifers (left pane), versus CEDcontrol which uses a sliding scale to prioritise the highest CED bulls to the females with the lowest body weight (right pane)

METHOD

A column in the data file with a descriptive header, e.g. "CEDcontrol" (calving ease control), or "BWTcontrol" (birthweight control), depending on what EBVs are available for ease of calving, has values entered as follows:

For each female candidate: Enter her current ***phenotype*** for body weight (or some such criterion of liability to calving difficulty).

For each male candidate: Enter -1 times the smallest weight of female that this bull should be allowed to mate, given its EBV for Calving Ease Direct (or some such EBV), according to the breeder's judgement.

With this setup, a mating between any bull and any cow gives a predicted progeny value for CEDcontrol that needs to be at least 0 to satisfy the breeder's desires in relation to calving ease for that female (see Table 1).

Table 1. Example calculation for male entries under data column CEDcontrol. Values for *Intercept* and *Slope* are calculated as shown below the table

Female body weight	Minimum Calving Ease Direct EBV chosen by breeder	Male entry for a bull of minimum EBV (*Intercept*+*Slope**EBV)	Predicted progeny value for CEDcontrol when using this bull
275Kg Heifer	+10	-400 + 12.5x10 = -275	0
300Kg Heifer	+8	-400 + 12.5x 8 = -300	0
400Kg Cow	0 (linear extrapolation)	-400 + 12.5x 0 = -400	0

Slope = $\frac{300-275}{10-8}$ = 12.5
Intercept = (2 x *Threshold*) -275 – (*Slope* x 10) = 0 -275 -125 = -400
Intercept = (2 x *Threshold*) -300 – (*Slope* x 8) = 0 -300 -100 = -400
… where *Threshold* = 0 in the example implementation (eg. last column, and in Figure 2).

The breeder only has to choose the four figures near the top left of Table 1: 275Kg, 300Kg, +10 and +8 (yellow shading). This represents the breeder's attitude to bull requirements for calving ease EBVs depending on female body weight. The third row (400Kg Cow) is only included for illustration. Notice that the fourth column (values = 0) is the average of the first and third columns, just as progeny predictions are the average of dam and sire EBVs.

Accordingly, what we enter for each bull in column CEDcontrol is *Intercept*+*Slope**CED where CED is the bull's Calving Ease Direct EBV. *Slope* and *Intercept* are calculated from the four figures chosen by the breeder, plus the target Progeny value threshold (*Thresh* = 0 here, but a different value can be chosen for cosmetic reasons).

The four driving figures (275Kg, 300Kg, +10 and +8 here) should cover the bodyweight region where calving ease is an issue. If there is little benefit from using high calving ease bulls over heavy cows, this is not critical, as the breeder can use a trait management tool in a way that gives no reward for high calving ease in heavy cows, eg. by only avoiding matings below the 0 threshold.

Once the CEDcontrol column has been made, the mate selection analysis is run with a constraint on progeny CEDcontrol to be at least 0 for all matings, eg. using 'Set minimum value at boundary' (see Kinghorn and Kinghorn 2019), as in Figure 2.

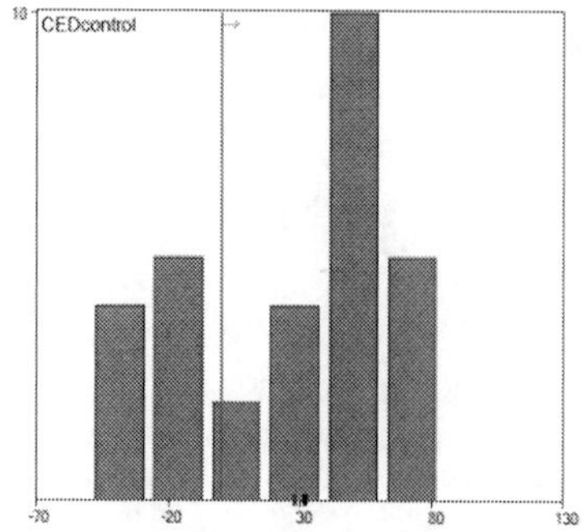

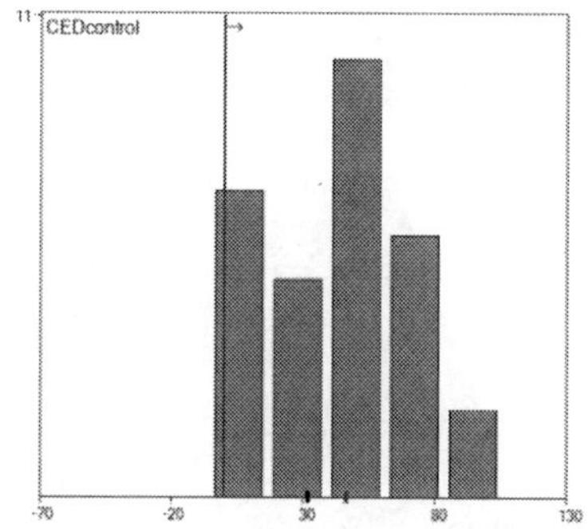

Figure 2. Setting CEDcontol boundary at zero to ensure that all matings conform to the breeder's calving ease policy with respect to female body weights. Left: Boundary not invoked (9 matings do not conform). Right: Boundary invoked (all matings conform). This is a small example with just 30 matings

RESULTS AND DISCUSSION

There are several main factors affecting calving difficulty including calf size, pelvic area of the cow, breed, parity of the calving, sex of the calf, gestation length, the season of the calving (Mekonnen and Moges 2016). Developing heifers on a low nutrient diet has clearly demonstrated an increase in dystocia. This is primarily due to poor skeletal growth and, therefore, smaller pelvic areas. While some studies have found that heifers of lighter weight have an increased risk of dystocia (Erb. *et al.* 1985; Naazie *et al.* 1989), other research has demonstrated that after calf size, the most important phenotypic predictor of dystocia is pre-calving pelvic area (Johnson *et al.* 1988). Heifers with a pelvic area of less than 140 cm^2 have increased incidence of dystocia compared to their above-average contemporaries. Larger heifers have larger pelvic areas, but they also have larger calves. Selecting large heifers for replacements may have little effect on dystocia unless pelvic areas are also known.

This paper has adopted female body weight as an indicator of liability to calving difficulties. Of course, the current method does not alleviate the situation by increasing the body weight of small heifers, but by aiming for them to have calves of smaller size, and/or whatever other attributes of calves lead to improved calving ease. This means that the observed impact of calf size on dystocia is also indicative of the value of the current method. It may be that some other trait or index of traits will be more diagnostic for the scenario in question than simple female body weight. The method proposed can use any such predictor.

High ewe liveweight and condition score during pregnancy may help indicate the risk of dystocia in sheep (Horton *et al.* 2017), such that the method proposed may be of some value in that species.

In a simple beef cattle example, a Trait Management tool (Kinghorn and Kinghorn 2019) was used to manage the progeny distribution of CEDcontrol, by setting a minimum boundary at 0 for predicted progeny merit, so that all matings satisfy the breeders desires (Figure 2).

The minimum boundary can be changed upwards from 0 to give even more overall emphasis on Calving Ease. This is a dynamic policy with smaller heifers always attracting more attention, whatever threshold is set. For example, to get 10 in the right-hand column of the top row of Table 1, still using the EBV = +10 bull, we would use a 295Kg heifer (as the average of 295 and -275 is 10). This means we would now afford a 295Kg heifer as much CED priority as we previously did for a 275Kg heifer. The breeder must then judge if the extra calving ease attained is worth the likely compromises seen in other issues, such as progeny merit for the selection index and inbreeding.

It is possible to scale this approach differently, so that a +1 progeny CEDcontrol value represents an increase of +1 in EBV units, setting the bar higher by that amount of EBV. Alternatively, if there is good predicted relationship between EBV and % calving difficulties, a breeder could operate directly at the level of predicted % calving difficulties.

REFERENCES

Horton B.J., Corkrey R. and Hinch G.N. (2017) *Anim. Prod. Sci.* **58:** 1125

Johnson S.K., Deutscher G.H. and Parkhurst A. (1988) *J. Anim. Sci.* **66**: 1081.

Kinghorn B.P. and Kinghorn A.J. (2019) 'Instructions for Matesel'. http://www.matesel.com Accessed 26 April 2019.

Mekonnen M. and Moges N. (2016) *Eur. J. Biol. Sci.* **8**: 91.

Naazie A, Makarechian M.M. and Berg RT. (1989) *J. Anim. Sci.* **67**: 3243.

WETHER TRIALS AND THEIR ROLE IN MERINO BLOODLINE EVALUATION

K.L. Egerton-Warburton[1], K.D. Atkins[1,2], L.M. Stephen[3] and S.I. Mortimer[3]

[1]NSW Department of Primary Industries, Orange, NSW, 2800 Australia
[2]Current address, Shoal Bay, NSW, 2315 Australia
[3]NSW Department of Primary Industries, Armidale, NSW, 2351 Australia

SUMMARY

Combined analyses of wether trial data have provided commercial Merino producers with reliable estimates of differences among bloodlines, based on the performance of their client flocks. The evolution of wether comparisons as a vehicle to obtain information on bloodline differences is briefly described along with the substantial changes in the genetic evaluation environment since the inception of the combined analyses. A future combined analysis of wether and ewe productivity trait data with genomic flock profiling may well represent the next step in the evolution of wether trials in Australia.

INTRODUCTION

Merino wether trials started as small scale commercial producer production competitions in several locations from the late 1970s. Simultaneously, resource flocks (such as D flock, Mortimer and Atkins 1989) showed that large differences existed among Merino studs and bloodlines that ram buyers could access when purchasing flock rams. Although a single wether trial provided little or no information on bloodline performance an innovative across-trial analysis was devised that combined all available data to produce reliable estimates of differences among studs, based on the performance of their client flocks (Hygate and Atkins 1988). At that time, there was no comparative across-stud performance information available in the public domain.

This paper will review the use of combined analyses of wether trial data to provide commercial Merino producers with information on Merino bloodline differences and briefly describe the evolution of wether trials as a vehicle to obtain information on bloodline differences. Given the changes in the genetic evaluation environment since the inception of the combined analyses, a future role of the combined wether trial analyses will be proposed.

THE BEGINNINGS

While resource flocks were demonstrating to commercial producers the large differences that existed between Merino bloodlines in wool production traits, wether trials became widespread and were used as a basis for regional breeding extension activities. In contrast to the resource flocks, wether trials had several practical advantages. Wether trials were located in a range of environments and were able to more cost-effectively collect wool production data on a wider range of bloodlines, albeit the bloodlines were represented by teams of wethers selected from commercial flocks. As importantly, the identity of the bloodline represented by each team was publicly reported, whereas non-disclosure agreements prevented this happening with the resource flocks. It became apparent that the data from wether trials could be useful in genetic evaluation of bloodline sources.

Following the pilot study of Hygate and Atkins (1988), the first attempt to comprehensively report on Merino bloodline performance was provided by Atkins *et al.* (1992). Their report used data from 48 wether trials, conducted in NSW between 1981 and 1991, and included robust estimates of performance in wool production and quality traits. Key features of the wether trials analysed were the wide distribution of trials across all regions of NSW, random sampling of teams of wethers and an average of at least 10 wethers per team. Regional variation in, and economic evaluations of bloodline performance, as well as guidance on interpretation and application of the results, were provided. The limitations of this form of bloodline evaluation were noted: potential for inaccurate description of the ram source; occurrence of non-random selection of wethers; and the historic nature of the data.

INFORMATION DELIVERED

The first across-trial analysis published by Hygate and Atkins (1988) reported on records of greasy fleece weight, fibre diameter (FD) and yield, and the derived trait of clean fleece weight (CFW). This initial publication was a 'proof of concept' using data from 12 wether trials across NSW, flagging a range of opportunities which were to become part of the future analysis and reporting of wether trial comparisons – now known as Merino Bloodline Performance.

Table 1 summarises the number of wether trials, teams and bloodlines represented in the across-trial analyses since 1992. Bodyweight and assessments of wool quality (inferred from wool type) were included in the analysis reported in 1992. Subsequently, stability traits (relative change with age in CFW and FD) were analysed and reported. The 2005 and later reports, plus supporting information, are accessible via the web (www.merinobloodlines.com.au).

Table 1. Summary of wether trials, bloodlines and teams represented in the Merino Bloodline Performance reports since inception

Year of published report	Number of wether trials contributing data	Bloodlines			Number of teams
		High and Medium Accuracy	Low Accuracy	Total	
1992	48	53	80	133	988
1995	54	61	83	144	1,110
1996	76	73	113	186	1,417
1998	67	75	117	192	1,184
2000	68	65	131	196	1,365
2005	58	71	95	166	1,182
2007	63*	137	85	222	1,087
2010	57*	145	123	268	1,285
2014	23*	71	1	72	922
2016	26*	77	0	77	457
2018	25	73	0	73	482

* Data from both ewe productivity trials and wether trials contributing to these reports.

Initially, economic analyses were reported using gross margins with different price periods selected to reflect a range of market scenarios (low to high micron premium; current versus long term average prices). Gross margins were reported on a per head and per dry sheep equivalent (DSE) basis to allow for differences in size and hence stocking rate.

Early attempts to model the whole farm economic impacts of differences between bloodlines were reported by Wilson *et al.* (1996). Their analysis, and that of Coelli *et al.* (2000), included the extrapolation from wether data to modelling of ewe enterprises. The 2010 analysis saw a change from gross margin to gross income, with gross income being calculated with a greater emphasis on income from meat (ratio of fleeces to surplus sheep sales of 2.5 to 1, compared with earlier reports where the ratio was 4:1) (Martin *et al.* 2010). Grassgro™ (Moore *et al.* 1997) has been used to model the financial performance of the bloodlines since 2014. Using base parameters for wether production systems at Bookham, Narrandera and Woolbrook (replaced by Armidale in 2018), the livestock production parameters were the outputs of the bloodline analysis. Three different price scenarios (median, 30th percentile and 70th percentile) can be simulated across the three environments.

Bloodline parameters and financial performance have also supported delivery of other activities such as the 'Merino Breeding & Selection' workshops (Hatcher and Bayley 1999), and the tool 'Bloodline benchmark' (Coelli *et al.* 1997). Other products that support the Merino Bloodline Performance analyses include the guidelines for the conduct of wether comparisons and on-farm genetic evaluations (Martin *et al.* 2005) and software (Sheep Wether Comparison – SWC) that supports collection and reporting of wether comparison results at individual sites, and facilitates provision of quality data to the across-trial analysis (Semple 2005).

CHANGES IN THE EXTERNAL ENVIRONMENT

Since the first wether trials were published, there have been significant changes to the Australian Sheep industry. The collapse of the reserve price scheme and a significant increase in value of surplus sheep has seen breeding objectives for a significant proportion of the sheep industry change to a more dual purpose (meat and wool) focus. This has generated increased interest in traits such as growth, carcase and reproductive performance, leading to the breeding ewe flock evaluations mentioned earlier. Sheep Genetics now runs MerinoSelect, the national genetic evaluation service (Brown *et al.* 2007) for Merino ram breeders, while a range of on-farm technologies have made the monitoring of animals and flock management easier.

The delivery network for wether trials has also changed significantly. The reduction in public sector extension by the state departments across the country has meant that the location and duration of wether trials is now largely in the hands of grower groups.

WHERE TO NEXT?

The need for wether trial information as the major source of across-flock differences in Merinos is less urgent today as more Merino studs participate in MerinoSelect. However, there are still many ram sources either not enrolled in MerinoSelect or that have inadequate or unreliable linkage with other flocks. A key question is "Do wether trials represent the only source of data in continuing to provide reliable and comparable bloodline differences for ram sources not available through MerinoSelect?"

Genomic flock profiles (Swan *et al.* 2018) are a relatively new source of data that provide information on the breeding value of flocks for various traits, including previously expensive or difficult to measure traits. Flock profiling combined with the range of phenotypes that can be recorded within wether trials, as well as the important forum for interactions between producers and service providers that wether trials promote, offer new opportunities for commercial evaluation of Merino bloodlines. For both adult CFW and FD there is good agreement between the genomic breeding values obtained from flock profiles of the single bloodline teams of the Peter Westblade Memorial Merino Challenge 2016-2018 (S. Martin, C. Wilson and T. Granleese, unpublished data) and the bloodline deviation estimates from an analysis of the 4 challenges conducted between 2010 and 2018 (Figure 1).

An innovative combined analysis of information from wether trials, ewe productivity trials and flock profiles can provide valuable and accurate information on across-stud differences in addition to that which is available in MerinoSelect.

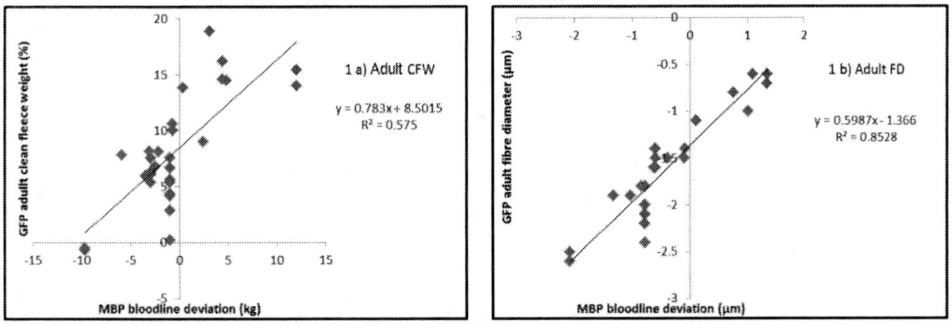

Figure 1. Relationship between mean bloodline deviations and genomic flock profile of single bloodline teams of the Peter Westblade Memorial Merino Challenge 2016-2018

ACKNOWLEDGMENTS

Merino Bloodline Performance and earlier reports were funded in part by Australian Wool Innovation and its predecessors. These reports were only possible due to the diligent efforts of the Committee members of the contributing wether trials in collecting data and making it available for further analysis. The efforts are acknowledged of current and past staff of the various state Departments of Agriculture, particularly NSW, in liaising with and supporting those committees in providing useable data for analysis, and in the conduct and reporting of the analyses.

REFERENCES

Atkins K.D., Semple S., Casey A.E. and Hygate L. (1992) 'Variation in production traits between Merino bloodlines – NSW wether comparisons 1981 -1991'. NSW Agriculture, Orange.

Brown D.J., Huisman A.E., Swan A.A., Graser H.-U., Woolaston R.R., Ball A.J., Atkins K.D. and Banks R.G. (2007) *Proc. Assoc. Advmt. Anim. Breed. Genet.* **17**: 187.

Coelli K.A., Atkins K.D., Casey A.E. and Semple S.J. (2000) 'Merino bloodlines: The comparisons: 1988-99'. Agnote DAI/52 (2nd ed.), NSW Agriculture, Orange.

Coelli K., Atkins K., Semple S. and Casey A. (1997) *Proc. Assoc. Advmt. Anim. Breed. Genet.* **12**: 365.

Hatcher S. and Bayley D. (1999) 'Merino Breeding and Selection – a commercial focus', NSW Agriculture and The Woolmark Company.

Hygate L. and Atkins K.D. (1988) *Proc. Assoc. Adv. Anim. Breed. Genet.* **7**: 414.

Martin S.J., Graham R.P., Langford C.M. and Lollback M.W. (2005) 'Designing and conducting Merino Wether Comparisons and On-farm Genetic Evaluations', (2nd ed.), NSW Department of Primary Industries, Orange.

Martin S.J., Atkins K.D., Semple S.J., Sladek M.A., Thackeray R.H., Staines J.M., Casey A.E., Graham R.P. and Russell A.J. (2010), 'Merino bloodlines: the comparisons 1999-2010', Primefact 930, Department of Industry & Investment NSW, Orange.

Moore AD, Donnelly JR and Freer M (1997) *Agric. Systems* **55**: 535.

Mortimer S.I. and Atkins K.D. (1989) *Aust. J. Agric. Res.* **40**: 433.

Semple S.J. (2005) NSW DPI Sheep Wether Comparison Database User Manual, NSW Department of Primary Industries, Orange.

Swan A.A., Gurman P.M., Boerner V., Brown D.J., Clark S. Gore K., Granleese T. and van der Werf J.H.J. (2018) *Proc. Wld Congr. Genet. Appl. Livest. Prod.* Vol. Electronic Poster Session - Theory to Application **3**: 346.

Wilson, J.M., Cummins, L.J. and Morley, F.H.W. (1996) *Proc. Aust. Soc. Anim. Prod.* **16**: 399.

GENOTYPE BY ENVIRONMENT INTERACTION IN AUSTRALIAN MATERNAL AND TERMINAL SHEEP

L. Li[1,3], A.A. Swan[1,3], D.J. Brown[1,3] and J.H.J. van der Werf[2,3]

[1]Animal Genetics & Breeding Unit*, University of New England, Armidale, NSW, 2351 Australia
[2]School of Environmental & Rural Science, University of New England, Armidale, NSW, 2351
[3]Cooperative Research Centre for Sheep Industry Innovation, Armidale, NSW, 2351

SUMMARY

Genotype by environment interactions (GxE) for economically important traits in Australian maternal and terminal sheep were investigated by both sire by flock interactions and multi-trait methods for performance data observed in eight flocks (sites) across Australia. The traits included growth traits: birth weight (bwt), weaning weight (wwt), post-weaning weight (pwt); carcass composition traits: carcass eye muscle depth (cemd), carcass fat (ccfat), lean meat yield (lmy) and dressing percentage (dressperc), and meat quality traits: intra-muscular fat (imf) and shear force (sf5). Univariate analyses showed that variation between genetic groups was relatively large compared to direct genetic variance and that maternal effects were significant for growth traits. The estimates of heritability were low for growth traits (from 0.08 to 0.11), moderate for most carcass composition traits (except for lmy) and sf5 (from 0.24 to 0.26), and high for lmy (0.44) and imf (0.50). Significant sire by flock interactions were found for growth traits and sf5. The average genetic correlations over pairs of flocks for each trait were 0.35 (bwt), 0.44 (wwt), 0.43 (pwt), 0.78 (cemd), 0.70 (ccfat), 0.77 (dressperc), 0.83 (lmy), 0.91 (imf) and 0.72 (sf5), respectively. Both the interaction term and multi-trait methods demonstrate that significant GxE existed for growth traits. The industry genetic evaluation should account for GxE for these traits.

INTRODUCTION

The Australian sheep industry has generated substantial gains through use of Australian Sheep Breeding Values and Indexes generated by Sheep Genetics (Swan, 2017). Where they are significant, genotype by environment interactions (GxE) result in changes in ranking across environments, with potential effects on selection response. Therefore, it is important to understand the magnitude of GxE for traits included in Australian sheep breeding programs. To date there have been no studies reporting GxE in meat quality and carcass traits in the terminal and maternal sheep breeds in Australia.

A well-structured distribution of genotypes across environments is crucial to detect GxE effects. The Sheep CRC Information Nucleus (INF, van der Werf *et al.* 2010) is an ideal resource to study GxE because a large number of sires were progeny tested at eight research flocks that represent the diversity of Australian sheep production environments. An extensive measurement program of meat quality traits was undertaken on individual animals at these eight flocks over a five-year period. In this study, these data were used to investigate the magnitude of GxE quantified by fitting a sire by flock interaction term and multi-trait methods for some economically important traits in terminal and maternal sheep breeds.

MATERIALS AND METHODS

Animals and data. The Sheep CRC IN flocks represented three sire breed types, Merino, Maternal and Terminal in the initial experimental design at eight research flocks. In this study performance data from progeny of Maternal and Terminal sire breed types mated to Merino or Border Leicester

* A joint venture of NSW Department of Primary Industries and the University of New England

x Merino dams were combined to conduct genetic analyses. The research flocks included Armidale, NSW (IN01); Trangie, NSW (IN02); Cowra, NSW (IN03); Rutherglen, Vic. (IN04); Hamilton, Vic. (IN05); Struan, SA (IN06); Turretfield, SA (IN07); and Katanning, WA (IN08). The measurement program was run for five years with animals born between 2007 and 2011. Pedigree and performance data of nine traits were extracted from the IN database. The traits analysed included three growth traits: birthweight (bwt, kg), weaning weight (wwt, kg) and post-weaning weight (pwt, kg); four carcass composition traits: carcass eye muscle depth at C site (cemd, mm), carcass fat at C site (ccfat, mm), dressing percentage (dressperc, %) and lean meat yield (lmy, %); two meat quality traits: intramuscular fat (imf, %) and shear force at five days aging at 3–4 °C (sf5, Newtons). All carcass composition and meat quality traits were measured on meat samples post-slaughter. A summary of the numbers of records and sires represented at each flock for each trait is shown in Table 1. The growth traits had the most records, followed by carcass composition traits and meat quality traits. Correspondingly, the total number of sires used differed across traits, varying from 263 for imf to 510 for bwt. However, the average number of sires used across pairs of flocks was not substantially different across traits, ranging from 67 to 82.

Statistical analyses. Univariate analyses were used to estimate variance components and heritabilities for each trait. Fixed effects included contemporary group (cg), birth type, rearing type, age of measurement (in days) fitted as a covariate and dam age fitted as linear and quadratic covariates for all traits with the exception that rearing type and age of measurement were not fitted for bwt. Hot carcass weight was included as a linear covariate for meat quality and carcass composition traits except dressperc and lmy. Contemporary group definitions were based on management group, flock, year, sex, breed type and date of measurement, with numbers of cg ranging from 369 (wwt) to 994 (sf5) across traits. Random effects included a genetic group effect (ranging from 124 to 159 genetic groups across traits, representing the original breeds and strains within breeds of the base animals), a direct genetic effect of animal, and sire x flock-year interaction (SF) for each trait. Random maternal effects (representing both maternal genetic and maternal permanent environmental effects) were fitted to growth traits only.

The genetic correlations of animal genetic effects between flocks which modelled each trait in the different flocks as different traits were estimated by two alternative models. The first model used pairwise bivariate analyses, with 28 analyses of all combinations of the eight flocks. The second model was the factor analytic model in which all data was used simultaneously to estimate all genetic correlations in a single analysis with heterogeneous residual variance fitted at the flock level. Both bivariate and factor analytic models used the same fixed and random effects as those used in the univariate analyses for each trait, but excluding the random SF effect. The random sire × flock, rather than a direct genetic of animal effect, was modelled with a factor analytic covariance structure (FA) in the factor analytic model. All analyses were conducted using software ASReml (Gilmour *et al.* 2009) with REML procedures.

RESULTS AND DISCUSSION

The summary statistics, phenotypic variance and ratios of variances are shown in Table 1. Significant values for the ratio of genetic group variances to direct genetic variances were found for all growth traits, decreasing from 4.73 (bwt) to 2.32 (pwt), showing that genetic groups account for a large proportion of the genetic variation for growth traits. The estimates of heritability were low for growth traits (from 0.08 for bwt to 0.11 for pwt), moderate for most carcass composition traits (except for lmy) and sf5 (from 0.24 to 0.26), and high for lmy (0.44) and imf (0.50). The heritability estimates for growth traits were lower than the weighted means of 0.15 (bwt), 0.18 (wwt) and 0.21 (pwt) reviewed by Safari et al. (2005) for meat breeds. Brown *et al.* (2016) also reported slightly higher heritability estimates for bwt (0.15) and pwt (0.15-0.16) for both maternal and terminal breeds, but with a similar estimate for wwt. This could be due to SF effects which were significant for these traits (variance ratio estimates ranging from 0.02 to 0.03) fitted in the models to account for the GxE across flocks. Negligible SF effects were observed for all carcass composition traits and imf. The heritability estimates for carcass composition traits were similar to those reported in the previous preliminary study by Mortimer et al. (2010) based on 2007-2008 INF data, except for the slightly higher estimate for lmy in this study (0.44 vs 0.34). For meat quality traits, Mortimer et al. (2014) reported similar estimates of heritabilities for imf (0.48) and sf5 (0.27) from INF data, along with a similar magnitude estimate of the SF effect for sf5 (variance ratio of 0.03). Maternal permanent environment effects were significant for all growth traits and decreased as age increased, from 0.34 for bwt to 0.21 for pwt. This trend was consistent with previous findings (sum of maternal genetic and permanent environment effects) but with slightly lower estimates for growth traits by Brown *et al.* (2016).

Table 1. Number of records (N), total number of sires with progeny data (tot_sire) across all flocks, average number of sires with progeny data across pairs of flock combinations (ave_sire), number of genetic groups (N_gg), number of contemory groups (N_cg), mean trait value (mean), estimates of phenotypic variance (σ^2_p), ratio of genetic group to additive genetic variance (b^2), direct heritability (h^2), sire by flock effect (s^2), maternal environmental effect (c^2) as a proportion of phenotypic variance, average genetic correlations of across pairs of flock combinations by factor analytic model (rg_fa) and by bivariate analyses (rg_bi) with standard errors in subscript for each trait

	bwt	wwt	pwt	cemd	ccfat	dressperc	lmy	imf	sf5
Unit	kg	kg	kg	mm	mm	kg	%	%	Newtons
N	16190	13144	12373	8996	8793	9483	7272	7016	7174
tot_sire	510	439	426	425	424	425	278	263	278
ave_sire	82	80	82	79	79	79	74	67	74
N_gg	159	159	139	137	138	138	124	125	125
N_cg	742	369	419	633	628	608	568	545	994
mean	4.9	29.4	32.9	31.3	4.6	46.1	58.1	4.2	26.9
σ^2_p	$0.71_{0.01}$	$17.08_{0.23}$	$19.88_{0.28}$	$10.76_{0.18}$	$3.81_{0.07}$	$5.15_{0.08}$	$5.61_{0.12}$	$0.61_{0.01}$	$48.41_{0.95}$
b^2	$4.73_{1.85}$	$3.96_{1.65}$	$2.32_{1.03}$	$0.15_{0.11}$	$0.20_{0.12}$	$0.06_{0.07}$	$0.49_{0.20}$	$0.15_{0.10}$	$0.12_{0.09}$
h^2	$0.08_{0.02}$	$0.09_{0.02}$	$0.11_{0.02}$	$0.25_{0.03}$	$0.27_{0.03}$	$0.24_{0.03}$	$0.44_{0.04}$	$0.50_{0.04}$	$0.26_{0.03}$
s^2	$0.03_{0.01}$	$0.02_{0.01}$	$0.03_{0.01}$	$0.01_{0.01}$	$0.01_{0.01}$	$0.01_{0.01}$	$0.01_{0.01}$	0	$0.03_{0.01}$
c^2	$0.34_{0.01}$	$0.24_{0.01}$	$0.21_{0.01}$	-	-	-	-	-	-
rg_fa	$0.35_{0.26}$	$0.44_{0.34}$	$0.43_{0.26}$	$0.78_{0.18}$	$0.70_{0.14}$	$0.77_{0.20}$	$0.83_{0.14}$	$0.91_{0.10}$	$0.72_{0.18}$
rg_bi	$0.41_{0.45}$	$0.41_{0.59}$	$0.41_{0.46}$	$0.82_{0.39}$	$0.61_{0.34}$	$0.75_{0.34}$	$0.77_{0.26}$	$0.82_{0.25}$	$0.74_{0.35}$

The average genetic correlations over all pairs of flocks from both the factor analytic model (rg_fa) and a series of bivariate analyses (rg_bi) are shown in Table 1. Similar magnitudes of genetic correlations were found from both approaches for all traits. However, the standard errors of genetic correlations from the factor analytic model were much smaller (from 0.10 to 0.34) than those from the bivariate analyses (from 0.25 to 0.59) across all traits, demonstrating that the factor analytic model is a more reliable and parsimonious approach to analyse eight flocks data simultaneously in this study. The results from the factor analytic model indicated low to moderate genetic correlations (between 0.35 and 0.44) for growth traits, and moderate to high genetic correlations (between 0.70 and 0.91) for both carcass composition and meat quality traits. These results are consistent with the significant level for sire by flock interaction term in the univariate analyses for most of the traits except for sf5. Although a significant SF effect was detected for sf5, a much higher average genetic correlation across flocks was found for sf5 than for the growth traits. The distributions of genetic correlations between each flock and other flocks for each trait is shown in Figure 1. What can be clearly seen in this figure is that low genetic correlations were generally found for each flock with other flocks for growth traits; imf had consistently high genetic correlations for each flock with other flocks; and for the carcass composition traits and sf5, most of the flocks had high genetic correlations with other flocks, but there was at least one flock that had only moderate genetic correlations with other flocks (e.g. dressperc and lmy for one and sf5 for three flocks).

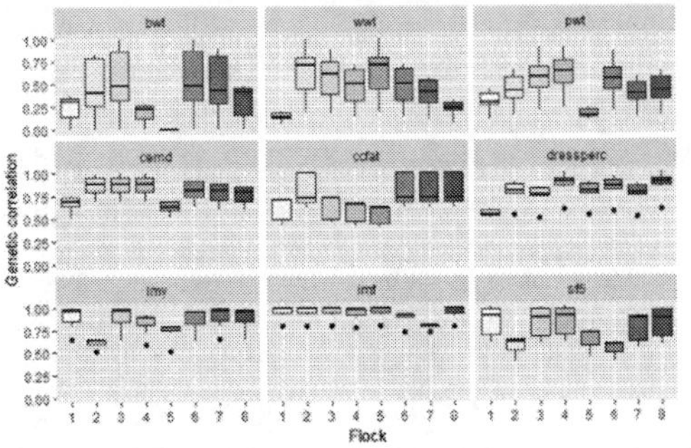

Figure 1. Genetic correlations of each flock with other flocks by using Factor analytic models

CONCLUSIONS

The results from both SF interaction and multi-trait models demonstrated that there were significant GxE for all growth traits (bwt, wwt and pwt) and negligible GxE were found in all flocks for imf and in most of the flocks for the carcass composition traits in maternal and terminal sheep. Our industry genetic evaluations should be able to account for these GxE effects by fitting a sire by environment interaction term in the models for these traits with significant GxE.

ACKNOWLEDGEMENTS

This work was partially funded by B.SGN.0127.

REFERENCES

Brown D.J., Fogarty N.M., Iker C.L., Ferguson D.M. Blache D. and Gaunt G.M. (2016) *Ani. Prod. Sci.* **56**: 767.

Gilmour A.R., Gogel B.J., Cullis B.R., and Thompson R. (2009) ASReml User Guide Release 3.0 VSN International Ltd, Hemel Hempstead, HP1 1ES, UK.

Mortimer S.I., van der Werf J.H.J., Jacob R.H., Pethick D.W., Pearce K.L., Warner R.D., Geesink G.H., Hocking Edwards J.E., Gardner G.E., Ponnampalam E.N., Kitessa S.M., Ball A.J., and Hopkins D.L. (2010) *Ani. Prod. Sci.* **50**: 1135.

Mortimer S.I., van der Werf J.H.J., Jacob R.H., Hopkins D.L., Pannier L., Pearce K.L., Gardner G.E., Warner R.D., Geesink G.H., Hocking Edwards J.E., Ponnampalam E.N., Ball A.J., Gilmour A.R. and Pethick D.W.(2014) *Meat Sci.* **96**: 1016.

Swan, A.A., Banks, R.G., Brown, D.J., and Chandler, H.R. (2017) *Proc. Assoc. Advmt. Anim. Breed Genet.* **22**: 365.

Safari, E., Fogarty, N. M., and Gilmour, A.R. (2005) *Liv. Prod. Sci.* **92**: 271.

Van der Werf J.H.J., Kinghorn B.P. and Banks R.G. (2010) *Ani. Prod. Sci.* **50**: 998.

SIMULATING GENOTYPIC MERIT WITH HIGH-ORDER EPISTATIC INTERACTIONS

A.B. Kinghorn[1] and B.P. Kinghorn[2]

[1]School of Biomedical Sciences, Li Ka Shing Faculty of Medicine, The University of Hong Kong
[2]School or Environmental and Rural Science, University of New England, Armidale, 2351 Australia

SUMMARY

The real map from genotype to phenotype is very complex indeed, and yet we use simple models to analyse it and simple models to simulate it. This paper illustrates a method to simulate phenotypes as a function of genotypes that aims to better emulate the underlying complexity involved, with multi-level epistatic interaction among all loci within large groups of loci. It is proposed that such simulated data will give a more realistic basis to test QTL detection, GWAS and genetic evaluation methods.

INTRODUCTION

We want to understand and exploit the relationship between genotype and phenotype. To do this we use simple models and methods that we hope will lead us to making good decisions. However, life is more complex than we can perceive, as it has not been designed, but has evolved in a random manner. How can we test the usefulness of these simple models? They might lead to what seems like good genetic progress, but do they miss something in the real complexity that alternative models and methods might capture for our benefit? In addition, our simple models often lead us to think that there are many hundreds of QTL affecting a trait, with relatively few QTL of large effect – could reality be that there are far fewer QTL that, because of their complex interactions, masquerade as many hundreds of QTL? If this were true and detectable, then we might take a different direction in QTL detection, GWAS analyses and genetic evaluations. Simulation can be used to test this. However, datasets that are simulated using the same or similar statistical models as will be used to analyse them are self-fulfilling and not appropriate. And of course, the real model is too complex for us to know and use. Instead we need a tractable approach that emulates the high complexity of true genotype-phenotype relationships, including the high-order epistatic interactions that are evident when gazing at a biochemical pathway chart.

The NK model (Kauffman and Levin 1987) is a theoretical fitness model that provides an objective function relating a sequence (genotype) to fitness score (phenotype). Each locus interacts with a given number of other loci that are either neighbours or randomly determined. Each locus is given an individual fitness score based on the loci with which it interacts. The individual loci fitness scores are summed to give a sequence's total fitness. This model is useful in that an NK fitness landscape's complexity can be tuned by altering the number of interactions at each locus. Cooper and Podlich added an extra layer of interaction to the standard NK model by introducing the concept of environmental dependent gene expression to simulate gene-by-environment interactions (Cooper and Podlich 2002). Although the NK model is useful, it has limitations in representing some biological systems. At higher interaction values that are biologically relevant the landscape descends into a chaotic surface on which additive adaptation is essentially not possible.

To solve this problem, Kinghorn and Tanner proposed an approach where the effects of groups of interacting loci ("phenotypic contributors") are added sequentially and in accordance with natural selection (Kinghorn and Tanner 2017), similarly to how gene networks probably evolved over time (Amoutzias *et al.* 2004). This approach is based on method for simulating the response surface of ligand/target molecule affinity as a function of DNA aptamer sequence (Kinghorn and Tanner 2017). We have used a similar approach to model SNP data from many genomes.

MATERIALS AND METHODS

The Selective Phenome Growth Adapted NK Model (SPANK) method of Kinghorn and Tanner (2017) operates on single DNA sequences (DNA Aptamers, typically 30 to 100 bases long). Our method follows the SPANK method quite closely, presented here briefly, in our context:

- N is the number of QTL
- PC_i is the ith Phenotypic Contributor, this being a vector of indicator variables {0,1} that point to loci involved in generating value for that PC. A key concept is that the genotypic merit for a haploid is the sum of many PCs – many components of genetic merit that contribute to expression of phenotype.
- $\varphi_{i,s}$ is the value of PC_i for sequence or haplotype s.
- n_{PC} is the number of PCs. This is unbounded.
- K is the maximum number of loci that can be involved in determining a PC. All levels from 1 to K can be involved, but only one level per PC.
- k_i This is the actual number of loci involved in determining PC_i.

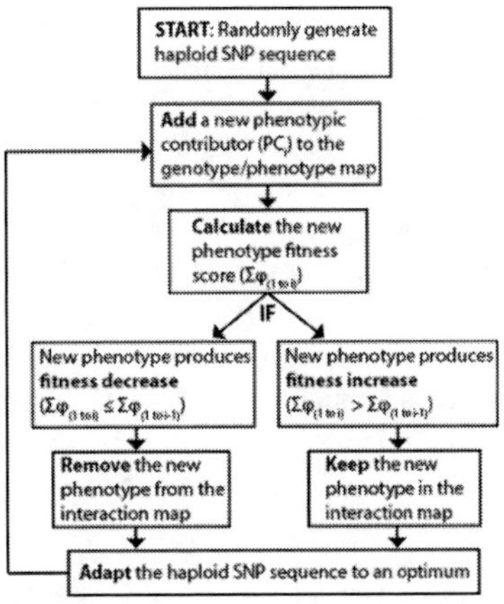

Figure 1. The SPANK method

There are three main parts to the method:

1. Generating the Genotype/Phenotype map (Figures 1, 2).
2. Analysing the SPANK Genotype/Phenotype map and comparing it to a randomly generated interaction map. To analyse the fitness landscapes we find 100,000 local optima and calculate their Hamming distance from the highest scoring optimum (Figure 3). The parameters used to drive the method can be changed to arrive at what is judged to be an appropriate fitness landscape, as indicated by such analysis.
3. For an implementation phase, the adopted Genotype/Phenotype map is used to generate phenotypes for the genotypes that are simulated into a real or simulated population.

The method follows Figure 1. A single haploid sequence is generated. This is the current Lead Sequence, which will direct the genotype/phenotype map evolution. The phenotypic contributors that make up the genotype/phenotype map will be formed around this lead sequence such that the lead sequence will be an optimum. To add a new phenotype to the interaction map, k_i is uniformly sampled from {1 to K}, and k_i loci randomly sampled from {1 to N}. The Lead Sequence alleles at these loci are used to determine its φi value, which is taken from a matrix of previously randomly generated φi values. For the PC_i to be accepted there must be an increase in the average merit across all prior PCs. The fitness score of the new phenotype ($\Sigma_{(1\text{-}i)}$) is then calculated and if it is greater than the fitness score of the old phenotype ($\Sigma\varphi_{(1\text{-}(i\text{-}1))}$) then the new PC_i is accepted, else it is rejected. Finally, noting that the lead sequence does not necessarily have the best genotype for a newly added PC, it is adapted using allelic substitution until it is at a fitness peak before the cycle is repeated. This allelic substitution proceeds by selecting the fittest 1-step mutant neighbour in sequence space and continuing the allelic substitution until no fitter 1-step mutant neighbours can be found.

RESULTS

To make a small illustrative example, the SPANK method was invoked with N=20 loci, n_{PC}=20 phenotypic contributors, and K= a maximum of 10 loci interacting to generate a PC. The resultant epistatic map is shown in Figure 2A. For comparison, an epistatic map was generated with random interactions that were not selected using a classical NK model, with 6 interactions per phenotypic contributor. In Table 1, a selection of the 20 phenotypic contributors (1, 2, 3 and 20) from the epistatic map (Figure 2) are shown. The interacting loci and their φ value for two test sequences are displayed. φ values were previously generated constants that are functions of the alleles mapped by the PC. The φ values for each *PC* are averaged to give the haploid fitness for each of the two sequences.

Our aim is to develop fitness landscapes that are of high order complexity, yet are reasonably smooth and not chaotic, to the extent that a practitioner might expect in real populations. The measure of landscape smoothness we have chosen is the Hamming distance from the fittest optimum (Figure 3). For each landscape, 100,000 sequences are chosen at random and from these sequences random mutational walks uphill are taken until a local optimum is reached for each starting sequence. The Hamming distance from each of these local optima to the fittest recorded optimum is calculated. It can be observed that for the SPANK generated fitness landscape (Figure 3A) there is a stronger relationship between fitness and distance from the fittest optimum. Additionally, the line of best fit shows that for the SPANK generated fitness landscape a greater number of random uphill walkers reach the fittest optimum, indicating a smoother landscape. For the random landscape (Figure 3B) there is a much weaker trend of having higher scoring optima closer to the fittest optimum. For the random landscape just 2709 random walkers reached the fittest optimum, whereas 16,328 random walkers reached the fittest optimum for the SPANK landscape.

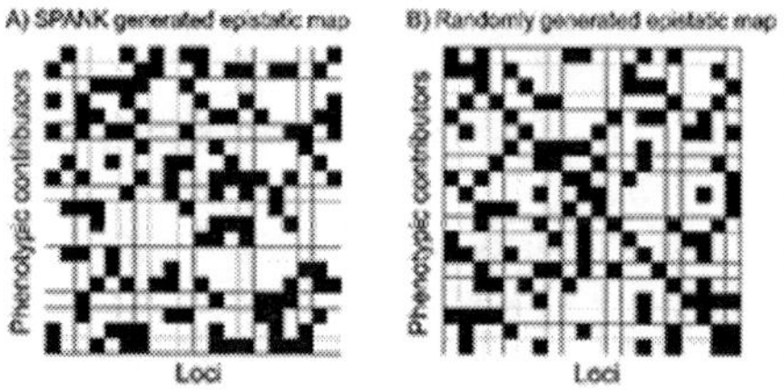

Figure 2. Epistatic maps generated A) by SPANK and B) at random without optimisation. The random map is a classical NK model with 6 interactions per phenotypic contributor. Each row is a phenotypic contributor and each column is a locus. Along each phenotypic contributor the interacting loci are denoted as dark shaded squares

DISCUSSION

As with the simple models we currently use to detect and exploit genotype-phenotype relationships, SPANK is not a model of the underlying biology. However, it does make a big step in the direction of emulating the biological complexity involved – permitting the involvement of multiple players (multiple loci and alleles) in each contribution of genotype to phenotype.

PC	k_{PC}	TS	Phenotypic contributors (PC) or Allele (TS)	φ	Haploid Fitness (Average of $PC_{(1\text{-}20)}$)
1	6		01000101011000000001		
		1	11221212222122111221	f(122221) = 0.7843	
		2	21221121112211222212	f(111122) = 0.8342	
2	9		10000011001011011010		
		1	11221212222122111221	f(112222112) = 0.9534	
		2	21221121112211222212	f(221211221) = 0.6934	
⇩			⇩	⇩	
20	10		10010110000111011100		
		1	11221212222122111221	f(1221122112) = 0.8133	
		2	21221121112211222212	f(2212211222) = 0.2643	
		1	11221211222122111221		Fitness = 0.8830
		2	21221121112211222212		Fitness = 0.7985

Table 1. Results from a small illustrative example. Using the Epistatic map in Figure 2, *N* is 20 loci and n_{PC} *is* 20. For phenotypic contributors (*PC*), 1 denotes a locus involved in the *PC*, else 0. For the two test sequences (*TS*) shown, 1 and 2 are the two alleles at each locus. φ is the value component for the *TS* under the prevailing *PC*, these being randomly generated constants.

Notice that f(122221) = 0.7843 appears twice, by chance

The functions in the φ column of Table 1 that allocate value to genotypes are determined by allele pattern jointly across loci – for example f(122221) = 0.7843 appears twice in the table, for different sets of loci. This departs from what we might model biologically. An alternative would be to require specific genotypes over many fixed loci, but this would suffer from such complexes being rare to occur and rare to transmit. In conjunction with the selective adaptation steps described, the current approach leads to sensible patterns of fitness across genotypes (Figure 3), without eg "witch's hat" peaks of extreme fitness (Kinghorn and Tanner 2017).

For implementation, attention has to be paid to diploidy and its effect on the expression of single-locus dominance as well as epistasis. For the latter, it is possible to assume dominance of epistasis by stipulating that a *PC* function is expressed if each locus is represented by either one or two of the enabling alleles. In a similar manner we could assume recessive inheritance of epistasis, or a mixture.

In addition, single locus effects need to be addressed. The method proposed can handle that by allowing k=1, which is not represented in Table 1, or indeed by using a classic approach to generate these components. Sampling of k from a Poisson or adapted Gamma distribution might give a presumed sensible weighting to the different levels of epistatic interaction, including k=1 for PCs involving no interactions. Additive and dominance single-locus effects could be conventionally simulated separately for each locus, then, for an interaction set involving k loci, the overall effect taken as the average across the single locus effects multiplied by the φ function shown in Table 1. This would diversify the single-locus effects from the relatively narrow sampling the method provides, and increase diversity of effects for higher-order interacting groups of loci.

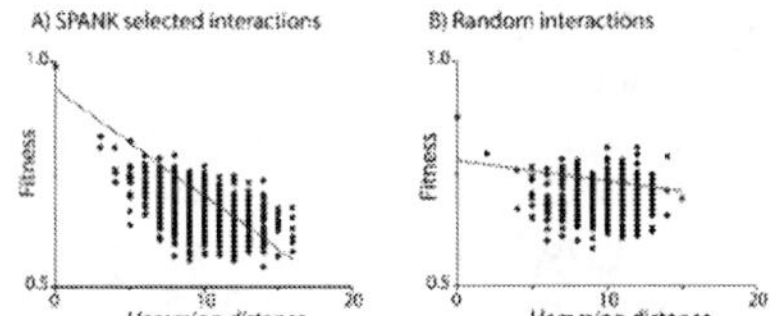

Figure 3. Hamming distance to fittest optimum. A) SPANK generated landscape B) Randomly generated landscape. The number of interacting loci for each landscape is matched. There is much superimposition of points, especially at the fittest optima (see text)

The SPANK model aims to mimic the complexity of genetic systems not from a top down approach, but from a bottom up approach that facilitates the emergence of complex interactions without devolving into chaos. Having a system that mimics gene interactions, in which the resultant interactions are explicitly recorded, we can evaluate the extent to which simple additive models exploit these interactions despite no specific fit to accommodate them. Many questions about the impacts of genetic interactions and our ability to detect and exploit them could be answered. Can a small number of complex interacting QTL masquerade as a large number of QTL? What number and strength of minimally interacting QTL are required to deviate observed causality from major QTL? Where is that missing heritability? Future work might be directed towards answering these and other questions regarding genetic interactions. This paper has only outlined and illustrated an approach to simulating phenotypes. If properly applied, the resulting genotype to phenotype map might approach the complexity of reality. This in turn could help provide insights to what we might be missing by using relatively simple models for QTL detection, GWAS and genetic evaluation.

REFERENCES

Kauffman S. and Levin S. (1987) *Journal of theoretical Biology* **128:** 11.

Cooper M. and Podlich D. W. (2002) *Complexity* **7**: 31.

Kinghorn A.B. and Tanner J.A. (2017) *Complexity* Article ID 6760852, 12 pages. https://doi.org/10.1155/2017/6760852.

Amoutzias G. D., Robertson D. L., Oliver S. G. and Bornberg-Bauer, E. (2004) *EMBO reports*, **5:** 274.

IMPACT OF AN APPROXIMATE INVERSE OF THE GENOMIC RELATIONSHIP MATRIX FOR SINGLE-STEP EVALUATION OF AUSTRALIAN MEAT SHEEP

K. Meyer and A.A. Swan

Animal Genetics & Breeding Unit*, University of New England, Armidale, NSW, 2351 Australia

SUMMARY

Common implementations of single-step genomic evaluation require the inverse of the genomic relationship matrix. Obtaining the inverse can become computationally prohibitive as its size increases. Stimulated by rapidly increasing numbers of genotyped animals, several procedures to approximate this inverse have been proposed. We examine the impact of two methods of approximation on predicted breeding values for a multi-breed population of Australian sheep. Results show that very high correlations with predictions using the full inverse can be achieved whilst reducing computational requirements. However, current levels of genotyping in our data were relatively low and results need to be validated as larger number of genotypes become available.

INTRODUCTION

The single-step procedure for joint genetic evaluation of genotyped and non-genotyped animals (ssGBLUP) has become routine in many livestock improvement schemes. In essence, it extends the classic breeding value model to include genomic information by replacing the pedigree based relationship matrix (**A**) with its counterpart (**H**) which combines both. Only $\mathbf{H}^{-1}$ is required in the mixed model equations (MME) to be solved. This can be formed directly, but does require the inverse of two matrices of size $n_2 \times n_2$, with n_2 the number of genotyped animals. The first is the inverse of the dense genomic relationship, **G**, which needs to be inverted explicitly. The second is the inverse of $\mathbf{A}_{22}$, the corresponding part of **A**, which can be obtained indirectly by exploiting partitioned matrix results (e.g. Strandén *et al.* 2017). Rapidly increasing numbers of genotyped animals have stimulated development of approximations for $\mathbf{G}^{-1}$. We examine the impact of two proposed schemes for a multi-breed set of sheep data, namely the 'algorithm for proven and young' (APY) sires (e.g. Misztal *et al.* 2014) and the use of the Woodbury matrix identity combined with a reduction in the number of principal components (PCs) considered, dubbed TBLUP (Mäntysaari *et al.* 2017).

MATERIAL AND METHODS

The APY inverse. Reorder and split **G** into a set of 'core' (or proven) animals and a set of 'non-core' (or young) animals, denoted by subscripts 'C' and 'N', respectively. This gives

$$\mathbf{G}^{-1} = \begin{bmatrix} \mathbf{G}_{CC}^{-1} + \mathbf{G}_{CC}^{-1}\mathbf{G}_{CN}\mathbf{G}^{NN}\mathbf{G}_{NC}\mathbf{G}_{CC}^{-1} & -\mathbf{G}_{CC}^{-1}\mathbf{G}_{CN}\mathbf{G}^{NN} \\ -\mathbf{G}^{NN}\mathbf{G}_{NC}\mathbf{G}_{CC}^{-1}\mathbf{G}_{NC} & \mathbf{G}^{NN} \end{bmatrix} \quad \text{for} \quad \mathbf{G} = \begin{bmatrix} \mathbf{G}_{CC} & \mathbf{G}_{CN} \\ \mathbf{G}_{NC} & \mathbf{G}_{NN} \end{bmatrix}$$

with $\mathbf{G}^{NN} = \left(\mathbf{G}_{NN} - \mathbf{G}_{NC}\mathbf{G}_{CC}^{-1}\mathbf{G}_{CN}\right)^{-1} = \mathbf{G}_{NN.C}^{-1}$, where $\mathbf{G}_{NN.C}$ is the matrix of relationships amongst non-core animals conditional on the core animals. For pedigree relationships, the diagonals of the corresponding function of **A** represent Mendelian sampling terms. Moreover, if non-core animals had no progeny, the matrix would be diagonal. Analogously, if non-core animals can be chosen so that $\mathbf{G}_{NN.C}$ is close to diagonal, a suitable approximation of $\mathbf{G}^{-1}$ can be obtained by substituting $\mathbf{D}_N = \text{Diag}\{\mathbf{G}_{NN.C}\}$ for it (Misztal *et al.* 2014). This gives an approximate inverse which is considerably sparser than $\mathbf{G}^{-1}$ and can reduce computational demands dramatically.

The TBLUP inverse. Consider **G** of form $(\lambda/s)\mathbf{ZZ}' + \mathbf{B}$ with **Z** the $n_2 \times m$ matrix of m centered marker counts and s a scale factor. A common choice for **B** is $(1-\lambda)\mathbf{A}_{22} + \lambda\alpha\mathbf{J}$ for $\lambda < 1$, $\alpha \geq 0$ a

* A joint venture of NSW Department of Primary Industries and the University of New England

small constant and $\mathbf{J}$ a matrix with all elements equal to unity. The Woodbury identity gives

$$\mathbf{G}^{-1} = \mathbf{B}^{-1} - (\lambda/s)\mathbf{B}^{-1}\mathbf{Z}\left(\mathbf{I} + (\lambda/s)\mathbf{Z}'\mathbf{B}^{-1}\mathbf{Z}\right)^{-1}\mathbf{Z}'\mathbf{B}^{-1} = \mathbf{B}^{-1} - \mathbf{T}'\mathbf{T} \qquad \text{with } \mathbf{T} \text{ of size } m \times n_2.$$

Similarly, $\mathbf{B}^{-1} = (1-\lambda)^{-1}[\mathbf{A}_{22}^{-1} - \psi\mathbf{A}_{22}^{-1}\mathbf{J}\mathbf{A}_{22}^{-1}]$ with $\psi = \lambda\alpha/(1-\lambda+\lambda\alpha\mathbf{1}'\mathbf{A}_{22}^{-1}\mathbf{1})$. This can reduce computational requirements to obtain $\mathbf{G}^{-1}$ if m is substantially smaller than n_2. Further, let $(\lambda/s)\mathbf{Z}'\mathbf{B}^{-1}\mathbf{Z} = \mathbf{VEV}'$, where $\mathbf{E}$ denotes the diagonal matrix of eigenvalues and $\mathbf{V}$ the corresponding matrix of eigenvectors. An approximate inverse of $\mathbf{G}$ can then be obtained by considering the $r < m$ largest eigenvalues and corresponding eigenvectors only, i.e replacing $\mathbf{T}$ above with $\mathbf{T}_r = (\mathbf{E}_r + \mathbf{I}_r)^{-1/2}\mathbf{V}_r'\mathbf{Z}$, of size $r \times n_2$ (Mäntysaari *et al.* 2017).

Data and model. Data consisted of 1,206,908 records for eye muscle depth, recorded for Australian terminal sire sheep breeds between 1990 and 2018. These included Poll Dorset, Suffolk, White Suffolk and Texel as the main breed groups and 18 other, less numerous breeds. Breed differences were modeled by appropriately defined genetic group effects.

Data were pre-corrected for fixed effects of birth and rearing type, age, age of dam and body weight. The model of analysis comprised additive genetic effects (random) for 1,698,838 animals in the pedigree, 54,094 contemporary groups (fixed), 93 genetic groups (random) and 56,212 sire × flock-year (random) effects. Genotype information, comprised of marker counts for $m = 48{,}599$ SNPs, was available for 23,040 animals.

Analyses. The 'raw' genomic relationship matrix, was built using Method 1 of Van Raden (2008), $\mathbf{G}_M = \mathbf{ZZ}'/s$, centering marker counts by observed gene frequencies. $\mathbf{G}$ was then formed as the weighted average of $\mathbf{G}_M$ and $\mathbf{A}_{22}$ aligning the matrices as described by Vitezica *et al.* (2011), $\mathbf{G} = \lambda(\mathbf{G}_M + \alpha\mathbf{J}) + (1-\lambda)\mathbf{A}_{22}$ for $\alpha = 0.02497$, and arbitrarily chosen weighting factor of $\lambda = 0.95$.

Analyses considered APY core sizes from $n_C = 2.5$K to 20K (with K denoting a factor of 1000). Core animals were chosen either by picking genotyped animals at random (RND) or by selecting those with the most progeny (PRG). TBLUP type approximations of $\mathbf{G}^{-1}$ utilised the leading PCs explaining between 90% and 99% of total variation. Single-step BLUP analyses were carried for all approximations of $\mathbf{G}^{-1}$ and contrasted to a 'standard' ssGBLUP analysis with the 'full' $\mathbf{G}^{-1}$ (FULL). MME were solved iteratively using a preconditioned conjugate gradient (PCG) algorithm with simple, diagonal preconditioner. All calculations were carried out using WOMBAT (Meyer 2007).

Summary statistics calculated were correlations between predicted total breeding values (EBV), i.e. the sum of the predicted additive genetic effects and the appropriate portions of the predicted genetic group effects, from FULL and APY or TBLUP analyses. In addition, corresponding regression coefficients and ranges of differences in EBVs were examined.

RESULTS

Correlations between and regressions of EBVs from FULL on APY analyses are summarised in Table 1. As in various literature reports, there were only small differences between schemes to select core animals. Core sizes about 15K were required to ensure correlations for non-core animals to be close to 0.999. This is in line with results of Pocrnic *et al.* (2016a,b) who demonstrated for a number of livestock species that core sizes of 15K or less sufficed to achieve peak predictive accuracies. Based on simulations linking core and effective population size, the authors recommended a core size equal to the number of eigenvalues (of $\mathbf{G}$) explaining 98% of total variation. For $\mathbf{G}_M$ and $\mathbf{G}$ this was equal to 15,220 and 16,714, respectively. In comparison, for a multi-breed population of New Zealand sheep, 18.8K eigenvalues were needed to capture 98% of the variation among 47K genotypes (Nilforooshan and Lee 2019). Linear regressions of FULL on APY EBVs for core sizes of 10K or more were essentially unity (with corresponding intercepts close to zero) demonstrating that approximation of $\mathbf{G}^{-1}$ at sufficient core size did not distort distributions of EBVs markedly.

Table 1. Relationship between total predicted breeding values from single-step analyses using the 'full' inverse of the genomic relationship matrix and its APY approximation

Type[a]	Sel.[b]	Correlation				Regression coefficient			
		2.5[c]	5	10	15	2.5	5	10	15
NOG	RND	0.9993	0.9997	0.9999	1.0000	1.0021	1.0013	1.0009	1.0000
	PRG	0.9992	0.9997	0.9999	1.0000	0.9980	0.9999	1.0006	1.0002
NOC	RND	0.9644	0.9831	0.9953	0.9986	0.9974	1.0038	1.0040	1.0007
	PRG	0.9636	0.9833	0.9953	0.9988	0.9789	1.0063	1.0074	1.0048
COR	RND	0.9941	0.9983	0.9996	0.9999	0.9849	0.9921	1.0006	1.0003
	PRG	0.9991	0.9991	0.9997	0.9999	0.9854	0.9956	0.9985	1.0004

[a] NOG: non-genotyped, NOC: non-core and COR: core animals [b] Selection of core animals: RND random, PRG most progeny [c] Number of core animals; in thousand

Table 2 shows the numbers of non-zero elements in $\mathbf{H}^{-1}$ for different APY approximation of $\mathbf{G}^{-1}$ and their effects on the number of iterates required to solve the MME. In comparison, corresponding numbers for FULL, were 271 million elements and 611 iterates. Use of APY tended to increase the number of iterates required somewhat, especially when selecting core animals with most progeny. A similar increase over the standard ssGBLUP has been reported by others (Strandén *et al.* 2017; Mäntysaari *et al.* 2017).

For $n_2 = 23{,}040$ genotyped animals and $m = 48{,}599$ SNPs considered, there was no computational advantage for the Woodbury inverse of $\mathbf{G}$. Moreover, the number of non-zero eigenvalues of $(\lambda/s)\mathbf{Z}'\mathbf{B}^{-1}\mathbf{Z}$ was limited to n_2. As shown in Table 3, sufficient PCs – just over 15K – to explain about 97% of total variation were required to yield correlations between TBLUP and FULL EBVs for genotyped animals of 0.999. Corresponding regression coefficients (not shown) were again close to unity. As for APY, there was a slight trend for the number of iterates to increase with less approximation, i.e. more PCs considered.

DISCUSSION

Approximation of $\mathbf{G}^{-1}$ via APY is widely used and has made ssGBLUP for very large numbers of genotypes feasible. For instance, Lourenco *et al.* (2018) described the APY implementation for American Angus cattle with 450K genotyped animals, and Masuda *et al.* (2017) reported on dairy analyses with 720K genotypes. There has been concern, though mainly anecdotal, that APY would work less well for multi-breed populations or at least require larger core sizes than for single breeds. A simulation study by Vandenplas *et al.* (2018) demonstrated good performance of APY for crossbred data when the core, of size equal to the number of eigenvalues explaining 98 to 99% of variation in $\mathbf{G}$, included animals from all breed compositions. Dealing with a beef cattle population involving 41

Table 2. Number of non-zero elements in $\mathbf{H}^{-1}$ (half-stored) for different APY schemes and number of iterates required to solve the corresponding mixed model equations

Select.[a]	Number of non-zero elements[b]					Number of PCG iterates				
	2.5[c]	5	10	15	20	2.5	5	10	15	20
RND	164	189	228	255	269	644	629	639	641	632
PRG	154	181	228	261	270	627	650	713	779	745

[a] Selection of core animals: RND random, PRG most progeny [b] In millions [c] Number of core animals; in 1000

Table 3. Correlations between total predicted breeding values from single-step analyses using the 'full' inverse of the genomic relationship matrix and its TBLUP approximation

	Proportion of variation explained					
	90%	95%	96%	97%	98%	99%
No. of eigenvalues	9,946	13,077	13,990	15,094	16,502	18,908
No. of PCG iterates	614	629	636	641	663	686
Non-genotyped animals	0.9999	1.0000	1.0000	1.0000	1.0000	1.0000
Genotyped animals	0.9950	0.9982	0.9987	0.9991	0.9995	0.9998

breeds, Mäntysaari *et al.* (2017) recommended TBLUP as a well defined and automatic approach to approximate $\mathbf{G}^{-1}$ for any population structure. Our results suggest that approximation of $\mathbf{G}^{-1}$ using either APY or TBLUP can result in predicted breeding values which are virtually identical to those obtained inverting **G** directly, whilst offering the scope for reducing computational requirements. Details will depend on the implementation of ssGBLUP and have not been considered in this study; see Mäntysaari *et al.* (2017) for some discussion of respective strategies and timings. A suitable APY core size or number of PCs to be used for TBLUP was identified to be about 15K. This fell well within the range of corresponding values reported in the literature for single breed studies. However, current levels of genotyping for our data were relatively modest and, moreover, the distribution of genotypes over breeds was very uneven. It remains to be seen whether such levels of approximation will be representative as the number of genotypes increases, especially for the minor breed groups.

CONCLUSIONS

Techniques available to approximate the inverse of the genomic relationship matrix in single step genomic evaluation can yield predicted breeding values for multi-breed sheep that are highly correlated with those obtained using a full inverse. Future work will need to re-evaluate suitable levels of approximation as numbers and breed diversity of genotyped animals increase.

ACKNOWLEDGEMENTS

This work was supported by Meat and Livestock Australia grant L.GEN.1704.

REFERENCES

Lourenco D.L., Tsurata S., Fragomeni B.O., Masuda Y., Aguilar I., Legarra A., Miller S., Moser D. and Misztal I. (2018) In *Proc. Eleventh World Congr. Genet. Appl. Livest. Prod.* Auckland, February 11–16. Paper No. 495.

Mäntysaari E.A., Evans R.D. and Strandén I. (2017) *J. Anim. Sci.* **95**:4728.

Masuda Y., Misztal I., Legarra A., Tsuruta S., Lourenco D.A.L., Fragomeni B.O. and Aguilar I. (2017) *J. Anim. Sci.* **95**:49.

Meyer K. (2007) *J. Zhejiang Univ. SCIENCE B* **8**:815.

Misztal I., Legarra A. and Aguilar I. (2014) *J. Dairy Sci.* **97**:3943.

Nilforooshan M.A. and Lee M. (2019) *J. Anim. Sci.* **97**:1090.

Pocrnic I., Lourenco D.A.L., Masuda Y., Legarra A. and Misztal I. (2016a) *Genetics* **203**.

Pocrnic I., Lourenco D.A.L., Masuda Y. and Misztal I. (2016b) *Genet. Sel. Evol.* **48**:82.

Strandén I., Matilainen K., Aamand G. and Mäntysaari E.A. (2017) *J. Anim. Breed. Genet.* **134**:264.

Van Raden P.M. (2008) *J. Dairy Sci.* **91**:4414.

Vandenplas J., Calus M.P.L. and ten Napel J. (2018) *J. Anim. Sci.* **96**:2060.

Vitezica Z.G., Aguilar I., Misztal I. and Legarra A. (2011) *Genet. Res.* **93**:357.

AN EVALUATION OF 'DEFLATION' TO IMPROVE CONVERGENCE RATES FOR SINGLE-STEP GENOMIC EVALUATION WITH THE HYBRID MODEL

K. Meyer and A.A. Swan

Animal Genetics & Breeding Unit*, University of New England, Armidale, NSW, 2351 Australia

SUMMARY

Single step genomic evaluation fitting a 'hybrid' model which combines marker effects for individuals with genotypes with breeding values for non-genotyped animals can readily accommodate large numbers of genotyped animals. However, iterative solution of the pertaining mixed model equations via a preconditioned gradient scheme has been reported to be afflicted by much slower convergence rates than the standard breeding value model. 'Deflation' of the coefficient matrix has been proposed as a second preconditioning step and shown to dramatically reduce numbers of iterations and computing time required. We describe its application for a set of sheep data. Results indicate that assignment of marker effects to subdomains in moderately sized chunks together with a separate treatment of genetic group effects could reduce total computing times by about a third.

INTRODUCTION

The single-step procedure for joint genetic evaluation of genotyped and non-genotyped animals has become routine in many livestock improvement schemes. Many implementations rely on extending the classic breeding value model (BVM) by combining the pedigree-based relationship matrix with estimates of genomic relationships. An equivalent alternative is the so-called hybrid model (HM) which fits marker effects instead of breeding values for genotyped animals (Fernando *et al.* 2016). This does not require the inverse of the genomic relationship matrix and thus readily accommodates large numbers of genotyped animals. However, initial experience with a preconditioned conjugate gradient (PCG) algorithm to solve the pertaining mixed model equations (MME) has been that convergence rates tended to be slow and that many iterates could be required. Recently, Vandenplas *et al.* (2018) showed that a second level of preconditioning – through a 'deflation' of the coefficient matrix in the MME – could dramatically improve convergence rates and demonstrated its effectiveness for a large, multi-trait analysis of dairy field data. This paper examines the scope of the resulting, deflated preconditioned gradient (DPCG) solver for a practical sheep data set.

BACKGROUND

Let $\mathbf{Cx} = \mathbf{r}$ represent the MME to be solved, with $\mathbf{C}$ (of size $N \times N$) the coefficient matrix, $\mathbf{x}$ the vector of effects and $\mathbf{r}$ the vector of right hand sides. A widely used iterative method to solve for $\mathbf{x}$ is the conjugate gradient (CG) algorithm. Its convergence rate is heavily influenced by the condition number, of $\mathbf{C}$, $\kappa(\mathbf{C})$, i.e. the ratio of its largest to its smallest eigenvalue. Convergence rates can be improved if $\kappa(\mathbf{C})$ can be reduced. An extensively used method to achieve this is to 'pre-condition' the MME, i.e. to solve $\mathbf{M}^{-1}\mathbf{Cx} = \mathbf{M}^{-1}\mathbf{r}$ instead. Choice of the preconditioning matrix $\mathbf{M}$ usually represents a compromise between $\mathbf{M}$ being close to $\mathbf{C}$ (so that $\mathbf{M}^{-1}\mathbf{C}$ is close to an identity matrix) and requirements for storing or inverting $\mathbf{M}$. Simple, effective choices are (block-) diagonal matrices where $\mathbf{M}$ contains the diagonals (or small diagonal blocks) of $\mathbf{C}$.

Deflation has been advocated as a method to eliminate 'unfavourable' eigenvalues of a matrix by projection on a suitable subspace. Let $\mathbf{P}$ denote a matrix comprised of S linearly independent columns (of size N) which form a subspace of $\mathbf{C}$ so that $\mathbf{CP} = \mathbf{PT}$ and $\mathbf{T}$ is a non-singular matrix of order S. For $\mathbf{VP} = \mathbf{I}$ (where $\mathbf{I}$ is an identity matrix), Householder (1961) showed that the deflated matrix $\mathbf{B} = \mathbf{C} - \mathbf{PTV}$ has S zero eigenvalues and the remaining eigenvalues of $\mathbf{B}$ are those of $\mathbf{C}$ that are not

* A joint venture of NSW Department of Primary Industries and the University of New England

eigenvalues of **T**. Hence, assuming **C** is non-singular, **B** has rank $N - S$. Similarly, the eigenvectors of **B** are those of **C** that correspond to their common eigenvalues. In other words, "deflation of an eigenspace cancels the eigenvalues without affecting the rest of the spectrum" (Frank and Vuik 2001).

Use of a deflation preconditioner for CG and PCG algorithms has been considered by various authors in a range of fields (e.g Tang *et al.* 2009; Jönsthövel *et al.* 2012). Combining deflation with the 'standard' preconditioner yields the DPCG, a two-level preconditioning scheme particularly suited to ill-conditioned systems of equations. It involves solving $\mathbf{M}^{-1}\mathbf{PCx} = \mathbf{M}^{-1}\mathbf{Pr}$ with $\mathbf{P} = \mathbf{I} - \mathbf{CS}(\mathbf{S}'\mathbf{CS})^{-1}\mathbf{S}'$ aimed at reducing $\kappa(\mathbf{C})$ and **S** a matrix of size $N \times S$ which defines the deflation subspace (Frank and Vuik 2001). This requires the choice of **S**. Loosely speaking, the closer the deflation vectors (i.e. columns of **S**) approximate the 'unfavourable' eigenvectors of **C** the more effective deflation is likely to be. However, as for **M** it involves trade-offs between improvements in convergence and extra computational requirements. A simple strategy is to divide the space of **C** in correspondence to non-overlapping subsets of equations, referred to as subdomains (Frank and Vuik 2001). Let the i–th element of **x** belong to the j–th domain ($j = 1$ to S). This gives a matrix **S** with ij–th element equal to unity while the remaining elements are equal to zero, i.e. each row of **S** has only one non-zero element. At the extreme, fitting subdomains for individual, single effects is analogous to 'absorbing' the pertaining equations in the mixed model.

MATERIAL AND METHODS

Data consisted of 1,206,908 measurements for eye muscle depth recorded for Australian terminal sire sheep breeds between 1990 and 2018. Data were pre-corrected for fixed effects other than contemporary groups. There were 1,698,838 animals in the pedigree and genotype information, comprised of marker counts for 48,599 SNPs, was available for 23,040 animals. Invoking the HM, additive genetic effects were fitted for non-genotyped animals and marker effects modelled those of genotyped individuals. For simplicity, additional polygenic effects were assumed to be absent. In addition, the model included 54,094 contemporary groups (fixed), 93 genetic groups (random) and 56,212 sire × flock-year (random) effects.

MME were built and solved using either PCG or DPCG with independent subdomains as described above, using a diagonal preconditioner, $\mathbf{M} = \text{Diag}\{\mathbf{C}\}$, throughout. Solutions were assumed to have converged when $\alpha\sqrt{(\mathbf{x}_k - \mathbf{x}_{k-1})'(\mathbf{x}_k - \mathbf{x}_{k-1})/\mathbf{x}_k'\mathbf{x}_k} < 10^{-7}$, with $\mathbf{x}_k$ denoting the vector of solutions from the k–th iterate and α the step size parameter in the (D)PCG algorithm. Analyses were carried out considering all markers and reduced marker panels. To select the latter a simple GWAS was performed fitting markers as fixed covariables, one at a time. Subsets, of size m, were then selected to include those with p-values less than 0.5, 0.2, 0.1 and 0.05. Following Vandenplas *et al.* (2018), single

Table 1. Numbers of iterates required to solve the mixed model equations for different deflation subdomain ('chunk') sizes and marker subsets

p[a]	m[b]	No. of iterates							Correlation[c]	
		–	200[d]	100	50	20	10	5	NOG[e]	GEN
–	48599	3961	2722	2222	1741	1188	859	612	–	–
0.50	28875	3348	2525	2091	1682	1167	840	599	1.000	0.995
0.20	13318	2565	2118	1871	1559	1126	832	598	0.999	0.977
0.10	7858	2159	1833	1654	1416	1085	824	606	0.998	0.962
0.05	4680	1756	1560	1461	1293	1025	820	619	0.997	0.943

[a] Minimum p value for marker subset selection [b] Number of markers [c] Correlation of total breeding values from analyses using all and a subset of markers [d] Number of markers per 'chunk' [e] NOG non-genotyped, GEN genotyped

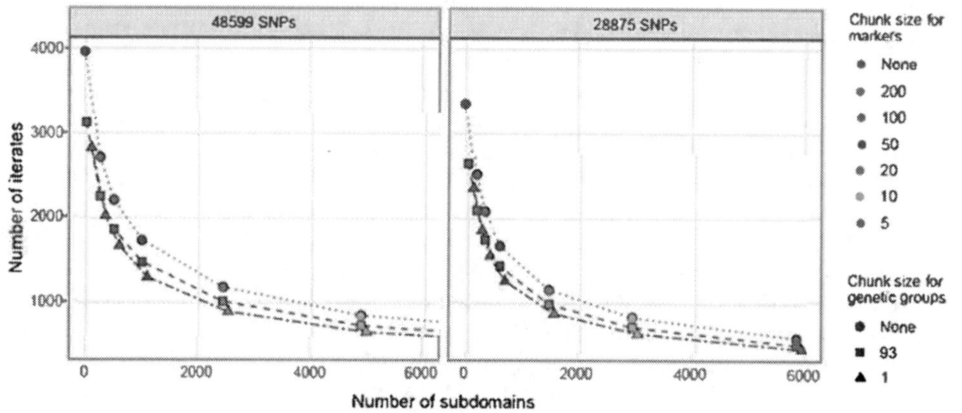

Figure 1. Numbers of iterates required for different deflation schemes and chunk sizes

domains were allocated to fixed effects and to all random effects other than marker effects. Equations for marker effects were divided into subdomains by selecting subsequent chunks (of equations) of size 5, 10, 20, 50, 100 or 200 to investigate the effect of chunk size on efficacy of deflation. This is referred to as scheme A. Scheme B was similar, but fitted separate subdomains for genetic group effects, with chunk sizes of 1 or 93. Computations were carried out under Linux on a shared machine with 512GB of RAM and 28 Intel Xeon CPU E5-2697 cores, rated at 2.6Gh using up to 28 threads.

RESULTS AND DISCUSSION

Numbers of iterates required to solve the MME for deflation scheme A are summarised in Table 1. For comparison, a corresponding analysis fitting the BVM and standard PCG (not shown) converged in 691 iterates. As reported by Vandenplas *et al.* (2018), deflation dramatically improved convergence, but small chunk sizes – and thus many subdomains – were required to achieve rates similar to those fitting the BVM. Reducing the number of markers decreased the number of iterates, especially for the larger chunk sizes (or no deflation), as well as reducing computations per iterate that were proportional to the number of markers. While correlations between predicted breeding values from analyses using the full and reduced marker sets for genotyped animals were less than 0.99 when markers with p-values less than 0.5 were eliminated, marker selection often affects the accuracy of evaluation considerably less, i.e. there is likely more scope for marker reduction than these correlations suggest. E.g., Saatchi and Garrick (2016) proposed a reduced panel for beef cattle comprising about 2,300 markers to capture most of the predictive performance of the full 50K panel.

Figure 1 illustrates the relationship between numbers of iterates required and deflation subdomains. Patterns for the other marker subsets were similar. Clearly, as emphasized by Frank and Vuik (2001), the efficacy of deflation increases with the number of subdomains employed. However, as S increases additional reductions in numbers of iterates achieved decrease. Our model of analysis fitted genetic groups as an additional random effect. This is known to affect convergence rates unfavourably – investigations for the BVM found that it almost doubled the number of iterates needed (Meyer *et al.* 2015). Additional analyses (not shown) identified a similar pattern for our data for the HM with standard PCG. Hence, scheme B attempted to counteract the detrimental effects of fitting genetic groups by defining additional subdomains. As shown in Figure 1 this yielded further reductions in the number of iterates required, the more so the larger chunk size for deflation of equations for marker effects. Even adding a single subdomain for all genetic groups (chunk size of 93) proved highly effective. Similarly, applying DPCG for the BVM, fitting a single subdomain for genetic groups (in addition to two subdomains comprising all fixed and all other random effects, respectively) reduced the number of iterates required from 691 to 536.

While DPCG has the scope to dramatically improve convergence rates and its implementation is straightforward, deflation incurs additional computational cost per iterate and for set-up steps which need to be balanced against reductions in numbers of iterates and additional memory requirements. Figure 2 shows total, elapsed computing times for different analyses. Matrices **CS** and $(\mathbf{S'CS})^{-1}$ only need to be computed once but the computational burden increases with S and S^2, respectively, and storage for large numbers may become prohibitive. For our data, values of S greater than about 2,000 (using all markers) tended to increase total computing times, primarily due to these overheads. Overall, moderate deflation for markers, involving chunks of 20 to 100 SNPs, paired with assigning genetic groups to individual subdomains appeared to yield a reasonable compromise between improvements in convergence behaviour and additional computations for deflation. Our implementation relied on in-core storage of **CS** and $(\mathbf{S'CS})^{-1}$ and the data part of **C**, but involved only limited optimisation of the computations associated with deflation. Values for 'iteration on data', out of core storage or improved parallel processing may differ; see Vandenplas *et al.* (2018) for some timings and discussion.

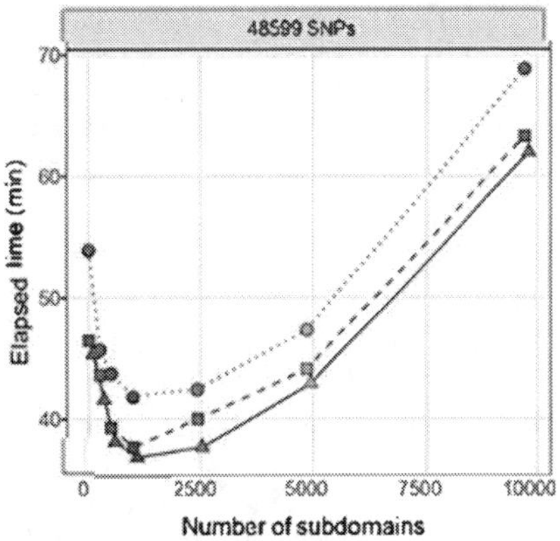

Figure 2. Total computing times[a]

[a] See Figure 1 for legend

As demonstrated for genetic groups, deflation assigning additional, separate subdomains to random effects other than markers was found to be advantageous. Further analyses (not shown) identified extra improvements in convergence rates when defining subdomains for groups of additive genetic effects for non-genotyped animals. Moreover, deflation also proved capable of improving convergence rates for the BVM. Further work will need to examine the efficacy of DPCG for multivariate analyses involving many traits and models fitting maternal effects, and to improve its implementation.

CONCLUSIONS

Deflation of the coefficient matrix in the mixed model equation reduces its condition number and thus improves convergence rates of an iterative solution scheme employing a conjugate gradient algorithm. It appears to be a valuable addition to our toolkit for genomic evaluation.

ACKNOWLEDGEMENTS

This work was supported by Meat and Livestock Australia grant L.GEN.1704.

REFERENCES

Fernando R.L., Cheng H., Golden B.L. and Garrick D.J. (2016) *Genet. Sel. Evol.* **48**:96.

Frank J. and Vuik C. (2001) *SIAM J. Sci. Comput.* **23**:442.

Householder A.S. (1961) *J. Soc. Indust. Appl. Math.* **9**:89.

Jönsthövel T.B., Van Gijzen M.B., MacLachlan S., Vuik C. and Scarpas A. (2012) *Comput. Mech.* **50**:321.

Meyer K., Swan A.A. and Tier B. (2015) *Proc. Ass. Advan. Anim. Breed. Genet.* **21**: Paper no. 18.

Saatchi M. and Garrick D.J. (2016) In *Plant and Animal Genome XXIV Conference*. San Diego, CA, January 9–13, 2016. Abstr.

Tang J.M., Nabben R., Vuik C. and Erlangga Y.A. (2009) *J. Sci. Comput.* **39**:340.

Vandenplas J., Eding H., Calus M.P.L. and Vuik C. (2018) *Genet. Sel. Evol.* **50**:51.

ALTERNATIVE IMPLEMENTATIONS OF PRECONDITIONED CONJUGATE GRADIENT ALGORITHMS FOR SOLVING MIXED MODEL EQUATIONS

D.J. Garrick[1,2], B.L. Golden[2] and D.P. Garrick[1,2]

[1]AL Rae Centre of Genetics and Breeding, Massey University, Hamilton, 3240 New Zealand
[2]Theta Solutions LLC, Atascadero, CA, 93422 USA

SUMMARY

Mixed model equations encountered in pedigree and genomic analyses are typically solved using an iterative preconditioned conjugate gradient algorithm. That algorithm requires a preconditioning matrix chosen to improve the condition number of the problem. Convergence is very fast when an appropriate preconditioning matrix is used, but some equations fail to converge unless an effective preconditioner can be found, and that is not always straightforward, especially in genomic analyses. Some preconditioning and absorption options are compared in the context of a national cattle evaluation for growth traits using a multi-trait single-step marker effects model. It is demonstrated that computing time is largely determined by the number of iterations required to obtain convergence, rather than the complexity of the equations or preconditioning. Further, a reliable convergence statistic for general applications remains problematic.

INTRODUCTION

Mixed linear models that include fixed effects other than the mean, and random effects other than the residuals, are fundamental to theoretical and applied aspects of animal breeding. Most genetic improvement programs rely on routine multiple-trait prediction that involves finding the solution to sets of simultaneous equations we refer to as the mixed model equations (e.g. Henderson, 1975) that are typically large, sparse, symmetric and positive semi definite. Early applications of national evaluation programs explicitly formed every contribution to the left- and right-hand sides of the equations, frequently after absorbing fixed effects such as herd-year-season, and then solved the resultant sparse set of equations using Gauss-Seidel (GS) iteration (e.g. Van Vleck and Dwyer 1985). Later applications avoided the accumulation of every element of the left- and right-hand sides and instead used iteration on data (Schaeffer and Kennedy 1986) to recreate matrix and vector elements as required. Gauss-Seidel iteration was replaced by the sometimes problematic but typically much faster approach of preconditioned conjugate gradient (PCG) (Berger *et al.* 1989; Stranden *et al.* 1999; Tsuruta *et al.* 2001).

A nice property of GS iteration is that every iteration results in a solution that when multiplied by the left-hand side coefficient matrix will more closely agree with the right hand side vector. However, GS can be slow to converge, and convergence tends to slow down with every subsequent iteration. In contrast, PCG tends to converge quite quickly, but in finite arithmetic the system is prone to rounding errors and to loss of conjugacy that can result in successive iterations being poorer fits than previous iterations. Performance is sensitive to the condition number of the equations, which can be dramatically influenced by the choice of preconditioner matrix. Calculating the perfect preconditioner matrix for a given problem is more effort than solving the equations. Finally, it is hard to know exactly when to stop iterating and accept the current iteration as a practical solution to the mixed model equations.

The adoption of single-step models for national evaluation that include both genotyped and non-genotyped animals in the same evaluation has created some additional challenges in obtaining PCG solutions. First, some submatrices of the mixed model equations are no longer sparse, and second, the equations are more likely to lose conjugacy than mixed model equations based on pedigree relationship matrices, at least when historically used diagonal or block preconditioners are applied. The objective of this study was to compare the performance of some alternate PCG implementations in the context of a multiple-trait single-step national cattle evaluation.

MATERIALS AND METHODS

The American Hereford Association runs its genetic evaluation on a weekly basis that includes performance and genotypic data along with pedigree records comprising about 2.5 million US and Canadian Hereford cattle. The complete evaluation comprises nine multiple-trait single-step marker effects models (Fernando *et al.* 2016) to publish 16 different EPD (Golden *et al.* 2018). Mixed model equations are solved using PCG, then the PCG solutions are used to seed parallel Markov chain Monte Carlo analyses using single-site Gibbs sampling to estimate prediction error variances (PEV) to calculate reliabilities, and PEV for contrasts between groups of one or more animals (Garrick *et al.* 2018). This paper reports the PCG solving performance for the multiple trait growth model. The model equations for each correlated trait in that analysis are

$$y_B = J_B j_B + X_B b_B + P_B p_B + Z_B a_B + M_B m_B + Z_B^n u_B^n + Z_B^g S_B \alpha_B + e_B$$
$$y_W = J_W j_W + X_W b_W + P_W p_W + Z_W a_W + M_W m_W + Z_W^n u_W^n + Z_W^g S_W \alpha_W + e_W$$
$$y_G = J_G j_G + X_G b_G + Z_G a_G + Z_G^n u_G^n + Z_G^g S_G \alpha_G + e_G$$

where y_i is a vector of phenotypic observations on B=birth weight, W=weaning weight, or G=post weaning gain, j_i is a fixed covariate accounting for the difference in expected value between genotyped and non-genotyped founders for each trait, b_i are all the other fixed effects, p_i are the random permanent environmental effects of the dam for birth or weaning weight, a_i are the random additional polygenic effects of each trait, m_i are the random maternal genetic effects of birth or weaning weight, u_i^n are the direct breeding values for non-genotyped animals for each trait, α_i are the random marker or SNP effects for each trait, and e_i are the random residual effects for each trait. The J_i matrices are formed from a vector of 1's corresponding to genotyped individuals and an imputed value for non-genotyped animals, X_i, P_i, Z_i, M_i, are incidence matrices for fixed effects, maternal permanent environmental effects, direct genetic effects, and direct maternal effects, respectively, Z_i^n, and Z_i^g are direct effect incidence matrices for non-genotyped and genotyped individuals with phenotypes, and S_i are marker matrices for centred SNP covariates for genotyped animals. The variance-covariance matrices and their inverses for this single-step marker effects model and its mixed model equations are in Fernando *et al.* (2016) and Garrick *et al.* (2018).

Two approaches to characterise convergence during PCG iteration are the two-norm of the preconditioned residual divided by the number of effects (which we denote the iteration *residual*), and the two-norm of the raw residual, divided by the two-norm of the right-hand side, which we denote as *cr* (following Lidauer *et al.* 2015). That is, for solving equations denoted by coefficient matrix, solution and right-hand side as $Cs = r$, based on the preconditioned equations $P^{-1}Cs = P^{-1}r$, the vector of raw residuals at iteration k is $\varepsilon^k = r - C\hat{s}^k$, which is used every round of iteration to compute the $residual = \varepsilon^{k\prime}P^{-1}\varepsilon^k/\text{length}(\varepsilon^k)$, and $cr = \sqrt{\varepsilon^{k\prime}\varepsilon^k/\text{r}'\text{r}}$, for all effects, or separately for each effect in the mixed model equations (i.e., j_i, b_i, p_i, a_i, m_i, u_i, and α_i).

Two options were compared for the preconditioning matrix, the simplest representing the inverse of the diagonal elements of the mixed model equations (i.e. diagonal preconditioning), and the other replacing the preconditioner elements for the fixed effects by the actual inverse of the submatrix of the mixed model equations for fixed effects, namely $(X'R^{-1}X)^{-1}$, either separately for each trait, or with one block for all three traits.

Two options for forming the mixed model equations were compared, one which explicitly fitted all the effects other than the random residual effects shown in the model equation above, and a reduced order set of equations in which fixed effects, b_i, for all three traits had been absorbed. The absorbed equations can be represented by striking out the rows and columns of the mixed model equations corresponding to the fixed effects to be absorbed, then subtracting some terms from the coefficient matrix and right-hand side to eliminate the absorbed equations. For the simplest mixed model equations represented by the model equation $y = Xb + Zu + e$ with $var(u) = G$ and $var(e) = R$, the complete mixed model equations would have order defined by the number of fixed effects plus the number of random effects and be given by

$\begin{bmatrix} X'R^{-1}X & X'R^{-1}Z \\ Z'R^{-1}X & Z'R^{-1}Z+G^{-1} \end{bmatrix}\begin{bmatrix} b \\ u \end{bmatrix} = \begin{bmatrix} X'R^{-1}y \\ Z'R^{-1}y \end{bmatrix}$, whereas the absorbed equations would have order defined by the number of random effects as in the equations

$$[Z'R^{-1}Z + G^{-1} - Z'R^{-1}X(X'R^{-1}X)^{-1}X'R^{-1}Z][u] = [Z'R^{-1}y - Z'R^{-1}X(X'R^{-1}X)^{-1}X'R^{-1}y].$$

RESULTS AND DISCUSSION

The number of iterations and computing times per iteration for BOLT PCG software on a 256Gb RAM Ubuntu server using one 12 Gb Titan V graphics processing unit are shown in Table 1 for the complete and absorbed sets of mixed model equations for various stopping criteria. Correlations between solutions for each factor from different approaches all exceeded 0.99 if not 0.999.

Table 1. Numbers of PCG iterations to achieve alternative stopping criteria in the North American Hereford multiple-trait single-step growth analysis using block or diagonal preconditioning of full or absorbed equations

Mixed Model Equations	Preconditioner	Stopping Criteria Change in residual					Time per iter
		1e-10	1e-11	1e-12	1e-13	cr 1e-5	
Complete[1]	Diagonal	1,768	1,826	4,373	5,435	2,617	0.17s
Complete[2]	Diagonal	1,575	3,227	4,627	6,182	2,624	0.15s
Complete[2]	Block	2,483	2,858	2,858	2,858	2,502	0.15s
Absorbed	Diagonal	2,123	3,289	3,386	6,965	8,641	0.21s

[1]Separate submatrix blocks for *J* factor and *X* factor for each trait
[2]Single submatrix block for *J* factors for B,W,G and another for *X* factors for B,W,G

The total computing time for PCG solution of the multi-trait single-step marker effects model varied from 4 minutes to 24 minutes but was influenced to a much greater extent by the number of iterations (1,575 to 8,641) required for convergence than by the computing time per iteration (0.15 to 0.21 s). The absorbed equations if formed explicitly are much less sparse than the complete set of mixed model equations, but the computing effort was little affected by the absorption of effects as the matrix multiplications were done in parts. This is not surprising as easily shown by denoting the coefficient matrix for the full equations to solve as $\begin{bmatrix} S^{-1} & T \\ T' & Q \end{bmatrix}$, where S^{-1} represents the fixed effects block diagonal partition to be absorbed, Q represents the block diagonal partition for all the other effects, and T represents the block off-diagonal partition between the effects being absorbed and the remaining effects, then the left-hand-side of the absorbed equations can be represented as $[Q - T'ST]$. Each iteration of PCG involves multiplying the coefficient matrix by a work vector, denoted w, as in $w' = [w_b' \;\; w_u]$, which for the complete equations requires computing $S^{-1}w_b$, Tw_u, $T'w_b$, and Qw_u, whereas for the absorbed equations it would requires computing Qw_u and $T'STw_u$. The latter term can be computed in parts as $T'(S(Tw_u))$, first involving the matrix-vector product Tw_u, then pre-multiplying this vector by S then pre-multiplying that product by T'. The only difference in effort between applying the PCG algorithm to the full or the absorbed equations is the computation of the matrix product involving S^{-1} rather than the matrix product involving S. In many mixed model equations, the sparsity and complexity of S^{-1} is similar to that of S, for example for $(X'R^{-1}X)^{-1}$ and $X'R^{-1}X$. Computation of the matrix-vector products in the full equations can be done in parallel, whereas the part equations requires the multiplications to be undertaken serially, involving the product of the first matrix-vector as the vector used in the second matrix-vector multiplication.

Changes in the number of iterations required to meet a given stopping criterion occur due to rounding errors and loss of conjugacy even when there is no change to the elements of the mixed model equations, or to the method of preconditioning, as shown by comparing rows 1 and 2 of Table 1 when the complete mixed model equations were partitioned into submatrices by factor and trait compared to when the factors for J were pooled across traits into one submatrix, and the factors for X were pooled across trait into another submatrix.

Using a block diagonal structure rather than a diagonal matrix for preconditioning fixed effects was initially slower but reached convergence much faster for higher convergence thresholds.

Changes in the number of iterations by absorbing the fixed effects did not have a consistent effect on the number of iterations. This is partly because the process of absorption reduces the two-norm in the denominator of the *cr* criterion, making the same tolerance (i.e. cr <1e-5) much more strict than in the complete mixed model equations.

Changes in the number of iterations by changing stopping criteria (from *residual* to *cr*) or the tolerance of the stopping criteria, resulted in reranking of the performance of the algorithms. The *residual* statistic is not a good stopping criterion because it tends to bounce around from iteration to iteration, but can occasionally achieve very small changes between iterations that result in apparent convergence that is not supported by the *cr* statistic. However, the *cr* statistic is sensitive to parameterisation of the mixed model equations, as shown by the effect of absorption, which also makes that criteria problematic for routine use.

CONCLUSIONS

The results demonstrate that uniformly appropriate convergence criteria for PCG systems are challenging to identify. Minor changes to the manner in which the mixed model equations are parameterised can have considerable influence on performance and run time, most notably by influencing the number of iterations required to achieve a given definition of convergence. Alternative blocking structures, preconditioning matrices, and parameterisation of models can notably influence results.

REFERENCES

Berger P.J., Luecke G.R. and Hoekstra J.A. (1989) J. Dairy Sci. 72:514.

Fernando R.L., Cheng H., Golden B.L., and Garrick D.J. (2016) Genet. Sel. Evol. 48:96.

Garrick D.J., Cheng H., Golden B.L. and Fernando R.L. (2017) Proc. Assoc. Advmt. Anim. Breed. Genet. 22:269.

Garrick D.P., Golden B.L. and Garrick D.J. (2018) WCGALP 2018:974.

Golden B.L., Spangler M.L., Snelling W.M. and Garrick D.J. (2018) Proc. 11th Beef Impr. Fed. Genet. Pred. Workshop 11:14.

Henderson C.R. (1973) J. Anim. Sci. Issue Symp. 70:23.

Lidauer M., Matilainen K., Mantysaari E., Pitkanen T., Taskinen M. and Stranden I. (2015) http://www.luke.fi/mix99

Schaeffer L.R. and Kennedy B.W. (1986) J. Dairy Sci. 69:575.

Stranden I. and Lidaeur M. (1999) J. Dairy Sci. 82:2779.

Tsuruta S., Misztal I. and Stranden I. (2001) J. Anim. Sci. 79:1166.

Van Vleck L.D. and Dwyer D.J. (1985) J. Dairy Sci. 68:1006.

ADJUSTING THE GENOMIC RELATIONSHIP MATRIX FOR BREED DIFFERENCES IN SINGLE STEP GENOMIC BLUP ANALYSES

P.M. Gurman, K.L. Bunter, V. Boerner, A.A. Swan and D.J. Brown

Animal Genetics & Breeding Unit*, University of New England, Armidale, NSW, 2351 Australia

SUMMARY

The genomic relationship matrix (GRM) routinely constructed for single-step genomic BLUP analyses is known to contain breed structure, observable via principal component analysis, while the pedigree relationship matrix uses coefficients that are constant between known relatives regardless of breed or genetic group membership. This paper explores the effect of using allele frequencies for each breed or genetic group when calculating the GRM to reduce breed or genetic group structures in the GRM in the presence of pedigree based genetic groups fitted as random effects. We investigated the effect of using a breed-adjusted GRM on estimated breeding values, showing cross-validation results, genetic trends and estimated breeding value accuracies. Cross-validation results across breed showed a slight increase in EBV accuracy using a breed-adjusted GRM, 0.220 ± 0.068 compared to a non-adjusted GRM, 0.206 ± 0.071. Genetic trends calculated from estimated breeding values (EBVs) using a breed-adjusted GRM were more closely aligned to those estimated using a pedigree-only model compared to a non-adjusted GRM. These results show that using a single set of allele frequencies in a GRM with a diverse number of breeds can result in biased breeding values and biased genetic trends relative to those obtained from pedigree model including breed groups.

INTRODUCTION

With the transition of routine genetic evaluations from pedigree- or genomic blending-based approaches, to single-step (Legarra *et al.* 2014), the alignment of the GRM to the pedigree-based numerator relationship matrix (NRM) has become a focus of research interest when genetic groups are present and included in the model as separate random effects. This research focus is, in part, due to the impact that any misalignment can have on genetic trends (Meyer *et al.* 2018). Scalar adjustment parameters have been suggested (Vitezica *et al.* 2011; Christensen 2012) to align the NRM and GRM, while leaving the general structure inherent in the GRM intact. The 'metafounders' framework (Legarra *et al.* 2015; Garcia-Baccino *et al.* 2017) was suggested as a method for modifying the NRM to be in better alignment with the GRM and in doing so, replacing genetic groups (Westell *et al.* 1988) that are currently used for managing missing pedigree. While the metafounders framework is a promising method for handling misalignment of the NRM with the GRM and genetic groups, it is challenging to implement in routine analyses. That approach assumes that each metafounder has genotyped animals in their forward pedigree, which may not occur in practice, and may require modifications to the rules currently used to assign animals to metafounders or genetic groups. An alternative method is to align the GRM to the NRM by removing breed/genetic group structure from the GRM, as described by Makgahlela *et al.* (2013). This paper aims to examine the impact of using a breed-adjusted GRM (hereby BGRM) in routine single-step genomic BLUP analyses on cross-validation correlations and regression slopes of evaluations on adjusted phenotypes and genetic trends compared to a standard non-adjusted GRM (hereby SGRM).

---

* A joint venture of NSW Department of Primary Industries and the University of New England

MATERIALS AND METHODS

A SGRM using the method of Yang *et al.* (2010) can be constructed as $\boldsymbol{W} = \frac{\boldsymbol{M} - 2\boldsymbol{p}}{\sqrt{2\boldsymbol{p}(1-\boldsymbol{p})}}$; $\boldsymbol{G} = \frac{\boldsymbol{WW'}}{m}$, where $\boldsymbol{M}$ is the marker matrix of dimensions animals by markers, $\boldsymbol{p}$ is the allele frequency of the animals in $\boldsymbol{M}$ and m is the number of markers in $\boldsymbol{M}$. The above equation was extended by Makgahlela *et al.* (2013) for allele frequencies that vary by breed (BGRM). A breed proportion matrix, $\boldsymbol{Q}$, was calculated from genotypes using BreedComp (Boerner *et al.* 2018), and the allele frequency for each breed in Q was calculated as F, allowing the method of Yang *et al.* (2010) to be extended for each column in the Q matrix. The expected allele frequency for each animal based on its breed proportion is then estimated as $\boldsymbol{P} = \boldsymbol{QF}$, and thus $\boldsymbol{W}_{ij} = \frac{\boldsymbol{M}_{ij} - 2\boldsymbol{P}_{ij}}{\sqrt{2\boldsymbol{P}_{ij}(1-\boldsymbol{P}_{ij})}}$ and $\boldsymbol{G} = \boldsymbol{WW'}/m$.

To examine the differences in EBVs using a pedigree-only relationship matrix or these two GRM construction methods in single step analyses utilising pedigree based genetic groups implemented as random effects, multi-trait BLUP analyses were performed on maternal reproduction data from sheep. The EBVs (including genetic group estimates) from these analyses were compared via genetic trends and EBV correlations for animals born after 2013. The data consisted of approximately 2.4 million animals in the pedigree, with 11,761 of these genotyped and phenotypes collected on up to 15 traits. Reproduction traits that were included in the analysis were: fertility of yearling (ycon) and adult (con) ewes, litter size of yearling (yls) and adult (ls) ewes, rearing ability of yearling (yera) and adult (era) ewes, and maternal behaviour score of adult ewes (mbs). Other traits included in this analysis were: post-weaning eye muscle depth (pemd), post-weaning carcase fat (pcf), post-weaning scrotal circumference (psc), yearling scrotal circumference (ysc), pre-joining weight of post-weaning (pwt) and adult (awt) ewes, and pre-joining condition score of yearling (ycs) and adult (cs) ewes. The number of phenotypes per trait varied, ranging from 595,978 (ls) to 1,746 records (ycs). The genotypes represented a variety of breeds, dominated by Border Leicester, Coopworth, Corriedale and crossbred animals, including Border Leicester-Merino cross sheep. BLUP analyses were performed using each of these three relationship matrices assuming common variance components, and included random effects for genetic groups. Further model details can be found in Bunter et al. (2019). Forward cross-validation was performed. Phenotypes for animals born after 2013 were removed from the analysis and breeding values were estimated for these animals from the remaining phenotypes. This year of birth was chosen to ensure sufficient reproduction records were included in the validation set, though some traits still had few validation phenotypes. Phenotypes recorded after 2013 were then adjusted for the relevant fixed effects to calculate correlations with the estimated breeding values, with phenotypes re-scaled by the square root of the heritability. For each trait, adjusted phenotypes were regressed on the EBVs; slopes less than one indicate over-prediction (i.e. bias) and slopes above one indicate under-prediction. The mean and standard deviation of the correlations and regression slopes across all 15 traits was calculated, weighted by the number of animals with phenotypes included in the validation set. Traits with fewer than 300 observations (n=4) were not included in these means.

RESULTS AND DISCUSSION

Comparing the breeding values of all 15 traits for animals born after 2013 estimated using the SGRM with those from a pedigree only analysis, the mean correlation was 0.988. The minimum correlation was 0.977, while the maximum was 0.997. The same correlations calculated using a MGRM were 0.996, with a minimum correlation of 0.993 and a maximum of 0.999. Within Border Leicester sheep, the mean correlation for a SGRM and a MGRM changed from 0.973 to 0.992, respectively, within Coopworth sheep the mean correlation changed from 0.978 to 0.999 and within Corriedale sheep the correlation increased from 0.983 to 0.999. The genetic trends for the four traits showing the lowest correlations between EBVs from pedigree and SGRM models are presented in Figure 1. These correlations and genetic trends show that using the BGRM produced breeding values and trends that were closer to those previously observed using the NRM and breed group effects.

The mean correlations from forward cross-validation estimated using NRM, SGRM and BGRM were 0.196 ± 0.0824 , 0.206 ± 0.071, and 0.220 ± 0.068, respectively. The mean regression slopes estimated using NRM, SGRM and BGRM were 0.910 ± 0.447, 1.004 ± 0.418 and 0.987 ± 0.370, respectively. The correlations and regression slopes by traits with sufficient data to make inference are presented in Table 1. These results indicate that the BGRM resulted in slightly higher cross-validation accuracies at the expense of a negligible increase in bias over the SGRM EBVs. A stronger bias was found in the NRM EBVs than for either of the single-step analyses. While the standard deviations were large across traits for both accuracies and biases, together these set of results suggest that correcting the relationship values of the GRM for breed can produce higher accuracy and lower bias in EBVs.

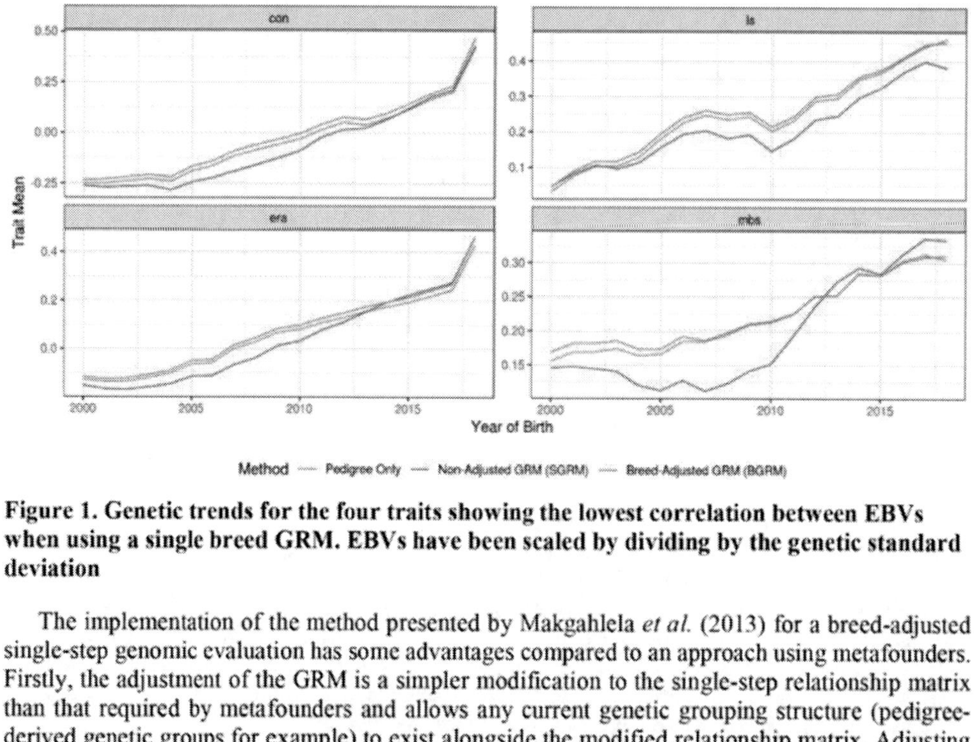

Figure 1. Genetic trends for the four traits showing the lowest correlation between EBVs when using a single breed GRM. EBVs have been scaled by dividing by the genetic standard deviation

The implementation of the method presented by Makgahlela *et al.* (2013) for a breed-adjusted single-step genomic evaluation has some advantages compared to an approach using metafounders. Firstly, the adjustment of the GRM is a simpler modification to the single-step relationship matrix than that required by metafounders and allows any current genetic grouping structure (pedigree-derived genetic groups for example) to exist alongside the modified relationship matrix. Adjusting the GRM is also simpler when breeds or genetic groups have no genotyped animals in their pedigree. Implementing a BGRM in a single-step analysis requires that genetic groups are also fitted in the model as they have for NRM based analyses. These genetic groups need to align with the breeds that were used in the construction of the GRM. There are situations where the implicit breed structure in the GRM has advantages, e.g. predicting breeding values for animals without pedigree across genetic groups, with metafounders allowing this structure to be imposed over the whole NRM.

Genetic groups or metafounders both require the assignment of animals into pre-defined group structures. Methods for creating the most parsimonious grouping structures require further investigation, minimising the number of groups required while maintaining enough groups for predictive purposes. The addition of genotypes can aid in this process.

CONCLUSIONS

In this paper, we show that the method presented by Makgahlela *et al.* (2013) reduces the breed structure implicit in a GRM constructed from multiple breeds, resulting in a GRM that is numerically more similar to the NRM. This change results in genetic trends that align closer with those seen from pedigree-only models. The BGRM resulted in slightly higher average cross-validation accuracies with similar biases, and less biased than pedigree alone, compared to BLUPs performed using a GRM constructed from a single set of allele frequencies.

Table 1. Table of forward cross-validation accuracies obtained from BLUP models using an NRM (r_NRM), a single-breed GRM (r_SGRM) and a multi-breed GRM (r_BGRM) and the corresponding biases, b_NRM, b_SGRM and b_BGRM. 'n' indicates the number of animals in the validation set

Trait	n	r_NRM	r_SGRM	r_BGRM	b_NRM	b_SGRM	b_BGRM
ycon	618	0.10	0.14	0.16	0.59	0.83	0.95
con	885	0.11	0.12	0.18	0.53	0.62	0.87
yls	627	0.18	0.20	0.17	0.79	0.84	0.78
ls	1,801	0.14	0.14	0.16	0.69	0.74	0.79
yera	377	0.22	0.21	0.25	1.84	2.05	2.05
era	1,583	0.34	0.27	0.29	2.00	1.90	1.72
pemd	3,476	0.15	0.17	0.18	0.74	0.83	0.79
pcf	3,467	0.21	0.24	0.25	0.94	1.13	1.10
pwt	431	0.25	0.27	0.26	0.80	0.86	0.83
awt	943	0.32	0.34	0.35	0.71	0.81	0.77
cs	545	0.34	0.34	0.34	1.08	1.20	1.16

Abbreviations: 'ycon' and 'con': fertility of yearling and adult ewes, respectively, 'yls' and 'ls': little size of yearling and adult ewes, respectively, 'yera' and 'era': rearing ability of yearling and adult ewes, respectively, 'pwt' and 'awt': pre-joining weight of post-weaning and adult ewes, respectively, 'cs': pre-joining condition score of adult ewes.

REFERENCES

Boerner V., Wittenburg D. (2018) *Front. Genet.* **9**:185.

Bunter K.L., Swan A.A., Gurman P.M., Boerner V., McMillan A.J., Brown D.J. (2019) *Proc. Assoc. Advmt. Anim. Breed. Genet.* **23**:(in press)

Christensen O.F. (2012) *Genet. Sel. Evol.* **44**:37.

Garcia-Baccino C.A., Legarra A., Christensen O.F., Misztal I., Pocrnic I., Vitezica Z.G., Cantet R.J.C. (2017) *Genet. Sel. Evol.* **49**:34.

Legarra A., Christensen O.F., Aguilar I., Misztal I. (2014) *Livest. Sci.* **166**:54.

Legarra A., Christensen O.F., Vitezica Z.G., Aguilar I., Misztal I. (2015) *Genetics* **200**:455.

Makgahlela M.L., Strandén I., Nielsen U.S., Sillanpää M.J., Mäntysaari E.A. (2013) *J. Dairy Sci.* **96**:5364.

Meyer K., Tier B., Swan A. (2018) *Genet. Sel. Evol.* **50**:39.

Vitezica Z.G., Aguilar I., Misztal I., Legarra A. (2011) *Genet. Res.* **93**:357.

Westell R.A., Quaas R.L., Van Vleck L.D. (1988) *J. Dairy Sci.* **71**:1310.

Yang J., Benyamin B., McEvoy B.P., Gordon S., Henders A.K., Nyholt D.R., Madden P.A., Heath A.C., Martin N.G., Montgomery G.W., Goddard M.E., Visscher P.M. (2010) *Nat. Genet.* **42**:565.

GENETIC CONTROL OF FERTILITY TRAITS ACROSS SPECIES: VARIANCE IN TROPICAL BEEF HEIFERS' AGE AT PUBERTY EXPLAINED BY GENES CONTROLING AGE AT MENARCHE IN WOMEN

R. Costilla, C. Warburton and B.J. Hayes

Centre for Animal Science, Queensland Alliance for Agriculture and Food Innovation, The University of Queensland, Brisbane, 4072 Australia

SUMMARY

Fertility traits are of paramount importance for humans and cattle. In cattle, they are one of the main profit drivers in the industry. Using data from genome-wide association studies (GWAS) from both species, we estimated the effect of genes associated with age at menarche in women (AaM) in the variance of age at puberty (AaP) in tropically adapted beef heifers. We found that variants within 100kb of AaM bovine orthologous genes explained 11.2% of the additive genetic variance of heifers AaP in the biggest cohort analysed. This represented about twice the variance explained by random gene-sets of similar size and number of SNPs ($P<0.2$). Our work suggests some potential of cross-species analyses to increase the cattle industry's productivity.

INTRODUCTION

Thanks to the recent advances in biomedical technology, the genetic basis of fertility in humans is better known now than ever. For instance, the biggest GWAS for female fertility to date with ~370,000 women, Day *et al.* (2017), reported hundreds of genomic loci associated with AaM in women, a female complex trait that is a milestone in pubertal development. An interesting question is, whether we can use the information coming out of the extremely powerful GWAS in humans to improve genomic predictions for related traits in cattle?

Given the evidence for genetic control of complex traits across mammalian species (Pryce *et al.* 2011; Bouwman *et al.* 2018), we hypothesised that genetic factors contributing to variation between individuals for age at puberty/age at menarche will be shared across humans and cattle. In humans, the heritability of AaM was estimated to be 0.32 (0.03) (Day *et al.* 2017). In cattle, AaP has been shown to be moderately to highly heritable in tropically adapted breeds (Johnston *et al.* 2009; Corbet *et al.* 2018) with heritabilities ranging from 0.22 (0.07) to 0.57 (0.12) for Santa Gertrudis and Brahman breeds respectively. Using bovine orthologous of genes associated with AaM, we estimated their contribution to the additive genetic variance of age at puberty (AaP) in heifers.

MATERIALS AND METHODS

Animals, genotypes and phenotypes. We used published data from several heifer populations: Beef Cooperative Research Centre for Beef Technology Brahman and Tropical Composite (CRC BB and CRC TC, respectively) and the Queensland Smart Futures population (Smart Futures). These herds contained heifers from several tropical beef breeds and were genotyped with the BovineSNP50 (CRC BB and CRC TC) and Geneseek GGP-LD array (Smart Futures). The Smart Futures heifers consisted of animals from three breeds: Brahman (979), Santa Gertrudis (1813) and Droughtmaster (914). Complete details for these animals and genotypes have been published elsewhere (Johnston *et al.* 2009; Corbet *et al.* 2018). In total, we used 3695, 960 and 868 animals from the Smart Futures, CRC TC and CRC BB herds. Genotypes were imputed twice up to 728,785 SNPs using Fimpute (Sargolzaei *et al.* 2014) and then to whole genome sequence using Minimac3 (Das *et al.* 2016). The phenotypes were age in days at first corpus luteum (AGECL) and corpus luteum score (CLscore) at ~600 days for the CRC and Smart Futures cohorts, respectively. The AGECL is a count variable and CLscore is an ordinal variable ranging from 0 "infantile tract" to 5 "pregnancy > 10 weeks". These two heifers AaP phenotypes, CLscore and AGECL, exhibit a very high genetic correlation (-0.83(0.09), Engle *et al.* 2019).

Bovine orthologous AaM genes. Using coding variation (nonsynonymous SNPs), associated expression in neural tissues (eQTL) and chromatin interaction data (Hi-C), Day *et al.* (2017) implicated 233 protein-coding genes in the regulation of AaM in women. We mapped these genes to the UMD3.1 bovine genome using Biomart Ensemb 94 and filtered them out by conservation status (orthology confidence=1 and gene identity > 60%), rendering a total of 205 highly conserved orthologous AaM genes in the bovine genome. Then, we located variants (SNPs and INDELs) in or around + 100kb using imputed sequence data from the CRC BB, CRC TC and SMF cohorts.

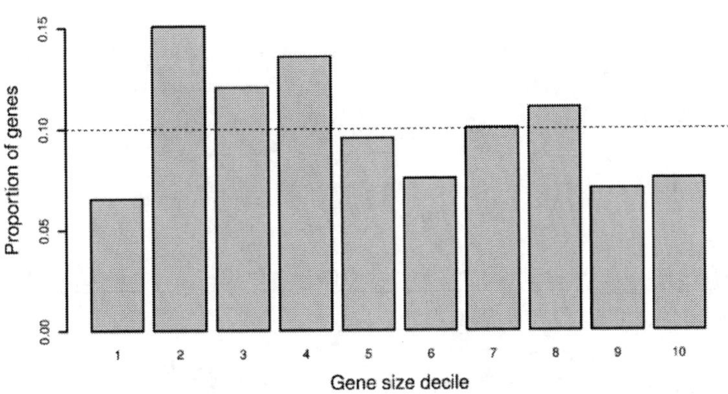

Figure 1. Gene size distribution (deciles) for bovine orthologous genes for age at menarche (AaM) in women

Statistical analysis. We estimated the variance of heifers' AaP explained by orthologous AaM genes using a model with two genomic relationship matrices (GRMs) constructed from the imputed to sequence genotypes described before. The first GRM is constructed from variants in or within 100kb of AaM genes and the second GRM from the remaining variants in the bovine genome. The model included additional continuous and categorical covariates as follows:

$$y = 1_n\mu + age + pc1 + pc2 + cgroup + g_1 + g_2 + \varepsilon$$

where y is a vector of phenotypes, μ the overall mean, 1_n is a vector of 1s, *age* is a vector with the heifers' age fitted as a continuous covariate, *pc1* and *pc2* the first and the second principal components (derived from the GRM), *cgroup* is a vector of contemporary groups that includes with herd, year, and season and is fitted as categorical covariate. g_1 and g_2 are vectors of random effects for the variants in or within 100kb of AaM genes and the remaining ones in the bovine genome with $g_1 \sim N(0, G_1\sigma^2_{g1})$ and $g_2 \sim N(0, G_2\sigma^2_{g2})$. ε is a vector of random residuals distributed $\varepsilon \sim N(0, \sigma^2_\varepsilon)$. G_1 and G_2 denote the corresponding GRM matrices constructed following the first method of VanRaden (2008) and σ^2_{g1}, σ^2_{g2}, σ^2_ε the corresponding genetic and error variances. We fitted the model separately for each cohort using GCTA (Yang *et al.* 2011).

In order to provide an appropriate comparison for the AaM genes, we also estimated the variance explained by 100 random gene-sets of similar length and SNP number, e.g. we performed a stratified random sampling by quantiles of gene size and number of SNPs, and ran a randomized permutation test for the percentage of AaP variance explained by AaM genes.

RESULTS AND DISCUSSION

Out of a total of 28.9 million imputed to sequence variants across all cohorts, there were 339,669 variants within +/- 100kb from 205 bovine orthologous AaM genes. The number of variants varied slightly within individual cohorts. Note also that in terms of gene physical size, AaM genes are over-represented in the lower deciles and thus tend to be smaller in size than the rest of protein-coding genes in the bovine genome (Figure 1).

Variants in AaM genes explained 2.5% phenotypic (11.2% genetic) variance of heifers AaP in the biggest cohort, Smart Futures (Table 1). This represented about twice the mean variance explained by variants in random gene-sets (1.2% phenotypic and 5.6% genetic) that had on average 379,325 variants. This result however did not reach significance in the randomized permutation test ($P<0.2$) (Figure 2). With regard to the CRC cohorts, variants in AaM genes explained negligible percentages when compared with variants in random gene-sets.

Table 1. SNP based heritability (h^2) partition for cohorts included in the meta-analysis

	Cohort					
	Smart Futures		CRC TC		CRC BB	
Component	h^2	se	h^2	se	h^2	se
AaM genes: V(G1)/Vp	0.025	0.019	0.005	0.059	0.015	0.058
Remaining: V(G2)/Vp	0.195	0.035	0.393	0.103	0.446	0.108
Overall: V(G1)+V(G2)/Vp	0.220	0.031	0.398	0.085	0.461	0.092
V(G1)/Vp for random gene-sets*	0.012	0.001	0.034	0.009	0.022	0.009

*Mean for 100 gene-sets (379,325 variants on average).

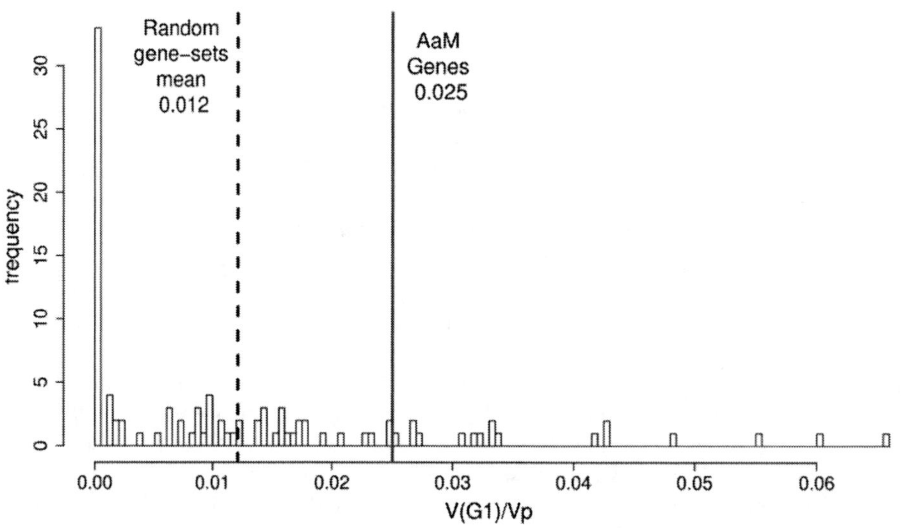

Figure 2. Randomised permutation test results for the Smart Futures cohort. Variance in heifers age at puberty (AaP) explained by age at menarche (AaM) genes (red line, 339,669 variants), and random gene-sets of similar size to AaM genes. Dotted lined displays the mean for 100 random gene-sets (379,325 variants on average)

Note that overall h^2 estimates by cohort: 0.220(0.031), 0.398(0.085), and 0.461(0.092) for Smart Futures, CRC TC and CRC BB, respectively, are consistent with previous estimates from published studies (Johnston *et al.* 2009; Corbet *et al.* 2018). In terms of individual genes, there were four genes in the AaM set (*ZNF654, LEPROT, CCDC40, CLUAP1*) that reached significance ($P<10^{-4}$) in the meta-analysis of AaP GWAS across the three cohorts. In humans, these genes are also associated with haemoglobin concentration (*ZNF654*), morbid obesity (*LEPROT*), blood protein levels (*CCDC40*), vital capacity and leukocyte count (*CLUAP1*) (Stelzer *et al.* 2016).

Taken together these results suggest that women's AaM genes are also associated with a similar phenotype in a different species, in this case fertility phenotypes in tropically adapted beef heifers. Importantly, however, is the issue of power for this complex trait as a large number of animals was required to pick up this signal, e.g. association was only presented in the biggest cohort with 3695 animals. An interesting extension would be to combined both CRC cohorts (Brahmans and composites) and performed the analyses presented here on this combined dataset.

CONCLUSIONS

Variants in AaM genes explained 2.5% phenotypic (11.2% genetic) variance of tropical beef heifers' AaP in the biggest cohort analysed here. This is about twice the variance explained by similar random gene-sets, although this result is not statistically significant ($P<0.2$), and the variance explained in the other cohorts was not different from zero. Some genes affecting AaM were also significant for AaP in heifers ($P<1 \times 10^{-4}$). Our work highlights the potential of cross-species analyses to increase the industry's productivity. Further research in terms of inclusion of variants in AaM genes in genomic prediction models is needed to achieve this potential.

ACKNOWLEDGEMENTS

We gratefully acknowledge the contribution of Drs. Nicholas J. Corbet, Jack M. Allen, Alan R. Laing, Geoffrey Fordyce, Russel Lyons, Michael R. McGowan, and Brian M. Burns who collected the data used in this study.

REFERENCES

Bouwman, A. C., Daetwyler H. D., Chamberlain A. J., Ponce C. H., Sargolzaei M., Schenkel F. S. *et al.* (2018) *Nat. Genet.* **50**: 362.

Corbet N. J., Allen J. M., Laing A. R., Fordyce G., McGowan M. R. and Burns B. M. (2018) *Anim. Prod. Sci.* **58**: 1735.

Das S., Forer L., Schonherr S., Sidore C., Locke A. E., *et al.* (2016) *Nat. Genet.* ***48***: 1284.

Day F. R., Thompson D. J., Helgason H., Chasman D.I., Finucane H., Sulem P.*et al.* (2017) *Nat. Genet.* **49**: 834.

Engle B. N., Corbet N. J., Allen J. M., Laing A. R., Fordyce G., McGowan M. R., Burns B. M., Lyons R. E. and Hayes B. J. (2019) *J. Anim. Sci.* **97**: 90.

Johnston D. J., Barwick S. A., Corbet N. J., Fordyce G., Holroyd R.G., Williams P.J. and Burrow H. M. (2009) *Anim. Prod. Sci.* **49**: 399.

Pryce, J. E., Hayes B. J., Bolormaa S. and Goddard M. E. (2011) *Genetics* **187**: 981.

Sargolzaei M., Chesnais J. P., and Schenkel F. S. (2014) *BMC Genomics*. **15**: 478.

Stelzer G., Rosen R., Plaschkes I., Zimmerman S., Twik M., Fishilevich S., *et al.* (2016) *Curr. Prot. in Bioinfo.* **54**: 1.30.1.

VanRaden P. M. (2008) *J. Dairy Sci.* **91**: 4414.

Yang J., Lee S. H., Goddard M. E., and Visscher P. M. (2011) *Am. J. Hum. Genet.* **88**: 76.

SIGNATURES OF SELECTION IN ADMIXED DAIRY CATTLE OF KENYA

H. Aliloo[1], R. Mrode[2,3], A.M. Okeyo[2] and J.P. Gibson[1]

[1]University of New England, Armidale, NSW 2351, Australia
[2]International Livestock Research Institute, Nairobi, Kenya
[3]Scotland's Rural College, Edinburgh, United Kingdom

SUMMARY

Small holder dairy farmers in Kenya rear crossbred cattle to combine the environmental adaptation features of indigenous populations with the high milk yield potential of exotic dairy breeds. The identification of signatures of selection in Kenyan admixed cattle could lead to a better understanding of the genetic structure of adaptation and productivity in challenging environmental conditions. Here, we examined the genome of the admixed cattle populations of Kenya for candidate regions under adaptive selection. We employed a haplotype based method, integrated extended haplotype homozygosity score (iHS), and scanned the genome of 1,475 admixed cattle using 521,362 SNPs. The local ancestry of the admixed cattle were inferred and used to identify the admixed cattle with more than 3 generations of crossing. This improved the power in detection of signatures of selection and after removing recently admixed animals, we identified 16 candidate regions and 8 candidate genes across 7 autosomes. Investigation of the candidate genes showed that several are involved in feed efficiency and disease resistance pathways that are important for adaptation under small-holder production systems. If substantiated, this information could be integrated into breeding programs aiming to improve dairy cattle productivity and adaptation in East Africa.

INTRODUCTION

The crossbred dairy cattle in Kenya consist of an admixed population resulting from around 50 years of crossing and inter-se matings of African indigenous cattle to several exotic dairy breeds, mainly from Friesian, Holstein, Ayrshire and related red dairy breeds, and Jersey. These animals are kept by smallholder dairy farmers, typically in herds of size 1 to 5 cows, and produce about 80% of the total milk in Kenya. The majority of Kenyan crossbred dairy cattle are bred via natural mating and only a small proportion of matings are made by AI to imported and locally bred purebred dairy bulls. Very few animals have pedigree records and there is no systematic genetic evaluation systems or breeding programs to support farmers. The identification of footprints of selection in admixed cattle through the use of molecular markers such as single nucleotide polymorphism (SNP) can lead to a better understanding of the genetic structure underlying adaptation and productivity in challenging environmental conditions. Genomic regions with selection advantage can be incorporated in breeding strategies to select animals that are well suited in such environments and production systems. In this study we scanned the genome of the Kenyan admixed cattle by applying an intra-population haplotype-based method (iHS) for signatures of post-admixture selection. We aimed to detect genomic regions responsible for adaptation and productivity under the challenging environment of East Africa. The local ancestry of individual loci are inferred to find the crossover events across the admixed genome and to assign each crossbred animal to a generation of crossing since the ancestral crossing happened.

MATERIALS AND METHODS

The genotypic data included 1,475 crossbred cattle sampled in Kenya between 2010 and 2014 and genotyped for 777,962 SNP markers using Illumina BovineHD BeadChip (Illumina, San Diego, CA). Routine QC was applied to genotypes and this resulted to 521,362 SNPs on 1,475 crossbred animals distributed over 29 autosomes based on the UMD3.1 bovine reference genome.

Local ancestry and crossing-overs in crossbred cattle. The local ancestry of the crossbred cattle was inferred at individual SNP level using samples from 3 groups of ancestral populations including *Bos indicus* (IND) African *Bos taurus* (AFT) and European *Bos taurus* (EUT) by LAMD-LD software (Baran *et al.* 2012). The local ancestry inferences were used to calculate the average number of crossover events across each crossbred genome by first counting the number of transitions from either IND or AFT ancestry to EUT ancestry and vice versa, and then standardizing it by chromosome length. A recombination rate of 1 cM = 1 Mb across the whole genome and 1 crossover per Morgan per generation after crossing was assumed to assign each crossbred animal to an approximate generation since the ancestral crossing (indigenous × taurine) happened. A minimum of 4 generations of crossing was used to remove the impact of recent admixture on selection of signature analysis. This was also to keep only animals for which selection has had enough time to leave its footprint on their genome.

Detection of footprints of selection. The integrated extended haplotype homozygosity score (iHS) was used as an intra-population measure of the extent of haplotype homozygosity in crossbreds (Voight *et al.* 2006). We used R software *rehh* package (Gautier *et al.* 2017) to calculate iHS and then we transformed these values into *p-values* according to Gautier and Naves (2011). The *qvalue* package in R software was then used to correct *p-values* for multiple testing by calculation of a false discovery rate and generating the corresponding *q-values*. A candidate region for selection was defined by first identifying SNPs with a *q-value* <0.1 and then searching within the 500 Kb interval downstream and upstream (1 Mb window) of the identified SNP for SNPs with a *p-value* $<10^{-3}$. Genes with at least 1 SNP with a *q-value* <0.1 found within them were deemed as candidate genes under selection.

RESULTS AND DISCUSSION

The haplotypes from the 3 ancestral groups, IND, AFT and EUT, were used to infer the local ancestries of the admixed cattle at individual loci level. The majority of haplotypes in the admixed cattle were found to have originated from EUT ancestor (≈0.73) while IND and AFT ancestral populations contributed smaller proportions of admixed haplotypes (≈0.24 and ≈0.03, respectively). The local ancestry inferences were further used to calculate the genome-wide average number of crossover events on haplotypes carrying the lowest number of crossovers between the two haplotypes of each individual for each chromosome (Figure 1). For most of the admixed cattle, the number of recent crossovers per Morgan was found to be relatively small (<3). This suggested that the admixed cattle in East Africa are mainly recent crosses of indigenous cattle with exotic breeds. Only 55 animals had an average of more than 3 crossovers per Morgan, which is approximately equivalent to 4 or more generations of inter-se mating after the original cross to an exotic or indigenous ancestor (Figure 1).

Selection needs time to leave its footprints on the genome and if there is not enough time since the most recent admixture, the detection analysis is underpowered. Including recently admixed animals in the analysis adds noise to the detection of signatures of selection and potentially masks the footprints that would have otherwise been detected. We found evidence for this in our results (not shown). When we included all admixed cattle in calculation of iHS, no candidate region at a FDR threshold of 0.1 was detected. However, removing crossbreds with a genomic average crossover frequency of less than 3 per Morgan identified 16 candidate regions across 7 autosomes at the same FDR shown in Figure 2.

The details of the 16 identified candidate regions from the iHS analysis of the filtered admixed cattle are in Table 2. The size of these candidate regions ranged from only 112.25 Kb on chromosome 12 up to 683 Kb on chromosome 7 and collectively encompassed 8 candidate genes. Chromosome 7 had the highest number of candidate regions for selection among all chromosomes and it contained 3 candidate genes. Chromosome 3 contained 2 candidate genes while chromosomes 6, 11 and 12 each had one candidate gene. The ancestry of all candidate regions in chromosome 3 was dominated by EUT while for chromosomes 6, 7 and 12 that had more than 1 candidate region, the dominant ancestry was either IND or EUT (Table 1).

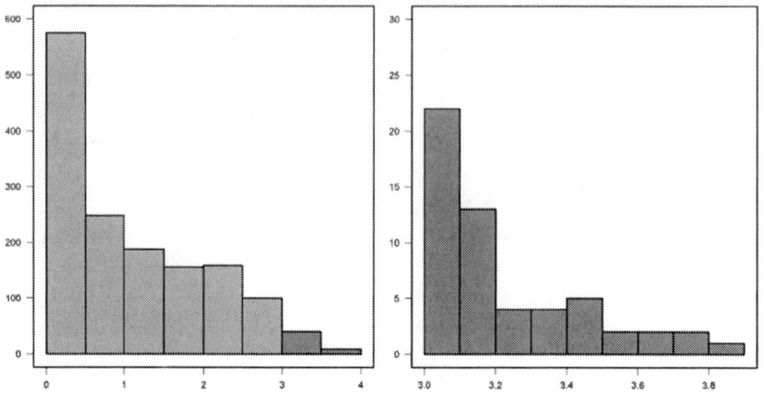

Figure 1. Average number of crossover events per Morgan in all admixed cattle (left) and in those with more than 3 crossovers per Morgan (right and green)

The *S100A10* gene is located on chromosome 3 and encodes a protein which regulates several cellular processes such as cell cycle progression and differentiation. It has been found as a candidate gene for residual feed intake in Angus cattle (Al-Husseini *et al.* 2013) through a single SNP genome-wide association study. Given that feed efficiency is a very important factor in low input smallholder production systems, it could be justified why this gene has been the target of selection in the African environment. Furthermore, the candidate region harbouring *S100A10* shows a dominant EUT ancestry in our study, suggesting possible EUT contribution to feed efficiency in the admixed cattle.

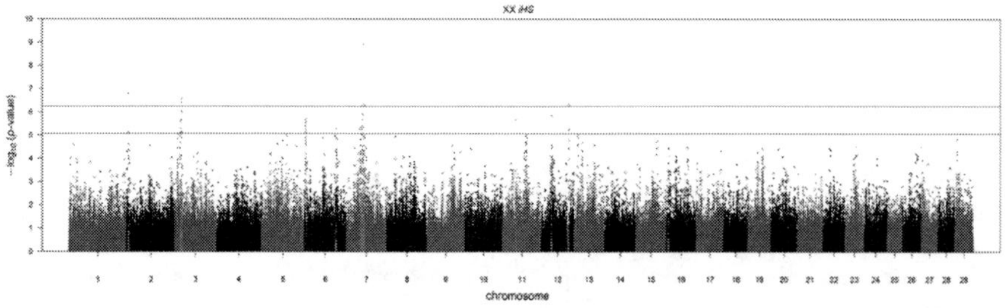

Figure 2. Manhattan plots of *p-values* for genome-wide iHS within the crossbred population. The red and blue horizontal lines correspond to false discovery rates at 5% and 10%, respectively

We identified *NLRP3* gene in a candidate region on chromosome 7 with a dominant IND ancestry. This gene encodes a pyrin-like protein and it plays a role in the regulation of inflammation, the immune response, and apoptosis. *NLRP3* has been found to be a candidate gene for Crohn's disease (Villani *et al.* 2009) and Johne's disease (Scanu *et al.* 2007; Mallikarjunappa *et al.* 2018) in human and livestock populations, respectively. The selection sweep harbouring this gene is of IND ancestry, suggesting that the IND ancestors may have contributed a version of *NLRP3* conferring resistance to local disease or other environmental challenges. Another candidate region on chromosome 7 harbours the gene *LYPD8*, which has been reported to be differentially expressed between cows with versus without subclinical mastitis (Song *et al.* 2016) and it provides defence against gram negative bacteria in the colon of non-ruminants. This region is of EUT origin, suggesting possible EUT contribution to disease resistance in the crossbred population.

Table 1. The details of the identified candidate regions from iHS analysis

Chromosome	Region (Mb)	Top SNP *q-value*	Dominant ancestry	Candidate genes
2	5.46 – 6.00	0.0378	IND	–
3	9.58 – 9.80	0.0995	EUT	–
3	17.18 – 17.70	0.0861	EUT	–
3	18.80 – 19.29	0.0578	EUT	*S100A10*
3	22.07 – 22.71	0.0390	EUT	*ACP6*
6	4.91 – 5.29	0.0578	IND	–
6	90.70 – 91.12	0.0861	EUT	*MTHFD2L*
7	38.55 – 38.92	0.0861	IND	–
7	41.40 – 42.00	0.0390	IND	*BTNL9, NLRP3*
7	43.84 – 44.16	0.0861	EUT	*LYPD8*
7	46.56 – 46.99	0.0006	EUT	–
7	49.91 – 50.25	0.0390	IND	–
11	36.81 – 37.13	0.0578	IND	*ACYP2*
12	28.64 – 29.05	0.0578	IND	–
12	76.82 – 76.93	0.0390	EUT	*CLDN10*
16	4.52 – 4.89	0.0995	IND	–

CONCLUSIONS

This study provides evidence that the genome of the admixed cattle in Kenya may have been shaped by adaptive selection in response to the challenging environment in which they exist. If our findings can be substantiated, the information might be used in breeding programmes to enhance productivity and adaptation traits in smallholder dairy systems of Kenya.

REFERENCES

Al-Husseini W., Gondro C., Quinn K., Herd R.M., Gibson J.P. and Chen Y. (2014) *Anim. Genet.* **45**: 12.

Baran Y., Pasaniuc B., Sankararaman S., Torgerson D.G., Gignoux C., Eng C., Rodriguez-Cintron W,. Chapela R., Ford J.G., Avila P.C., Rodriguez-Santana J., Burchard E.G. and Halperin E. (2012) *Bioinformatics* **28**: 1359.

Gautier M. and Naves M. (2011) *Mol. Ecol.* **20**: 3128.

Gautier M., Klassmann A. and Vitalis R. (2017) *Mol. Ecol. Resour.* **17**: 78.

Mallikarjunappa S., Sargolzaei M., Brito L.F., Meade K.G., Karrow N.A. and Pant S.D. (2018) *J. Dairy Sci.* **101**: 7280.

Scanu A.M., Bull T.J., Cannas S., Sanderson J.D., Sechi L.A., Dettori G., Zanetti S. and Hermon-Taylor J. (2007) *J. Clin. Microbiol.* **12**: 3883.

Song M., He Y., Zhou H., Zhang Y., Li X. and Yu Y. (2016) *Sci. Rep.* **6**: 29390.

Villani A.C., Lemire M., Fortin G., Louis E., Silverberg M.S., Collette C., Baba N., Libioulle C., Belaiche J., Bitton A., Gaudet D., Cohen A., Langelier D., Fortin P.R., Wither J.E., Sarfati M., Rutgeerts P., Rioux J.D., Vermeire S., Hudson T.J. and Franchimont D. (2009) *Nat. Genet.* **41**: 71.

Voight B.F., Kudaravalli S., Wen X. and Pritchard J.K. (2006) *PLoS Biol.* **4**: e5350.

GENOME-WIDE ASSOCIATION STUDY OF CARCASE AND EATING QUALITY TRAITS IN AUSTRALIAN ANGUS BEEF CATTLE

W.M.S.P. Weerasinghe[1], B.J. Crook[1], S.A. Clark[2], N. Moghaddar[2] and A.I. Byrne[3]

[1]Agricultural Business Research Institute, UNE, Armidale, NSW, 2350 Australia
[2]School of Environmental and Rural Science, UNE, Armidale, NSW, 2351 Australia
[3]Angus Australia, 86 Glen Innes Road, Locked Bag 11, Armidale, NSW, 2350 Australia

SUMMARY

Eating quality traits are important determinants of consumer satisfaction and are considered as traits of economic importance for genetic improvement in the Australian beef industry. In this study, the genotypic and phenotypic data of 3,454 Angus cattle were analysed to identify genomic regions that potentially influence carcase traits, especially those related to eating quality. A genome-wide association study revealed 3, 5, 1 and 13 significant SNPs associated with carcase weight (CWT), carcase eye muscle area (EMA), Meat Standards Australia Index (MSA_I) and ossification score (OSS) respectively. They were located across chromosomes 3, 7, 13 and 21 and accounted for 2%, 4%, 6% and 12% of the total genetic variance for CWT, EMA, MSA_I and OSS, respectively. No significant SNPs were evident for MSA marble score (MSA_M). Results of this study may have potential practical application in the design of marker SNP chips and improving the accuracy of genomic prediction for carcase and eating quality traits in Angus beef cattle.

INTRODUCTION

Expectations of eating quality are a primary determinant of purchasing decisions made by consumers of Australian beef products. Consequently, Meat and Livestock Australia (MLA) developed the Meat Standards Australia (MSA) grading system to provide consumers with a level of assurance as to the eating quality of beef products (Watson *et al.* 2008). The current MSA Index, denoted by a single number score, represents a standard national measure that allows beef carcases to be ranked according to predicted eating quality and potential merit (McGilchrist *et al.* 2019). The Index is a weighted average of the predicted eating quality of 39 carcase cuts based on parameters collected by accredited MSA graders and of relevance to consumer preferences for tenderness, juiciness, flavour and overall perceptions of meat products (McGilchrist *et al.* 2019). The moderate heritability reported for MSA Index in both Angus and Brahman breeds (Jeyaruban *et al.* 2017) demonstrates a level of genetic control, suggesting improvements in MSA Index may be possible via selective breeding.

While most beef carcase and eating quality traits demonstrate a level of genetic control, less is known about the structure of these traits at the genomic level. Furthermore, phenotypic information on these traits requires slaughter at ages of maturity that allow market specifications to be met, which means that assessment of genetic merit for these traits is delayed. Genome-wide association study (GWAS) of these traits might therefore have practical application in the design of marker SNP chips as well as improving the accuracy of genomic prediction for these traits, especially of young candidate animals. Several such studies using SNP arrays have been reported for carcase traits in beef cattle breeds (Koohmaraie *et al.* 2006; Saatchi *et al.* 2014; Sudrajad *et al.* 2016).

The objectives of the present study were to investigate the presence of significant genomic regions in association with carcase and eating quality traits in Australian Angus beef cattle, and to quantify the amount of total genetic variation explained by such informative SNPs.

MATERIALS AND METHODS

Phenotypic data used in this study were derived from the performance extracts for Angus Australia as used in the March 2019 Angus BREEDPLAN analysis. Carcase trait records included: hot carcase weight (CWT), eye muscle area (EMA) and ossification score (OSS), the latter being an assessment of physiological maturity and indicative of eating quality (AUS-MEAT 2019). Eating quality traits were represented by two traits of relevance in the MSA grading system: MSA marble score (MSA_M) and MSA Index (MSA_I). Slaughter-based contemporary groups were constructed according to standard BREEDPLAN procedures (Graser *et al.* 2005) with criteria including herd, year, sex and prior performance contemporary group, plus slaughter group and slaughter date. Single animal groups were excluded.

Genomic data for animals with carcase and eating quality phenotypes was supplied by Angus Australia. The reference population for the genotype imputation consisted of 11,226 animals genotyped with a number of 50k arrays (LDMAX_SNPMap, ZM2_SNPMap, GSTP_SNPMap, ZOE-50K). Quality control (QC) was applied where only autosomal SNPs and the SNPs with a call rate higher than a 0.6 GeneCall score were kept. Further QC was undertaken using Plink v1.90b3.42 (Chang *et al.* 2015), filtering out those SNPs with minor allele frequency (MAF) < 0.01, deviation from Hardy Weinberg equilibrium ($P<10^{-6}$), and those SNPs with more than 5% missing genotypes. Only animals that had a valid genotype on more than 95% of SNPs were kept in the analysis. A final data set containing 37,974 SNPs for 3,454 animals was available for GWAS. Although the majority of these animals originated from the Angus Sire Benchmarking Program (Banks 2011), this was not an essential criterion *per se* for this study. Individuals required at least a CWT record and genotypes, within a contemporary group of at least two animals, for inclusion.

GWAS analysis of SNP effects and significance was conducted for each carcase and eating quality trait using the program GCTA (Yang *et al.* 2011) and the following linear regression model:

$$\boldsymbol{y} = \boldsymbol{Xb} + \boldsymbol{Za} + \boldsymbol{e}$$

where $\boldsymbol{y}$ is a vector of phenotypes, $\boldsymbol{b}$ is a vector of fixed effects including contemporary group, linear regression of age and SNP effect, $\boldsymbol{a}$ is a vector of random additive genetic effects and $\boldsymbol{e}$ is a vector of random residual effects. $\boldsymbol{X}$ and $\boldsymbol{Z}$ are incidence matrices relating fixed effects and additive genetic effects to phenotype, respectively. The additive genetic effects were assumed to be normally distributed as: $a \sim N(0, \boldsymbol{G}\sigma_a^2)$, where $\boldsymbol{G}$ is a genomic relationship matrix based on the 50k SNP genotypes, and σ_a^2 is the additive genetic variance. Significant SNPs were identified using a Bonferroni correction with α=0.05 and $-\log10$ (p)=5.88. Significant SNPs present in the same genomic regions were subjected to joint multivariate regression analysis using GCTA to identify the most informative SNPs for the particular trait.

The variances explained by all SNPs and the heritability were estimated using the restricted maximum likelihood analysis with GTCA including the genomic relationship matrix (GREML). Individual SNP variances were calculated as $2pq\alpha^2$, where p and q are allele frequencies and α is SNP effect, once SNPs were confirmed as being in Hardy-Weinberg equilibrium.

RESULTS AND DISCUSSION

Table 1 provides descriptive statistics for the carcase and eating quality traits of the 3,454 animals included in the GWAS.

There were 12 significant SNPs on chromosome 13 associated with OSS after Bonferroni correction (Figure 1). Only one SNP remained significant after multivariate regression analysis, reflecting that all 12 SNPs refer to the same QTL due to high LD between them. A second significant SNP for OSS was evident on chromosome 21 (Table 2). Similar outcomes were evident in the GWAS results for EMA and CWT, with 5 and 3 significant SNPs on chromosome 7 respectively after Bonferroni correction, and reducing to one significant SNP for each trait after multivariate regression analysis (Table 2).

Table 1. Descriptive statistics for carcase and eating quality traits

Trait	No of Animals	Mean	SD	Minimum	Maximum	Heritability
Carcase traits						
CWT (kg)	3,454	420.24	75.45	167.60	571.50	0.49 ± 0.03
EMA (cm^2)	2,954	89.42	10.89	57.00	128.00	0.47 ± 0.03
OSS (score)	2,704	150.97	17.54	100.00	280.00	0.29 ± 0.04
Eating quality traits						
MSA_M (score)	2,963	500.04	117.17	100.00	1030.00	0.40 ± 0.03
MSA_I (score)	2,658	64.88	1.78	59.15	70.48	0.40 ± 0.04

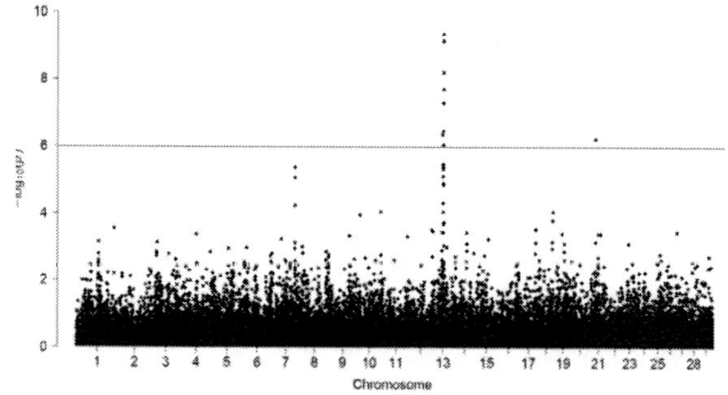

Figure 1. Manhattan plot of -log10 (p) from the Angus cattle GWAS of OSS. The horizontal reference line indicates the genome-wise significance levels (– log10 (p))

In terms of the two eating quality traits, only one SNP remained significant for MSA_I and no significant SNPs remained for MSA_M after Bonferroni correction (Table 2). Manhattan plots for both eating quality traits did suggest that several regions across the genome may warrant further detailed investigation.

Table 2. Significant SNPs and estimates of variance for the carcase and eating quality traits*

Trait	Chr	Mb	P-values	V(G)	%V(snps)
CWT	7	93	1.64E-07	451.1	2
EMA	7	93	3.42E-11	31.10	4
OSS	13	41	4.47E-07	54.04	8
	21	22	7.35E-07	53.21	4
MSA_M	-	-	-	4058.00	-
MSA_I	3	13	1.06E-07	0.74	6

* Chr = Chromosome; Mb = Mega base pairs position; V(G) = total genetic variance =; V(snps) = percentage of total genetic variance explained by significant SNPs.

The variance components and heritability derived for each trait in the current study are similar to those reported by Jeyaruban *et al.* (2017). This is not surprising given the current data extract includes the subset used in the former study. Given the high proportion of base females represented as dams in this data extract, differences in variance components may reflect differences in how relationships were modelled. The former study used pedigree information whereas the present study used realised relationships via the G matrix.

Sudrajad *et al.* (2016) identified six SNPs distributed across chromosome 4, 6, 27, 10, 9 and 20 as having significant associations with carcase weight, eye muscle area, fat depth and marble score in a commercial population of Hanwoo cattle. In the present study of Australian Angus cattle, the significant SNPs identified for CWT, EMA and MSA_I after Bonferroni correction explained 2%, 4% and 6% of total genetic variance respectively (Table 2). The two significant SNPs identified for OSS (one on each of chromosomes 13 and 21) explained 12% of total genetic variance for the trait. This is a substantial proportion of the genetic variance, encompassing a relatively small number of SNPs.

Chromosome 7 (93Mb position) has been reported previously in association with certain growth and carcase traits in beef cattle. Saatchi *et al.* (2014) reported an association with weight traits and eye muscle area in American Angus, as well as Hereford and a number of other breeds, while Koohmaraie *et al.* (2006) identified the calpastatin gene on chromosome 7 (98 Mb position) in association with meat tenderness. The significant SNPs on chromosome 13 associated with ossification in the present study may perhaps reflect a QTL related to physiological maturity and/or calcium metabolism, given that certain SNPs on chromosome 13 have shown significant associations with lean meat yield and milk yield traits in Holstein Friesian cattle (Doran *et al.* 2014).

CONCLUSIONS

In conclusion, this study identified significant SNPs in the bovine genome associated with eating quality traits for Angus cattle, supported by results from previous studies. Outcomes of the study suggest that significant markers might be added to SNP arrays used for developing Angus-specific SNP panels. Inclusion of these trait-specific markers in genetic evaluation models might also improve the accuracy of prediction of breeding values for such traits.

ACKNOWLEDGMENTS

The authors acknowledge the contributions of Angus Australia and its members who provided access to the data used in this study.

REFERENCES

AUS-MEAT. (2019) *Handbook of Australian Beef Processing*. AUS-MEAT Ltd. Murarrie, Qld.

Banks R.G. (2011) *Proc. Assoc. Advmt. Anim. Breed. Genet.* **19**: 399.

Chang C. C., Chow C. C., Tellier L. C., Vattikuti S., Purcell S. M., and Lee J. J. (2015) *Giga Sci* **4**:7. doi:10.1186/s13742-015-0047-8.

Doran A.G., Berry D.P. and Creevey C.J. (2014) *BMC Genomics* **15**: 837.

Graser H.U., Tier B., Johnston D.J. and Barwick S.A. (2005) *Aust. J. Exp. Agric.* **45**: 913.

Jeyaruban M. G., Johnston D. J. and Walmsley B.J. (2017) *Proc. Assoc. Advmt. Anim. Breed. Genet.* **22**: 313.

Koohmaraie M., Shackelford S.D., White S.N., Wheeler T.L., Smith T.P.L., Casas E., Chase C.C., Jr., Riley D.G. and Johnson D.D. (2006) *J. Anim. Sci.* **84**: 520.

McGilchrist P., Polkinghorne R.J., Ball A.J. and Thompson J.M. (2019) *Animal* **13**: 1750.

Saatchi M., Schnabel R.D., Taylor J.F. and Garrick D.J. (2014) *BMC Genomics* **15**: 442.

Sudrajad P., Sharma A., Dang C., Kim J., Kim K.S., Lee J., Kim S. and Lee S. (2016) *Asian-Austr. J. Anim. Sci.* **29**: 1541.

Watson R., Polkinghorne R. and Thompson J.M. (2008) *Aust. J. Exp. Agric.* **48**: 1368.

Yang J., Lee S.H., Goddard M.E. and Visscher P.M. (2011) *Americ. J. Hum. Genet.* **88**: 76.

MOLECULAR INVESTIGATION OF SEVERAL EMERGING INHERITED DISEASES IN CATTLE AND SHEEP

S.A. Woolley[1], E.R. Tsimnadis[1], R.L. Tulloch[1], P. Hughes[1], B. Hopkins[1], S.E. Hayes[1], M.R. Shariflou[1], A. Bauer[2], I.M. Häfliger[2], V. Jagannathan[2], C. Drögemüller[2], T. Leeb[2], M.S. Khatkar[1], C.E. Willet[3], B.A. O'Rourke[4], I. Tammen[1]

[1]The University of Sydney, Faculty of Science, Sydney School of Veterinary Science, Camden, NSW, 2570 Australia
[2]The University of Bern, Institute of Genetics, Vetsuisse Faculty, Bern, Switzerland
[3]The University of Sydney, Sydney Informatics Hub, Core Research Facilities, Sydney, NSW, 2006 Australia
[4]The Elizabeth Macarthur Agricultural Institute, NSW Department of Primary Industries, Menangle, NSW, 2568 Australia

SUMMARY

Emerging inherited diseases can cause numerous issues for producers, including productivity loss, profit loss and animal welfare problems. Current collaborative efforts between the University of Sydney and the Elizabeth Macarthur Agricultural Institute, NSW Department of Primary Industries has resulted in the ongoing investigation of several inherited diseases using both SNP-based homozygosity mapping and whole genome sequencing approaches to identify positional candidate genes and likely causal variants. This paper serves as a brief update for eight of the investigated inherited diseases in cattle and sheep, with these studies aiming to identify positional candidate genes and causal variants to facilitate the improved management of at-risk populations for each inherited disease investigated.

INTRODUCTION

The advancement of livestock breeding has allowed for desirable traits and elite genetics to be disseminated throughout livestock populations within relatively short periods of time. Small effective population sizes and inbreeding poses a risk for the inheritance of deleterious alleles in homozygous form and can contribute to the increased observation of animals with recessive inherited diseases (Charlier *et al.* 2008; Groeneveld *et al.* 2010), especially when considering closed herds or flocks. The reporting of inherited diseases within Australian livestock is limited due to either misdiagnosis of a prospective inherited disease or concern for reputation damage and profit losses. Detailed clinical and phenotypic descriptions of suspected recessive inherited diseases is imperative to future molecular investigations. Without consistent reporting and detailed phenotype information, the molecular characterisation of emerging inherited diseases can be delayed due to resource loss or lack of key information such as pedigree data and clinical descriptions. This can therefore impact on the monitoring and management of the inherited disease in at-risk populations, especially if detailed pedigrees are unknown when genotyping tests become available (Man *et al.* 2007).

Collaborative projects between researchers at the University of Sydney and the Elizabeth Macarthur Agricultural Institute, NSW Department of Primary Industries (EMAI) has enabled the investigation of several emerging recessive inherited diseases in livestock. With an increasing number of suspected inherited disease cases being investigated, the use of SNP-chip based homozygosity mapping and whole genome sequencing approaches is becoming routine in identifying positional candidate genes, causal variants and for facilitating the development of genotyping tests for inherited diseases with little pedigree information or phenotypic descriptions. This paper serves as an update for eight of the emerging inherited diseases with a suspected recessive mode of inheritance currently under investigation by the University of Sydney and EMAI. These emerging inherited diseases include: cardiomyopathy and woolly haircoat syndrome (CWH) in Hereford cattle, congenital mandibular prognathia (CMP) in Droughtmaster cattle, Niemann-Pick type C disease (NPC) in Angus cattle, new variants of ichthyosis fetalis (IF) in Hereford and Shorthorn cattle, the previously reported brachygnathia, cardiomegaly and renal hypoplasia syndrome (BCRHS) in Merino sheep (Shariflou *et al.* 2013), cervicothoracic vertebral subluxation (CVS) in Merino sheep, ovine dermatosparaxis (OD) in Merino sheep, and pulmonary hypoplasia with anasarca (PHA) in Persian sheep. The aim for these studies was to identify positional candidate genes and likely causal variants to facilitate improved management of at-risk populations for each inherited disease investigated.

MATERIALS AND METHODS

Analysis of SNP genotype data for carrier and affected animals (Table 1) using sliding windows of 25, 50 and 100 SNPs to identify runs of homozygosity (ROH) was previously conducted (Table 1) for affected animals using the bovine UMD3.1 genome assembly and the ovine Oarv1.0 genome assembly (Woolley *et al.* 2017). ROH were analysed using PLINK (Purcell *et al.* 2007) and were considered to be regions of interest if these regions were shared by all of the affected animals only. These regions were scanned for positional candidate genes based on gene function and comparative genomics methods.

Table 1. Number of affected and carrier DNA samples submitted for SNP chip genotyping and regions of homozygosity, including species specific OMIA ID

Disease	*OMIA ID*[1]	*Breed*	*Affected/ Carrier*	*SNP chip*
Cardiomyopathy and woolly haircoat syndrome	000161-9913	Poll Hereford	2/0	SNP80[2]
Congenital mandibular prognathia	-	Droughtmaster	9/4	SNP80[2]
Ichthyosis fetalis	000547-9913	Hereford	1/3	SNP80[2]
Niemann-Pick disease	-	Angus/Angus X	2/2	SNP80[2]
Cervicothoracic vertebral subluxation	000077-9940	Merino	14/2	SNP50[3]
Pulmonary hypoplasia with anasarca	000493-9940	Persian	5/5	SNP50[3]

[1]OMIA http://omia.angis.org.au, - indicates no species specific OMIA ID. [2]SNP80 = GeneSeek® Genomic Profiler Bovine HD Chip 80K chip (Neogen, NE, USA). [3]SNP50 = Illumina® OvineSNP50 Genotyping BeadChip (CA, USA).

Sanger sequencing for inherited diseases with identified positional candidate genes commenced but was cost and labour intensive. Whole genome sequencing (WGS) was conducted for affected animals for CMP, BCRHS, CVS and PHA (Woolley *et al.* 2017) with 150bp paired-end reads at an expected coverage of 20X or 30X (Table 2). Sequence reads were aligned with BWA-mem (Li 2013) to either the bosTau8 or oviAri3 reference genome assemblies and analysed for novel genetic variants using a modified GATK best practice pipeline (McKenna *et al.* 2010; DePristo *et al.* 2010). Large structural variant calling was completed using DELLY (version 0.7.6), LUMPY-sv (version 0.2.12) and LUMPY SVtyper (Rausch *et al.* 2012; Layer *et al.* 2014). WGS data generated at the University of Bern similarly applied standard bioinformatics pipelines using software and steps to process fastq files into bam and GVCF files in accordance to the latest 1000 Bulls processing guidelines (www.1000bullgenomes.com). For variant filtering, control genomes from other samples that were sequenced during this study were used according to species and breed, and for the Shorthorn IF and OD samples, 341 control genomes of various cattle breeds and 16 control genomes of various sheep breeds were used to identify novel variants for affected animals only. Genetic variants were annotated using SnpEff for predicted effects and filtered using SnpSift (Cingolani *et al.* 2012). To predict the functional effects of candidate causal variants, both SnpEff and SIFT (Kumar *et al.* 2009; Cingolani *et al.* 2012) were used to assess whether candidate disease-causing variants were deleterious to protein function.

RESULTS AND DISCUSSION

As previously identified, homozygosity mapping was able to successfully reveal and/or exclude positional candidate genes for all of the inherited diseases investigated, with a likely causal variant in a positional candidate gene identified for NPC through Sanger sequencing of affected animals (Shariflou *et al.* 2013; Woolley *et al.* 2017). Affected samples for BCRHS, CMP, CWH, IF, CVS, OD and PHA were re-sequenced using WGS (Table 2) as either homozygosity mapping did not identify positional candidate genes of interest or Sanger sequencing of affected animals did not identify causal variants within candidate positional candidate genes. Preliminary quality control analysis of the WGS data was positive (Woolley *et al.* 2017), however WGS for CWH in Poll Hereford cattle and IF in Hereford cattle was unsuccessful due to inadequate DNA quality. Further investigation of other positional candidate genes and genomic regions of interest based on SNP genotyping data will be required for CWH and IF.

After application of filtering parameters on samples that were whole genome sequenced, numerous genetic variants that were homozygous for the alternate allele in the affected animal(s) only were identified either across the genome or within previously identified ROH (Table 2) (Woolley *et al.* 2017).

Table 2. Variants identified in affected animals for which each animal was homozygous alternate to the reference sequence

Disease	*Breed*	*Affected/ Carrier*	*No. homozygous alternate variants*	*Likely causal variant identified*
Brachygnathia, cardiomegaly and renal hypoplasia syndrome	Merino	1	215[1]	Yes
Cervicothoracic vertebral subluxation	Merino	2	Ongoing	Ongoing
Ovine dermatosparaxis	Merino	1	1864[2]	Yes
Pulmonary hypoplasia with anasarca	Persian	2/1	333[1,3]	Under validation
Congenital mandibular prognathia	Droughtmaster	2	5780[4]	Under validation
Ichthyosis fetalis	Shorthorn	1	298[2]	Yes

[1]Filtered for low, moderate and high impact with known dbsnps included.[2]Private homozygous alternate and heterozygous protein-changing variants with a moderate or high predicted impact.[3]At least one animal was homozygous alternate.[4]Includes SNPs and small indels.

Further manual filtering based on the predicted variant impact on protein function revealed candidate causal variants for BCRHS, PHA, CMP and IF in Shorthorn cattle (Table 2). A candidate causal variant with possible heterogeneity was identified for OD in Merino sheep and requires greater sample numbers to facilitate further validation. Genotyping assays were developed for these five inherited diseases, with preliminary validation results showing variant segregation with disease in related herds or flocks. The development of these genotyping assays has allowed for producers to facilitate forward planning breeding management strategies.

Despite small sample sizes, poor phenotypic descriptions and challenging sample types, candidate causal variants have been successfully identified through the combined use of genome wide SNP genotyping, homozygosity mapping and WGS. These approaches have successfully identified candidate genes and causal mutations in a range of recessive inherited diseases in cattle, including ichthyosis fetalis in Chianina cattle (Charlier *et al.* 2008). The reporting of the inherited diseases investigated in these studies has enabled for better screening and preliminary management and has showcased the ability to identify candidate causal variants using modern genomic technologies.

CONCLUSIONS

Despite the challenges surrounding insufficient sample numbers and poorly defined phenotypes, the results from these studies indicate that candidate causal variants can be identified by utilising targeted approaches. The identification of likely causal variants for BCRHS, OD with possible genetic heterogeneity, PHA, CMP, NPC and IF in Shorthorn cattle, has enabled for the development of genotyping assays that are able to successfully discriminate between homozygous wildtype, heterozygous and homozygous alternate genotypes. These assays are being used as a preliminary screen for related or founder herds or flocks and would prove to be a useful tool for screening wider populations to gain a more holistic understanding of population allele frequencies and future breed management strategies.

ACKNOWLEDGMENTS

The authors acknowledge and thank the producers and veterinarians for the submission of samples to EMAI and the University of Sydney. The authors thank the genetics laboratory staff at EMAI for their assistance. The authors acknowledge the Sydney Informatics Hub, a Core Research Facility at the University of Sydney for facilitating access to the High Performance Computer Artemis, where computational analysis in this study was performed. The University of Sydney Faculty of Veterinary Science provided research support for 3 honours projects (NN 2013, ET 2014 & SW 2015), and funding support from the Animal Welfare, FS Quiney and Dorothy Minchin Bequests. The Australian Wool Education Trust is acknowledged for the support of a research scholarship for RT (2016) and PH (2018) and the Australian Government for the Australian Government Research Training Program (RTP) Scholarship for SW. The WGS is funded by a Faculty of Veterinary Science - DPI collaborative compact fund and research grant CRSII3_160738 from the Swiss National Science Foundation to TL.

REFERENCES

Auwera G.A., Carneiro M.O., Hartl C., *et al.* (2013) *Curr Protoc Bioinformatics* **43**: 11.0.1
Charlier C., Coppieters W., Rollin F., *et al.* (2008) *Nat. Genet.* **40**: 449.
Cingolani P., Platts A., Wang le L., *et al.* (2012) *Fly (Austin)* **6**: 80.
DePristo M.A., Banks E., Poplin R., *et al.* (2011) *Nat. Genet.* **43**.
Groeneveld L.F., Lenstra J.A., Eding H., *et al.* (2010) *Anim. Genet.* **41**: 6.
Kumar P., Henikoff S. and Ng P.C. (2009) *Nat Protoc* **4**: 1073.
Layer R.M., Chiang C., Quinlan A.R. and Hall I.M. (2014) *Genome Biol.* **15**: R84.
Li H. (2013) arXiv:1303.3997v1 [q-bio.GN]
Man W.Y.N., Nicholas F.W. and James J.W. (2007) *Prev. Vet. Med.* **78**: 262.
McKenna A., Hanna M., Banks E., *et al.* (2010) *Genome Res.* **20**: 1297.
Purcell S., Neale B., Todd-Brown K., *et al.* (2007) *Am. J. Hum. Genet.* **81**.
Rausch T., Zichner T., Schlattl A., *et al.* (2012) *Bioinformatics* **28**: i333.
Shariflou M.R., Wade C.M., Kijas J., *et al.* (2013) *Animal Genet.* **44**: 231.
Woolley S.A., Tsimnadis E.R., Nowak N. *et al.* (2017) *Proc. Assoc. Advmt. Anim. Breed. Genet.* 15.

THE POPULATION GENOMIC SIGNATURE OF ENVIRONMENTAL SELECTION IN CHICKENS FROM MALAWI, SOUTH AFRICA AND ZIMBABWE

K. Hadebe[1], E.F. Dzomba[2] and F.C. Muchadeyi[1]

[1]Agricultural Research Council, Biotechnology Platform, Private Bag X5, Onderstepoort, 0110 South Africa
[2]Discipline of Genetics, School of Life Sciences, University of KwaZulu-Natal, Private Bag X01, Scottsville, 3209 South Africa

SUMMARY

Indigenous chickens in Africa are found across heterogeneous landscapes, and heritable adaptive variations across environmental gradients suggest local adaptation. The direction of adaptive differentiation is still underestimated and may have negative impact on the conservation programs. This study examined 60K genotyping data from 311 village chickens from Zimbabwe, Malawi and South Africa, and conserved flocks (Venda, Naked Neck, Potchefstroom Koekoek, and Ovambo) to identify runs of homozygosity (ROH) and selection signatures using ROH islands and association of SNPs with bioclimatic and geographic variables. Overall, 5537 ROH were detected, with short segments more prevalent across all populations. Larger windows (>40 Mb) were found in the South Africa and Zimbabwe flocks only, suggesting less genetic diversity. Thirty-three ROH islands (50% of population) were only found in Naked Neck, Potchefstroom Koekoek and Venda and were located in 4352 genes. Two SNPs Gga_rs14045047 (chromosome 12) and Gga_rs13560712 (chromosome 6) were associated with 7 variables and longitude, altitude, BIO8, BIO17 was common for both. This suggests their importance and complexity of genetic adaptation. This study identifies regions potentially under selection pressure of production system and environmental adaptation and provides baseline for identifying populations adapted to local environment.

INTRODUCTION

Indigenous chickens in Africa have an extended geographic distribution across agro-ecological zones and production systems, thriving in environments with limited resources due to their unique adaptive traits. After an initial description and characterisation of the extensively raised village chickens populations from Zimbabwe and Malawi (Muchadeyi *et al.* 2007) and South Africa (Mtileni *et al.* 2011), A detailed population genetic studies using the Illumina 60K SNP BeadChip were completed (Khanyile *et al.* 2015a; 2015b). These studies observed genetic divergence with sufficiently strong geographic barrier of village genetic groups among African countries (Muchadeyi *et al.* 2007), differentiation of South African conservation populations to founder village populations due to reproductive isolation (Mtileni *et al.* 2011; Khanyile *et al.* 2015a; 2015b), regions with high linkage disequilibrium suggestive of selection of signatures (Khanyile *et al.* 2015a). Understanding the role of natural and artificial selection in the shaping diversity may provide new insights into the genetic mechanisms underlying their adaptation to their production environment. ROH have been used in livestock genomic studies, confirming the correlation between shared ROH and genomic regions putatively under selection (ROH island) (Mastrangelo *et al.* 2018). In addition, signatures of past climatic trends have played large roles in shaping genetic structure of livestock species. Therefore, the objectives of our study were to detect runs of homozygosity (ROH) and detect key bioclimatic and geographic factors that drive adaptive differentiation and assess their association with SNPs using landscape genomics approach.

MATERIALS AND METHODS

Genomic data and quality control. Illumina chicken iSelect SNP60 Beadchip genotype data (SNP= 57636) of 311 chickens from different regions of Malawi, South Africa and Zimbabwe was used and has been previously described (Khanyile *et al.* (2015a; 2015b). Briefly, 135 village chickens were from three Zimbabwean agro-ecological zones (AEZ, AEZ1 = 92, AEZ3 = 34, and AEZ5 = 10) and 30 chickens were sampled from Malawi. South African village chickens (SAFIELD = 76) were ecotypes from Limpopo (n = 15), Eastern Cape (n = 26) and Northern Cape (n = 35) provinces. In addition, four conserved flocks (n = 70, Venda (VD = 20), Naked Neck (NN = 20), Potchefstroom Koekoek (PK = 20) and Ovambo (OV = 10) at the Agricultural Research Council Poultry Breeding Resource, Irene Pretoria, South Africa. Genotypes with a failed call rate of > 0.95, minor allele frequency of > 0.05 and Hardy-Weinberg equilibrium > 1^{-5} were used in this study. Accordingly, 46160 SNPs from 290 individuals were used for further analyses.

Runs of homozygosity (ROH) and ROH islands. Runs of homozygosity (ROH) was defined per animal as 1) 50 or more consecutive homozygous SNPs, 2) a minimum physical length of 1 Mb to exclude short ROH deriving from LD, 3) a density of 50 Kb/SNP and 4) maximum of 3 heterozygous calls within ROH using *detectRuns* (Biscari *et al.* 2018). ROH islands were defined by ROHs that occurred in 50% of the individuals.

Environmental contribution to genetic structure and selection. Bioclimatic variables over the 30-year period (1970 to 2000) were available from the WorldClim version 2 (Fick and Hijmas 2017) using the GPS coordinates (latitude and longitude) at each district level for each individual. Districts level was used as individual farm data were not available for all countries. In Zimbabwe, Temperature (°C) and precipitation (mm) variables included annual mean temperature (BIO1), Mean Diurnal Range (BIO2), Isothermality (BIO3), Temperature Seasonality (BIO4), Maximum Temperature of Warmest Month (BIO5), Minimum Temperature of Coldest Month (BIO6), Temperature Annual Range (BIO7), Mean Temperature of Wettest Quarter (BIO8), Mean Temperature of Driest Quarter (BIO9), Mean Temperature of Warmest Quarter (BIO10), Mean Temperature of Coldest Quarter (BIO11), Annual Precipitation (BIO12), Precipitation of Wettest Month (BIO13), Precipitation of Driest Month (BIO14), Precipitation Seasonality (BIO15), Precipitation of Wettest Quarter (BIO16), Precipitation of Driest Quarter (BIO17), Precipitation of Warmest Quarter (BIO18), Precipitation of Coldest Quarter (BIO19).. To prevent overestimation on the contribution to the genetic structure, correlation analysis was performed on the bioclimatic (BIO1-BIO19), geographic (longitude, latitude and latitude) variables using ggcorr in *GGally* package (Schloerke *et al.* 2013). Redundancy analysis (RDA) detected the contribution of the variables on the spatial genetic structure, using *vegan* package (Oksanen *et al.* 2015). Association analysis using latent factor mixed model (lfmm) was then performed using the *LEA* package (Frichot *et al.* 2013). Parameters included 10,000 sweeps, 5,000 burn-in sweeps, 10 repetitions and 6 latent factors (Khanyile *et al.* 2015). SNPs with a false discovery rate of P< 0.001 were considered as significantly associated.

RESULTS AND DISCUSSION

Runs of homozygosity (ROH) and ROH islands. A total of 5537 ROHs were detected across the 7 chicken populations. The frequency of ROHs and their length-distribution differed across populations (Table 1). In all populations, shorter segments of between 1 to 10Mb predominated the homozygosity present and accounted for approximately 82% of all ROH detected suggestive of more ancient relatedness, inbreeding and long-term selection within these populations. OV had the least number of ROH for all categories. Zimbabwe had the highest number of segments larger than 10Mb suggestive of the more likely that recent inbreeding occurred within a pedigree (Khanyile *et al.*, 2015b) which remains unaccounted for in village populations, due to lack or recording system. The increase of homozygous regions in NN, PK, and VD could be a consequence of inbreeeding, bottleneck effect and the decline in effective population size because they have been a closed populations since 25 years ago (Mtileni *et al.* 2011). Across all populations, the mean ROH length was 2.34 Mb and the longest segment was 50.09Mb in length (1994 SNPs) which was found on chromosome 3 in Zimbabwe population. The number of ROH per chromosome decreased with chromosome length and was greater for chromosome 1 (796 ROH) and lower for micro-chromosomes including chromosome 23 (23 ROH). High level of homozygosity in chromosome 1 was consistent with the presence of high number haploblocks (Khanyile *et al.* 2015b), which could be due to differences in recombination rates, genetic drift and selection across the different geographical distribution.

Table 1. Number of runs of homozygosity (*n*ROH) and length (in Mb) categorised by ROH length class (ROH_{1-5Mb}, ROH_{5-10Mb}, $ROH_{10-20\ Mb}$, $ROH_{20-40\ Mb}$, and $ROH_{>40\ Mb}$)

Class	NN	OV	PK	VD	SAFIELD	MALAWI	ZIMBABWE
1-5Mb	765	178	750	767	649	309	1102
5-10Mb	125	10	77	142	104	38	230
10-20Mb	31	1	12	46	35	21	88
20-40Mb	4	0	2	7	5	6	31
>40Mb	0	0	0	0	1	0	1
Total	925	189	841	962	794	374	1452

Thirty-three ROH islands, which indicate regions of strong selection were evident across the genome of NN (n = 5), PK (n = 7) and VD (n = 21) only. ROH islands were not found on other populations observed to have highly admixed individuals (Khanyile *et al.* 2015a). The longest ROH island was observed in VD on chromosome 7 (46.65Mb), while the shortest one was observed on chromosome 4 (1.73Mb). Within all of the ROH islands reported, we identified from 4352 genes (827 NN, 324 PK, 3202 VD). Functions of the genes varied and included metabolic, cardiac muscle and vascular smooth muscle contraction and signaling pathways,

Landscape genomics. Landscape genomics studies in indigenous livestock have gained momentum in past years. Highly correlated (r> 0.90) bioclimatic and geographic variables and those that did not explain genomic variation using RDA were removed, whilst annual mean temperature (BIO1), mean diurnal range (BIO2), isothermality (BIO3), temperature seasonality (BIO4), temperature annual range (BIO7), mean temperature of wettest quarter (BIO8), mean temperature of driest quarter (BIO9), mean temperature of warmest quarter (BIO10), precipitation of wettest month (BIO13), precipitation of driest quarter (BIO17), altitude, longitude, latitude were retained for LFMM analysis. RDA 1 and RDA 2 explained 3.82% and 1.23% of the variance, respectively. BIO3, BIO9, BIO1 and altitude explained most of the genetic variation (P <0.001) and BIO8 explained the least. RDA showed evidence of population structuring, consistent of the population structure described in Khanyile *et al.* (2015). Village populations clustered together despite the geographic distances between the countries potentially due to similar production environments. Reproductive isolation and sharing of production environment of the conservation flocks resulted in the populations clustering together despite different genetic backgrounds. Overall, a total of 3090 SNPs (6.69%) were associated with one or more variables, whilst 1888 SNPs were associated with a specific variable. BIO2 has been associated with thermoregulation and was the most significant variable in 283 SNPs suggestive of their role in adaptation to diurnal environmental conditions and BIO4 had the least (n= 184). SNP Gga_rs14045047 on chromosome 12 was associated with BIO1, BIO2, BIO8, BIO10, BIO17, altitude and longitude. Association of BIO3, BIO7, BIO8, BIO13, BIO17, altitude and longitude with Gga_rs13560712 on chromosome 6 indicates their complexity in shaping the genetic diversity. Five SNPs (Gga_rs13705188 on chromosome 5, Gga_rs14836389 on chromosome 1, Gga_rs14091452 on chromosome 15, Gga_rs15243798 on chromosome 27, Gga_rs14142376 on chromosome 2) were only associated with longitude and latitude suggestive of role in the local adaptation. Genes were related to pathogen and disease defence and adaptive phenotypic traits including weight and fast growth.

CONCLUSIONS

The existence of ROH and islands demonstrated the role of the production systems in increasing homozygosity in specific regions of the genome. In low input village production systems in the sampled regions, climate ranges from heat and drought, pose selection pressure. Although different regions were identified between the two analysis, gene functions overlapped showing the complexity of response to production and environmental pressure.

ACKNOWLDEGMENTS

Agricultural Research Council Biotechnology Platform funded the study.

REFERENCES

Biscarini, F., Cozzi, P., Gaspa, G., Marras, G. (2018) https://github.com/bioinformatics-ptp/detectRUNS/tree/master/detectRUNS.

Frichot E., Schoville S.D., Bouchard G. and François O. (2013) *Mol. Biol. Evol* **30**: 1687.

Fick, S.E., and Hijmans, R.J. (2017) *International Journal of Climatology*.

Khanyile, K., Dzomba, E.F., and Muchadeyi, F.C. (2015a) *Frontiers in Genetics*. **6**:1.

Khanyile, K., Dzomba, E.F., and Muchadeyi, F.C. (2015b) *Genetics of Molecular Research* **14**:12276:.

Mastrangelo, S., Sardina, M.T., Tolone, M, Di Gerlando, R., Sutera, A.M., Fontanesi, L., and Portolano, B. (2018) *Animal*. **12**: 2480

Mtileni, B.J., Muchadeyi, F.C., Maiwashe, A., Groeneveld, E., Groeneveld ,L.F., Dzama K. and Weigend, S. (2011) *J. Anim. Breed. Genet.* **128**: 209.

Muchadeyi, F. C., Eding, H., Wollny, C.B.A., Groeneveld, E., Makuza, S.M., Shamseldin, R., Simianer, H. and Weigend, S. (2007) *Animal Genetics* **38**: 332.

Oksanen. J., Blanchet, F.G., Kindt, R., Legendre, P., Minchin, P.R., O'Hara R.B, Simpson, G.L., Solymos, P., Stevens, M.H.H., Wagner, H. (2015). http://cran.r-project.org, https://github.com/vegandevs/vegan.

Schloerke, B., Crowley, J., Cook, D., Hofmann, H., Wickham, H., Briatte, F., Marbach, M. (2013). http://CRAN.R-project. org/package=GGally.

ASSESSMENT OF GENOMIC PREDICTION ACCURACY FOR MEAT QUALITY TRAITS IN HANWOO CATTLE

M. Bedhane[1], J.H.J. van der Werf[1], M. Al Kalaldeh[1], D. Lim[2], B. Park[2], M.N. Park[2], R.S. Hee[2] and S.A. Clark[1]

[1]School of Environmental and Rural Science, University of New England, Armidale, NSW, 2351 Australia
[2]National Institute of Animal Science, RDA, Republic of Korea
[3]Hanwoo Genetic Improvement Center, NAGI, Republic of Korea

SUMMARY

In beef cattle, genomic selection has promising benefits for the improvement of carcass traits such as meat quality, because estimated breeding values can be obtained without sacrificing the selection candidates. The objective of this study was to assess genomic prediction accuracy for meat quality traits in Hanwoo beef cattle. Genomic and phenotypic data from 2,110 Hanwoo steers were used to predict genomic estimated breeding values for marbling score, meat texture and meat colour. The accuracy of the genomic breeding value was assessed by using cross-validation for two scenarios; 1) when the reference population of animals with phenotype and genotype included family members and 2) when family members were excluded. The mean cross-validation accuracy of genomic predictions for marbling score were 0.32 and 0.46 for the distantly and closely related validations set, respectively. These accuracies were 0.28 and 0.39 for meat texture and 0.19 to 0.31 for meat colour. The results indicated that the accuracy of prediction was affected by the heritability of the trait and the degree of relationship between reference and test population. These results were based on a small sample size and should be validated with a larger data set.

INTRODUCTION

Genomic prediction uses DNA information to produce genomically enhanced estimation of breeding values (GBV) and it is increasingly applied in breeding programs for livestock species. The Genome-wide SNP based genomic prediction has the most benefit for traits that are difficult to measure, expensive to record or that are measured late in an animal's life compared to pedigree-based estimates of breeding value (Meuwissen *et al.* 2001). Thus, genomic information can be applied to select young animals for meat quality traits without sacrificing the selection candidates, which is an important advantage of genomic selection in beef cattle. Prediction accuracy of GBV is an important parameter in designing breeding programs with genomic selection. The accuracy of genomic prediction mainly depends on the size and the diversity of the reference population, the heritability of the trait, the linkage disequilibrium (LD) between SNP and QTL, and the methods that will be used for prediction (Daetwyler *et al.* 2012). The accuracy also depends on the relationship between the reference population and the target animals to be predicted (Clark *et al.* 2012). The accuracy of GBVs should be validated before implementing a genomic selection-breeding program and the most common way to assess GBV accuracy is using cross-validation.

Several genomic prediction studies have been reported for meat quality and carcass traits on various beef cattle around the world (Chen *et al.* 2015; Magalhães *et al.* 2019). However, few have included indigenous Korean beef cattle (Hanwoo) and prediction accuracies may differ between breeds due to the effective population size (N_e) differences among breeds. As a result, there is no comprehensive study on assessment of genomic prediction accuracy for marbling score, meat texture and meat colour in Hanwoo cattle. Therefore, the objective of this study was to assess genomic prediction accuracy for meat quality traits in Hanwoo cattle.

MATERIALS AND METHODS

Data structure and quality control. Phenotypic data from 2110 Hanwoo steers were used and all individuals were slaughtered at the same age (24 months). Details of feeding, management practices and traits measurements are reported elsewhere (Bhuiyan *et al.* 2018). Marbling score (MS) was assessed and scored (1 to 9 scale). Similarly, meat colour (MC) was assessed and graded from very light red (grade 3) to dark red (grade 7), and meat texture (MT) was evaluated on a scale from very fine (grade 1) to coarse (grade 3). All animals with phenotypic data were genotyped with the 50k SNP Chip (Illumina Bovine SNP50 BeadChip; Illumina, San Diego, CA). SNPs that had a minor allele frequency (MAF) less than 1% were removed as well as those with p-values for Hardy-Weinberg equilibrium (HWE) less than 0.1%. Finally, 40197 SNPs passed the quality control thresholds and were used for the analysis.

Statistical model and data analysis. Genomic best linear unbiased prediction (GBLUP) was used to predict the breeding value for each trait. The genomic relationship matrix (G) (Yang *et al.* 2010) was used in a univariate linear mixed model to estimate the GBV and heritability. The model was: $\mathbf{y} = \mathbf{Xb} + \mathbf{Zu} + \mathbf{e}$, with $\mathbf{b}$ being a fixed effect of kill group and u was a random additive genetic effect of the animal with var($\mathbf{u}$)=G. ASReml version 4.1(Gilmour *et al.* 2015) was used for the data analysis.

Cross-validation. Two 10-fold cross-validation (CV) scenarios were used. In the first scenario, 2110 steers were divided into 10 folds using random sampling of individuals (RCV). Each of the folds (n=211, 10%) was used as validation whereas the other folds (n=1,899, 90%) were used as the reference population. In the RCV scenario, there was a relatively close relationship between validation and the reference population, because half-sibs of animals in the validation set could be present in the reference population. In the second CV scenario, the 2110 steers were divided into ten folds based on family-based sampling techniques (FCV). Steers in every ten subsets came from 25 sires and the number of steers in each validation data set was varied from 179 to 238. Thus, in the FCV scenario, the validation steers did not have any siblings in the corresponding reference population, indicating that there was a relatively distant relationship between validation and reference population. Finally, the accuracy of GBV was assessed using the Pearson product-moment correlation between GBV and corrected phenotypic value (*yc*) divided by the square root of heritability, where *yc* was the phenotypic value corrected for the kill batch effect. The bias in the variance of the estimated breeding values was measured through the regression coefficient (slope) of the corrected phenotypes on the estimated predicted breeding values. $Prediction\ accuracy = \frac{R_{corr}\,(GBV,y_c}{\sqrt{h2}}$ $Prediction\ accuracy = \frac{R_{corr}\,(GBV,y_c)}{\sqrt{h2}}$.

Summary statistics for meat quality traits Mean, minimum (Min), maximum (Max), standard deviation (SD) and coefficient of variation (CV%) are shown in Table 1.

Table 1. Summary statistics for the three meat quality traits in the 2110 Hanwoo steers

Traits	Sample size	Min	Mean	SD	Max	CV%
Marbling score	2110	1	3.23	1.50	9	46.4
Meat texture	2110	1	1.65	0.50	3	30.3
Meat colour	2110	3	4.8	0.55	7	11.5

RESULTS AND DISCUSSION

Assessment of genomic prediction accuracy. The estimated heritabilities for MS, MT and MC are shown in Table 2, with traits with higher heritability having higher prediction accuracy (Figure 1). In the RCV scenario, MS had the highest (0.46) prediction accuracy, with accuracies being 0.39 for MT and 0.31 for MC. In the FCV scenario, the accuracy of genomic prediction for MC was lower (0.19) compared with 0.32 for MS and 0.28 for MT. As shown in Table 2, the RCV scenario was more accurate and less biased than the FCV scenario for all studied traits.

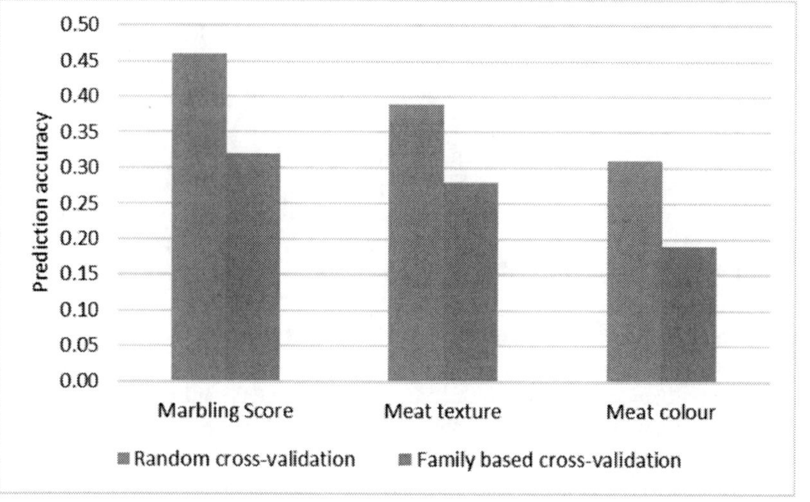

Figure 1. Accuracies of genomic prediction for the three studied traits with random and family-based cross-validations

Table 2. Slope of genomic predictions and heritability for the studied traits

Traits	MS	MT	MC
Slope RCV	0.92±0.11	0.90±0.07	0.77±0.16
Slope FCV	0.88±0.11	0.88±0.15	0.70±0.24
Heritability	0.46±0.05	0.30±0.05	0.15±0.04

In the current study, the empirical accuracy based on FCV were 14, 11 and 12% lower than those based on RCV for MS, MT and MC traits, respectively. A similar study in chicken for the traits associated with growth showed that FCV yielded lower genomic prediction accuracy than RCV (Liu *et al.* 2017). In our study, the accuracy of genomic prediction increased with increasing the relationship between validation and reference population. Similarly, (Clark *et al.* 2012) found that the prediction accuracy was improved as the degree of relationship between the validation and reference population increased. The way of a data splitting strategy for cross-validation affects prediction accuracy. For instance, the RCV does not consider the data structure such as age, family and relatedness, while FCV increases relationships within a group but decreases between groups. Thus, the genetic distance of the reference population from the group of selection candidates determines the accuracy of GBV.

Reports on the accuracy of genomic prediction for beef cattle are limited and are usually based on small data sets. A previous study in Hanwoo cattle showed that the genomic prediction accuracy for IMF varied from 0.37 to 0.45 based on different GRMs (Choi *et al.* 2017) using 778 genotyped Hanwoo steers. The study used 5- fold family-based cross-validation techniques and sampled 706 and 72 steers into reference and validation data sets, respectively. The genomic prediction accuracies for meat quality traits have also been studied in other beef cattle breeds. Chen *et al.* (2015) reported an accuracy of 0.37 for genomic prediction of marbling score in Angus cattle using 543 genotyped steers. A recent study (Magalhães *et al.* 2019) reported a prediction accuracy of 0.40 for a trait associated with meat colour in Nellore cattle using 5000 genotyped animals. In that study, the animals were divided into two groups for reference and validation sets based on year of birth and animals born in the last year were used as a validation population.

In general, it is difficult to compare the accuracies from different studies because of differences in trait heritabilities, training and validation set sizes, data splitting strategies to reference and validation, and statistical methods to estimate marker effects. Likewise (Luan *et al.* 2009; Daetwyler *et al.* 2012) established that the data splitting strategies to reference and validation affected prediction accuracy.

Furthermore, different breeds have different population structure and vary in diversity. In a less diverse population with small effective population size (Ne), animals share large chromosome segments, which lead to relatively high prediction accuracy. In the current study, moderate (0.31 to 0.46) prediction accuracies were found in the RCV scenario for the studied meat quality traits. The small sample size could affect the prediction accuracy in our study. Therefore, the obtained prediction accuracy in the current study should be confirmed with a large sample size prior to starting the intended breeding program in Hanwoo cattle.

CONCLUSIONS

Genomic predictions for meat quality traits in beef cattle are potentially valuable because it can be applied early in life and do not require potential selection candidates to be sacrificed. Our study shows that marbling score and meat texture traits had higher genomic prediction accuracy, suggesting that selection for these traits may improve meat quality in Hanwoo cattle. The accuracy of genomic prediction was affected by the heritability of the studied traits and the method of sampling the training and validation sets, which affected the degree of relationship between validation and reference populations. Overall, the current estimated genomic prediction accuracy could be affected by the small sample size used in the study and should be confirmed with large sample sizes.

REFERENCES

Bhuiyan M. S., Lim, D., Park M., Lee S., Kim Y., Gondro C., et al. (2018) *Frontiers in genetics* **9**.

Chen L., Vinsky M. and Li, C. (2015) *Animal Genet.* **46**: 55.

Choi T., Lim D., Park B., Sharma A., Kim J.-J., Kim, S. and Lee S. H (2017) *J. Anim. Sci.* **30**: 907.

Clark S. A., Hickey J. M., Daetwyler H. D. and van der Werf J. H. (2012) *GSE* **44**: 4.

Daetwyler H. D., Swan A. A., van der Werf J. H. and Hayes B. J. (2012) *GSE* **44**: 33.

Gilmour A., Gogel, B. Cullis B., Welham S. and Thompson R. (2015). ASReml user guide V. 4.1

Liu T., Luo C., Wang J., Ma J., Shu D., Lund M. S., Su G. and Qu H. (2017) *PLOS ONE* **12**: e0173620.

Luan T., Woolliams J. A., Lien S., Kent M., Svendsen M. and Meuwissen T. H. (2009) *Genet.* **183**: 119.

Magalhães A. F. B., Schenkel F. S., Garcia D. A., Gordo D. G. M., Tonussi R. L., Espigolan et al. (2019) *Meat Sci.* **148**: 32.

Meuwissen T. H., Hayes B. J. and Goddard M. E. (2001) *Genet.* **157**: 1819.

Yang J., Benyamin B., McEvoy B. P., Gordon S., Henders A. K., Nyholt D. R., Madden P. A. et al. (2010) *Nature Genet.* **42**: 565.

GENOMIC PREDICTIONS FOR FERTILITY TRAITS IN TROPICAL BEEF CATTLE FROM A MULTI-BREED, CROSSBRED AND COMPOSITE REFERENCE POPULATION

B.J. Hayes, G. Fordyce and S. Landmark

Queensland Alliance for Agriculture and Food Innovation, The University of Queensland, St Lucia, QLD, 4067 Australia

SUMMARY

Cow fertility is a major driver of profitability in Northern beef herds. Cow fertility has been difficult to select for, and the availability of genomic estimated breeding values (GEBV) would enable more rapid gains to be made. Ideally GEBV would be from a multi-breed genomic evaluation, given the wide range of breeds, composites and crossbreds used in Northern Australia. With this ultimate goal in mind, 14,552 heifers in 54 herds across Northern Australia were genotyped and phenotyped for CLscore (presence or absence of a corpus luteum at approximately 600 days), a proxy trait for age at puberty, a trait in turn correlated with cow lifetime productivity. Genomic heritabilities estimated from the data set were 0.32, 0.42, 0.22 and 0.25 for weight, hip height, body condition score and CLscore respectively. The accuracy of GEBV in nine validation herds (where accuracy was the correlation of GEBV for CLscore and the actual CLscore for the heifers within a herd, representing a wide range of breed composition were 0.30, 0.50, 0.25 and 0.40 for weight, hip height, body condition score and CLscore respectively. For CLscore, this accuracy suggests gains for fertility could be made through selection on GEBV. The data set analysed here represents approximately half the data that will be collected in the Northern Genomics Project.

INTRODUCTION

Cow fertility is a key driver of productivity and profitability of beef production in northern Australia (Taylor and Rudder 1986; Fordyce 2012; Johnstone *et al.* 2014). Genomic estimated breeding values (GEBV) for cow fertility would enable more rapid genetic gains for these traits. However, accurate genomic evaluations for low heritability traits such as fertility require large reference populations (e.g. Goddard and Hayes 2009), with thousands of cows measured for both the traits of interest and genotyped for genome wide markers. Assembling such large reference populations for each breed used in Northern Australia is likely to be challenging, as the cattle population consists of many breeds, crossbreds and composites. Cattle populations include high proportion *Bos indicus* breeds (e.g. Brahman), stabilised composites (e.g. Droughtmaster and Santa Gertrudis), adapted *Bos taurus* breeds, and many composites. An alternative to constructing reference populations within each breed is to use multi-breed genomic evaluations, where the reference set includes cows from across Northern Australia.

Here we test the accuracy of GEBV for fertility and other traits from using such a reference set, in this case consisting of 14,552 heifers (reference and validation) from 54 commercial properties across Northern Australia.

MATERIALS AND METHODS

Animals and Phenotypes. Fifty-four collaborator herds from across Northern Australia are participating in the Northern Genomics project. The data set includes crossbred and, in some cases, purebred Angus, Belmont Red, Brahman, Charolais, Droughtmaster, Shorthorn, Limousin, Santa Gertrudis, Boran and Wagyu heifers. The fertility trait measured on the heifers to date is cycling or not cycling by approximately 600 days (CLscore) assessed by ovarian scanning, as described by (Corbet *et al.* 2018). To maximise genetic variation, the trait is actually measured when an estimated 50% of heifers are pubertal, ie, at 1.0-2.5 years of age. As an alternative to CLscore, CLrate was also measured, where 1 = Acyclic, 2 = Dominant follicle 10mm or less, 3 = Dominant follicle greater than 10mm, 4 = Corpus luteum is present, and 5= Cow is pregnant (Burns *et al.* 2016). Weight, body condition score, hip height, fly lesions, and tick scores were also collected at the time of scanning. Tail hairs have currently been taken from all heifers for genotyping.

Genotypes. All heifers were genotyped with the 35K tropBeef SNP array by Neogen, Australasia. Genotypes were imputed up to 728,785 SNP (Bovine HD array) using the Fimpute software (Sargolzaei, *et al.* 2014), and a panel of 3,140 cattle of relevant breeds genotyped for the Bovine HD array.

Statistical Analysis. We first estimated breed proportions of each heifer for each of the 12 breeds known to be in the data set (using the 35K array data only). Previously, a separate large data set consisting of only purebred cattle was used to estimate SNP effects for breed composition. A GBLUP model was fitted, where the phenotype was 1 if the animal was of that breed and 0 if not (Dodds *et al.* 2014). The effects of each SNP for the proportion of each breed was then derived by back-solving for the SNP effects (Yang *et al.* 2011), and the resulting prediction equations for each breed were used to estimate breed proportions in the heifers. Then the model fitted to the CLscore, CLrate, height and weight data was

$$y = \mu + cohort + year + het + breedprop + animal + \text{error}$$

where **y** is a vector of trait records (CLscore,CLrate,weight, hip height or body condition score, μ is the population mean, **cohort** is the property+yeardrop+paddock that the heifers were in prior to mustering for trait recording, year is the year of recording, **het** is the heterozygosity of each heifer as measured by the proportion of markers that were heterozygous (to capture heterosis effects), fitted as a liner effect **breedprop** is a series of 12 covariates (11 breeds and *Bos indicus* content), measuring the proportion of each breed in the heifers as described above, and **animal** is a vector of random effects $\sim N(0, \mathbf{G}\sigma_g^2)$, with G the genomic relationship matrix among all heifers (Yang *et al.* 2011) and σ_g^2 the genetic variance captured by the SNP markers, and **error** is a vector of random deviations $\sim N(0, \mathbf{I}\sigma_e^2)$. Variance components were estimated in GCTA (Yang *et al.* 2011), and the heritability of the traits (actually the proportion of phenotypic variance captured by the SNP) was estimated as $h^2 = \widehat{\sigma_g^2}/(\widehat{\sigma_g^2} + \widehat{\sigma_e^2})$.

The accuracy of GEBV was evaluated by dropping out 9 herds at random (but these 9 herds had to have at least two-year cohorts in the data set). The breed composition within the 9 herds (2,205 heifers) ranged from purebred Brahman to crossbreds of *Bos taurus* breeds. There were 12,347 heifers in the reference population. GEBV were predicted for the heifers in the 9 excluded herds, then the GEBV were correlated with the actual phenotypes (adjusted for fixed effects) of the heifers within each herd. This correlation was divided by the square root of the heritability of the trait to get the accuracy of genomic prediction. Accuracy was calculated either dropping out all of the data from the 9 herds, or just the last year drop. The latter approach was taken to assess the improvement in accuracy when a herd has some data in the reference set.

Ultimately for multi-breed evaluations, head to head comparisons of breeds in the same herd/environment are necessary. We assessed how many head to head comparisons as $\sum_i^n X_i' X_i$ where for each of n cohorts in the data set, X_i is a matrix of breed proportions in cohort i, of dimensions number of heifers in the cohort x number of breeds (12).

RESULTS AND DISCUSSION

The heritability of the traits estimated from the genomic data was moderate (for CLscore, CLrate and body condition score), and higher for weight and hip height (Table 1).

Table 1. Trait genomic heritabilities and standard errors

Trait	Heritability	Standard error
Weight	0.32	0.02
Hip height	0.43	0.02
Body condition score	0.22	0.02
CLscore	0.25	0.01
CLrate	0.22	0.01

Heritabilities were consistent with previous estimates for these traits in tropical beef cattle data sets derived from pedigree (eg Corbet *et al.* 2018). Accuracies of GEBV in the 9 validation herds were moderate (Figure 1). Accuracies of GEBV were higher when the validation herds had a cohort in the dataset.

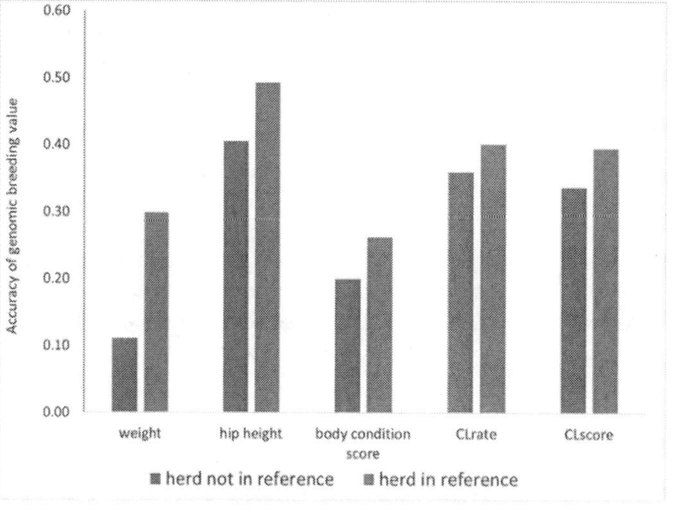

Figure 1. Accuracy of GEBV in nine validation herds. The data for these herds was either completely removed from the reference population ("herd not in reference") or the last cohort of heifer data was removed from the reference and used as the validation set ("herd in reference)

The number of head to head comparisons possible from the data set, which enables estimates of breed effect, reasonable for Angus versus Brahman, Brahman versus Droughtmaster and Brahman versus Santa Getrudis, but was lower for other breed combinations, Table 2. This suggests the data set will contribute to multi-breed genomic evaluations for many, but not all breeds used in Northern Australia.

CONCLUSIONS

The results of this preliminary study, both in terms of genomic heritabilities, and accuracy of GEBV are promising. Heritability's of the traits measured on the 14,552 heifers phenotypes and genotyped to date are consistent with heritability previously reported for tropical beef cattle, based on pedigree and data. Accuracies of GEBV, including for the fertility traits CLscore and CLrate were moderate, but of sufficient magnitude to suggest genetic gains could be made by selecting for GEBV for these traits. The utility of GEBV are enhanced by the fact that they work to some extent across breeds (the validation set included herds with *Bos indicus*, *Bos indicus* x *Bos taurus* and *Bos taurus* cattle) The heifer data here represents approximately half the data that will be collected in the Northern Genomics Project (which aims to genotype and phenotype 30,000 heifers from the 54 collaborating herds). Additional traits will include heifer rebreed success and follow up pregnancy tests for a number of years. Given the results reported here, the complete data set should enable reasonably accurate GEBV for several fertility traits related to cow lifetime productivity, especially when this data is combined with other data sets, for example in BREEDPLAN.

Table 2. Number of head to head breed comparisons in the data set, where each cell represents the number of genomes for a breed being compared to the number of genomes of the other breed. Empty cells indicate no comparisons for that breed combination

	Angus	Belmont Red	Brahman	Charolais	Drought-master	Hereford	Limousin	Santa Gertrudis	Shorthorn	Wagyu
Angus										
Belmont Red	11									
Brahman	315	37								
Charolais	26		165							
Droughtmaster	120	17	603	42						
Hereford	52		94	16	59					
Limousin	25		87	14	31	15				
Santa Gertrudis	116	12	311	34	144	37	31			
Shorthorn	36		79	12	58	16	12	45		
Wagyu	12		40		27			19		
Boran			51		13					

ACKNOWLEDGEMENTS

The authors gratefully acknowledge the participation and efforts of the 54 Northern herds in the Northern Genomics Project, and assistance from various DAF staff at various stages of the project. We are grateful for funding from the MLA Donor Company and University of Queensland through project P.PSH.P0833.

REFERENCES

Burns, B.M., (2016) Queensland Government Smart Futures Research Partnerships Program (2012-2015). University of Queensland.

Corbet N.J., Allen J.M., Laing A.R., Fordyce G., McGowan M.R., and Burns B.M. (2017) *Anim. Prod. Sci.* **58**: 1735.

Dodds K.G., Auvray B., Newman S.A. and McEwan J.C. (2014) *BMC Genetics* **16**: 15.

Fordyce G (2012) Beef CRC exit Report..

Johnston D.J., Barwick S.A., Fordyce G., Holroyd R.G., Williams P.J., Corbet N.J. and Grant T. (2013) *Anim. Prod. Sci.* **54**: 1.

Sargolzaei M., Chesnais J.P. and Schenkel F.S. (2014) *BMC Genomics*. **15**:478.

Taylor, W.J., and T. H. Rudder. (1986). Proceedings of the Australian Society of Animal Production 16:379-382.

Yang J, Lee SH, Goddard ME, Visscher PM (2011) Amer. Jour. Human. Genet. **88**:76-82.

SELECTION OF REFERENCE CANDIDATES FOR WHOLE GENOME SEQUENCING IN AN AUSTRALIAN WAGYU POPULATION

R.A. McEwin[1], M.L. Hebart[1], H. Oakey[2], R. Tearle[1], J. Grose[3], G.I. Popplewell[4] and W.S. Pitchford[1]

[1]Davies Research Centre, School of Animal and Veterinary Sciences, University of Adelaide, SA, 5371 Australia
[2]Biometry Hub, School of Agriculture, Food and Wine, University of Adelaide, SA, 5064 Australia
[3]3D Genetics Pty LTD, 939 Pukawidgi Rd, Bukkulla, NSW, 2360 Australia
[4]Popplewell Genetics, 33 Tom Schmidt Court, Mount Samson, Qld, 4520 Australia

SUMMARY

Denser genotypes on individuals has the potential to enhance genetic gain through accuracy of genomic selection. However denser genotypes, such as whole genome sequencing, on large numbers of animals is costly. This can be overcome by selecting suitable reference candidates for denser genotyping to describe the population, allowing for accurate imputation of the unselected candidates (target population). Two methods to select reference candidates were compared: the MCA method which utilises a pedigree based relationship matrix, and the MCG method which utilises a genomic relationship matrix. In a Wagyu population, the MCG method gave slightly superior imputation accuracies in the target population across differing reference population sizes as well as explaining 5% more of the genetic variance in the population when 100 candidates were selected. Similarity between chosen candidates was high between the two methods having selected 71 animals in common out of 100 with a high rank correlation of 0.82.

INTRODUCTION

Whole genome sequencing presents as an opportunity to capture more information about the genetic structure of a population which can be utilised in breeding program and mating decisions through genomic selection methodologies. However, it is costly with sequencing to 30x coverage costing approximately $1000 per sample. Through the use of imputation, high density genotyping does not need to be carried out population wide as "filling in the blanks" of sparsely genotyped individuals to higher densities can be completed using inferred haplotypes. One method of genomic selection is genomic best linear unbiased prediction (G-BLUP; Clark and van der Werf 2013) which utilises a relationship matrix calculated from a genotyped set of individuals. The resulting relationship matrix can be utilised to determine which individuals are the best candidates to describe variation in the population (i.e. form the reference population) and therefore suitable for genome sequencing to achieve high imputation accuracies. This is the aim of the commercial Wagyu breeding program behind this study and while other methods exist, e.g. Bickhart *et al.* (2015), the convenience of using a relationship matrix, having been already constructed for genomic evaluations, was appealing.

MATERIALS AND METHODS

Selection of candidates was carried out using two methods described by Yu *et al.* (2014). The first, denoted the MCA method, selects candidates for whole genome sequencing by minimising the genetic variation of the target population, relative to the selected pool, in order to improve their imputation accuracy. This method utilises Wrights numerator relationship matrix (A) such that;

$$\mathbf{A}_{11}^{*} = \mathbf{A}_{11} - \mathbf{A}_{12}\mathbf{A}_{22}^{-1}\mathbf{A}_{21}$$

where the 1 subscript denotes the set of target animals and 2 subscript denotes the set of animals selected to be sequenced. Diag($\mathbf{A}_{11}^{*}$) are the residual variances that are expected to remain if sequence data were to be obtained from the selected individuals and used to predict/impute genotypes of the target set. Animals were selected using an iterative process. **A** was constructed using an Australian Full-Blood Wagyu pedigree comprised of 10,549 individuals with a depth of up to 9 generations from the current generation using the R package *pedigreemm* (Bates and Vazquez 2014).

The second method (MCG) is akin to MCA but utilises a genomic relationship matrix (G) in place of A. G was constructed as per VanRaden (2007) method 2, utilising genotype information on 5,334 individuals genotyped with 30K GGP-LD (Neogen: GeneSeek Operations) or Bovine VersaSNP 50K (Weatherbys Scientific) chips. Animals genotyped on the Versa SNP were imputed to 30K from the approximate ~10K overlap between the chips, due to the significantly larger reference population available (4940 vs. 394), using Fimpute 2.2 (Sargolzaei *et al.* 2014). After imputation, SNPs were retained that had a minor allele frequency greater than or equal to 0.05 before building the GRM. All genotyped animals were present in the pedigree resulting in 5,334 animal overlap between the numerator (A) and genomic (G) relationship matrices.

Imputation accuracy, described here as the correlation between true and imputed genotypes (r), was calculated for the 4,940 individuals genotyped on the 30K chip by masking their true genotypes to a ~10K density. Seven rounds total of single replicate genotype imputation (Fimpute 2.2) was then carried out using 4 reference population sizes (100, 50, 25, 10) of animals selected for whole genome sequencing from the 2 methods (MCA or MCG respectively).

RESULTS AND DISCUSSION

The degree of similarity between the MCA and MCG methods was very high with MCA selecting 71/100 individuals that were selected by MCG. Of the animals that were selected by both methods, they were ranked very similarly with a strong positive rank correlation of 0.82 (Figure 1). This is a stronger relationship than previously reported Yu *et al.* (2014), however with approximately half of the animals in the pedigree having been genotyped and the target population also being the potential selection pool it is less surprising the lists are similar. The MCG method did account for slightly more genetic variance reaching 35% when 100 animals were selected compared to 30% accounted for using the MCA method. The first 20 selected animals accounted for 19% and 21% of the genetic variance for the MCG and MCA method respectively with each additional animal there after contributing less information (Figure 2).

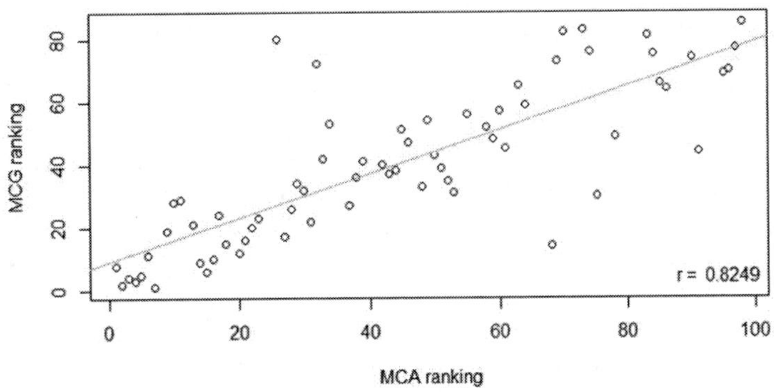

Figure 1. Correlation between ranks of candidates selected for whole genome sequencing using the MCA or MCG methods respectively

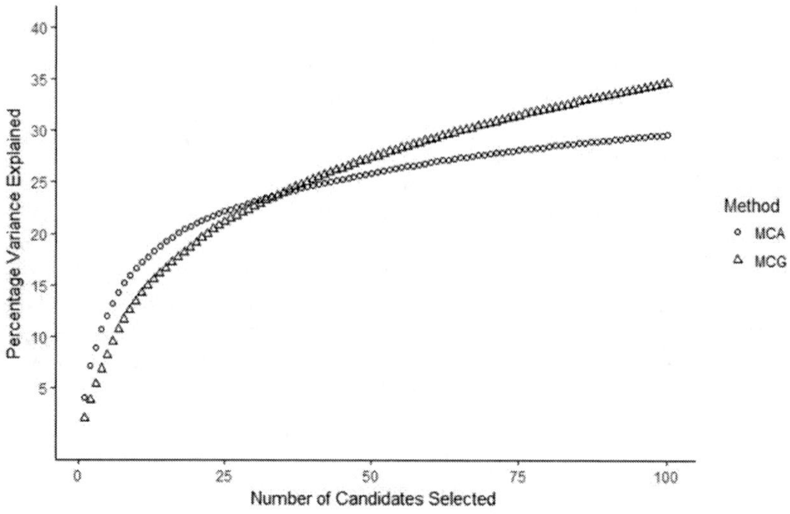

Figure 2. Diagonal values of A* representing the percentage of genetic variance explained for each additional selected candidate for whole genome sequencing using the MCG method (top) or MCA method (bottom)

A common logic is to identify animals that have a higher number of descendants, i.e. are considered influential, to be selected for sequenicng. For 100 genotyped sires (with effective progeny numbers of 1 to 437, mean = 47, in this population) the amount of genetic variation accounted for was 30%, equivalent to the MCA method but lower than MCG method.

Imputation accuracy was calculated for both the MCG and MCA method. Larger reference populations gave the highest imputation accuracies which is to be expected. For animals selected using the MCG method, selecting 100 animals was fairly comparable to selecting 50 animals with a noticeable drop in the mean accuracy from ~0.96 to 0.94 and 0.83 when 25 and 10 animals are selected respectively (Table 1). For MCA, out of the 100 selected animals, only 75 were genotyped and so could be used to calculate imputation accuracy. For comparisons sake, only reference populations of 50, 25 and 10 were constructed. MCA selected animals who weren't genotyped were generally lowly ranked, however a few non-genotyped candidates were present in the higher ranks. Therefore the MCA reference populations are not "perfectly" ranked as per the MCA method. Higher ranked animals that were not genotyped were replaced by the next available ranked animal until the desired number was sampled. The mean accuracies for MCA at 50, 25 and 10 reference animals were comparable to MCG although MCG was slightly superior. Noticeably though MCA did have a much higher minimum accuracy of 0.67 compared to 0.55 for MCG which indicates a narrower spread of imputation accuracies giving more successful imputation overall (Table 1).

Proc. Assoc. Advmt. Anim. Breed. Genet. 23:286-289

Table 1. Imputation accuracy calculated for sparse 11K genotypes imputed to 30K using differing reference populations of different sizes selected from two methods

MCG					MCA			
Ref size	100	50	25	10	-*	50	25	10
Min	0.5495	0.5482	0.5474	0.5101	-	0.6724	0.5317	0.5139
Mean	0.9756	0.9659	0.9395	0.8331	-	0.9627	0.9271	0.834
Max	0.9996	0.9992	0.9968	0.9725	-	0.9995	0.9978	0.9863

* For MCA, out of the 100 selected animals, only 75 were genotyped and so could be used to calculate imputation accuracy. For comparisons sake, only reference populations of 50, 25 and 10 were constructed. For MCA the next available candidate was selected if no genotype was available and so reference populations do not display perfect ranking but can be used as an example.

Both the MCA and MCG method assumed that all potential selection candidates had DNA available for sequencing and in a commercial pedigree this is not always the case. This fact became partially evident in the imputation study where not all MCA selected candidates had genotypes to form the reference. This is an important consideration and both methods could be easily modified to account for this. Within an iteration, the animal that is selected is logically the one that reduces the residual genetic variance of the target population i.e. Diag($\mathbf{A}_{11}^{*}$), the most. Multiplying each candidates impact on the residual by a simple vector of 0 (no DNA available) or 1 (DNA available) would ensure that only candidate animals with DNA are selected. This would also prevent bias when selecting sequence candidates to form the reference if you were just to remove animals with no DNA from the analysis all together.

CONCLUSIONS

For a full-blood Australian Wagyu herd their appeared to be little difference between the MCG and MCA methods for selection of candidates for whole genome sequencing. Both methods accounted for greater than 30% of the genetic of the target population when selecting 100 candidates and had comparable imputation accuracies up to 30K. Given the large volume of genotypes and deep, complete pedigree, either method would be suitable to select whole genome sequencing candidates to form the reference population for imputation. At the commercial level, the MCG method was selected to sample 73 candidates for sequencing due to the higher likely hood of selecting candidates with DNA sources (hair or semen) available in the first instance and the need to QC each individual hair sample in store prior to DNA extraction if semen was not available.

REFERENCES

Bates D. and Vazquez A. (2014). pedigreemm: Pedigree-based mixed-effects models. R package version 0.3-3.https://CRAN.Rproject.org/package=pedigreemm

Bickhart DM., Hutchison DJ., Null DJ., VanRaden PM. and Cole JB. (2015) *J. Dairy Sci.* **99**:5526.

Clark S., van der Werf J. (2013) In: Genome-Wide Association Studies and Genomic Prediction. Methods in Molecular Biology (Methods and Protocols), pp. 321- 330, editor Gondro C., van der Werf J. and Hayes B. Humana Press, Totowa, NJ

Neogen: GeneSeek Operations. https://genomics.neogen.com/pdf/ag151_ggp_ts.pdf

Sargolzaei M., Chesnais J. and Schenkel F. (2014) *BMC Genomics* **15**:478

VanRaden PM (2008) *J. Dairy Sci.* **91**: 4414.

Weatherbys Scientific. 2017, https://weatherbysscientific.com/versasnp/

Yu X., Woollimas J. and Meuwissen T. (2014) *Gen. Sel. Evol.* **46**: 46.

THE ACCURACY OF GENOTYPE IMPUTATION IN SELECTED SOUTH AFRICAN SHEEP BREEDS FROM AUSTRALIAN REFERENCE PANELS

C.L. Nel[1], K.P. Gore[2], A.A. Swan[2], S.W.P. Cloete[1,3], J.H.J. van der Werf [4] and K. Dzama[1]

[1]Department of Animal Science, University of Stellenbosch, Stellenbosch, WC, 7602 South Africa
[2]Animal Genetics & Breeding Unit*, University of New England, Armidale, NSW, 2351 Australia
[3]Directorate Animal Sciences, Western Cape Department of Agriculture, Elsenburg, 7607 South Africa
[4]School of Environmental and Rural Science, University of New England, Armidale, NSW, 2351 Australia

SUMMARY

The cost of genotyping is becoming increasingly affordable but remains an influential factor for determining the SNP-density at which genotyping can proceed. Compared to Australian breeding programs, the South African wool sheep industry represents parallel objectives within similar environments but presently lacks the necessary infrastructure to exploit modern technologies such as genomic selection. The aim of the study was to determine the feasibility of across country imputation as an alternative to high density genotyping on a local basis. Following imputation from a 15k to 50k density, mean accuracy levels of 0.87 and 0.85 were observed in the Merino and Dohne Merino breeds, while the highest levels of accuracy of 0.88 and 0.90 was observed in the Dorper and White Dorper breeds, respectively. The extent of genetic relationships was considered amongst the key factors that limit the ability to impute at an accuracy above 90%, but the observed results suggest that across country imputation could remain useful. Imputation from reference panels genotyped at densities higher than 50k and research into across country prediction is recommended.

INTRODUCTION

Genomic prediction and Genome Wide Association Studies (GWAS) depend on the size of the reference population as well as the density at which informative individuals were genotyped. Even though medium and high density genotyping options are becoming more affordable, cost remains a restricting criterion for the choice of a genotyping platform. Economic restrictions are likely to be more severe within a developing infrastructure as is currently experienced in South Africa (Van Marle-Köster and Visser 2018). There could be potential to exploit similarities in South African and Australian *ovine* breeds and environments through the compilation of genotypic resources. Imputation of un-typed markers of animals genotyped at a lower density has proven a reliable and affordable alternative to widespread genotyping on high density platforms (Browning and Browning 2007; Berry and Kearney 2011; Hickey *et al.* 2011; Huang *et al.* 2012; Moghaddar *et al.* 2015). The objective of this study is thus to investigate the potential of across country imputation of South African datasets from Australian reference populations from low (15k) to medium (50k) densities.

MATERIALS AND METHODS

Data Structure and Distribution. The South African sample set was selected from multiple breeds within respective resource flocks (Schoeman *et al.* 2010) as well as a smaller proportion of animals originating from the industry sector. Genotyping of the South African (SA) sample set was performed with the OvineSNP50 (Illumina Inc., CA, USA) beadchip at GeneSeek Inc. (Lincoln, NE,

* A joint venture of NSW Department of Primary Industries and the University of New England

USA) and subjected to quality control measures (> 0.25 GenCall score, > 0.5 GenTrain score, > 0.01 MAF, > 0.95 call rate, > 0.95 sample call rate). Following imputation of randomly missing SNPs, 986 samples with 50095 SNPs remained available for further analysis. Animals were grouped by breed type, namely Merino (552), Dohne Merino (60), Dorper (59), South African Mutton Merino (57), Dormer (42), Meatmaster (39) and White Dorper (27) while the hardy native breeds Damara (30), Pedi (29) and Afrikaner-type (13) animals were grouped together as 'Indigenous' (72). The Australian reference set constituted a database of ~ 84 000 samples from multiple breeds that serve as respective reference populations in genomic prediction programs. The major proportions of the dataset were classified as Merino, maternal (Border Leicester and Coopworth) and terminal (Poll Dorset and White Suffolk) groups. The same OvineSNP50 genotyping platform was used in generating the Australian database and 48 599 SNPs were available for analysis following quality control.

Design. All 986 SA samples were subset to ~ 15k SNPs using Illumina map information to simulate a commercial 15k beadchip. Analysis proceeded by the subsequent imputation back up to the 50k density using an Australian reference. The accuracy of imputation was evaluated by Pearson correlation coefficients between the imputed and observed SNP genotypes. To reduce computation time and increase accuracy (Moghaddar *et al.* 2015), the Australian reference set was screened by assigning an animal in the sample set with the top 50 highest values of animals in the reference set according to a genomic relationship matrix (GRM) that included all the animals in the study. Thus, animals from the reference set not meeting this criterion for any of the animals in the sample set were not used for imputation.

Software. Genotype imputation was performed using FIMPUTE (V2.2) (Sargolzaei *et al.* 2014). The program assumes a level of relatedness between all individuals and phases reference sets with overlapping sliding windows that is shrunk in proceeding increments with each chromosome sweep. The initial larger window sizes aim to capture the long-range haplotypes expected from highly related individuals, while the subsequent sweeps aim to capture relationships between more distant individuals. The inclusion of pedigree information is an optional addition to FIMPUTE, but it was not supplied in the current analysis. Summary statistics and visual analyses were performed in R (R Core Team 2016, Vienna, Austria).

RESULTS AND DISCUSSION

The accuracy of imputation varied considerably both between and within breeds. The accuracy of imputation for indigenous breed group was very low (mean = 0.68) and is not represented in subsequent figures and tables. A low accuracy is to be expected considering their heterogeneous nature and poor representation within the Australian reference set. Moreover, concerns have been raised surrounding an underrepresentation of indigenous breeds in the design of commercial bead chips (Sandenbergh *et al.* 2016). Table 1 shows the summary statistics for imputation accuracy (correlation coefficients) for the remaining breed groups.

The accuracy for Merino samples was moderate, as Pearson's correlations ranged from 0.82 to 0.90. This is considerably lower than correlation coefficients of 0.93 to 0.96 reported by Moghaddar *et al.* (2015) for 1,000 purebred Merinos imputed from smaller proportions of the same reference set. Hayes *et al.* (2012) reported considerably lower values of accuracy (71%) for imputing Merino samples from 5k to 50k densities, but with a reference set confined to ~ 5000 animals. Moderate accuracies were also observed for Dohne Merino and the South African Mutton Merino (SAMM) individuals, while the imputation accuracy for Dormers and Meatmasters were below 0.80.

Table 1. Summary statistics for the imputation accuracy of all South African breed groups in the sample set

	Merino	Dohne Merino	Dorper	SAMM	Dormer	Meat -master	White Dorper
(n)	552	60	59	57	42	39	27
Min.	0.82	0.82	0.85	0.81	0.76	0.72	0.87
1st Quartile	0.86	0.85	0.87	0.83	0.78	0.74	0.89
Mean	0.87	0.85	0.88	0.85	0.79	0.75	0.90
3rd Quartile	0.87	0.86	0.89	0.86	0.79	0.76	0.91
Max.	0.90	0.88	0.90	0.87	0.81	0.78	0.92

The Dorper and White Dorper breeds achieved moderately high to high imputation accuracies. The Dorper originates from South Africa, and it is possible that the animals that represent them in the Australian database have not drifted extensively from the ancestral lines or share relatively recent parental links. Considering the size of the Australian reference set and the large proportion of Merinos included, it could be considered somewhat unexpected that none of the Merino test samples attained an imputation accuracy of above 0.90. However, the number of reference samples available as well as their relatedness to the sample population is considered essential factors in the accurate phasing of haplotypes for the imputation of un-typed markers.

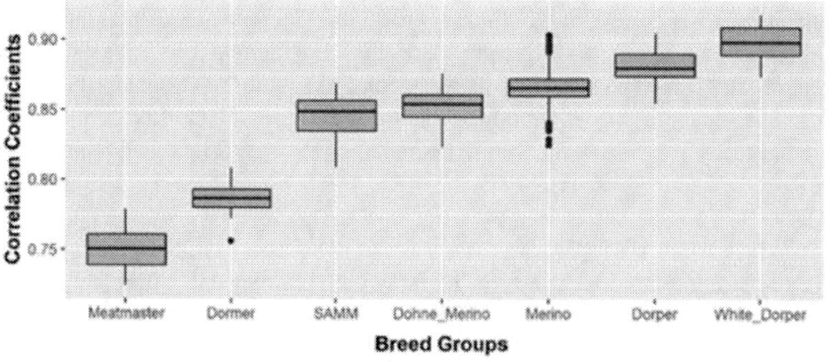

Figure 1. Box plots for the imputation accuracies for all South African breeds in the sample set

Analyses that characterize haplotypes using population linkage disequilibrium (LD) based methods do not utilize a pedigree, but indirectly capture patterns associated with identity by descent (IBD), the accuracy of which is complemented by the indication that there is little benefit in including pedigree data if the reference set is large enough (Browning and Yu 2009; Larmer *et al.* 2014; Moghaddar *et al.* 2015). Hayes *et al.* (2012) proposes that haplotypes are not necessarily shared across breeds and that 50k genotyping platforms do not capture LD to an adequate level for across breed application. With markedly less family linkage, the proportion of genomic regions possibly considered IBD should be markedly smaller when attempting across country imputation. Thus, a similar argument to that proposed by Hayes *et al.* (2012) could be extended to the current results, despite the current study being within breed analysis. It is possible the denser 500k platform could provide improved phasing of the reference set that is less dependent on long range haplotypes and more appropriate for capturing linkage disequilibrium observable over distant populations.

CONCLUSIONS

Genotype imputation of un-typed markers in a population depends on the representation of that population within a reference set. There is little benefit in the addition of genetically dissimilar animals. Across country imputation will likely be limited by a lack of direct genetic links, but moderately high levels of accuracy can still be achieved within breeds. Research into across country genomic prediction for shared breeds is recommended.

REFERENCES

Berry D.P. and Kearney J.F. (2011) *Animal* **5**: 1162.
Browning S.R. and Browning B.L. (2007) *Am. J. Hum. Genet.* **81**: 1084.
Browning B.L. and Yu Z. (2009) *Am. J. Hum. Genet.* **85**:847.
Hayes B.J., Bowman P.J., Daetwyler H.D., Kijas J.W. and van der Werf J.H.J. (2012) *Anim. Genet.* **43**: 72.
Hickey J. M., Kinghorn B. P., Tier B., Wilson J. F., Dunstan N., and van der Werf J. H. J. (2011) *Genet. Sel. Evol.* **43**: 1.
Huang Y., Hickey J.M., Cleveland M.A. and Maltecca C. (2012) *Genet. Sel. Evol.* **44**: 1.
Larmer S.G., Sargolzaei M. and Schenkel F.S. (2014) *J. Dairy Sci.* **97**: 3128.
Moghaddar N., Gore K.P., Daetwyler H.D., Hayes B.J. and van der Werf J. H. J. (2015) *Genet. Sel. Evol.* **47**: 1.
Sandenbergh L., Cloete S., Roodt-Wilding R., Snyman M. A. and Bester-van der Merwe A. E. (2016) *S. Afr. J. Anim. Sci.* **46**: 2011.
Sargolzaei M., Chesnais J.P. and Schenkel F.S. (2014) *BMC Genomics* **15**: 478.
Schoeman S.J., Cloete S.W.P. and Olivier J.J. (2010) *Livest. Sci.* **130**: 70.
Van Marle-Köster E. and Visser C. (2018) *S. Afr. J. Anim. Sci.* **48**: 808.

APPLICATION OF GENOMIC SELECTION TO VIETNAMESE HOUSEHOLD DAIRY HERDS

N.N. Bang[1,2], B.J. Hayes[1], I.A.S. Randhawa[1], R.E. Lyons[1], J.B. Gaughan[1], N.V. Chanh[3], N.X. Trach[2], N.D. Khang[3] and D.M. McNeill[1]

[1]The University of Queensland, Gatton QLD 4343 and St Lucia QLD, 4067 Australia
[2]Vietnam National University of Agriculture, Hanoi, 131000 Vietnam
[3]Nong Lam University, Ho Chi Minh, 700000 Vietnam

SUMMARY

Household dairy farms (HDFs) account for most of the demand for animal breeding support in Vietnam, as they comprise 97% of the national herd. However, most do not have individual cow pedigrees or production data. Consequently, neither pedigree-based nor genomic selection (GS) methods have been used in Vietnam. The aim of this project was to establish a milk production database and assess the accuracy of GS for production traits using only a single test-day measurement (average of pm + am milking). Phenotypic data included milk yield (MILK, kg/d), milk dry matter (DM%), fat (FAT%), and protein (PRO%) contents of 345 lactating cows from 4 dairy regions, with 8 HDFs per region. The cows were genotyped using the Bovine 50K chip. GBLUP was used to estimate genomic heritability (h^2) and evaluate the accuracy of GS per trait. Moderate heritabilities and accuracies of GS were detected for FAT% ($h^2 = 0.45$, accuracy = 0.28), PRO% ($h^2 = 0.21$, accuracy = 0.23), and DM% ($h^2 = 0.18$, accuracy = 0.48). However, the heritability for MILK was very low (0.01) and the standard errors for all heritabilities and GS accuracies were high. These data suggest the potential for a single test-day to assess Vietnamese dairy cows for milk solid content, but not milk yield, using GS.

INTRODUCTION

The dairy industry in Vietnam is characterized by approximately 500,000 household dairy farms (HDFs) (Nguyen *et al.* 2016). The HDFs account for 97% of the national dairy herd (Trach 2017), and supply >80% of fresh milk production (Vinamilk 2017). Genotypes commonly used are European breeds (predominantly Holstein Friesian but also Jersey) crossed with tropically adapted breeds (Red Sindhi and Sahiwal) and local breeds (Yellow and Lai Sind) (Hayley 2010; Lam *et al.* 2010). Improving the genetic potential of dairy cows for milk production in HDFs is necessary to improve the national supply of fresh milk. However, a national breeding program for household dairy herds is yet to be officially implemented, even for basic traits such as milk yield, fat, or protein. A lack of individual cow pedigree and phenotypic data is the main reason for this.

Genomic selection (Meuwissen *et al.* 2001) is a recently proven method which is now widely used globally in dairy cow selection. This could be a suitable tool for dairy selection in Vietnam as it enables the selection of animals based on genomic markers, most commonly single nucleotide polymorphisms, or SNPs, without the need for pedigrees. However, whilst genomic selection does not require a pedigree, it does require a large number of animals with phenotypic data to allow the development of accurate prediction equations. These data are not easily obtained in Vietnamese HDFs. Such data is expensive to collect in terms of money, time, and labour, as it requires manually separating, weighing, and sampling milk from each cow at each milking time. Consequently, we aimed to estimate the genomic heritability for key milk production traits and to assess the accuracy of genomic selection on these traits using only a single test-day measurement for each.

MATERIALS AND METHODS

Phenotype data. From August to October 2017, data from 345 lactating cows located on 32 HDFs, 8 from each of four main dairy regions in Vietnam were recorded: Lam Dong – a south high-altitude province; Ho Chi Minh – a south low-altitude city; Son La – a north high-altitude province, and Ha Nam – a north low-altitude province. Each farm was visited twice to correspond with a milking in an afternoon and following morning. At these visits, individual cow age, number of lactations and days in milk were obtained by asking the farmer and/or checking their record books where possible, and tail hair was sampled from each lactating cow. The mean ± SD obtained for age (years), number of lactations, and days in milk of these cows were 4.5 years ± 1.7, 2.3 lactations ± 1.4, and 191.4 days ± 120, respectively. A single day milk yield (MILK, kg/d) for each cow was obtained by weighing and summing the afternoon and the following morning milk yields. Milk samples for each cow were also collected at each milking, analysed at the Food Chemistry Lab (Vietnam National University of Agriculture) and averaged for milk dry matter (DM%, which is the percentage of all milk constituents excluding water), milk fat (FAT%), and milk protein (PRO%) contents. These data were used to calculate the yield of milk dry matter (DM, kg/d), fat (FAT, kg/d), protein (PRO, kg/d), and energy-corrected milk (ECM, kg/d) using the equation of Tyrrell and Reid (1965).

Genotype data. Hair samples were genotyped by Neogen Australasia, The University of Queensland, Gatton. DNA was extracted from the samples using Sbeadex Livestock Kits (LGC Limited, 2017), and then genotyped using the GGP Bovine 50K chip, which assays 48,268 SNPs (Neogen GeneSeek Operations, 2018).

Quality control. R Software (R Core Team, 2016) was used for all data processing. The quality control on the genotype data removed 3,313 SNPs, which were either mitochondrial SNPs or unmapped SNPs, 5478 SNPs with call rates lower than 95%, 1980 SNPs with minor allele frequency lower than 95%, and 633 SNPs with a heterozygosity deviating ±3 SD from the SNPs' heterozygosity mean. One sample with a call-rate less than 95% was removed, in addition to three cows with heterozygosity deviating ±3 SD from the samples' heterozygosity mean, 14 cows from fours farms with less than five lactating cows. Three cows had missing phenotypic data. The final data set for analysis comprised 323 cows from 28 farms, genotyped for 36864 SNPs.

Genomic heritability and genomic breeding values. Univariate animal linear mixed models with common environmental effects (Mrode and Thompson 2013) were fitted using the GBLUP method in the R "Sommer package" and (Covarrubias-Pazaran 2019). The matrix notation describing the model was: $y = X\beta + Z\alpha + Wc + e$, where: y was the vector of the traits observed, β was the vector of fixed effects (age, lactations, days in milk, days in milk squared), α was the vector of random additive genetic animal effects [$\alpha \sim N(0, A\sigma^2_a)$], wherein A was the genomic relationship matrix derived from the SNPs, c was the vector of the random environmental farm effect (28 farms) [$c \sim N(0, I\sigma^2_c)$], e was the vector of residual random effects [$e \sim N(0, I\sigma^2_e)$], and X, Z, and W were the incidence matrices of the fixed effects, random additive genetic effects, and random environmental effects, respectively. Animal, random environmental and residual effects were assumed to be independently distributed. Heritability (h^2) was estimated as the ratio of the additive genetic variance to total phenotypic variance [$h^2 = \sigma^2_a / (\sigma^2_a + \sigma^2_e)$] (Falconer and Mackay 1996).

Accuracy of genomic selection. Due to the relatively small data-set, a 10-fold cross-validation approach was applied (Kang *et al.* 2017). Briefly, the entire data set, of 323 samples, were randomly partitioned into 10 subsets of equal size. Nine were used as a training set to determine genomic estimated breeding values (GEBV) for the retained validation set (10%). This process was repeated 10 times so that each subset was used only once as the validation set. The accuracy of genomic selection for each trait was determined by: Accuracy = $r(GEBV, y)/h$, where r was the correlation between GEBV and the original phenotype (y) of each validation set and h was the square root of genomic heritability of the trait.

RESULTS AND DISCUSSION

MILK of the average household dairy cow in Vietnam was 17.87 kg/d (Table 1), which was much higher than other published survey estimates in Vietnam (14.0 – 16.0 kg/d) (Lam *et al.* 2010; Vu *et al.* 2016) but much lower than surveys in other developed countries such as Australia (22.9 kg/d) (DataGene 2017) or Asian countries such as Korea (27.8 kg/d) (Cho *et al.* 2013).

Table 1. Descriptive statistics for milk production traits in Vietnamese dairy cows

Trait	n	Mean	SD	Median	Minimum	Maximum	IQR
MILK (kg/d)	321	17.87	6.28	17.8	5.3	36.65	8.75
DM%	323	12.22	1.22	12.1	9.46	16.36	1.63
FAT%	323	3.65	0.78	3.55	1.98	5.97	0.92
PRO%	323	3.30	0.48	3.25	2.29	5.64	0.61
ECM (kg/d)	321	16.75	5.23	16.44	5.31	33.23	7.43
DM (kg/d)	321	2.14	0.67	2.11	0.75	4.26	0.88
FAT (kg/d)	321	0.63	0.20	0.61	0.14	1.31	0.28
PRO (kg/d)	321	0.58	0.19	0.56	0.19	1.20	0.27

n: number of observations; SD: standard deviation; IQR: interquartile range; MILK: milk yield; DM: milk dry matter; FAT: milk fat; PRO: milk protein; ECM: energy corrected milk.

Genomic heritability estimates (Table 2) for DM%, FAT%, PRO%, DM, FAT, and PRO in our study ranged from 0.12 (PRO) to 0.45 (FAT%), which were moderate and similar to other comparable studies that used a far greater number of cows (Kim *et al.* 2009; Toghiani 2012; Cho *et al.* 2013). These studies presented heritability for FAT% ranging from 0.15 to 0.36, PRO% from 0.07 to 0.50, FAT from 0.28 to 0.52, and PRO from 0.26 to 0.34. However, in the current study it should also be realised that the standard errors for these milk solid traits, except FAT%, were high. These high standard errors of the heritabilities could be because the single test-day measurements in our study were derived from the cows at wide ranges of lactations and days in milk, whereas the heritabilities for milk productions traits change widely throughout a lactation (Kim *et al.* 2009).

The heritability for MILK in our study was also lower than expected (0.01) and with a high standard error (13 times the mean) when compared with other studies (0.15 to 0.46, Kim *et al.* 2009). The low heritability for MILK is likely due to high environmental and residual variances or measurement errors and so indicates a larger sample size would be required for a more acceptable estimate for that trait. Similarly, the low heritability for MILK was also the reason for the low heritability for ECM (0.08), as these was calculated from MILK.

The accuracies of GEBV from GBLUP for DM%, FAT%, and PRO% were moderate (0.23 to 0.48) and significantly different from zero, as the mean of accuracies for these traits were at least almost twice their standard errors. However, the accuracy for MILK in our study was inflated by its very low heritability (0.01) to become an unrealistically high number (>1). The accuracies of GS for other traits were unstable with moderate means (0.11 to 0.34), but with very high standard error (0.7 to 2 times the mean). To avoid inflated accuracy resulting from close family relationships between training and test animals, partitioning animal into training and validation sets should be based on family so that the highly related animals were in the same validation set (Pszczola *et al.* 2012). However, due to the lack of pedigree data, cows in our study were just randomly partitioned into training and validation sets and this could be a bias source in our GS accuracies.

Table 2. Estimates of additive genetic variance (σ^2_a), random environmental variance (σ^2_c), residual variance (σ^2_e), heritability (h^2), correlation between GEBV and phenotype, and accuracy of genomic selection of milk production traits using univariate models

Trait	σ^2_a	σ^2_c	σ^2_e	h2 ± SE	Correlation ± SE	Accuracy ± SE
MILK (kg/d)	0.16	9.84	18.35	0.01 ± 0.13	0.15 ± 0.08	1.60 ± 0.82
DM%	0.17	0.27	0.75	0.18 ± 0.14	0.21 ± 0.06	0.48 ± 0.14
FAT%	0.21	0.04	0.25	0.45 ± 0.15	0.19 ± 0.05	0.28 ± 0.07
PRO%	0.03	0.02	0.12	0.21 ± 0.14	0.11 ± 0.05	0.23 ± 0.12
ECM (kg/d)	1.09	7.51	12.78	0.08 ± 0.14	0.06 ± 0.11	0.21 ± 0.41
DM (kg/d)	0.04	0.11	0.21	0.15 ± 0.14	0.13 ± 0.09	0.34 ± 0.24
FAT (kg/d)	0.005	0.009	0.019	0.19 ± 0.15	0.05 ± 0.08	0.11 ± 0.18
PRO (kg/d)	0.002	0.012	0.016	0.12 ± 0.14	0.08 ± 0.08	0.24 ± 0.25

Abbreviations of traits as in Table 1; SE: standard error

CONCLUSIONS

This study suggests that genomic selection using the GGP Bovine 50K chip and a single test day measurement could potentially be applied in Vietnam to the milk solid traits DM%, FAT%, and CP%, but not to milk yield traits. However, a larger sample size is recommended to confirm these findings. The very low estimation of the heritability of MILK could give misleading results when the accuracy of genomic selection is assessed.

REFERENCES

Cho C., Cho K., Choy Y., Choi J., Choi T., Park B. and Lee S. (2013) *J. Ani. Sci. Tec.* **55**: 7.
Covarrubias-Pazaran, G. (2019) 'Solving Mixed Model Equations in R'. CRAN.
DataGene (2017) 'Australian Dairy Herd Improvement Report 2016'. DataGene Limited.
Falconer D.S. and Mackay T.F.C. (1996) 'Introduction to quantitative genetics. Longman.
Hayley R.A.B.A. (2010) Master Thesis, The Ohio State University.
Kang H., Zhou L., Liu J., Kang H., Zhou L. and Liu J. (2017) *Fro. Agri. Sci. Engi.* **4**: 268.
Kim B.W., Lee D., Jeon J.T. and Lee J.G. (2009) *Asian-Aust. J. Anim.Sci.* **22**: 923.
Lam V., Wredle E.K., Thao N.T., Man N.V. and Svennersten S.K. (2010) *Afr. J. Agr. Re.* **5**: 2668.
LGC Limited (2017) Sbeadex livestock kit.
Meuwissen T. H., Hayes B. J. and Goddard M. E. (2001) *Genetics* **157**:1819.
Mrode A. R. and Thompson R. (2013) 'Linear Models for the Prediction of Animal Breeding Values' 3rd Ed. CABI.
Neogen GeneSeek Operations (2018) GeneSeek Genomic Profiler Bovine 50K.
Nguyen M.H., Cesaro J.D., Pham D.K., Hoang V.Q. and Duteurtre G. (2016) *J. Anim. Sci. Tech.* **61**: 12.
Pszczola M. Strabel T. Mulder H. A. Calus M. P. L. (2012) *J. Dairy Sci.* **95**: 389.
R Core Team (2019) 'R: A language and environment for statistical computing'.
Toghiani S. (2012) *Archiv Tierzucht* **55**: 458.
Trach N. X. (2017) *Improving dairy cattle health and production in Vietnam.*
Tyrrell H. F. and Reid J. T. (1965) *J. Dairy Sci.* **48**:1215.
Vinamilk (2017) Vinamilk invested many high-tech dairy farms.
Vu N. H., Lambertz C. and Gauly M. (2016) *Asian-Aust. J Anim. Sci.* **10**: 290.

GENOMIC TOOLS FOR USE IN THE NEW ZEALAND DEER INDUSTRY

K.G. Dodds, S-A.N. Newman, S.M. Clarke, R. Brauning, A.S. Hess, T.P. Bilton, J.F. Ward, A.J. Chappell, J.C. McEwan, T.C. Van Stijn, M. Bates and S.J. Rowe

AgResearch, Invermay Agricultural Centre, Mosgiel, New Zealand

SUMMARY

The New Zealand deer industry has recently adopted genotyping-by-sequencing (GBS) as a tool for parentage analysis. For cost reasons, sequencing is performed on many individuals at once with low sequencing read depth supporting the genotypes. It is important to account for the partial information provided by these low depth reads and to account for the high genetic diversity between breeds present in the population in any analysis. The genomic information provided by the more than 70,000 markers scored can also be used for additional purposes such as inbreeding and relatedness estimation, plus gender and breed prediction. The data provides a platform for genome wide association studies and genomic selection, which are being developed for this industry. These results provide evidence that GBS is a useful technique for genomic studies.

INTRODUCTION

The New Zealand (NZ) deer industry has been using DNA-marker testing since the early 1990s. This has been primarily for parentage assignment, as deer behaviour prevents manual recording of pedigree at birth. DNA markers have also been used to provide information about breed. The primary breeds are wapiti or elk (*Cervus canadensis*) and red deer (*Cervus elaphus*), which are regarded as distinct species, but there is also interest in estimating the components of differing European origin (Eastern or Western) in red deer. Initially a small panel of isozymes was used for breed discrimination. This was subsequently replaced by a microsatellite panel. Since 2017 genotyping-by-sequencing (GBS) has been used as the marker system. We show how recently developed methods for low-depth GBS data are being used in the New Zealand deer industry for parentage, breed prediction and gender assignment and consider how this GBS resource can be used for gene discovery and genomic selection.

MATERIALS AND METHODS

Animals. The Invermay, AgResearch deer herd is used to illustrate the use of genomics in the NZ deer industry. The 2018 cohort consisted of 554 genotyped calves, 621 potential dams and 46 potential sires. Industry-wide data used here refers to deer genotyped by GenomNZ (https://www.agresearch.co.nz/genomnz) using GBS, first used for the 2016 calf-drop and their parents. This industry GBS dataset currently contains ~80,000 animals.

GBS genotypes. The animals were genotyped by GBS using the methods described by Dodds *et al.* (2015). The resulting sequence reads from a set of animals likely to represent much of the genetic variation were adapter-trimmed and then UNEAK (Lu *et al.*, 2013) was used to detect variants (without the use of a reference genome). These variants were placed into a catalog which was used to report counts of reference and alternate alleles for each variant and sample (including any subsequently sequenced samples) using TagDigger (Clark and Sacks 2016). Each new set of GBS count data is compared against any previous results for the same sample, by comparing the relatedness, estimated taking into account the read depths (Dodds *et al.* 2015), between a pair of results for the same sample with the mean self-relatedness of those samples. Differences greater than 0.4 are reported for checking. Accepted results are then appended to the file of previous results. There is a corresponding comparison made during downstream analysis for any pairs of samples that have come from the same animal.

Population structure and breed prediction. The genetic structure of the population was portrayed as the principal components of the genomic relationship matrix (GRM), which in turn was calculated using the method of Dodds *et al.* (2015) which is based on VanRaden's (2008) first method, but accounts for the read depth in a genotype call. The GRM was calculated for a random sample of approximately 5000 deer from NZ commercial samples supplemented with NZ and overseas samples of reputedly pure 'breed' (wapiti/elk, English red and Eastern European red, denoted 'Wapiti', 'English' and "Eastern', respectively) standards. Breed prediction was undertaken by regressing the observed proportions of A alleles at each SNP for an animal on each breed's allele frequency (Kuehn *et al.* 2011). The breed allele frequencies were calculated from the breed standards.

Gender prediction. Gender is predicted using the method of Bilton *et al.* (2019) using the proportion of Y chromosome SNPs with reads and the heterozygosity of X chromosome SNPs. There were 15 SNPs located on the Y chromosome and 1006 SNPs located on the non-pseudoautosomal region of the X which passed the criteria given in Bilton *et al.* (2019).

Parentage analysis. Parentage assignment is based on the methods of Dodds *et al.* (2019) with the highest related potential sire and dam were assigned provided they achieved the chosen thresholds. The thresholds used for assigning parentage were 0.3 for estimated relatedness (from the GRM), 0.015 for parent-offspring excess (raw minus expected, where expected rate is calculated for the given read depths and offspring genotype) mismatch rate (EMM) and 0.03 for trio EMM.

RESULTS AND DISCUSSION

The GBS process resulted in a catalog of calls for 74,798 SNPs. After filtering SNPs for a Hardy-Weinberg disequilibrium coefficient (proportion of animals observed as homozygous for the A allele minus the squared A allele frequency; calculated using the Industry dataset) greater than -0.05 and a minor allele frequency (calculated for Invermay dataset) greater than 0.01, there were 66,824 SNPs remaining. These SNPs had a 76.9% call rate and mean read depth of 3.37 in the Invermay dataset.

Population structure and breed prediction. The first two principal components of an analysis of 6269 deer (109 breed standards, 1,211 Invermay herd deer, 4949 randomly chosen) is shown in Figure 1. The first component explains 80% of the variance and reflects the large genetic difference between wapiti and red deer which are at opposite ends of this axis. The second component explains 14% of the variance, and English and Eastern deer occur at the opposite ends of this axis. The Invermay deer mainly occur in a continuum between these two red deer types, with a few plotting part-way towards the Wapiti group, suggesting some wapiti ancestry in those animals.

The Invermay animals were predicted to be an average of 58% Eastern, 39% English and 3% Wapiti. The range in predicted breed percentages in the progeny were 4-91% Eastern, 7-96% English and 0-19% Wapiti. An estimated breed proportion could be used for a national across-breed genetic evaluation, but proportions estimated by different methods (marker systems and pedigree) need to be consistent.

Gender prediction. The results of the gender prediction for the Invermay herd are shown in Figure 2. The mean read depth of the sex chromosome SNPs in this herd was 2.97. The parents matched their recorded gender (apart from one uncertain), as expected. For the calves, three that were recorded as male were predicted as female, while one recorded as female was predicted as male. One of these recorded males was subsequently corrected to female, but the other inconsistencies could not be checked (sold or died). The gender test can be made at the same time as a parentage analysis and provides a check for assigned gender which can be difficult to assign in the field with 100% accuracy in young calves.

Parentage analysis. Both parents were assigned for 535 calves, 17 calves were assigned to a dam only, one was assigned a sire only and one was not assigned either parent (excluded based on the trio EMM which was 0.05). A seemingly low threshold (0.3) was used for assigning parentage to accommodate variations in breed structure and the fact that the GRM used allele frequencies estimated from the same dataset which can depress relatedness estimates (Yang *et al.* 2010). Only four of the final sire or dam assignments were with estimated relatedness below 0.4. One assignment had relatedness to both parents less than 0.4 and in this case the sire was predominantly (81%) Eastern while the dam was predominantly (82%) English.

A GRM of the dam only progeny, visualised using the heatmap function in R (Figure 3), suggested that two or perhaps three different sires were involved. Such information could help to find additional sires to include in the analysis.

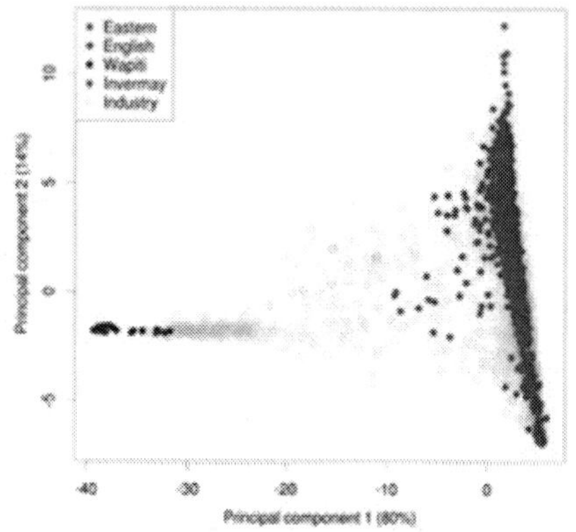

Figure 1. Principal components plot of the Invermay herd, breed standards and a random industry set of 5000 deer

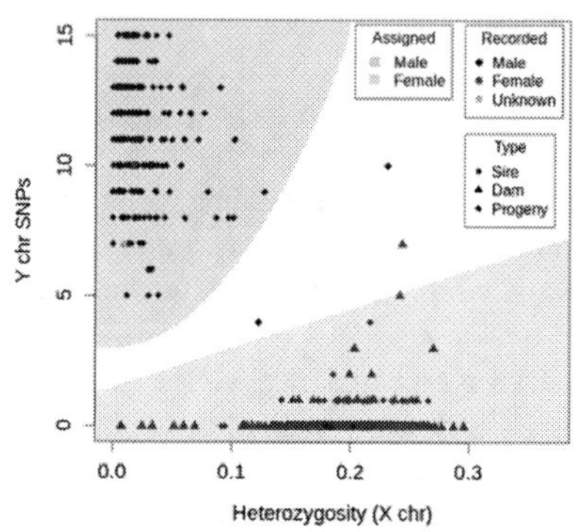

Figure 2. Gender plot with number of Y chromosome SNPs with reads plotted against heterozygosity of X chromosome SNPs. The lower and upper shaded areas are predicted as females and males, respectively

A by-product of calculating a GRM is that estimates of inbreeding are available (self-relatedness minus 1). The distribution of these estimates is shown in Figure 4. As is the case of most genomic estimators of inbreeding, values outside of [0,1] are possible. The progeny with both parents assigned with relatedness less than 0.4 had estimated inbreeding of -0.3, reflecting the high genetic separation between its parents. Reporting inbreeding estimates to breeders will alert them to issues with their breeding programme if high estimates are present, however care is needed to help breeders understand these values compared to pedigree-based calculations which are always within [0,1].

Future directions. The use of GBS in the NZ deer industry has enhanced the information available to breeders compared with that from marker systems previously used. Parentage, breed and estimated inbreeding results are returned to the breeders, but there is no systematic way of returning genomic relationships to breeders or service providers to allow enhanced breeding plan designs (e.g. optimal contributions). Further opportunities are available, such as the use of this genomic information for genome-wide association studies and genomic selection. Some initial investigations have been made by Rowe *et al.* (2017), including the consideration of calculating appropriate GRMs with GBS data in a multi-breed context. These or other methods need to be tested for their feasibility in the full industry dataset, and the data from non-genotyped animals included in the analysis. As the deer industry is much smaller than the dairy, beef and dual-purpose sheep industries, it will need to learn from the use of genomic information in those industries to allow affordable implementation.

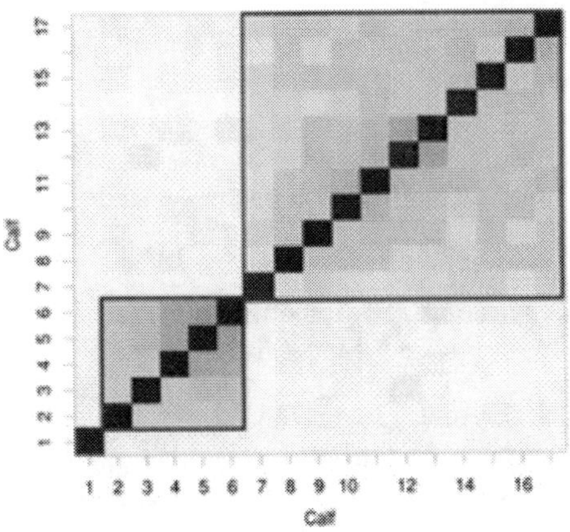

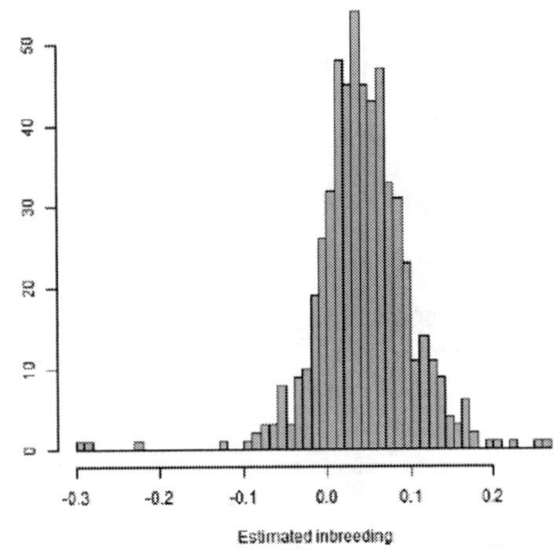

Figure 3. Heatmap of the relatedness between calves with only a dam assigned. The red boxes group potential sire groups

Figure 4. Distribution of estimated inbreeding in the Invermay progeny

ACKNOWLEDGEMENTS

This work was funded by DEEResearch and the Ministry of Business, Innovation and Employment (NZ), through the "Genomics for Production & Security in a Biological Economy" programme.

REFERENCES

Bilton T.P., Chappell A.J., Clarke S.M., Brauning R., Dodds K.G., McEwan J.C. and Rowe S.J. (2019) *Anim. Genet.* (in press).

Clark L.V. and Sacks E.J. (2016) *Source Code Biol. Med.* **11**:1.

Dodds K.G., McEwan J.C., Brauning R., Anderson R.A., Van Stijn T.C., Kristjánsson T. and Clarke S.M. (2015) *BMC Genomics* **16**:1047.

Dodds K.G., McEwan J.C., Brauning R., Van Stijn T.C., Rowe S.J., McEwan K.M. and Clarke S.M. (2019) *bioRxiv* 582585.

Kuehn L.A., Keele J.W., Bennett G.L., McDaneld T.G., Smith T.P.L., Snelling W.M., Sonstegard T.S. and Thallman R.M. (2011) *J. Anim Sci.* **89**:1742.

Lu F., Lipka A.E., Glaubitz J., Elshire R., Cherney J.H., Casler M.D., Buckler E.S. and Costich D.E. (2013) *PLoS Genet.* **9**:e1003215.

Rowe S.J., Dodds K.G., Ward J.F., Asher G.W., Brauning R., McEwan J.C. and Hess A. (2018) *Proc. 11th World Congress Genet. Appl. Livest. Prod.*:432

VanRaden P.M. (2008) *J. Dairy Sci.* **91**:4414.

Yang J., Benyamin B., McEvoy B.P., Gordon S., Henders A.K., Nyholt D.R., Madden P.A., Heath A.C., Martin N.G., Montgomery G.W., Goddard M.E. and Visscher P.M. (2010) *Nat. Genet.* **42**:565.

GWAS FOR METHANE YIELD, RESIDUAL FEED INTAKE AND LIVEWEIGHT IN NEW ZEALAND SHEEP

M.K. Hess[1], P.L. Johnson[1], K. Knowler[1], S.M. Hickey[2], A.S. Hess[1], J.C. McEwan[1] and S.J. Rowe[1]

[1]AgResearch Ltd., Invermay Agricultural Centre, Mosgiel, 9092 New Zealand
[2]AgResearch Ltd., Ruakura Agricultural Centre, Hamilton, 3214 New Zealand

SUMMARY

Identification of genomic regions associated with environmentally important traits, such as methane yield and residual feed intake, has the potential to improve genomic selection models for these traits, which are typically difficult and/or expensive to measure. Methane Yield, Residual Feed Intake and Liveweight phenotypes were available on 965 composite ewe lambs that had gone through an individual feed intake facility between 5 and 10 months of age. Our aim was to estimate heritabilities and identify genomic regions associated with these traits. Bayesian genomic models showed moderate heritability estimates between 0.38 and 0.44 for all traits. A genome wide association study failed to identify any large-effect QTL for any of the traits, consistent with these being highly polygenic traits. Future studies will explore the relationships between these environmentally important traits and other production traits and assess prediction accuracies.

INTRODUCTION

Animals that have a lower environmental footprint will be a vital part of future livestock production. Methane Yield and Residual Feed Intake (RFI) are environmentally important traits that are typically difficult and expensive to measure on large numbers of animals, therefore identifying causative mutations or indicator traits may facilitate selection on these traits. Genomic selection is a promising approach (Rowe *et al.* 2014) and incorporation of QTL or their tagging SNPs into current selection models may further improve the potential of using genomics in these traits.

Dual-purpose composite ewe lambs (Dodds *et al.* 2014) were put through an individual feed intake facility between 5 and 10 months of age, and gas emissions, including methane, were measured using portable accumulation chambers (PAC). Bayesian Genome Wide Association Studies were run for RFI, Methane Yield (CH4Yield) and Liveweight at 8 months (Liveweight) to identify genomic regions that are associated with these environmentally important traits.

MATERIALS AND METHODS

This study utilises a series of feed intake trials at AgResearch's Invermay campus (Elmes *et al.* 2014; Johnson *et al.* 2016; Johnson *et al.* 2018). Five cohorts of approximately 200 lambs were put through an individual feed intake facility for ~42 consecutive days between 5 and 10 months of age, where they were fed a Lucerne pellet diet. Gas emissions, including methane, were measured twice for each individual using PAC, ~14 days apart, as described in Jonker *et al.* (2018). The 965 lambs in this study came from 102 sires (range = 1-27 offspring per sire, median = 8).

Traits. Methane Yields, expressed as methane volume divided by total gas volume, were averaged for each individual to get the phenotype of CH4Yield. RFI was calculated as described in Johnson *et al.* (2018), whereby energy intake was corrected for metabolic mid-weight, average daily gain, trial, cohort, and pen. Liveweight was recorded when the lamb was 8 months of age.

Genotypes. Animals were genotyped on a variety of sheep SNP chips, ranging in density from 6,000 to over 500,000 markers. These SNPs were imputed to ~48,000 SNPs present on the 50,000

SNP panel using a reference population of 16,320 New Zealand sheep using FImpute (Sargolzaei *et al.* 2014).

Statistical Analyses. Univariate Bayesian selection models for CH4Yield, RFI, and Liveweight were run using GenSel v4.90 (Fernando and Garrick 2009) with a chain length of 51,000, including burn-in of 1,000. Model equations were:

$$y = \mu + brr + aod + bdev + cg + SNPs$$

where y is the trait of interest: CH4Yield (n=959), RFI (n=962) or Liveweight (n=963); brr and aod are the fixed class effects for birth rear rank and age of dam, respectively; bdev is the fixed covariate for birth date deviation; cg is the contemporary group: flock*birthYear for CH4Yield and RFI, and flock.rbyrmx for Liveweight. Heritability estimates were obtained by fitting SNPs in a BayesC model assuming all SNPs were associated with the trait ($\pi = 0$). GWAS were performed by fitting SNPs in a BayesB model assuming ~5% of SNPs are associated with the trait ($\pi = 0.95$).

RESULTS AND DISCUSSION

Heritabilities. Moderate heritability estimates were obtained for all traits (Table 1). These estimates are consistent with other estimates for these traits in the same population (Pickering *et al.* 2012; Pinares-Patiño *et al.* 2013; Johnson *et al.* 2018; Jonker *et al.* 2018). The heritability for methane yield is a little higher than has previously been published in an expanded New Zealand dataset (Jonker *et al.* 2018), however this could be due to averaging the two measurements for CH4Yield rather than fitting a repeatability model. The two measurements were averaged because GenSel cannot fit random effects apart from SNP effects (Fernando & Garrick 2009).

Table 1. Genetic and phenotypic variance and heritability estimates for Methane Yield, Residual Feed Intake and Liveweight in ewe lambs

Trait	Genetic Variance	Phenotypic Variance	Heritability
Methane Yield	0.004	0.009	0.41
Residual Feed Intake	0.65	1.71	0.38
Liveweight (kg)	9.2	20.8	0.44

Genome wide association studies. All traits were found to be highly polygenic and the highest peak explained less than 0.6% of the total genetic variance (Figure 1). Within the 15 windows that explained the most variation in each trait, there was at least one window with a high posterior probability of inclusion in the model (>90%; larger points in Figure 1), suggesting that there may be causal mutations in these genomic regions that have a small impact on these phenotypes.

Selection for environmentally important traits. Although no large QTL were identified for CH4Yield or RFI, it is still possible to make genetic progress in these traits, due to their moderately high heritability estimates. 181 sheep were from the Methane Yield selection lines, which have been successfully bred for high and low Methane Yield using pedigree-based selection (Rowe *et al.* 2019). This flock is a valuable resource for uncovering relationships between Methane Yield and environmentally and economically important traits. Incorporation of additional information into evaluations, such as rumen microbial community profiles; has the potential to further improve our ability to select elite animals for these traits (Kittelmann *et al.* 2014).

CONCLUSIONS

This study confirms that both methane yield and residual feed intake are heritable but polygenic traits in New Zealand sheep; therefore, marker assisted selection for a limited number of markers is unlikely to be as successful as genomic selection for making genetic progress in these traits. Future studies will evaluate genetic and phenotypic relationships between these environmentally important traits and production traits, including carcass quality. Ongoing research is assessing prediction accuracy for these traits and evaluating the impact of including rumen microbial profiles.

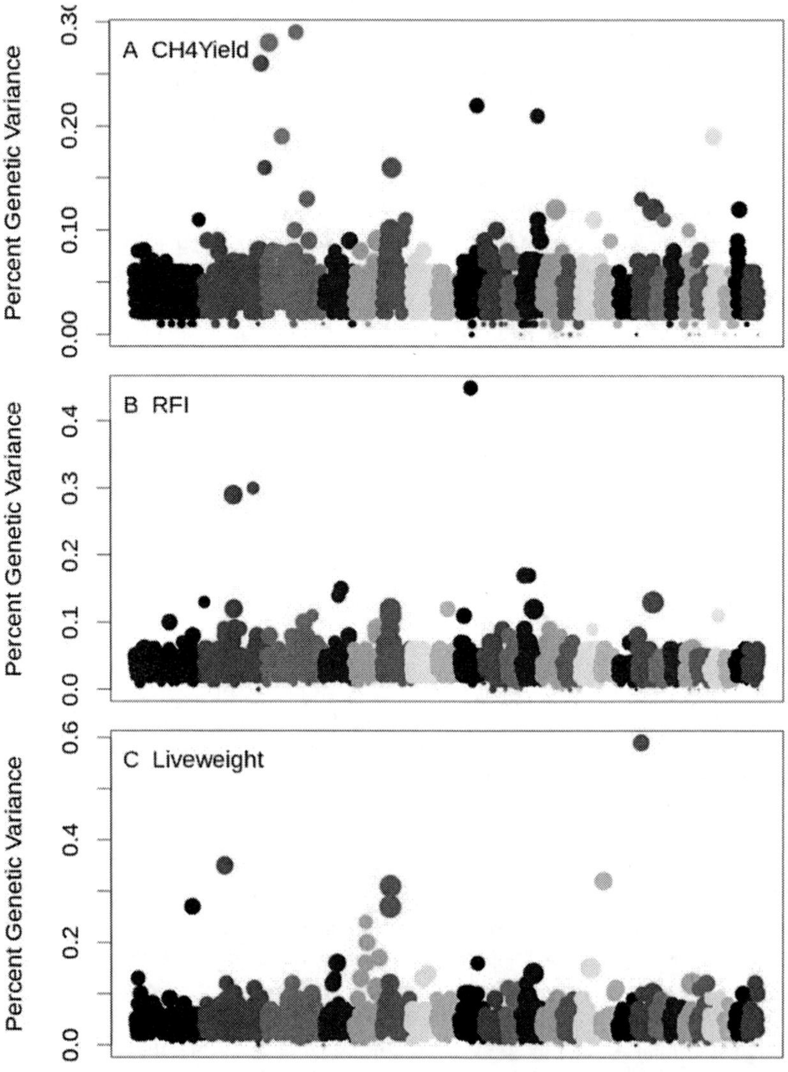

Figure 1. Genome wide association study of Methane Yield (A), Residual Feed Intake (B) and Liveweight (C) in ewe lambs whereby the size of the point is relative to the proportion of iterations in which the 1 Mb window explains some of the overall genetic variance

ACKNOWLEDGEMENTS

This project was made possible by funding and/or animals provided by Beef + Lamb New Zealand Genetics, the Ministry for Business, Innovation and Employment, Beef + Lamb New Zealand, the Pastoral Greenhouse Gas Research Consortium, Sustainable Land Management and Climate Change, the New Zealand Agricultural Greenhouse Gas Research Centre, the Global Partnerships in Livestock Emissions Research, and the many farm and support staff.

REFERENCES

Dodds K., Auvray B., Lee M., Newman S. & McEwan J. (2014) *Proc. 10th Wld Congr. Genet. Appl. Livest. Prod.* Vancouver, Canada.

Elmes S., Bain W., Greer G., Hickey S., Young E., Pickering N., Rowe S., Knowler K., Pinares-Patiño C. and McEwan, J. (2014) *Proc. NZ Soc. Anim. Prod.*. Napier, New Zealand.

Fernando, R. L. & Garrick, D. J. (2009) GenSel - User Manual for a Portfolio of Genomic Selection Related Analyses. *Second edition. Iowa State University. http://bigs.ansci.iastate.edu/bigsgui/help.html. Accessed 27 June 2016.*

Johnson P.L., Miller S.P. and Knowler K. (2016) *Proc. NZ Soc. Anim. Prod.*. Adelaide, Australia.

Johnson P. L., Knowler K., Wing J., Hickey S. and Johnstone, P. (2018) *Proc. 11th Wld Congr. Genet. Appl. Livest. Prod.* Auckland, New Zealand.

Jonker A., Hickey S. M., Rowe S. J., Janssen P. H., Shackell G. H., Elmes S., Bain W. E., Wing J., Greer G. J., Bryson B., MacLean S., Dodds K. G., Pinares-Patiño C. S., Young E. A., Knowler K., Pickering N. K. and McEwan J. C. (2018) *J. Anim. Sci.* **96;** 3031.

Kittelmann S., Pinares-Patiño C. S., Seedorf, H., Kirk M. R., Ganesh S., McEwan J. C. and Janssen P. H. (2014) *PLoS One,* 9, e103171.

Pickering N. K., Dodds K. G., Blair H. T., Hickson R. E., Johnson P. L. and McEwan J. C. (2012) *J. Anim. Sci.* **90:** 1411.

Pinares-Patiño C., Hickey S., Young E., Dodds K., MacLean S., Molano G., Sandoval E., Kjestrup H., Harland R. and Hunt C. (2013) *Animal,* **7:** 316.

Rowe S., McEwan J., Hickey S., Anderson R., Hyndman D., Young E., Baird H., Dodds K., Pinares-Patiño C. and Pickering N. (2014) *Proc. 10th Wld Congr. Genet. Appl. Livest. Prod.*. Vancouver, Canada.

Rowe S.J., Hickey S.M., Jonker A., Hess M.K., Janssen P.H., Johnson T., Bryson B., Knowler K., Pinares-Patiño C.S., Bain W.E., Elmes S., Young E., Wing J., Waller E., Pickering N. and McEwan J.C. (2019) Selection for divergent methane yield in New Zealand sheep - a ten-year perspective. *Assoc. Advmt Anim. Breed. Genet.* **23**,

Sargolzaei M., Chesnais J. P. and Schenkel F. S. (2014). *BMC Genomics,* 15, 478.

SELECTION FOR DIVERGENT METHANE YIELD IN NEW ZEALAND SHEEP – A TEN-YEAR PERSPECTIVE

S.J. Rowe[1], S.M. Hickey[2], A. Jonker[3], M.K. Hess[1], P. Janssen[3], T. Johnson[1], B. Bryson[4], K. Knowler[1], C. Pinares-Patino[3], W. Bain[1], S. Elmes[1], E. Young[1], J. Wing[1], E. Waller[1], N. Pickering[1] and J.C. McEwan[1]

[1]AgResearch, Puddle Alley, Mosgiel, New Zealand
[2]AgResearch, Ruakura, Hamilton, New Zealand
[3]AgResearch, Grasslands, Palmerston North, New Zealand
[4]Woodlands Research Centre, Invercargill, New Zealand

SUMMARY

A flock of 200 breeding ewes (originally selected from extremes of 1,000 genetically diverse animals from national progeny test flocks) have been selected for divergent methane emissions over a ten-year period. Sheep were ranked for breeding using measures from respiration chambers. Over this period, a number of proxies have been investigated and effects of selection on methane emissions, production traits, feed intake, carcass and milk quality have been evaluated. The lines differ on average by 10-12% for methane emissions. Low methane animals appear to be economically favourable, grow more wool, have smaller rumens, are leaner, have different microbiomes and differ in fatty acid profiles in muscle.

INTRODUCTION

New Zealand is heavily reliant on pastoral based agriculture. Grazing livestock, however, are responsible for 80% of methane emissions and around 1/3rd of the total NZ greenhouse gas emissions (Steinfeld 2006). Maternal sheep production is reliant on feeding and maintaining ~18.5 million breeding ewes through the winter months and successfully rearing at least one lamb. Sheep breeders can obtain breeding values for their stock (Newman 2009), expressed as $ gross profit per breeding ewe. The sustainability and therefore profitability of this system, however, is facing a new threat as awareness grows of the magnitude and impact of ruminant methane emissions on the environment. Strategies, such as carbon taxes on livestock production, have been put forward to protect the environment and to maintain global food security. Independent breeding strategies exist for increased production and for reduced methane emissions but, to date there has been no data to show whether these breeding objectives might be synergistic, neutral or antagonistic.

Ten years ago, a divergent flock of sheep was created to evaluate the effects of selection for methane on other breeding objectives. Here we describe the main results and describe the flock divergence for methane and other traits over the ten-year period.

MATERIALS AND METHODS

This report summarizes the creation of the methane yield selection lines and their subsequent development. Sheep were selected from central progeny test flocks (Maclean *et al.* 2006). Initial extremes of methane yield were selected using born 2007, then 2009, 2010 and 2011 from 4 research progeny test flocks (A, B, C, D). One thousand ewes in total were screened with the top and bottom 100 ewe lambs retained based on methane yield. Ram lambs were screened from 96 born 2009 animals from flock A. The lines were closed in 2012 and currently all sires used since 2012 were born in flock 'GHG' (the methane yield selection lines). Methane measures were performed in respiration chambers with 48-hour measures repeated after 10-14 days. These are described fully in Pinares-Patino *et al.* (2013). At least 96 male and female lambs were measured annually to select the next generation.

RESULTS AND DISCUSSION

Figure 1 shows changes in breeding value for methane yield over time. The GHG selection lines have diverged every year and do not overlap. Table 1 shows that the average methane yield measured in multiple flocks is 16 g CH_4 /kg dry matter intake (DMI). Currently the average of the GHG selection lines differ by approximately 12%.

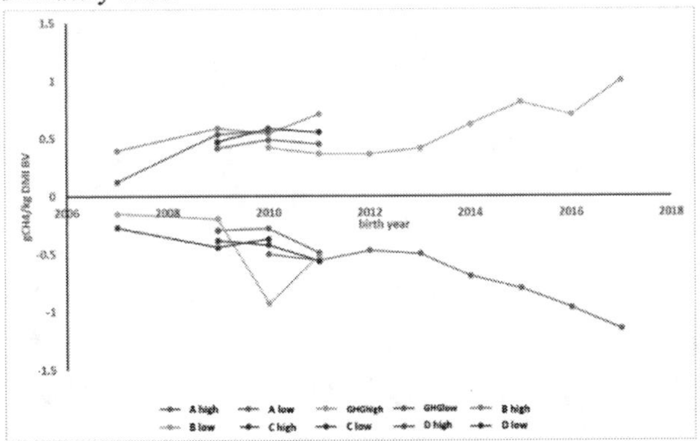

Figure 1. Graph of estimated methane yield research BVs of foundation line ewes by flock (A, B, C, D) and birth year (circles) and GHG selection line progeny

Table 1. Genetic and fixed effect estimates from respiration chamber (RC) measurements in sheep <15 mo

Trait	Mean	s.d.	Total variance	σ_p	direct h^2 (s.e.)	maternal h^2 (s.e.)	Repeatability (s.e.) 2 Day	14 Day
BW, kg	45.9	8.00	23.09	4.80	0.35 (0.05)	0.07 (0.03)		0.89 (0.004)
CH_4, g/d	24	8.28	7.91	2.81	0.23 (0.04	0.05 (0.02)	0.92 (0.003)	0.65 (0.01)
CO_2, g/d	1066	99.5	8926.7	94.5	0.34 (0.05)	0.03 (0.02)	0.94 (0.002)	0.76 (0.01)
CH_4+CO_2, mol/d	25.64	3.7	5.47	2.34	0.33 (0.05)	0.03 (0.02)	0.94 (0.002)	0.76 (0.01)
CH_4/(CH_4+CO_2)	0.059	0.006	0.00002	0.005	0.17 (0.03)	0.03 (0.02)	0.91 (0.003)	0.43 (0.02)
CH_4 yield, g/kg DMI	16	1.42	1.92	1.39	0.13 (0.02)	0.02 (0.02)	0.85 (0.005)	0.38 (0.02)
DMI, kg	1.573	0.255	0.019	0.140	0.39 (0.05)	0.05 (0.03)	0.97 (0.001)	0.83 (0.01)

Table 2 gives trends for breeding values for standard production traits in the lines over time. Although genetic correlations between methane yield and maternal and production traits have been shown to be generally neutral (Rowe *et al.* 2019 in press), in general, predicted breeding values are favourable in the low methane selection line. The general production index in 2018 was $13.20 gross margin greater per ewe without including any financial value associated with reduced methane. These differences were driven by greater fleece weights, increased growth, lean yield and greater parasite resistance. Given the narrow genetic base and limited numbers in the population (each year 5 rams are used in each line of 100 ewes), founder effects cannot be disregarded.

Table 2. Mean breeding values by year for GHG flock low and high selection line progeny surviving to at least 4 months of age

Estimated Breeding Value $	2012 high	2012 Low	2014 high	2014 low	2016 high	2016 low	2018 high	2018 low	2014 -18 Δ (l-h)	2014 -18 ~pval diff.
Weaning weight	1.56	2.21	1.80	2.76	1.84	2.46	2.17	2.72	0.53	0.008
Weaning weight maternal	0.96	1.11	0.78	1.54	0.65	1.62	0.58	1.71	0.91	0.000001
Liveweight 8 months	3.16	3.95	3.44	4.67	3.72	4.70	4.48	5.31	0.60	0.10
Carcass weight	1.09	1.28	1.13	1.62	1.17	1.54	1.43	1.66	0.23	0.04
Adult ewe weight	2.61	1.65	3.34	1.72	3.72	0.90	3.85	1.76	-2.31	0.0001
Lamb fleece weight	0.054	0.057	0.028	0.066	0.036	0.077	0.028	0.082	0.05	0.00002
Fleece weight 12 months	0.36	0.40	0.20	0.47	0.25	0.53	0.20	0.56	0.30	0.00001
Ewe fleece weight	0.30	0.34	0.17	0.41	0.21	0.45	0.16	0.49	0.27	0.000004
Survival	0.023	0.020	0.021	0.030	0.019	0.034	0.019	0.039	0.01	0.003
Survival maternal	0.005	-0.005	-0.001	0.000	-0.008	-0.003	0.003	-0.004	-0.002	0.31
Number lambs born	0.20	0.17	0.20	0.17	0.22	0.23	0.22	0.20	-0.040	0.06
Adult faecal egg count %	-14	-28	-8	-36	-15	-32	-9	-32	-25	0.00003
Summer Faecal egg count %	5	-3	9	-8	4	-1	7	2	-12	0.001
Autumn Faecal egg count %	5	-8	7	-17	0	-11	6	-11	-20	0.00001
Shoulder lean yield	-0.01	0.12	0.02	0.16	0.01	0.18	0.04	0.19	0.14	0.00001
Hindquarter lean yield	-0.04	0.14	0.01	0.18	0.03	0.22	0.05	0.24	0.17	0.00007
Lean leg yield	-0.03	0.05	0.00	0.05	0.01	0.07	0.02	0.08	0.05	0.0005
Fat yield	0.17	-0.19	0.06	-0.16	0.05	-0.16	0.05	-0.18	-0.22	0.000001
Lamb dag score	-0.01	0.14	-0.02	-0.08	-0.09	0.16	-0.14	0.11	0.17	0.02
Adult dag score	-0.04	0.19	-0.19	0.03	-0.27	0.24	-0.23	0.25	0.41	0.0003
*Dual purpose Index	1413	1960	1239	2615	1221	2804	1491	2811	1239	0.0001
**Methane Yield research BV	0.38	-0.45	0.63	-0.78	0.64	-0.75	0.92	-1.09	-1.71	0.000001

*Dual purpose index is a weighted combination of al traits except methane, **Research methane BV selecting for low methane yield (g CH_4/kg DMI).

Further investigations into physiological differences between the lines, however, have shown that animals with lowered methane emissions have fundamental physiological differences from their high emitting counterparts. These include 20% smaller rumens (Goopy *et al.* 2013, Bain *et al.* 2014), different microbial fermentation profiles (Kittelmann *et al.*, 2014) and a higher ratio of propionate to butyrate supplied to the animal as an energy source (Jonker *et al.* 2017, Pinares-Patino *et al.* 2011). There is also preliminary evidence that these changes are also associated with a leaner animal (Elmes *et al.* 2014). Furthermore, preliminary analyses on fatty acid profiles in meat suggest differences in intra-muscular fat, feed intake, feeding behaviour and feed efficiency (T. Johnson, personal communication).

CONCLUSIONS

Methane yield has been shown to be heritable and therefore under host control. Breeding for lowered methane emissions has been successfully shown to be a permanent and cumulative strategy for the mitigation of methane in sheep. This strategy, however, has resulted in physiological changes affecting the rumen, feeding behaviour, outputs from the rumen and resulting body composition. These changes appear to be economically favourable, however given the limited size of the flock involved these results require validation on a much larger scale. Including methane as part of the national breeding objectives would enable the selection of animals that are low emitters whilst efficient for production. This indicates that breeding is a credible strategy for the mitigation of greenhouse gases from livestock. This is particularly pertinent when considering the targets set by the Paris agreement, one of the first of which, is to reduce global greenhouse gases by 30% by 2030.

ACKNOWLEDGEMENTS

This work was funded by the New Zealand Agricultural Greenhouse Gas Centre (NZAGRC) and the Pastoral Greenhouse gas research consortium (PGgRc). Progeny test flocks were funded by Beef and Lamb Genetics New Zealand.

REFERENCES

Bain W.E., Bezuidenhout L., Jopson N.B., Pinares-Patiño C. and McEwan J.C. (2014) *Proc. 10th Wld Congr. Genet. Appl. Livest. Prod.*

Elmes S.N., Bain W.E., Greer G.J., Hickey S.M., Young E.A., Pickering N.K., Knowler K.J., Pinares-Patiño C.S. and McEwan J.C. (2014) *Proc. NZ Soc. Anim. Prod.*

Kittelmann S., Pinares-Patiño C.S., Seedorf H., et al. (2014) *PLoS One* 9, e103171.

McEwan J.C., Morris C.A., Fennessy P.F., Greer G.J., Bain W.E. and Hickey.S.M. (2001) *Anim. Sci.* **73**: 241

McLean N.J., Jopson N.B.,.Campbell A.W., Knowler K., Behrent M., Cruickshank G., Logan C.M.,.Muir P.D.,.Wilson T. and McEwan J.C.(2016) *Proc. NZ Soc.Anim. Prod.***66**: 368.

Newman S.A.N., McEwan J.C. and Young M.J. (2009) *Assoc, Advmt. Anim, Breed. Genet.* **18**: 624.

Pinares-Patiño C.S., Hickey S.M., Young E.A., Dodds K.G., MacLean S., Molano G., Sandoval E., Kjestrup H., Harland R., Hunt C., Pickering N.K., and McEwan J.C. (2013) *Animal.* **2**: 316.

VISUAL MARBLE SCORE AS A PREDICTOR OF INTRAMUSCULAR FAT FOR THE GENETIC IMPROVEMENT OF EATING QUALITY IN LAMB

S.Z.Y. Guy[1,3], P. McGilchrist[2,3] and D.J. Brown[1,3]

[1]Animal Genetics & Breeding Unit*, University of New England, Armidale, NSW, 2351 Australia
[2]School of Environment and Rural Science, University of New England, Armidale, NSW, 2351
[3]Advanced Livestock Measurement Technologies Project, Meat & Livestock Australia, North Sydney, NSW, 2060 Australia

SUMMARY

Marbling refers to the small flecks of fat deposits between muscle fibres and is used as a subjective measure of intramuscular fat (IMF) – a key determinant of eating quality in red meat. In lamb, there is limited literature describing visual scoring guides and the trait correlations with other carcase traits. The objectives of this study were to establish a visual scoring system for marbling in lamb, estimate genetic parameters for the trait and estimate relationships with other eating quality and carcase traits. A 5-point visual marble score guide was constructed, which was highly correlated to the corresponding IMF of each sample ($r = 0.99$). To estimate genetic parameters, 1,120 loin samples were scored for visual marbling, with an average score of 3.01 (± 0.68, SD). On a phenotypic level, a 1 unit score increase was associated with a significant increase in IMF by 0.83 ± 0.04% ($p < 2e^{-16}$). The heritability estimate for visual marble score was 0.28 ± 0.09, and there was a high genetic correlation between visual marble score and IMF ($r_g = 0.93 \pm 0.08$). While more data are required for better genetic parameter estimates, these results indicate that visual marble score is an accurate phenotypic and genetic predictor of IMF in lamb. Therefore, there is potential for the use of visual marble scoring in lamb for the genetic improvement of eating quality in the interim period before a more rapid and accurate technology is commercially available to measure IMF.

INTRODUCTION

Marbling refers to the small flecks of fat deposits between muscle fibres. Visual marbling is used in the beef industry as a subjective measure of intramuscular fat (IMF), and is commonly accepted as a key determinant of eating quality in red meat. Measures of IMF obtained using chemical analysis of loin samples (using soxhlet extraction or near-infra red) are currently used as a selection criteria in Sheep Genetics eating quality indexes. However, this is a time-consuming and costly process.

In beef, marbling is visually scored during chiller assessment on the cut surface made between the 12th and 13th ribs. Burrow *et al.* (2001) summarised within-breed heritability estimates for a visual marbling scoring system in beef cattle, which ranged from 0.26 to 0.93. In addition to this, Reverter *et al.* (2003) report a close to unity genetic correlation between IMF and visual marble score in beef. However, no such studies exist for lamb as a cut surface is not available to grade lamb carcasses during processing, and lamb is historically not known to express the variation in visual marbling as seen in beef. Therefore, no such marble score system currently exists for lamb meat during carcase grading.

There is limited literature available on visual marble scoring in lamb. The trait, scored from 1 to 5, has reported heritability estimates ranging from 0.31 to 0.40 (Johnson *et al.* 2015a, 2015b; Brito *et al.* 2017). However, details of the visual marble scoring system used and its correlations with IMF were not provided in those studies. Therefore, the objectives of this study are to establish a visual scoring system for marbling in lamb, estimate genetic parameters for the trait and estimate relationships with other eating quality and carcase traits.

---

* A joint venture of NSW Department of Primary Industries and the University of New England

MATERIALS AND METHODS

Data. This study examined carcase data from 836 lambs slaughtered from the 2017-drop MLA Resource Flock (RF) and 284 lambs from 4 different commercial ram breeding flocks. Table 1 outlines the carcase measures collected, which included hot carcase weight (HCWT), carcase eye muscle depth (CEMD), fat measured hot at the Girth Rib (GR) site (110 mm from the midline between the 12th & 13th rib; GR) and fat measured cold at the C-site (45mm from the midline between the 12th & 13th rib; Csite). Traits that reflected eating quality analysed in this study were IMF and shear force aged at five days (SF5). All these traits were measured in accordance to the Information Nucleus Flock (INF) operations manual (Sheep CRC 2009).

Table 1. Summary of carcase traits measured on lambs from a commercial ram breeder and the 2017-drop MLA Resource Flock (n = 1,120)

Trait	Abbreviation	Mean	Standard deviation	Range	Coefficient of Variation
Hot carcase weight	HCWT	24.8	3.3	13.6 - 38.6	0.13
Eye muscle depth	CEMD	33.4	4.7	20 - 49	0.14
GR fat	GR	16.7	4.5	4 - 30	0.27
C-site fat	CSite	4.6	2.3	1 - 17	0.50
Intramuscular fat	IMF	4.8	1.1	2.6 - 9.8	0.22
Shear force at day 5	SF5	35.1	9.7	16.4 - 80.9	0.28

Visual Marble Score Guide. A loin sample for each lamb was butterflied and prepared as per protocol for assessment of retail colour. A visual marbling 5-point scale guide was constructed without knowledge of the IMF content. The aim was to produce a scale, where a score of 1 corresponded to no marbling and 5 corresponded to high marbling. Bloomed loin samples were scored for visual marbling by an experienced assessor.

Analysis. The phenotypic association between visual marble score and chemical IMF was firstly assessed by including visual marble score as a linear covariate in a linear regression model for IMF. Model selection was conducted using stepwise linear regression.

Genetic parameters for all carcase traits were then estimated with REML in ASReml (Gilmour *et al.* 2009) using a series of bivariate analyses. Fixed effects included birth type, rearing type, and the covariates of age at measure, age of dam (linear and quadratic) and HCWT. Random effects included additive genetic effect, breed-based genetic group (35 groups) and contemporary group. Contemporary group was defined as a combination of breed, flock, management group, sex, date of measurement and kill group. Genetic correlations between traits were estimated using a series of bivariate analyses.

RESULTS AND DISCUSSION

Visual marble score guide. Figure 1a is the 5-point scale visual marble score guide that was constructed. The 5 samples used for the visual marble score guide were highly correlated with their corresponding chemical IMF ($r = 0.99$) (Figure 1b). Using a simple linear regression model, a one unit increase in the visual marble score guide corresponded to a predicted increase in IMF by 0.94 ± 0.07% ($p = 0.001$, $R^2 = 0.98$, RMSE = 0.23). This indicates that the samples used for the subjective score guide accurately reflected the objective measure of IMF.

a)

a)

b)

Figure 1. a) The 5-point scale visual marble score guide for lamb and b) the relationship between samples used and chemical intramuscular fat (IMF) percentage

The average visual marble score was 3.01, with a minimum of 1 and maximum of 5, and a standard deviation of 0.68. A 1 unit increase in marble score was predicted to correspond to a significant increase in IMF by 0.83 ± 0.04% (p < 2e-16). The removal of visual marble score decreased the variability explained in the model, from 42% to 18%, and an increase in RMSE from 0.81 to 0.96. Therefore, visual marble score is a significant phenotypic predictor of IMF.

Genetic analysis. The heritability estimate of visual marble score was 0.28 ± 0.09 ($\hat{\sigma}^2_a = 0.11 \pm 0.03$, $\hat{\sigma}^2_p = 0.38 \pm 0.02$, $\hat{\sigma}^2_e = 0.28 \pm 0.03$). Taking into account standard errors, this aligns with estimates reported in lamb of 0.31 ± 0.03 by Brito *et al.* 2017, 0.32 ± 0.10 by Johnson *et al.* 2015b and 0.40 ± 0.06 by Johnson *et al.* 2015a. Visual marble score was very highly genetically correlated with IMF ($r_g = 0.93 \pm 0.08$). Therefore, there is potential for genetic gains in visual marbling, and selection for increased marbling is predicted to also increase IMF.

Genetic correlation estimates between visual marble score, IMF and other carcase traits are presented in Table 2. The genetic correlations for visual marble score and IMF were consistent in direction and magnitude for SF5, HCWT and GR. However, estimates did not overlap when taking into account standard errors for HCWT and CSite. More data are required to reduce standard errors and to obtain better genetic parameter estimates.

Table 2. Genetic correlation estimates (± SE) between intramuscular fat (IMF), visual marble score and other carcase traits*

	SF5	HCWT	CEMD	GR	CSite
IMF	-0.45 ± 0.07	0.77 ± 0.03	-0.19 ± 0.08	0.28 ± 0.06	0.26 ± 0.06
Visual marble score	-0.41 ± 0.22	0.96 ± 0.02	0.01 ± 0.24	0.31 ± 0.17	-0.15 ± 0.18

*SF5: shear force at day 5; HCWT: hot carcase weight; CEMD: eye muscle depth; GR: fat at girth rib; Csite: fat at C-site

Taking into account standard errors, genetic correlation estimates for IMF align with those previously reported for SF, CEMD, GR and Csite, but not for HCWT (Mortimer *et al.* 2014, 2018). Genetic correlation estimates for visual marble score reported by Brito *et al.* (2017) also align for

SF5, CEMD and GR, but not for HCWT. This may have due to the variation in HCWT, which was larger in this current study compared to the other studies.

Visual marble score is currently being used as a proxy for IMF in New Zealand sheep genetic evaluation (Johnson *et al.* 2018). However, while higher marbling in pasture-fed lambs was reported to be associated with higher IMF, marbling score did not affect eating quality in New Zealand lambs (Young *et al.* 2009), possibly due to a small range in IMF. To our knowledge, there is currently no literature available on investigations of selection for eating quality through marble score in Australian lambs.

CONCLUSIONS

While the subjective scoring of lamb loins may not be viable for grading of lamb carcases in a commercial environment, this study indicates that visual marble score is an accurate phenotypic and genetic predictor of IMF in lamb. Therefore, there is potential for the use of visual marble scoring in lamb for the genetic improvement of eating quality in the interim period before a more rapid and accurate technology is commercially available to measure IMF.

ACKNOWLEDGEMENTS

This work was conducted as part of the Advanced measurement technologies for globally competitive Australian meat (ALMTech) project, funded by the Australian Government Department of Agriculture and Water Resources as part of its Rural R&D for Profit programme, in partnership with Research and Development Corporations, Commercial Companies, State Departments and Universities. The authors would also like to thank the Sheep CRC for Sheep Industry Innovation, Meat and Livestock Australia, the Australian Meat Processor Corporation, and the contributions of research staff and industry members involved in the data collection at UNE.

REFERENCES

Brito L.F., McEwan J.C., Miller S., Bain W., Lee M., Dodds K., Newman S.-A., Pickering N., Schenkel F.S. and Clarke S. (2017). *Small Rumin. Res.* **154**:81.

Burrow H.M., Moore S.S., Johnston D.J., Barendse W. and Bindon B.M. (2001) *Aust. J. Exp. Agr.* **41**:893.

Gilmour A.R., Gogel B.J., Cullis B.R. and Thompson R. (2009) 'ASReml User Guide Release 3.0' VSN International Ltd, Hemel Hempstead, HP1 1ES, UK.

Johnson P.L., Charteris A., Tate M.L. and Dodds K.G. (2015a). *Proc. NZSAP.* **75**:111.

Johnson P.L., Young E.A., Russenklau D., Anderson T. and Dodds K.G. (2015b) *Proc. NZSAP.* **75**:106.

Johnson P.L., Agnew M.P., Shorten P.R., Reis M.M., Cullen N. and Craigie C.R. (2018). *Proc. WCGALP.* **11**: 611.

Mortimer S.I., van der Werf J.H.J., Jacob R.H., Hopkins D.L., Pannier L., Pearce K.L., Gardner G.E., Warner R.D., Geesink G.H., Hocking Edwards J.E., Ponnampalam E.N., Ball A.J., Gilmour A.R. and Pethick D.W. (2014) *Meat Sci.* **96** :1016.

Mortimer S.I., Fogarty N.M., van der Werf J.H.J., Brown D.J., Swan A.A., Jacob R.H., Geesink G.H., Hopkins D.L., Hocking Edwards J.E., Ponnampalam E.N.,Warner R.D., Pearce K.L. and Pethick D.W. (2018) *J. Anim. Sci.* **96** :3582.

Reverter A., Johnston D.J., Perry D., Goddard M.E. and Burrow H.M. (2003) *Aust. J. Agric. Res.* **54**: 119.

Sheep CRC (2009). Information Nucleus Operation Manual.

Young E.A., Bain W.E., McClean N.J., Campbell A.W. and Johnson P.L. (2009). *Proc. NZSAP.* **69**:104.

THE GENETIC RELATIONSHIPS BETWEEN INTRAMUSCULAR FAT MEASURED IN FOUR DIFFERENT LAMB MUSCLES

S.Z.Y. Guy[1,4], S.F. Walkom[1,4], F. Anderson[2,4], G.E. Gardner[2,4], P. McGilchrist[3,4] and D.J. Brown[1,4]

[1]Animal Genetics & Breeding Unit*, University of New England, Armidale, NSW, 2351 Australia
[2]School of Veterinary and Life Sciences, Murdoch University, Murdoch, WA, 6150 Australia
[3]School of Environmental and Rural Science, University of New England, Armidale, NSW, 2351 Australia
[4]Advanced Livestock Measurement Technologies Project, Meat and Livestock Australia, North Sydney, NSW, 2060 Australia

SUMMARY

Intramuscular fat percentage (IMF) is a key determinant of eating quality in red meat. Measures of IMF from the short loin muscle (*M. longissimus lumborum*, LL) are currently used as selection criteria in Sheep Genetics eating quality indexes. To understand how routine selection on the short loin impacts IMF across the whole carcase, this pilot study examines IMF data collected from three additional muscles from the fore quarter (*Muscularis supraspinatus*, SS) and hind quarter (*Musculus semimembranosus*, SM; *Muscularis semitendinosus*, ST) of the carcase. The heritability of IMF was relatively high and consistent across the SS, LL and ST muscles, and lower in the SM. The genetic correlation estimate between IMF measured in the different muscles were all positive, ranging from 0.49 ± 0.13 to 0.97 ± 0.10. Therefore, IMF measurements from the short loin, which is currently being used as selection criteria for eating quality, will be a useful indicator for IMF across muscles from other parts of the carcase. Further, the genetic selection to increase IMF in one muscle should result in an increase in IMF in the other muscles, although at differing rates.

INTRODUCTION

Intramuscular fat (IMF) is a key determinant of eating quality in red meat as it has been found to have a positive influence on flavour, juiciness and tenderness (Hopkins *et al.* 2006; Pannier *et al.* 2014). Currently used as a selection criteria in Sheep Genetics eating quality indexes (Swan *et al.* 2015), IMF is extracted from the short loin using near-infra red technology. Most research has focused around IMF measured in the loin muscle, with very little research on other muscles.

Pre-adjusted IMF phenotypes measured on different muscles have been found to have moderately positive phenotypic correlations, ranging from 0.24 to 0.68 (Anderson *et al.* 2015). However, there are no reports on genetic relationship between IMF across different muscles. The objective of this pilot study was to estimate genetic correlations between IMF in different muscles from the fore-quarter, saddle (or loin) and hind-quarter sections of the lamb carcase.

MATERIALS AND METHODS

Data. Data was collected on 400 lambs slaughtered from the 2011-drop of the Information Nucleus Flock from the Katanning site (Fogarty *et al.* 2007) and 1,111 lambs slaughtered from the 2017-drop from the MLA Resource Flock (900 from the Kirby site and 211 from the Katanning site). Lambs were slaughtered at an average age of 280 ± 44 (± SD) days and an average hot carcase weight of 23.7 ± 3.3 kg. There were no common sires between the 2017 and 2011 drop lambs, with 64 sires in common across the two sites in the 2017 drop. In addition to the standard carcase traits measured

* A joint venture of NSW Department of Primary Industries and the University of New England

(including IMF measured on the loin), three additional muscles were measured for IMF. In total, there were four muscles that were measured for IMF from the fore, saddle and hind section of the carcass:

- Fore-quarter: *Muscularis supraspinatus* (SS)
- Saddle: *M. longissimus lumborum* (LL)
- Hind-quarter: *Musculus semimembranosus* (SM) and *Muscularis semitendinosus* (ST).

All carcase traits, including IMF, were measured according to the Information Nucleus Flock operations manual (Sheep CRC 2009). The collection of IMF from additional muscles was as described in Anderson *et al.* 2015. Due to carcass imperfections and muscle trimming, IMF measures could not always be obtained for all muscles and or carcasses. Table 1 provides a summary of the number of samples, means and standard deviation for each muscle within each flock. Across the four muscles, IMF ranged from 2.16% to 11.79%, with an average of 4.71%.

Table 1. Summary of intramuscular fat (%) records available in muscles* from the fore-quarter, saddle and hind-quarter, sampled from lambs slaughtered from the Katanning 2011-drop, Katanning 2017-drop and the Kirby 2017-drop (*n* = 1,383)

			Fore-quarter	Saddle (loin)	Hind-quarter	
Site	Drop		SS	LL	SM	ST
Katanning	2011	Count	337	344	341	338
		Mean (*SD*)	5.04 (1.10)	4.36 (0.84)	3.69 (0.78)	4.87 (1.18)
Katanning	2017	Count	134	199	194	199
		Mean (*SD*)	6.29 (1.40)	5.00 (0.98)	4.07 (1.07)	5.63 (1.20)
Kirby	2017	Count	187	837	761	830
		Mean (*SD*)	5.44 (0.86)	4.95 (1.10)	3.92 (0.58)	5.25 (1.07)
Overall		Count	658	1380	1296	1367
		Mean (*SD*)	5.40 (1.21)	4.56 (1.05)	3.89 (0.74)	5.20 (1.14)

*SS: *Muscularis supraspinatus*; LL: *M. longissimus lumborum*; SM: *Musculus semimembranosus*; ST: *Muscularis semitendinosus*

Analysis. The IMF traits were analysed using a multivariate sire model in ASReml (Gilmour *et al.* 2009). An animal model was explored but due to the small number of records, a sire model was preferred. Fixed effects included birth type, rearing type, age, age of dam, age of dam squared, sire breed, dam breed and hot carcase weight. Contemporary group was defined by breed, flock, management group, sex, date of measurement and kill group. Maternal effects and genetic groups were not tested as there was insufficient data. Therefore, the genetic components estimated from this genetically diverse resource population are expected to be larger than estimates reported in literature (Walkom and Brown 2016). This was further exacerbated by the subsampling of the Australian sheep population in this study.

RESULTS AND DISCUSSION

The variance component and heritability for the IMF traits are presented in Table 2. These heritabilities are higher than presented in literature due to the inability to completely take into account breed and maternal effects in this subset of a genetically diverse reference population (Walkom and Brown 2016). Nevertheless, the heritability estimate for IMF measured in the LL (0.60 ± 0.10) reflects the estimate of 0.48 ± 0.05 reported by Mortimer *et al.* (2014), which included genetic groups in their analysis of LL samples from the same sheep resource population. The genetic and phenotypic variation was lowest in the SM muscle. Meanwhile, the SS and ST muscles exhibited the greatest genetic and phenotypic variation, which exhibited more than double the variation observed in the SM muscle.

Table 2. Genetic parameter estimates ± SE for intramuscular fat traits measured in four muscles* from the fore-quarter, saddle and hind-quarter (*n* = 1,383)

	Fore-quarter	Saddle		Hind-quarter
Genetic parameter estimate	SS	LL	SM	ST
Phenotypic variance	1.00 ± 0.07	0.85 ± 0.04	0.43 ± 0.02	1.10 ± 0.05
Residual variance	0.76 ± 0.05	0.73 ± 0.03	0.41 ± 0.02	0.89 ± 0.04
Sire variance	0.24 ± 0.06	0.13± 0.03	0.03 ± 0.01	0.18 ± 0.03
Heritability	0.96 ± 0.19	0.60 ± 0.10	0.25 ± 0.08	0.68 ± 0.11

*SS: *Muscularis supraspinatus*; LL: *M. longissimus lumborum*; SM: *Musculus semimembranosus*; ST: *Muscularis semitendinosus*

The phenotypic correlations between IMF traits from the multivariable analysis were all positive (Table 3). The genetic correlations were also positive and stronger, ranging from 0.49 ± 0.13 to 0.97 ± 0.10. Therefore, IMF measurements from the LL muscle (short loin) will be a useful indicator for IMF across muscles from other parts of the carcase. Further, the genetic selection to increase IMF in one muscle should result in an increase in IMF in the other muscles, although at differing rates.

Table 3. Genetic correlations (below diagonal) and phenotypic correlations (above diagonal) for IMF measured in four muscles* from the fore-quarter, saddle and hind-quarter (*n* = 1,383)

	Fore-quarter	Saddle	Hind-quarter	
	SS	LL	SM	ST
SS		0.30 ± 0.03	0.30 ± 0.03	0.37 ± 0.04
LL	0.68 ± 0.11		0.44 ± 0.02	0.53 ± 0.02
SM	0.76 ± 0.17	0.97 ± 0.10		0.34 ± 0.03
ST	0.49 ± 0.13	0.70 ± 0.08	0.71 ± 0.13	

*SS: *Muscularis supraspinatus*; LL: *M. longissimus lumborum*; SM: *Musculus semimembranosus*; ST: *Muscularis semitendinosus*

Although more data will improve the accuracy of these estimates by reducing standard errors, this pilot study demonstrates that there are no detrimental consequences on the eating quality of the entire carcase when selecting on only measurements taken from the loin. These results also suggest that IMF should be recorded in the SS muscle, as this was the muscle that exhibits the most genetic variability. However, the muscle from which IMF samples are taken routinely should also consider the ease of sampling and the financial value of muscle.

Variation in the functional requirements of muscles leads to differences in muscle fibre type, the proportion of oxidative fibres and in turn the levels of triglycerides and IMF in the muscle (Hocquette *et al.* 2010). Muscles associated with posture tend to have more oxidative fibers and higher IMF (Picard *et al.* 2002; Anderson *et al.* 2015), which is reflected in the lower mean and heritability observed for the SM.

CONCLUSIONS

Intramuscular fat percentage (IMF) is a key determinant of eating quality in red meat. The analysis of IMF measures from four different muscles from 1,383 lambs suggests that the heritability of IMF was relatively high and consistent across the SS, LL and ST muscles, and lower in the SM. There were moderate to high genetic correlations between IMF across the four muscles. Therefore, IMF measurements from the short loin (LL), which is currently being used as selection criteria for eating quality, will be a useful indicator for IMF across muscles from other parts of the carcase.

ACKNOWLEDGEMENTS

This work was conducted as part of the Advanced measurement technologies for globally competitive Australian meat (ALMTech) project, funded by the Australian Government Department of Agriculture and Water Resources as part of its Rural R&D for Profit programme. The authors would also like to thank the Sheep CRC for Sheep Industry Innovation, Meat and Livestock Australia and the Australian Meat Processor Corporation, and the contributions of the research staff and resources provided at the Information Nucleus Flock/ Resource Flock research sites, employed by Murdoch University and UNE.

REFERENCES

Anderson F., Williams A., Pethick D.W. and Gardner G.E. (2015). *Animal* **9**:1239.

Fogarty N.M., Banks R.G., van de Werf J.H.J., Ball A.J. and Gibson J.P. (2007). *Proc. Assoc. Advmt. Anim. Breed. Genet.* **17**: 29.

Gilmour A.R., Gogel B.J., Cullis B.R. and Thompson R. (2009) 'ASReml User Guide Release 3.0' VSN International Ltd, Hemel Hempstead, HP1 1ES, UK.

Hopkins D., Hegarty R., Walker P. and Pethick D.W. (2006) *Aus. J. Exp. Agri.* **46**:879.

Hocquette J., Gondret F., Baéza E., Médale F., Jurie C. and Pethick D.W. (2010) *Animal* **4**:303.

Pannier L., Pethick D.W., Geesink G. H. and Ball A.J. (2014). *Meat Sci.* **96**: 1068.

Picard B., Lefaucheur L., Berri C. and Duclos M.J. (2002) *Repro. Nutri. Develop.* **42**:415.

Mortimer S.I., van der Werf J.H.J., Jacob R. H., Hopkins D.L., Pannier L., Pearce K.L., Gardner G.E., Warner R.D., Geesink G.H., Hocking Edwards J.E., Ponnampalam E.N., Ball A.J., Gilmour A.R. and Pethick D.W. (2014). *Meat Sci.* **96**: 1016.

Sheep CRC (2009). Information Nucleus Operation Manual.

Swan A.A., Pleasants T. and Pethick D.W. (2015). *Proc. Assoc. Advmt. Anim. Breed. Genet.* **21**: 29.

Walkom S.F. and Brown D.J. (2016). *Ani. Prod. Sci.* **57**: 20.

GENETIC PARAMETERS FOR PRIMAL CUT WEIGHTS IN PIGS

N. R. Sarker, B.J. Walmsley and S. Hermesch

Animal Genetics & Breeding Unit*, University of New England, Armidale, NSW, 2351 Australia

SUMMARY

A study was conducted to estimate genetic parameters for phenotypes describing primal cuts recorded on 2,077 pigs with pedigrees comprising 5,011 animals over three generations. The four main primal cuts considered were: shoulder, leg, loin and belly. These were analysed as primal weights, or as a percentage of cold carcase weight. Heritabilities ranged from 0.10 (± 0.04) to 0.24 (± 0.06) for primal cut weights and from 0.12 (± 0.05) to 0.24 (± 0.06) for primal cut percentages. The loin primal was the least heritable. The genetic correlations between primal weights and primal percentages for the same primals ranged from 0.97 to 1.00. The genetic correlations among primal cut weights ranged from -0.45 to 0.07 which were similar to those found among the primal cut percentages (-0.63 to 0.06). The genetic correlations between the shoulder and leg primal with the belly primal were negative. The strongest negative genetic correlation was found between the leg and belly primals (-0.45 for weight trait and -0.63 for percentage trait). The leg weight was genetically uncorrelated with loin weight, suggesting that selection for high leg weights would not result in high loin weights. The phenotypic correlation between loin and belly was negative but the genetic correlation was not significant. Genetic correlations between the loin and other primal cut weights were weaker in comparison to the genetic correlations between the belly and other primal cuts. The genetic correlations amongst primal cut weight traits were similar to those found among the primal cut percentages. Incorporating these genetic parameters into a pig breeding program could help to increase the total economic return from pig carcases but would need to be done in association with other traits that impact pig production.

INTRODUCTION

Pork producers and retailers could make better marketing decisions if quantitative information on primal cut yield per carcase were available. A primal cut is a piece of meat initially separated from the carcase of an animal during butchering. Until now, the weight of each primal cut has not been considered in the price Australian farmers receive for pigs. Producers or wholesalers are currently paid on the basis of hot standard carcase weight (HSCW) and back fat. Total carcase weight and leanness do not provide complete information about carcase market value. The economic return per carcase could be determined by the market value of each of its primal cuts and in turn increase returns to farmers (Hermesch 2008). The Australian pig industry lacks classification systems to measure variation in primal cut weights in commercial abattoirs that is required to quantify the economic benefits of higher saleable meat yield for a given carcase weight. Therefore, determining the weight of primal cuts is an important area of interest worth exploring (Lisiak et al. 2015). This study aimed to estimate the genetic parameters of primal cut weights and primal cut percentages in pigs and to determine the relationships between these traits.

MATERIALS AND METHODS

Animals. Data on primal cut traits were combined with pedigree and performance records from 2,077 pig carcases recorded in 2012. The pedigree was extended back three generations for parameter estimation which comprised 5,011 animals including 523 sires and 2411 dams. The subset of pigs

* A joint venture of NSW Department of Primary Industries and the University of New England

with primal cut records was represented by 98 sires and 1135 dams. Pigs were recorded on one farm and represented three different breeds. There were 25 contemporary groups defined by the week of slaughter, which were fully nested within five different grow-out facilities.

Traits. The four main primal cuts weighed (kg), and expressed also as percentages (relative to cold carcase weight, %) were shoulder (S_WT and S_P), leg (L_WT and L_P), loin (LN_WT and LN_P) and belly (B_WT and B_P). All primal cut weights were recorded on one side of the carcase only. Each primal cut weight was multiplied by two to express them on a per carcase basis.

Statistical analysis. Descriptive statistics for carcase measurement and Pearson correlations were calculated using the statistical package SAS. The (SAS) procedure GLM was used to test for the significance of fixed effects for each trait. All genetic parameter estimates were obtained under an animal model using ASReml (Gilmour *et al.* 2015). Two mixed linear animal models were used: for primal cut weight: $Y_{ijkmn} = \mu + b_i + c_j + cw_k + an_m + pe_n + e_{ijkmn}$ and for primal cut percentage: $Y_{ijmn} = \mu + b_i + c_j + an_m + pe_n + e_{ijmn}$ where, Y = observation for of a trait, μ= overall mean, b_i= fixed effect of the i_{th} breed, c_j = fixed effect of the j_{th} contemporary group, cw_k = linear covariate (cold carcase weight), an_m = random effect of the m_{th} animal, pe_n = common litter effect of n_{th} litter and e_{ijkmn} =error. The common litter effect was only retained in the models for the loin and belly primals, because it was not significant for the shoulder and leg primals.

RESULT AND DISCUSSION

Descriptive statistics. Animals were on average slaughtered at 168 days of age with an average hot carcass weight of 79.0 kg. Considerable variation was observed in all four primal cuts (Table 1). The coefficients of variation (CV) were higher for the loin (18%) and belly (17%) cuts which may be due inconsistencies during the butchering process.

Table 1. Descriptive statistics for the weights and percentages of four primal cuts

Trait		N	Mean	SD	CV	Min	Max
Shoulder weight (kg)	S_WT	2064	22.7	2.56	11.3	12.0	32.2
Leg weight (kg)	L_WT	2068	23.2	2.40	10.3	14.6	31.5
Loin weight (kg)	LN_WT	2073	12.9	2.30	17.8	6.8	30.0
Belly weight (kg)	B_WT	2038	10.2	1.75	17.1	5.2	16.2
Shoulder percentage (%)	S_P	2061	32.8	1.71	5.2	21.9	41.8
Leg percentage (%)	L_P	2065	33.6	1.53	4.6	22.1	41.6
Loin percentage (%)	LN_P	2070	18.7	2.45	13.2	9.4	35.7
Belly percentage (%)	B_P	2035	14.7	1.67	11.3	8.7	22.6

N: number of pigs, SD: standard deviation, CV: coefficient of variation, Min and Max: minimum and maximum

Model fit. The total variation accounted for by the GLM model (R^2; Table 2) was high for all primal cut weights due to the adjustment for cold carcase weight. The R^2 values were 0.86 and 0.87 for S_WT and L_WT and slightly lower at 0.62 and 0.77 for LN_WT and B_WT, respectively. As expected, when compared with the R^2 values for the primal weights those for the primal percentages were lower without carcass weight adjustment (0.10 to 0.28) and when adjusted for cold carcase weight (0.32 to 0.48) because part of the variance in primal cut percentages is masked by the variance in cold carcase weight which is part of the percentage trait.

Heritabilities. The heritabilities for primal cuts were low to moderate, ranging from 0.10 to 0.24 for cut weights and 0.12 to 0.24 for percentages with or without adjustment for cold carcase weight (Table 2). The loin and belly primal traits were less heritable than the other primals. Adjusting primal cut percentage traits for cold carcase weight did not affect heritability estimates significantly and the phenotypic variances were only slightly reduced. The reduction in additive genetic variation resulting from this adjustment supports the simpler model for primal cut percentages. Higher heritabilities for primal cut weight traits compared to the current study have been observed previously (Newcom *et al.* 2002; Van Wijk *et al.* 2005). However, primal cuts definition differed between studies and it is difficult to directly compare heritability estimates. Nonetheless, Newcom *et al.* (2002) estimated heritabilities of 0.60, 0.61, 0.24 and 0.23 for ham, loin, belly and shoulder weights of Yorkshire and Duroc breeds, whereas Van Wijk *et al.* (2005) reported heritabilities of 0.40 and 0.29 for leg and loin weights. Mérour *et al.* (2009) reported higher heritabilities for loin and leg (0.43 and 0.46) than belly and shoulder (0.35 and 0.23).

Table 2. Estimates of heritability (h^2) and common litter effect (c^2) with standard errors (se) along with variance components for primal cut traits[a]

Traits[a]	R^2	h^2	c^2	σ_{pe}^2	σ_a^2	σ_p^2
S_WT	0.86	0.24 (0.06)	-	-	0.44	1.82
L_WT	0.87	0.23 (0.06)	-	-	0.35	1.53
LN_WT	0.62	0.10 (0.04)	0.09 (0.03)	0.28	0.30	3.07
B_WT	0.77	0.09 (0.04)	0.16 (0.04)	0.18	0.10	1.16
S_P	0.12	0.24 (0.06)		-	0.64	2.67
S_P (adjusted for CW)	0.34	0.23 (0.06)		-	0.60	2.60
L_P	0.10	0.23 (0.06)		-	0.49	2.16
L_P (adjusted for CW)	0.32	0.24 (0.05)		-	0.48	2.01
LN_P	0.15	0.12 (0.05)	0.07 (0.03)	0.38	0.61	5.23
LN_P (adjusted for CW)	0.36	0.12 (0.05)	0.07 (0.03)	0.36	0.62	5.19
B_P	0.28	0.16 (0.06)	0.15 (0.04)	0.31	0.32	2.05
B_P (adjusted for CW)	0.48	0.14 (0.05)	0.13 (0.04)	0.25	0.27	1.91

[a] For trait abbreviations see Table 1; CW: cold carcase weight; R^2: coefficient of determination; σ_{pe}^2: variance due to common litter effect; σ_a^2: additive genetic variance and σ_p^2: phenotypic variance

Correlations. The genetic correlations between primal weights and primal percentages for the same primal ranged from 0.97 to 1.00 (Table 3). These very high correlations indicate that the weight and percentage traits for the same primal are essentially the same trait. The genetic correlations among primal cut weights ranged from -0.45 to 0.07 which were similar to those found among the primal cut percentages (-0.63 to 0.06). The genetic correlations the shoulder and leg primals have with the belly primal were negative. The strongest negative genetic correlation was found between the leg and belly primals (-0.45 for weight trait and -0.63 for percentage trait), suggesting that selection for longer pigs would be associated with lower leg weights. The leg weight was genetically uncorrelated with loin weight, which means selection for high leg weights would not result in high loin weights. This result was also found by Mérour *et al.* (2009). In comparison, Van Wijk *et al.* (2005) reported positive genetic correlations (0.22 to 0.58) between leg and loin weights. The phenotypic correlation between loin and belly primals was negative but the genetic correlation was not significant. In comparison, a number of studies (Newcom *et al.* 2002; Mérour *et al.* 2009) reported moderate to high negative genetic correlations (-0.54 to -0.57) between loin and belly primals. Genetic correlations between the loin and other primal cut weights were weaker in comparison to the genetic correlations the belly primal has with the other primal cuts. The genetic correlations amongst the primal cut weight traits were similar to those found among the primal cut percentage traits.

Table 3. Estimates of phenotypic (above diagonal) and genetic correlations (below diagonal) with standard error (se) among primal cut weight and percentage traits

Traits	S_WT	L_WT	LN_WT	B_WT	S_P	L_P	LN_P	B_P
S_WT		38(2)	8(2)	21(2)	90(0)	4(3)	-18(3)	-9(3)
L_WT	22(17)		16(2)	18(2)	5(3)	88(1)	-5(3)	-16(3)
LN_WT	1(27)	-4(28)		-14(3)	-14(2)	-3(2)	99(0)	-47(2)
B_WT	-10(29)	-45(31)	7(34)		-7(2)	-14(2)	-54(2)	99(0)
S_P	97(2)	-8(19)	5(25)	-11(27)		7(2)	-16(2)	-11(2)
L-P	-18(18)	97(2)	-7(26)	-54(23)	-19(18)		-5(2)	-20(2)
LN_P	-17(24)	-21(25)	100(0)	-4(34)	-3(24)	-10(25)		-42(2)
B_P	-27(22)	-63(19)	-0(31)	100(0)	-33(20)	-63(18)	6(29)	

All correlations and se were multiplied by 100

CONCLUSION

Overall, these results suggest that the size of the most valuable primals (loin and belly) could be improved through breeding, which may also reduce the size of the least valuable primals (shoulder and leg). This could in-turn increase the total economic value of pig carcases. The primal weight traits were genetically highly correlated with the percentage traits, so the primal cut weight and percentage traits are basically same trait. The actual primal weight is needed to calculate primal percentages so breeders need only use primal weights adjusted for cold carcass weight. In breeding programs, these primal traits need to be used with respect to other production and welfare traits in order to change overall profit per carcase without causing detrimental changes in the pig.

ACKNOWLEDGEMENTS

Data recording was funded by Australian Pork Limited (Project 1025). The first author was supported by an Australian Government RTP International Scholarship.

REFERENCES

Gilmour A. R., Gogel B.J., Cullis B.R., Welham S. J., Thompson R. (2015) 'ASReml User Guide Release 4.1 Functional Specification' VSN International Ltd., Hemel Hempstead, HP1 1ES, UK.

Hermesch S. (2008) *Anim.* **2:** 1178.

Lisiak D., Duziński K., Janiszewski P., Borzuta K. and Knecht D. (2015) *Anim. Prod. Sci.* **55:** 1044.

Mérour I., Hermesch S., Jones R.M., and Tribout T. (2009) *Proc. Assoc. Advmt. Anim. Breed. Genet.* **18:** 195.

Newcom D.W., Baas T.J., Mabry J.W. and Goodwin R.N. (2002) *Proc. World Congr. Genet. App. Livest. Breed.* **2002:** 19.

Wijk V.H.J., Arts D.J.G., Matthews J.O., Webster M., Ducro B.J., and Knol E.F. (2005) *J. Anim. Sci.* **83:** 324.

INVESTIGATING RELATIONSHIP BETWEEN TRAITS ASSOCIATED WITH EATING QUALITY AND MARKET END POINT

S.J. Lee, M.L. Hebart and W.S. Pitchford

Davies Research Centre, School of Animal and Veterinary Sciences, University of Adelaide, Roseworthy, SA, 5371 Australia

SUMMARY

BREEDPLAN reports estimated breeding values (EBVs) for many traits, but with the exception of carcass weight and rib fat, there are no EBVs specifically for the inputs into the Meat Standards Australia (MSA) Index that producers can use to make genetic progress in eating quality. Further it is not known how selection using BREEDPLAN EBVs influences the MSA Index and if these relationships are the same for different market endpoints. The motivation behind this study was to examine the extent to which MSA Index of commercial animals is related to EBVs of sires.

INTRODUCTION

With the development of Meat Standards Australia (MSA, Polkinghorne *et al.* 2008) and MSA Index, the interest by the industry to improve eating quality through genetic selection has been heightened. In investigating genetics underlying eating quality and carcass traits, Reverter *et al.* (2003), noted that it was important to determine whether there were significant genotype by environment (GxE) interactions for finishing systems (pasture- vs. feedlot-finished). Reverter *et al.* (2003) and Johnston *et al.* (2003) reported on genetic parameters for temperate cattle breeds for feedlot- vs. pasture-finished for a range of growth, body composition, carcass and meat quality traits. They reported generally increasing additive genetic variance with increasing carcass weight end points but minimal GxE and subsequent re-ranking of sires. In a study of 1.7M carcass records from 37,637 lots (slaughter groups) from the MSA database for cattle from nine processing plants in southern Australia from 2010-2013 Hebart *et al.* (2016) investigated the relationship between carcass end point defined either by weight or marbling and phenotypic variance. Hebart *et al.* (2016) found that increased lot mean carcass weight was associated with increasing phenotypic variance in carcass weight. Furthermore, higher lot mean MSA Marbling and carcass weight was associated with increased phenotypic variance in MSA Marbling. How eating quality traits respond to selection is a function of the selection intensity, heritability and the phenotypic variance of the traits. Currently producers can select for increased intramuscular fat (IMF) using BREEDPLAN estimated breeding values (EBVs) to improve marbling and in turn increase MSA Index. Differences in the phenotypic variance have the potential to change the magnitude of the regression coefficient for MSA Marbling on BREEDPLAN IMF EBV and, therefore, of the relationship between IMF EBV and MSA Index. The regression coefficient is calculated as:

$$b_{MSA,EBV} = r_G \text{x} \frac{\sqrt{h^2} \text{ x } \sigma_P}{\sigma_{EBV}}$$

which is a function of the genetic correlation (r_G) between the traits (could be same trait at different endpoints), the heritability (h^2) of the trait, variation in the carcass trait (σ_P) and the variation in EBV (σ_{EBV}). Since the genetic correlation between traits, the heritability and the variation in EBV are likely to remain constant any scale effect observed in the variation of carcass traits is likely to have the greatest effect on the regression coefficient estimate. The motivation behind this study is to examine the extent to which MSA Index is related to estimated breeding values (EBVs) of sires for different market end points.

MATERIALS AND METHODS

Data from 12 industry or research data sets totalling 6,997 animals from four breeds (Angus, Charolais, Hereford and Limousin) and 433 sires have been included for analysis. The datasets are Maternal Productivity (MP – Vasse) (Pitchford *et al.* 2017), 4 x Regional Combinations (RC – NSW, RC – SA, RC – WA, RC – Vic.) animals (McKiernan *et al.* 2005), Rockdale (Herd *et al.* 2017) and Trangie (Arthur *et al.* 2005). In addition, three Beef Information Nucleus datasets (BIN, Angus BIN, Charolais BIN, and Hereford BIN) and data from two Team Te Mania herds (Central West NSW with calves born in early spring, and western Victorian autumn calving) were included. The datasets contain a range of growth paths (slow vs. fast), finishing regimes (Short feedlot <200 days, Long feedlot >200 days and Pasture) and carcass end point (200-500kg carcass weight) included in the analysis.

The carcass traits measured included hot standard carcass weight (HSCW, kg), rib fat (Rib, mm), intramuscular fat (IMF %, measured in the laboratory), MSA marbling (Marb), and MSA Index. Carcass traits (IMF, MSA marbling, MSA Index, Rib, and HSCW) were regressed on BREEDPLAN sire EBVs (IMF EBV, Rib EBV, 600 day weight EBV) after taking into account contemporary groups (a concatenation of dataset, management group and kill date), appropriate genetic "line" effects (high IMF, high yield, high RFI etc.) and management (Pasture, Short-fed, Long-fed) for each dataset. Sire BREEDPLAN EBVs were standardised by subtracting the mean sire EBV of a breed and dataset group within each breed within each dataset to allow for between breed comparisons and to account for EBVs being estimated at different times for each dataset. A general linear model was fitted in ASReml (Gilmour *et al.* 2009) which included dataset contemporary groups as fixed effects, standardised sire EBVs and interactions between finishing system, breed, dataset and the standardised sire EBVs to determine if there was a significant difference in the magnitude and or direction of the relationships between carcass traits and sire EBVs.

Sire variance components were estimated in ASReml (Gilmour *et al.* 2009) for each of the 12 datasets to determine whether the genetic variance in MSA index and its input traits changes with carcass weight. The same fixed effects as used for the regression analysis (excluding the sire EBVs and interactions) were fitted. Sire was included as a random effect in the benchmark model. Additional random effects were tested as interactions with sire: finish by sire, breed by sire and dataset by sire were included in separate models with separate sire variance components for finish regime, breed and dataset estimated. The log likelihood ratio test statistic was calculated to determine if the additional random terms significantly improved the model.

RESULTS AND DISCUSSION

The regressions for all carcass traits regressed on their associated EBV were significant. There was a significant interaction demonstrating a different regression coefficient between the finish systems for all regressions. In all cases the regression coefficients were greater for the Long-fed cattle than the short and pasture which tended to be similar to each other. For Rib Fat on sire Rib EBV and IMF% regressed on sire IMF EBV the Long-finished coefficients were significantly greater than the Short and Pasture finished regression coefficients (Table 1). This difference was the greatest when IMF% was regressed on IMF EBV where there was a 6.5-fold difference between the Pasture and Long feeding regimes. Moreover, the effect of selecting for improved IMF EBV was almost 5 fold greater in long grain finished cattle than short for MSA marbling (Table 1). For every 1 % increase in sire IMF EBV the increase in MSA marbling was 36.7 MSA marbling scores in Long-fed cattle relative to 7.6 in Short-fed cattle. MSA index was most closely related to the IMF EBV with an increase in IMF EBV being associated with a significant increase in MSA Index with the Long finish almost 3 times greater than Pasture finished (Table 2). A 1 % increase in sire IMF EBV was worth 0.28 MSA Index points under a Long feedlot finishing regime relative to a 0.10 unit increase under Pasture

(Table 2). Similar results were observed for sire Rib and carcass weight EBVs where Long finished regression coefficients were significantly higher than pasture finished. There were no significant ($P<0.05$) differences between the breeds in their relationship between MSA Index and sire EBVs.

Including sire by finishing system and estimating separate sire variance components for each finishing system (i.e. placing a G structure on the data) resulted in a significant improvement to the model for all traits based on the likelihood ratio test statistic. For almost all traits the sire variance under a Long-fed finishing regime was significantly greater than both Short and Pasture (Table 3). The exception was MSA Index where the sire variance of the Short-fed cattle was higher than both Long and Pasture fed cattle. For HSCW, Rib Fat, and MSA marbling, the sire variance for Long-fed animals was between 4 and 6-fold higher than pasture finished cattle. The difference for IMF was even larger however there were fewer animals with IMF measured. The sire variances were larger than those estimated by Reverter *et al.* (2003) for temperate beef breeds.

Table 1. Finishing system regression coefficients for carcass traits on BREEDPLAN sire EBVs (± standard errors)

Finish	HSCW on CWT EBV	Rib Fat on Rib EBV	IMF% on IMF EBV	MSA Marbling on IMF EBV
Long	$0.72^{a} \pm 0.03$	$0.75^{a} \pm 0.04$	$1.08^{a} \pm 0.04$	$36.7^{a} \pm 1.9$
Short	$0.55^{b} \pm 0.06$	$0.21^{b} \pm 0.08$	$0.17^{b} \pm 0.13$	$7.6^{b} \pm 2.4$
Pasture	$0.48^{b} \pm 0.06$	$0.31^{b} \pm 0.09$	$0.15^{b} \pm 0.13$	$8.9^{b} \pm 2.8$
P-Value	0.009	<0.001	<0.001	<0.001

Different superscripts indicate significantly different regression coefficients between finishing systems.

Table 2. Finishing system regression coefficients for MSA Index on BREEDPLAN sire EBVs (± standard errors)

Finish	600D Wt EBV	CWT EBV	Rib EBV	IMF EBV
Long	0.012 ± 0.003	0.014 ± 0.002	0.003 ± 0.020	0.34 ± 0.03
Short	0.009 ± 0.003	0.010 ± 0.004	-0.085 ± 0.033	0.29 ± 0.06
Pasture	0.005 ± 0.003	0.002 ± 0.004	-0.023 ± 0.036	0.12 ± 0.07
P-Value	0.308	0.035	0.046	0.020

Table 3. Sire variances for each finishing system (± standard errors)

	HSCW	Rib Fat	IMF	Marbling	Index
Long	159 ± 27	3.03 ±0.56	1.63 ± 0.27	1619 ± 286	0.34 ± 0.07
Short	82 ± 21	2.05 ± 0.42	0.22 ± 0.07	588 ± 112	0.63 ± 0.11
Pasture	39 ± 12	0.49 ± 0.19	0.04 ± 0.03	352 ± 107	0.07 ± 0.04

It was hypothesised that despite genetic correlations for marbling between various end points and finishing regimes being close to 1 the regression may change substantially depending on market end point. For example, where there is low variance, the regression coefficient of MSA marbling on IMF EBV is expected to be lower, in contrast where there is higher variance the regression coefficient is expected to be higher. There appears to be systematic differences (increases) in variance of traits of interest such as MSA Marbling and IMF for heavier carcass weights (associated with Long fed feedlot) or faster growth paths. This highlights the importance of considering target market end point weight when reporting estimating breeding values.

CONCLUSIONS

This work has quantified relationships between carcass traits and sire BREEDPLAN EBVs with the regressions for all carcass traits regressed on their associated EBV being significant. Importantly, there was a significant interaction demonstrating a different regression coefficient between the finishing systems for regressions with greater regression coefficients observed for the Long-fed cattle than the Short and Pasture which tended to be similar to each other. At a commercial level, this will have major effects on the increase in MSA Index expected through genetic improvement for traits linked with eating quality depending on market end point.

ACKNOWLEDGEMENTS

The authors gratefully acknowledge the funding of the project "Addressing key issues affecting compliance rates of pasture-fed cattle in southern Australia MLA" (B.SBP.0110) by Meat and Livestock Australia. The authors also acknowledge the provision of datasets from this project from breeders, breed societies and other Meat and Livestock Australia funded projects.

REFERENCES

Arthur P.F., Herd R.M., Wilkins J.F. and Archer J.A. (2005) *Aust. J. Exp. Agric.* **45**: 985.

Gilmour A.R., Gogel B.J., Cullis B.R. and Thompson R. (2009) *ASReml User Guide Release 3.0.* VSN International.

Hebart M.L., Lee S.J. and Pitchford W.S. (2016) *Proc. Aust. Soc. Anim. Prod.* **31**: 1299.

Herd R.M., Arthur P.F., Bottema C.D.K, Edgar A.R., Geesink D.S., Lines D.S., Piper J.P., Siddell J.M., Thompson J.M. and Pitchford W.S. (2018) *Anim. Prod. Sci.* **58**: 164.

Johnston D.J., Reverter A., Burrow H.M., Oddy V.H. and Robinson D.L. (2003) *Aust. J. Exp. Agric.* **54**: 107.

McKiernan W.A., Wilkins J.F., Barwick S.A., Tudor G.D., McIntyre B.L., Graham J.F., Deland M.P.B. and Davies L. (2005) *Aust. J. Exp. Agric.* **45**:.959.

Pitchford W.S., Accioly J.M., Banks R.G., Barnes A.L., Barwick S.A., Copping K.J., Deland M.P.B., Donoghue K.A., Edwards N., Hebart M.L., Herd R.M., Jones F.M., Laurence M., Lee S.J., McKiernan W.A., Parnell P.F., Speijers E.J., Tudor G.D. and Graham J.F. (2018) *Anim. Prod. Sci.* **58**: 20.

Polkinghorne R., Thompson J.M., Watson R., Gee A. and Porter M. (2008) *Aust. J. Exp. Agric.* **48**: 351.

Reverter A. Johnston D.J., Perry D., Goddard M.E. and Burrow H.M. (2003) *Aust. J. Exp. Agric.* **54**: 119.

THE HERITABILITY OF CONGENITAL ENTROPION IN DUAL-PURPOSE NEW ZEALAND SHEEP

K.M. McRae, S.M. Clarke, K.G. Dodds, S.J. Rowe, S-A. Newman, H.J. Baird and J.C. McEwan

AgResearch, Invermay Agricultural Centre, Mosgiel, New Zealand

SUMMARY

Congenital entropion, a condition in which one or more eyelids are inverted, resulting in contact between the eyelashes and the cornea, was recorded in three South Island progeny test flocks between 2000 and 2018. A total of 42,535 lambs were scored for entropion (as 0, 1 or 2, where the score reflected the number of eyes affected) within 24 hours of birth. The overall incidence rate for entropion was 6.5%. The incidence of entropion varied between flock ($P < 0.0001$) and year ($P < 0.0001$), ranging from 1% to 15%. The heritability of entropion was estimated to be 0.18 (± 0.01), indicating that the incidence of entropion within flocks can be reduced through selective breeding.

INTRODUCTION

Entropion is a condition in which one or more eyelids are inverted, resulting in contact between the eyelashes or external hair and the cornea, which can lead to blindness. Many mammalian species are affected by congenital entropion, including humans (Vallabhanath and Carter 2000), domestic animals (Priester 1972; Glaze 2005) and livestock (Warwick and Berry 1962; Allbaugh and Davidson 2009; Donnelly *et al.* 2014; Mészáros *et al.* 2015).

In sheep, the incidence of congenital entropion is variable. Reported frequencies of entropion range from 1% to 11% (Warwick and Berry 1962; Green *et al.* 1995; Claine *et al.* 2013; Greber *et al.* 2013). Congenital entropion is usually present at birth or occurs soon afterward and can be detected by either examination of the eye, or observation of excessive ocular discharge, conjunctivitis or keratitis (Boileau and Gilmour 2012). Treatment is relatively straightforward, however if left untreated, the contact between the eyelashes and the cornea can lead to blindness (Moore and Whitley 1984).

There is evidence of both between-breed and within-breed variation in the occurrence of entropion in sheep. In intensively reared lambs in south west England, Charollais and Texel sired lambs had an increased risk of entropion compared to Suffolk sired lambs (Green *et al.* 1995). In a separate study in France, Ile de France and crossbred lambs were significantly more affected by entropion than purebred Texel lambs (Claine *et al.* 2013). Heritability estimates for entropion range between 0.08 and 0.21 in purebred (Columbia, Polypay, Rambouillet, Suffolk, and Targhee) and crossbred sheep in the U.S. (Sakul and Kellom 1997), however, the incidence and heritability of entropion in New Zealand lambs has not been reported.

MATERIALS AND METHODS

Animals and data. Animals were managed in accordance with the provisions of the New Zealand Animal Welfare Act 1999, and the New Zealand Codes of Welfare developed under sections 68–79 of the Act.

Congenital entropion was recorded on all lambs born in three genetically linked progeny test flocks, run at the Woodlands Research Station in Southland (flocks A and C), and Invermay Research Farm in Otago (flock B). Recording began in 2000 in flock A, 2002 in flock C, and 2011 in flock B, and continued until 2017 (flocks A and C) or 2018 (flock B). Dams were composites of the main dual-purpose sheep breeds used in New Zealand, including Romney, Coopworth and Perendale, and sires were from a mixture of dual-purpose and terminal breeds.

A total of 42,535 lambs from 574 sires were scored for the presence or absence of entropion (EYE) within 24 hours of birth. EYE was initially scored on a 0-5 scale for each animal, depending on the number of in-turned eyelids and the presence or absence of infection, but this was reduced to a 0/1/2 score in 2014, where the score reflected the number of eyes affected. Subsequently, all scores were converted to a 0/1/2 scale.

Statistical Analysis. All pedigree and phenotypic records were obtained from Sheep Improvement Limited (SIL), the New Zealand sheep genetic evaluation database. Contemporary group (CG) was defined as flock, birth year, sex and weaning mob, and records were removed if the contemporary group contained less than five observations or had a mean incidence of entropion of zero. The resulting dataset consisted of 37,208 animals (Table 1).

Heritability of EYE was examined firstly using the reported values transformed using the formula $EYEt = EYE/\sqrt{[EYEm*(2\text{-}EYEm)]}$, where m is the mean incidence rate within the CG where phenotypic score is being adjusted, and secondly reporting it as a binary (presence/absence) trait. Fixed effects were determined using the GLM procedure in SAS (SAS Institute Inc., Cary NC, USA). The final model included fixed effects of contemporary group, birthday deviation from the mean of the contemporary group (BDEV), birth-rearing rank (BRR) and age of dam (AOD). Heritability estimates were obtained by running a univariate analysis using ASReml (Gilmour et al. 2015).

RESULTS AND DISCUSSION

Of the 37,208 animals with records remaining after data cleaning, 2,409 lambs had congenital entropion, giving an overall incidence rate over 19 years of 6.5%. The incidence of entropion varied between flock ($P < 0.0001$) and year ($P < 0.0001$), ranging from 0.01 to 0.13 (Table 1).

Heritability estimates (± standard error) for entropion adjusted for incidence rate per contemporary group (EYEt) and reported as a binary trait (EYEb) were 0.18 (± 0.01) and 0.19 (± 0.01), respectively. This is in line with a previous study that examined the heritability of entropion in purebred and crossbred U.S. sheep, which gave an overall estimate of 0.15 (ranging from 0.08-0.21). While entropion has not been reported to impact upon lamb growth (Claine *et al.* 2013), reduction in flock incidence will alleviate welfare concerns. This can therefore be achieved through scoring of lambs at birth for congenital entropion, and the use of the scores in selective breeding programs.

Table 1. Incidence of congenital entropion (% = incidence) between 2000 and 2018 in lambs at birth in three pedigree-recorded flocks. EYE score reflects the number of eyes affected (0 = unaffected; 1= one eye affected; 2 = both eyes affected)

	EYE Flock A				EYE Flock B				EYE Flock C			
Year	*0*	*1*	*2*	*%*	*0*	*1*	*2*	*%*	*0*	*1*	*2*	*%*
2000	1,393	58	43	7%								
2001	1,105	53	45	8%								
2002	1,039	44	25	6%	615	10	18	4%				
2003	1,010	67	78	13%	679	61	55	15%				
2004	1,024	48	35	7%	589	7	15	4%				
2005	1,145	66	44	9%	937	9	13	2%				
2006	1,089	63	44	9%	761	31	32	8%				
2007	1,162	71	43	9%	1,013	47	14	6%				
2008	1,150	56	35	7%	479	11	8	4%				
2009	1,116	89	59	12%	1,070	33	36	6%				
2010	973	30	42	7%	930	18	23	4%				
2011	1,157	72	39	9%	1,178	33	26	5%	43	3	3	12%
2012	1,214	21	47	5%	997	21	10	3%	110	0	2	2%
2013	1,092	26	30	5%	866	35	20	6%	272	15	10	8%
2014	1,499	45	20	4%	1,022	50	26	7%	314	2	4	2%
2015	653	9	10	3%	930	34	17	5%	70	0	2	3%
2016	183	1	1	1%	1,241	16	11	2%	110	3	1	4%
2017	821	15	12	3%	1,005	52	34	8%	70	0	1	1%
2018					673	34	17	7%				
Overall	18,825	834	652	7%	14,985	502	375	6%	989	23	23	4%

This study provides the first estimate of the heritability of congenital entropion in dual-purpose New Zealand lambs.

ACKNOWLEDGEMENT

The flocks involved in this study were from Beef and Lamb New Zealand Genetics (via Ovita), FarmIQ, and Pastoral Greenhouse Gas Research Consortium (PGgRc) funded progeny test programs. The authors gratefully acknowledge the AgResearch Animal Genomics field staff, and the staff at both the Woodlands and Invermay Research Stations for their assistance in collecting the data.

REFERENCES

Allbaugh, R.A. and Davidson, H.J. (2009) *Vet. Ophthalmol.* **12**: 115.

Boileau, M.J. and Gilmour, M.A. (2012) In 'Sheep and Goat Medicine' 2nd ed. W.B. Saunders, Saint Louis.

Claine F., Raes M., Leemans J., Muylkens B. and Kirschvink N. (2013) *Small Ruminant Res.* **111**: 1.

Donnelly K.S., Pearce J.W., Giuliano E.A., Fry P.R. and Middleton, J.R. (2014) *Vet. Ophthalmol.* **17**: 443.

Glaze M.B. (2005) *Clin. Tech. Small Anim. Pract.* **20**: 74.
Greber D., Doherr, M., Drögemüller, C. and Steiner, A. (2013) *Acta Vet. Scand.* **55**: 27.
Green L.E., Berriatua E. and Morgan, K.L. (1995) *Prev. Vet. Med.* **24**: 15.
Mészáros G., Stuckler M.P., Ferenčaković M. and Sölkne, J. (2015) *Poljorivreda* **21**: 48.
Moore C.P. and Whitley R.D. (1984) *Vet. Clin. North Am. Large Anim. Pract.* **6**: 641.
Priester W.A. (1972) *J. Am. Vet. Med. Assoc.* **160**: 1504.
Sakul H. and Kellom T.R. (1997) *Small Ruminant Res.* **23**: 187.
Vallabhanath P. and Carter S.R. (2000) *Curr. Opin. Ophthalmol.* **11**: 345.
Warwick B.L. and Berry R.O. (1962) *J Hered.* **53**: 10.

CORRECTING SAMPLING BIAS IN MICROSATELLITE MARKER TESTING FOR POLLEDNESS

N.K. Connors, R.G. Banks and D.J. Johnston

Animal Genetics & Breeding Unit*, University of New England, Armidale, NSW, 2351 Australia

SUMMARY

The poll microsatellite test has been available to Australia's beef industry for approximately 7 years and in that time, the bias in polled phenotyped animals submitted for testing from industry has influenced the accuracy of polled probability assignment to observed haplotypes. This article describes examples of observed mis-assigned haplotypes and their respective phenotypic observations, and the steps taken to correct the poll probabilities and resulting genotype estimations.

INTRODUCTION

The costs associated with carcase defects are largely attributed to damage from horned animals (Prayaga 2007). While dehorning is common practice to address these issues, questions remain regarding the animal's welfare, and breeding naturally polled animals provides a long term solution. The microsatellite DNA marker test for polledness was developed by the Beef Cooperative Research Centre and CSIRO (Henshall *et al.* 2011), and has been available to Australia's beef industry for approximately 7 years. In that time, samples submitted from industry have been biased towards polled submissions, due to a logical disinterest in testing horned animals. Prior to this study, the vast majority of phenotypes submitted to the test were unknown (>60%), over a quarter polled (27%) and the least horned (5%) and scurred (5%) (Connors *et al.* 2018). Given the number of potential haplotypes possible, there is no realistic option of a large enough reference population. As such the test uses all available industry data to estimate genotypes and an appropriate representation of different phenotypes should be present in the data so that microsatellite haplotypes can be assigned the appropriate poll probability based on the observed phenotypes. The bias in polled phenotypes has influenced the accuracy of the genotype estimations, such that haplotypes which should be assigned as horned, have been mis-assigned as polled due to only polled phenotypes being observed with this haplotype. Recently additional horned phenotypes were sourced for inclusion into the test to correct this sampling bias and to demonstrate the effect that these additional phenotypes have on haplotype poll probability assignment. This paper describes a number of haplotypes with mis-assigned poll probabilities, the resulting genotype estimations, and the effect of including additional horned phenotypes on the haplotypes' assignments.

MATERIALS AND METHODS

The microsatellite test estimates an animal's genotype as homozygous polled (PP), heterozygous polled (PH), or homozygous horned (HH), and detailed methodology has been discussed previously (Piper *et al.* 2014; Connors and Tier 2016; Connors *et al.* 2016). Briefly, samples submitted for testing are accompanied with a phenotype (i.e. horned, polled, scurred, or unknown). The test uses ten microsatellite markers to form haplotype pairs for each sample, where each haplotype is labelled with a unique number. Haplotypes are assigned as either horned or polled, providing each haplotype with a polled probability based on the following criteria: (i) observed in polled animals with homozygous haplotypes; (ii) observed within progeny-tested animals (i.e. phenotyped progeny); (iii) observed in horned animals; (iv) observed in polled or scurred animals, where the other haplotype is horned. If the

* A joint venture of NSW Department of Primary Industries and the University of New England

haplotypes are not observed in any of these situations, then they cannot be assigned as horned or polled.

Samples from 278 animals from four different breeds (Angus, Santa Gertrudis, Brahman, and Droughtmaster) were phenotyped at dehorning (with photographic records) and microsatellite genotyped. Angus samples were sourced from breeder/s, and all others originated from the Repronomics™ project (Johnston *et al.* 2017). Genotypes were compared with the phenotypes and agreement or mismatch was quantified. Where a mismatch between the genotype and phenotype occurred, the haplotypes were investigated for potential bias in phenotype observations.

RESULTS AND DISCUSSION

Of the 278 samples sent for genotyping, 45 samples had incomplete microsatellite results (less than 10 markers). Microsatellite genotypes were obtained from 231 animals, consisting of 5 scurred, 14 polled, and 212 horned animals. Genotype estimations from the poll test had complete concordance with 221 phenotypes, such that:

- 5 samples with ≥90% PP microsatellite call matching polled phenotype;
- 6 samples with ≥90% PH microsatellite call matching polled and scurred phenotypes;
- 177 samples with ≥90% HH microsatellite call matching horned phenotype;
- 33 samples with 70-90% HH microsatellite call matching horned phenotype;

Six samples had a mismatch with the phenotype result (shaded orange in Table 1), and another four had low probability genotype estimations (i.e. <70%) due to haplotype uncertainty (shaded blue in Table 1). Haplotypes with poll probability of 0.01 are high likelihood of being horned, and are associated with high number of horned phenotypes. Those with a poll probability of 0.99 are high likelihood of being polled, and are associated with high number of polled phenotypes. Deviation from either end towards the centre (i.e. 0.5) represents a level of uncertainty in the assignment of polled or horned, and is most often due to variation in phenotype observations. Haplotypes suspected of mis-assignment/low certainty are shaded grey in Table 1. Phenotypes associated with these haplotypes are counted, shown in Table 2.

Table 1. Microsatellite poll results from mis-assigned/uncertain haplotypes. Orange shading indicates mismatching genotypes and probability (e.g. 96% PH); blue shading indicates low probability genotypes; grey shading indicates mis-assigned/uncertain haplotype

Breed	Phenotype	Haplotype 1	Haplotype 2	Haplotype 1 poll probability	Haplotype 2 poll probability	PP	PH	HH
Santa Gertrudis	horned	19	660	0.01	0.97	0.01	0.96	0.03
Santa Gertrudis	horned	22	660	0.01	0.97	0.01	0.96	0.03
Droughtmaster	horned	87	1655	0.01	0.85	0.01	0.84	0.15
Santa Gertrudis	horned	3	463	0.01	0.69	0.01	0.68	0.31
Angus	scurred	8	166	0.99	0.99	0.98	0.02	0
Angus	scurred	6	999	0.99	0.92	0.91	0.09	0
Droughtmaster	horned	254	383	0.2	0.38	0.07	0.43	0.5
Brahman	horned	135	771	0.01	0.38	0	0.39	0.61
Santa Gertrudis	horned	3	745	0.01	0.3	0	0.31	0.69
Droughtmaster	horned	254	1587	0.2	0.15	0.03	0.29	0.69

The phenotype counts for haplotypes driving mismatched genotypes (orange in Table 2) overwhelmingly show a bias towards polled phenotypes along with a significant number of unknown phenotypes which are uninformative. These phenotype counts explain the haplotype's high poll probability assignment of each mismatching genotype highlighted orange in Table 1.

The phenotype counts of less certain haplotypes are shown in blue in Table 2. Each such haplotype has a mix of contradicting phenotypes. The inclusion of varied and contradicting phenotypes leads to probability uncertainty and thus, low probability genotype estimations.

Table 2. Phenotype counts for haplotypes with mis-assigned poll probabilities (before additional samples submission). Orange shading indicates mis-assigned haplotypes; blue shading indicates uncertain probability haplotypes

Haplotype	Unknown	Scurred	Horned	Polled	Total
660	24	1	0	9	36
1655	1	1	1	3	8
463	4	2	0	1	8
166	8	0	2	7	17
999	2	0	0	5	7
254	17	5	1	5	28
771	3	2	3	0	8
383	40	1	9	17	68
745	10	1	3	8	22

Inclusion of more consistent phenotype observations will improve the certainty of the haplotype probabilities. An example of this is shown in Figure 1. Haplotype 383 had a poll probability of 0.38 due to contradicting horned and polled phenotypes (Table 2). Incremental inclusion of over 20 horned phenotypes saw the poll probability drop to approximately 0.01.

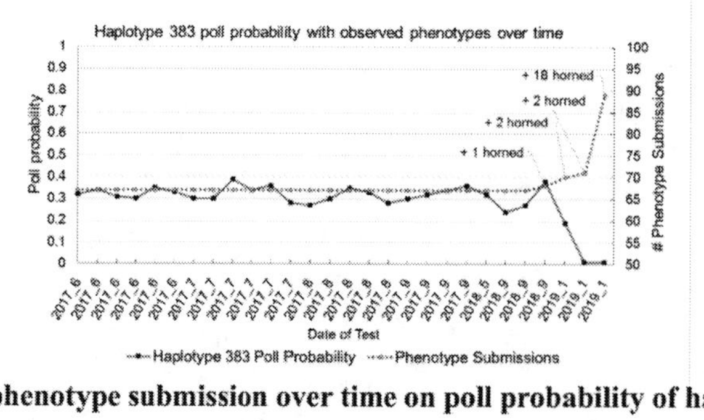

Figure. 1. Effect of phenotype submission over time on poll probability of haplotype 383

Table 3. Poll probability changes for less certain haplotypes (after additional horned submissions)

Haplotype	Poll probability before	Poll probability after	Poll probability change	No. horned additions
383	0.38	0.01	-0.37	22
771	0.38	0.31	-0.07	1
745	0.3	0.03	-0.27	5
254	0.2	0.1	-0.1	5

As a result of inclusion of more than 212 horned phenotypes, the less certain haplotypes from Table 2 have shifted poll probabilities significantly, shown in Table 3. These shifts towards zero poll probability are a direct result of the inclusion of horned phenotypes associated with these haplotypes. Unfortunately, further horned samples possessing haplotypes causing the mismatches in Table 1 could not be sourced; inclusion of further samples would be needed to adequately shift the poll probabilities of these haplotypes.

A shift in poll probability of some haplotypes may have happened historically at any point, and is a direct reflection of the reference data of the test. Reliable horned phenotypes are the most informative data as they exclude the possibility of being genetically polled. Submission of horned phenotypes is challenged in two major aspects. Firstly, it is difficult to ensure animals' phenotypes are accurate when (i) horns may be labelled as scurs and vice versa; (ii) horns may develop after the phenotyping time; and (iii) animals may be dehorned and mis-phenotyped polled. Secondly, data submission under commercial conditions makes submission of horned animals extremely unlikely; the cost of receiving a horned genotype result, when the horned phenotype is already known is unnecessary. Each of these aspects have likely impacted the observed sampling bias of the poll microsatellite test. As a result, some historical genotype predictions may be incorrect. This will likely become apparent using newly available technology, such as the commercial poll SNP test, which is now offered to the beef industry, where the microsatellite test will run in parallel. It is possible that the SNP test will provide a SNP result contradicting the microsatellite result, where the microsatellite haplotypes have been mis-assigned due to phenotype observations. It is impossible to know how many haplotypes have been affected, though reassuringly in this dataset, the microsatellite test had approximately 96% accuracy in relation to known phenotypes recorded.

CONCLUSIONS

This paper describes the effect of phenotyping bias on haplotype poll probabilities and resulting genotype estimations for the poll microsatellite test. This dataset had 96% genotype to phenotype concordance. The remaining four percent was demonstrated to be a result of haplotype mis-assignment due to associated phenotype observations, which can be corrected with additional horned phenotype submissions.

ACKNOWLEDGEMENTS

This research includes data generated by the BeefCRC. This research is supported by Meat and Livestock Australia (MLA) projects L.GEN.1803 and L.GEN.0174.

REFERENCES

Connors N.K. and Tier B. (2016) In 'Breeding Focus 2016 - Improving Welfare', pp. 59. editors S Hermesch, S Dominik. Animal Genetics and Breeding Unit, University of New England, Armidale, NSW.

Connors N.K., Tier B. and Johnson D.J. (2016) *Proc. Northern Beef Research Update Conference.* 196.

Connors N.K., Tier B. and Johnston D.J. (2018) *Proc. World Congr. Genet. App. Livest. Breed.* **11:** 344.

Henshall J.M., Piper E.K. and McDonald A. (2011) *Proc. Assoc. Advmt. Anim. Breed. Genet.* **19:** 79.

Johnston D.J., Grant T.P., Schatz T.J., Burns B.M., Fordyce G. and Lyons R.E. (2017) *Proc. Assoc. Advmt. Anim. Breed. Genet.* **22:** 385.

Piper E.K., Tier B. and Henshall J.M. (2014) *Proc. World Congr. Genet. App. Livest. Breed.* **10:** 237.

Prayaga K.C. (2007) *Aust. J. Agric. Res.* **58:** 1.

GENETIC ANALYSIS OF TAIL-BITING VICTIMS IN PIGS

S. Hermesch and S.Z.Y. Guy

Animal Genetics & Breeding Unit*, University of New England, Armidale, NSW, 2351 Australia

SUMMARY

Tail biting is a welfare concern in pigs for both victims of tail biting and tail biters. This study aimed to estimate genetic parameters for tail-biting victims using medication records routinely collected on farm. Medication records for 771 pigs were available from 2011 until 2017 and most pigs ($n = 459$) needed medication due to tail-biting injury. There were 10,335 pigs with growth and backfat records that had not been medicated during this time period. Three different health traits were analysed as binary traits, defined as medication due to tail-biting injury, overall medication and medication for any health issue other than tail biting. Linear and logistic sire models were used to estimate genetic parameters. Heritability estimates for tail-biting victims were 0.09 (± 0.02) and 0.25 (± 0.09) based on a linear and logistic sire model, respectively. Medication due to other sicknesses was not heritable indicating that heritabilities for overall medication reflected additive genetic effects for tail-biting victims. There were no genetic associations between being tail bitten and growth rate or backfat indicating that current selection emphasis for these performance traits does not affect tail-biting victims. These first genetic parameter estimates of being a tail-biting victim indicates opportunities to select pigs less prone to becoming a victim of tail biting.

INTRODUCTION

Tail biting is a behaviour in pigs that causes pain, injury and in severe cases mortality in victims of tail biting. Further, biters start tail biting because their own welfare is compromised. The causes of tail biting are multi-factorial and the prevalence of tail biting may depend on interactions between some factors of the environment and the animal (Sonoda *et al.* 2013). This makes it difficult to find solutions to reduce the incidence of tail biting. So far, information about genetic factors affecting tail biting is limited. Previous research has focussed on tail biters (Breuer *et al.* 2005). Only recently has the first information about genetic variation for the incidence of victims of tail biting been reported (Canario and Flatres-Grall 2018), where tail-biting injury was recorded as a binary trait observed when pigs were approximately 100 kg. Alternatively, medication records available on farms for veterinary purposes may be used to identify victims of tail biting.

This study aimed to estimate heritability for tail-biting victims in pigs using medication records and to estimate their genetic correlations with growth rate and backfat.

MATERIALS AND METHODS

Medication records were available from January 2011 until September 2017 for 771 Large White pigs. Most pigs required medication due to tail biting ($n = 459$ pigs). These medication data were combined with other performance data recorded on farm during the same time period. Three different health traits were defined according to the reason for medication: due to having a tail-bite injury, overall medication and due to sickness other than tail biting (Other sickness). For these health traits, any pigs that were medicated were defined as 1 (case) while non-treated pigs received a 0 (control) for these health traits. There were 10,335 pigs with performance data that had not been medicated. These non-medicated pigs were recorded for growth rate and backfat at an average age of 126 days and an average body weight of 85.7 kg. There were 326 medicated pigs and 179 tail-bite victims that

* A joint venture of NSW Department of Primary Industries and the University of New England

were also recorded for growth and backfat. The prevalence of tail-biting victims was estimated as the proportion of pigs recorded for growth and backfat which were also medicated for tail bite. Overall, pigs were the progeny of 180 sires and 1,082 dams.

The three health traits were analysed as binomial variables using generalized linear mixed models which were fitted on a sire level with a logistic link. Therefore, a logistic distribution was assumed for the underlying liability scale. In addition, variance components were estimated for health traits applying a linear sire model which was also used to estimate genetic correlations between health traits and growth or backfat. Genetic models for health traits as well as average daily gain and backfat included month of birth as contemporary group and sex of the animal fitted as fixed effects. The weight of the animal at recording was fitted as a linear covariate for backfat. Random common litter effect was fitted as an additional random effect for all traits. For sire models, additive genetic variance was calculated as four times the estimated sire variance. Further, the residual variance was specified as $\pi^2/3 \approx 3.29$ for logistic sire models. Genetic parameters were estimated using ASReml (Gilmour *et al.* 2009).

RESULTS AND DISCUSSION

Prevalence of tail biting. The prevalence of tail-biting victims based on medication records was 4.2% in this study. However, the prevalence observed in this study should only be regarded as an indication of the true prevalence of tail biting because both the number of tail-biting victims needing medication and the number of pigs not being a victim of tail biting were estimated from incomplete data. The prevalence of tail-biting victims was 6.6% and 10.8% in two different herds based on a binary trait to identify pigs with tail damage (Canario and Flatres-Grall 2018). These two prevalence scores are not directly comparable because only a proportion of pigs with tail damage require medication and a higher prevalence of tail damage score is expected.

Heritability estimates. Tail biting had a heritability of 0.09 (± 0.02) based on a linear sire model (Table 1). In comparison, the heritability estimate of tail-biting victims was higher (0.25 ± 0.09) based on a logistic sire model (Table 2). A higher heritability based on a logistic sire model in comparison to a linear sire model has been observed in other studies (Baeza-Rodriguez *et al.* 2017). In comparison, Canario and Flatres-Grall (2018) found a heritability of 0.06 (± 0.01) based on an animal model that also included social genetic effects. Jointly, these results indicate that the incidence of tail-biting victims has a genetic component that can be used for selective breeding.

Table 1. Phenotypic (Vp) variances, heritability (h^2) and common litter effect (c^2) estimates (standard errors, se) for tail biting, health and performance traits fitting linear models

Trait	Vp	h^2	se	c^2	se
Tail biting[1]	0.0377	0.09	0.02	0.11	0.01
Overall medication[1]	0.0602	0.07	0.02	0.08	0.01
Other sickness[1]	0.0276	0.00	0.01	0.04	0.01
Growth rate[2]	2668	0.22	0.03	0.12	0.01
Backfat[2]	1.74	0.23	0.02	0.07	0.01

[1] a linear sire model was fitted; [2] a linear animal model was fitted

No genetic variation was evident in the health traits defined by sickness other than tail biting indicating that heritability found for overall medication was predominantly due to tail biting incidence. Medication records were explored in detail by Guy *et al.* (2019) who used a subset of the data presented in this study. Alternative approaches to derive pseudo identifications for pigs without performance records from weaning records were explored. Information about litters weaned each week was used to derive pseudo pedigree for pigs that were expected to be weaned from each litter each week. Heritabilities for medication incidence from a logistic sire model were similar for both approaches which defined controls based on performance-tested pigs (reduced-control: 0.06 ± 0.04) or based on pigs weaned per litter (full-control: 0.04 ± 0.03).

Estimates of common litter effects were 0.11 (± 0.01) and 0.14 (± 0.03) for tail-biting victims based on the linear and logistic sire model, respectively. Litter mates are likely to be housed in the same pen post weaning which may have contributed to these significant common litter effects for tail-biting victims.

Table 2. Phenotypic (Vp) variances, heritability (h^2) and common litter effect (c^2) estimates (standard errors, se) for tail biting and health traits fitting logistic sire models

Trait	Vp	h^2	se	c^2	se
Tail biting	4.14	0.25	0.09	0.14	0.03
Medication	3.84	0.13	0.06	0.11	0.03
Other sickness	3.61	0.02	0.06	0.08	0.05

Genetic correlations. Estimates of genetic correlations between tail biting and growth rate or backfat were not significantly different from zero (Table 3). Other correlations between tail biting and growth rate were lowly negative demonstrating that higher prevalence of tail biting was associated with lower growth rate at the residual, common litter and phenotypic level. These negative non-genetic associations between tail biting and growth rate were not found for backfat. Further, estimates of genetic and non-genetic associations between overall medication and growth rate or backfat were like associations between tail biting and growth rate or backfat. Genetic correlations were also not significantly different from zero indicating that selection for higher growth rate and lower backfat will not adversely affect tail-biting victims or overall medication.

Table 3. Genetic (r_g), common litter (r_c), residual (r_r) and phenotypic (r_p) correlations (with standard errors) between tail biting or overall medication and growth rate or backfat fitting linear sire models

Trait	r_g (se)	r_c (se)	r_r (se)	r_p (se)
		Tail biting		
Growth rate	0.03 (0.18)	-0.14 (0.06)	-0.06 (0.01)	-0.07 (0.01)
Backfat	-0.09 (0.19)	-0.09 (0.07)	-0.01 (0.01)	-0.02 (0.01)
		Overall medication		
Growth rate	0.03 (0.19)	-0.11 (0.07)	-0.10 (0.01)	-0.09 (0.01)
Backfat	-0.01 (0.19)	-0.09 (0.08)	-0.01 (0.01)	-0.02 (0.01)

Medication due to other sickness was not heritable and genetic correlations with other traits could therefore not be estimated. No information was found in the literature regarding genetic associations between being a victim of tail biting and growth rate or backfat. Estimates of genetic correlations between tail biters and growth or backfat presented by Breuer *et al.* (2005) are not comparable because the behaviour of tail biting is different to the behaviour of a tail-biting victim.

Selection strategies. Medication records were used in this study to identify victims of tail biting in pigs. This measurement of tail-biting prevalence does not capture all victims of tail biting because only a proportion of tail-biting victims require medication. Therefore, a binary score identifying tail damage as was used by Canario and Flatres-Grall (2018) may be a better measure of tail-biting victims because the prevalence of such a score is expected to be higher than the prevalence based on medication records. A higher prevalence of a score identifying tail-biting victims results in a higher variance for the binary trait. Overall, it is recommended that tail damage of pigs is recorded when pigs are performance tested for weight or fat depth in order to verify these initial heritability estimates available for tail-biting victims.

Tail biting leads to economic losses because tail-bitten pigs are at higher risk of infections, carcase condemnation, reduced weight gain and increased medication and labour costs (review by Valros and Heinonen 2015). Often these cost components are difficult to quantify and information about medication records provides information about additional medication and labour costs.

The prevalence of tail biting is high when an outbreak of tail biting occurs. Generally, tail biting is not observed continuously and the overall prevalence of tail-biting victims is low. This is desirable of course, however, a low prevalence implies that variance for tail-biting victims is low which in turn limits opportunities for genetic improvement. Therefore, selection criteria that can be recorded easily on all pigs to reduce biting behaviour and prevalence of tail biting are desirable. First indications that social genetic effects for growth are associated with multiple biting behaviours including tail biting were presented by Camerlink *et al.* (2015) and should be investigated further. Social genetic effects for prevalence of tail-biting victims were found by Canario and Flatres-Grall (2018). However, estimating these social genetic effects for tail-biting victims directly is difficult due to the low prevalence and binary nature of this trait. Therefore, investigating social genetic effects for growth as an indirect selection criterion for tail biting in pigs may be a more feasible alternative. This approach also requires information of tail-biting victims and recording a simple (binary) score to identify tail-biting victims should be priority.

CONCLUSIONS

Being a tail-biting victim, identified by medication records, was heritable. No genetic associations were found between tail-biting victims and growth rate or backfat. Simple (binary) scores to identify victims of tail biting based on medication records or observations of tail damage of pigs on farms should be considered as new welfare traits in pig breeding programs.

ACKNOWLEDGEMENTS

This project was funded by Australian Pork Limited under project (APL 2016/094). Provision of data by Gatton piggery at University of Queensland is gratefully acknowledged.

REFERENCES

Baeza Rodriguez J.J., Vega Murillob V.E., Rios Utrera A., Martinez Velazquez G., Arechavaleta Velasco M.E. and Montano Bermudez M. (2017) *Rev. Mex. Cienc. Pecu.* **8:** 249.

Breuer K., Sutcliffe M.E.M., Mercer J.T., Rance K.A., O'Connell N.E., Sneddon I.A. and Edwards S.A. (2005) *Livest. Prod. Sci.* **93**: 87.

Camerlink I., Ursinus W.W., Bijma P., Kemp B. and Bolhuis J.E. (2015) *Behav. Genet.* **45**: 117.

Canario L. and Flatres-Grall L. (2018) *Proc. World Congress on Genet. Appl. to Livest. Prod.*, **11:** 572.

Gilmour A.R., Gogel B.J., Cullis B.R. and Thompson R. (2009) ASReml User Guide Release 3.0. VSN International Ltd, Hemel Hempstead, HPI IES, UK.

Guy S.Z.Y., Li L., Thomson P.C. and Hermesch S. (2019) *J. Anim. Sci.* **97:** 1076.

Sonoda L.T., Fels M., Oczak M., Vranken E., Ismayilova G., Guarino M., Viazzi S., Bahr C., Berckmans S. and Hartung J. (2013) *Berl. Munch. Tierarztl. Wochenschr.* **126**: 104.

Valros A. and Heinonen M. (2015) *Porcine Health Manag.* **1**: 2.

GENOMIC PREDICTION OF METABOLIC PROFILES IN DAIRY COWS

T.D.W. Luke[1,2], T.T.T. Nguyen[1], S. Rochfort[1,2], W.J. Wales[3], C. Richardson[1,2], M. Abdelsayed[4] and J.E. Pryce[1,2]

[1]Agriculture Victoria Research, AgriBio, Centre for AgriBioscience, Bundoora, Victoria, 3083 Australia
[2]School of Applied Systems Biology, La Trobe University, Bundoora, Victoria, 3083 Australia
[3]Agriculture Victoria Research, Ellinbank Centre, 1301 Hazeldean Rd., Ellinbank, Victoria, 3820 Australia
[4]Datagene Ltd., AgriBio, Centre for AgriBioscience, Bundoora, Victoria, 3083 Australia

SUMMARY

Improving animal health and resilience is an increasingly important breeding objective for all livestock industries. In this study we estimated genetic parameters of serum metabolic profiles in early lactation dairy cows. A single serum sample was taken from 1,393 cows, located on 14 farms in south eastern Australia, within 30 days after calving. Sera were analysed for biomarkers of energy balance (β hydroxybutyrate (BHBA) and non-esterified fatty acids (NEFA)), mineral status (Ca and Mg), protein nutrition (urea and albumin) and immune status (globulins and albumin to globulin ratio (A:G)). After editing, 47,162 single nucleotide polymorphism marker genotypes were used for estimating genomic heritabilities and breeding values (gEBV) for these traits in ASReml. Heritabilities were low for BHBA, NEFA, Ca, Mg and urea (0.09, 0.18, 0.07, 0.19 and 0.18, respectively), and moderate to high for albumin, globulins and A:G (0.27, 0.46 and 0.41, respectively). The accuracy of genomic predictions was assessed by (1) calculating empirical accuracy using 5-fold cross validation, and (2) calculating theoretical accuracy using the prediction error variance obtained from ASReml. Empirical accuracies ranged from 0.20 to 0.40, being higher for traits with higher heritabilities. Theoretical accuracies were higher than respective empirical accuracies (0.31 – 0.51), but the results of the 2 methods were in excellent agreement ($R^2 = 0.89$). While increasing the size of the reference population should theoretically improve accuracies, our results suggest that genomic prediction may allow identification of healthier cows that are less susceptible to diseases in early lactation.

INTRODUCTION

Most disease events affecting dairy cows occur in the first 30 days after calving (LeBlanc *et al.* 2006) and many of these diseases are associated with metabolic disorders such as ketosis and hypocalcaemia (Ospina *et al.* 2010). While heritability estimates of metabolic disorders are generally low (Uribe *et al.* 1995), sufficient genetic variance exists to suggest that improvements in metabolic stability can be achieved through genetic selection.

One way of assessing the metabolic health of cattle is serum metabolic profiling, which employs well-established epidemiological associations between the concentrations of several metabolites in serum, and the presence of both subclinical and clinical metabolic disorders (Payne *et al.* 1970). These metabolites include those associated with energy balance (BHBA and NEFA), mineral status (Ca and Mg), protein nutrition (urea and albumin) and immune status (globulins and albumin to globulin ratio). While extremely valuable, these phenotypes are costly and invasive to collect, making them impractical for traditional large-scale genetic evaluations. Genomic selection offers exciting potential for achieving genetic improvement in such difficult to measure and lowly heritable traits, by using data obtained from relatively small genotyped reference populations with high quality phenotypic data.

The objectives of this study were to (1) estimate the genetic parameters of serum biomarkers of health in early lactation dairy cows using data collected from a genotyped female reference population, and (2) estimate the accuracy of genomic predictions of serum biomarker concentrations. If sufficiently accurate, genomic selection for metabolic stability offers the potential to provide permanent and incremental improvements in dairy cow health and welfare, thereby increasing farm profitability and sustainability.

MATERIALS AND METHODS

Phenotypes. A single serum sample was taken from of 1,393 Holstein-Friesian cows from 14 farms in south eastern Australia between August 2017 and October 2018, according to the protocol described in Luke *et al.* (2019). All animals had been calved 30 days or less at the time of sampling. Sera were analysed for biomarkers of energy balance (BHBA and NEFA), mineral status (Ca and Mg), protein nutrition (urea and albumin) and immune status (globulins) by Regional Laboratory Services (Benalla, Victoria, Australia). Descriptive statistics of phenotypes are shown in Table 1.

Genotypes. Genotypes for the 1,393 animals used in this study were provided by DataGene Ltd. (Victoria, Australia). After editing, 47,162 single nucleotide polymorphism (**SNP**) markers were available for genomic analyses. A genomic relationship matrix (**GRM**) was constructed according to Yang *et al.* (2010).

Genetic parameters. Variance components were estimated for each trait using univariate linear mixed animal models in ASReml (Gilmour *et al.* 2015). In matrix notation, the model used was $\mathbf{y} = \mathbf{Xb} + \mathbf{Zu} + \mathbf{e}$ (Model 1), where **y** is a vector of metabolite concentrations (BHBA, NEFA, Ca, Mg, urea, albumin, globulins), **b** is a vector of fixed effects of DIM, herd, parity, age and sample collection date, **u** is a vector of random genetic effects, and **e** is a vector of the random residual effects; and **X** and **Z** are incidence matrices for **b** and **u** respectively. It is assumed that var(**u**) = $\mathbf{GRM}\ \sigma_u^2$, and var(**e**) = $\mathbf{I}\sigma_e^2$. Estimated variance components were then used to calculate the genomic heritability of each biomarker.

Genomic predictions. Genomic estimated breeding values (**GEBV**) were predicted using genomic best linear unbiased prediction (**gBLUP**), using variance components estimated from the univariate model (Model 1). The accuracy of genomic predictions was assessed in 2 ways. Firstly, empirical accuracy was calculated using 5-fold cross validation. This involved randomly dividing the reference population into 5 equally sized groups or folds. Data from 1 fold (approximately 20% of the reference population) were set aside as a validation set, and data from the remaining 4 folds (approximately 80% of the reference population) formed the training set for model development. The resulting model was then used to predict GEBVs for animals in the validation set. This was repeated 5 times so that all animals appeared in the testing set once. Empirical accuracy was then calculated as the Pearson's correlation between the predicted GEBVs and actual phenotype values, corrected for the fixed effects described in Model 1. Predicted accuracies of the true breeding values were calculated by dividing the empirical accuracies by the square root of the heritability of the trait. Secondly, theoretical accuracy was calculated as

$$\mathrm{r}_i = \sqrt{1 - \frac{SE_i^2}{\sigma_g^2\ GRM_{ii}}}$$

where SE_i SE_i is the standard error of GEBV of individual *i*, and $\sigma_g^2\sigma_g^2$ is the genetic variance of each trait estimated from Model 1, adjusted for inbreeding by multiplying by the corresponding diagonal elements in the GRM for each individual (GRM_{ii}).

RESULTS AND DISCUSSION

Estimated heritabilities for all traits, obtained from Model 1, are shown in Table 1. Heritability estimates were low for serum BHBA, NEFA, Ca, Mg and urea at 0.09 0.18, 0.07 0.19 and 0.18, respectively. Heritabilities of albumin, globulins and A:G were higher at 0.27, 0.46 and 0.41, respectively. Standard errors for all heritabilities were low (0.04 - 0.06).

Heritability estimates were consistent with the literature for NEFA (Oikonomou *et al.* 2008), Mg (Tsiamadis *et al.* 2016), albumin, globulins and A:G (Cecchinato *et al.* 2018). We could find no reports of the heritability of serum urea concentration in the literature, however our results are consistent with the reported heritability of milk urea nitrogen (Mitchell *et al.* 2005), the concentration of which is linearly correlated with serum urea.

The heritability of serum BHBA in our dataset was 0.09 ± 0.04, which is in excellent agreement with the findings of Weigel *et al.* (2017) (0.093 ± 0.045), slightly lower than those of van der Drift *et al.* (2012) (0.17 ± 0.06), and considerably lower than those of Oikonomou *et al.* (2008) and Cecchinato *et al.* (2018) (0.40 ± 0.06 and 0.37 ± 0.14, respectively). Oikonomou *et al.* (2008) demonstrated that the heritability of BHBA concentration is highest in immediately post calving and decreases rapidly over the first 7 weeks of lactation. In our study only 209 cows were in the first week of lactation at the time of sampling, and we expect that adding more data from animals in this highest risk period could improve heritabilities. Similarly, the heritability of Ca in our dataset was 0.07 ± 0.04, significantly lower than reported by Tsiamadis *et al.* (2016) who found that the heritability of serum Ca at days 1, 2, 4 and 8 post-partum ranged from 0.23 (± 0.02) to 0.32 (± 0.03). Serum Ca concentrations drop in the 12 to 24 hours immediately post-calving before rapidly returning to normal physiological levels once homeostatic mechanism are restored, and it is likely that our low heritability estimate is the result of having sampled only 14 cows in this period of highest phenotypic variability. These results demonstrate the importance of careful trait definition when investigating the genetic parameters of health traits in the transition period.

Table 1. Number of samples (n), phenotypic means (μ) and standard deviations (σ), estimated genomic heritabilities (± standard errors), empirical reliabilities, and theoretical reliabilities of serum metabolic profiles

Phenotype	n	μ	σ	h^2	r_e	r_t
BHBA	1393	0.48	0.22	0.09 ± 0.04	0.29	0.34
NEFA	1393	0.55	0.33	0.18 ± 0.05	0.36	0.41
Ca	1327	2.31	0.18	0.07 ± 0.04	0.20	0.31
Mg	1294	0.98	0.14	0.19 ± 0.06	0.28	0.41
Urea	1393	5.24	0.17	0.18 ± 0.05	0.30	0.41
Albumin	1294	32.8	2.95	0.27 ± 0.06	0.38	0.44
Globulin	1294	38.4	6.04	0.46 ± 0.06	0.40	0.51
A:G	1294	0.88	0.17	0.41 ± 0.06	0.40	0.49

Accuracies of genomic predictions resulting from univariate models are shown in Table 1. Empirical accuracies of the true breeding values were low to moderate (0.20 and 0.40), with more heritable traits having higher prediction accuracies. Theoretical accuracies, calculated from the standard errors estimated from Model 1, were higher than respective empirical accuracies, but the results of the 2 methods were in excellent agreement ($R^2 = 0.89$). Although low, our results are consistent with a small female reference population and low to moderate trait reliabilities (Gonzalez-Recio *et al.*

2014). We expect that increasing the size of the reference population and refining trait definitions to maximise heritabilities should improve genomic prediction accuracies. Given the cost and logistical challenges of blood sampling large numbers of cows, one method for dramatically increasing the number of phenotypes may be to use mid-infrared spectroscopy of milk to predict serum biomarker concentrations. Other high throughput metabolomic methods such as nuclear magnetic reasonance spectroscopy may also offer potential for the discovery of novel biomarkers of health in milk and serum, which could help to further improve the genomic prediction accuracies.

CONCLUSIONS

Our results show that genetic variance exists in the concentration of biomarkers of energy balance, protein nutrition, micromineral status and immune status in early lactation dairy cows. Genomic prediction accuracies were low, and while increasing the size of the reference population should theoretically improve accuracies, our results suggest that genomic prediction may allow identification of healthier cows that are less susceptible to diseases in early lactation.

ACKNOWLEDGEMENTS

The authors thank the farmers involved and acknowledge DairyBio (Agriculture Victoria, Dairy Australia and Gardiner Foundation) for funding this study.

REFERENCES

Cecchinato A., Bobbo T., Ruegg P.L., Gallo L., Bittante G. and Pegolo S. (2018) *J Dairy Sci.* **101**: 11108.

Gilmour A.R., Gogel B.J., Cullis W.S.J. and R. Thompson. (2015) ASReml. 4.1 ed. VSN International Ltd, Hemel Hempstead, HP1 1ES, UK.

Gonzalez-Recio O., Coffey M. P. and Pryce J. E. (2014) *J Dairy Sci.* **97**:7905.

LeBlanc S.J., Lissemore K.D., Kelton D.F., Duffield T.F. and Leslie K.E. (2006) *J Dairy Sci.* **89**: 1267.

Luke T.D.W., Rochfort S., Wales W.J., Bonfatti V., Marett L. and Pryce J. E. (2019) *J Dairy Sci.* **102**: 1747.

Mitchell R.G., Rogers G.W., Dechow C.D., Vallimont J.E., Cooper J.B., Sander-Nielsen U., and Clay J.S. (2005) *J Dairy Sci.* **88**: 4434.

Oikonomou G., Valergakis G.E., Arsenos G., Roubies N., and Banos G. (2008) *J Dairy Sci.* **91**: 2814.

Ospina P., Nydam D., Stokol T., and Overton T. (2010) *J Dairy Sci.* **93**: 1596.

Payne J., Dew S.M., Manston R., and Faulks M. (1970) *Vet. Rec.* **87**: 150.

Tsiamadis V., Banos G., Panousis N., Kritsepi-Konstantinou M., Arsenos G. and Valergakis G.E. (2016) *J Dairy Sci.* **99**: 5535.

Uribe H.A., Kennedy B.W., Martin S.W. and Kelton D.F. (1995) *J Dairy Sci.* **78**: 421.

van der Drift S.G., van Hulzen K.J., Teweldemedhn T.G., Jorritsma R., Nielen M. and Heuven H.C. (2012) *J Dairy Sci.* **95**: 6781.

Weigel K.A., Pralle R.S., Adams H., Cho K., Do C., and White H.M. (2017) *J Anim. Breed. Genet.* **134**: 275.

Yang J., Benyamin B., McEvoy B.P., Gordon S., Henders A.K., Nyholt D.R., Madden P.A., Heath A.C., Martin N.G., Montgomery G.W., Goddard M.E., and Visscher P.M. (2010) *Nat. Gen.* **42**: 565.

POLL TESTING EFFICIENCY, ACCURACY AND TRENDS IN AUSTRALIAN CATTLE

I.A.S. Randhawa[1], M.R. McGowan[1], L.R. Porto-Neto[2], B.J. Hayes[3], K.M. Schutt[4] and R.E. Lyons[1,4]

[1]School of Veterinary Science, University of Queensland, Gatton, QLD, 4343 Australia
[2]Agriculture and Food, CSIRO, St Lucia, QLD, 4067 Australia
[3]Centre for Animal Science, QAAFI, University of Queensland, St Lucia, QLD, 4072 Australia
[4]Neogen Australasia, University of Queensland Gatton campus, Gatton, QLD, 4343 Australia

SUMMARY

Poll testing is becoming common practice in Australia because it helps early prediction of head-phenotypes in calves to avoid dehorning and disbudding. It further improves cattle welfare by selecting breeding animals which are not carriers of horn alleles to avoid horned or scur calves, which may undergo physical dehorning. Current testing assays are limited to some breeds and often give inconclusive outputs as "Not Determined" or "No Results", because of ascertainment bias and marker failures. This study presents comparison of previously used poll testing assays (microsatellite and SNP based) with an optimized poll test (OPT) and poll allele distribution in beef breeds harbouring the Celtic and Friesian mutations.

INTRODUCTION

Cattle (*Bos taurus* and *Bos indicus*) species naturally evolved as horned and grow horns of different shapes and sizes as a unique phenotypic diversity between the breeds (Ajmone-Marsan *et al.* 2010). Modern cattle have further evolved their head-status as, *horn:* permanent pointy appendages attached to the skull, *scur:* pseudo horns loosely attached to the head-skin, or *poll:* complete absence of horn and scur (Wiener *et al.* 2015). Current management of horns in cattle production systems poses both welfare and economic challenges. Presence of horns poses potential hazards for other animals (injuries, damaged hides and bruised carcass), buildings, equipment and transport, and farm workers. Growth of horns can be avoided by physically dehorning. However management practices to remove horns or stop their growth remain invasive and painful for the animals (Knierim *et al.* 2015). Surgical dehorning affects growth and increases risks of infection and subsequently causes production loss and mortality, while there are also risks to workers and increased labour costs (Bunter *et al.* 2013). Genetically, the presence of horn is a qualitative trait which has been mapped on chromosome 1 (Mariasegaram *et al.* 2012). Although the genetic mechanisms underpinning *horn*, *scur* and *poll* status remain to be fully understood, inheritance of the conditions suggest that *poll* is the dominant gene, i.e., PP (polled) and pp (horned), and Pp animals generally present as poll or scurs (Capitan *et al.* 2011; Tetens *et al.* 2015).

Commercial DNA diagnostics for *poll* status are rapidly increasing and is routinely practiced by cattle farmers in Australia for informed and strategic breeding plans to reduce dehorning and disbudding. Microsatellite (MSAT) markers were used to establish the first-generation of poll testing assays (Mariasegaram *et al.* 2012). A total of 14 microsatellite markers have shown strong associations with polledness across different populations. In the poll-haplotype diagnostic test, 8 MSATs were initially used. However, the updated haplotype test contains 10 MSATs (Piper *et al.* 2014). MSAT single marker and haplotype assays were generally successful in Brahman and *Bos taurus* breeds respectively. The second generation of *poll* testing is single nucleotide polymorphism (SNP) based, and is rapidly replacing MSATs as SNP genotyping technologies become more accessible and cost effective. SNP testing has only become available with sequencing of chromosome 1 which has identified genetic heterogeneity across breeds linking the polledness with 4 distinct insertion-deletions at the poll locus, called, Celtic (Pc), Friesian (Pf), Mongolian (Pm) and Guarani (Pg) (Medugorac *et al.* 2012; Rothammer *et al.* 2014; Wiedemar *et al.* 2014; Wiener *et al.* 2015; Utsunomiya *et al.* 2019). Notably, Pc and Pf alleles are prevalent in most of the naturally polled cattle. Predictions in SNP-based diagnostic assays rely on genetic linkage between the contiguous markers in the *poll* locus harbouring Pc and Pf (Medugorac *et al.* 2012; Rothammer *et al.* 2014; Wiedemar *et al.* 2014). Up to 10 SNPs with strong LD with the known *poll* alleles (Pc and Pf) are available in various cattle breeds and the current poll testing (CPT) assays include 5-8 SNPs.

Poll testing assays help horn management in cattle herds by early predictions of head-status. However, some breeds are disadvantaged because of ascertainment bias, marker types and other factors relating to these diagnostic tools (Connors *et al.* 2018). Here, we have investigated the efficiency and limitations of available assays which use MSATs or SNPs and propose an optimized poll testing (OPT) to efficiently diagnose the presence of Pc and Pf across 10 breeds of cattle.

MATERIALS AND METHODS

Animal ethics approval for tail-hair samples, head phenotypes, genotyping and sequencing were obtained (AEC # SVS/301/18). Genomic data of 37,694 animals across ten breeds (Table 1) was used to compare the available poll test results using MSATs (n=20,534) and SNPs (n=18,793) based assays, and with the proposed SNPs-based OPT (n=18,793). To assess the phenotypic concordance, information about their head-status (*horn*, *scur*, and *poll*) from 6,930 (out of 18,793) registered animals of 8 breeds in Australia (excluding Angus and Wagyu) were obtained from the BREEDPLAN database (http://breedplan.une.edu.au/index.php). Hair samples of Brahman (n=60 out of 2691) were used from available stocks for targeted DNA sequencing. In addition, collection of hair samples and assessing of head-status of Droughtmaster (n=84) cattle from UQ's research herd were performed for phenotypic and genetic concordance for validation.

First, we compared the efficiency of available MSAT and CPT assay based predictions using available poll test results on different samples, because most animals were tested with either MSATs or SNPs based markers. Second, 10 SNPs in the poll region were investigated for genotyping failures, monomorphism and overall informativeness to develop the optimized poll testing (OPT). Third, OPT based predictions were evaluated for phenotypic concordance with BREEDPLAN data (available for 6,930 registered animals only) and finally, validated by UQ-herd.

Table 1. Breed samples tested with microsatellite (MSAT) and SNP-based assays

Breeds	Samples	MSAT tested	SNP tested	Tested by both
Angus	1630	28	1602	0
Brahman	7009	4532	2691	214
Brangus	754	745	37	28
Charolais	3148	2666	900	418
Droughtmaster	2611	2223	708	220
Hereford	6424	3485	3341	402
Limousin	2193	2124	207	138
Santa Gertrudis	4427	4306	136	15
Shorthorn	316	224	121	29
Wagyu	9182	201	9050	69
Total	37,694	20,534	18,793	1,533

RESULTS AND DISCUSSIONS

We found that MSAT-based testing failed to predict genotypes (HH, PH or PP) for *horn* or *poll* conditions in 11.7% of tests performed across all breeds combined and these were reported as Not Determined (Figure 1). The CPT diagnostic test significantly reduced the frequency of "No Results" being reported, with the notable exception of Zebu (*Bos indicus*) and their cross-bred cattle which were still constrained by No Result issues (Figure 1). Out of 18,793 SNP-tested samples, 5.48% had failed to identify an unambiguous genotype with CPT and hence were predicted as No Result with the majority of these no results being Brahman (18.3%) and Brangus (22.2%).

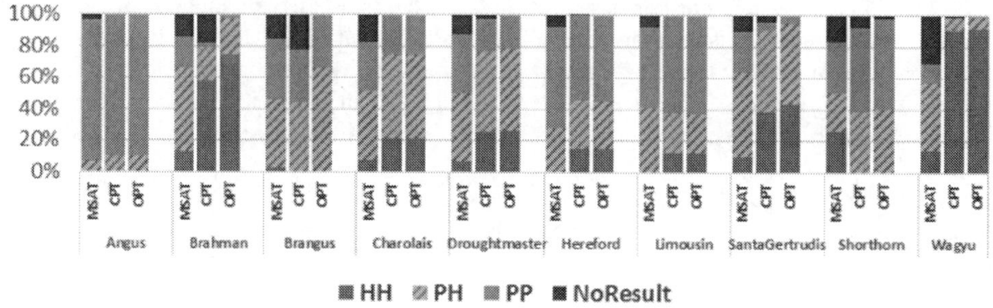

Figure 1. Comparison of poll testing by microsatellites (MSAT) and SNP-based current poll test (CPT) and optimized poll test (OPT) in 10 breeds of Australia

Utility of 10 SNPs was investigated for prediction of Pc and Pf allele prevalence in different breeds. Initially, a single SNP (rs800947704) was found to be failing in genotyping assays in 8 of the 10 breeds (n=662 out of 16,828), especially in Brahman which had a 14.4% failure rate. Targeted sequencing of a 1,098 bp (1,654,527-1,655,625) fragment surrounding rs800947704 in Brahman samples showed that the probe region (within 50bp of target SNP) was unstable causing genotyping failure. Hence the SNP was rejected as a useful marker. Further investigation found that 4 other SNPs were not reliable for accurate predictions, of which 2 SNPs (rs798116945 and rs800767839) were highly monomorphic and 2 SNPs (rs799187101 and rs799920960) were not in complete LD. The No Results predictions were caused by issues with 1 or more of these 5 SNPs, indicating that they were unsuitable for the poll testing assays in Zebu cattle. The other 5 SNPs passed the inclusion criteria for the OPT predictions and were evaluated using 18,589 samples of European, Zebu and their cross-bred populations. Previously successful predictions (n=18,019) were found to remain unchanged (100%) using OPT relative to the original prediction. Of the previously unsuccessful (No Results, n=570) samples, 569 (99.8%) were effectively classified into one of the head-status predictions. Overall allele frequencies were found as H = 0.57, Pc = 0.40 and Pf = 0.03 (Figure 2). Genotype distributions (HH = 40.4, HP = 32.3, PP = 27.3) were different than phenotypic rates (*Horn* = 42.7, *Scur* = 6.2, *Poll* = 51.1), predominantly because many heterozygous animals (HP) are *poll* (Table 2). Numbers of OPT-based genotypes (and associated phenotypes as a %) in the UQ herd were; HH =15 (*horn* 100%), HP = 45 (*scur* 51% and *poll* 49%) and PP = 24 (*poll* 100%).

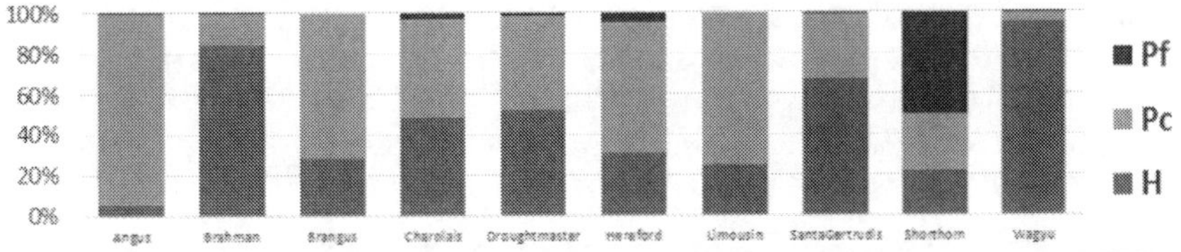

Figure 2. Frequency (%) of *horn* (H) and *poll* (Pc & Pf) alleles in 10 breeds (n=18,793)

Table 2. Concordance between OPT genotypes and BREEDPLAN phenotypes in 8 breeds

OPT genotypes	Number tested	Phenotypic concordance (%) with head-status		
		Horn	*Scur*	*Poll*
HH	2800	94.8 %	3.10 %	2.10 %
HPc	2121	13.5 %	15.7 %	70.8 %
HPf	120	2.50 %	5.80 %	91.7 %
PcPc	1595	0.75 %	0.25 %	99.0 %
PcPf	267	0.37 %	0.37 %	99.2 %
PfPf	27	-	-	100 %

OPT-based genotypes have shown high concordance with known head-status, except for HPc and HPf that can result in *scur* phenotypes, with indications in literature pointing to the probability that sex (female) and sex hormones (steer) sway heterozygotes to be *poll* (Randhawa *et al.* 2019). It is very unlikely that HH animals can be either *scur* (3.1%) or *poll* (2.1%). However, inaccuracies with phenotypic recording are common (Connors *et al.* 2018). Overall, using the OPT can effectively resolve MSAT and CPT limitations to accurately predict true *poll* conditions (over 99%) in *Bos taurus*, *Bos indicus* and cross-bred beef cattle. We continue to investigate the genetics of the *scur*.

ACKNOWLEDGEMENTS

This study is funded by the Meat and Livestock Australia, project L.GEN.1713.

REFERENCES

Ajmone-Marsan P., Garcia J.F. and Lenstra J.A. (2010) *Evol Anthropol* **19**: 148.

Bunter K.L., Johnston D.J., Wolcott M.L. and Fordyce G. (2013) *Anim. Prod. Sci.* **54**:.25.

Capitan A., Grohs C., Weiss B., Rossignol M.-N., Reversé P. and Eggen A. (2011) *PLoS One* **6**:.e22242.

Connors N.K., Tier B. and Johnston D.J. (2018) Current status of Australia's diagnostic poll haplotype test. *Wld Congr. Genet. Appl..Livest. Prod.* 1**1:** 344.

Knierim U., Irrgang N. and Roth B.A. (2015) *Livestock Sci.* **179**:.29.

Mariasegaram M., Harrison B.E., Bolton J.A., Tier B., Henshall J.M., Barendse W. and Prayaga K.C. (2012) *Anim. Genet.* **43**:.683.

Medugorac I., Seichter D., Graf A., Russ I., Blum H., Göpel K.H., Rothammer S., Förster M. and Krebs S. (2012) *PLoS One* **7**:.e39477.

Piper E.K., Tier B. and Henshall J.M. (2014) *Wld Congr. Genet. Appl. Livest. Prod.* **10**: 1.

Randhawa I.A.S., Lyons R.E., Hayes B.J., Porto-Neto L.R. and McGowan M.R. (2019) *Proc. Assoc. Advmt. Anim. Breed. Genet.* **23**: .

Rothammer S., Capitan A., Mullaart E., Seichter D., Russ I. and Medugorac I. (2014) *Genet. Sel. Evol.* **46**: 44.

Tetens J., Wiedemar N., Menoud A., Thaller G. and Drögemüller C. (2015) *Anim. Genet.* **46**: 224.

Utsunomiya Y.T., Torrecilha R.B.P., Milanesi M., Paulan S.d.C., Utsunomiya A.T.H. and Garcia J.F. (2019) *Anim. Genet.* **50**: 187.

Wiedemar N., Tetens J., Jagannathan V., Menoud A., Neuenschwander S., Bruggmann R., Thaller G. and Drögemüller C. (2014) *PLoS One* **9**: e93435.

Wiener D.J., Wiedemar N., Welle M.M. and Drögemüller C. (2015) *PLoS One* **10**: e0127691.

DETERMINING THE GENE EXPRESSION PROFILES OF 17 CANDIDATE GENES FOR HOST RESISTANCE TO TICKS IN SOUTH AFRICAN BEEF CATTLE

B. Dube[1], K.J. Marima[1,2], C.M. Marufu[3], F. Muchadeyi[4], N.O. Mapholi[5], N.N. Jonsson[6] and K. Dzama[2]

[1]Agricultural Research Council-Animal Production, P. Bag X2, Irene, 0062, South Africa
[2]Department of Animal Sciences, University of Stellenbosch, P. Bag X1, Matieland 7602, South Africa
[3]Faculty of Veterinary Sciences, University of Stellenbosch, Onderstepoort 0110
[4]Department of Life and Consumer Sciences, University of South Africa, Florida, 1710, South Africa
[5]Biotechnology Platform, Agricultural Research Council-Onderstepoort Veterinary Institute, Onderstepoort 0110
[6]School of Veterinary Medicine, Glasgow University, Glasgow, Scotland

SUMMARY

Beef production is under threat from tick infestation problems, which have so far not been successfully controlled because of shortcomings in chemical and vaccine usage. The variation in resistance to ticks among breeds provides an opportunity to determine the mechanisms that underlie resistance to ticks. Brahman, Nguni and Angus animals were used to study gene expression following artificial infestation with *Rhipicephalus* ticks. Skin biopsies were collected, and RNA extracted for gene expression analyses. Variation in gene expression was observed in genes involved in discouraging long-term supply of blood meal to the tick and those associated with immune responses.

INTRODUCTION

Cattle ticks pose the risk of inflicting deleterious effects on production traits by hindering the growth and weight gain, productivity, fertility, as well as the meat quality of cattle (Marufu *et al.* 2011). The profitability of the beef cattle industry may be compromised as many beef enterprises maximise their profit margins by concentrating more on fertility and a high weaning weight off the veld (Mapholi 2014). Current tick control methods include grazing practices and use of acaricides and vaccines, which have however not been successful in completely eradicating ticks. The widespread use of acaricides to control tick burdens places strong evolutionary pressure towards the emergence of new chemical-resistant strains of ticks, faster than new chemicals can be produced (Gasbarre *et al.* 2009). Ticks also mutate the targeted epitopes into unfamiliar forms and nullify the effect of a particular vaccine. There is also increasing public concern about chemical residues in animal products and the environment (Mapholi *et al.* 2014). A relatively simple and cheap method of reducing the effects of parasite infestation would be use of genetically tick-resistant animals. Resistance to ticks among cattle breeds is variable (Muchenje *et al.* 2008) and this presents an opportunity to exploit the host's resistance to ticks in developing more cost-effective and sustainable tick control programs. Tick bites trigger immune responses in the host animal's body by releasing specific proteins that fight infection at the site, suggesting that response to tick infestation may be under genetic control (Marufu *et al.* 2014). Thus, a better insight into the mechanism of resistance to ticks may be achieved by identifying the genes expressed as a result of tick infestation. The objective of the study was therefore to evaluate the genetic expression differences in different cattle breeds in response to infestation by two different tick species.

MATERIALS AND METHODS

Thirty-six cattle, comprising of 12 Nguni bulls, 12 Brahman bulls, six Angus heifers and six bulls aged between 12 and 15 months were artificially infested with unfed *Rhipicephalus* tick larvae. Half of the animals per breed were infested with *R. microplus*, while the other half were infested with *R. decoloratus*. The Angus groups were further divided in terms of sex, with three animals of each sex being infested with *R. microplus* and the remaining three being infested with *R. decoloratus* larvae. Skin biopsies were collected pre-infestation and the animals' mid-back area was shaved and a calico bag was attached, after which the tick larvae were placed inside. Twelve hours post-infestation, the bags were opened and skin biopsy samples were collected from the tick bite sites. The biopsies were preserved in 5 ml RNAlater® RNA stabilization Reagent (Qiagen) and stored at -80°C. About 50-100 mg of each biopsy sample was used for RNA extraction, which was conducted following the TRIzol® Reagent protocol. Samples showed separation of the 28S and 18S bands with partial smearing after running the 1% agarose gel. Purity test was done using the NanoDrop spectrophotometer to ensure that all samples had 260/280 values $\geq$ 1.70. Samples which were below this value were then repurified.

cDNA was synthesised using equal amounts of total RNA and the RT^2 First Strand Kit was used according to manufacturer's protocol. To obtain optimal results, 400ng of total RNA per sample was used to obtain a total cDNA volume of 30µl. Then, genetic analysis was done, where threshold cycle (C_T) values generated were used to calculate the expression levels of a panel of 17 candidate genes using the RT^2 Profiler PCR Array Data Analysis Webportal (SABioscience - Qiagen). The panel of genes included cytokines (*TLR5, TLR7, TLR 9, TRAF6, CD14*), chemokines and their receptors (*CCR1, CCL2, CCL5*), toll-like receptors (*IL-1β, CXCL8, IL-10, TNF*) and other candidate genes (*BDA20, OGN, TBP, LUM, B2M*). The fold change value of each gene, normalised against the reference genes Ribosomal protein, large, P0 (*RPLP0*), 18S ribosomal RNA (*RN18S1*), Glyceraldehyde-3-phosphate dehydrogenase (*GAPDH*) and Beta-actin-like (*LOC616410*). Fold change was calculated using the $\Delta\Delta C_T$ method described by Livak and Schmittgen (2001). An analysis of variance for two-way factorial designs was used to test the interaction between the main effects, breed and tick species, for each of the genes. The primers for each of the genes of interest were custom designed by Qiagen using forward and reverse primer sequences associated with the GenBank and UniGene reference sequence numbers.

Ninety-six-well RT^2 Profiler PCR arrays were used for the real-time PCR analyses and facilitated high-throughput focused expression analysis on the genes of interest. Each plate enabled the analyses of four samples at a time to generate amplification data for 17 genes of interest and four reference genes per sample. The gene expression profiles of selected genes were examined using the ABI 7500 real-time PCR thermocycler. A PCR components mix was prepared in a 5ml tube for each sample according to manufacturer's protocol. The arrays were also fitted with primers designed to amplify three Qiagen recommended quality control parameters, namely Bovine Genomic DNA Control (BGDC), Reverse Transcription Control (RTC) and Positive PCR Control (PPC).

RESULTS AND DISCUSSION

Four reference genes, namely *RPLP0, RN18S1, GAPDH* and *LOC616410* were chosen to normalise the data. The average C_T values for the reference genes were 24.153, 15.717 and 25,399 for *RPLP0, RN18S1* and *GAPDH*, respectively. There was no significant interaction between the main effects, breed and tick species, observed for any of the genes, which may suggest similar responses to both tick species' infestations by these breeds. While the expression of most of the genes did not differ significantly according to breed, the expression profiles of genes *TRAF6, TBP, LUM* and *B2M* were significantly different among breeds. There were differences between the Nguni and Angus in the expression levels of *TBP* and *TRAF6* ($P < 0.05$), as well as between the Brahman and Angus in the expression levels of *LUM* and B2M ($P < 0.01$). Increases in the expression levels of six genes (*CCL2, CCL26, CD14, OGN, LUM,* and *B2M*) post-infestation for all breed × tick species treatment groups were observed. Five genes (*CCR1, TLR5, TRAF6, TBP, BDA20*) increased expression or remained approximately equal after infestation with ticks for all groups. Mixed results were obtained in the breed × tick species groups for expression levels for the genes *IL1-β, TLR7* and *TLR9*, while the expression levels of three genes (*CXCL8, IL10, TNF-α*) decreased or remained the same after tick challenge in all breed × tick species groups.

The results of this study were broadly consistent with previous work (Wang *et al.* 2007; Piper 2010). The genes encoding the extracellular matrix constituents, most importantly, *LUM* and *B2M*, were upregulated at much higher levels in the high (Brahman) and intermediate (Nguni) resistance breeds than the genes involved in immune system regulation and inflammatory responses. This was in agreement with the observation by Piper *et al.* (2010), where there was upregulation of genes encoding constituents of the extracellular matrix in the tick-resistant Brahman in comparison to the susceptible Holstein-Friesian cattle. Kongsuwan *et al.* (2010) attributed resistance to ticks to the epidermal permeability barrier of the skin, which is associated with the heightened expression of these genes in the tick-resistant Brahman cattle. The genes, *LUM, B2M* and *TBP* induced resistance to ticks by promoting continued cellular regeneration, tissue repair and detoxification of the tick bite site, instead of initiating host immune responses. This activated the mechanism required to discourage long term supply of blood meal to the tick. These genes, except *TBP*, were upregulated in all treatment combination groups, excluding the Angus-*R. microplus* group.

The highest upregulation values were detected for *LUM* in the Brahman treatment groups and Nguni-*R. microplus*. As a gene that encodes a member of the small leucine-rich proteoglycan (Weizmann Institute of Science 2016a), *LUM* serves in conjunction with *OGN* to induce immune responses. The gene *OGN* similarly presented higher upregulation values than the rest of the genes of interest. Both *LUM* and *OGN* are capable of regulating fibril organisation and circumferential growth as well as epithelial cell migration in the process of tissue repair at the tick bite site (Weizmann Institute of Science 2016a). The significantly high expression level of *LUM* in the Brahman more than the Angus suggested that the Brahman had a stronger capacity to prevent tick feeding through continuous tissue repair than the Angus. This was true for both tick species. The results suggest that *LUM* can be used as a biomarker for resistance to both *R. microplus* and *R. decoloratus* tick species.

Unlike *LUM*, the significant differences in the expression levels of *TBP* and *B2M* in different treatment groups were unexpected. *TBP* is a component of the RNA polymerase III; hence it was expected to behave like a housekeeping gene exhibiting stable expression levels in all treatment combinations to facilitate continued cell growth. While *TBP* was upregulated in most treatment groups, the gene displayed a downregulated but stable expression level in Angus-*R. microplus* group. This may be attributed to the stressful conditions inflicted by the tick infestations, which might have resulted in the regulatory protein *Maf1* repressing RNA polymerase III activity (Vannini *et al.* 2010). The *B2M* gene is a component of the MHC class I that is responsible for presenting peptide antigens (including tick antigens) to the immune system, while simultaneously forming amyloid fibrils in pathological challenges (Weizmann Institute of Science 2016b). Therefore, the significantly low *B2M* expression levels produced by the Angus animals may imply that this breed's nucleated cells had a poor capacity to detect the tick antigens to prompt host immune responses.

CONCLUSIONS

The differences in the expression profiles of different genes in breeds of different levels of resistance to ticks may provide an insight into the mechanism of resistance to ticks. Genes that show variation in responses to tick infestation among breeds are involved in discouraging long term supply of blood meal to the tick, although there was some variation in the genes associated with immune responses. The gene *LUM* may be used as biomarker for resistance to ticks. Given that resistance to ticks is a polygenic trait, deep sequencing may reveal more genes associated with this trait. Further studies should be conducted to investigate the association between skin permeability, genes expressed and resistance to ticks.

REFERENCES

Gasbarre L.C., Smith L.L., Lichtenfels J.R. and Pilitt P.A. (2009) *Vet. Parasitol.* **166**: 281.

Kongsuwan K., Piper E.K., Bagnall N.H., Ryan K., Moolhuijzen P., Bellgard M., Lew A., Jackson L. and Jonsson N.N. (2008) In 'Animal Genomics for Animal Health' pp. 77-88, editor M.-H. Pinard, Developments in biologicals, Basel.

Livak K.J. and Schmittgen T.D. (2001) *Methods.* **25**, 402.

Mapholi N.O., Marufu M.C., Maiwashe A., Banga C.B., Muchenje V., Macneil M.D., Chimonyo M. and Dzama K. (2014) *Ticks Tick-Borne Dis.* **5**: 475.

Marufu M.C., Dzama K. and Chimonyo M. (2014) *Exp. Appl. Acarol.* **62**: 241.

Marufu M.C., Qokweni, L., Chimonyo M. and Dzama K. (2011) *Ticks Tick-Borne Dis.* **2**: 172.

Muchenje V., Dzama K., Chimonyo M., Raats J.G and Strydom P.E. (2208) *Animal.* **2**: 298.

Piper E.K., Jackson L., Bielefeldt-Ohmann H., Gondro C., Lew-Tabor A.E. and Jonsson N.N. (2010) *Int. J. Parasitol.* **40**: 431.

Vannini A., Ringel R., Kusser A.G., Berninghausen O., Kassavetis G.A. and Cramer P. (2010) *Cell.* **143**: 59.

Wang Y.H., Reverter A., Kemp D., McWilliam S.M., Ingham A., Davis C., Moore R.J. and Lehnert S. (2007) *Aust. J. Exp. Agric.* **47**:1397.

Weizmann Institute of Science. 2016a. LUM Gene. http://www.genecards.org/cgi-bin/carddisp.pl?gene=LUM [2016, November 04].

Weizmann Institute of Science. 2016b. B2M Gene. http://www.genecards.org/cgi-bin/carddisp.pl?gene=B2M [2016, November 04].

PRODUCTION AND POLLEDNESS: GENETIC CORRELATIONS BETWEEN TARGET TRAITS IN BEEF CATTLE

I.A.S. Randhawa[1], L.R. Porto-Neto[2], M.R. McGowan[1], B.J. Hayes[3], K.M. Schutt[4] and R.E. Lyons[1,4]

[1]School of Veterinary Science, University of Queensland, Gatton, QLD, 4343 Australia
[2]Agriculture and Food, CSIRO, St Lucia, QLD, 4067 Australia
[3]Centre for Animal Science, QAAFI, University of Queensland, St Lucia, QLD, 4072 Australia
[4]Neogen Australasia, University of Queensland Gatton campus, Gatton, QLD, 4343 Australia

SUMMARY

Keeping horns or physically removing them pose economic and welfare risks, therefore, producing naturally hornless (polled) animals would make livestock production more humane and sustainable. The cattle industry is rapidly breeding polled cattle with the aid of advanced genomic technologies. However, some reluctance has been noticed due to perceived trade-offs associating polled animals with increased inbreeding and loss of production for various traits. Estimated breeding values (EBVs) of 243,330 animals, born in last 70 years, from three beef breeds (Brahman, Droughtmaster and Hereford) were obtained from BREEDPLAN. We have compared eight economically important traits for production (birth weight, mature cow weight carcase weight, retail beef yield, intramuscular fat and milk yield) and reproduction (scrotal size and days to calving). At various levels of EBVs accuracy (60%, 75%) a few significant differences of small effect sizes were found in no consistent direction of either horn or poll cohorts. Overall, we conclude that polledness had no detrimental effects on target traits of beef cattle.

INTRODUCTION

Many modern cattle are naturally horned, which pose risks to animals and workers, and management practices to remove horns are expensive, painful and unsafe (Bunter *et al.* 2013; Thompson *et al.* 2017). Alternatively, with growing support of consumers, genetic polledness is being progressively adopted, as a welfare-oriented and an effective management approach, to breed hornless (polled) cattle. Poll cattle have a long-history being kept in colder regions for easy confinement of cattle, however, commercial adoption of genetically polled cattle can sometimes face resistance because of a few perceived trade-offs associating polledness with increased inbreeding and loss of production for various traits (Schafberg and Swalve 2015). In dairy cattle, the frequency of polled bulls is so low that including this trait as selection criteria generally results in higher inbreeding and thus slower genetic improvement (Gaspa *et al.* 2015; Windig *et al.* 2015; Scheper *et al.* 2016). However no significant differences were found between horn and poll cattle in dairy traits (Onaciu *et al.* 2012) at the population level.

In beef cattle, the prevalence of natural mating and higher proportions of males in the herd suggest the need for better horn management by adapting to poll breeding. Randhawa *et al.* (2019a) noted that horn appearance and agonistic behaviour were generally male centric. Some beef cattle breeds have already achieved fixation of polledness, e.g., Angus, however, many beef breeds grow horns and entail impact assessment for poll breeding schemes. The impact of polledness on production and fertility traits of different breeds and cross-bred cattle have generally shown no significant difference for several beef traits, such as; live weight, growth rate, carcass weight and quality, dystocia, fertility and mortality rates (Frisch *et al.* 1980; Stookey and Goonewardene 1996; Kommisrud and Steine 1997; Goonewardene *et al.* 1999).

For comparison of genetic merit between animals, evaluation of genetic effects of a trait is more practical by substituting estimated breeding value (EBV) for phenotypic values. EBV is a tool of genetic evaluation between animals for a particular trait by accounting for heritability and fixed effects. EBVs for a quantitative trait capture the aggregate additive genetic value by using phenotype of an animal together with phenotypes of its relatives (Henderson 1975). EBVs denote that how an animal's genetics is different than the genetic base, for example breed averages. Accuracy of EBV predictions increases as more information become available for animal's direct performance, pedigree and progeny. BREEDPLAN (http://breedplan.une.edu.au/index.php) is an advanced genetic evaluation system, implemented for national beef recording scheme in Australia to compute EBVs, which can be used to highlight the genetic differences in various beef traits between various head-status cohorts. The aim of this study is to compare the genetic merit of naturally horned and polled animals for eight economically important traits for three breeds of Australian beef cattle.

MATERIALS AND METHODS

There were 243,330 animals from the Brahman (BRH), Droughtmaster (DRM) and Hereford (HFD) breeds from the BREEDPLAN database included in this study (Table 1). Animals were born between 1950 and 2018 and were classified into head-status cohorts as; horn (101,287), scur (5,297) and poll (131,792). BREEDPLAN EBVs are classified for interpreting accuracy, such that less than 50% = preliminary, 50-74% = medium, 75-90% = medium-high, and above 90% = high accuracy estimates of the animal's true breeding value. EBV records were obtained for eight traits where the EBV accuracy $\geq$ 60%. The number of EBVs for each trait and cohort are given in Table 1. Total number of samples for each breed at various accuracy of EBVs thresholds (%) and birth years were;

EBV 60% and born 1950 - 2018 (BRH: 50,392, DRM: 4,545, and HFD: 188,393) EBV 75% and born 2000 - 2018 (BRH: 4,210, DRM: 365, and HFD: 14,788).

In addition, a subset of 5,586 animals (BRH: 2,476, DRM: 323, and HFD: 2,787) had genomic horn and poll predictions obtained from the recently developed optimised poll testing (OPT) (Randhawa *et al.* 2019b). Samples with phenotype-genotype discrepancy (n = 374) were excluded, which were previously deemed as phenotyping and data recording errors (Randhawa *et al.* 2019b).

Table 1. List of eight traits and number of breed-wise samples for head-status for EBVs (60% accuracy)

Traits	Acronym (unit)	Brahman (N)			Droughtmaster (N)			Hereford (N)		
		Horn	*Scur*	*Poll*	*Horn*	*Scur*	*Poll*	*Horn*	*Scur*	*Poll*
Birth Weight	BW (kg)	27664	872	2818	826	295	2902	58642	3723	124066
Mature Cow Weight	MCW (kg)	41620	1054	4726	722	301	3248	28250	2021	47013
Carcase Weight	CW (kg)	18228	572	1607	417	131	1546	15632	1108	24842
Retail Beef Yield	RBY (%)	313	-	3	18	7	56	1340	164	3182
Intra Muscular Fat	IMF (%)	271	35	40	3	2	35	2277	215	5593
Milk Yield	MY (kg)	7667	127	652	760	186	2510	27965	1332	46819
Scrotal Size	SS (cm)	13213	377	1400	1025	520	3000	17318	2289	30873
Days to Calving	DTC (days)	3351	187	379	-	-	-	60	14	117

For each trait, four comparisons were made between the phenotype-based cohorts within breeds to screen the impacts of levels of medium (60%) and medium high (75%) EBV accuracy, birth years and poll test genotype-based composition of cohorts. Data analyses were conducted using the R program (R Core Team 2018). Because highest number of samples with EBVs at medium level accuracies represent extensively the available herds of beef cattle, therefore, summary statistics of Mean±SD were computed between the three cohorts (horned, scurred and polled) at 60% accuracy. The descriptive statistics including ANOVA, p-values by Tukey multiple comparisons of means (95%

family-wise confidence level) and pairwise comparisons using t-tests with pooled SD, and effect size (Cohen's *d*) were computed and probed for the poll-*vs*-horn cohorts.

RESULTS AND DISCUSSION

Table 2 shows the distribution of EBVs of 8 traits between within-breed cohorts of horn-status and provides an overview of significance levels and effect sizes. Of the eight traits, desirability for breeding differ for higher (MCW, CW, RBY, MY, IMF, SS) and lower (DTC) EBV values. Note that because BW is the major genetic cause of calving difficulty, small or moderate BWs are more favourable. Our results were computed for poll-*vs*-horn comparisons for trait-wise EBVs at medium to medium-high accuracies, as the number of animals with high EBVs accuracies (≥ 90%) were too low, e.g., BRH: 44, DRM: 11 and HFD: 593. At medium accuracy of 60%, comparisons of mean EBVs of several traits between poll and horn were highly significant ($p \leq .001$), however, the effect size were small ($d \sim 0.2$). As we increased the accuracies to 75%, there were a very few significant differences (Table 2).

Table 2. Descriptive statistics, effect sizes and statistical significance (t-test) of eight traits in three breeds of Australian beef cattle

Breed	Trait	Avg.§	Mean ±SD within cohorts at 60%¶			d^ and p° values between Poll-Horn		
			Horn	*Scur*	*Poll*	60%¶	2K†,75%	OPT‡
BRH	BW	2.5	2.37±1.95	1.73±1.85	1.83±1.77	-0.29 ***	-0.24 *	0.09
	MCW	41	32.4±22.6	33.9±22.3	30.2±22.4	-0.10 ***	-0.01	0.09
	CW	22	17.0±9.82	16.6±9.10	15.8±9.34	-0.12 ***	-0.41	-0.02
	RBY	0.6	0.13±0.86	-	-0.10±0.46	-0.33	-	0.22 **
	IMF	-0.1	-0.02±0.29	-0.13±0.27	-0.10±0.25	-0.29 *	-1.15	0.07
	MY	-1.0	-0.72±2.83	-1.40±2.91	-0.65±2.85	0.02 *	-0.19	0.27***
	SS	0.7	0.71±1.28	1.23±1.42	0.95±1.17	0.20 ***	0.18 **	0.12***
	DTC	-0.9	-5.09±7.40	-5.28±6.72	-5.41±7.54	-0.04	0.45 *	-0.01
DRM	BW	0	-0.43±1.56	-0.22±1.34	-0.23±1.34	0.13 **	0.32 *	0.01
	MCW	25	25.3±17.8	22.4±18.4	23.0±20.3	-0.12 *	0.05	0.24
	CW	14	14.8±6.22	14.4±6.35	13.9±6.18	-0.16 *	-0.21	0.01
	RBY	0.6	0.76±1.09	0.69±0.63	0.68±0.70	-0.09	-	-0.36
	IMF	0.0	0.20±0.87	0.00±0.00	0.01±0.48	-0.26	-	0.04
	MY	4.0	6.00±3.52	5.32±3.49	5.17±3.38	-0.24 ***	-0.34	0.11
	SS	1.3	1.15±1.03	1.36±1.12	1.36±1.04	0.20 ***	0.58	0.35 *
HFD	BW	4.4	4.25±2.20	4.7±1.97	4.12±1.97	-0.07 ***	-0.24 ***	-0.20***
	MCW	68	56.3±22.7	66.9±22.6	58.8±22.7	0.11 ***	0.08	0.14 *
	CW	50	32.0±14.8	42.9±16.0	35.7±15.9	0.24 ***	0.06	0.54***
	RBY	0.8	0.95±0.91	0.79±0.87	0.72±0.89	-0.25 ***	-0.78 *	-0.37***
	IMF	0.4	0.15±0.61	0.43±0.67	0.23±0.68	0.13 ***	0.00	0.54***
	MY	16	9.86±5.11	10.9±5.93	8.69±5.78	-0.21 ***	-0.29 ***	0.09 *
	SS	2.0	1.37±0.90	1.79±1.02	1.69±0.97	0.34 ***	0.27 ***	0.65***
	DTC	-2.7	-1.10±2.14	-2.07±3.13	-1.92±2.57	-0.34	-0.71	-0.62***

§ Avg. is breed averages of each trait EBVs for the 2017 born calves (Source: BREEDPLAN).

¶ EBVs were used at accuracies ≥ 60% and 75% thresholds.

† EBVs were used from animals born between 2000 and 2018.

‡ EBVs were used from animals which were also genotyped with OPT (optimized poll testing).

^ Cohen's *d* represents effect size in pair-wise trait comparison (Sawilowsky 2009), and interpreted as; *d* 0.01: very small, *d* 0.20: small, *d* 0.50: medium, *d* 0.80: large, *d* 1.20: very large, *d* 2.0: huge.

° Significance differences between Poll and Horn cohorts were calculated by t-test and *p*-values results are denoted by $p \leq 0.001$: ***, $p \leq 0.01$: **, and $p \leq 0.05$: *.

Genetic merit for most traits in the 3 breeds experienced significant changes within the last two decades. Therefore, by using cohorts born since 2000 and EBVs' accuracy ≥ 75% might have reliably found very few significant differences of small effects. For instance, the effect of head-status on BW (kg) was significant across 3 breeds, however small effect sizes suggesting that on average poll animals were -0.47kg (BRH), 0.45kg (DRM) and -0.48kg (HFD) different than horn animals at birth. Another significant difference was noted for SS (cm), higher SS is associated with increased semen production, and results showed that poll cohorts were better by 0.24cm (BRH), 0.68cm (DRM) and 0.28cm (HFD). DTC (days) is another important trait, measured from female introduced to bull until subsequent calving and is mainly affected by the time taken to conceive. DRM are not recorded for DTC, while poll BRH and HFD showed 3.94 (p=0.05) and -1.92 DTC, respectively. Although, BRH showed significant difference for DTC, however, opposite trends in BRH and HFD suggested that the polledness may not be directly involved. Our results by using OPT genotypes to classify poll and horn cohorts were consistent, except for HFD which may have been affected by relative very small cohort-size of horned animals. Overall, our results coincide with previous findings (Frisch *et al.* 1980; Stookey and Goonewardene 1996; Kommisrud and Steine 1997; Goonewardene *et al.* 1999). A few significant differences of small effects were found in some beef traits for horned animals however the claims were not sustained with EBV estimates at high accuracies (≥ 75%). On the other hand, the polled animals were consistently significantly better for fertility traits (SS) than the horned animals in three breeds. This study concludes that poll and horn animals have equal genetic potential for production, carcass and fertility traits.

ACKNOWLEDGEMENTS

Meat and Livestock Australia financially supported the project (L.GEN.1713).

REFERENCES

Bunter K.L., Johnston D.J., Wolcott M.L. and Fordyce G. (2013) *Anim. Prod. Sci.* **54**: 25.

Frisch J.E., Nishimura H., Cousins K.J. and Turner H.G. (1980) *Anim. Sci.* **31**: 119.

Gaspa G., Veerkamp R.F., Calus M.P.L. and Windig J.J. (2015) *Livest. Sci.* **179**: 86.

Goonewardene L.A., Price M.A., Liu M.F., Berg R.T. and Erichsen C.M. (1999) *Can J Anim Sci* **79**: 383.

Henderson C.R. (1975) *J. Dairy Sci.* **58**: 1731.

Kommisrud E. and Steine T. (1997) *Reproduction in Domestic Animals* **32**: 221.

Onaciu G., Pentelescu O. and Jurco E. (2012) *ABAH Bioflux* **4**: 58.

R Core Team (2018) R: A language and environment for statistical computing. R Foundation for Statistical Computing, Vienna, Austria.

Randhawa I.A.S., Lyons R.E., Hayes B.J., Porto-Neto L.R. and McGowan M.R. (2019a) *Proc. Assoc. Advmt. Anim. Breed. Genet.* **23**:(in press).

Randhawa I.A.S., McGowan M.R., Porto-Neto L.R., Hayes B.J., Schutt K.M. and Lyons R.E. (2019b) *Proc. Assoc. Advmt. Anim. Breed. Genet.* **23**:(in press).

Sawilowsky S.S. (2009) *Journal of Modern Applied Statistical Methods* **8**:467.

Schafberg R. and Swalve H.H. (2015) *Livest. Sci.* **179**: 54.

Scheper C., Wensch-Dorendorf M., Yin T., Dressel H., Swalve H. and König S. (2016) *Genet. Sel. Evol.* **48**: 50.

Stookey J.M. and Goonewardene L.A. (1996) *Can J Anim Sci* **76**: 1.

Thompson N.M., Widmar N.O., Schutz M.M., Cole J.B. and Wolf C.A. (2017) *J. Dairy Sci.* **100**: 4941.

Windig J.J., Hoving-Bolink R.A. and Veerkamp R.F. (2015) *Livest. Sci.* **179**: 96.

GENETIC CORRELATIONS BETWEEN DAYS TO CALVING AND OTHER MALE AND FEMALE REPRODUCTION TRAITS IN BRAHMAN CATTLE

D.J. Johnston and K.L. Moore

Animal Genetics & Breeding Unit*, University of New England, Armidale, NSW, 2351 Australia

SUMMARY

Heritabilities and genetic correlations for male and female reproduction traits were estimated for Brahman cattle raised in northern Australia. The traits included the female reproduction traits of days to calving (DC), age at puberty (AP) and lactation anoestrous interval (LAI). Days to calving using repeat records (DCr) was further considered as separate DC traits for first (DC1) and second parity (DC2) records, as well as a simple binary trait for calving rate (CR). Male reproduction traits included scrotal circumference (SC) and percent normal sperm (PNS) measured in young bulls. The heritability estimates for DCr, CR, DC1, DC2, AP, LAI, SC and PNS, were 0.09, 0.10, 0.09, 0.15, 0.47, 0.40, 0.44 and 0.15, respectively. Genetic correlations between DC1 and AP, LAI, SC and PNS were 0.62, 0.52, -0.32 and -0.66, respectively. For DC2, the genetic correlation with DC1, AP, LAI, SC and PNS were 0.46, 0.56, 1.0, -0.29 and -0.71, respectively. The study has shown that the various reproduction traits were heritable. The 0.46 genetic correlation between DC1 and DC2 suggests they should be considered as separate traits in genetic evaluation and this would allow fitting different genetic correlations with important component traits. Improvement of the genetic evaluation will increase accuracies of female reproduction EBVs and allow more genetic progress in tropical beef breeds in northern Australia.

INTRODUCTION

Reproduction is a key profit driver in many northern production systems. However little or no genetic progress has occurred in the tropical beef breeds due to low levels of recording and difficulty in recording the traits. Days to calving EBV has been used in BREEDPLAN since the 1990s and several breeds and individual breeders have shown significant improvements. Research by Beef CRC showed early-in-life female reproduction traits were moderately heritable (Johnston *et al.* 2009) and could be used in selection to improve lifetime reproductive performance. These new traits have recently been included in BREEDPLAN multiple-trait evaluations of northern breeds. The recently completed Repronomics project (Johnston *et al.* 2017) measured large numbers of Brahman females for these key traits and therefore the aim of this work is to re-estimate the genetic parameters from the additional records to inform the genetic evaluation systems and industry recording.

MATERIALS AND METHODS

Data used were from a February 2019 extract of Australia Brahman Breeders' Association database and included a large amount of female reproduction data submitted from the Repronomics project. Traits used in the study included female reproduction traits: days to calving (DC), heifer age at puberty (AP) and 1st calf-cow lactation anoestrous interval (LAI), and male traits: scrotal circumference (SC) and percent normal sperm (PNS). For this study the days to calving (repeat records, DCr) were separated into traits for first (DC1) and second parity (DC2) records and were further simplified into a binary trait for calving rate (CR).

Adjusted phenotypes and contemporary groups (CG) were obtained from a full BREEDPLAN evaluation for each of the traits. Adjustment methods, DC and CG definitions were as defined by

* A joint venture of NSW Department of Primary Industries and the University of New England

Graser *et al.* (2005), and for PNS by Jeyaruban and Johnston (2017). AP and LAI were adjusted for significant experimental design effects. The CR phenotype for individuals was constructed from the DC records by assigning a value of 0 for those females that did not calve (i.e. received a penalty DC record) and 1 for those that calved (i.e. valid DC record).

Statistical analyses. Univariate REML analyses were performed for each trait using linear animal model (ASReml, Gilmour *et al.* 2009) and included 3 generations of pedigree where available. The exception was CR, it was analysed using the logit model and EBVs predicted from the underlying scale (CR_{under}) were transformed to the observed calving percent scale (CP_{obs}) using the following equation:

$$CP_{obs} = (\Phi(\tau 1 + CR_{under}) - \Phi(\tau 1)) * 100$$

where, Φ is the cumulative density function, $\tau 1$ is a threshold (-0.553). The CP_{obs} EBV were regressed against DC EBV to obtain the linear relationship.

A bivariate analysis was used to estimate the genetic correlation between DC1 and DC2. The resultant genetic correlation was significantly less than 1 (see Table 2), so DC was considered as the two separate traits for estimation of correlations with the other male and female reproduction traits. Estimates of genetic correlations for DC2 with other traits resulted in variances and heritabilities that were close to the univariate estimates of DC2 (not presented) and as such there was no need to include DC1 in the estimation of DC2 with other traits.

RESULTS AND DISCUSSION

Number of records and raw trait statistics are presented in Table 1. Days to calving from repeat records (DCr) totalled 29,269 records and for the two separate traits of DC1 (N=19,668) and DC2 (9,601).

Variance components and heritabilities from univariate analyses are presented in Table 1. Heritabilities for DC were low as expected, whereas the component traits were moderately heritable, and in agreement with previous estimates using sub-sets of these data (Johnston *et al.* 2009, 2014a, Corbet *et al.* 2013, Jeyaruban and Johnston 2017).

Table 1. Numbers of records, raw statistics for adjusted phenotypes, variance components (V_a=additive genetic, V_e=residual, V_{pe}=permanent environment, V_p=phenotypic) and estimated heritability (h^2) and standard error (in brackets) for days to calving (DCr, DC1, DC2), calving rate (CR), heifer age at puberty (AP), anoestrous interval in first-calf cows (LAI), scrotal circumference (SC) and percent normal sperm (PNS) in Brahman

Trait	N	Mean	std	V_a	V_e	V_{pe}	V_p	h^2
DCr (d)	29,269	368.1	58.2	222.5	1,216.2	1,120.8	2,562.2	0.09 (0.01)
CR	29,269	0.72	0.45	0.39#	3.29	0.08	3.76	0.10 (0.02)
DC1 (d)	19,668	366.4	63.1	231.2	2,403.0	-	2,634.2	0.09 (0.01)
DC2 (d)	9,601	371.5	48.2	268.2	1,472.8	-	1,741.0	0.15 (0.02)
AP (d)	2,021	632.0	112.4	5,462.0	6,120.5	-	11,582.0	0.47 (0.06)
LAI (d)	1,420	126.1	90.6	2,763.1	4,176.5	-	6,939.5	0.40 (0.07)
SC (cm)	33,983	26.9	4.2	2.90	3.68	-	6.59	0.44 (0.02)
PNS (%)	3,023	67.7	25.3	86.1	480.0	-	566.1	0.15 (0.05)

on underlying scale and repeat records

The relationship between the EBVs of DCr and CP on the observed scale for 672 sires with 10 or more daughters are plotted in Figure 1. The linear relationship was significant (P<0.0001) with a regression coefficient of b = -1.098 %/d and an R^2 = 0.73. This simple analysis and Figure 1 suggests these are highly correlated traits but not exactly the same trait. This is not surprising because the DC trait not only captures all of the CR trait but also includes differences in calving date.

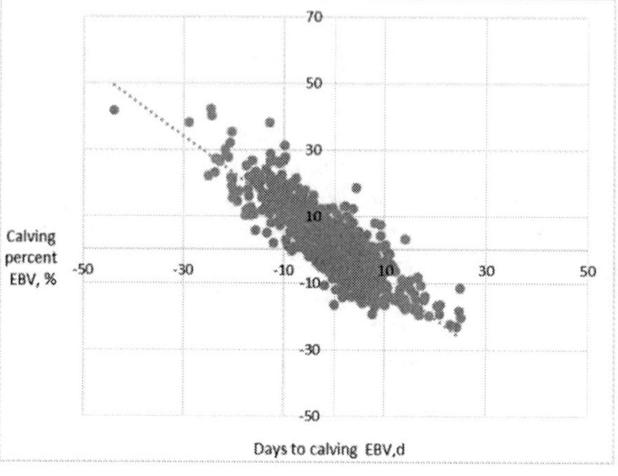

Figure 1. Calving percent EBV (observed scale) versus days to calving EBV for Brahman sires (N=672) with 10 or more daughters with records

Genetic correlations from bivariate analyses are presented in Table 2. The estimate of the genetic correlation between DC1 and DC2 was 0.46. This shows these are not the same trait and re-ranking of sires could occur for daughter's reproduction performance at these two stages. This estimate is not surprising given the different physiological state of the females with regard to lactation status at the two measurement times, and it suggests a repeatability model is not the most suitable method for handling these records.

Genetic correlations between DC1 and the other female traits were positive for AP and LAI, showing shorter DC1 was genetically related to younger AP and shorter LAI. Likewise, the negative correlations for the male traits indicate a shorter DC1 was associated with larger SC and higher PNS. These estimates reflect the -0.71 genetic correlation between AP and PNS. Correlations are similar to earlier estimates from Johnston *et al.* (2014b).

Table 2. Estimated genetic correlations (standard error) between DC1 and DC2 and other reproduction traits in Brahman

Traits#	DC2	AP	LAI	SC	PNS
DC1	0.46 (0.10)	0.62 (0.14)	0.52 (0.19)	-0.32 (0.07)	-0.66 (0.22)
DC2		0.56 (0.14)	1.0*(0.06)	-0.29 (0.07)	-0.71 (0.22)
AP			0.29 (0.12)	-0.48 (0.10)	-0.71 (0.23)
LAI				-0.29 (0.11)	-0.64 (0.23)

see Table 1 for trait names *estimate at bounds

The same pattern of correlations existed between DC2 and the traits, however LAI was larger in magnitude with the correlation going to the bounds, showing this is essentially the same trait, but more heritable. This is likely due to more precision of measurement of LAI compared to DC. PNS was also highly correlated with DC2 and is in agreement with estimates using subset of the data reported by Jeyaruban and Johnston (2017) and reflects the -0.64 between LAI and PNS.

DC1 was most associated with traits related to puberty (e.g. AP), whereas DC2, where the cows are lactating at mating, was very highly correlated with LAI. Male traits offer advantage to genetic evaluation of female reproduction traits, if they are genetically correlated, because they can be measured at a young age in bulls (i.e. before selection) and provide an early prediction of genetic differences in future daughters. Results show PNS was highly correlated with both DC1 and DC2. SC showed similar correlations but at much lower in magnitude but the trait was much more heritable than PNS.

CONCLUSIONS

Improving the genetic evaluation of reproduction traits, especially in northern Australia breeds will have large payoffs for commercial production. DC has been used as an easy to record trait in the genetic evaluation of several tropical beef breeds. However these results confirm that more heritable measures in both males and females can be used as correlated traits in the genetic evaluation of female reproduction. DC is strongly associated with CR and the regression coefficient of -1d DC EBV = +1% CP EBV provides easy to use benchmark.

Changes should be considered to the definition of traits used in BREEDPLAN. The current DC trait could be easily modified to separate it into two traits for DC1 and DC2, and this would provide the added benefit of being able to include the appropriate genetic correlations with correlated traits. However consideration is needed on how these changes would impact on traits in the breeding objective. Also the correlated traits are more costly to record and therefore should be the focus of recording in genomic reference populations to ensure the added value is fully captured. Additional records are required for AP, LAI, and PNS to further reduce the standard errors of these correlation estimates, and to allow the development of genetic evaluation of reproduction traits in other tropical breeds.

ACKNOWLEDGEMENTS

The authors thanks MLA for funding the study under MLA L.GEN.0174 and data from MLA B.NBP.0759. Thanks to the collaborating organisations in Repronomics project and the large team of people involved in designing, running and collecting data in the project. We acknowledge Australian Brahman Breeders' Association and their members for providing access to their database.

REFERENCES

Corbet N.J., Burns B.M., Johnston D.J., Wolcott M.L., Corbet D.H., Venus B.K., Li Y., McGowan M.R. and Holroyd R.G. (2013) *Anim. Prod. Sci.* **53**: 101.

Gilmour A.R., Gogel B.J., Cullis B.R., Thompson R. (2009) ASReml user guide, release 3.0. VSN International Ltd, Hemel Hempstead, UK.

Graser H.U., Tier B., Johnston D.J., Barwick S.A. (2005) *Aust. J. Expt. Ag.* **45**: 913.

Jeyaruban M.G. and Johnston, D.J. (2017) *Proc. Assoc. Advmt. Anim. Breed. Genet.* **22**: 305.

Johnston D.J., Barwick S.A., Corbet N.J., Fordyce G., Holroyd R.G., Williams P.J., Burrow H.M. (2009) *Anim. Prod. Sci.* **49**: 399.

Johnston D.J., Barwick S.A., Fordyce G., Holroyd R.G., Williams P.J., Corbet N.J. (2014a) *Anim. Prod. Sci.* **54**: 1.

Johnston D.J., Corbet N.J., Barwick S.A., Wolcott M.L., Holroyd R.G. (2014b) *Anim. Prod. Sci.* **54**: 74.

Johnston D.J., Grant T.P., Schatz T.J., Burns B.M., Fordyce G., and Lyons R.E. (2017) *Proc. Assoc. Advmt. Anim. Breed. Genet.* **22**: 385.

GENOME-WIDE ASSOCIATION STUDIES FOR BODY WEIGHT AND AVERAGE DAILY FEED INTAKE DURING THE FEEDLOT TEST PERIOD

J.A. Torres-Vázquez, J.H.J. van der Werf and S.A. Clark

School of Environmental and Rural Science, University of New England, Armidale, 2351 Australia

SUMMARY

Feed efficiency component traits such as body weight (BW) and daily feed intake are economically relevant traits in beef cattle breeding programs. The objectives of this work were to identify genomic regions associated with BW and average daily feed intake (ADFI) during the feedlot period, and to evaluate whether these genetic variants for each trait were consistent over the 70-day test period. Data on 2070 Angus steers were used to estimate (co)variance components using the genomic relationship matrix (gREML) fitted in ASReml. For the studied traits, a two-trait repeatability (TT-REPM) and a two-trait random regression (TT-RRM) models were performed. SNP-effects for the TT-REPM and TT-RRM were estimated using a post analysis back-solving approach using the genomic estimated breeding values from each model respectively. For each trait, results were validated with single-trait animal models (ST-ANIMs) at the beginning and at the end of the test period using single-SNP regression in the GCTA software. Results from the genome-wide association studies (GWAS) using TT-REPM and TT-RRM were similar to the conventional approach using the ST-ANIMs. For all models, the variants rs43350564 and rs109326204 presented the strongest association with BW and ADFI, respectively. The identified SNP effects remained constant throughout the feedlot test period and could be useful for understanding the biology of feed efficiency. Further studies with more data and possibly with longer feed lot test periods are needed to investigate the effect of genomic regions for feed efficiency traits over the feedlot trajectory.

INTRODUCTION

Feed efficiency, commonly referred to as the conversion of feed into useable animal products, and their component traits such as body weight (BW) and average daily feed intake (ADFI) are economically relevant traits in beef cattle breeding programs. Identifying single nucleotide polymorphisms (SNPs) as genetic markers linked to quantitative trait loci (QTL) associated with BW and ADFI may aid in unravelling the biology underlying feed efficiency. These genetic markers for QTL can be identified through genome-wide association studies (GWAS). In beef cattle, GWAS for BW and ADFI have been addressed using the average measurements during the feedlot test period (Bolormaa *et al.* 2011).

For both traits, previous studies have documented that pedigree-based genetic parameters change during the trajectory of the feedlot period (Torres-Vázquez *et al.* 2018), and therefore it is expected that QTL-effects may also change over time. The objective of this study was to identify genomic regions associated with body weight and average daily feed intake during the feedlot period, and to evaluate whether these genetic variants for both traits were consistent over the 70-day test period.

MATERIALS AND METHODS

The phenotypic data included BW and feed intake measures from 2220 Angus steers collected from 2013 to 2017 at Tullimba Research Feedlot (30°20'S, 151°10'E, altitude 560 m), NSW, Australia. On entry to the feedlot, steers ranged from 500-600 days of age with an average weight of 578 kg. Initially steers were conditioned for 21 days and fed for an additional 70 days over which time all data was collected. Steers were weighed 6 times over the 70-day test period (fortnightly). Daily feed intake measurements were averaged over 14-day periods to align them with the BW measurements to create average daily feed intake (ADFI). Duplicated and incomplete records were discarded (see Torres-Vázquez *et al.* 2018). The final data file consisted of 2,070 Angus steers. The pedigree file included an historical file with 14,662 animals with 1,454 sires and 7,835 dams; with 191 sires and 1,782 dams having progeny with phenotypic records. Contemporary groups, as defined by BREEDPLAN, included the concatenation of herd, year of birth, birth type (single or twin), breeder-defined management group, observation date and age (Graser *et al.* 2005).

Animals with phenotypes were genotyped with a range of low-density marker chips. These genotypes were imputed to higher density based on a reference of 7626 animals genotyped using the Illumina Bovine 50K v2 (54609 SNP). Quality control of the SNP markers was performed to eliminate SNP with a call rate less than 90% and minor allele frequency less than 1%. The remaining 39,136 SNPs passed the quality control measures and were acceptable for the analyses. Low density genotypes were then imputed to 39,136 SNP using FImpute (Sargolzaei *et al.* 2011). The genomic relationship matrix (GRM) was subsequently created using the GCTA software (Yang *et al.* 2011) from the imputed genotypes.

To analyze the genomic associations between traits a two-trait repeatability (TT-REPM) and a two-trait random regression (TT-RRM) model were undertaken. (Co)variance components for each analysis were estimated using ASReml incorporating the genomic relationship matrix (GRM (Gilmour *et al.* 2009). The most suitable fit of the models were assessed based on the log likelihood (LogL), Akaike's information criterion (AIC), and the Bayesian information criterion (BIC).

Several researchers have documented the equivalences between snpBLUP and gBLUP for genomic selection (Strandén and Garrick, 2009; Gondro, 2015). Therefore, SNP effects for the two-trait models were obtained following the methodology described in Gondro (2015), where:

$$\hat{u}_i = \left(\frac{1}{d} W\right) GRM^{-1}\, GEBV_i$$

where $\hat{\mathbf{u}}_i$ is a vector of the predicting SNP marker effects for the *i*th individual; d represent a scalar of the deviation effects calculated as $2 * \sum(p * q)$; **W** represent the SNP marker matrix corrected for the allele frequency differences (**M** – 2 * (p – 0.5)). **M** is the matrix of marker genotypes coded as: 1 for the heterozygous genotype, 0 and 2 for the genotype which is homozygous for the first and second allele, respectively; GRM^{-1} represent the inverse of the GRM; and $\mathbf{GEBV}_i$ is a vector of genomic estimated breeding values (GEBVs) obtained from the gREML model. For this approach, p-values of 0.05 were estimated based on a t-distribution calculated as the probability value of the 95th percentile of the GEBV distribution.

To validate the SNP-effects calculated for the TT-REPM and for the TT-RRM at days 5 and 70, GWAS were conducted with single-trait animal models (ST-ANIMs) based on data evaluated at the beginning and at the end of the test period (days 1 and 70, respectively) using single-SNP regression in GCTA (Yang *et al.* 2011).

RESULTS AND DISCUSSION

The TT-RRM had the highest LogL, and smallest value for AIC and BIC, showing the best fit for this model (Table 1). In general, high genomic heritability estimates were observed for BW compared to ADFI. The TT-RRM yielded the highest range for genomic heritability estimates with higher genomic heritability estimates for BW. As expected, repeatability estimates for both traits increased across the feed lot test period, and these estimates were higher for BW compared to ADFI, suggesting that measurement errors are more relevant for the accuracy of ADFI. Our genomic heritability estimates for BW and ADFI followed the same pattern as the pedigree-based estimates reported by Torres-Vázquez *et al.* (2018). Using the TT-REPM, the genetic correlation was of 0.69 ± 0.05. However, with the TT-RRM, this correlation increased from 0.63 at day 1 up to 0.75 at day 82. This was a slight reduction in the difference observed in the pedigree analysis (0.56 to 0.82) by Torres-Vázquez *et al.* (2018). Nevertheless, the increase in correlation potentially indicates that the genes that cause variation in the two traits are more similar over time.

Table 1. Measures of goodness of fit, genomic heritability (h^2) and repeatability (rep) for the two-trait models

Model / Trait	n	Log L	AIC	BIC	h^2	rep
TT-REPM, BW	9	-26.5	100,071.0	100,143.5	0.46 ± 0.04	0.88 ± 0.01
TT-REPM, ADFI	9	-26.5	100,071.0	100,143.5	0.26 ± 0.03	0.59 ± 0.01
TT-RRM, BW	23	7,187.8	94,421.5	94,606.3	From 0.42 to 0.53	From 0.92 to 0.94
TT-RRM, ADFI	23	7,187.8	94,421.5	94,606.3	From 0.36 to 0.23	From 0.68 to 0.70

SNP-effects obtained by back-solving the TT-REPM and the TT-RRM at days 5 and 70 followed the same pattern as those yielded by the ST-ANIM using GCTA (Figure 1). For each trait, only one SNP exceeded the significance level.

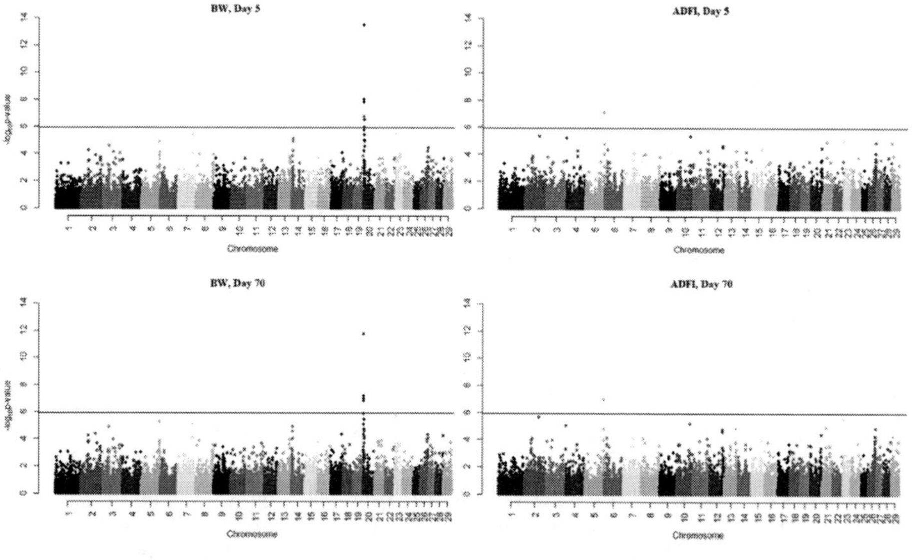

Figure 1. Manhattan plot for the genomic associations of BW and ADFI at days 5 and 70, using the back-solving approach with the two-trait random regression model

The most strongly associated marker (rs43350564) for BW was located on chromosome 20 at the position 4,618,689. This SNP was located 269 bp downstream from *ERGIC1* (*endoplasmic reticulum-golgi intermediate compartment 1).* This gene is potentially involved in increased protein turnover and has been previously associated with increases in multiple liveweight measures in American Simmental, Red Angus and Gelbvieh (Saatchi *et al.* 2014). The most significant SNP (rs109326204) associated with ADFI was located on chromosome 5 at the position 120,378,417. This SNP is located in the CELSR1 gene, which has been associated with decreases in body mass in mice (Zerbino *et al.* 2018).

The methodology implemented in this work to obtain the SNP effects, is easy to apply without transforming phenotypes but it has some limitations. Given the increasing genetic correlation together with the low accuracies, the GEBVs obtained with the TT-RRM were highly correlated between days (>0.987). This yielded very similar SNP effects in each trait with low probabilities of identifying accurately other genetic markers along the test period. Besides, the implemented methodology was sensitive to several factors. For example, the number of animals with phenotypic data collected with the greatest precision, imputation quality, and the genomic accuracy of the trait. In addition, in the presence of small SNP effects for a trait, further samples may be necessary to detect them.

CONCLUSIONS

The TT-RRM showed that the genetic parameters tended to change over the feedlot test period. With this model, the genomic correlation estimates increased over the whole trajectory from 0.63 to day 1 to 0.75 at the end of the period. In this work, SNP effects obtained by the TT-REPM and TT-RRM yielded similar results to conventional GWAS approaches. Two SNPs were identified by GWAS that may be useful for understanding the biology of feed efficiency in beef cattle. Further studies are necessary to investigate the change of genomic regions including more samples in longer feedlot test period for feed efficiency traits.

ACKNOWLEDGEMENTS

The authors thank the Angus Society of Australia for their support of this research and access to the data for this study through the Angus Sire Benchmarking Project (ASBP). Meat and Livestock Australia (MLA) have provided co-funding support for the ASBP (PSH.0528). Generous support has also been provided to the ASBP by numerous Angus bull breeders, co-operator herds, supply chain partners, technicians and research organizations.

REFERENCES

Bolormaa S., Hayes B.J., Savin K., Hawken R., Barendse W., Arthur P.F., Herd R.M. and Goddard M. E. (2011) *J. Anim. Sci.* **89**: 1684.

Gilmour A.R., Gogel B., Cullis B., Thompson R., and Butler D. (2009) VSN International Ltd, Hemel Hempstead, UK.

Gondro C. (2015) 'Primer to analysis of genomic data using R'. Springer. Cham.

Graser H.-U., Tier B., Johnston D.J. and Barwick S.A. (2005) *Aust. J. Agric. Res.* **45**: 913.

Sargolzaei M., Chesnais J. and Schenkel F. (2011) *J. Dairy Sci.* **94**: 421.

Saatchi M., Schnabel R.D., Taylor J.F. and Garrick D.J. (2014) *BMC Genomics* **15**: 1004.

Strandén I. and Garrick D. (2009) *J. Dairy Sci.* **92**: 2971.

Torres-Vázquez J.A., van der Werf J. and Clark S.A. (2018) *Proc. World Congress on Genetics Applied to Livestock Production.* Feed Intake and Efficiency **2**:372.

Yang J., Lee S.H., Goddard M.E. and Visscher P.M. (2011) *Am. J. Hum. Genet.* **88**: 76.

Zerbino D.R, Achuthan P., Akanni W., Amode M.R., Barrell D., Bhai J., Billis K., Cummins C., Gall A., García Giron C., Gil L., Gordon L., Haggerty L., Haskell E., Hourlier T., Izuogu O.G., Janacek S.H., Juettemann T., Kiang To J., Laird M.R., Lavidas I., Liu Z., Loveland J.E., Maurel T., McLaren W., Moore B., Mudge J., Murphy D.N., Newman V., Nuhn M., Ogeh D., Ong C.K., Parker A., Patricio M., Riat H.S., Schuilenburg H., Sheppard D., Sparrow H., Taylor K., Thormann A., Vullo A., Walts B., Zadissa A., Frankish A., Hunt S.E., Kostadima M., Langridge N., Martin F.J., Muffato M., Perry E., Ruffier M., Staines D.M., Trevanion S.J., Aken B.L., Cunningham F., Yates A. and Flicek P.F. (2018) *Nucleic Acids Research.* **46**: D754.

COEFFICIENT OF VARIATION AS A MEASURE OF GENERAL RESILIENCE FOR YEARLING WEIGHT IN NELLORE CATTLE

D.C.B. Scalez[1,2], A. Reverter[2], B.O. Fragomeni[3], L.R. Porto-Neto[2], L.H.S. Iung[1], L.G. Albuquerque[1,4] and R. Carvalheiro[1,4]

[1]Department of Animal Science, School of Agricultural and Veterinarian Sciences, Sao Paulo State University (UNESP), Jaboticabal, SP, Brazil
[2]CSIRO Agriculture & Food, 306 Carmody Rd., St. Lucia, Brisbane, QLD, 4067 Australia
[3]University of Connecticut, Mansfield, USA
[4]National Council for Scientific and Technological Development (CNPq), Brasília, DF, Brazil

SUMMARY

Resilience can be defined as the capacity of the animal to be minimally affected by disturbances or to rapidly return to the state pertained before exposure to a disturbance (Colditz and Hine 2016). As indicators for general resilience have not yet been defined, our aim is to investigate the potential of the coefficient of variation (CV) as a measure of general resilience for yearling weight (YW). Using 138,590 Nellore cattle, sired by 560 Nellore bulls, we computed the CV based on within-sire progeny groups (PGs) that comprised of at least 10 progenies from the same sex, farm and year of birth. From this, we generated 5 datasets based on the size of the PG: maximum of 20, 30, 50, 100 and no limit. A two-trait single-step GBLUP model was adopted (mean YW and CV), considering the genotypes of the sires and the pedigree information relating to a given PG with its sire. Smaller groups resulted in higher estimates of heritability for both traits. Moreover, estimates of genetic correlations were positive yet of low magnitude, and closer to zero for PG with a maximum size of 20. We conclude that the use of the CV combined with the grouping, offers an opportunity to select animals that have high genomic estimated breeding values for YW with reduced CV.

INTRODUCTION

One of the current challenges of livestock production is to achieve successful intensification of production, without detrimental effects on animals, which requires healthy and easy-to-manage animals (Elgersma *et al.* 2017). Although highly important, few studies have investigated general resilience, a feature that can be defined as the capacity of the animal to be minimally affected by disturbances (Colditz and Hine 2016).

According to Berghof *et al.* (2019), indicators for general resilience to environmental disturbances have not yet been defined, and measuring this variable is difficult. Most studies regarding to resilience have been conducted at experimental level, which does not represent the reality of the production system. Also, most of these studies have focused on disease resilience and, although these studies can provide useful information in physiology, the results may not be representative of resilience under non-disease conditions.

Here, we propose to use coefficient of variation (CV) as an alternative to evaluate general resilience, based on within-family data across environments (sex-farm-year).

MATERIALS AND METHODS

Data for YW from 138,590 cattle, born between 1986 and 2016 and sired by 560 bulls, were extracted from the Alliance Nellore database. The number of progenies per sire averaged 247 and ranged from 10 to 12,612. Cattle were raised on pasture in herds from Brazil and Paraguay, and YW was measured at an average age of 533 days (ranging from 338 to 627 days).

The CV was considered as a measure of general resilience, which was computed based on within-sire progeny groups (PGs). We took the assumption that a PG size of 10 would be sufficient to estimate the mean and the CV, then each individual PG was comprised of at least 10 progenies from the same sire, sex, farm and year of birth. From this, we generated 5 datasets based on the size of the PG, considering the growth rate (age and weight) during the regrouping process, making the groups more homogeneous: (*i*) TS_20: PGs with more than 20 observations were splitted into other groups, respecting a minimum of 10 and a maximum of 20; (*ii*) TS_30: PGs with more than 30 observations were divided into groups with maximum size of 30; (*iii*) TS_50: PGs with more than 50 records were divided into groups with maximum of 50; (*iv*) TS_100: PGs with more than 100 were splitted respecting the maximum of 100; (*v*) No_TS: no limits were established, i.e. there was no regrouping.

Genotypic information from 560 sires genotyped with the Illumina® BovineHD chip was used. In the quality control of genotypes, non-autosomal SNPs, SNPs with minor allele frequency lower than 0.02, p-value for Hardy-Weinberg equilibrium test less than 10^{-5} and call-rate lower than 0.98 were removed, so that 405,442 SNPs remained for the analyses. All genotyped bulls had a call rate higher than 0.90, passing the quality control.

A two-trait single-step GBLUP animal model was adopted for the average YW and CV within PGs as the phenotypes, considering the genotypes of the sires and the pedigree information relating a given PG with its sire. Sex and year were used to create contemporary groups (CGs), fitted as fixed effects. In addition, the size of the PG (linear), the average age of the PG (linear and quadratic) and the heterozygosity (HET) of the sires were also included in the model as covariates.

RESULTS AND DISCUSSION

A summary of the number of PGs generated, mean and standard deviations for each dataset used in the bivariate analyses is reported in Table 1.

Table 1. Summary statistics of yearling weight (YW, kg) and coefficient of variation (CV, %) of progeny groups, and correlation estimates between sire's heterozygosity (HET) with YW and CV, in Nellore cattle

		YW		CV		Pearson correlation with HET			
Groups*	N	Mean	SD	Mean	SD	YW	P-value	CV	P-value
TS_20	10,290	300	45.9	7.08	2.62	0.123	0.004	-0.079	0.060
TS_30	8,951	300	44.6	7.80	2.44	0.120	0.004	-0.100	0.020
TS_50	8,459	300	44.3	8.11	2.33	0.118	0.005	-0.111	0.008
TS_100	8,341	300	44.1	8.19	2.29	0.116	0.006	-0.135	0.001
No_TS	8,327	300	44.1	8.20	2.29	0.116	0.006	-0.135	0.001

*TS_20: target size 20; TS_30: target size 30; TS_50: target size 50; TS_100: target size 100; No_TS: with no regrouping.

Estimates of correlation between sire's HET and the mean for YW, although low, were positive (Table 1). The opposite tendency was observed for the CV, being negative and more pronounced as the size of the PGs increased. Even though the estimates are discrete for both traits, the behavior of the estimates is desirable, i.e. the greater the heterozygosity the greater the YW and the lower the CV. Heterozygosity also has the potential to be used in mate selection in order to maximize heterozygosity in the offspring (de Cara *et al.* 2011). This could be achieved through the selection of parents that are opposite homozygotes for either as many loci as possible or for the relevant alleles for the trait of interest (Iversen *et al.* 2019).

Table 2 shows parameter estimates from the bivariate analyses and the different datasets. Smaller groups yielded higher heritability estimates for both traits. Moreover, although they are of low magnitude, estimates of genetic correlations were smaller for TS_20. Therefore, the creation of more groups by sire (within family) and consequently making the sizes of the groups more homogeneous, appears to be a sensible approach. In addition, despite being a non-favorable correlation, the use of the CV combined with the grouping, demonstrates that there is a chance of selecting animals that have high genomic estimated breeding values for YW and with reduced CV.

Table 2. Estimates of direct additive genetic variance (σ_a^2)σ_a^2), heritability (h^2h^2), and correlation (r)r) for yearling weight (YW) and coefficient of variation (CV) of progeny at each dataset in Nellore cattle

Groups*	YW		CV		$r_{yw,cv}$
	σ_a^2	h^2	σ_a^2	h^2	
TS_20	637.59	0.476	4.0201	0.556	0.0956
TS_30	509.39	0.411	2.5401	0.420	0.1339
TS_50	485.99	0.400	2.2209	0.399	0.1214
TS_100	476.98	0.396	2.0248	0.378	0.1497
No_TS	485.24	0.402	2.0006	0.376	0.1483

*TS_20: target size 20; TS_30: target size 30; TS_50: target size 50; TS_100: target size 100; No_TS: with no regrouping.

In Figure 1, animals that presented GEBVs above 1 standard deviation for YW and below 1 standard deviation for CV (16 sires) for the TS_20 dataset, are highlighted in blue. Selecting these sires would assist making progress towards both traits simultaneously: high and consistent growth.

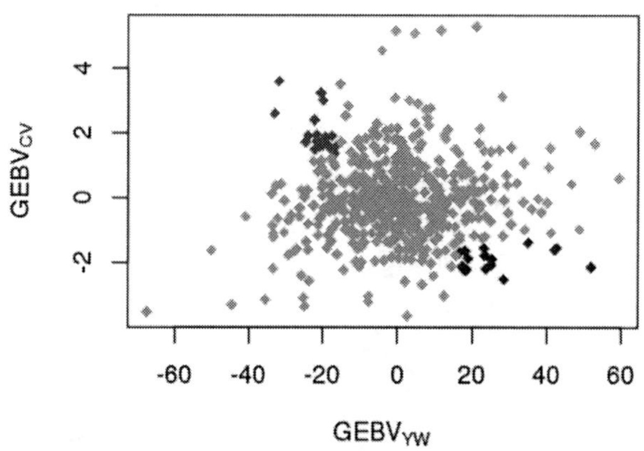

Figure 1. Scatter plot between genomic estimated breeding values for yearling weight ($GEBV_{YW}$) and for coefficient of variation ($GEBV_{CV}$) for all 560 Nellore sires. The blue lozenges represent the animals with favourable GEBVs for both traits, and the red lozenges represent animals with unfavourable GEBVs for both traits

The real-life nature of the data (within-family data) made this study particularly challenging, because bulls were used in different intensities through artificial insemination, presenting different sizes of progeny, and also some groups were in the same environment (sex-farm-year). While originally large in size, limiting the minimum size of the group (required to compute CV with some confidence) caused the exclusion of a lot of data, so further strategies are warranted.

CONCLUSIONS

TS_20 presented the highest heritability for YW and CV, and the smallest correlation between them, showing that the use of CV combined to the grouping strategy is feasible for studies considering within-family data, making possible the selection for weight and uniformity simultaneously. These are preliminary results of an ongoing study indicating that the use of CV is one alternative to select animals for resilience. Further research is warranted to test new variables and new strategies to assess general resilience.

ACKNOWLEDGMENTS

To São Paulo Research Foundation (FAPESP) grant 2018/13749-3.

REFERENCES

Berghof T.V.L., Poppe H.W.M. and Mulder H. A. (2019) *Front. Genet.* 1–15.

Colditz and Hine. (2016) *Animal Prod. Sci.* **56**: 1961.

de Cara M.A.R, Fernández J., Toro M.A. and Villanueva B. (2011) *J. Anim. Breed. Genet.* **128**: 456.

Elgersma G.G., de Jong G., van der Linde R. and Mulder H.A. (2017) *J. Dairy Sci.* **101**: 1240.

Iversen M.W., Nordbø Ø., Gjerlaug-Enger E., Grindflek E., Lopes M.S. and Meuwissen T. (2019) *Genet. Sel. Evol.* **51**: 1.

INCREASES IN ACCURACY OF FEMALE REPRODUCTION GENETIC EVALUATIONS FOR BEEF BREEDS IN NORTHERN AUSTRALIA

K.L. Moore[1], C.J. Girard[1], T.P. Grant[2] and D.J. Johnston[1]

[1]Animal Genetics & Breeding Unit*, University of New England, Armidale, NSW, 2351 Australia
[2]Queensland Department of Agriculture and Fisheries, Toowoomba, QLD, 4350 Australia

SUMMARY

Female reproduction traits influence profitability of beef enterprises, but genetic improvement has been limited. This study assessed the impact of including phenotypic reference data and genotypes on the genetic evaluations for three northern Australian beef breeds with different recording and data structures. For young (2016/17) animals, accuracy for days to calving estimated breeding value (EBV) increased 14.7% for Brahman (BB: well-recorded and genotyped), 6.2% for Santa Gertrudis (SG: limited data and small number of strategic genotypes) and 6.0% for Droughtmaster (DM: limited data and not genotyped). With these accuracy increases there is potential to increase the rates of genetic gain for key female reproduction traits.

INTRODUCTION

Female reproduction is an important driver in the production and profitability of beef enterprises. Genetic progress for these traits is typically limited as reproduction traits are difficult to measure, measured late in life, are sex limited and often have low heritabilities (Cammack *et al.* 2009). These issues make reproduction ideal to benefit from genomic selection and single-step genomic selection is ideal as non-genotyped and genotyped animals can be analysed together. A key female reproduction trait is age at puberty (AP), which can be determined from the age at first corpus luteum (CL). Johnston *et al.* (2009) estimated AP to be highly heritable in both BB (h^2=0.57) and Tropical Composite (h^2=0.52) heifers. Not only do cows need to reach puberty quickly, but they also need to be able to return to cycling quickly after calving, to produce a calf in the annual management cycle. Using real-time ultrasound ovarian scans to detect the first CL in the mating period, Johnston *et al.* (2014) estimated lactation anoestrous interval (LAI) to be moderate to highly heritable for both BB (h^2=0.51) and Tropical Composite (h^2=0.26) cows. The aim of this paper was to assess the improvement of EBV accuracies for female reproduction traits (days to calving (DC), AP and LAI) when reference data and single-step genomic selection was included into BREEDPLAN genetic evaluations for three tropically adapted northern Australian beef breeds with different recording and data structures.

MATERIALS AND METHODS

Three breeds (BB, SG and DM) were extensively recorded over five years as part of the Repronomics™ project in northern Australia (Johnston *et al.* 2017). Using real-time ultrasound, regular ovarian scans were undertaken to accurately identify the age that heifers become pubertal (AP) and when lactating cows first cycled after their first calving (LAI). Repronomics herds were fully BREEDPLAN recorded with data submitted for BREEDPLAN genetic evaluations. The reference dataset for this study included records from the Repronomics herds as well as additional BB AP and LAI phenotypes collected as part of the Beef CRC (Johnston *et al.* 2009, 2014).

The BB BREEDPLAN evaluation has recently implemented single-step genomic selection and this will soon be implemented for SG. There were 14,821 BB and 3,464 SG animals genotyped with approximately 40K SNPs available for single-step analysis, see Connors *et al.* (2017) for details of

* A joint venture of NSW Department of Primary Industries and the University of New England

the genomic pipeline. To evaluate the benefits of including the fertility reference data and single-step genomic selection, BREEDPLAN data from February 2019 extracts for each breed was used in three separate genetic evaluations using the BREEDPLAN methodology (Johnston *et al.* 2018). All evaluations were multi-trait and provided EBVs and accuracies for all traits, with results for three female reproduction EBVs reported in this study; DC, AP and LAI. In all analyses the number of animals remained the same with only the level of recording and inclusion of genomic information changing across runs. Table 1 outlines the information available for the three genetic evaluations considered.

Table 1. Number of days to calving (DC, days), age at puberty (AP, days) and lactation anoestrous interval (LAI, days) records available for genetic evaluations with different levels (GE1, GE2, GE3) of data included for Brahman, Santa Gertrudis and Droughtmaster

Breed	Brahman			Santa Gertrudis			Droughtmaster		
Trait	DC	AP	LAI	DC	AP	LAI	DC	AP	LAI
GE1#	54,154	0	0	34,704	0	0	1,682	0	0
GE2/GE3	1,398	2,020	1,403	238	216	115	808	627	481

\# GE1 = no phenotypic reference data or genomic selection included; GE2 = phenotypic reference data included but no genomics; GE3 = both the phenotypic reference data and single-step genomics included

RESULTS AND DISCUSSION

The breeds considered represent three different recording and data structures. BB was the largest breed numerically with 449,620 animals included in genetic evaluations and had the highest level of recording and genotyping in the study. For all subsets of animals and traits, increasing the information in the genetic evaluation increased EBV accuracy. Table 2 records the average accuracy for reproduction traits for three data scenarios and three subsets of animals. Despite AP and LAI having substantially fewer records available, the average accuracy was not much lower than the DC accuracy and this was due to the higher heritabilities for AP and LAI.

Table 2. Average accuracy of days to calving (DC, days), age at puberty (AP, days) and lactation anoestrous interval (LAI, days) from genetic evaluations with different levels (GE1, GE2, GE3) of data included for Brahman

Animal subset	Genotyped n=14,821			16/17 born n=24,555			Genotyped 16/17 born n=3,062		
Dataset	GE1#	GE2	GE3	GE1	GE2	GE3	GE1	GE2	GE3
DC	35.4	41.2	49.7	25.7	28.4	33.3	30.1	35.1	44.8
AP	18.2	34.3	44.9	13.1	20.2	26.8	14.5	28.9	41.3
LAI	24.4	32.6	40.9	17.6	20.8	25.7	19.5	25.4	35.2

\# See Table 1 for descriptions

Including phenotypic reference data and genomic selection (GE3, Table 2) significantly increased accuracy for AP and LAI. Despite only a small increase in the number of DC records, DC accuracy increased by 14.3% for genotyped animals. AP and LAI are both highly heritable and strongly correlated with DC (Johnston *et al.* 2019), and 41% of the increase in DC accuracy of genotyped animals was attributed to correlated responses with AP and LAI, and the remaining increase the result of genomic selection. Approximately 20% of the genotyped animals were also young animals born in 2016/2017. With smaller numbers and no phenotypes themselves at the time of analysis, the average accuracy for these animals were lower than other subsets. However, the increase in accuracy observed after including the reference data and genomics was similar to that for all genotyped animals. Compared to all young animals, the genotyped young animals tended to have slightly higher accuracy in the base evaluation (likely due to selective genotyping) but they also had approximately twice the improvement in accuracy when phenotypic reference data and genotypes were added to the genetic evaluation; with EBV accuracy for genotyped young animals increased by 14.7% (GE3) compared to the base genetic evaluation (GE1).

Both SG and DM had similar sized evaluations, with 280,596 and 232,551 animals, respectively. However, they have different levels of recording and genotyping for the female reproduction traits analysed for this study. SG have relatively few AP and LAI records available in the reference dataset but were well-recorded for DC. DC is currently a research EBV for DM and has smaller numbers of industry records available, with Repronomics DC records contributing significantly to the total number of DC records for the breed. For the SG breed, the genotyping strategy has focused on genotyping well-recorded and influential industry animals, particularly those with high DC EBV accuracy. This targeted genotyping strategy was evident with genotyped animals having higher EBV accuracies compared to BB in the base evaluation (Table 3). With the smaller genotyped reference population the impact of including reference phenotypic data and genomic selection was smaller for SG than BB, with a 2.7% increase in DC accuracy for genotyped animals. Similar trends were found for the accuracy of sires in the evaluation with DC accuracy increasing by 1.9% when reference data and genotypes were added. The greatest increases in accuracy for the breed were observed for the young 2016/2017 animals; DC accuracy increased 6.2% when reference data and genotypes were included but only 9% of this was due to genomic selection. Despite having small numbers of records in the reference data, non-trivial increases were observed for these young animals as the reference animals were highly influential and well connected to the wider SG population. Across all subsets of SG animals the impact of including genomic selection was small. For genomic selection to have an impact on female reproduction traits more animals need to be genotyped that also have female reproduction traits recorded, as well as increasing the number of AP and LAI phenotypes in the reference population.

Table 3. Average accuracy of days to calving (DC, days), age at puberty (AP, days) and lactation anoestrous interval (LAI, days) from genetic evaluations with different levels (GE1, GE2, GE3) of data included for Santa Gertrudis

Animal subset	Genotyped n=3,464			Sires n=5,211			16/17 born n=12,282		
Dataset	GE1[#]	GE2	GE3	GE1	GE2	GE3	GE1	GE2	GE3
DC	55.6	57.5	58.3	44.6	46.4	46.5	31.7	37.4	37.9
AP	28.6	33.6	35.0	16.7	19.6	19.8	13.1	21.9	22.7
LAI	32.7	35.4	36.5	19.2	21.2	21.4	15.1	20.4	21.0

[#] See Table 1 for descriptions

DC is currently a research trait for DM and there are fewer records. With no CRC data, all AP and LAI records came from the Repronomics project. There are insufficient genotypes to date to enable single-step to be implemented. Although, comparable in overall size to the SG breed, DM's have limited phenotypic recording for female reproduction, and this was evident in the lower trait accuracies (Table 4) compared to other breeds. Including the reference phenotypic data resulted in large increases in accuracy for Repronomics animals where the data was collected, showing the power of recording to lift EBV accuracy for high heritability traits. The DC accuracy for sires increased by 4% and 6% for young 2016/17 born animals when the reference data was added. Despite having lower starting accuracies in the base evaluation, the increase in accuracy as a result of including reference data was comparable to the other breeds.

Table 4. Average accuracy of days to calving (DC, days), age at puberty (AP, days) and lactation anoestrous interval (LAI, days) from genetic evaluations with different levels (GE1, GE2, GE3) of data included for Droughtmaster

Animal subset	'Repronomics' n=2,846		Sires n=2,534		16/17 born n=11,347	
Dataset	GE1[#]	GE2	GE1[#]	GE2	GE1[#]	GE2
DC	18.8	45.5	32.3	36.3	15.0	21.0
AP	3.0	46.4	4.2	15.6	1.4	15.5
LAI	3.0	37.3	4.2	12.1	1.4	10.6

[#] See Table 1 for descriptions

Including the reference phenotypic data and single-step increased the number of animals with accuracies ≥40%. For DC in the base evaluation, 13.8, 37.0 and 3.1% of animals had DC accuracy ≥40% for BB, SG and DM, respectively. In the evaluations including all the available information this increased to 29.5, 47.0 and 12.9% of animals, respectively for BB, SG and DM.

CONCLUSIONS

The inclusion of intensively recorded female reproduction reference phenotypic data and genotypes increased the accuracies for DC, AP and LAI. Only DC is reported to industry, but the new traits are highly heritable and correlated to DC and results showed they contributed significantly to the increase in DC EBV accuracy. The magnitude of accuracy increase depended on the volume of records and the data structure. Results for SG showed that where industry and reference data are closely related, the increase in EBV accuracy from a small reference data set can still be beneficial. The increase in accuracy for genotyped SG was smaller than that for genotyped BB animals. This was due to the smaller number of genotypes and small phenotypic reference dataset but also because the genotyped animals were from industry and had DC recorded, so they already had higher levels of accuracy. The increase in accuracy when selecting young bulls has the potential to significantly increase the rates of genetic improvement for female reproduction traits and thus improve the overall production and profitability of the beef industry.

ACKNOWLEDGEMENTS

The authors acknowledge funding from MLA (project B.NBP.0759). Thank you to the collaborating organisations and the large team involved in designing, running and collecting data. We acknowledge the three breed societies and their members for access to their database.

REFERENCES

Cammack K.M., Thomas M.G. and Enns R.M. (2009) *The Prof. Anim. Sci.* **25**: 517.

Connors N.K., Cook J., Girard C.J., Tier B., Gore K.P., Johnston D.J. and Ferdosi M.H. (2017) *Proc. Assoc. Advmt. Anim. Breed. Genet.* **22**: 317.

Johnston D.J., Barwick, S.A., Corbet N.J., Fordyce G., Holroyd R.G., Williams P.J. and Burrow H.M (2009) *Anim. Prod. Sci.* **49**: 399.

Johnston D.J., Barwick, S.A., Fordyce G., Holroyd R.G., Williams P.J., Corbet N.J. and Grant T.P (2014) *Anim. Prod. Sci.* **54**: 1.

Johnston D.J., Ferdosi M.H., Connors N.K., Boerner V., Cook J., Girard C.J., Swan A.A. and Tier B. (2018) *Proc. 11th WCGALP, New Zealand* 269.

Johnston D.J., Grant T.P., Schatz T.J., Burns B.M., Fordyce B.M. and Lyons R.E. (2017) *Proc. Assoc. Advmt. Anim. Breed. Genet.* **22**: 385.

Johnston D.J. and Moore K.L. (2019) *Proc. Assoc. Advmt. Anim. Breed. Genet.* **23**.

SHARING MULTIBREED COW DATA WITH NEW ZEALAND TO IMPROVE PREDICTION FOR AUSTRALIAN CROSSBREED COWS FOR MILK YIELD TRAITS

M. Haile-Mariam[1], I.M. MacLeod[1], M. Khansefid[1], C. Schrooten[2], E. O'Connor[3], G. de Jong[2], H.D. Daetwyler[1,4] and J.E. Pryce[1,4]

[1]AgriBio Centre for AgriBioscience, Agriculture Victoria, Bundoora, VIC, 3083 Australia
[2]CRV, 6800 AL Arnhem, the Netherlands
[3]CRV Ambreed, Hamilton, 3216 New Zealand
[4]School of Applied Systems Biology, La Trobe University, Bundoora, VIC, 3083 Australia

SUMMARY

Ways to improve accuracy of genomic prediction (GP) for Australian (AU) crossbred cows by using data of about 33,000 cows from New Zealand (NZ), where crossbreeds are the dominant breed group (BG), and AU data were assessed. Accuracy of GP for validation cows was tested using single trait and multi-trait models, with data from different BGs considered as correlated traits. When data of the different BGs were considered as separate traits, the genetic correlations for milk yield (MY) were higher compared to that for fat yield (FY). The lowest correlations for all traits were between pure Holstein (H) and Jersey (J) as expected, and among the milk yield traits the lowest correlations were for FY. The estimated heritability and genetic correlations using the high-density SNP chip were slightly higher than 50K chip. Accuracy of GP using the NZ reference set (RS) was not better than AU reference. For MY, the accuracy of GP for AU crossbreed cows was like that observed for pure breed H cows. However, for FY and protein yield (PY), the accuracy of GP was lower in HJ (F_1) and HHJ (back cross to H) crosses. The joint NZ-AU RS resulted in 1 to 5% increase in accuracy for FY and PY of mainly crossbred cows.

INTRODUCTION

A joint project to improve accuracy of GP by sharing cow data in the pasture-based dairy systems of NZ and AU has been established by Agriculture Research Victoria and CRV (cattle breeding company in The Netherlands). A recent analysis showed that reliability (i.e. squared accuracy) of GP for milk traits for NZ validation bulls can be increased by 4 to 7% by including about 60,000 AU cows to a RS that included all NZ animals (Haile-Mariam *et al.* 2019). The benefit of adding NZ cows to AU RS is expected to be low for AU pure breed prediction because the number of genotyped NZ cows is relatively small. However, the number of crossbred cows from NZ is more than that from AU and this could be used to improve accuracy GP for AU crossbreed cows and possibly even for purebred Jersey for which the AU RS is small. Several studies have shown that the accuracy of GP from multibreed RS is not better than single-breed RS particularly when the breeds are distantly related (Calus *et al.* 2018). The inclusion of crossbred animals in the RS could improve the accuracy GP for crossbreds which was reported to be lower than those observed for pure breeds (Khansefid *et al.* 2019) and for all animals by improving the links between the pure breeds.

Data from several breeds for GP have been used in joint analyses in several ways including by considering the same trait recorded in different breeds as correlated traits in multi-trait (MT) model (Calus *et al.* 2018; Karoui *et al.* 2012) or by fitting breed as fixed effect in univariate model (Uni). In the MT model, the marker effects could be assumed to be different in different breed groups (BGs) where performance in J and H and their different crosses are treated as different but correlated traits. Using milk yield traits as response variable, the objectives of this study were: 1) to estimate genomic correlation (r_g) between the same trait measured in different BGs; 2) to assess the accuracy of GP for AU crossbred and purebred validation cows using NZ and AU cows as RS.

MATERIALS AND METHODS

Performance data of about 33,000 NZ genotyped cows and their contemporaries were obtained from NZ and included in the May 2018 genetic evaluation of DataGene for AU dairy cattle. In addition to NZ cows, there were close to 60,000 AU cows in the dataset. All NZ cows and most AU cows were genotyped with low density SNP chips (~ 10K SNP). These genotypes were imputed first to Bovine 50K SNP chip and then to High Density (HD) 800K SNP panels. After edit, in total the HD genotype set included 633,374 SNP and the 50K chip included 40,850 SNPs. The HD and 50K genotypes were used to create genomic relationship matrices (GRMs). The GRM that included J, H and crossbreeds was calculated for NZ and AU reference and validation cows (Table 1) separately and jointly following Yang et al. (2010). The number of cows included in the RS (born before 2011) and cows used for validation (born after 2010) is shown in Table 1. The response variable which were DRP for milk yield traits were analysed using MTG2 (Lee and van der Werf 2016). When all data were considered as the same trait, BG was fitted as fixed effect and in the multi-trait model data of each BG was considered as separate trait. In addition to the 5-trait (BGs) in NZ and 4-trait model in AU, the data from each country were analysed assuming a 3-trait model by combining the back crosses (i.e. HHJ or JJH) into their respective pure BG.

Adjusted accuracy was calculated as correlation between direct genomic breeding values (DGVs) and DRP, divided by the accuracy of the DRP of the validation cows. To ensure that the accuracies were less affected by high relationship among AU reference and validation cows, a cow was included in the validation set if its genomic relationship to the average of the top 10 cows in the RS (Clark *et al.* 2012) was below 0.25. As a result of this, no J cows were used for validation.

Table 1. Number of NZ and AU cows in reference set by breed group (BG: Holstein [H], back cross to H [HHJ], F_1 [HJ], back cross to J [JJH] and Jersey [J]) which their records were considered as different traits (5, 4 or 3 traits), and the number of AU validation cows

Breed group	NZ reference set			AU reference set			AU validation set
	Number	5-Trait	3-Trait	Number	4-Trait	3-Trait	
H	8624	Trait 1	Trait 1	21633	Trait 1	Trait 1	4944
HHJ	10125	Trait 2	Trait 1	1401	Trait 2	Trait 1	965
HJ	8675	Trait 3	Trait 2	1308	Trait 3	Trait 2	344
JJH	1481	Trait 4	Trait 3	-	-	-	-
J	3915	Trait 5	Trait 3	5905	Trait 4	Trait 3	-

RESULTS AND DISCUSSION

Tables 2 and 3 show the proportion of variance explained by the GRM (genomic h^2) in NZ and AU cows for MY and FY when the HD SNP chip was used. The genomic h^2 was the lowest for PY when using NZ data where they varied from 0.14 to 0.18 and 0.15 to 0.19 using the 50K and HD SNP chip, respectively. In the AU data, genomic h^2 for PY were only slightly lower than or similar to that for FY. In all cases the HD SNP chip explained about 2 to 5% more variance than 50K SNP chip (results not shown). Genetic correlations (r_g) among the BGs, when each BG was considered as traits, were lower for FY than for MY (Table 2 and 3). The pattern of r_g for PY was more similar to MY than to FY in NZ data but similar to FY in AU data. Differences in r_g between SNP chips were small, but in general the HD SNP chip showed higher correlations among the BGs than the 50K SNP chip (Table 2 and 3). As expected, r_g had higher standard errors (up to 0.10) than genomic h^2 (up to 0.04). Although the genomic h^2 were higher when AU cow data were used, the standard errors of the genomic h^2 and the r_g, particularly those involving crossbred BG, were higher in AU than NZ cows. Overall the use of 50K SNP chip "correctly" estimated the r_g to be the lowest between J and

H, whereas the HD SNP chip estimated the lowest correlation to be between H and the HJJ (Table 2), though the differences were not significant given the standard errors which were up to 0.15 in AU data. The observation that the HD SNP chip explained more variance than the 50K SNP chip agrees with van den Berg *et al.* (2016), where they found that adding selected sequence variants increased h^2 compared to the 50K SNP chip. Lower r_g between breeds when BGs are considered as traits for FY compared to MY in this study also agrees with other studies (Calus *et al.* 2018; van den Berg *et al.* 2016). Overall that our r_g estimates even between the two pure breeds (H and J) are higher than most literature estimates (van den Berg *et al.* 2016) may be due to some level of crossbreeding between J and H in NZ and AU (de Roos *et al.* 2008; Pryce *et al.* 2011) several generations back or due to similarity in production environment (i.e. pasture-based). The 3-trait model based on NZ and AU data sometimes showed the lowest r_g to be between HJ and J rather than between J and H which was unexpected. This may be due the small sample size and possibly some errors in the BG classification.

Table 2. Genomic h^2 in NZ reference cows of high-density SNP chip (HD) for milk and fat yield on the diagonal (in bold) and genetic correlations between breed groups for milk and fat yields using HD (above diagonal) and 50K SNP chip (below diagonal) in 5-trait model

Breed group	Milk					Fat				
	H	HHJ	HJ	JJH	J	H	HHJ	HJ	HJJ	J
H	**0.30**	0.97	0.85	0.73	0.77	**0.26**	0.97	0.76	0.42	0.49
HHJ	0.96	**0.30**	0.93	0.85	0.85	0.98	**0.23**	0.86	0.65	0.57
HJ	0.83	0.91	**0.31**	0.95	0.87	0.78	0.86	**0.22**	0.85	0.80
HJJ	0.77	0.84	0.92	**0.36**	0.86	0.41	0.63	0.84	**0.24**	0.90
J	0.72	0.82	0.86	0.89	**0.4**	0.47	0.55	0.78	0.88	**0.27**

Table 3. Genomic h^2 in AU reference cows of high-density SNP chip (HD) for milk and fat yield on the diagonal (in bold) and genetic correlations between breed groups for milk and fat yields using HD (above diagonal) and 50K SNP chip (below diagonal) in 4-trait model

Breed group	Milk				Fat			
	H	HHJ	HJ	J	H	HHJ	HJ	J
H	**0.34**	0.96	0.81	0.88	**0.23**	0.96	0.66	0.57
HHJ	0.94	**0.37**	0.88	0.97	0.92	**0.24**	0.65	0.57
HJ	0.78	0.78	**0.40**	0.78	0.63	0.56	**0.35**	0.55
J	0.75	0.91	0.72	**0.43**	0.43	0.58	0.50	**0.26**

The accuracy of GP for HJ and HHJ was higher than H for MY when NZ cows were used as RS (Table 4) because the crossbred cows dominate the set (Table 1). When using AU RS only, accuracy of GP was lower for crosses compared to H for FY (Table 2 and 3) and PY (results not shown) where r_g between the BGs were also lower. The use of NZ cows as a RS is expected to have less contribution for GP of PY because the r_g between performance in NZ and AU is lower (0.60 in H and 0.70 in J, Haile-Mariam *et al.* 2019) compared to both MY and FY and this will likely reduce the benefit of adding NZ RS to improve GP. However, the use of AU+NZ RS increased adjusted accuracy by 1 to 5% (Table 4). Table 4 also shows that considering performance of cows of different BGs in MT or Uni model has little benefit on the accuracy.

Table 4. Adjusted accuracy as correlation between direct genomic breeding values (DGVs) and DRP, divided by the accuracy of the DRP for validation Australian (AU) cows using New Zealand or AU cows in the reference, assuming data of cows from different breed groups to be the same trait (Uni.) or different (multi-traits) models from HD GBLUP

Trait	Breed group	New Zealand			Australia			AU+NZ
		Uni.	3-Trait	5-Trait	Uni.	3-Trait	4-Trait	Uni.
Milk	HJ	0.46	0.45	0.46	0.61	0.61	0.60	0.61
	HHJ	0.43	0.44	0.42	0.59	0.59	0.58	0.60
	H	0.33	0.33	0.35	0.58	0.58	0.57	0.59
Fat	HJ	0.26	0.24	0.22	0.43	0.42	0.42	0.47
	HHJ	0.28	0.26	0.23	0.46	0.46	0.45	0.48
	H	0.29	0.28	0.28	0.55	0.55	0.55	0.56
Protein	HJ	0.36	0.36	0.34	0.40	0.42	0.40	0.44
	HHJ	0.23	0.24	0.24	0.36	0.36	0.36	0.37
	H	0.34	0.33	0.34	0.54	0.54	0.54	0.55

CONCLUSIONS

Although the NZ reference did not provide better GP accuracy for AU crossbreed cows than AU RS, the joint use of AU and NZ RS increased GP for FY in HJ and HHJ cows and for PY in HJ only. In the case of MY accuracy of GP in crosses and H was similar, so adding NZ cows was not beneficial.

ACKNOWLEDGEMENTS

We thank DataGene and DairyNZ for data and Drs Konstantinov and Nieuwhof of DataGene for the DRPs. This study is funded by DairyBio (Agriculture Victoria and Dairy Australia) and CRV.

REFERENCES

Calus M.P.L., Goddard M.E., Wientjes Y.C.J., Bowman P.J. and B.J. Hayes (2018) *J. Dairy Sci.* **101**:4279.

Clark S.A., Hickey J.M., Daetwyler H.D. and van der Werf J.H. (2012) *Genet. Sel. Evol.* **44**:4.

de Roos, A.P.W., Hayes, B.J., Spelman, R.S. and Goddard M.E. (2008) *Genetics* **179**:1503.

Haile-Mariam M., MacLeod I.M., Schrooten C., de Jong G, O'Connor E., Daetwyler H. D. and Pryce J.E. (2019) Submitted to *J. Dairy Sci.*

Khansefid M., MacLeod I.M., Haile-Mariam M., Schrooten C., de Jong G, O'Connor E., Daetwyler H. D., Pryce J.E. and Goddard M.E. (2019) *Proc. Assoc. Advmt. Anim. Breed. Genet.* **23**:(in press).

Karoui S., Carabano M.J., Diaz C. and Legarra A. (2012) *Genet. Sel. Evol.* **44**:39.

Lee, S. H, and van der Werf J. H. (2016) *Bioinformatics*. **1;32**(9):1420.

Pryce J. E., B. Gredler, S. Bolormaa, P.J. Bowman, C. Egger-Danner, C. Fuerst, R. Emmerling, J. Sölkner, M.E. Goddard and B.J. Hayes. (2011) *J. Dairy Sci.* **94**:2625.

Van den Berg I., D. Boichard and M. S. Lund. (2016) *Genet Sel. Evol.* **48**:83.

Yang J, Benyamin B., McEvoy B., Gordon S., Henders A., Nyholt D., Madden P., Heath A., Martin N., Montgomery G., Goddard M.E and Visscher P.M. (2010) *Nature Genet.* **42**:565.

ASSESSING THE VALUE OF WHOLE GENOME SEQUENCE DATA IN SELECTING FOR AGE AT PUBERTY IN TROPICALLY ADAPTED BEEF HEIFERS

C. Warburton and B.J. Hayes

Queensland Alliance for Agriculture and Food Innovation, The University of Queensland, St Lucia QLD, 4067 Australia

SUMMARY

Age at puberty (AP) has been shown to be heritable in tropically adapted beef heifers, and is associated with lifetime productivity, but it is a difficult and expensive trait to measure. This study investigated whether whole genome sequence (WGS) genome wide association study (GWAS) results could be used to improve the accuracy of selection for AP by using various methods of SNP inclusion and different densities of SNP panels. These results suggest that the most benefit of WGS SNP inclusion would be made in lower density marker panels, with the 6K plus WGS analyses having prediction accuracies equivalent to the 50K base analysis. The ability to use a less expensive, lower density marker panel to make selection decisions will have a financial benefit to producers and warrants further investigation. Further research is required to determine the best technique to select WGS SNP and the most appropriate method to include these SNP into prediction models.

INTRODUCTION

Age at puberty (AP) has been shown to be moderately heritable in tropically adapted beef populations and is favourably correlated to female lifetime reproductive capacity (Johnston *et al.* 2009; Zhang *et al.* 2013; Johnston *et al.* 2014; Farah *et al.* 2016). AP has also been shown to be heritable using genomic information, however, the accuracy of selection using these methods has been low (Zhang *et al.* 2013; Engle *et al.* 2019; Hayes *et al.* 2019).

One possibility for improving accuracy of genomic predictions is to use (imputed) whole genome sequence (WGS). To date, the use of WGS data in genomic predictions within livestock species has shown modest improvements in selection accuracy (0%-5%), and there is much interest in developing novel techniques to best utilise this data (Raymond *et al.* 2018). The aim of this study was to investigate methods to incorporate WGS data into the genomic prediction for AP in a multi-breed population of tropically adapted beef heifers, to improve the accuracy of selection.

MATERIALS AND METHODS

Animals and Phenotypes. Fertility records used in this study were obtained from two research herds, the Northern Breeding Project research herds from the Cooperative Research Centre for Beef Genetic Technologies (Beef CRC) and the Queensland Smart Futures (SMF) population.

Briefly, 868 Brahman heifers and 960 Tropical Composite heifers with both a phenotype for AP and genotype data were obtained from the Beef CRC. In this study, AP was defined as age, in days, at first *corpus luteum*, obtained by ultrasound scanning heifers every 4 to 6 weeks (Johnston *et al.* 2009). Detailed herd structure, management and data recording have been outlined in Johnston *et al.* (2009).

A total of 3,695 reproductive maturity scores (a proxy trait for AP; measured at 600 days by ultrasound and is a 0 to 5 score) were obtained from the SMF database on heifers from 3 breeds, Brahman, Santa Gertrudis and Droughtmaster (Burns *et al.* 2016). Full information on herd structure, management and data recording can be found in Burns *et al.* (2016).

The SMF results presented in this paper have been analysed across breeds to determine if multi-breed genomic predictions could be viable for use in industry data. However, it must be noted that the genomic estimated breeding values (GEBV's) shown in these results are not true multi-breed GEBV's as heifers of each breed were managed separately and there were no mixed breed cohorts analysed in this data.

Genotypes. Beef CRC heifers were genotyped with the BovineSNP50 BeadChip (Illumina, SanDiego, CA) and SMF heifers were genotyped with the 24,121 SNP from the Geneseek GGP-LD array. Full details on genotype quality control are described in Hayes *et al.* (2019). Genotypes were imputed up to 728,785 SNP (Bovine HD array) using the Fimpute software (Sargolzaei *et al.* 2014), and a panel of 1500 cattle of relevant breeds genotyped for the Bovine HD array. All genotypes were then imputed to 23 million whole genome sequence variant genotypes using the 1000 bull genomes Run6 data base (Hayes *et al.* 2019) using Eagle phasing and Minimac3 for imputation.

Statistical analysis. Three datasets, Brahman (Beef CRC), Tropical Composite (Beef CRC) and SMF (Brahman, Santa Gertrudis and Droughtmaster) were used in these analyses. The analysis proceeded in two steps: 1) Identify SNP associated with AP in the imputed sequence data by within breed GWAS analysis in the Beef CRC animals, then 2) Test the accuracy of genomic predictions when these SNPs are added to base SNP panels in the SMF data.

The final models for each analysis included contemporary group fitted as a covariate, which was defined as herd, year and season in the SMF dataset. In the SMF dataset age at AP measurement was also included as a covariate. In the Brahman analysis, age of dam was fitted as a covariate and in the Tropical Composite analyses zebu percentage was fitted as a continuous covariate. Animal was fitted as a random effect in all models.

Two strategies were used to identify SNPs associated with AP in the GWAS:

- TOP GWAS (SNP significance threshold 5.0e-06) - all SNP from the WGS GWAS that met the significance threshold from either breed were included in each analysis.
- TOP META (SNP significance threshold 5.0e-07) - Meta-analyses were conducted on the output from the WGS GWAS of the combined Brahman and Tropical Composite populations using the program Metal (Willer *et al.* 2010) and the SNP that met the significance threshold were included in each analysis.

The numbers of significant SNP from the WGS data for each analysis and each SNP selection strategy are shown in Table 1.

Genomic predictions in the SMF data were conducted using 3 different density of base SNP panels, 6K (BovineLD array), 50K (BovineSNP50 BeadChip) and 800K (BovineHD array). A GBLUP approach was used. Genome-wide complex trait analysis (GCTA) was used to construct genomic relationship matrices (GRM) and perform genomic predictions for each of the datasets for each SNP density, see Yang *et al.* (2011) for more detail.

Significant, unique (not already included in base marker panels) SNP from the sequence GWAS were incorporated into each analysis using one of two methods; first, by adding the significant WGS SNP into the GRM for each analysis (6K plus WGS SNP, 50K plus WGS SNP or 800K plus WGS SNP) or secondly, by using a multi GRM method where the base GRM remained the same but a second GRM, with only the WGS SNP, was added and analysed simultaneously. Any significant WGS SNP that were already included on marker panels were excluded from the WGS GRM but remained in the base GRM in the MGRM analyses. The GEBV from each GRM in the MGRM analyses were added together to calculate total GEBV which was used to calculate prediction accuracy (6K MGRM, 50K MGRM or 800K MGRM).

Five way cross validation within the SMF data set was used to determine the accuracy of prediction of GEBV, where each dataset was randomly split five times and four fifths of the data (reference) was used to predict the GEBV of the last fifth (validation) and the validation animals were then used to calculate the correlation between their predicted GEBV and phenotype adjusted for the model fixed effects. The prediction accuracy was the correlation of the GEBV and the phenotype divided by the square root of the heritability of AP in the SMF 800K analysis, h^2=0.196.

RESULTS AND DISCUSSION

Results in Table 1 show that accuracy was improved more by using a higher density SNP panel than through the addition of WGS SNP in the TOP GWAS analyses. One reason for this may be due to the high level of SNP redundancy that may be occurring with this SNP selection strategy. Of the 165/172 SNP used from the WGS data in the TOP GWAS analysis, a large proportion of SNP occurred on just 3 chromosomes (results not shown), chromosome 1 n=44, chromosome 14 n=77 and chromosome 21 n=18, total number of significant SNP on these 3 chromosomes is 139. There is a probability that a number of these SNP are in close proximity to a single SNP of large effect and, due to linkage disequilibrium, these SNP may appear significant in a GWAS due to this association. Therefore, the actual number of effective SNP that are being used for selection in the TOP GWAS analysis may be lower than the 165/172 shown, which may explain the limited improvement in accuracy seen in Table 1.

Table 1. Prediction Accuracy for TOP GWAS and TOP META analyses in SMF data

Analysis	TOP GWAS		TOP META	
	Prediction accuracy *(s.e)*	No. sig. WGS SNP	Prediction accuracy *(s.e)*	No. sig. WGS SNP
6K	0.36 *(0.04)*		0.36 *(0.04)*	
6K plus WGS SNP	0.37 *(0.05)*	172	0.40 *(0.05)*	1591
6K MGRM	0.37 *(0.04)*	172	0.40 *(0.05)*	1591
50K	0.41 *(0.05)*		0.41 *(0.05)*	
50K plus WGS SNP	0.41 *(0.05)*	172	0.42 *(0.05)*	1587
50K MGRM	0.41 *(0.05)*	172	0.43 *(0.06)*	1587
800K	0.42 *(0.05)*		0.42 *(0.05)*	
800K plus WGS SNP	0.42 *(0.05)*	165	0.42 *(0.05)*	1502
800K MGRM	0.42 *(0.05)*	165	0.44 *(0.05)*	1502

The prediction accuracy of GEBV for TOP META analyses were also improved through the use of higher density SNP panels. In contrast to the TOP GWAS results, the addition of the significant WGS TOP META SNP did result in small improvements in prediction accuracy within each of the analyses, although the improvement is not significant. The inclusion of WGS META SNP into the 6K analysis improved the prediction accuracy of this analysis so that it became equivalent to the 50K analysis. The 6K marker panel is more cost effective for producers than the higher density panels, therefore, if equivalent prediction accuracies can be made from the 6K panel with the use of WGS SNP the financial benefit to producers would be significant.

It is evident that there are many more significant WGS SNP being used in the TOP META analysis, in comparison to the TOP GWAS analysis, which may explain the small improvement in accuracy. Similar to the TOP GWAS results, a large proportion of SNP discovered in the TOP META analysis existed on a single chromosome, 14 (results not shown), n=~1,400 (depending upon the analysis). More research needs to be done to determine the most effective way to select WGS SNP and reduce this potential redundancy.

The MGRM analyses in the TOP META SNP selection strategy resulted in slight improvements in prediction accuracy (though not significant), in comparison to the single GRM analyses, in the 50K and 800K analyses. In the MGRM analysis the WGS SNP are being fitted in their own GRM and, as a result, their effect is less regressed. As these SNP have been selected for having a significant effect upon the AP phenotype from a meta-analysis, it can be argued that fitting these SNP into a single, large GRM may regress their effect by too great an extent. More research is required.

CONCLUSIONS

While the results presented in this paper are not conclusive, there is an indication to suggest that improved methods of WGS SNP selection may be used to improve GEBV prediction accuracy particularly for the less dense marker panels. The inclusion of 1,591 WGS META SNP into the 6K analysis was able to improve the prediction accuracy for puberty to a similar level as the 50K base analysis, which would be a much more cost-effective genotyping solution for producers.

Further research is warranted into appropriate methods to select WGS SNP that are able to explain variation in the AP trait in multi-breed tropically adapted beef populations and the best way to incorporate these SNP into future genomic analysis. More AP phenotypes will be required to improve the accurate detection of WGS SNP that can explain variation in AP across a number of tropically adapted breeds.

ACKNOWLEDGEMENTS

We gratefully acknowledge the Beef CRC and the scientists and technicians who pioneered the traits used in this paper, and the huge effort that went into designing and conducting those experiments. We also gratefully acknowledge the contribution of Dr Brian Burns, Nicholas Corbet, Jack Allen, Alan Laing and Geoffry Fordyce for the SMF data set used in this paper.

REFERENCES

Burns, B.M., Corbet N.J., Allen J.M., A. Laing and M. T. Sullivan (2016).Queensland Government Smart Futures Research Partnerships Program (2012-2015).

Engle, B. N., N. J. Corbet, Allen J.M., Laing A.R., Fordyce G, McGowan M.R., Burns B.M., Lyons R.E. and Hayes B.J. (2019) *J. Anim. Sci.* **97**: 90.

Farah M. M., Swan A.A., Fortes M.R.S, Fonseca R., Moore S.S. and Kelly M.J. (2016) *Anim. Genet.* **47**: 3.

Hayes B. J., Corbet N.J., Allen J.M., Laing A.R., Fordyce G., Lyons R., McGowan M.R. and Burns B.M. (2019) *J. Anim. Sci.* **97**: 55.

Hayes B.J. and Daetwyler H.D. (2019) *Annu. Rev. Anim. Biosci.* **7**: 89.

Johnston D., Barwick S.A., Corbet N., Fordyce G., Holroyd R.G., Williams P. and Burrow H. M. (2009) *Anim. Prod. Sci.* **49**: 399.

Johnston D., Barwick S.A., Fordyce G., Holroyd R.G., Williams P., Corbet N. and Grant T. (2014). *Anim. Prod. Sci.* **54**: 1.

Raymond B.,A. Bouwman C., Schrooten C., Houwing-Duistermaat J. and Veerkamp R. F. (2018) *Genet. Sel. Evol.* **50**: 27.

Sargolzaei, M., Chesnais J. and Schenkel F. (2014) *BMC Genomics* **15**: 478.

Willer C.J., Li Y. and Abecasis G. R. (2010) *Bioinformatics* **26**: 2190.

Yang J., Lee S.H., Goddard M.E. and Visscher P.M. (2011) *Am. J. Hum. Genet.* **88**: 76.

Zhang Y.D., Johnston D.J., Bolormaa S., Hawken R.J. and Tier B. (2013) *Anim. Prod. Sci.* **54**: 16.

INCREASING THE ACCURACY OF GENOMIC PREDICTION IN CROSSBRED DAIRY CATTLE

M. Khansefid[1], M.E. Goddard[1,2], M. Haile-Mariam[1], C. Schrooten[3], G. de Jong[3], E. O'Connor[4], J.E. Pryce[1,5], H.D. Daetwyler[1,5] and I.M. MacLeod[1]

[1]AgriBio Centre for AgriBioscience, Agriculture Victoria, Bundoora, VIC, 3083 Australia
[2]Faculty of Veterinary and Agricultural Sciences, University of Melbourne, Parkville, VIC, 3010 Australia
[3]CRV, 6800 AL Arnhem, the Netherlands
[4]CRV Ambreed, Hamilton, 3216 New Zealand
[5]School of Applied Systems Biology, La Trobe University, Bundoora, VIC, 3083 Australia

SUMMARY

This study assessed the accuracy of genomic prediction for crossbred dairy cows (mixed crosses between Holstein and Jersey) when purebreds and crossbreds were combined in a single mixed-breed reference set. The reference population consisted of 36,695 bulls and cows. There were six validation breed groups including crossbred cows (from New Zealand and Australia) as well as purebred cows (Holstein or Jersey cows from New Zealand). The effect of using genotypes of different marker densities (50K or HD) and different analytical models (GBLUP or emBayesR) on the accuracy and bias of genomic predictions was studied. The results showed that on average for milk traits (milk, fat and protein yields), the accuracies increased using HD genotypes compared to 50K genotypes, regardless of the prediction model. However, emBayesR outperformed GBLUP in all validation populations with the highest increase observed for Australian crossbreds when HD genotypes were used. Additionally, the bias of genomic prediction was reduced when using HD compared to 50K genotypes in both GBLUP and emBayesR models.

INTRODUCTION

Genomic prediction (GP) within breeds is generally very accurate using the standard 50K SNP panel when the linkage disequilibrium (LD) between markers and the causal mutations is preserved over long distances (e.g. using 50K in a purebred Holstein reference to predict into young Holstein bulls). However, using a single breed reference population for GP in crossbreds generally has low reliability, similar to the low accuracy for across breed prediction (e.g. Kemper *et al.* 2015). This is likely due in part to the fact that LD decays faster in crossbreds compared to purebreds so that markers that accurately predict QTL effects in Holsteins may not always be in LD with the same causal mutation allele in the crossbred. This is particularly the case when crossbreeding occurs over several generations as is common in the New Zealand dairy industry (New Zealand Dairy Statistics, www.dairynz.co.nz/dairystatistics).

The use of a mixed breed reference population to increase the reference population size, can potentially increase the accuracy of GP if the markers segregating across breeds have the same LD phase with the causal mutation alleles (Kemper *et al.* 2015). Moreover, inclusion of crossbreds in the reference population should also help to find the most predictive markers closest to causal mutations because LD would be preserved over shorter distances. This also helps to limit the number of multiple SNPs in high LD with QTLs.

Therefore, the aim of this study was to increase the accuracy of GP in crossbred dairy cattle using a mixed breed and crossbred reference population, increasing the density of markers (HD versus 50K) and using models which tend to calculate individual SNP effects (Bayesian) rather than haplotype effects (GBLUP). The dairy crossbreds were cows that had varying proportions of Holstein and Jersey (approximately 50%:50% cross = "HJ", approximately 75% Holstein = "HHJ" and approximately 75% Jersey = "HJJ"). The accuracy of GP in the crossbreds was also compared with purebred Holstein "H" and Jersey "J".

MATERIALS AND METHODS

Animals. The reference set consisted of 7,463 purebred bulls mainly from New Zealand and the Netherlands (953 Red H, 5,409 H and 1,101 J) as well as 29,232 purebred and crossbred cows from New Zealand (NZ) (7,623 H, 9,262 HHJ, 7,807 HJ, 1,157 HJJ and 3,383 J). There were five NZ cow validation populations: 1,002 H, 863 HHJ, 868 HJ, 324 HJJ and 532 J. An Australian (AU) cow validation set of 344 HJ was included to demonstrate GP in a less related group.

Relatedness between validation and reference. The validation sets were selected to reduce high relationships with the reference: no sires or half-sib brothers of validation cows were included in the reference. It has been demonstrated that the strength of the top 10 genomic relationships between validation and reference animals ($\text{Rel.}_{\text{Top10}}$) gives a good indicator of the relative accuracy of GP (Clark *et al.* 2012). Therefore, this is reported for each validation population.

Phenotypes. Milk, fat, and protein yields were analysed separately but the results are reported as the average across three traits. The phenotypes for CRV bulls were de-regressed proofs (DRP) on the Australian scale, derived from international MACE (2018) breeding values (Liu 2009). The NZ and AU cow phenotypes were also DRP which were processed together by DataGene (2018) using test day records and correcting for known fixed effects as for the official Australian dairy cattle evaluations (https://datagene.com.au/).

Genotypes. Two sets of imputed genotypes were used in GP: the standard Illumina 50K SNP panel (40,850 SNP) and Illumina HD 800k SNP panel (633,375 SNP), where the latter included an additional custom set of ~ 1200 variants. In the HD genotype set, one of each pair of SNP in LD $r^2 > 0.95$ was pruned out leaving 316,396 SNP. The majority of genotypes were first imputed from low density chips (~ 10k SNPs) up to 50K and then imputed from 50K to HD using FImpute (Sargolzaei *et al.* 2014). The SNPs with minor allele frequency (MAF) < 0.002 were removed.

Models. The GBLUP (Meuwissen *et al.* 2001) analysis used the following model (with MTG2 software: Lee and Van der Werf 2016):

$$\mathbf{y} = \mathbf{Xb} + \mathbf{Zu} + \mathbf{e} \qquad (1)$$

where, **y** is the vector of phenotypes (MY, PY or FY DRP) for the animals in the reference, **X** is a design matrix allocating phenotypes to fixed effects (sex and breed group), **b** is the vector of fixed effect solutions, **Z** is a design matrix allocating records to individual additive genetic values in **u**, $\mathbf{u} \sim N(0, \mathbf{G}\sigma^2_g)$ is a vector of genomic breeding values (GEBVs) in which σ^2_g is the additive genetic variance and **G** is the GRM constructed from animal genotypes (50K or pruned HD), and $\mathbf{e} \sim (0, \mathbf{E}\sigma^2_e)$ is a vector of random residual effects in which σ^2_e is the error variance and **E** is a diagonal matrix constructed as $\text{diag}(1/w_i)$ where w_i is the weighting coefficient for each animal. Weighting coefficients were calculated differently for cows and bulls following Equation 5 and 6 of Garrick *et al.* (2009), with heritability h^2=0.33, repeatability t=0.56 and proportion of variance not explained by markers is c=0.2.

We also analysed the data with "emBayesR" (Wang *et al.* 2016: in-house software):

$$\mathbf{y} = \mathbf{Xb} + \mathbf{Wv} + \mathbf{e} \qquad (2)$$

where, **y**, **X**, **b** and **e** are as for equation 1, **v** is the vector of SNP effects (50K or pruned HD), **W** is a design matrix of SNP marker genotypes (50K or pruned HD). In emBayesR model, the initial EM (Expectation-Maximisation) phase was set for a maximum of 1,500 iteration with the convergence parameter set as 1×10^{-7} and the BayesR phase was set to complete 5,000 iterations. For each trait, the emBayesR model was run in 5 independent replicated analyses (MCMC chains) to check for convergence and the results were averaged across the 5 chains. The accuracy of GP for each validation breed group was defined as the Pearson's correlation coefficient between GEBVs and DRPs ($r_{GEBV,DRP}$). The bias of GP was assessed by calculating the regression coefficient of DRP on GEBVs ($b_{DRP,GEBV}$) (no bias $b_{DRP,GEBV} = 1$).

RESULTS AND DISCUSSION

The accuracy and bias of GEBVs in each of the six validation breed groups are shown in Figure 1, where the values are averaged across MY, FY, and PY.

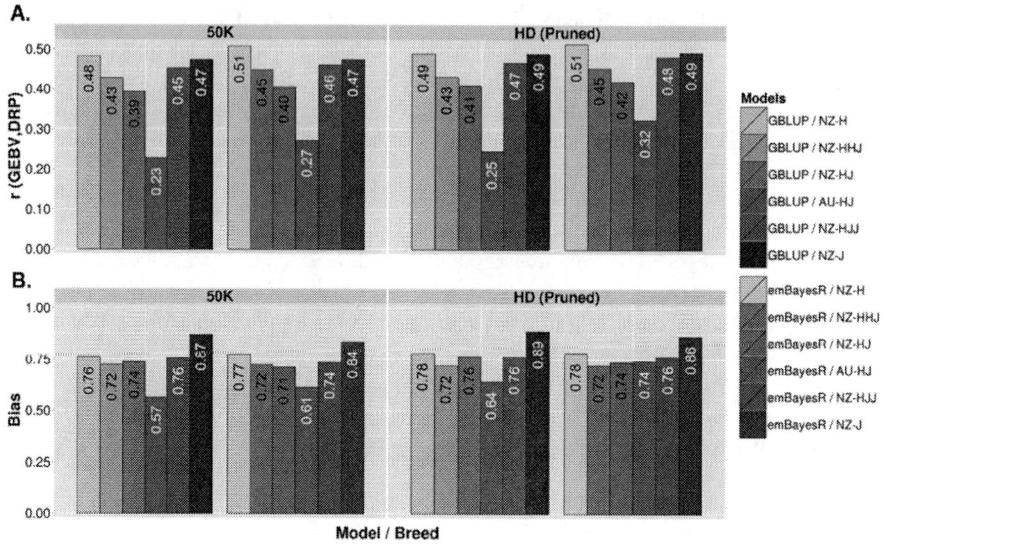

Figure 1. The accuracy (A) and bias (B) of GP in validation breed groups using different marker density genotypes and analytical models

Crossbred vs. purebred. Regardless of SNP density and method used, the accuracy of GP in purebred H was highest followed by pure J cows. It is not surprising that H was the most accurate because the H breed dominated the reference population. Furthermore, the relationships between the different validation sets and the reference may also partly explain the results. The $Rel._{Top10}$ in purebred H and J cows was 0.250 and 0.345, respectively, but was generally lower in crossbred cows: HHJ=0.209, NZ-HJ=0.207, AU-HJ=0.195 and HJJ=0.282. Therefore, this may be partly contributing to the observed lower in accuracy of GP for: crossbreds compared to purebreds, as well as HJ compared to other crosses. The lower accuracy observed in crossbreds could also be partly due to the lower reliability of some crossbred phenotypes compared to purebreds and potentially, genotype imputation in crossbreds may be less accurate than for purebreds. Although pure J validation had the strongest relationships with the reference, the accuracy of GP in pure J was slightly lower than H. This may occur because the proportion of J in the reference is very low compared to H, therefore if some QTL segregate only in J they may not be accurately predicted. The GP in pure J and H was generally less biased than crossbreds. However, the bias across the different crossbred validations was almost the same. Although all validations showed some bias, $b_{DRP,GEBV}$ lower than 1 is a common observation in dairy cattle (Khansefid *et al.* 2014).

50K vs. pruned HD genotypes. There was a consistent increase in accuracy when using pruned HD instead of 50K genotypes, regardless of validation breed group and prediction method (on average ~ 2%; from 0.41 to 0.43). Additionally, the bias of GP was reduced when using denser genotypes (on average ~ 3%; from 0.73 to 0.76). This suggests that increasing the marker density enables more precise estimates of QTL effects because markers tend to be closer and in stronger LD with the causal variants. Moreover, in HD genotypes the markers tend to have the same LD phase with the causal mutation alleles across different breeds. Therefore, for the validation breed group AU-HJ, in which the cows were least related to the reference, the amount of gain from using denser markers was expected to be greatest. However, the amount of gain in AU-HJ accuracy compared to other validation sets was greater only when emBayesR was used in GP. This suggests that to obtain the most benefit from increased marker density, the Bayesian model works better than GBLUP because it provides a more precise estimate of QTL effects. In AU-HJ using HD genotypes instead of 50K genotypes, also reduced the bias of predictions more than other validation breed groups.

GBLUP vs. emBayesR. The accuracy of GP was increased using emBayesR instead of GBLUP in all validation breed groups regardless of marker density (on average ~ 2%; from 0.42 to 0.44). This is likely because the genetic architecture of the milk traits is better modelled by the Bayesian sparse mixture model compared to the quasi-infinitesimal GBLUP model (Goddard *et al.* 2016). In GBLUP the effect of causal QTLs tends to be spread across many markers that are in LD with the causal mutations and all effects come from the same normal distribution. However, in emBayesR the SNP effects are estimated more precisely because we allow a mixture distribution of SNP effects where some may be small medium or large, and a proportion of SNP may have no effect on the trait. This Bayesian model would therefore be expected to show the most benefit when the validation animals are less related to the reference group. Using emBayesR instead of GBLUP did not have a large effect on the bias of GP, except in AU-HJ where the bias of prediction reduced.

CONCLUSIONS

The accuracy of GP in crossbreds was lower than purebreds. Using HD instead of 50K genotypes and emBayesR instead of GBLUP increased the accuracy and reduced the bias of genomic predictions.

ACKNOWLEDGEMENTS

We thank DataGene (particularly Drs K Konstantinov and G Nieuwhof), CRV and DairyNZ staff for processing and provision of cow data. The authors acknowledge financial support from CRV and DairyBio (a joint venture between Agriculture Victoria and Dairy Australia).

REFERENCES

Clark S.A., Hickey J.M., Daetwyler H.D. and van der Werf J.H. (2012) *Genet. Sel. Evol.* **44**: 4.

Kemper K.E., Reich C.M., Bowman P.J., Vander Jagt C.J., Chamberlain A.J., Mason B.A., Hayes B.J. and Goddard, M.E. (2015) *Genet. Sel. Evol.* **47**: 29.

Khansefid M., Pryce J.E., Bolormaa S., Miller S.P., Wang Z., Li C. and Goddard M.E. (2014) *J. Animal Sci.* **92**: 3270.

Garrick D.J., Taylor J.F. and Fernando R.L. (2009) *Genet. Sel. Evol.* **41**: 55.

Goddard M.E., Kemper K.E., MacLeod I.M., Chamberlain A.J. and Hayes B.J. (2016) *Proc R Soc Lond [Biol]* **283**: 20160569.

Meuwissen T.H., Hayes B J. and Goddard M.E. (2001) *Genetics* **157**: 1819.

Lee S.H. and van der Werf J.H.J. (2016) *Bioinformatics* **32**: 1420.

Liu Z. (2009) Deregressing MACE proofs for genomic evaluations, Proteje Meeting, Brussels, Belgium.

Sargolzaei M., Chesnais J.P. and Schenkel F.S. (2014) *BMC Genomics* **15**: 478.

Wang T., Chen Y.P.P., Bowman P.J., Goddard M.E. and Hayes B.J. (2016) *BMC Genomics* **17**:744.

GENOTYPE PANEL REQUIREMENTS FOR INCLUSION INTO BREEDPLAN SINGLE STEP EVALUATIONS

N.K. Connors and M.H. Ferdosi

Animal Genetics & Breeding Unit*, University of New England, Armidale, NSW, 2351 Australia

SUMMARY

Increased demand for genomic data driven by the transition of BREEDPLAN to single-step genomic BLUP, has seen an increase in the numbers of genotyping providers and SNP panels. To assess the suitability of new panels, AGBU has developed a standardised procedure to ensure high quality genomic data for inclusion into BREEDPLAN.

INTRODUCTION

The transition of Australia's national beef genetic evaluation, BREEDPLAN, to Single-Step Genomic Best Linear Unbiased Prediction (ssGBLUP) in 2017 (Johnston *et al.* 2018) has driven increased demand for genomic data. The resultant growth in genotyping has stimulated the rapid introduction of new commercial genotyping companies and new single nucleotide polymorphism (SNP) panels. Given that the end use for the majority of the beef industry's genotyping is incorporation into BREEDPLAN, it is necessary to ensure that SNP panels offered to breeders/breed societies are compatible with the genomic pipeline quality control (QC) requirements (Connors *et al.* 2017) used for building the genomic relationship matrix for BREEDPLAN single-step evaluations. These QC checks can only be performed if the genotypes are compatible and as such, new genotypes and new panels require analysis prior to inclusion into the BREEDPLAN genomic pipeline. The Animal Genetics and Breeding Unit (AGBU) has developed a set of industry standards for genotype panels, along with a process of analysing new SNP panel products and validating their compatibility for the BREEDPLAN genomic pipeline. This paper describes the analyses and validation process for new SNP panels and their genotypes.

MATERIALS AND METHODS

The fundamental requirements for BREEDPLAN genomics compatibility are consistent format, SNP quality and informativity. These requirements ensure accuracy and consistency across all genomic records included in the national genomic evaluations. These standards are currently in consultation with Meat and Livestock Australia, industry partners and genotype providers. Firstly, genotypes are required to be sent in a particular format derived from Illumina Genotyping Exports with specifications designated by AGBU. This consistent format ensures all genotypes are processed in the correct order, enables automated analysis of the new SNP panels, and ensures genomic pipeline QC can be performed. Analyses of new SNP panels requires access to the panel map file, which provides SNP names, SNP location (chromosome and base pair position), and allele coding (e.g. manufacture strand/customer strand) in a particular genome assembly. The map file must also be in a particular consistent format as designated by AGBU. The map file provides an overview of the panel, including SNP density and chromosome coverage.

Analysis of SNPs on new panels enables a comparison with existing panels, and more importantly, the consensus panels developed at AGBU (AGBU 6k, AGBU 150k) used for the BREEDPLAN genomic pipeline to combine SNP data from panels of varying densities. AGBU 6k consists of a set of approximately 6000 SNPs common across SNP panels and is available in the public domain (Boichard

* A joint venture of NSW Department of Primary Industries and the University of New England

et al. 2012); AGBU 150k is a set of approximately 150,000 SNPs used to format all genotypes to a common consensus panel (Connors *et al.* 2017). Analysis of the new panel map file in relation to the consensus panels indicates its compatibility with the BREEDPLAN genomic pipeline. For example, genotypes from panels missing the common 6000 SNPs (AGBU 6k) cannot be compared with existing genotypes and will be incompatible with BREEDPLAN.

AGBU has developed automated reporting processes to analyse new SNP panels, providing a detailed breakdown including the number of SNPs (overall and per chromosome), minimum and maximum distance between SNPs, first and last position of SNPs, mean and standard deviation of positions per chromosome. A common SNPs analysis is performed by creating a SNP overlap matrix between the new panel and a subset of the most informative existing panels, including the AGBU consensus panels. These statistics are presented in an automated report in both tabular and graphical representations, enabling rapid identification of potential issues. Automated reports can be made available to relevant industry and commercial bodies considering the use of the analysed new SNP panel.

RESULTS AND DISCUSSION

To date, AGBU has analysed more than 50 different SNP panels, enabling a detailed understanding of the requirements for BREEDPLAN compatibility. Based on this experience, AGBU has formulated a set of requirements for new panels to ensure compatibility for the genomic pipeline and inclusion into BREEDPLAN single-step evaluations, forming the basis of new panel assessment Levels 1-3.

Currently Level 1 assessments are enforced for BREEDPLAN inclusion, such that genotypes from a new panel not meeting these requirements will be excluded from the evaluation until such time as the requirements are met. Levels 2 and 3 assessments, along with the formation of reference populations, are proposed for future implementation and are currently being negotiated with industry bodies to determine funding structures and accountability.

Level 1 assessment – Panel format, quality and content. New panel inclusion requirements are as follows:

- Provide AGBU with the aim and/or target breed/s of a newly developed panel;
- Genotypes and map files must be formatted as per specified guidelines (as supplied by AGBU);
- The genome assembly used for the panel map file must be provided;
- Panel must contain at least 90 percent of SNPs in the AGBU 6k consensus panel;
- Panel must have at least 10000 SNPs in common with the autosomal region of the AGBU 150k consensus panel;
- There must be at least 200 SNPs on the non-autosomal region of the X chromosome;
- Confidential or patented SNPs and any restrictions regarding their use must be provided;
- The raw genotypes are to be supplied in Illumina AB format (guidelines as supplied by AGBU) and must not be imputed or manipulated (e.g. removal of chromosome X, Y or mitochondrial SNPs);
- Genotypes must possess GenCall (GC) scores. If GC scores are not provided, a statement of quality assurance is required. AGBU will not accept any responsibility of quality regarding issues related to SNPs without GC scores;
- Companies which do not use GC as quality control (e.g. Affymetrix) must provide AGBU with the formula to convert their QC to GC. Additionally, they should provide a statement regarding the concordance of their panels with Illumina panels;
- SNP names should not contain a prefix or suffix. If prefixes or suffixes are present, specific recommendations to AGBU must be included on the SNPs involved;
- There must be no SNP duplication in the map file; i.e. SNPs with same position and chromosome but different name, or same SNP name and different positions. If important SNPs must be tested on the

panel more than once, they must have a suffix 'dup' and the reason for duplication must be provided.

In addition to meeting the above criteria for panel inclusion, AGBU also recommends the following:

- At least 20 individuals should be genotyped with both an existing panel of equal or higher density and the new panel, to check quality and concordance;
- Suggested inclusion of SNPs in mitochondrial regions on new panel;
- Suggested inclusion of SNPs on the non-autosomal region of Y chromosome on new panel.

These BREEDPLAN inclusion requirements and recommended features are communicated to genotyping companies in relation to the design of new panels, and/or in response to inclusion/exclusion of genotypes from new panels.

Level 2 assessment – Heterozygosity. The Level 2 heterozygosity assessment is proposed for implementation in future, dependent on ongoing negotiations with industry and commercial bodies. This assessment is not currently enforced for new panels to be included into BREEDPLAN. Genotypes for at least 200 individuals per breed are required. Individuals should be purebred according to software used for the BREEDPLAN genomic pipeline (Boerner 2017; Boerner and Wittenburg 2018) and therefore representative of the breed population, forming a reference dataset for that breed. Individuals can be genotyped either with the new panel, or with a panel of high enough density (e.g. Illumina Bovine HD (777k)) such that there are two panels (one of which is the new panel) sharing at least 95 percent of SNPs in common for comparison. The heterozygosity and allele frequencies of the genotypes are assessed to check for the following requirements:

- The heterozygosity over all SNPs for the entire breed's population must be greater than 30 percent;
- The heterozygosity for pure breed animals must be less than 50 percent;
- Major allele frequencies for each individual must not exceed 75 percent;
- The histogram for SNPs with minor allele frequency (MAF) above 0.05 should be reasonably distributed.

Importantly, all SNPs included in BREEDPLAN single-step evaluations will require at least 1000 genotypes with that SNP (i.e. 1000 individuals), to ensure high imputation accuracy. If less than 1000 genotypes are available (e.g. because they're included in a new SNP panel) these SNPs will not be used until they can be further validated. Thus the new panels require more stringent assessment of those SNPs in common with other panels.

Level 3 assessment – Imputation accuracy. The Level 3 imputation accuracy assessment is proposed for implementation in future, dependent on ongoing negotiations with industry and commercial bodies. This assessment is not currently enforced for new panels to be included into BREEDPLAN. The aim of this assessment is to investigate how well genotypes from existing panels can be imputed to the new panel, and vice versa. Genotypes of at least 2000 individuals with high density (e.g. 777k) per breed are required. Individuals should be suitably representative of the average breed population, such that they form a reference dataset for that breed. This assessment will be performed by extracting a subset of SNPs based on the new panel's map file from 1000 individuals and imputing up to 777k, to other panels, and to SNPs used for building the GRM. The remaining 1000 individuals will be used as a reference set for imputation. Furthermore, if the new panel is of higher density (i.e. >50k), imputation accuracy of low density panels (i.e. <20K) will also be assessed.

Reference populations. Testing SNP panels for heterozygosity and imputation accuracy requires reference populations. The creation of reference populations is proposed for implementation in the near future to ensure ongoing robustness of genomics in BREEDPLAN. Genotypes for at least 200 individuals per breed are required. Individuals should be purebred, according to BREEDPLAN genomic pipeline requirements (Boerner 2017; Connors *et al.* 2017) and thus suitably representative of the breed's population. Reference individuals can be used to test concordance between panels, and ongoing research needs for BREEDPLAN single-step evaluations.

These reference populations should be dynamic, continually supplemented and updated yearly, to ensure genetic trends in the population are captured. The significant cost involved in establishing reference populations and maintaining them requires strategic negotiations with appropriate industry bodies and commercial parties to determine funding structures and accountability. These strategic discussions are currently underway and will determine the expected time for implementation of assessment Levels 2-3 and reference populations.

CONCLUSIONS

This paper describes industry standards developed by AGBU for the inclusion of genotypes from new SNP panels into the BREEDPLAN single step evaluations. These requirements ensure accuracy and consistency across genomic records from various different genotyping platforms. Increasing numbers of new SNP panels are being introduced, requiring SNP panel analyses to determine BREEDPLAN compatibility. AGBU has developed an automated analysis process, which provides reports on new SNP panels for interested commercial and industry partners. This assessment process benefits industry by ensuring animals are not genotyped with panels that are not compatible with BREEDPLAN, and ensures genotyping quality is maintained.

ACKNOWLEDGEMENTS

This research includes data generated by the Beef CRC. This research is supported by Meat and Livestock Australia (MLA) project L.GEN.0174.

REFERENCES

Boerner V. (2017) *Proc. Assoc. Advmt. Anim. Breed. Genet.* **22:** 97.

Boerner V., Wittenburg D. (2018) *Frontiers in Genetics* **9:** 185.

Boichard D., Chung H., Dassonneville R., David X., Eggen A., Fritz S., Gietzen K.J., Hayes B.J., Lawley C.T., Sonstegard T.S., Van Tassell C.P., VanRaden P.M., Viaud-Martinez K.A., Wiggans G.R., for the Bovine L.D.C. (2012) *PLOS ONE* **7:** e34130.

Connors N.K., Cook J., Girard C.J., Tier B., Gore K.P., Johnston D.J., Ferdosi M.H. (2017) *Proc. Assoc. Advmt. Anim. Breed. Genet.* **22:** 317.

Johnston D.J., Ferdosi M.H., Connors N.K., Boerner V., Cook J., Girard C., Swan A., Tier B. (2018) *Proc. World Congr. Genet. App. Livest. Breed.* . **11:** 269.

FINDING THE OPTIMAL REFERENCE POPULATION FOR GENOMIC PREDICTION OF AUSTRALIAN RED DAIRY CATTLE

I. van den Berg[1], I.M. MacLeod[1] and J.E. Pryce[1,2]

[1]Agriculture Victoria Research, AgriBio, 5 Ring Road, Bundoora, Victoria, 3083 Australia
[2]School of Applied Systems Biology, La Trobe University, Bundoora, Victoria, 3083 Australia

SUMMARY

Genomic prediction for breeds with a small population size, such as the Australian Red, is challenging, because reliability depends on the size of the reference population and its relatedness to the animals evaluated. Our objective was to find the optimal reference population for Australian Red, comparing within breed and multi breed prediction for milk yield, fat yield, protein yield and somatic cell count.

Our results show that while multi breed prediction can result in higher accuracies than within breed prediction, adding fewer animals that are more closely related to the validation population can result in a higher reliability than adding a much larger number of individuals that are more distantly related.

INTRODUCTION

Genomic prediction for breeds with a relatively small population size, such as Australian Red cattle, is challenging, because the reliability of prediction is dependent on the size of the reference population (Goddard 2009). Sharing reference populations across breeds or countries may increase the size of the reference population, though this has only been advantageous for closely related breeds, such as the Nordic Red cattle breeds (Brøndum *et al.* 2011). Australian Red cattle are influenced by several Red dairy breeds, including Scandinavian Red cattle breeds, Ayrshire, Shorthorn, Illawarra and Red and White Holstein (http://www.aussiereds.com.au).

Multi breed prediction often analyses the same trait in different breeds as a single trait with a breed effect to account for differences across breeds. Not all QTL impact the expression of quantitative traits in the same way across breeds (Raven *et al.* 2014) and there may be QTL by breed interactions resulting in different effects of QTL for different breeds. Therefore, it may be appropriate to fit the same trait in different breeds as multiple correlated traits (Olson *et al.* 2012).

Because linkage disequilibrium is maintained over much shorter distances across breeds than within breed (de Roos *et al.* 2008), prediction reliability is expected to decrease faster across breeds than within a breed when the distance between causal mutations and prediction markers increases (van den Berg *et al.* 2016). Consequently, the standard 50K SNP chip may not be dense enough for accurate prediction from Holstein to Australian Red, and variants close to causal mutations could potentially result in a higher reliability.

The objective of this study was to find the optimal reference population for Australian Red dairy cattle. Within and multi breed reference populations were compared, with multi breed populations containing either a low number of Holstein animals that are relatively closely related to Australian Red cattle based on a genomic relationship matrix between Holstein and Australian Red cattle, or larger numbers of more distant Holstein and Jersey individuals, used a single trait model or a multi trait model that fitted the same trait in different breeds as multiple correlated traits.

MATERIALS AND METHODS

We calculated the reliability of genomic prediction in Australian Red bulls for different reference populations. The reference population contained up to 3,248 Holstein bulls, 48,386 Holstein cows,

807 Jersey bulls, 8,734 Jersey cows and 3,041 Australian Red cows. Genome-wide complex trait analysis (GCTA) (Yang *et al.* 2011) was used to first construct a genomic relationship matrix of the full reference population and perform a principal component analysis (PCA). In total, 10 reference populations were used. The largest reference population contained 3,041 Australian Red cows, 51,634 Holstein and 9,541 Jersey individuals, and the smallest only the Australian Red cows. Additional reference populations contained the Australian Red cows and either all Holstein individuals or only Holsteins with a value for the first principal component (PC1) above a certain threshold. Figure 1 shows the first two principal components of the PCA, and indicates the groups used to construct different reference populations. The number of individuals in each of these seven subsets is shown in Table 1. The validation population contained 280 Australian Red bulls. Deregressed proofs (DRP) for milk (MY), fat (FY) and protein yield (PY) and somatic cell count (SCC) were calculated following Garrick *et al.* (2009) and used as phenotypes.

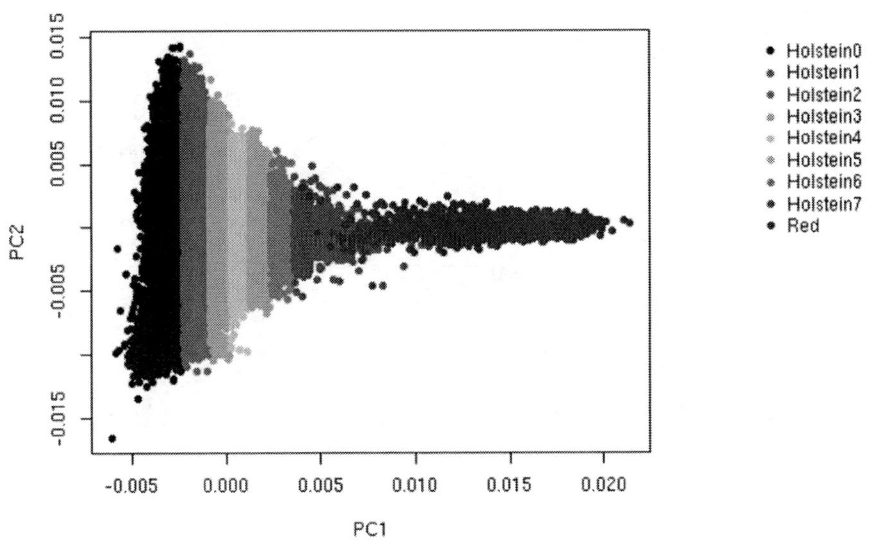

Figure 1. First two principal components (PC1 and PC2) of the genomic relationship matrix of the multibreed reference population containing Holstein and Australian Red individuals. Different colours show different subsets of animals that are used to construct different reference populations

Table 1. Number of Holstein and Australian Red (Red) individuals in different reference populations based on the first principal component

Breed	H1-7+R	H2-7+R	H3-7+R	H4-7+R	H5-7+R	H6-7+R	H7+R
Holstein	39,788	29,809	19,835	9,880	4,915	2,436	1,197
Red	3,041	3,041	3,041	3,041	3,041	3,041	3,041

Genotypes were available for the Illumina BovineSNP50 chip (50K, real or imputed). Because the LD between QTL and prediction markers on the 50K chip may not be conserved across breeds, we also analysed genotypes on a custom chip with 46,516 imputed sequence variants selected by Xiang *et al.* (2019) that are expected to be enriched for dairy trait QTL (XT).

For each of the reference populations, we used the GBLUP model as implemented in MTG2 (Lee and van der Werf 2016) to predict GEBV of the validation population. The reliability of genomic prediction was calculated as the squared correlation between DRP and GEBV divided by the average reliability of individuals in the validation population. The model either considered the same trait in different breeds as a single trait, fitting a breed effect to correct for breed differences (ST-GBLUP), or fitted the same trait in different breeds as different, correlated traits, using a multi trait model (MT-GBLUP).

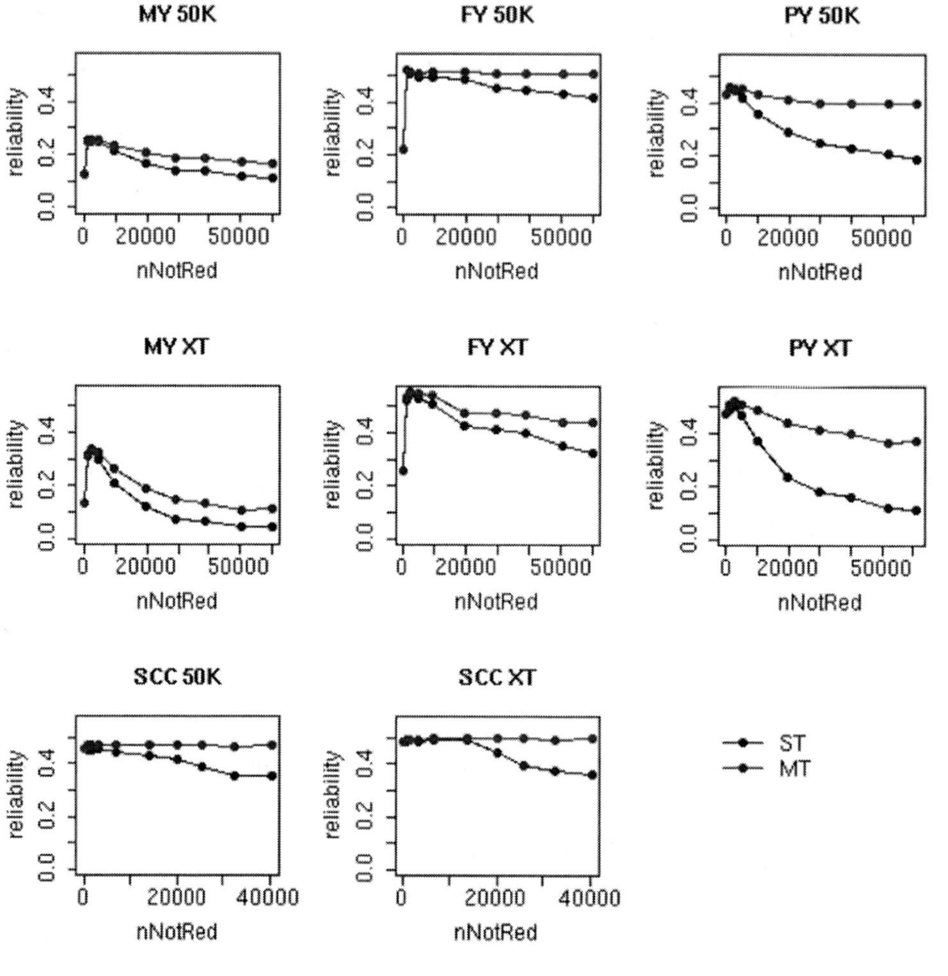

Figure 2. Reliability of genomic prediction as a function of the number of Holstein and Jersey individuals in the reference population (nNotRed) for milk yield (MY), fat yield (FY), protein yield (PY) and somatic cell count (SCC), using variants on the 50K SNP chip (50K) or selected sequence variants (XT). For the multi breed reference populations, the same trait in different breeds was analysed using a single trait model fitting a breed effect (ST) or a multi trait model considering the trait as multiple correlated traits in different breeds (MT)

RESULTS AND DISCUSSION

Figure 2 shows the reliability as a function of the composition of the reference population. The overall pattern was similar for all traits: the highest accuracies were obtained using a multi breed reference population with a limited number of Holstein individuals that are relatively closely related to the Australian Red. population.

For all traits tested, the highest reliability was obtained with the MT model and a multi breed reference population. The XT variants only led to a small difference in reliability compared to the 50K variants. For MY, FY and PY, the reference population resulting in the highest reliability contained around 2,400 Holsteins (with reliabilities of 0.34, 0.56, 0.52 for MY, FY and PY, respectively), while for SCC, the highest reliability (0.50) was obtained with 13,822 Holsteins. Adding Jerseys to full reference population containing all Holstein individuals resulted in a similar reliability as obtained without the Jerseys.

Except for FY, the reliability obtained with the full multi breed reference population was lower than the reliability obtained with the within breed reference population. The decrease in reliability when adding larger numbers of Holstein individuals to the reference population was larger with the ST model than with the MT model.

The GBLUP prediction models assume all variants are equally important to predict the trait. Models that can allocate higher importance to variants linked to causal mutations, such as Bayesian variable selection models or a weighted GBLUP, may result in higher and be less prone to the decrease in reliability we observed when adding larger numbers of Holstein individuals to the reference populations.

CONCLUSIONS

Our results show that while multi breed prediction can result in higher accuracies than within breed prediction, adding fewer animals that are more closely related to the validation population can result in a higher reliability than adding a much larger number of individuals that are more distantly related. To implement genomic prediction in Australian Red cattle, an international reference population containing other Red breeds is likely to lead to a higher reliability than a multi breed Australian reference population.

ACKNOWLEDGEMENTS

We acknowledge Bolormaa Sunduimijid for providing the imputed sequence variants, DataGene for providing access to data used in this study, and our partners in the 1000 Bulls Genomes Project for access to the reference genomes.

REFERENCES

Brøndum R.F., Rius-Vilarrasa E., Strandén I., Su G., Guldbrandtsen B., Fikse W.F. and Lund M.S. (2011) *J. Dairy Sci.* **94**: 4700.

de Roos A.P.W., Hayes B.J., Spelman R.J. and Goddard M.E. (2008) *Genetics* **179**: 1503.

Garrick D.J., Taylor J. F. and Fernando R.L. (2009) *Genet Sel Evol* **41**:44.

Goddard M.E. (2009) *Genetica* **136**: 245.

Lee S.H. and van der Werf J.H.J. (2016) *Bioinformatics* **32**: 1420.

Olsen K.M., VanRaden P.M., and Tooker M.E. (2012) *J. Dairy Sci.* **95**: 5378.

Raven, L-A., Cocks B.G. and Hayes B.J. (2014) *BMC Genomics* **15**: 62.

van den Berg I., Boichard D., Guldbrandtsen B. and Lund M.S. (2016) *G3* **6**: 2553.

Xiang R., van den Berg I., MacLeod I.M., Hayes B.J., Prowse-Wilkins C.P., Wang M., Bolormaa S., Liu Z., Rochfort S.J., Reich C.M., Mason B.A., Vander Jagt C.J., Daetwyler H.D., Chamberlain A.J. and Goddard M.E. (2019) *Submitted*

Yang J., Lee S.H., Goddard M.E. and Visscher P.M. (2011) *Am. J. Hum. Genet.* **88**: 76.

APPLYING NEXT GENERATION PHENOTYPING STRATEGIES FOR GENETIC GAIN IN DAIRY CATTLE

J.E. Pryce[1,2], T.T.T. Nguyen[1], P.N. Ho[1], T.D.W. Luke[1], S. Rochfort[1], W.J. Wales[3], P. Moate[3], L.C. Marett[3], G. Nieuwhof[1,4], M. Abdelsayed[4], M. Axford[4], M. Shaffer[4], M. Haile-Mariam[1]

[1]Agriculture Victoria Research, AgriBio, Bundoora, Victoria, 3083 Australia
[2]School of Applied Systems Biology, La Trobe University, Bundoora, Victoria, 3083 Australia
[3]Agriculture Victoria Research, Ellinbank, Victoria, 3821 Australia
[4]DataGene Ltd, Agribio, Bundoora, Victoria, 3082 Australia

SUMMARY

Genomic selection is changing how we selectively breed animals and, more recently, for the traits we select. In addition to providing genomic breeding values for traits that were traditionally evaluated through progeny-testing of Australian dairy cattle, genomic breeding values have already been provided for two novel traits. Feed Saved, and Heat Tolerance, were released in 2015 and 2017, respectively. Our focus for dairy cattle breeding is now on traits associated with animal health, fertility and impacts on the environment. This is being achieved by directly selecting measurable phenotypic traits, or indirectly using tools such as mid-infrared spectral data and automated sensor devices to identify predictors of these traits. Greater collaboration between scientific disciplines and countries is likely to facilitate development of data-sets that will serve as better reference populations for genomic selection of new traits into the future.

INTRODUCTION

Genomic selection has transformed worldwide livestock and plant breeding. While genomic selection has changed how we select, it has not substantially changed the traits we select for. Having said this, there are recent examples of traits that are now being selected for that would not be possible without genomic selection. In this paper we describe two examples of how genomic selection has enabled the next generation of breeding values for dairy cattle. In addition, we will explore new opportunities that leverage off advances in phenotyping.

GENOMIC REFERENCE POPULATIONS

For most traits evaluated in dairy breeding, the genomic reference population is usually composed of bulls with large daughter groups. Often published breeding values include information from the animal's pedigree, including progeny and ancestors, in addition to the genomic component. However, for expensive or difficult to measure traits, it is not cost effective to phenotype large daughter groups. Instead the reference population can be genotyped females that have the desired phenotype measured directly. The Australian **G**enomic **Info**rmation Nucleus (Ginfo) started in 2013 with around 100 herds and 30,000 cows and has contributed to the increase of reliabilities of genomic ABVs and played a key role in development of genomic breeding values of novel traits.

TECHNOLOGICAL ADVANCES

The use of fully automated phenotyping in animal breeding is still in its infancy. Many precision farming technologies, such as pedometers, automatic temperature devices, automated oestrus detection, daily body condition scoring and bodyweight scales are becoming more common on modern dairy farms (Egger-Danner *et al.* 2014). To make substantial advances in low and moderately heritable traits, it is important to measure phenotypes on a large number (>10,000) of animals. Generating quality phenotypes from large amounts of data is a challenge that requires expertise in interpretation of data and how it can be used for selection. It is important to consider the effect of any new breeding objective on other traits in the breeding goal.

Sensors. The use of sensor technology alongside genomic selection could lead to a further improvement in the prediction of complex traits, such as fertility, as the data obtained is likely to be more objective than other sources and potentially provide new information. For example, Talukder *et al.* (2015) compared gold standard progesterone-evaluated oestrus detection, infrared thermography, heat and rumination tags (e.g. Hi Tag, SCR Engineers, Israel) and visual assessment of mounting indicators. The mounting indicators had 100% positive predictive values, while prediction using thermography was poor. The sensor tags performed reasonably well with 70% positive predictive values.

Mid-infrared spectroscopy (MIR). Mid-infrared spectroscopy involves passing a beam of light through a milk sample to provide data in the form of spectra (absorbance or reflectance at specific wavelengths). Farmers currently receive regular reports from their herd test centres with information on milk volume and fat and protein concentration generated from MIR. Analysis of milk MIR has been used to predict other milk characteristics such as milk fatty acids, milk protein composition, milk coagulation properties, milk acidity, mineral composition and ketone bodies with reasonable accuracy (De Marchi *et al.* 2014).

Mid-infrared prediction equations are already showing promise to aid management decisions regarding complex traits. A good example is beta-hydroxy-butyrate (BHB) concentration, where most MIR prediction equations are calculated using the concentration of BHB in milk (Grelet *et al.* 2016). The BHB concentration in milk can also be used to predict the BHB concentration in blood (Luke *et al.* 2019). This is important as sub-clinical ketosis in dairy cattle is often diagnosed using of the concentration of BHB in blood, hence using MIR in milk to predict metabolites measured in blood is likely to be a suitable practical approach to manage metabolic disease. The way in which MIR can be used for selection purposes may differ from management purposes i.e. it may not enhance genomic prediction. However, it has already been shown to be a powerful tool to identify genetic variants associated with milk composition (Benedet *et al.* 2019).

Multi-omics. There may also be opportunities to use information from multiple sources. Examples include direct measurements, the metagenome (e.g. rumen, reproductive etc), the proteome/metabolome (protein and metabolite structure and function) and functional genomic assays (e.g., methylation, transcriptomics etc.). When these techniques are used in conjunction with sequencing technologies, causal variants can be identified, which should lead to better responses to selection. Ultimately, multi-omics approaches could enhance selection of existing and novel traits.

FEED SAVED AND GREENHOUSE GAS EMISSIONS

The Feed Saved Australian Breeding Value (ABV) comprises the energy required for maintenance, through liveweight breeding values calculated from conformation scores (Haile-Mariam *et al*, 2014) and residual feed intake (RFI) as a genomic prediction evaluated in heifers and cows. The reference population for RFI comprises around 2000 Australian heifers and cows and European Holstein cows (Pryce *et al.* 2015). The current reliability of Feed Saved is 35-40%.

Maintenance of this breeding value requires additional data from cows that are contemporaries of the current national population. One way to increase the accuracy of genomic breeding values is to increase the size of the reference population through large international collaborations. An example of this is the global dry matter initiative (gDMI) where a reference population of around 10,000 cows (Berry *et al.* 2014). More recently, the Efficient Dairy Genome Project, which is a multi-national research initiative led by Canada and it currently (April 2019) has collated feed intake records from 4,779 cows (T. Chud, personal communication). Each partner has free access to the database that includes feed intake, production and liveweight phenotypes in addition to pedigree and genomic information.

Technological advances to measure feed intake in commercial cows are occurring rapidly. For example, in dairy cattle in confined systems, cameras are being used to estimate volume changes along a feed lane before and after feeding to estimate the change in volume of feed (Bloch *et al.* 2019). In grazing environments, bite meters can be used to measure feeding behaviour discriminating between time spent biting, chewing and ruminating. It is more challenging to measure the volume of each bite and the nutrient concentrations of the pasture eaten. If this can be overcome, then individual feed intake records might become more common, especially if the sensors have multiple functions, so that farmers are motivated to purchase them.

In recent times, there has been a push to share phenotypes on individual cow methane emissions. Methane production is an expensive phenotype to measure and again, international cooperation is an attractive way to develop a dataset that is large enough for genomic prediction. There are many ways to measure methane emissions, some of which measure the total methane emitted by an individual cow in a day (Deighton *et al.* 2014). Others measure the methane emitted only at certain times or locations (Hegarty 2013). Therefore, there has been a requirement to develop statistical ways to combine heterogeneous data (Haas *et al.* 2018). In addition to multi-country reference populations, there may also be a benefit to bringing together data on different breeds, such as beef breeds.

HEAT TOLERANCE

Worldwide, heat stress is a concern for many livestock production systems as it affects animal welfare and reduces productivity. In December 2017 genomic estimated breeding values (GEBV) for heat tolerance in dairy cattle were released for the first time in Australia. The dataset was constructed by merging herd-test production records with weather station data. Heat tolerance phenotypes were defined as the rates of decline in milk, fat and protein yield after a heat stress event (i.e. temperature-humidity index exceeds 60) and were estimated using a reaction norm model (Nguyen *et al.* 2016). The GEBV has been validated using divergent lines managed in controlled hot weather events (Garner *et al.* 2016). Although the reliability of heat tolerance is moderate (on average around 38%), it is expected that this will improve as the size of reference populations are increased. The genetic trend for heat tolerance has worsened, the genetic correlation with the Australian national selection index (Balanced Performance Index; BPI) is -0.20. Including heat tolerance in the BPI could improve farm profitability.

There are other ways that heat impacts dairy cows. For example, Dahl *et al.* (2016) stated that heat stress reduces dry matter intake, which in turn reduces yield and compromises immune function and if heat events are experienced in late gestation calf survival and performance is affected. Therefore, there is a need for further research on the impacts of heat stress on other traits to develop a multi-faceted heat tolerance breeding value.

HEALTH TRAITS

Most genomic breeding values associated with health of dairy cattle have either been developed using records of "clinical cases" collected from farms, or by using predictor traits. For example, Abdelsayed *et al* (2017) obtained clinical disease data from >150k cows in 90 Ginfo herds, concluding that many health traits have sufficient genetic variation for selection purposes.

The problem with farmer recorded data is that it is often inconsistent, incomplete, or sparse and generally only works well when electronic record keeping is mandatory for other purposes. However, there are opportunities to improve the reliability of genomic breeding values through the use of predictor traits, such as conformation traits, e.g. udder conformation for mastitis resistance and feet and leg traits for lameness. Technological advances in phenotyping, described earlier, could offer potential solutions for genetic selection.

FERTILITY

Currently, most fertility breeding values around the world consider calving, mating and pregnancy data usually recorded by farmers. More extensive use of mid-infrared (MIR) spectroscopy (generated through machines used in routine commercial herd-testing), advanced phenotyping (using sensor technology etc) and genes identified to explain some of the genetic variation in fertility are under study and to expected deliver more precise genomic breeding values of fertility by getting closer to the biology of this complex trait.

CONCLUSIONS

New technologies will generate large amounts of data that can be used for selection purposes and it is expected these will improve the way we select for current and future breeding objectives. As the emphasis of genetic evaluations changes from increasing output to reducing production costs and environmental footprints and improving animal welfare, access to quality data will be a challenge. This can be met by collaboration including with international partners and with farmers and research working in other disciplines to ensure expensive data is used for many purposes.

ACKNOWLEDGEMENT

Agriculture Victoria staff acknowledge Dairy Australia, Gardiner Foundation and Agriculture Victoria for funding through DairyBio and DairyFeedbase.

REFERENCES

Abdelsayed M., Haile-Mariam M. and Pryce J. E. (2017) *J. Dairy Sci.* **100**: 9643.

Benedet A., Ho P.N., Xiang R., Bolormaa S., de March M., Goddard M.E. and Pryce J.E. (2019) *J. Dairy Sci.* **102**: 7189.

Berry D., Coffey M., Pryce J., de Haas Y., Løvendahl P., Krattenmacher N., Crowley J., Wang Z., Spurlock D. and Weigel K. (2014) *J. Dairy Sci.* **97**: 3894.

Bloch V., Levit H. and Halachmi I. (2019) *J. Dairy Res.* **86**: 34.

Dahl G. E., Tao S. and Monteiro A.P.A. (2016) *J. Dairy Sci.* **99**: 3193.

De Marchi M., Toffanin V., Cassandro M. and Penasa M. (2014) *J. Dairy Sci.* **97**: 1171.

Deighton M.H., Williams S.R.O., Hannah M.C., Eckard R.J., Boland T.M., Wales W.J. and Moate, P.J. (2014) *Anim. Feed Sci. and Tech.* **197**: 47.

Egger-Danner C., Cole J., Pryce J., Gengler N., Heringstad B., Bradley A. and Stock K. (2014). *Animal*: **1**.

Garner J.B., Douglas M.L., Williams, S.R.O., Wales W.J., Marett L.C., Nguyen T.T.T., Reich C.M. and Hayes B.J. (2016) *Sci. Rep.* **6**: 34114.

Grelet C., Bastin C., Gelé M., Davière J.-B., Johan M., Werner A., Reding R., Pierna J.F., Colinet F. and Dardenne P. (2016) *J. Dairy Sci.* **99**: 4816.

Haas Y.d., Wall E., Garnsworthy P.C., Kuhla B., Negussie E. and Lassen J. (2018) P*roc. of the World Congr. on Genet. App. to Livest. Prod.*

Haile-Mariam M., Gonzalez-Recio O. and Pryce J. (2014) *J. Dairy Sci.* **97**: 3173.

Hegarty R.S. (2013). *Animal* **7**: 401.

Luke T.D.W., Rochfort S., Wales W.J., Bonfatti V., Marett L. and Pryce J.E. (2019) *J. Dairy Sci.* **102**: 1747.

Nguyen T.T., Bowman P.J., Haile-Mariam M., Pryce J.E. and Hayes B.J. (2016) *J. Dairy Sci.* **99**:2849.

Pryce J., Gonzalez-Recio O., Nieuwhof G., Wales W., Coffey M., Hayes B. and Goddard M. (2015) *J. Dairy Sci.* **98**: 7340.

Talukder S., Thomson P.C., Kerrisk K.L., Clark C.E.F. and Celi P. (2015) *Therio.* **83**: 739.

THE GENETIC ANALYSIS OF ADULT BIRD PERFORMANCE TOGETHER WITH SLAUGHTER TRAITS IN OSTRICHES

S.W.P. Cloete[1,2], A. Engelbrecht[3], A.R. Gilmour[4], M.F. Schou[5], Z. Brand[3] and C.K. Cornwallis[5]

[1]Department of Animal Sciences, University of Stellenbosch, Stellenbosch, 7602 South Africa
[2]Directorate Animal Sciences: Elsenburg, Western Cape Department of Agriculture, Elsenburg, 7607 South Africa
[3]Directorate Animal Sciences: Oudtshoorn, Western Cape Department of Agriculture, Oudtshoorn, 6620 South Africa
[4]11 Holman Way, Orange, NSW, 2800 Australia
[5]Department of Biology, Lund University, Lund, Sweden

SUMMARY

Genetic analyses on economically important ostrich traits have currently only been done within trait complexes, such as breeding or slaughter traits. This study resolved the issue by investigating genetic correlations across the complexes of production traits in adult birds (i.e. egg and chick production as well as adult live weight) and slaughter traits in immature birds (slaughter weight, skin size, hair follicle score and nodule size score). All traits were heritable and variable indicating that responses to selection may be possible. Heritability estimates and genetic correlations within trait complexes were consistent with previously derived parameters. Two-trait analyses on single traits from each complex found generally negative genetic correlations of reproduction with slaughter weight and skin size, with the correlation between egg production and skin size being significant at -0.41. The genetic correlation between slaughter weight and adult weight was high and positive at 0.81, as expected when comparing the same trait measured at different life stages. Size dependent slaughter traits (skin size and nodule size) were also positively correlated to adult weight. These results are discussed in relation to ostrich production.

INTRODUCTION

Up to 70% of global commercial ostrich products originate from South Africa (Brand and Jordaan 2011). It is therefore not surprising that the literature on scientific ostrich breeding is also heavily dependent on studies of South African genetic resources (see Cloete *et al.* 2008b for a review). So far, most analyses have been conducted within trait complexes, e.g. live weight up to slaughter age (Bunter and Cloete 2004; Engelbrecht 2013), adult weight and reproduction (Cloete *et al.* 2008a), feather traits (Brand and Cloete 2015), chick survival (Wang *et al.* 2011) as well as slaughter traits (Engelbrecht 2013). As a result, little is known about the genetic correlations among traits of these trait complexes. This study therefore investigates genetic parameters for traits in the adult animal trait complex (Cloete *et al.* 2008a) and the slaughter bird trait complex (Engelbrecht 2013) and estimates genetic correlations between them.

MATERIALS AND METHODS

The study utilised data from the ostrich resource population maintained at the Oudtshoorn Research Farm of the Western Cape Department of Agriculture, which has been well-documented (Bunter and Cloete 2004; Cloete *et al.* 2008a; 2008b; Engelbrecht 2013). Only data of the South African Black strain on the farm were used. Data in the adult bird complex included repeated egg production, chick production (total egg and chick numbers over a breeding season) and adult live weight records, as described in detail by Cloete *et al.* (2008a). Data in the slaughter trait complex were slaughter weight and skin size, as well as the subjectively assessed traits of nodule size and hair follicle scores. All these traits were reviewed by Engelbrecht *et al.* (2009) and subsequently assessed in a genetic analysis conducted by Engelbrecht (2013). The number of records used varied from 1079 for skin size to 6292 for adult weight (Table 1).

Genetic (co)variance components and ratios were derived from a three-trait repeatability model for adult birds and a four-trait animal model for slaughter traits in ASREML (Gilmour *et al.* 2016). For adult production traits, fixed effects included production year and animal age, sex for adult weight and the length of the breeding season in days as linear covariates for the reproduction traits. Year of hatch, sex and age at measurement were modelled for slaughter traits. Additive animal, animal permanent environment and service sire (for egg and chick production) were fitted as random for adult birds and additive animal for slaughter traits. Further analyses involved two-trait combinations of each slaughter trait with each adult bird trait, using parameters previously derived as priors. Animals with records were 1391 for adult weight, 678 females with reproduction records and 721 service sires mated to at least one female.

RESULTS AND DISCUSSION

The descriptive statistics reported in Table 1 were consistent with those of previous studies on adult birds (Cloete *et al.* 2008a) and slaughter birds (Engelbrecht 2013). Likewise, coefficients of variation were previously above 50% for reproduction traits, below 20% for live weight traits, below 10% for skin size and between 20 and 50% for subjective skin quality traits.

Table 1. Descriptive statistics for traits in the adult animal and slaughter animal trait complexes

Complex and trait	Number of observations	Mean	SD	CV	Range
Adult traits at 5.4 (SD = 3.2) years:					
Egg production (n)	3023	44.0	25.2	57.3	0 – 121
Chick production (n)	3023	21.0	17.5	83.3	0 – 90
Adult weight (kg)	6292	120.1	15.2	12.7	68 – 178
Slaughter traits at 364 (SD = 71) days:					
Slaughter weight (kg)	4085	92.4	17.1	18.5	40 – 148
Skin size (dm^2)	1079	140.8	9.3	6.6	104 – 170
Nodule size (n)	1749	4.49	1.22	27.2	1 – 9
Hair follicles (n)	1771	3.77	1.73	45.9	1 – 9

Heritability estimates from the three-trait model amounted to 0.16 for egg production, 0.11 for chick production and 0.37 for adult live weight (Table 2). Genetically, egg production and chick production were the same trait (r_g = 0.99), while the genetic correlations of reproduction traits with adult live weight were below 0.10 and not significant (P > 0.05).

Table 2. The observed phenotypic variance (σ^2_p) and (co)variance ratios for the traits in the adult animal complex. Significant (P < 0.05) correlations are denoted by an asterisk

Components and traits	Trait: Egg production	Chick production	Adult weight
σ^2_p	498.7	269.2	196.0
(Co)variance ratios: Heritability in bold on the diagonal, with genetic correlations below and phenotypic correlations above the diagonal (± standard error)			
Egg production (n)	**0.16 ± 0.04**	0.74* ± 0.01	0.06 ± 0.03
Chick production (n)	0.99* ± 0.03	**0.11 ± 0.04**	0.04 ± 0.03
Adult weight (kg)	-0.01 ± 0.15	0.06 ± 0.17	**0.37 ± 0.04**

Animal permanent environmental effects ranged between 0.15 ± 0.04 for egg production and 0.30 ± 0.04 for adult weight (data not shown). Likewise, service sire effects amounted to 0.045 ± 0.011 for egg production and 0.046 ± 0.011 for chick production. These results were generally consistent with our previous study on the same resource population involving the same traits (Cloete *et al.* 2008a) as well as with results reviewed from the literature (Cloete *et al.* 2008b).

Heritability estimates from the four-trait model for slaughter traits were quite consistent in magnitude, ranging from 0.33 for skin size to 0.38 for slaughter weight (Table 3). Genetic correlations between slaughter weight, skin size and nodule size were positive, while those involving hair follicle score were variable in sign and not significant ($P > 0.05$). The heritability of slaughter weight compared well with those previously reported for live weight at an age close to the anticipated slaughter date (Bunter and Cloete 2004; Engelbrecht *et al.* 2009; 2011; 2013; Engelbrecht 2013). Genetic parameters involving skin size and subjectively assessed leather quality were also consistent with previous studies (Engelbrecht *et al.* 2009; Engelbrecht 2013).

Table 3. The observed phenotypic variance ($\sigma 2_p$) and (co)variance ratios for the traits in the slaughter animal complex. Significant ($P < 0.05$) correlations are denoted by an asterisk

Component and traits	Trait			
	Slaughter weight (kg)	Skin size (dm^2)	Nodule size score (n)	Hair follicle score (n)
σ^2_p	165.3	61.8	1.16	2.53
(Co)variance ratios: Heritability in bold on the diagonal, with genetic correlations below and phenotypic correlations above the diagonal				
Slaughter weight (kg)	**0.38 ± 0.04**	0.69* ± 0.02	0.33* ± 0.02	0.04 ± 0.03
Skin size (dm^2)	0.88* ± 0.05	**0.33 ± 0.06**	0.45* ± 0.03	0.03 ± 0.03
Nodule size score (n)	0.37* ± 0.10	0.55* ± 0.10	**0.36 ± 0.06**	0.14* ± 0.03
Hair follicle score (n)	0.06 ± 0.12	-0.17 ± 0.13	0.19 ± 0.13	**0.37 ± 0.06**

Genetic correlations of reproduction traits with slaughter weight and skin size were consistently negative in sign and mostly not significant ($P > 0.05$; Table 4). The exception was for the correlation of egg production with skin size which was more than double the corresponding standard error. If it is considered that growth is a key trait to ensure early slaughter and a reduced feed cost (Cloete *et al.* 2008b), these correlations are potentially unfavourable. On the other hand, it could be argued that the unbridled improvement of size could result in an increased maintenance need, as well as heavy animals that are difficult to handle, thereby compromising animal welfare and human occupational health and safety. Further research into the management of breeding programs, in light of these relationships, is therefore needed. Reproduction traits were not genetically correlated to the subjectively assessed skin quality traits. Genetic correlations of adult live weight with slaughter weight and skin size were positive and high. Adult live weight was positively related to nodule size but was uncorrelated with hair follicle score.

It was notable that some of the genetic and animal permanent environmental variation in the reproduction traits partitioned towards service sire. This was evident in analyses involving two-trait combinations of slaughter traits with adult reproduction traits, but most pronounced for analyses involving chick production. Heritability estimates and animal permanent environmental effects amounted to 0.09 to 0.18 respectively in these analyses, while service sires effects amounted to approximately 0.11. These results support our previous contention that the joint analysis of egg and chick production has assisted with the partitioning of additive, animal permanent environmental and service sire variances for these traits (Cloete *et al.* 2008a). Analysing slaughter weight with adult weight also resulted in a slight repartitioning of variances in the latter trait and resulted in a heritability of 0.41 and an animal permanent environmental variance ratio of 0.27. Except for estimates for service sire variances for chick production, it was impossible to demonstrate significance ($P < 0.05$) for these repartitioned variances. Further research on this phenomenon as more data accrue is therefore also required.

Table 4. Genetic correlations of the traits in the slaughter animal complex with those in the adult animal complex. Significant ($P < 0.05$) correlations are denoted by an asterisk

Adult animal traits	Slaughter traits			
	Slaughter weight (kg)	Skin size (dm^2)	Nodule size score (n)	Hair follicle score (n)
Egg production (n)	-0.15 ± 0.15	-0.41 ± 0.20*	-0.10 ± 0.17	0.15 ± 0.17
Chick production (n)	-0.21 ± 0.18	-0.43 ± 0.24	0.05 ± 0.20	0.28 ± 0.18
Adult weight (kg)	0.81 ± 0.05*	0.65 ± 0.12*	0.31 ± 0.11*	0.11 ± 0.11

CONCLUSIONS AND RECOMMENDATIONS

This study confirmed that all traits in the slaughter and adult animal complexes were variable and heritable, as was also reported in previous studies. Genetic correlations within trait complexes were also consistent with previous results. Genetic correlations across trait complexes suggested that correlations of reproduction traits with quantitative slaughter traits were possibly unfavourable, although only significant for the estimate involving egg production and skin size. Further research on these correlations is needed. Live weight expressed in adult animals was genetically highly correlated with slaughter weight, as could be expected for the same trait recorded at different life stages. In line with this, previously determined size-dependent slaughter traits, namely skin size and nodule size score, were also genetically related to adult size. Given the profound effect of animal size on maintenance requirements in other species, it is important to update these results as data accrue to optimise the economic efficiency of selection strategies in ostriches. More accurate genetic correlations between trait complexes are needed for the design of breeding programmes to ensure a balanced breeding strategy for ostriches.

REFERENCES

Brand T.S. and Jordaan J.W. (2011) *Appl. Anim. Husb. Rural Develop.* **4**: 1.

Brand Z. and Cloete S.W.P. (2015) *S. Afr. J. Anim. Sci.* **45**: 20.

Cloete S.W.P., Brand Z., Bunter K.L. and Malecki I.A. (2008a) *Austr. J. Exp. Agric.* **48**: 1314.

Cloete S.W.P., Engelbrecht A., Olivier J.J. and Bunter K.L. (2008b) *Austr. J. Exp. Agric.* **48**: 1247.

Engelbrecht A. (2013) PhD dissertation, Stellenbosch University, South Africa.

Engelbrecht A., Cloete S.W.P., Bunter K.L. & van Wyk J.B. (2009) *Proc. Assoc. Advmt. Anim. Breed. Genet.* **18**: 548.

Engelbrecht A., Cloete S.W.P., Bunter K.L. and van Wyk J.B. (2011) *Proc. Assoc. Advmt. Anim. Breed. Genet.* **19**: 183.

Engelbrecht A., Cloete S.W.P., Bunter K.L., van Wyk J.B. and Hoffman L.C. (2013) *Proc. Assoc. Advmt. Anim. Breed. Genet.* **20**: 381.

Engelbrecht A., Hoffman L.C., Cloete S.W.P. and Van Schalkwyk, S.J. (2009) *Anim. Prod. Sci.* **49**: 549.

Gilmour A.R., Gogel B.J., Cullis B.R., Welham S.J. and Thompson R. (2016) ASREML - User Guide Release 4.1. VSN International Ltd, Hemel Hempstead, HP11ES, UK.

Wang M.D., Cloete S.W.P. and Dzama, K. (2011) *Proc. Assoc. Advmt. Anim. Breed. Genet.* **19**: 486.

TRAIT DEVELOPMENT FOR *APIS MELLIFERA* IN COMMERCIAL BEEKEEPING IN NEW ZEALAND

G.E.L. Petersen[1], P.F. Fennessy[1] and P.K. Dearden[2,3]

[1]AbacusBio Ltd, Dunedin, Otago, New Zealand
[2]Department of Biochemistry, University of Otago, Dunedin, Otago, New Zealand
[3]Genomics Aotearoa, Dunedin, Otago, New Zealand

SUMMARY

Honeybee populations have been modified for centuries by selection and culling, but traditional selection criteria are no longer sufficient to address the needs of modern beekeeping and to counter threats such as spread of disease. While evaluating selected honeybee traits for their relevance and measurability in commercial beekeeping and their presumed heritability and their scale of variation, we encountered a dichotomy in the requirements of small-scale hobbyist beekeepers and large-scale commercial beekeeping operations. A number of traditional traits feasible for selection under commercial conditions could be identified, eight out of which can be considered high-priority traits in the design of an industry-wide honeybee breeding objective: honey production, gentleness, colony strength, brood viability, wintering ability and disease resistance. However, the costs of hive evaluations are often prohibitive to implementation of breeding and selection schemes in commercial operations. This can be overcome with the deployment of remote beehive monitoring equipment that provides continuous observations on colony status in conjunction with Machine Learning tools to evaluate change in trait expression in different environmental conditions. Simultaneously, image analysis and hive telemetry provide opportunities for the definition of novel traits such as nectar reactivity or the pattern of honey deposition. Using these recent technological advances, bee breeding can be made accessible to large-scale commercial beekeepers as well as dedicated small-scale queen breeders.

INTRODUCTION

Since domestication, century-long breeding programmes have made dramatic changes to most livestock species, creating fit-for-purpose breeds adapted to their respective management systems which perform well across a range of environments (van der Werf *et al.* 2009). The development of specialised breeds has resulted in high within-species genetic and phenotypic diversity, making it a natural practice for farmers to choose a breed that suits the particular production systems.

These long-term developments are largely absent in beekeeping, the only exceptions being the establishment of the "Buckfast" bee, a hybrid of several honeybee sub-species (Brother Adam 1987) and the accompanying breed regulations (Gemeinschaft der Europäischen Buckfastimker e.V. 2016). Sustainable genetic improvement systems can only be established with strong support from commercial beekeepers, who tend to manage large parts of the national honeybee populations but are slow to adopt modern animal breeding methods. For the design of data-driven and economically focused genetic honeybee breeding schemes, traits need to be selected carefully to ensure that they are not only valuable parts of the breeding objective, but also feasible selection criteria in a commercial environment.

MATERIALS AND METHODS

A list of traditional honeybee traits was compiled and then grouped into areas that contribute to beekeeping profitability (see Table 1). While some of these traits had been advocated for up to a century (Armbruster 1919; Brother Adam 1987) and/or are currently being used by European breed associations, they had not been evaluated for their suitability within modern commercial honeybee populations.

Table 1. Traditional Honeybee traits with potential for incorporation into bee breeding schemes, with relevance and measurability

Area	Trait	Unit of measurement	Relevance	Measurability*
production	Honey production	kg / hive / season	✓	✓
	Wax production	kg / hive / season	×	×
workability	Gentleness	subjective score (1-5)	✓	✓
	Docility / Calmness	subjective score (1-5)	✓	✓
	Swarming urge	attempts / season	✓	✓
strength	Brood strength	No of full frames of brood	✓	✓
	Colony strength	No of full bee spaces	✓	✓
	Spring growth	rate of growth in spring	✓	✓
health	Brood viability	percentage	✓	✓
	Disease resistance	*variable*		
queen	Q: laying pattern	pass / fail	✓	✓
health	Q: laying capability	laying rate in eggs / day	✓	×
	Q: longevity	weeks grafting to failure	✓	✓
robustness	Wintering index	% of surviving bees	✓	✓

*Grey tick marks indicate traits that can be measured in a queen breeding operation but cannot be readily measured in most commercial operations due to management strategies.

Traits were evaluated with regards to their relevance to commercial beekeeping and their practical measurability in the field, based on the published literature as well as discussions with commercial beekeepers. Trait heritabilities were compiled from the scientific literature.

RESULTS AND DISCUSSION

Only a small number of traits had been previously been investigated for their genetic parameters and were generally found to be of medium to high heritability (see Table 2), with the exception of swarming urge, which was found to have low heritability.

Table 2. Heritability estimates for selected honeybee traits. Estimates marked with ‡ are for Africanised honeybees (hybrids between African and European subspecies of *A. mellifera*, common beekeeping in South America)

Area	Trait	Heritability (Standard Error)
production	Honey production	0.27 (0.06) (Brascamp *et al.* 2016); 0.54 (0.18) (Bar-Cohen *et al.* 1978)
workability	Gentleness	0.37 (0.06) (Brascamp *et al.* 2016)
	Docility / Calmness	0.38 (0.05) (Brascamp *et al.* 2016)
	Swarming urge	0.06 (0.04) (Brascamp *et al.* 2016)
strength	Brood strength	0.10 (0.10) (Bar-Cohen *et al.* 1978)
	Colony strength	0.49 (0.44) (Koffler *et al.* 2017)

Most traits in Table 1 were found to be relevant to commercial beekeeping operations. However, evaluations under commercial conditions and with non-destructive methods were found to be too expensive to be feasible for honey production or pollination companies and require specialised queen breeders to evaluate their stock (see grey ticks under "Measurability" in Table 1). While evaluation costs can be prohibitive for commercial operators, the same is not true for dedicated queen breeders, who can expect to recover the costs of queen evaluation and selection in returns from the sale of elite breeding stock. When establishing an industry-wide honeybee breeding programme, both of these levels need to be taken into account, since the success of elite queen breeders (or academically-driven breeding programmes established by universities) hinges on the continuous adoption of their improved stock by commercial operators (Ibrahim *et al.* 2007).

These findings suggest that while there are a number of feasible traits for the development of economically sustainable honeybee breeding schemes, there is a need in the beekeeping industry for the development and deployment of low-cost alternatives to the hands-on and visual inspection / evaluation of honeybee colonies. Machine vision tools can be used to rapidly evaluate brood-related traits such as worker brood viability and brood pattern / queen laying pattern. Novel phenotyping technologies exist (although the hardware is often still in development) and could facilitate the establishment of industry-wide genetic improvement schemes by bridging the gap between elite queen breeders and the commercial beekeeping operators that are essentially their clients.

Table 3. Honeybee traits for commercial honeybee breeding that could benefit from novel phenotyping strategies

Area	Trait	would benefit from novel phenotyping technology
production	Pattern of honey deposition	✓
workability	Swarming attempts	✓
strength	Colony strength	✓
health	Worker brood viability	✓
	Wintering Index	✓
queen health	Queen: laying pattern	✓
pollination ability	Spring population growth	✓
	Flight temperature	✓

There are a number of feasible honeybee characteristics that could form the basis of a bee breeding scheme for commercial beekeeping. However, some of the relevant traits are currently not feasible in commercial operations because the costs associated with recording are too high.

Nevertheless, breeding of improved stock would be an efficient and permanent way to address the challenges that honeybee breeders and commercial beekeepers are facing today, and recent advances in science and technology allow for innovative solutions to be developed. Technology surrounding data collection and analysis both at the hive level and at honey extraction is leaping forward, with more and more automated systems breaking into the market (e.g. www.arnia.com, www.hivemind.co.nz). This creates an opportunity for the development of a cutting-edge honeybee genetic improvement programme in collaboration with commercial beekeepers.

Honey yield and temperament are the traits currently most modified by queen breeders and are also the ones that should be treated as paramount in the definition of a New Zealand honeybee breeding objective. They are highly relevant to the beekeeper, relatively easy to record and have been shown to be heritable. A third highly relevant trait, winter survival, needs to be investigated further, as there are currently no estimates of genetic parameters for this trait available. However, since low winter survival presents a crucial issue to beekeepers worldwide, it should be included in breeding programs from the beginning. New Zealand beekeepers currently export upwards of 25,000 packages of live bees containing queens every year (New Zealand Ministry for Primary Industries 2018), mainly to

Canada, which continues to experience annual winter losses of ~30%. Doubts have been voiced on the suitability of New Zealand queen genetics for the harsher Canadian winters (Harpur *et al.* 2015) and although New Zealand is currently not experiencing high colony mortality over winter, inclusion of winter survival as a key trait would be a valuable step towards future-proofing the beekeeping industry and making New Zealand genetics desirable overseas.

Additional traits that should be prioritised are colony strength, brood viability and disease resistance. However, some of these high-priority traits can be expected to be costly to measure and novel phenotyping methodologies such as machine vision (image analysis supported by artificial intelligence) or remote hive monitoring / hive telemetry systems will need to be developed allow measurement to be practiced.

CONCLUSION

Novel automated phenotyping systems can support modern honeybee breeding programmes to support large-scale commercial beekeeping industries such as in New Zealand. These programmes can be expected to have a positive long-term impact on both domestic bee productivity and health as well as the survival and overall quality of bees exported to e.g. North America by being able to incorporate standardised data from all around the globe.

REFERENCES

Bar-Cohen R., Alpern G. and Bar-Anan, R. (1978). *Apidologie* **9**: 95.

Brascamp E.W., Willam A., Boigenzahn C., Bijma P. and Veerkamp R.F. (2016) *Apidologie* **47**: 739.

Brother Adam. (1987). *Breeding the Honeybee*. Northern Bee Books.

Gemeinschaft der Europäischen Buckfastimker e.V. (2016).

Harpur B.A., Chapman N.C., Krimus L., Maciukiewicz P., Sandhu V., Sood K., Lim J., Rinderer T.E., Allsopp M.H., Oldroyd B.P. and Zayed A. (2015) *Insectes Sociaux* **62**: 479.

Ibrahim A., Reuter G.S. and Spivak M. (2007) *Apidologie* **38**: 67.

Koffler S., de Matos Peixoto Kleinert A. and Jaffé R. (2017) *Conservation Genetics* **18**: 689.

New Zealand Ministry for Primary Industries. (2018) https://mpi.govt.nz/document-vault/5302.

van der Werf J.H.J., Graser H.-U., Frankham R. and Gondro C. (2009). *Adaptation and Fitness in Animal Populations*. Dordrecht: Springer Netherlands.

A METHOD FOR DEVELOPING A BREEDING OBJECTIVE TRAIT FROM MULTIPLE COMPONENTS USING THE EXAMPLE OF IMMUNE COMPETENCE IN AUSTRALIAN ANGUS CATTLE

S. Dominik[1], L. Porto-Neto[2], C.J. Duff[3], A.I. Byrne[3], B. Hine[1], A. Ingham, I.G. Colditz[1] and A. Reverter[2]

[1]CSIRO Agriculture and Food, Armidale, NSW, 2350 Australia
[2]CSIRO Agriculture and Food, St Lucia, QLD, 4067 Australia
[3]Angus Australia, Armidale, NSW, 2350 Australia

SUMMARY

Traits that are being recorded in livestock improvement programs might not be suitable breeding objective traits themselves, which is an important aspect for the consideration of novel traits in breeding programs. Here we demonstrate, using the example of immune competence in cattle, how multiple novel traits can be reduced to a single breeding objective trait. It was demonstrated that it is possible to achieve a high heritability for the novel single breeding objective trait and maximise the genetic correlation with one of the major production traits, here final weight. An approach as described here would maximise the genetic gain in the novel trait.

INTRODUCTION

In order to respond to future livestock industry needs, novel traits are being developed to promote sustainable livestock production. One such desirable attribute of the animal is immune competence (IC) (Wilkie *et al.* 1999), which has demonstrated health benefits in dairy cattle (Thompson-Crispi *et al.* 2012; Aleri *et al.* 2019). A protocol for measuring IC has been developed in Australian Angus Cattle (Hine *et al.* 2019). Immune competence has two components: cell-mediated (Cell-IR) and antibody-mediated immune responses (Ab-IR). These represent two aspects of adaptive immune responses that help control infectious disease. However, in a breeding objective context, it would be easier to use immune competence, as a combination of Ab-IR and Cell-IR, as a single breeding objective trait. The aim of this study was to combine the two immune response traits into a single breeding objective trait, here IC, so that the heritability for IC is high and the correlation of IC with final weight (FW), one of the key profit drivers, is maximised to allow for the highest possible genetic gain in the novel trait through direct and correlated response.

MATERIALS AND METHODS

Data. A protocol has been developed to measure Cell-IR and Ab-IR in commercial beef herds (Hine *et al.* 2019). Immune response phenotypes were recorded on 1,149 Angus cattle from the Angus Sire Benchmarking Program. Animals originated from five different herds and were born across three years. Link sires were used to provide connections between herds and birth cohorts. Not all animals within a herd could be tested in one day, and up to 7 test cohorts exist within a herd. Immune response phenotypes were assessed during the yard weaning period. On the day of weaning (Day 0) cattle were vaccinated with a multi-valent clostridial vaccine containing tetanus toxoid antigen (Zoetis, Australia). The ability to mount an Ab-IR was measured as production of tetanus toxoid specific IgG1 antibody in blood between day 8 and day 21 post-vaccination. The actual sampling day was dependent on the specific herd and their prior clostridial vaccination history. Vaccination history was identical for animals tested within herd and test cohort. The antibody concentration measured in blood represents a cumulative response to the vaccination given at day 0 and to any vaccinations administered previously. An in-house indirect ELISA method was used to measure antibody levels (Aleri *et al.* 2015). The Ab-IR was recorded as optical density values (OD) and for analysis the OD values were square root transformed.

Cell-mediated immune response (Cell-IR) was assessed as delayed type hypersensitivity (DTH) by measuring changes in skin fold thickness in response to intradermal injection of the clostridial vaccine (Ultravac 7 in 1 clostridial and leptospira vaccine (Zoetis)) in the caudal fold of the tail. Testing day was consistent within herd and test cohort and was conducted around day 14 post vaccination aligning with blood collection for antibody testing. One side of the tail was injected with 100 µL of Ultravac 7 in 1 (test) and the other with 100 µL of saline (control). Skin thickness was measured in millimetres using callipers prior to injection (T0) and after 48 hours (T48). The magnitude of DTH responses was determined as the T48 test response in relation to the T48 control response (DTH T48 test/DTH T48 control). The DTH response at T0 (DTH T0 test/DTH T0 control) was fitted as a covariate in the linear model. The Cell-IR variable and covariate were log transformed prior to analysis to ensure normality.

The two immune response traits, Cell-IR and Ab-IR, were both multiplied by 100 for analysis. Cattle were finished through a feedlot after backgrounding at pasture for approximately nine months and final weight, the weight when animals were sent to the feedlot at approximately 600 days (FW), was also used for analysis. Fixed effects included contemporary group (herd, birth year, test cohort) for all traits. For Cell-IR and Ab-IR, age at testing was fitted as a covariate. Age at the measurement of FW was fitted as covariate for FW. To prove the hypothesis, this study only animals with all data for phenotypes, fixed effects and covariates were included in the study, which resulted in a data set with 851 animals (all male) and 2,128 animals in the pedigree.

Analysis. Variance components and heritabilities were estimated using VCE 6.0.2 (Kovac *et al.* 2010) and genetic and phenotypic correlations were estimated for FW, Cell-IR and Ab-IR. Here we explored whether the two immune response traits could be combined into a single IC trait as the relevant single breeding objective trait in a breeding program. The hypothesis was that the two traits could be combined such that the heritability for IC is high and the correlation between IC with FW is most strongly negative. This is not a realistic example as we would not attempt to maximise an unfavourable correlation, but the data set offered the highest number of records for FW and a strong correlation with IC, which helps to prove the hypothesis. Immune competence was calculated using the following function: $IC = \alpha \times$ Cell-IR $+ (1- \alpha) \times$ Ab-IR, with $\alpha = 0$ to 1. For each α, ranging from 0 to 1, a bivariate analysis was run for IC and FW. Breeding values (EBV) and heritabilities were estimated for IC along with genetic correlations between IC and FW.

RESULTS AND DISCUSSION

The summary statistics for immune response and FW traits are shown in Table 1. The amount of phenotypic variation in the immune response traits was expected since they have not been traits of direct selection. However, at this point there are no results how that variation relates to variation in disease protection. The negative minimum of Cell-IR indicates that in some animals the skin thickness after challenge reduced compared to Day 0 as is seen in other studies. Final weights ranged from 476kg to 880kg across contemporary groups/properties. Within properties there is much less variation, highlighting the need to fit contemporary group.

Table 1. Descriptive statistics of cell-mediated and antibody-mediated immune response (Cell-IR and Ab-IR, multiplied by 100) and final weight (kg)

	Minimum	Maximum	Mean ± Standard deviation
Cell-IR*	-2.30	56.08	24.22 ± 9.01
Ab-IR*	14.90	143.26	78.36 ± 25.08
FW	476.00	880.00	559.87 ± 88.37

*Cell-IR was log transformed and Ab-IR square root transformed

Heritabilities for Cell-IR and Ab-IR were moderate (Table 2) and are in line with previous estimates from all 1,149 animals (Hine *et al.* 2019). The genetic correlation between the immune response traits was moderately positive, which confirmed previous estimates from the full data set (Hine *et al.* 2019). Genetic correlations of Cell-IR and Ab-IR with FW are negative, possibly indicating that high immune response diverts energy resources from growth.

Table 2. Heritabilities (diagonal, bold) and genetic correlations (below diagonal)

	Cell-IR	Ab-IR	FW
Cell-IR	**0.33 ± 0.11**		
Ab-IR	0.40 ± 0.22	**0.30 ± 0.10**	
FW	-0.27 ± 0.21	-0.50 ± 0.19	0.48 ± 0.12

Table 3 outlines the results from the bivariate analyses of IC and FW, where IC is a function of the weighted component traits Cell-IR and Ab-IR. At $\alpha = 0.0$ IC is the same as Ab-IR, at $\alpha = 1.0$ IC is the same as Cell-IR. The heritability of IC is moderate and the genetic correlation with FW is most strongly negative at $\alpha = 0.3$. Genetic gains for IC could be maximised through selection on IC and the highly correlated trait FW, however, because the correlation is negative, FW would be reduced. Alternatively, at $\alpha = 1.0$, the negative correlation with FW would be minimised, but would result in IC being only a representation of Cell-IR.

Table 3. Heritability of immune competence (IC) and genetic correlation with final weight (FW) at different weightings (α) to combine cell-mediated and antibody-mediated immune response traits; standard errors (se) in brackets

α	h^2IC (se)	r_g (se)
0.0	0.299 (0.107)	-0.500 (0.223)
0.1	0.306 (0.106)	-0.503 (0.210)
0.2	0.315 (0.103)	-0.505 (0.214)
0.3	0.325 (0.111)	-0.506 (0.198)
0.4	0.338 (0.111)	-0.505 (0.206)
0.5	0.354 (0.114)	-0.500 (0.195)
0.6	0.371 (0.112)	-0.488 (0.206)
0.7	0.384 (0.114)	-0.465 (0.193)
0.8	0.384 (0.116)	-0.425 (0.207)
0.9	0.362 (0.119)	-0.362 (0.211)
1.0	0.322 (0.109)	-0.274 (0.210)

Figure 1 shows, animals can be differentiated based on the EBV for IC. As can be expected at α=0.3, EBV for IC are a closer reflection of the EBV for Ab-IR than for Cell-IR. Ideally, animals with high EBV for IC would reflect high EBV for both Ab-IR and Cell-IR, because both types of responses are required to effectively control environmental pathogens. In addition to the approach presented here, other ways to define IC as breeding objective trait to allow equal emphasis on both component traits of IC need to be explored.

Results using the example of IC are instructive. The small amount of variation observed in r_g for values of α between 0.1 and 0.6 suggest that for this trait there is considerable latitude to vary weighting on Ab-IR and Cell-IR with little variation in the penalty to FW. Ab-IR and Cell-IR also influence other drivers of profitability (Hine *et al.* 2016), highlighting the need for a more comprehensive method for incorporating Ab-IR and Cell-IR as the novel multi-component trait IC.

Figure 1. Estimated breeding values for Cell-IR, Ab-IR and IC at α=0.3, sorted from highest to lowest IC EBV

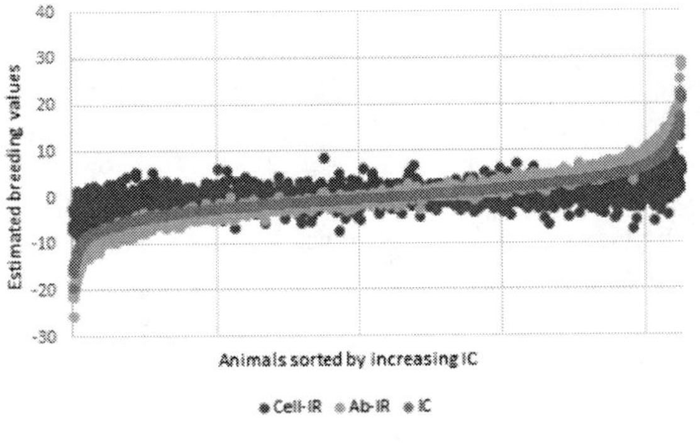

CONCLUSION

Novel traits provide an opportunity to extend traditional livestock breeding objectives to ensure the industry's future sustainability. Strategically defining the breeding objective trait can assist in incorporating novel trait in breeding programs.

ACKNOWLEDGEMENT

The authors would like to gratefully acknowledge the funding for this study by MLA. We would also like to thank the staff of CSIRO Agriculture and Food and Angus Australia who ensured an accurate and seamless data collection.

REFERENCES

Aleri J.W., Hine B.C., Pyman M.F., Mansell P.D., Wales W.J., Mallard B. and Fisher A.D. (2015) *J. Anim. Sci.* **93**: 3713.

Aleri J.W., Hine B.C., Pyman M.F., Mansell P.D., Wales W.J., Mallard B., Stevenson M.A. and Fisher A.D. (2019) *J. Dairy Sci.* **102**: 3282.

Hine B.C., Ingham A.B., Dominik S. and Colditz I.G. (2016) MLA Final report B.STU.0244.

Hine B.C., Bell A.M., Niemeyer D.D., Duff C.J., Dominik S., Ingham A.B. and Colditz I.G. (2019) *Anim. Sci.* Accepted.

Kovac M., Groeneveld E. and Garcia-Cortez A. (2010). VCE 6.0.2.

Mallard B.A., Wagter L.C., Ireland M.J. and Dekkers J.C. (1997) *Vet. Immunol. Immunopathol.* **60**: 61.

Rauw W.M., Kanis E., Noordhuizen-Stassen E.N. and Grommers F.J. (1998) *Livest. Prod. Sci.* **56**: 15.

Thompson-Crispi K.A., Hine B., Quinton M., Miglior F. and Mallard B.A. (2012) *J. Dairy Sci.* **95**: 3888.

SELECTIVE BREEDING FOR IMPROVED SURVIVAL TO JUVENILE PEARL OYSTER MORTALITY SYNDROME IN SILVER LIPPED PEARL OYSTER, *PINCTADA MAXIMA*

C. Massault[1], K. Zenger[1], J.M. Strugnell[1], J. Knauer[2], G. Firman[3], R. Barnard[4] and D.R. Jerry[1]

[1]Centre for Sustainable Tropical Fisheries and Aquaculture, College of Science and Engineering, James Cook University, Townsville, Australia
[2]Autore Pearls, Sydney, Australia
[3]Cygnet Bay Pearls, Broome, Australia
[4]RMB Aqua Pty Ltd, Perth, Australia

SUMMARY

Juvenile pearl oyster mortality syndrome (JPOMS) causes mass-mortality of young pearl oysters impacting the production and financial revenue of farms. The use of selective breeding to improve survival is an effective solution to reduce impact, especially in the case of a poorly understood disease such as JPOMS. Here, we investigate the potential of implementing a selective breeding program for increased survival to JPOMS in the silver lipped pearl oyster, *Pinctada maxima*. Simulation results show that significant increases in survival could be achieved if selective breeding was applied (8% increase in survival per generation). Attention should be paid to balance genetic gain (increase of survival) and diversity to limit inbreeding where individuals are selected on their family mean.

INTRODUCTION

The revenue from Australian pearl production has dramatically reduced from AU$ 144 million in 2010, to AU$ 70 million in 2017 (ABARES, 2018). Part of the decline can be explained by the recurrence of mass-mortality of juvenile stocks, which severely reduces the number of oysters that are cultured to pearl seeding sizes. Repeated observations of mass-mortality have been reported over the last decade, but the actual pathogenic agent/s causing this juvenile pearl oyster mortality syndrome (JPOMS) is yet to be identified. Due to the limited knowledge on the actual causative agent, no treatment exists to this day. While the in-depth knowledge of the disease is extremely important for the Australian pearl industry, the process of unravelling the causative agent of JPOMS is non-trivial and time consuming.

Selective breeding for disease resistance (or survival) is increasingly becoming a primary selection trait in aquaculture (Gjedrem 2012; Houston 2017) and offers a practical alternative when little information about the causative agent of the disease is known. In the last few years, hatchery technology has been refined in pearl farms which permits broodstock selection and, to some degree, the control of family contribution to progeny cohorts. The advantage of the use of selective breeding to increase survival is that improvement can still be achieved even though understanding of the disease is limited.

To explore the potential of selective breeding focused on JPOMS survival, we simulated a breeding program under various parameters (family size, family number, selection methods) for 20 years. The mode of selection used for JPOMS survival is based on breeding value for survival per half-sib family and can potentially lead to high level of inbreeding. This study focus on the impact of restriction of selected individuals per half-sib family on the genetic merit and inbreeding generated by such breeding program.

MATERIALS AND METHODS

The simulated breeding program was run for 20 years which corresponded to 10 generations of the target species, the silver lip pearl oyster (*Pinctada maxima*) which becomes sexually mature around 2 years of age. The base scenario uses 100 males mated to 200 females. All individuals in the base generation are unrelated. Each female produced 100 offspring (which corresponds to 100 offspring per full-sib family and 200 offspring per half-sib family). Survival is recorded as an average per family. Under the current farm setting, it is not possible to obtain survival data per individual. The oysters, at this stage, are usually attached in panel nets on long-lines suspended in the ocean for the grow-out period and it is difficult to keep track of individual performance throughout the entire grow out phase. However, average of survival per half-sib family, after genotyping individuals before and after a JPOMS event, can be determined. Half-sib families were ranked according to their family mean survival and a maximum of 10 males and 20 females per half-sib family were taken as selection candidates from the best families until a total of 250 males and 500 females were selected. Families had unequal sizes due to mass mortality, in particular at the beginning of the breeding program. The best ranking families were also the largest families. Therefore, the number of families represented among the selection candidates was variable and decreased as the breeding program continued and the mortality decreased. From these selection candidates, 100 males and 200 females were selected and mated, half-sib and full-sib mating were not allowed.

Unfortunately, at this stage, actual estimates for heritability of PJOMS are currently in the process of being calculated and are not yet known. Heritability of survival is usually low and varies between 0.06 and 0.16 in aquaculture (Gjedrem and Olesen, 2005) and a review of selective breeding for disease resistance in oyster by Degremont *et al.* (2014) shows that heritability is scarcely reported (expect for summer mortality in Pacific oysters), therefore, we chose a conservative heritability of 0.1 at the start of the breeding program for the subsequent simulations. Breeding program simulations were run to explore the impact of the number and size of families and the maximum number of males and females selected per family (see Table 1). For all scenarios, the disease killed 90% of the reared progeny in the first generation and then occurred every subsequent generation. All scenarios were replicated 100 times.

Selective breeding scenarios were assessed using the cumulative genetic gain of survival of PJOMS (designated as the proportion of survival), the accuracy of breeding values (calculated from survival family mean) and rate of inbreeding per generation. The genetic gain of generation t was the difference of the average survival family mean between generation t and generation t-1. The rate of inbreeding was calculated as follow:

$$\Delta F = \frac{F_t - F_{t-1}}{1 - F_{t-1}}$$

with F_t the average of inbreeding coefficient for a generation, t the current generation.

Table 1. Description of selection parameters used for the various simulated scenarios

Simulated scenario	# families	Half-sib families size	# sires / # dams selected / family
1	50	200	10 /20
2	150	200	10 /20
3	100	100	10 /20
4	100	300	10 /20
5	100	200	5/10
6	100	200	No restriction

RESULTS AND DISCUSSION

The cumulative genetic gain, accuray and rate of inbreeding in the base scenarios are shown in Figure 1 (black lines). After 10 generations of selection, survival had reached 75% and rate of inbreeding stayed under the acceptable limit of 1% per generation. This simulation showed that the South Sea pearl industry could benefit from implementing selective breeding for survival to JPOMS..

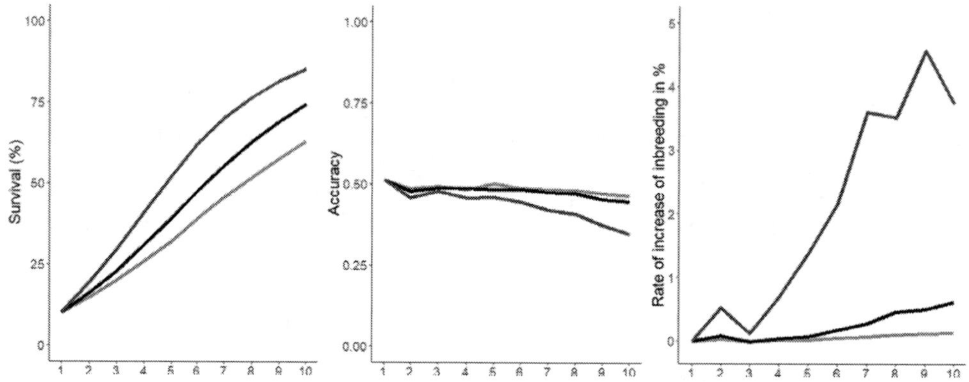

Figure 1. Proportion of survival (left), accuracy of breeding values (middle) and rate of increase of inbreeding (right) as a function of the number of sires and dams selected per families

There were no major differences in survival, accuracy of breeding program and rate of inbreeding when varying the number of families used or the family size. However, large impacts on genetic parameters were observed when varying the number of sires and dams selected per family (Figure 1). We observed an increase in survival as well as an increase in rate of inbreeding associated with an increased number of males and females selected per family. Both scenarios that included a restricted number of males and females per family as selection candidates had an acceptable rate of inbreeding (rate of inbreeding increase < 1%). However, decreasing the number of males and females selected per family as selection candidates also reduced the genetic gain with regard to survival (60% survival with 5 males and 10 females selected per family compared to 75% survival with 10 males and 20 females selected per family at generation 10). In the instance where there was no restriction on the maximum number of males and females selected per family (i.e. selecting whole families, red line in Figure 1), the survival rate reached 85% after 10 generations of selection, which is substantially higher than for the basic scenario (75%). However, this scenario also led to the highest rate of inbreeding (>4%), which will result in a higher number of matings between relatives at each generation and a likely increase in inbreeding depression in the long term.

The number of males and females per family that were used as selection candidates played a key role in the success of the simulated breeding program for pearl oysters. At the beginning of the breeding program, when survival was at 10%, families had few offspring survive. Therefore, to reach the goal of 250 male and 500 female selection candidates, a large number of families was required (around 20% of families under the base scenario with 5 males and 10 females maximum selected per family). With a maximum of 10 males and 20 females used per family, the average proportion of family selected was 10% (scenario 5), while with no restriction on the number of males and females used per family, selection candidates were coming from only 3% of the families (scenario 6). In the latest scenario, selection candidates were generated from only 3 families and therefore the number of related of individuals was very high and increase of inbreeding was therefore inevitable. Additionally, as selection on survival was performed, the proportion of survival increased and a larger number of offspring survived per family, which decreased the number of families selected per generation and increased the chances of mating among related individuals. This highlights the importance of balancing genetic gain and the maintenance of family-specific genetic diversity.

Survival to JPOMS under the base scenario would result in an increase of genetic gains of 7% per generation over 10 generations and by 6% at the first generation (from 10% at generation 1 to 16% at generation 2). These results are in line with those reviewed by Gjedrem (2012) for various species and diseases. Elston *et al.* (1987) reported a 73% survival increase between non-selected and selected individuals over a period of 4 to 10 generations of selection for resistance of *Bonamia ostrea* in European flat oyster (*Ostrea edulis)*, similar to the 64% increase in our simulation, over 10 generations.

The case of disease resistance to JOD in the eastern oyster *Crassostrea virginica* is of particular interest for this study as it demonstrated the benefit of selective breeding (increase of 85% survival in 2 generations) for a disease that primarily affects juveniles, like JPOMS, and that exhibits similar characteristics to it. Therefore, we can expect that the implementation selective breeding for survival to JPOMS will result in significant improvement of survival.

CONCLUSION

This study shows that implementing selective breeding for JPOMS in *P. maxima* could theoretically be very beneficial for the South Sea pearl industry and the predictions of genetic gain and rate of inbreeding for the basic scenario are in accordance to reported gains of selective breeding for disease resistance in other oyster species. However, attention should be paid to maintaining genetic diversity as well as improving survival, as the rate of genetic gain was also linked to higher inbreeding.

ACKNOWLEDGMENT

This research is founded by CRC-P58609 "Breeding for resilience of JPOMS" and in collaboration with Ellies Pearls, Clipper Pearls and Cygnet Bay Pearls.

REFERENCES

ABARES (2017) Australian Fisheries and Aquaculture Statistics 2017 report

Barber B., Davis C. and Crosb, M. (1998) *J. Shellfish Res.* **17**: 1171.

Degremont L., Garcia C. and Standish, K. A. (2015) *J. Invertebr. Pathol.* **131:** 226.

Gjedrem, T. and Olesen, I. (2005) Chapter 5. Basic statistical parameters in Selection and Breeding programs in Aquaculture, Berlin, Germany, Springer – p66-70.

Gjedrem T. (2015) *J. Mar. Sci. Eng.* **3**: 146.

Houston R. (2017) *R. Bras. Zootec.* **46**: 545.

SEQUENCING STRATEGY, IMPUTATION AND GENOMIC PREDICTION IN A LARGE PIG SEQUENCING STUDY

R. Ros-Freixedes[1,2], A. Whalen[1], A. Somavilla[1], S. Gonen[1], M. Battagin[1], M. Johnsson[1,3], G. Gorjanc[1], C.Y. Chen[4], W.O. Herring[4], A.J. Mileham[5] and J.M. Hickey[1]

[1]The Roslin Institute and Royal (Dick) School of Veterinary Studies, The University of Edinburgh, Easter Bush Campus, EH25 9RG Midlothian, Scotland, UK
[2]Departament de Ciència Animal, Universitat de Lleida-Agrotecnio Center, Lleida, Spain.
[3]Department of Animal Breeding and Genetics, Swedish University of Agricultural Sciences, SE-750 07, Uppsala, Sweden
[4]Genus plc, 100 Bluegrass Commons Blvd., Suite 2200, Hendersonville, TN, 37075 USA
[5]Genus plc, 1525 River Road, DeForest, WI, 53532 USA

SUMMARY

The use of whole-genome sequence data has great potential in livestock breeding programs but suitable sequencing strategies and imputation methods need to be developed to generate sequence information for a large number of individuals at an affordable cost. We describe the sequencing strategy that we followed in a study that sequenced more than 7,848 pigs from nine commercial lines, mostly at low coverage. Results demonstrate that the coupling of appropriate sequencing strategies and imputation methods such as hybrid peeling is a viable strategy for producing whole-genome sequence data for large livestock pedigreed populations, but it remains to be determined whether these large datasets can provide an increased accuracy of genomic predictions.

INTRODUCTION

The use of whole-genome sequence data has great potential in livestock breeding programs. It may increase the power of discovery of causative variants (Pasanuic *et al.* 2012; Daetwyler *et al.* 2014; Nicod *et al.* 2016) and may enable more accurate and persistent predictions of breeding values than marker arrays (Meuwissen and Goddard, 2010; Iheshiulor *et al.* 2016). To capture the full potential of sequence data in livestock, sequence and phenotype data are required on a large number of individuals, perhaps millions, to accurately estimate the effects of the large number of causative variants that underlie quantitative traits (Hickey *et al*, 2014).

Low-cost sequencing strategies combined with imputation can be utilised to generate the required amount of sequence information for a large number of individuals at an affordable cost (Brøndum *et al.* 2014; van Binsbergen *et al.* 2014; VanRaden *et al.* 2015; Pausch *et al.* 2017). Low-coverage sequencing (LCSeq) enables the sequencing of a larger number of animals, which provides four advantages: (1) higher variant discovery rates, particularly for low-frequency variants; (2) inclusion of rare haplotypes; (3) a more precise capture of the recombination events that have occurred in the population, which enables better definition of haplotypes and thus better imputation of these haplotypes into the individuals that carry them; and (4) more sequenced animals that are related, which improves the imputation of the sequence data to the whole population.

We first describe the sequencing strategy that we followed in a study that sequenced more than 7,848 pigs from nine commercial lines, mostly at low coverage (1x or 2x). Then, we demonstrate that the coupling of that sequencing strategies with the imputation method 'hybrid peeling' is a viable strategy for producing whole-genome sequence data for large livestock pedigreed populations. Finally, we test the benefit that these large datasets can provide an increased accuracy of genomic predictions.

MATERIALS AND METHODS

Sequencing strategy. We performed whole-genome sequencing of 7,848 individuals from nine commercial pig breeding lines (Genus PIC, Hendersonville, TN) with a total coverage of approximately 32,114x. Sequencing effort in each of the nine lines was proportional to population size. Approximately 2% (1.7-2.5%) of the pigs in each line were sequenced. Most pigs were sequenced at low coverage, with target coverage of 1 or 2x, but a subset of pigs were sequenced at higher coverage of 5x, 15x, or 30x. Thus, the average individual coverage was 4.1x, but the median coverage was 1.5x.

We selected the individuals and the coverage at which they were sequenced using a three-step strategy: (1) we first selected sires and dams that contributed most genotyped progeny in the pedigree (referred to as 'top sires and dams') to be respectively sequenced at 2x and 1x; (2) conditional on the first step, we used AlphaSeqOpt part 1 (Gonen *et al.* 2017) to identify the individuals whose haplotypes represented the greatest proportion of the population haplotypes (referred to as 'focal individuals') and to determine an optimal level of sequencing coverage between 0x and 30x for these individuals and their immediate ancestors (i.e., parents and grandparents) under a total cost constraint; and (3) conditional on the second step, we used the AlphaSeqOpt part 2 (Ros-Freixedes *et al.*, 2017) to identify individuals that carried haplotypes whose cumulative coverage was low (i.e., below 10x) and distributed 1x sequencing amongst those individuals so that the cumulative coverage on the haplotypes could be increased (i.e., at or above 10x). AlphaSeqOpt used haplotypes inferred from marker array genotypes (GGP-Porcine HD BeadChip; GeneSeek, Lincoln, NE), which were phased with AlphaPhase (Hickey *et al.* 2011) and imputed with AlphaImpute (Hickey *et al.*, 2012). The sequencing resources were split so that approximately 30% of the sequencing resources were used for sequencing the top sires at 2x, 15% for the top dams at 1x, 25% for the focal individuals and their immediate ancestors at variable coverage, and the remaining 30% for individuals that carried under-sequenced haplotypes at 1x.

Variant discovery. The reads were preprocessed using Trimmomatic (Bolger *et al.* 2014) to cut adapter sequences from the reads. Then the reads were aligned to the Sscrofa11.1 reference genome using the BWA-MEM algorithm (Li & Durbin 2009). Duplicates were marked with Picard (http://broadinstitute.github.io/picard). SNPs and short insertions and deletions (indels) were genotyped jointly for all samples using a pipeline based on the HaplotypeCaller tool from GATK 3.8 (DePristo *et al.* 2011). To avoid biases towards the reference allele introduced by GATK when applied on low-coverage sequence data we extracted the read counts supporting each allele directly from the aligned reads stored in the BAM files with a pile-up function using the pipeline described in (Ros-Freixedes *et al.* 2018). A total of 60 million SNPs were discovered across the nine lines.

Imputation of whole-genome sequence data. Most individuals in every population were genotyped using commercial marker arrays, with either 15,000 (LD) or 75,000 (HD) markers genome-wide. Imputation to whole-genome sequence was performed in each population separately using hybrid peeling, as implemented in AlphaPeel (Whalen *et al.* 2018) with the default settings. This method involves two stages: (1) multi-locus iterative peeling to estimate the segregation (the probability that each pair of grandparental gametes was co-inherited at a given locus) at the positions genotyped with the marker arrays; and (2) a modified single-locus iterative peeling step to impute the genotypes at each variant position discovered from the sequence data. This two-stage method reduces the computational cost of the imputation by estimating segregation of the markers from the array only and then approximating the segregation estimates at any other loci based on the estimates of the markers from the array that flank them. The accuracy loss of this approximation is negligible due to the limited number of recombinations in each chromosome and the high probability that nearby markers are inherited together. Multi-locus iterative peeling was performed on all available marker array data to estimate the segregation probabilities for each individual. The individuals genotyped with LD marker arrays were not imputed to HD prior to this step. The segregation probabilities were used for segregation-aware single-locus iterative peeling for the remaining segregating variants. The total number of pigs with imputed data across the nine lines ascends to around 350,000.

To assess imputation accuracy, we used 284 individuals from four of the nine populations who were sequenced at high coverage (15x or 30x). Of these, 37 belonged to a 20,000-individual population, 65 to a 35,000-individual population, 92 to a 70,000-individual population, and 90 to a 110,000-individual population. Many of these individuals sequenced at high coverage belonged to early generations of the pedigree of each population. Sequence data of the 284 individuals was completely masked, using a leave-one-out design. The imputed allele dosages were compared to those obtained with the complete data, considered as the 'true' values. For estimating the accuracies, we used 50,000 non-consecutive SNPs chosen randomly from chromosome 5.

Genomic prediction. Genomic prediction accuracy was tested in a single line with 30k pigs with imputed genotypes for 16 million of SNPs. Genomic predictions were performed using ridge regression as implemented in AlphaBayes software. The model was trained on 22,318 individuals and validated on 1,458 individuals. Genomic predictions were performed for nine synthetic traits with different heritability (0.1, 0.25, or 0.5) and with different number of QTN underlying their variation (100, 1,000, or 10,000 QTN), selected randomly from among all variants. The effect of the QTN was sampled from a normal distribution $N(0,1)$. Genomic predictions were performed using four sets of markers: the 57k markers from the array (HD), 248k variants preselected from the sequence data based on LD pruning (WGS_LD), around 183k variants preselected from the sequence data based on results of single-marker regression with a set of 13k individuals independent from the training and testing sets (WGS_SMR), or 67k variants preselected from the sequence data by keeping only every 200th variant (WGS_200th). Accuracy of the gEBV was estimated as the correlation between the gEBV and the synthetic phenotypes in the validation set.

RESULTS AND DISCUSSION

Imputation accuracy. The imputation accuracy in the real data was high for most of the tested individuals. The imputation accuracy achieved for each of the 284 tested individuals is shown in Figure 1. The average individual-wise dosage correlation was 0.94 but there was substantial variation with an asymmetrical distribution (median: 0.97; min: 0.11; max: 1; interquartile range: 0.94-0.98). Some of the oldest individuals that belonged to the earliest generations of the pedigree (some of the 106 individuals located in the first 20% of the pedigree) had lower imputation accuracy than individuals in the remainder of pedigree, who had consistently high imputation accuracy. This pattern was observed for all four populations. The imputation accuracy of the individuals in later generations (the 178 individuals after the first 20% of the pedigree) was higher, with an average dosage correlation of 0.97 and with much lower variability (median: 0.98; min: 0.69; max: 1; interquartile range: 0.96-0.99).

The marker array density of the individuals was confounded with the number of ancestors that were genotyped with marker arrays. The non-genotyped individuals (n=19) and approximately half of the individuals genotyped at HD (n=87 out of 157) belonged to early generations of the pedigree, which reduced the chances that they had ancestors with data and penalized the imputation accuracy for these two groups of individuals. On the contrary, most individuals genotyped at LD belonged to later generations (n=91 out of 108), ensuring that their ancestors had enough data to enable high imputation accuracies for the LD individuals. The average dosage correlation for the non-genotyped individuals was 0.81, for the HD individuals was 0.94, and for the LD individuals was 0.96. The average dosage correlation for the HD individuals in the earliest generations was lower (0.91) than for the HD individuals in later generations (0.97). For individuals in the later generations there were no significant differences between marker array densities and the average dosage correlation of both the HD and LD individuals was 0.97 and therefore no intermediate imputation steps were required for the LD individuals. There was no clear trend that population size affected imputation accuracy.

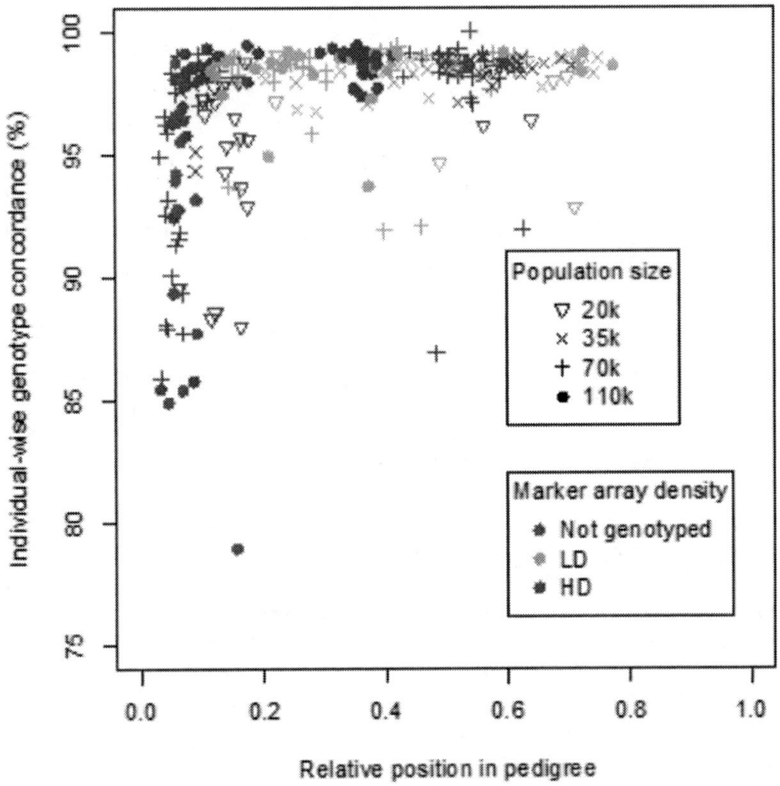

Figure 1. Imputation accuracy on relative position of the individual in the pedigree, marker array density, or population size

Genomic prediction. Sequence data can provide better prediction accuracy than marker arrays in some cases, but its advantage may depend on the genetic architecture of the trait. The genomic prediction accuracies for the nine synthetic traits are shown in Table 1. When a low number of QTN determine the phenotype, there may be sufficient statistical power to identify variants that underlie the genetic variation of the trait and prediction accuracy using those variants (WGS_SMR) is higher than with the markers from commercial marker arrays (HD). This is consistent with previous observations that adding one or a few markers with large effects as predictors can improve prediction accuracy of the marker arrays (Estany *et al.* 2017; Lopes *et al.* 2017; Nani *et al.* 2019; Al Kalaldeh *et al.* 2019). In such contexts, the information from markers with large effect could overcome the noise that arises from a higher number of markers with low effects. When the number of QTN is large, it became more difficult to identify these variants with single-marker regression and WGS_SMR performed worse than HD. In such cases, other sets of variants selected from the sequence data can be (marginally) more beneficial than the commercial marker arrays as they are not affected by ascertainment bias in the same way as commercial marker arrays.

Table 1. Prediction accuracies for nine synthetic traits

QTN	h^2	HD	WGS_LD	WGS_SMR	WGS_200th
100	0.1	0.370	0.367	0.389	0.368
	0.25	0.416	0.395	0.422	0.418
	0.5	0.625	0.615	0.626	0.626
1,000	0.1	0.373	0.345	0.356	0.370
	0.25	0.396	0.393	0.402	0.404
	0.5	0.620	0.594	0.597	0.620
10,000	0.1	0.430	0.411	0.395	0.430
	0.25	0.437	0.430	0.398	0.444
	0.5	0.657	0.644	0.617	0.658

In this test we did not observe an improvement in prediction accuracy using sequence data when the number of QTN was large, which is the case of many traits of economic interest in livestock. These results are partly due to the already high prediction accuracies obtained with the current implementation of genomic selection using commercial marker arrays. These results are in line with other studies that found no improvement or only small variations in genomic prediction when using sequence data, often by preselecting variants, compared to HD marker arrays (van Binsbergen *et al.* 2015; Calus *et al.* 2016; Veerkamp *et al.* 2016; van den Berg *et al.* 2017; VanRaden *et al.* 2017). However, these genomic prediction results are preliminary results for a single line. With a more complete set of sequenced individuals, it remains to be determined whether the results will improve due to: data from multiple breeds, enabling multi-breed training and a much larger training set; or genomic prediction methods that are more suited for exploiting sequence data at a large scale than ridge regression.

CONCLUSIONS

The coupling of an appropriate sequencing strategy and hybrid peeling is a powerful method for generating whole-genome sequence data in large pedigreed populations, as long as the individuals are connected to enough informative relatives with marker array or sequence data, and regardless of population size. It remains to be determined whether these large datasets can provide the leverage for increased accuracy of genomic predictions.

REFERENCES

Al Kalaldeh M., Gibson J., Duijvesteijn N., Daetwyler H.D., MacLeod I., Moghaddar N., Lee S.H. and van der Werf J.H.J. (2019) *Genet. Sel. Evol.* **51**: 32.

Bolger A.M., Lohse M. and Usadel B. (2014) **30**: 2114.

Brøndum R.F., Guldbrandsten B., Sahana G., Lund M.S. and Su G. (2014) *BMC Genomics* **15**: 728.

Calus M.P.L., Bouwman A.C., Schrooten C. and Veerkamp R.F. (2016). *Genet. Sel. Evol.* **48**: 49.

Daetwyler H.D., Capitan A., Pausch H., Stothard P., van Binsbergen R., R.F. Brøndum R.F., …and Hayes B.J. (2014) *Nat. Genet.* **46**: 858.

DePristo M.A, Banks E., Poplin R., Garimella K.V., Maguire J.R., Hartl C., …and Daly M.J. (2011) *Nat. Genet.* **43**: 491.

Estany J., Ros-Freixedes R., Tor M. and Pena R.N. (2017) *J. Anim. Sci.* **95**: 2261.

Gonen S., Ros-Freixedes R., Battagin M., Gorjanc G. and Hickey J.M. (2017) *Genet. Sel. Evol.* **49**: 47.

Hickey J.M., Kinghorn B.P., Tier B., Wilson J.F., Dunstan N. and van der Werf J.H.J. (2011) *Genet Sel Evol.* **43**: 12.

Hickey J.M., Kinghorn B.P., Tier B., van der Werf J.H.J. and Cleveland M.A. (2012) *Genet Sel Evol.* **44**: 9.

Hickey J.M., Gorjanc G., Cleveland M.A., Kranis A., Jenko J., Mésázros G., Woolliams J.A. and Perez-Enciso M. (2014) *Proc. 10th World Congress of Genetics Applied to Livestock Production*, 17-22 Aug, Vancouver, BC, Canada.

Iheshiulor O.O.M., Woolliams J.A., Yu X., Wellmann R. and Meuwissen T.H.E. (2016) *Genet. Sel. Evol.* **48**: 15.

Li H. and Durbin R. (2009) *Bioinformatics* **25**: 1754.

Lopes M.S., Bovenhuis H., van Son M., Nordbø Ø., Grindflek E.H., Knol E.F. and Bastiaansen J.W. (2017) *J. Anim. Sci.* **95**: 59.

Meuwissen T. and Goddard M. (2010) *Genetics* **185**: 623.

Nani J.P., Rezende F.M. and Peñagaricano F. (2019) *BMC Genomics* **20**: 258.

Nicod J., Davies R.W., Cai N., Hassett C., Goodstadt L., Cosgrove C., …and Flint J. (2016) *Nat. Genet.* **48**: 912.

Pasanuic B., Rohland N., McLaren P.J., Garimella K., Zaitlen N., Li H., …and Price A.L. (2012) *Nat. Genet.* **44**: 631.

Pausch H., MacLeod I.M., Fries R., Emmerling R., Bowman P.J., Daetwyler H.D. and Goddard M.E. (2017) *Genet. Sel. Evol.* **49**: 24.

Ros-Freixedes R., Battagin M., Johnson M., Gorjanc G., Mileham A.J., Rounsley S.D. and Hickey J.M. (2018) *Genet. Sel. Evol.* **50**: 64.

Ros-Freixedes R., Gonen S., Gorjanc G. and Hickey J.M. (2017) *Genet. Sel. Evol.* **49**: 78.

van Binsbergen R., Bink M.C.A.M., Calus M.P.L., van Eeuwijk F.A., Hayes B.J., Hulsegge I. and Veerkamp R.F. (2014) *Genet. Sel. Evol.* **46**: 41.

van Binsbergen R., Calus M.P.L., Bink M.C.A.M., van Eeuwijk F.A., Schrooten C. and Veerkamp R.F. (2015) *Genet. Sel. Evol.* **47**: 71.

van den Berg I., Bowman P.J., MacLeod I.M., Hayes B.J., Wang T., Bolormaa S. and Goddard M.E. (2017) *Genet. Sel. Evol.* **49**: 70.

VanRaden P.M., Sun C. and O'Connell J.R. (2015) *BMC Genet.* **16**: 82.

VanRaden P.M., Tooker M.E., O'Connell J.R., Cole J.B. and Bickhart D.M. (2017) *Genet. Sel. Evol.* **49**: 32.

Veerkamp R.F., Bouwman A.C., Schrooten C. and Calus M.P.L. (2016) *Genet. Sel. Evol.* **48**: 95.

Whalen A., Ros-Freixedes R., Wilson D.L., Gorjanc G. and Hickey J.M. (2018) *Genet Sel Evol.* **50**: 67.

GENOMIC PREDICTION AND CANDIDATE GENE DISCOVERY FOR DAIRY CATTLE TEMPERAMENT USING SEQUENCE DATA AND FUNCTIONAL BIOLOGY

I.M. MacLeod[1], P.J. Bowman[2], A.J. Chamberlain[1], C.J. Vander Jagt[1], H.D. Daetwyler[1,2], B.J. Hayes[3] and M.E. Goddard[1,4]

[1] Agriculture Victoria, Centre for AgriBioscience, Bundoora, VIC, 3083 Australia
[2] School of Applied Systems Biology, La Trobe University, Bundoora, VIC, 3083 Australia
[3]Queensland Alliance for Agriculture and Food Innovation, University of Queensland, St. Lucia, QLD, 4067 Australia
[4]Faculty of Veterinary and Agricultural Sciences, University of Melbourne, VIC, 3010 Australia

SUMMARY

Dairy cow temperament is a complex trait affecting both animal and human welfare. Using Bayesian methods, differential gene expression and sequence variant annotation, we increased the accuracy of genomic prediction for temperament compared to using only HD genotypes. Candidate genes for temperament overlapped with genes associated with human neuropsychiatric disorders. More generally, the results indicate that for complex traits, we could make further gains in the accuracy of genomic prediction from access to more specific knowledge of functional biology. This study demonstrates a practical approach to use imputed sequence genotypes and functional biology to improve the accuracy of genomic prediction.

INTRODUCTION

Since the time of cattle domestication some 10,000 years ago there has been continuous genetic selection for animals of docile temperament (excepting animals bred for combat). In dairy cattle, good temperament is critical for animal welfare as well as human safety because of the daily interaction between cattle and agricultural technicians carrying out tasks such as milking and semen collection. Dairy cattle temperament is a polygenic trait with low to moderate heritability (Visscher and Goddard 1995). Given the intensive selection pressure for docility, we hypothesise that a significant proportion of the segregating variants that affect temperament will be relatively rare and recent. If this is the case, it is likely that for candidate gene discovery and genomic prediction there would be an advantage in using sequence variants rather than high density (HD) SNP chips. The reason for this is that SNP on commercial arrays are chosen to be common variants and are therefore not in strong LD with rare variants which are much more common in sequence data.

This study had three aims: 1) to use sequence variants to improve the accuracy of genomic prediction for temperament, 2) to use differential gene expression and functional annotation as a biological prior to increase the accuracy of genomic prediction, 3) to discover candidate genes affecting dairy cattle temperament.

MATERIALS AND METHODS

Phenotypes & Genotypes. Australian dairy cow milking temperament is routinely scored by farmers on a scale of 1 to 5 (where 1 is good and 5 is bad) and phenotypes are processed by DataGene for use in national dairy cattle evaluation. For this study, DataGene provided temperament phenotypes pre-corrected for herd-year-season for Holstein (7,354), Jersey (3,224) and Australian Red (103) animals, including records on 7,343 cows, and 3,338 bulls with progeny test of ≥ 20 daughters. Phenotypes were expressed as trait deviations for cows and daughter trait deviations for bulls (mean=-0.20, SD=0.61, min=-2.25, max=3.48) as used for the national dairy cattle evaluations. DataGene also provided pedigree information. All animals had either real or imputed Illumina 800K BovineHD beadChip genotypes (HD). Subsequently, their genotypes were imputed to sequence variants in all gene coding regions (exons) as well as 5000 bp flanking all known genes. The combined HD and sequence data, "SEQ", was then pruned for SNP pairs in perfect linkage disequilibrium (LD, $r^2 > 0.99$) and for variants with minor allele frequency (MAF) < 0.002 (details in MacLeod *et al.* 2016). After filtering, 994,019 variants remained and the animal genotypes were then centred and scaled to a unit variance. The Australian Reds (all bulls) were used only for validation of genomic predictions. The reference set included all Holstein and Jersey animals.

Statistical models. The data was analysed using the BayesR and BayesRC methods described by Erbe *et al.* (2012) and MacLeod *et al.* (2016) respectively. Briefly the model fitted was:

Temperament = mean + breed-sex group + SNP effects + pedigree + error,

where pedigree was fitted to account for any polygenic genetic variance not explained by the combined SNP effects. To account for heterogeneous error variance associated with cow and bull phenotypes, the residuals were weighted following Garrick *et al.* (2009) and this was implemented in the Bayesian models as described in Kemper *et al.* (2015). Our Bayesian models fit SNP effects jointly as a mixture of four normal distributions with a mean of zero and variance: $\sigma^2_1=0\sigma^2_g$, $\sigma^2_2=0.0001\sigma^2_g$, $\sigma^2_3=0.001\sigma^2_g$ and $\sigma^2_4=0.01\sigma^2_g$, where σ^2_g is the additive genetic variance. All analyses were replicated with 5 MCMC chains, each with 40,000 iterations (20,000 burn-in). The accuracy of genomic prediction was estimated as the correlation between the genomic predictions and phenotypes, and bias was assessed as the regression coefficient of phenotypes on predictions.

The BayesRC approach is very similar to BayesR but incorporates prior biological knowledge in the model. For example, if one or more groups of variants are thought to be more enriched for QTL or causal variants, these can be allocated to a separate variant category *a priori*. In BayesRC, each category is then independently modelled as a mixture of the four BayesR distributions described above, but each starting with equal priors. If a category of variants is found to be enriched for causal variants in the data, this can improve the fit of the model.

Therefore, *a priori* we used independent differential gene expression data measured in 18 bovine tissues (Chamberlain *et al.* 2016), to identify 500 genes that were most highly differentially over-expressed in each of: caudal brain tissue, cerebral brain tissue and adrenal tissue. There was a strong overlap between the top 500 over-expressed genes in each of these three tissues, resulting in a unique set of 1006 genes that we refer to collectively as the "DE" gene set. To further inform the selection of variants for potentially enriched categories, we annotated all non-synonymous coding variants (NSC) associated with the DE genes as well as variants < 50 Kb up- and down-stream of DE genes (REG). We tested four BayesRC models, the first being "DE7" with 7 variant categories (of which 6 used functional annotation):

1) NSC in DE genes overlapping in both caudal and cerebral tissue (N=1617)
2) NSC in DE genes in either caudal or cerebral tissue (N=1447)
3) NSC in the remaining DE genes in adrenal tissue (N=1430)
4) REG flanking DE genes overlapping in both caudal and cerebral tissue (N= 30549)
5) REG flanking DE genes in either caudal or cerebral tissue (N= 28893)
6) REG flanking the remaining DE genes in adrenal tissue (N= 22151)
7) All remaining variants (N= 907932)

"DE2" was the second BayesRC model, where variants in categories 1 to 6 above were combined into one category, and remaining variants to a second category. The third and fourth models, "Random7" and "Random2", had variant categories that matched DE7 and DE2, except that the DE gene set was replaced with a random set of 1006 genes chosen from 24,580 known bovine genes. The BayesR model was run with SEQ or HD genotypes.

RESULTS AND DISCUSSION

The estimated heritability of temperament in the BayesR SEQ model was 0.1 which, although low, indicates that there is still important genetic variation for this trait. Previously, Visscher and Goddard (1995) estimated the heritability of Australian dairy cattle temperament to be 0,2 using only bull progeny test data and a sire model. More recent literature, in Holsteins, report similar heritability estimates to ours for farmer scored temperament (e.g. Stephansen *et al.* 2018). The accuracy of genomic prediction in the Australian Red validation set improved when sequence variants and HD SNP were combined (SEQ) in the BayesR model compared to HD only (Table 1). This may be a result of the sequence variants being in stronger linkage disequilibrium (LD) with causal variants and/or causal variants being included. If it is due to stronger LD, this could reflect the possibility that variants affecting temperament are rare because there has been strong selection pressure for docile temperament in dairy cattle since domestication. Previous studies in cattle for other traits have also shown small improvements from using selected subsets of sequence data compared to 50K or HD SNP genotypes (eg. Brøndum *et al.* 2015; MacLeod *et al.* 2016). However, use of full genome sequence has not yet shown consistent improvement compared to SNP chip genotypes (eg. Calus *et al.* 2016; van den Berg *et al.* 2017). We had therefore pre-selected a subset of sequence variants from gene coding regions and regions adjacent to genes, hoping to capture important missense or regulatory mutations for candidate genes.

In our study, the BayesRC DE7 and DE2 models showed a further small increase in the accuracy of prediction (Table 1). These two models used the same variants as BayesR SEQ, but used prior biology to identify categories of variants that were in or close to genes highly over-expressed in brain or adrenal tissue compared to 17 other tissues (DE genes). Additionally, the DE7 model incorporated a biological prior on variant annotation: non-synonymous coding variants and those that might be regulatory. In the BayesRC Random2 and Random7 models, we replaced the DE gene set with a random set of genes and used this as the prior to group variants into 2 or 7 categories. The accuracy of prediction in the Random2 and Random7 models was lower than the DE2 and DE7 models (Table 1). This lends support to our assumption that genes which are highly expressed in brain and/or adrenal tissue are more enriched for variants controlling dairy cow temperament. However, the level of enrichment for the different variant categories was not very high compared to the random models, suggesting that more specific prior biology is required to better inform the BayesRC model. The accuracy for the Random7 model was slightly lower than the HD. Although this is likely not significant, it could reflect the inclusion of some poorly imputed sequence variants that add noise to the prediction. This could be further tested by constructing random models multiple times. The bias of the predictions suggests a tendency to under-predict genomic breeding value but it is similar across the models.

Our Bayesian methods have previously been demonstrated to be a useful approach for fine mapping genes and mutations that affect complex traits (eg. MacLeod *et al.* 2016). Following our previous study, we used the Bayesian "posterior probability of a variant having a non-zero effect" to detect QTL regions and identify candidate genes. In the BayesR SEQ model, if there is very strong LD across a QTL region, the model will have difficulty distinguishing which variant to prioritise, so the posterior probability will be relatively low and spread across all variants in strong LD. Therefore, to locate candidate gene regions, we summed the posterior probability in windows of 20 SNP, sliding 10 SNP to the next window. We identified 11 known genes in or closest to the top 13 QTL regions genome-wide: NCOA7, GAD2, PDGFD, TMPRSS5, DRD2, IQSEC1, MAOB, PTPRF, SLC25A16, TMCO5A, SNRPB2. The first seven were highly differentially expressed in bovine brain and/or adrenal tissue (Chamberlain *et al.* 2016) in line with our assumption that the DE genes were more likely to be associated with cow temperament than other genes. Furthermore, 10 genes of these 11 overlap candidate genes or gene families associated with a range of human neuropsychiatric or neurodevelopmental disorders including: schizophrenia, autism, intellectual disability, post-traumatic stress and anxiety (eg. http://atgu.mgh.harvard.edu/~spurcell/genebook/genebook.cgi?user=guest&cmd=overview).

Table 1. Accuracy and bias of genomic prediction in 103 Australian Red bulls using a range of BayesR and BayesRC analytical models

Model[1]	Accuracy	Bias	Increase in accuracy vs. HD
BayesR HD	0.236	1.4	-
BayesR SEQ	0.269	1.6	3.4%
BayesRC DE7	0.289	1.6	5.3%
BayesRC DE2	0.282	1.7	4.6%
BayesRC Random7	0.221	1.3	-1.4%
BayesRC Random2	0.254	1.5	1.8%

[1] See Materials & Methods for acronyms

CONCLUSIONS

This study demonstrates a practical approach to exploiting sequence data and functional biology to improve the accuracy of genomic prediction and for causal gene discovery. It is likely that more specific functional biology would be beneficial for this approach.

ACKNOWLEDGEMENTS

Dairy farmers and DataGene are acknowledged for collection, processing & provision of phenotypes and pedigree data. We are grateful to Dr P Stothard (University of Alberta) for annotation of sequence variants and to the 1000 Bull Genomes Project for access to sequences.

REFERENCES

Brondum R., Guldbrandtsen B., Sahana G., Lund M. and Su G. (2014) *BMC Genomics* **15**: 728.

Calus M.P.L., Bouwman A.C., Schrooten C. and Veerkamp R.F. (2016) *Genet. Sel. Evol.* **48:** 49.

Chamberlain A.J., Vander Jagt C.J., Goddard M.E. and Hayes B.J. (2014) *Proceedings of the World Congress of Genetics Applied to Livestock Production* **Paper 180.**

Erbe M., Hayes B.J., Matukumalli L.K., Goswami S., Bowman P.J., Reich C.M., Mason B.A. and Goddard M.E. (2012) *J. Dairy Sci.* **95:** 4114.

Garrick D.J., Taylor J.F. and Fernando R.L. (2009) *Genet. Sel. Evol.* **41**: 44.

Kemper K.E., Reich C.M., Bowman P., vander Jagt C.J., Chamberlain A.J., Mason B.A., Hayes B.J. and Goddard M.E. (2015) *Genet. Sel. Evol.* **47**: 29.

MacLeod I.M., Bowman P.J., Vander Jagt C.J., Haile-Mariam M., Kemper K.E., Chamberlain A.J., Schrooten C., Hayes B.J. and Goddard M.E. (2016) *BMC Genomics* **17**: 144.

Stephansen R.S., Fogh A. and Norberg E. (2018) *J. Dairy Sci.* **101**: 11033.

van den Berg I., Bowman P.J., MacLeod I.M., Hayes B.J., Wang T., Bolormaa S. and Goddard M.E. (2017) *Genet. Sel. Evol.* **49**: 70.

Visscher P. and Goddard M. (1995) *J. Dairy Sci.* **78**: 205.

GENE NETWORK ANALYSIS FOR MARBLING DEVELOPMENT USING GENE EXPRESSION (RNA-SEQ) IN HANWOO

S. de las Heras-Saldana[1*], D. Lim[2], S.H. Lee[3] and J.H.J. van der Werf[1]

[1]School of Environmental and Rural Science, University of New England, Armidale, NSW, 2351 Australia
[2]Animal Genomics & Bioinformatics Division, National Institute of Animal Science, RDA, Republic of Korea
[3]Division of Animal and Dairy Science, Chungnam National University, Deajeon, South Korea

SUMMARY

This study aimed to use co-expression networks to reach a better understanding of the genetic mechanisms underlying marbling development and assess the effect of different diets on the transcriptomic profile and the resulting phenotype. We evaluated development of marbling in *Longissimus dorsi* muscle extracted from 45 Hanwoo steers at 8, 12, 18, 24 and 30 months of age. The effect of two different feeding conditions (high/low feeding) were evaluated on the gene expression in the relation with the marbling score. In the four groups according to the marbling score (MS) and the feeding treatment: HighFeed_HighMS, HighFeed_LowMS, LowFeed_HighMS, and LowFeed_LowMS we found 818, 928, 899, and 946 co-expressed genes respectively. The activity of modules changed significantly with the age indicating different expression profiles for genes involved in muscle growth (*PI3K-Akt signaling, Focal adhesion, ECM-receptor interaction*), metabolic regulation (*Biosynthesis of amino acids* and *Glutathione metabolism*) and lipid deposition (*Fatty acid metabolism, Regulation of lipolysis in adipocytes*) through the animal development. The effect of the feeding condition in animals that developed low MS showed the activation of pathways related to stress and maintenance of cell homeostasis under nutritional limitations, while pathways involved in fat deposition and lipid mobilization where observed under high feeding.

INTRODUCTION

In beef cattle, the evaluation of intramuscular fat, measured as marbling score, is an important indicator of meat quality and achieving higher levels of this trait is an economical incentive for producers. Improving the knowledge of the genes and pathways involved in the development of economically important traits can potentially help to improve management strategies, genomic selection, or molecular tools to improve beef production. There have been some studies that attempted to describe the mRNA (Lim *et al.* 2015) and miRNA (Seong *et al.* 2016) abundances in muscle samples from high and low phenotypes. However, all these studies were performed using tissue from animals at age of slaughter, showing the final phenotype (high or low marbling). The use of gene expression (RNA-seq) could be applied in the identification of markers for the onset of marbling at younger ages. Since differences in marbling are subtle at early ages, the analysis of co-expressed genes with the identification of networks could be more informative to explain the development of marbling and find markers than the analysis of differentially expressed genes alone. Differential expression analysis relied on big changes in expression between conditions, while in the co-expression analysis it is possible to identify the genes with similar expression profile which are more likely to be involved in the same metabolic pathway, have related function, may have been co-regulated and identifying them can assist in the finding of hubs or molecular targets (Russo *et al.* 2018).

MATERIALS AND METHODS

RNA sequences were obtained from previous study (Lim *et al.* in print). Briefly, 45 Hanwoo steers were grown on high (23 steers) and low (22 steers) energy diets from eight months of age until slaughter at 30 months. Muscle samples were taken from *Longissimus dorsi* at 8, 12, 18, and 24 by biopsy of the tissue, while muscle was sampled after slaughter at 30 months. The samples were sequenced in Illumina HiSeq200 to obtain paired-end reads of 100 bases pairs. Standard procedures were followed on the reads for quality control, cleaning, mapping and assembly (Lim *et al.* in print). Analysis was done for groups differentiated by diet (High vs Low Feed) and marble score at slaughter (High vs Low MS). The co-expression networks were performed separately for each sample group (HighFeed_HighMS, HighFeed_LowMS, LowFeed_HighMS and LowFeed_LowMS). For each group, we normalized the gene expression counts and filtered out the genes with low expression as well as the genes with low variance resulting in around 13,000 expressed genes. Pearson correlations were calculated on the logarithmic copies per million (lcpm) and the pairs of genes with a correlation ≥ 0.8 were selected for subsequent analysis. Similarities between these gene expression was used to identify modules of genes with similar expression profile by a dynamic tree cut. The biological role of the selected genes in each module was found through an over representation analysis to identify the pathways involved (adjusted P-value <0.05). The activity of the genes in each module, its activation or repression according to the age, was evaluated with a gene set enrichment analysis (GSE) using the R Package CEMItool (Russo *et al.* 2018). This analysis ranked the genes according to the correlation of their expression with the phenotypic class (high or low marbling) and determined whether the genes of each module tend to be at the top or bottom of the ranked list. The normalized enrichment score (NES) will be higher as the gene in the module is found in the ranked list; alternatively, the score value is negative if the genes are not found in the list. The score value could be zero if the set of genes are randomly distributed in the list. Finally, network graphs were made from the co-expression information and combined with information of protein-protein interaction extracted from the STRING v11.0 database to identify the hub genes (representing genes with interactions with multiple other genes).

RESULTS AND DISCUSSION

We found a similar number of genes selected for each of the four groups after filtering out low expression and variance, and keeping the genes that are highly correlated (Table 1). In general, most of the genes belong to module one, and there were "not-correlated" genes (NC) in each group. To assess the activity of the genes in each module and across ages we performed a gene set enrichment analysis (GSEA) for each group. The GSEA results showed variation in the normalized enrichment score indicating that the activity of the genes changed according to the age of the animals (Figure 1) suggesting that their expression have an effect in the growth of muscle and marbling. Every group presented modules with high activity at 30 months of age indicating also association with the final marbling phenotype. Genes with potential role in marbling development because of their activity at 30 months (red color) could be found in the modules M2, M3, and M5 for HighFeed_HighMS group; M3 in HighFeed_LowMS group; M1 and M3 in LowFeed_HighMS; while M2, M3, and M5 were identified in LowFeed_LowMS group.

Table 1. Number of genes and pathways represented for each sample group

Group	Genes	Pathways	M1	M2	M3	M4	M5	M6	NC
HighFeed_HighMS	818	47	172	160	152	122	101	49	62
HighFeed_LowMS	928	62	595	131	63	62	38	0	39
LowFeed_HighMS	899	78	495	150	69	61	0	0	124
LowFeed_LowMS	946	55	395	245	167	55	43	0	41

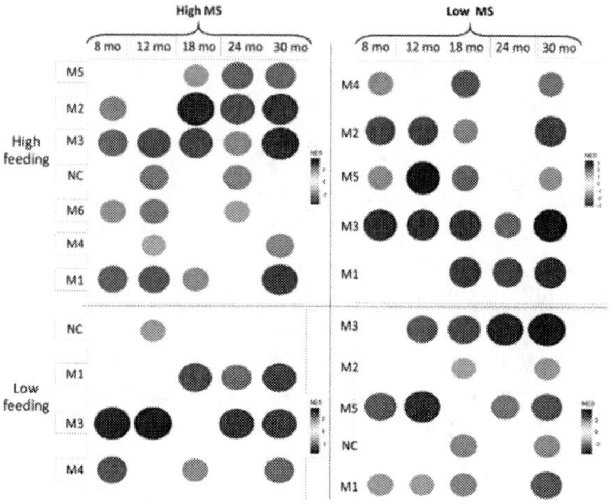

Figure 1. Gene set enrichment analysis showing the modules (M) and the normalized enrichment score (NES); NC= no correlated

We performed an over-representation analysis to determine which processes are mostly associated with de development of marbling in each group and we investigated if these pathways are affected by the feeding condition. To compare the pathways between groups, we selected 40 pathways with an important role in the muscle growth and the development of marbling (Figure 2).

We observed the presence of genes in multiple pathways showing that there is a cross-talk/interaction between them. Nine pathways reflected the conserved process involved in the morphology and growth of the skeletal muscle in all the groups: *PI3K-Akt signaling*, *Focal adhesion*, *ECM-receptor interaction*, *Carbon metabolism*, *Biosynthesis of amino acids*, *Glutathione metabolism*, *Fatty acid metabolism*, *Regulation of lipolysis in adipocytes*, and *Pentose phosphate*. Interestingly, there were also pathways represented exclusively in each group (except for HighFeed_LowMS). In the HighFeed_HighMS group the physiologic response under to the HighFeed diet activated important pathway involved in transport of glucose, body weight, fat deposition, and vasculature: *Apelin signaling*, *Rap1 signaling*, *FoxO signaling*, *Relaxin signaling*, and *Insulin resistance*. In skeletal muscle the entry of glucose is promoted by apelin which also affect the activity of FOXO1 gene (Hwangbo *et al.* 2017). The gene Rap1 have been reported in the control of body weight and metabolic regulation in mice (Yeung *et al.* 2013). The genes involved in pathways related to stress and low energy disposition (i.e. *HIF-1 signaling*, *Protein processing in endoplasmic reticulum*, *Alanine, aspartate and glutamate metabolism*) were identified in steers that developed low MS under low feeding conditions. Special attention was focused on the LowFeed_HighMS group since these represent animals with a high marbling score even under under low feeding conditions. In this group, the pathways *Starch and sucrose metabolism*, *Adipocytokine signaling*, *Pantothenate and CoA biosynthesis*, and *Oxidative phosphorylation* seems to have an important role in the regulation energy metabolism and deposition of intramuscular fat. The hub gene ADIPOR2 seems to have an important function in the *Adipocytokine signaling* (Figure 2). ADIPOR2 is an adipocytokine which expression affects lipid accumulation, the activity of PPAR-α signaling pathway (Ouchi *et al.* 2012; Cao 2014). Also in this pathway, the gene PGC-1α is a key regulator of the conversion of muscle fiber types from glycolysis (uses glycolysis as energy source) into oxidised (use fatty acid oxidation to produce energy) muscle fibers (Gu *et al.* 2019).

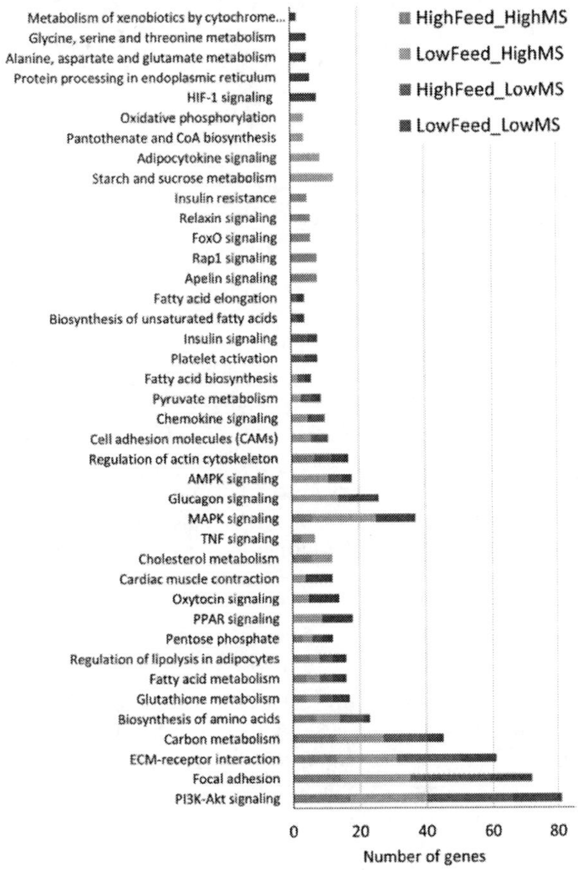

Figure 2. Number of genes involved in the pathways identified in each studied group

CONCLUSIONS

The co-expression analysis was shown to be a useful approach for the identification of processes and genes related to marbling development, particularly with the identification of gene modules that are associated with early age onset of marbling. The genes FOXO1, ADIPOR2, PGC-1α are promising markers to select for animals for high marbling.

REFERENCES

Cao H. (2014) *J. End.* **220**: T47.

Gu H., Li J., Ying F., Zuo B., Xu Z. (2019) *Mol. Biol. Rep.* **1**.

Hwangbo C., Wu J., Papangeli I., Adachi T., Sharma B. et al. (2017) *Sci. Trans. Med.* **9**: eaad4000.

Lim D., Chai H.H., Lee S.H., Cho Y.M., Choi J.-W., *et al.* (2015) *Asi-Aust. J. Anim. Sci.* **28**: 1075.

Ouchi N., Ohashi k., Shibata R., Murohara T. (2012) *Nagoya J. Med. Sci.* **74**: 19.

Russo P.S., Ferreira G.R., Cardozo L.E., Bürger M.C. *et al.* (2018) *BMC Bioinformatics.* **19**: 56.

Seong J., Yoon H., Kong H.S. (2016) *Genes & Genomics* **38**: 529.

Yeung F., Ramírez C.M., Mateos-Gomez P.A., Pinzaru A., Ceccarini G. *et al.* (2013) *Cell Rep.* **3**: 1847.

EXPLORING THE REGULATORY POTENTIAL OF LONG NON-CODING RNA IN BOVINE FEED EFFICIENCY THROUGH COEXPRESSION IN LIVER AND MUSCLE

P.A. Alexandre[1,2], A. Reverter[2], L.R. Porto-Neto[2], R.B. Berezin[1], J.B.S. Ferraz[1] and H. Fukumasu[1]

[1]College of Animal Sciences and Food Engineering, USP, Pirassununga, SP, Brazil
[2]CSIRO Agriculture & Food, St. Lucia, Brisbane, QLD, 4067 Australia

SUMMARY

Long non-coding RNA (lncRNA) control gene expression through a variety of processes, thus, can be central regulators of important traits in livestock production, such as feed efficiency (FE). With the aim to identify regulatory lncRNA in liver and muscle of bulls divergent for FE, we used RNAseq data to obtain a list of novel lncRNA in bovine. Then, lncRNA in liver and muscle of high and low FE bulls were tested for differential expression (DE) and their potential to regulate gene expression in those tissues assessed through a regulatory impact factor algorithm. A total of 102 lncRNA were DE in both tissues and 126 were considered regulators of gene expression. We were able to highlight at least three lncRNA with the potential to be further explored in our ongoing study. The variety of mechanisms by which lncRNA influences gene expression make these molecules of particular interest to be validated as central regulators of FE.

INTRODUCTION

The flow of information from DNA to protein synthesis comprises several steps in which gene expression can be controlled, and lncRNA (long non-coding RNAs) are involved in several of them. Structurally, lncRNA have low or no potential for protein-coding and low conservation among species. In contrast to protein-coding genes, lncRNA are more tissue-specific and have a lower level of expression, although they also present multiple exons, polyA tail, 5 'cap and may contain CpG islands in their promoter regions (Weikard *et al.* 2016).

Long non-coding RNAs are involved in the regulation of gene expression either by recruiting chromatin-modifying complexes, regulating DNA methylation levels, modulating allele-specific expression or functioning as guides, precursors or sequester of miRNAs. They can also act as sequester of transcription factors or as scaffolds of multiprotein complexes. They act in the alternative splicing, in enzymes activity and can either be exported and act in other cells. In gene networks, they act as regulators, being expressed in coordinated pattern with other genes. This variety of lncRNA mechanisms of action confer these molecules great potential to regulate important traits for animal production, such as feed efficiency (FE), a proxy for both increased productivity and reduced environmental impact.

In cattle, a catalogue of lncRNA in multiple tissues was published by Koufariotis *et al.* (2015), but there is little overlap between those and lncRNA found by other studies (Weikard *et al.* 2016). This makes the tissue/cell/time/condition-specific character of the lncRNA evident and highlights the need to study these molecules in the context of FE. The objective of this study was to identify lncRNAs with regulatory potential for FE in beef cattle that could be explored as biomarkers of this trait.

MATERIALS AND METHODS

Tissue Sampling and Bioinformatics Analyses. We used RNAseq data (Illumina HiSeq 2500, 100bp, paired-end) from our database which comprises 86 samples from liver, skeletal muscle, adrenal gland, hypothalamus and pituitary of high and low FE Nellore (*Bos indicus*) bulls. Refer to

Alexandre *et al.* (2019) for details of the experimental resource and bioinformatics pipeline protocol for sequence alignments. In brief, Cufflinks pipeline (Ghosh and Chan 2016) was used to generate an annotation file from which new transcripts were selected and filtered to exclude those smaller than 200bp. Then, transcripts presenting open reading frame greater than 300bp between a START and a STOP codon were excluded. Transcripts were tested for similarity to the UniProt/SwissProt database using BLAST+ and discarded if E-value $<10^{-6}$. A final check was performed to consider as novel lncRNA only transcripts presenting more than 1 exon. Cuffcompare was used to compare novel transcripts to the NCBI annotation file containing predicted genes by the GNOMON method. Novel lncRNA were also compared to the bovine NONCODE database using BLAST+ and were considered significant when E-value<10-6.

Differential Expression and Gene Network Inference. Using data from liver and muscle RNAseq, read counts were generated by HTseq and expression values were estimated as fragments per kilobase of transcript per million mapped reads (FPKM). Genes with average expression lower than 0.2 FPKM across all samples in the two tissues were discarded. Differential expression (DE) analysis was performed as described in Alexandre *et al.* (2019). Genes were considered DE when the difference between the expression in high FE and low FE groups were greater than ±2.57 SD from the mean, corresponding to a t-test $P < 0.01$. In order to identify lncRNA with regulatory potential, the regulatory impact factor (RIF) metric was calculated considering as targets the totality of 14,466 mRNAs expressed in liver and muscle (Alexandre *et al.* 2019), and the DE lncRNA as potential regulators. Those scores deviating ±2.57 SD from the mean (corresponding to $P < 0.01$) were considered significant. For gene network inference, we input all the mRNAs plus the DE lncRNA significant according to RIF into the PCIT algorithm in order to identify significant connections that would establish an edge in the reconstruction of the network. Only edges involving at least one DE lncRNA were considered for the network visualisation and analysis. We used Cytoscape (Shannon *et al.* 2003) to visualise the network and BINGO plugin (Maere *et al.* 2005) to perform functional enrichment, using the whole genome as background, hypergeometric test and Benjamini–Hochberg FDR correction ($P < 0.05$).

RESULTS AND DISCUSSION

We were able to initially identify 174,793 novel transcripts of which 174,431 (99.8%) presented length greater than 200bp, a criterion used to distinguish small ncRNA from lncRNA. The filter that most excluded transcripts was ORFs greater than 300nt, maintaining only 8.5% (14,904) of the transcripts initially identified. Then, 5,865 transcripts were excluded because they presented high similarity with known proteins and 12 were excluded because they presented coding potential. From the resulting 9,027 transcripts, only 4,603 presented more than one exon and were considered as novel lncRNA.

Among the 4,603 transcripts identified as lncRNA, the most represented class was u - intergenic transcripts (Figure 1a), which is in accordance with the literature. However, when we compared the lncRNA identified with predicted genes in the bovine genome annotation, we found that a large portion of them are, actually, new isoforms (class j) or overlap exon of a predicted gene (class o). A few of them (6%) overlap perfectly with predicted genes (Figure 1b).

A total of 2,677 (58%) lncRNA presented high similarity with NONCODEv5, suggesting that a large part of those have already been identified previously. While that percentage indicates a certain level of efficiency in the identification of lncRNA, it also reflects a deficit in the annotation of these molecules, especially considering differences between the subspecies *taurus* and *indicus*, as most of the experiments for characterization of lncRNA were performed in taurine breeds. Given the regulatory role of this class of RNAs and their apparent lack of conservation among species, it is expected that important differences exist between zebu and taurine cattle, especially regarding adaptive responses.

Differential expression analysis was performed using all the 2,688 expressed lncRNA and resulted in 58 DE lncRNA in liver and 46 in skeletal muscle (Figure 2), being 42 up-regulated and 60 down-regulated in high FE (P<0.01), as two DE transcripts in both tissues were concordant. Thirty-eight of the DE lncRNA were isoforms of known/predicted genes and one, TCONS00029353, draws attention for being DE in muscle and a non-coding isoform of TSC22D1, a gene stimulated by transforming growth factor beta (TGFB) that regulates the transcription of multiple genes. In fact, TGFB1 was recently pointed as a central regulator of FE in muscle (Alexandre *et al.* 2019), making that lncRNA of particular interest. Two transcripts completely overlapped predicted genes LOC101905381 and LOC104970784 which are indeed described as a non-coding RNA, but there is no literature discussing their potential function.

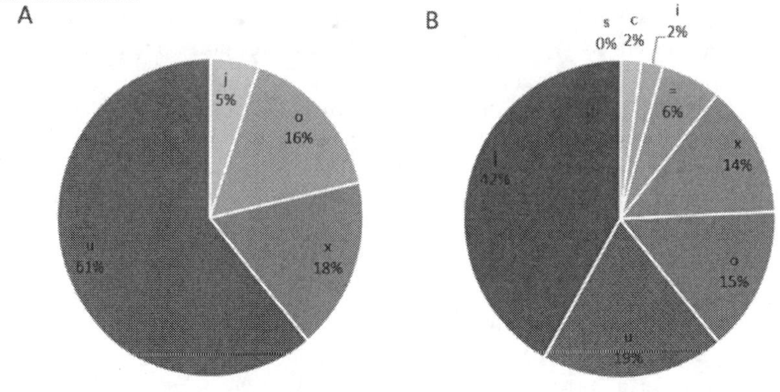

Figure 1. Distribution of the lncRNA identified according to their classification. A) Only annotation of known genes and B) Annotation including genes predicted by Gnomon (NCBI): = - overlap with known gene; c - contained in a known gene; i – overlap with known intron; j - new isoform; o - generic overlap with known exon; s - contains intron that overlaps known intron on the opposite string; u - intergenic transcript; x – overlap with known gene on opposite string

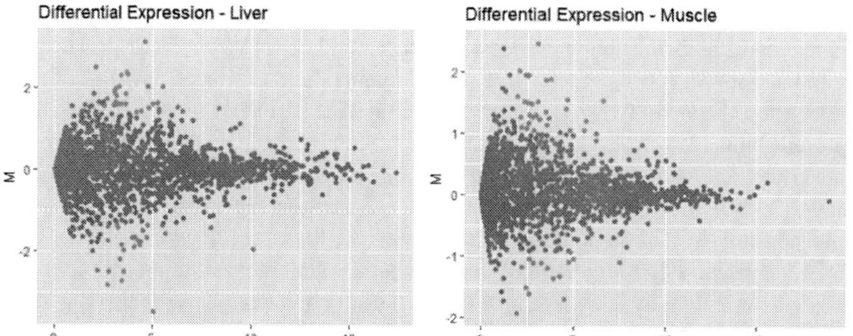

Figure 2. MA plots of 2,688 lncRNA expressed in liver and muscle between high and low feed efficiency (FE). Differential expression (M, y-axis) is plotted against average expression (A, x-axis). Pink dots represent statistically significant DE lncRNA (P<0.01)

A total of 8 lncRNA were considered regulators and a co-expression network containing their directly co-expressed mRNAs was created (Figure 3). Of these, two lncRNA are co-expressed with most of the genes in the network and the functional enrichment of their co-expressed genes showed that these transcripts are related to biological functions already previously associated with FE in these tissues (Alexandre *et al.* 2019), supporting their potential as regulators of FE. TCONS00030954 is an intergenic transcript up-regulated in the liver of high FE animals and their co-expressed genes are related to translation (adjusted P=3.8x10^{-7}) and gene expression (adjusted P=4.9x10^{-5}). Moreover, it is co-expressed with NR2F6, also recently pointed as a central regulator of FE in liver. TCONS00089983 overlaps PTMA gene in the opposite string and is also up-regulated in the muscle of high FE bulls. Its co-expressed genes are enriched mainly for protein metabolic process (adjusted P=1.1x10^{-2}).

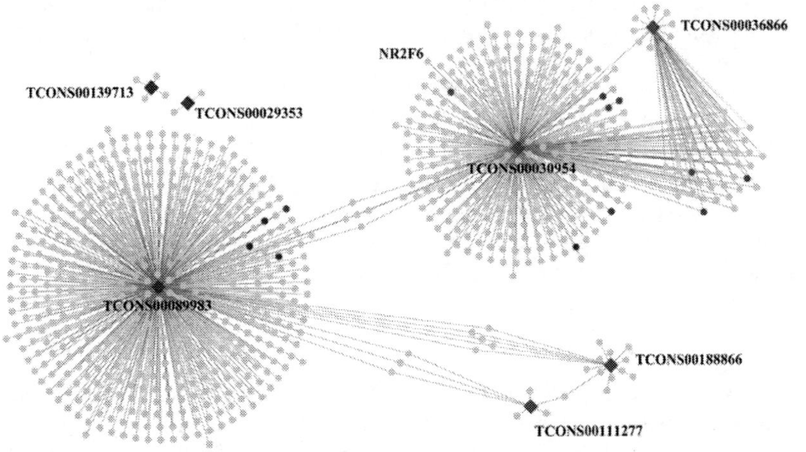

Figure 3. Gene co-expression network of eight lncRNA that are differentially expressed and with regulatory potential, and their 878 directly co-expressed mRNA. Ellipses represent mRNA, diamonds represent lncRNA; blue and red nodes represent differential expression in liver and muscle, respectively. Only coexpression correlations above 0.7 are represented

CONCLUSIONS

These are preliminary results of an ongoing study but at this point, we can already indicate that lncRNA have great potential as regulators of FE in beef cattle, in particular, three lncRNA that will be further explored regarding their functional effect as regulators of gene expression and their potential as biomarkers for selection purposes.

ACKNOWLEDGEMENTS

To São Paulo Research Foundation (FAPESP) Proc. 2015/22276-3 and 2014/07566-2.

REFERENCES

Alexandre P.A., Naval-Sanchez M., Porto-Neto L.R., Ferraz J.B.S., Reverter A. and Fukumasu H. (2019) *Front Genet.* **10**:3.

Ghosh S. and Chan C.K.K. (2016) *Methods Mol Biol.* **1374**:339.

Koufariotis L.T., Chen Y.P., Chamberlain A., Jagt C.V. and Hayes B.J. (2015) *PLoS ONE.* **10**:10.

Maere S., Heymans K. and Kuiper M. (2005) *Bioinformatics.* **21**:16.

Shannon P., Markiel A., Ozier O., Baliga N.S., Wang J.T., Ramage D. and Ideker T. (2003) *Genome Res.* **13**:11.

Weikard R., Demasius W. and Kuehn C. (2016) *Animal Genet.* **48**:1.

GENETIC IMPROVEMENT OF PASTURE INTAKE AND EFFICIENCY IN BEEF CATTLE: ARE WE THERE YET?

P.F. Arthur[1], R.M. Herd[2], P.L. Greenwood[2], K.A. Donoghue[3] and T. Bird-Gardiner[3]

[1]NSW Department of Primary Industries, Elizabeth Macarthur Agricultural Institute, Menangle, NSW, 2568
[2]NSW Department of Primary Industries, Livestock Industries Centre, Armidale, NSW, 2351
[3]NSW Department of Primary Industries, Agricultural Research Centre, Trangie, NSW, 2823

SUMMARY

Currently feed intake can be accurately and easily measured in cattle that are lot-fed prepared feeds, commonly using automatic feed intake recorders, and the data from these underpin the BREEDPLAN® feed efficiency trait, Net Feed Intake. However, a large proportion of beef production in most countries including Australia and New Zealand is either fully or partially pasture-based. There are now emerging technologies which have capability for automation, have reasonable level of accuracy in estimation of feed intake, and the ability to be deployed on pastures. With further development a pasture intake and efficiency trait can be a reality.

INTRODUCTION

Profitability of beef production depends on both inputs and outputs. Providing feed to cattle is the single largest input cost in most commercial beef production enterprises, and thus improving the efficiency of feed use will help reduce input costs. In beef cattle, genetic improvement of feed use is currently possible in Australia through the use of the BREEDPLAN® trait, Net Feed Intake (NFI), which is also known in the scientific literature as residual feed intake (Berry and Crowley 2013). An important factor in the development of any feed efficiency trait is an ability to accurately measure feed intake of each animal. Currently feed intake can be accurately and easily measured in cattle that are lot-fed prepared feeds, commonly using automatic feed intake recorders. However, a large proportion of beef production in most countries including Australia is either fully or partially pasture-based; as illustrated by Australian commercial labels – "grassfed beef" and "grainfed beef"; the latter referring to cattle raised on pasture and finished on grain-based diet in feedlots, and the national breeding cow herd is pasture-based. There is currently no individual animal feed efficiency measure for pasture-based cattle, mainly due to the challenge of measuring individual animal pasture intake accurately and easily. The objective of paper is to review currently available methods of measuring individual animal pasture intake and provide some of the relevant results to date with a view to stimulating interest in progressing development of strategies for a pasture intake and/or efficiency trait for genetic improvement in beef cattle.

PASTURE INTAKE MEASUREMENT TECHNOLOGIES

Unlike prepared rations, pasture intake by cattle is difficult to measure directly. There are currently a number of technologies available for estimating pasture intake by cattle and their suitability for use would depend, among other considerations on accuracy, cost and ease of use and potential for automation. These measurement technologies include:

Marker technologies. *Chromium oxide, other metallic oxides and non-metallic markers.* These markers have been used for many decades to estimate faecal output by grazing animals, and with knowledge of the composition and digestibility of the diet consumed, an estimate of pasture intake can be calculated (Mayes and Dove 2000). The technique requires daily dosing of animals with the marker, which can interrupt animal's normal grazing behaviour, or administration of a bolus of marker in an intra-ruminal controlled release devise (CRD) which releases a controlled dose of marker into the rumen over days and weeks, obviating the need for daily dosing. Both methods require animals to be mustered regularly, typically twice weekly, for collection of a faecal sample which must then be subsequently analysed for its' concentration of chemical marker. Concern over residues of metallic oxide makers in animals and the environment and non-registration of these markers for use in livestock currently limit their repeated use in individual animals and large numbers of animals.

n-alkanes and other plant chemical constituents. These have been proposed as alternative markers to metallic oxides and also offer the capacity to estimate the proportions of some pasture species and supplements consumed by grazing animals, as well as estimation of faecal output and feed intake (Mayes and Dove 2000).

Use of both groups of markers share the same limitations, being the need for dosing of animals, frequent mustering for collection of faecal samples, sampling of pasture species and supplements on offer, sample storage and preparation (typically, freezing, drying and grinding) and laboratory chemical analysis. Only the last step offers scope for automation. Experiments using markers to estimate faecal output and feed intake have typically been conducted over much shorter time periods (rarely more than 3 weeks) then recommended for cattle undergoing a feed efficiency test on complete mixed rations. Despite these limitations, markers have been used in Australia to estimate pasture intake by relatively large numbers of grazing beef cattle (Barlow *et al.* 1990; Dicker *et al.* 1998; Herd *et al.* 2004), and by sufficiently large numbers of sheep to estimate phenotypic and genetic parameters (Lee *et al.* 1995; Lee *et al.* 2002; Fogarty *et al.* 2006)

Wireless sensor technologies. Use of wireless sensors in estimating pasture intake has advantages including provision of an array of sensor-based data including locations, behaviours such as grazing and ruminating, and specific applications such as pasture intake over longer time periods in the commercial grazing environment, and in a timely manner, potentially in real-time, without the high labour inputs and need for repeated handling of cattle of other methods. They also have potential for commercial applications in addition to research. Development of simple initial algorithms from sensor data to classify behaviours that predicted 60% of measured variation in pasture intake by individual beef cattle has been achieved under experimental conditions (Greenwood *et al.* 2017). A range of sensor types including movement-based sensors such as accelerometers, magnetometers and gyroscopes within inertial measurement units, and acoustic and pressure sensors, attached to various on-animal locations using devices such as collars, ear tags, halters, nose-bands and leg straps have been used in studies aimed at extracting features that enable classification of cattle behaviours and other measurements from which estimates of intake may be made in intensive and extensive situations (Greenwood *et al.* 2014; Andriamandroso *et al.* 2016). Further research and developments are required to enable validation and refinement of algorithms, including more specific classification of pasture consumption events within grazing systems using sensors, development of new algorithms that include other sources of variation within and between grazed and browsed forages that impact on intake and selectivity, development of more commercially useful sensor devices, and deployment of sensor devices on large numbers of livestock in commercial grazing environments. In achieving these outcomes, issues that need to be considered include software development required for the collection of data from multiple types of sensors, the management and analyses of the very large volumes of data, determination of which sensing modalities are sufficient and/or necessary, management of the constrained power source, and generation of high-quality benchmark pasture intake and other data from which sensor data features, and behaviour classification and pasture intake algorithms can be established (Greenwood *et al.* 2014, 2017).

Emissions measurement technologies. Heat production (HP) is commonly used to assess the energy expenditure of an animal. This is based on the close relationship between HP and the process of oxidation of organic matter (feed) where oxygen (O_2) is consumed, and carbon dioxide (CO_2) and methane (CH_4) emitted (Blaxter 1962). Dry energy intake (DMI) can then be calculated by HP and a measure of the energy retained by the animal. The main challenge is accurately measuring the gases in the animal's production setting, and procedures used in the past include trachea fistulation (e.g. Flatt *et al.* 1958), isotope dilution technique (e.g. Whitelaw *et al.* 1972) and respiration pattern analysers (Young *et al.* 1975). The main drawbacks of these techniques are they cannot be used in the animal's production setting and that only a small number of animals can be measured at a time.

Sulphur hexafluoride (SF_6) tracer-gas technique. SF_6 can be used in the animal's production environment and has the potential of being used to measure a reasonable number of animals at a time. It has a minimal capacity for automation. It should be noted that while most studies have reported good accuracy for CH_4 measurement by SF_6, some studies have reported overestimation of individual animal CO_2 results, although animal rankings are maintained relative to respiration chamber measurements (Boadi *et al.* 2002; Pinares-Patiño *et al.* 2007).

Respiration chambers. These have the highest accuracy of measurement and can be used to measure large numbers of animals but cannot be used in the animal's production environment. Animals in respiration chambers usually consume less feed than they would in their normal production setting (Bickell *et al.* 2014; Herd *et al.* 2016).

In a beef cattle study, CH_4 and CO_2 production from 1,096 roughage fed beef cattle was measured in respiration chambers. Phenotypic and genetic correlations between DMI and CH_4 were 0.71 and 0.84, respectively (Donoghue *et al.* 2016). Similarly high correlations were obtained between DMI and CO_2 for the same cattle (Bird-Gardiner *et al.* 2018). While these results are promising, it should be noted that in the respiration chambers the cattle were fed 1.2 times their estimated maintenance requirements, which is lower than their expected *ad libitum* intake. These results therefore needed to be confirmed in cattle fed *ad libitum* in their production environment.

Short-term Breath Monitors. There are currently available equipment that use multiple short-term breath measurements, such as the GreenFeed Emission Monitors (GEM; C-Lock Inc., Rapid City, SD, USA). The GEM offers a good degree of automation, can be used in the animal's production environment, and can be used to measure large numbers of animals. This technology is new relative to the others and test procedures to improve the estimation of emissions are being optimised (e.g. Arthur *et al.* 2017).

In a study involving 119 heifers (in yards at Trangie) fed roughage diet *ad libitum* and 326 steers (in a feedlot near Armidale) fed feedlot diet *ad libitum* using GEM, Bird-Gardiner *et al.* 2017 reported a phenotypic correlation between DMI and CH_4 of 0.75 for the heifers and 0.62 for the steers. Phenotypic correlations between DMI and CO_2 of 0.84 for the heifers and 0.83 for the steers were reported by Arthur *et al.* (2018). Similar results were obtained in a Canadian study with heifers (Manafiazar *et al.* 2017). These results are encouraging given that of all the emissions measurement technologies, the GEM is easiest to deploy on pasture, and have been used in some pasture studies (Velasco *et al.* 2017; Gunter and Beck 2018) for purposes other than predicting pasture intake.

GENERAL CONSIDERATIONS

Irrespective of the technology used to estimate pasture intake, it will be necessary for samples of pastures consumed by animals to be taken at strategic times for analyses of potential digestibility and metabolisable energy (ME) content, to enable estimates of faecal output or energy metabolism to be converted to units of intake. Secondly, animal measurements such as liveweight, weight gain and fat scans taken while the animals are on pasture is necessary to relate intake to efficiency, and advances in automated liveweight recording will be very useful. While it may be logical to use several individual traits (e.g. CO_2 and CH_4) to produce an estimate of pasture intake, there is the possibility that using one or more of the traits directly may be a more efficient and cost-effective approach. An example may be a scenario where the genetic correlation between CO_2 and ME intake of cattle fed *ad libitum* in their production setting is greater than 0.85.

In conclusion, there is a need to capture pasture intake and efficiency in our beef genetic improvement system, given that most of the beef production systems in Australia and New Zealand are pasture based. There are emerging technologies which have capability for automation with reasonable levels of accuracy, and the ability to be deployed at pasture.

REFERENCES

Andriamandroso A.L.H., Bindelle J., Mercatoris B., and Lebeau, F. (2016) *Biotechnol. Agron. Soc. Environ.* **20**(S1): 273.

Arthur P.F., Barchia I.M., Weber, C., Bird-Gardiner T., Donoghue K.A., Herd R.M. and Hegarty R.S. (2017) *J. Anim. Sci.* **95**: 645.

Arthur P.F., Bird-Gardiner T., Barchia I.M., Donoghue K.A. and Herd R.M. (2018) *J. Anim. Sci.* **96**: 4859.

Barlow R., Ellis K.J., McLennan D.J., Costigan P. and Hirst P.H. (1990) *J. Agric. Sci. Camb.* **115**: 117.

Berry D.P. and Crowley J.J. (2013) *J. Anim. Sci.* **91**: 1594.

Bickell, S. L., Revell D.K., Toovey A.F. and Vercoe P.E. (2014) *J. Anim. Sci.* **92**: 2259.

Bird-Gardiner T., Arthur P.F., Barchia I.M., Donoghue K.A. and Herd R.M. (2017) *J. Anim. Sci.* **95**: 4391.

Bird-Gardiner T., Donoghue K.A., Arthur P.F. and Herd R.M. (2018) *Proc. WCGALP*, Poster 23.

Blaxter, K. L. (1962) The energy metabolism of ruminants. Hutchinson & Co. Ltd, London.

Boadi, D., Wittenberg K.M. and Kennedy, A.D. (2002) *Can. J. Anim. Sci.* **82**: 125.

Dicker R.W., Herd R.M., Lee G.J., Oddy V.H. and Ayres J.F. (1998) *Anim. Prod. Sci.* **22:** 355.

Donoghue, K. A., Bird-Gardiner T., Arthur P.F., Herd R.M. and Hegarty R.S. (2016) *J. Anim. Sci.* **94**: 1438.

Flatt W.P., Waldo D.R., Sykes J.E. and Moore L.A. (1958) *Eur. Ass. Anim. Prod.* No. 8: 101.

Fogarty N.M., Lee G.J., Ingham V.M., Gaunt G.M. and Cummins L.J. (2006) *Aust. J. Agric. Res.* **57:** 1037.

Greenwood P.L., Valencia P., Overs L., Paull D.P. and Purvis I.W. (2014) *Anim. Prod. Sci.* **54**: 1796.

Greenwood P.L., Paull D.R., McNally J., Kalinowski T., Ebert D., Little B., Smith D.V., Rahman A., Valencia P., Ingham A.B. and Bishop-Hurley G.J. (2017) *Crop Pasture Sci.* **68**: 1091.

Gunter S.A. and Beck M.R. (2018) *Transl. Anim. Sci.* **2**: 11.

Herd, R.M., Velazco J.I., Arthur P.F. and Hegarty R.S. (2016) *J. Anim. Sci.* **94**: 4882.

Herd R.M., Dicker R.W., Lee G.J., Johnston D.J., Hammond A.J. and Oddy V.H. (2004) *Anim. Prod. Aust.* **25**: 93.

Lee G.J., Atkins K.D. and Mortimer S.I. (1995) *Livest. Prod. Sci.* **41:** 143.

Lee G.J., Atkins K.D. and Swan A.A. (2002) *Livest. Prod. Sci.* **73:** 185.

Manafiazar, G., Zimmerman S., and Basarab J.A. (2017) *Can. J. Anim. Sci.* **97**: 118.

Mayes R.W. and Dove H. (2000) *Nutr. Res. Reviews* **13:** 107.

Pinares-Patiño C.S., D'Hour P., Jouany J-P. and Martin C. (2007) *Agric. Ecosyst. Environ.* **121**: 30.

Velazco J.I., Herd R.M., Cottle D.J. and Hegarty R.S. (2017). *Anim. Prod. Sci.* **57**: 627.

Whitelaw F.G., Brockway J.M. and R. S. Reid. (1972). *Q. J. Exp. Physiol. Cogn. Med. Sci.* **57**: 37.

Young B.A., Kerrigan, B. and Christopherson R.J. (1975). *Can. J. Anim. Sci.* **55**: 17.

THE EFFECT OF INCLUDING STATURE IN SIRE SELECTION ON THE LIVE WEIGHT, MILK YIELD, FERTILITY AND FEED EFFICIENCY OF HOLSTEIN COWS

C.J.C. Muller[1], S.W.P. Cloete[1,2] and M. Burger[2]

[1]Faculty of AgriSciences, University of Stellenbosch, Stellenbosch, 7602 South Africa
[2]Directorate Animal Sciences, Western Cape Department of Agriculture, Elsenburg, 7607 South Africa

SUMMARY

The efficiency of milk production is increasingly under investigation to improve the profit margins of dairy herds. Preliminary results on the live weight (LW), production, fertility and efficiency of first parity Holstein cows descended from sires selected for (i) estimated breeding values (EBVs) for combined fat and protein yield (Prod) or (ii) EBVs fat plus protein yield and the lowest (negative) stature values (Stat) in a pasture-based production system are presented in this paper. Progeny from Stat bulls weighed less while also producing less milk ($P<0.05$) in first lactation. Fertility traits did not differ in heifers and cows from Prod and Stat sires. Efficiency was higher ($P<0.05$) in the progeny of sires selected for production compared with the progeny of sires selected for production and stature.

INTRODUCTION

Feed efficiency in dairy cows is a complex trait as cows undergo lactation cycles marked by declines and increases in milk yield, body reserves and LW (Connor 2015; Roche *et al.* 2009). Breeding and selection programmes in dairy herds have always focused on production and conformation traits (Walsh *et al.* 2007). Because of genetic improvement, better housing, improved feeding and general overall management, the milk yield of dairy cows in most countries has doubled over the last 30 years (Capper *et al.* 2009). This has indirectly resulted in an increased dairy feed efficiency (VandeHaar and St-Pierre 2006). However, higher milk yields have resulted in cows becoming larger and heavier. Berry *et al.* (2003) found that genetic correlations between LW at different stages of the lactation and total lactation milk production were close to zero although becoming positive (0.01 to 0.39) after adjusting LW for body condition score. Furthermore, to perform well in cattle showing, Holstein breeders put a strong emphasis on size. These factors had a positive effect on the LW and body size of dairy cows. This is problematic for cows in housing systems designed and built earlier. For pasture-based dairying systems, farmers suggest that larger cows have difficulty in walking large distances to and from pastures especially on uneven and hilly terrain. Maintenance requirements of cows also increase with size. It is expected that efficiency would be reduced when milk yield does not follow an increase in LW.

Hansen *et al.* (1999) in the USA reported results from a long-term study starting in 1966 comparing large and small Holstein cows in a total mixed ration feeding system. Sires were selected on predicted transmitting ability for stature, strength and body depth using an index consisting of (0.5 x stature) + (0.25 x strength) + (0.25 x body depth). Live weight and body measurements of cows differed significantly between sire groups but there was a large overlap of body size across the two genetic lines. Milk yield did not differ between sire lines resulting in poorer efficiency measures for large cows. Reproduction did not differ between genetic lines ($P>0.05$) although all traits numerically favoured the small line.

No research has been conducted in South Africa on the effect of sire selection regarding cow size. The aim of the study was to compare the effect of sire selection based on (i) EBVs for fat and protein yield (Prod) and (ii) yield (Prod) + lowest negative stature on the LW, production, some efficiency measures and reproductive performance of Holstein cows in a pasture-based system.

MATERIALS AND METHODS

Data. The study was conducted at the Elsenburg Research Farm of the Western Cape Department of Agriculture. The area has a typical Mediterranean climate with short, cool, wet winters and long, warm, dry summers with an average annual rainfall of about 630 mm. Two groups of five Holstein bulls each were selected annually for artificial insemination based on (i) EBVs for combined fat and protein yield (Prod) and (ii) within the top 25 bulls ranked for EBV for fat and protein yield, sires with the lowest negative values for stature (Stat). At the start of the trial, 120 Holstein cows were divided into two groups based on EBV for milk yield and stature scored by a trained mating programme technician. Bulls were matched to cows within each group using a commercial mating programme to control inbreeding. Heifers born from the two sire groups were reared similarly to first calving. After calving, cows were put in a kikuyu-ryegrass pasture-based production system supplemented with a commercial concentrate mix at 7 kg/cow/day. Fresh drinking water was freely available at all times. Walking distance to the pasture varied from 1-3 km/day. Cows were machine-milked twice a day.

The milk yield of cows at the evening and following morning's milking was recorded according to standard milk recording procedures. At each recording event, milk samples were collected at both the evening and morning milking sessions. Samples were combined and analysed at the milk testing laboratory of the National Milk Recording Scheme for fat, protein and lactose content.

Heifers were weighed at birth and thereafter once a month until first calving. Body size measurements (girth circumference, shoulder height, body length, depth and width) of heifers were recorded at fixed stages during the rearing period until first calving. The LWs of cows were recorded within one week before and after calving. During the lactation period cows were weighed after each milking using a walk-over scale. Insemination dates and pregnancy check results by rectal palpation by a veterinarian were recorded for both heifers and cows. The following efficiency measures were estimated:

1. Component values (CV):	((fat yield x 6)+(protein yield x 13))/2
2. Alternative Kleiber ratio: (Hurley *et al.* 2014)	(daily milk yield x ((0.0929 x fat%) + (0.0547 x protein%) + (0.0395 x lactose%)) / metabolic live weight ($LW^{0.75}$)
3. FCM/LW: (Gaines *et al.*, 1940)	4% fat corrected milk yield (FCM)/ live weight
4. MY/LW:	Lactation milk yield / live weight
5. DMi/day: (NRC, 2001)	DM intake (kg/day) = (0.372 x FCM) + (0.0968 x $LW^{0.75}$) x $(1 - e^{(-0.192 \times (\text{Week of lactation} + 3.67))})$
6. MY/DMi	Milk yield(kg)/kg DMi

Statistical analysis. Analysis of variance was performed, using cows as random replicates, using GLM Procedure of SAS software (Version 9.4; SAS Institute Inc, Cary, USA) to test the effect of breed on production parameters and efficiency measures. Third order polynomial trend lines were fitted for body size measurements of heifers.

RESULTS AND DISCUSSION

Body size measurements for heifers from different sire groups did not differ (P>0.05) until 22 months of age when growth rate curves showed a divergence (Figure 1). While shoulder height of heifers from Prod and Stat sires did not differ (P>0.05), chest circumference differed (P<0.05), being 188 vs. 192 cm, respectively. At first calving heifers from Stat sires weighed less (P<0.05) than Prod group heifers, i.e. 610±65 vs. 647±60 kg, respectively.

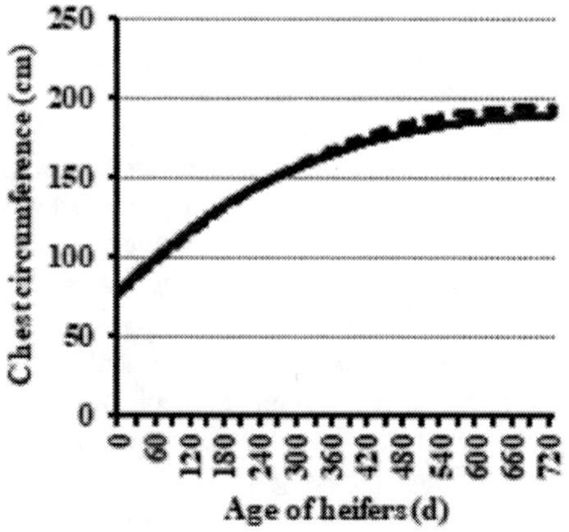

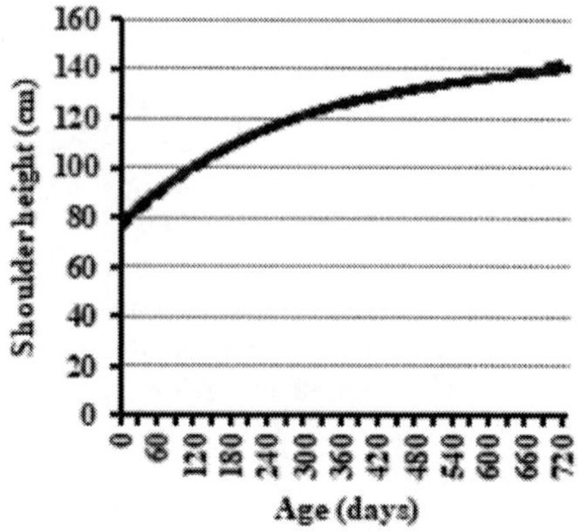

Figure 1. The chest circumference and shoulder height of heifers from production (---) and stature (—) sires from birth to 24 months of age

While the mean LW of cows from Prod and Stat sires differed (P<0.05), a considerable overlap was observed (Table 1) similar to the results reported by Hansen *et al.* (1999). Graham *et al.* (1991) showed that New Zealand sires produced first lactation cows similar in LW to Canadian sires, i.e. 533 vs. 528 kg, respectively. Test-day milk yield of Stat cows was less (P<0.05) than that of cows from Prod sires. While the estimated dry matter (DM) intake of Stat cows was lower, efficiency measures were also compromised (P<0.05) relative to progeny of Prod sires.

Table 1. Means (±SEM) for production parameters and efficiency measures for first parity progeny of Holstein sires selected for production (Prod) and production + stature (Stat) in a pasture-based system (FCM: fat corrected milk yield; CV: component value; KR: Kleiber ratio; MY: milk yield; LW: live weight; DMi: dry matter intake; ^{a,b}Values differ at P<0.05)

Production parameters	Sire selection		Efficiency measures	Sire selection	
	Prod	Stat		Prod	Stat
Cows (Test days)	36(199)	24(132)	LW (kg)	546^{a}±4.4	520^{b}±5.0
MY (kg/d)	23.0^{a}±0.4	20.8^{b}±0.4	CV (kg)	7.51^{a}±0.13	6.61^{b}±0.14
FCM (kg/d)	23.0^{a}±0.4	20.1^{b}±0.4	KR (kg)	0.153^{a}±0.036	0.139^{b}±0.032
Fat (%)	4.01^{a}±0.05	3.80^{b}±0.05	FCM/LW (kg)	0.042^{a}±0.001	0.039^{b}±0.001
Protein (%)	3.19±0.02	3.16±0.02	MY/LW (kg)	0.042±0.001	0.040±0.001
Lactose (%)	4.90^{a}±0.01	4.78^{b}±0.01	DMi (kg/d)	18.0^{a}±0.2	16.9^{b}±0.2
Fat (kg/d)	0.92^{a}±0.02	0.79^{b}±0.02	MY/DMi(kg)	1.29±0.02	1.24±0.02
Protein (kg/d)	0.73^{a}±0.01	0.65^{b}±0.01			

Fertility traits in heifers did not differ (P>0.05) between sire lines although favouring the Stat line, i.e. age at first calving was 24.1 vs. 23.6 months while the proportion of heifers calving down before 24 months of age was 0.58 and 0.73 for Prod and Stat sires, respectively. Hansen *et al.* (1999) showed better fertility for small cows in comparison to large cows.

The body weight of cows is associated with size and is the most practical measurement in dairy cows (Arango & Plasse 2002). However, Berry *et al.* (2004) showed moderate to strong genetic correlations between stature and milk yield and stature and average body weight of 0.42 and 0.63 respectively. These genetic correlations may reflect past emphasis on milk yield together with increased cow stature, body width, depth and angularity in Holstein breeding programmes. This has resulted in taller, wider and deeper cows that tend to be more angular with less body condition. Moderate to strong genetic correlations of most of the body-related type traits with average body weight reflect an element of the size of animals. As cows are often not routinely weighed, body-related traits could be used to predict live weight. The USA Holstein Association (2017) has developed an efficiency equation using PTA for milk, fat and protein yields together with a body weight composite trait which includes stature, strength, body depth, rump width and dairy form.

A larger source of sires for selection may have resulted in reducing the difference in milk yield potential while reducing body size. Currently no EBVs for the LW of the daughters of bulls are available. The difficulty in recording the LW of cows may be overcome by using automatic milk recording systems which also record cow LW after each milking. Determining the actual feed intake of cows would make it possible to compare sire groups on residual feed intake.

CONCLUSION

This study reported differences in the growth rates and production performance of the progeny of sires selected for production and production plus stature in a pasture-based system. Cows from Stat bulls weighed less than cows from Prod bulls while also producing less milk during first lactation. This reduces efficiency measures for Stat sires progeny.

ACKNOWLEDGEMENTS

This research was funded by the Western Cape Agricultural Research Trust and the Western Cape Department of Agriculture. The inputs of the dairy management team at the Elsenburg Research Dairy are greatly appreciated.

REFERENCES

Arango J. and Plasse D. (2002) *7th World. Congr. Genet. Appl. Livest. Prod.* Aug 19-23, 557.

Berry D.P., Buckley F., Dillon P., Evans R.D. and Veerkamp R.F. (2004) *Irish J. Agric. Food Res.* **43**: 161.

Berry D.P., Buckley F., Dillon P., Evans R.D., Rath M. and Veerkamp R.F. (2003) *J. Dairy Sci.* **86**: 2193.

Capper J.L., Cady R.A. and Bauman D.E. (2009) *J. Anim. Sci.* **87**: 2160.

Connor E.E. (2015) *Animal* **9:** 395.

Gaines W.L., Rhode C.S. and Cash J.G. (1940). *J. Dairy Sci.* **23**: 1031.

Graham N.J., Burnside E.B., Gibson J.P., Rapitta A.E. and McBride B.W. (1991) *Can. J. Anim. Sci.* **2**: 293.

Hansen L.B., Cole J.B., Marx G.D. and Seykora A.J. (1999) *Adv. Dairy Techn.***11**: 39.

Hurley A.M., McParland S., Kennedy E., Lewis E., McDonovan M., López-Villalobos N. and Berry, D.P. (2014). *Proc. 10th World. Congr. Gen. Appl. Livest. Prod.* Vanvouver, BC, Canada.

Roche J.R., Friggens N.C., Kay J.K., Fisher M.W., Stafford K.J. and Berry, D.P. (2009) *J. Dairy Sci.* **92**: 5769.

USA Holstein Association (2017) www.holsteinusa.com/pdf/Upcoming_Changes_aug17.pdf.

VandeHaar M.J. and St-Pierre N. (2006) *J. Dairy Sci.* **89**: 1280.

Walsh S., Buckley F., Berry D.P., Rath M., Pierce K., Byrne N. and Dillon P. (2007) *J. Dairy Sci.* **90**: 5769.

GENETIC PARAMETERS OF MILK LACTATION CURVE TRAITS OF THAI DAIRY CATTLE

S. Pangmao[1], P.C. Thomson[2] and M.S. Khatkar[1]

[1]Sydney School of Veterinary Science, University of Sydney, 425 Werombi Road, Camden, NSW, 2570 Australia

[2]School of Life and Environmental Sciences, University of Sydney, 425 Werombi Road, Camden, NSW, 2570 Australia

SUMMARY

The aim of this study was to estimate genetic parameters of lactation traits, viz. 305-day milk yield, peak milk yield, days to peak milk yield and persistency in Thai dairy cattle from three government farms. Milk yield and peak milk yield have moderate heritability suggesting these can be improved by selective breeding while days to peak milk yield and persistency have low heritability which can be improved by adjusting environmental management such as feed and farm management. The genetic correlations of most of these traits across lactations were high suggesting that selection based on the first lactation will improve the first three lactations in these herds.

INTRODUCTION

Lactation curve is important for prediction of milk yield and for feeding management in lactating cows. The inclusion of accurate description of lactation curve can be very useful for breed improvement programs. The test interval method (TIM) is an empirical-based method and is one of the standard methods approved by the International Committee for Animal Recording (ICAR) for calculating 305-day milk yield (ICAR 2019, Schaeffer and Jamrozik 1996). In addition, there are many mathematical models that have been used to describe lactation curves. The Wood model (Wood 1967) has been widely used (Dijkstra *et al.* 2010) due to its simple form and ability to capture key features of the lactation curve (Tozer and Huffaker 1999) and more robust compared to other models when fitting lactation curve to irregular and infrequent test-day sampling regimes (TDSR) (McGill *et al.* 2013). This study used data from three dairy cattle research centres recorded by the Department of Livestock Development of Thailand. The objectives of this study were to estimate the genetic parameters, including heritability, genetic and phenotypic correlations of lactation curve traits across lactations of Thai dairy cattle in three government research farms.

MATERIALS AND METHODS

Description of data and animal management. The data from three Thai government dairy cattle farms provided by the Bureau of Animal Husbandry and Genetic Improvement (BAHGI), Department of Livestock Development (DLD) in Thailand were used for this study. Two farms are located in Chiang Mai province and Nakhon Ratchasima province raised both Holstein-Friesian (HF) and crossbred HF while the third farm located in Sakon Nakhon province raised only crossbred HF. Upgrading local or zebu breeds with HF semen or natural mating with HF bulls has been used to improve productivity and maintain tropical insect and disease tolerance in these herds.

All animals in the three farms were raised under the same guidelines given by BAHGI. Nevertheless, some aspects were different such as feeding and health management because of the differences of locations, weather conditions, feed resources, farm machinery and diseases prevalence. Fresh Ruzi and Napier grass were fed during the rainy season while in the dry season, Ruzi silage and hay were fed (nutrient components of roughage and concentrate feed using in dairy cattle farm as described in Pangmao *et al.* (2017)).

Data and statistical analysis. The data comprised of 25,789 monthly test day milk yield records of 1,468 cattle from lactation 1 to 3 of HF and crossbred HF dairy cattle who calved between 1990 and 2015. The total number of animals, sires and dams in the pedigree file were 4,753; 287 and 1,237, respectively. The records of cows with less than three test day milk yield records were excluded for further analysis. Five breed groups of animals were classified based on the percentage of HF blood as 1 (≤ 75), 2 (>75 – 87.5), 3 (>87.5 – 93.75), 4 (>93.75 – 99.99) and 5 (100), as calculated from the pedigree information. The calving months were grouped into three seasons, namely winter (November to February), summer (March to June) and rainy (July to October). The traits examined were accumulated 305-day milk yield calculated by using the adapted test interval method (Sargent *et al.* 1968) and Wood model (Wood 1967) in the form of non-linear model (MT and MW), peak milk yield (PY), days to peak milk yield (DP) and persistency (PS). The 305-day milk yield calculated by using the adjusted test interval method as follows:

$$MT = I_0 M_1 + I_1 * \left(\frac{M_1 + M_2}{2}\right) + I_2 * \left(\frac{M_2 + M_3}{2}\right) + \ldots + I_{n-1} * \left(\frac{M_{n-1} + M_n}{2}\right) + \frac{I_n M_n}{2}$$

where $M_1, M_2, \ldots, M_n$ are the milk yields (kg) recorded in the 24 hours of the recording day and given to one decimal place, $I_1, I_2, \ldots, I_{n-1}$ are the intervals in days between recording dates, I_0 is the interval in days between the lactation period start date and the first recording date, and I_n is the interval in days between the last recording date and the end of the lactation period. The nonlinear Wood model is as follows:

$$W(t; k, b, c) = \exp(k + b\log_e t - ct)$$

where $W(t)$ is the theoretical or expected milk yield at time t, k is a scalar factor, b is the rate of increase prior to the peak and c is the rate of decrease after the peak. Cumulative milk yield to day T (say day 305) was then obtained as $MW = \int_0^T W(t)\,dt$. Fitting of the Wood model was conducted using the *nlme* library in R (Pinheiro *et al.*, 2018) and calculation of cumulative milk yield through use of the *pgamma* function in R. The parameters derived from Wood model were used to calculate lactation curve characteristics namely peak milk yield ($PY = e^k(b/c)^b e^{-b}$), days to peak milk yield ($DP = b/c$) and persistency ($PS = -(b+1)\log_e c$).

Estimates of genetic parameters were obtained with an animal model, formulated as a linear mixed model fitted to each of the five traits. Univariate models were used to estimate variance components and heritability estimates of each trait, while bivariate models were used to estimate covariance components and correlations estimates between a trait in two different lactations. Model fitting was conducted using ASReml-R (Butler *et al.* 2009). The univariate model used in the analysis is as follows:

$$y_{ijkl} = \mu + \text{Herd}_i + \text{YSOC}_j + \text{BG}_k + \text{Anim}_l + e_{ijkl}$$

where y_{ijkl} is an observation of trait on animal l; μ is the overall mean; Herd_i is the fixed effect of herd (level, 1-3); YSOC_j is the fixed effect of year and season of calving (level, 1-78); BG_k is the fixed effect of breed group (level, 1-5); Anim_l is the random animal effect; and e_{ijkl} is the random residual effect.

RESULTS AND DISCUSSION

The descriptive summary of milk yield and lactation characteristic traits for lactation 1 to 3 is shown in Table 1. Both accumulated milk yields (MT and MW) were similar within all the three lactations. The lowest MW in lactation 1 compared with lactation 2 and 3 (3,348 ± 1282 kg vs 3,685 ± 1,503 kg and 3,564 ± 1,317 kg, respectively) was in agreement with Hossein-Zadeh (2014) (9,186 kg vs 10,386 kg and 10,000 kg, respectively) which may be due to partition of nutrition for growth and milk production in lactation 1 cows. The highest PS in lactation 1 was in agreement with Gengler (1996) who reported higher persistency in first lactation than the other lactations.

Table 1. Descriptive summaries of the lactation curve traits from lactation 1 to 3 of Thai dairy cattle (mean ± standard deviation (number of observations))

Traits[a]	lactation 1	lactation 2	lactation 3
MT	3,379 ± 1,409 (1,301)	3,682 ± 1,604 (949)	3,599 ± 1,534 (665)
MW	3,348 ± 1,282 (1,210)	3,685 ± 1,503 (906)	3,564 ± 1,317 (636)
PY	14.95 ± 4.97 (1,210)	17.52 ± 6.48 (906)	17.49 ± 6.09 (633)
DP	49 ± 26 (1,209)	39 ± 18 (904)	41 ± 17 (638)
PS	6.68 ± 0.48 (1,209)	6.43 ± 0.39 (906)	6.43 ± 0.41 (638)

[a]MT: milk yield calculated by test interval method; MW: milk yield calculated by Wood model; PY: peak milk yield; DP: days to peak milk yield and PS: persistency

The estimated heritabilities for lactation curve traits of lactation 1 to 3 are shown in Table 2. In general, the heritability estimates of all traits from first parity cows were greater than or equal to those from parities 2 and 3. The heritability estimate of MW was higher in lactation 1 compared to lactation 2 and 3 (0.21 vs 0.01 and 0.08, respectively) while for MT, heritability is similar between lactation 1 (0.19) and 3 (0.19) but lower in lactation 2 (0.12). The heritability estimates of MT and MW in lactation 1 was similar to Mohammed *et al.* (2013) (0.24) but lower compared to the reported in other studies (0.35: König *et al.* 2005 and 0.34: Sarakul *et al.* 2011). The lower milk yield heritability in this study compared to other studies may be due to the low estimate of additive variance and high estimate of residual variance suggesting this trait was highly affected by the environmental factors such as farm and feed management, hot and humid tropical environment. In addition, the different size of the data and the models used for analysis also might have an effect on estimation of variance components and heritability estimates. Nonetheless, milk yield calculated from both methods show the potential of improvement by selection program. Lactation curve traits' heritability estimates (PY, DP and PS) were low to moderate ranging from 0.00 to 0.23 for lactation 1 to 3. The heritability of PY and DP in lactation 1 was similar to Chegini *et al.* (2015) (0.23 vs 0.26 and 0.10 vs 0.10) although PS was higher (0.14 vs 0.05). In general, the heritability estimates of lactation curve traits for all lactations were low except the moderate heritability of PY in lactation 1 and 3 (0.23 and 0.17 respectively) which means only PY can be improved by selective breeding while DP and PS trait may be improved with better management practices.

Table 2. Estimates of heritability (± standard errors) of the lactation curve traits from lactation 1 to 3 of Thai dairy cattle

Traits[a]	lactation 1	lactation 2	lactation 3
MT	0.19 ± 0.06	0.12 ± 0.07	0.19 ± 0.10
MW	0.21 ± 0.06	0.01 ± 0.05	0.08 ± 0.09
PY	0.23 ± 0.06	0.09 ± 0.07	0.17 ± 0.10
DP	0.10 ± 0.05	0.01 ± 0.05	0.00 ± 0.00
PS	0.14 ± 0.06	0.04 ± 0.06	0.00 ± 0.00

[a]MT: milk yield calculated by test interval method; MW: milk yield calculated by Wood model; PY: peak milk yield; DP: days to peak milk yield and PS: persistency

Table 3 shows the genetic and phenotypic correlation estimates of lactation curve characteristics between lactation 1 to 3. Most of the genetic correlation estimates of the traits were high (0.75 to 0.99). The high genetic correlation estimates of all traits between two different lactations suggested that the selection of animal for first lactation curve traits in these herds will improve traits in the second and third lactation as well, although the phenotypic correlations for most of the traits were low and negative for PS.

Table 3. Genetic and phenotypic correlation (in parenthesis) (± standard errors) of lactation curve traits of Thai dairy cattle between traits in two different lactations

Traits[a]	lactation 1 vs lactation 2	lactation 1 vs lactation 3	lactation 2 vs lactation 3
MT	0.96 ± 0.11 (0.32 ± 0.05)	0.92 ± 0.16 (0.28 ± 0.05)	0.97 ± 0.26 (0.53 ± 0.05)
MW	0.99 ± 0.06 (0.26 ± 0.05)	0.87 ± 0.11 (0.29 ± 0.05)	0.94 ± 0.07 (0.15 ± 0.06)
PY	0.99 ± 0.06 (0.40 ± 0.04)	0.75 ± 0.12 (0.26 ± 0.05)	0.94 ± 0.11 (0.32 ± 0.06)
DP	0.92 ± 1.07 (0.08 ± 0.05)	NA	NA
PS	0.85 ± 0.74 (0.06 ± 0.05)	NA (-0.02 ± 0.06)	NA (-0.02 ± 0.06)

[a]MT: milk yield calculated by test interval method; MW: milk yield calculated by Wood model; PY: peak milk yield; DP: days to peak milk yield and PS: persistency

CONCLUSIONS

The results from this study suggested that two methods of calculation of accumulated 305-day milk yield were quite similar with moderate heritability estimates for lactation 1 which means that milk yield in these herds can be improved by selective breeding. However, only peak milk yield will respond to selective breeding for lactation curve traits of cows in lactation 1. The high genetic correlation estimates of most lactation traits of the first lactation with the second and third lactation suggested that the selection of animals based on the first lactation could increase the overall milk yield in these three government farms in Thailand.

ACKNOWLEDGEMENTS

The authors are grateful to the Department of Livestock Development of Thailand for providing the data. SP was supported by the Royal Thai government through a postgraduate fellowship.

REFERENCES

Butler, D. G., Cullis, B. R., Gilmour, A. R. and Gogel, B. J. (2009) ASReml-R Reference Manual. Queensland Department of Primary Industries and Fisheries: Brisbane.

Chegini, A., Shadparvar, A. A. and Ghavi Hossein-Zadeh, N. (2015) *Iran. J. Appl. Anim. Sci.* **5:** 61.

Dijkstra, J., Lopez, S., Bannink, A., Dhanoa, M. S., Kebreab, E., Odongo, N. E., Fathi Nasri, M. H., Behera, U. K., Hernandez-Ferrer, D. and France, J. (2010) *J. Agric. Sci.* **148:** 249.

Gengler, N. (1996) *Interbull Bulletin* **12:** 87.

Hossein-Zadeh, N. G. (2014) *J. Agric. Sci.* **152:** 309.

ICAR, 2019. International Committee for Animal Recording [Online]. Available at: https://www.icar.org/index.php/icar-recording-guidelines/ (verified 8 April 2019)

König, S., Chongkasikit, N. and Langholz, H. J. (2005) *Arch. Anim. Breed.* **48:** 233.

McGill, D. M., Thomson, P. C., Mulder, H. A. and Lievaart, J. J. (2013) *Proc. Assoc. Advmt. Anim. Breed.* **20:** 74.

Mohammed, E., Tumwasorn, S., Sopannarath, P. and Prasanpanich, S. (2013) *ANRES* **47:** 228.

Pangmao, S., Thomson, P. C. and Khatkar, M. S. (2017) *Anim. Prod. Sci.* **57:** 1242.

Pinheiro, J., Bates, D., DebRoy, S., Sarkar, D. and Team, a. R. C. (2018).

Sarakul, M., Koonawootrittriron, S., Elzo, M. and Suwanasopee, T. (2011) *Asian-Aust. J. Anim. Sci.* **24:** 1031.

Sargent, F. D., Lytton, V. H. and Wall, O. G., Jr. (1968) *J. Dairy. Sci.* **51:** 170.

Schaeffer, L. R. and Jamrozik, J. (1996) *J. Dairy. Sci.* **79:** 2044.

Tozer, P. R. and Huffaker, R. G., (1999) *Aust. J. Agric. Res.* **50:** 431.

Wood, P. D. P. (1967) *Nature* **216:** 164.

GENETIC VARIATION AND ESTIMATING BREEDING VALUES FOR SMALL-HOLDER CROSSBRED DAIRY CATTLE IN INDIA

M. Al kalaldeh[1], Y. Gaundare[2], M. Swaminathan[2], S. Joshi[2], H. Aliloo[1], E.M. Strucken[1], V. Ducrocq[3] and J.P. Gibson[1]

[1]University of New England, School of Environmental and Rural Science, Armidale, NSW, 2350 Australia
[2]BAIF Development Research Foundation, Central Research Station, Uruli-Kanchan, Pune, Maharashtra, India
[3]INRA 78352 Jouy-en-Josas Cedex, France

SUMMARY

We report the results of the first large scale milk recording and genetic evaluation for crossbred cows in a smallholder dairy production system in India. Preliminary results represented 8,144 smallholder crossbred cows with a total of 140,214 daily milk records, of which 2,946 animals were genotyped with the GGP Bovine and Illumina SNP assays. Data were adjusted for fixed effects and analysed with a random regression (RR) model with the 1st degree Legendre polynomial and heterogeneous variance. Heritabilities of milk yield ranged from 0.14 to 0.22 throughout the lactation period, with an average value of 0.19. Genomic Estimated breeding values (GEBV) for the genotyped animals including the smallholder crossbred cows and the bulls and dams from the BAIF bull stud ranged from +1.9 to -1.4 kg/day. The moderate heritability of the milk yield found in our results together with the wide range of GEBV, indicate that a good response to genomic selection for milk yield can be expected for smallholder dairy farms in India.

INTRODUCTION

Genomic selection is now widely applied in the dairy industry, which has resulted in a rapid rate of genetic gain for dairy cattle breeds in developed countries. In India, the world's largest milk producer, the majority of milk production is undertaken by smallholders, who mostly keep one to three cows. These cows are mainly multi-generation crosses between indigenous Indian cattle and exotic dairy breeds, which have no pedigree information. Consequently, implementing traditional genetic evaluation for these smallholder cows has not been possible. Use of genomic selection based on genome-wide single nucleotide polymorphisms (SNPs) and phenotype recording in smallholder herds offers a route to implementing genetic improvement in such populations. The objective of this study was to estimate genetic parameters for milk yield and the genomic estimated breeding values (GEBV) for a large genetic gains project initiated by BAIF, a large NGO serving smallholder farmers in India.

MATERIALS AND METHODS

A mixed model analysis using random regression to estimate genetic parameters and GEBV was performed on animals with genotypes. These included 2,946 smallholder crossbred cows with milk records plus 653 bulls and cows from the BAIF bull stud. 2,389 smallholder crossbred cows were genotyped with the GGP Bovine 50K array and 557 crossbred cows were genotyped with the Illumina 780k BovineHD array. 496 animals from the BAIF bull stud animals were genotyped with the Illumina BovineSNP50 array and the rest of the bulls (157) were genotyped with the GGP Bovine 50K array. SNPs were excluded if they had a call rate (CR) < 0.90 and if they are located on the X and Y chromosomes. Individual samples with CR < 0.90 were excluded. Furthermore, an individual sample for which the correlation between its genotype and that of another sample > 0.98 was removed.

The genotypes of all animals across the different arrays were then combined to impute the medium density genotypes up to the high-density using 2,961 reference animals. The reference set for imputation included 1,309 indigenous plus crossbred cattle, 968 Holsteins and 684 Jerseys all genotyped on the Illumina BovineHD array (Illumina, San Diego, CA). The imputation was performed using FImpute v2.2 (Sargolzaei *et al* 2014) and resulted in 737,073 SNP genotypes across 29 Bos *taurus* autosomes. A prior analysis using data for all animals with phenotypes (8,144 cows with 140,214 milk records) corrected milk records for fixed effects. Cows were included in the analysis when they had at least five monthly records in at least one lactation and there was a minimum of 10 cows per AI distribution centre. Milk records were corrected for fixed effects, including parity, season, centre, the interaction of the season by centre, the average lactation curve of the population modelled by a 3rd-order Legendre polynomial, the lactation curves for parities modelled by the 3rd order Legendre polynomials, and the lactation curves for centres modelled by the 3rd order Legendre polynomials. In matrix notation, the fitted random regression model was:

$$y = Xb + Z_1 a + Z_2 p + Z_3 h + e,$$

where y is a vector of corrected milk records, b is a vector of fixed effects. The genetic model included fixed effects for the three environments (high, medium and low production environments, based on the estimates for the centre and the herd in which each cow sits), five crossbreed groups (0-20%, 20-40%, 40-60%, 60-80% and 80-100% exotic dairy breed as estimated from SNP genotypes), and the interaction between environment and breed composition; a, p, and m are vectors of animal additive genetic, permanent environmental, and household effects; X, Z_1, Z_2, and Z_3 are incidence matrices of fixed, additive genetic, permanent environmental, and household effects; and e contains the residual effects. Legendre polynomials of the 1st order (α_0 and α_1) were fitted to the animal additive genetic, animal permanent environmental and farm random effects. Days of lactation used to estimate Legendre polynomials ranged from 8 to 340. Residual variance was assumed to be homogenous for test-day records within, but heterogeneous between eight lactation period classes: 8-49, 50-91, 92-133, 134-175, 176-217, 218-259, 260-301, and 302-340 days. Variance components and GEBV were estimated using restricted maximum likelihood in Wombat software (Meyer 2007).

RESULTS AND DISCUSSION

Figures 1 and 2 show the estimated lactation curves for parities (the first 7 parities) and centres (81 centres) within the six states. The variation in lactation curves between Centres is much greater than that between lactations, perhaps reflecting the smaller number of cows per Centre (ranging 10 to 224) but likely also the substantial differences in true lactation shape between the different environments represented by different Centres. Some estimated lactation curves have implausible shapes. For example, those with extreme curvature and peak yields in mid lactation seen for several Centres in Punjab, Maharashtra and Jharkhand states. Unusual shapes of curves could result if cows are not evenly balanced across the lactation. However, examination of the most extreme curves did not indicate that this was a major factor.

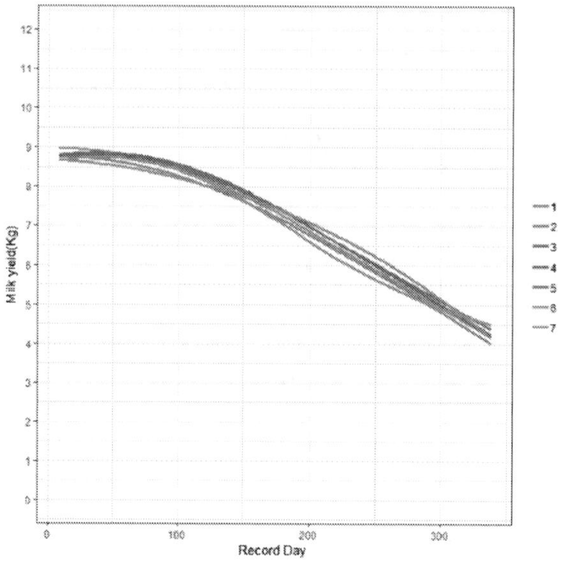

Figure 1. Estimated lactation curves for the first seven parities

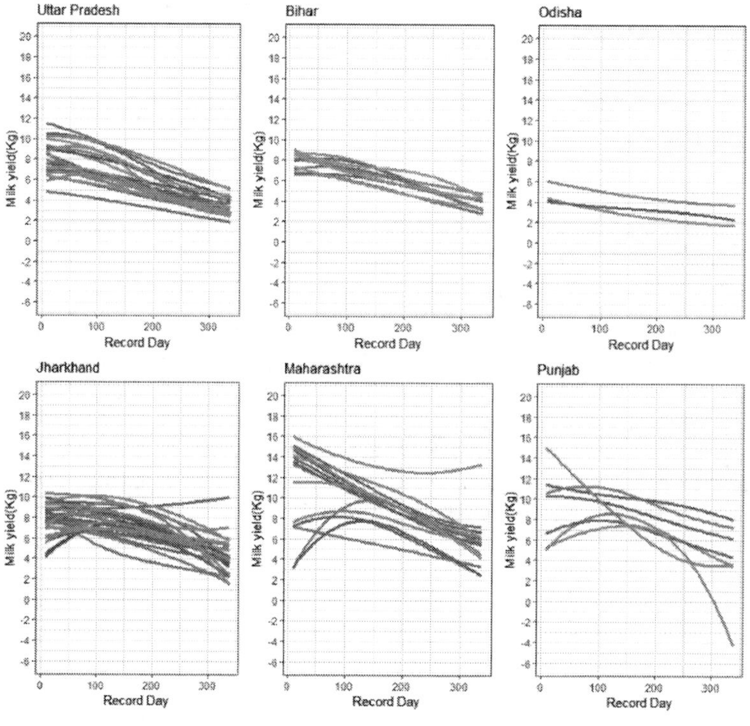

Figure 2. Estimated lactation curves for centres in each state

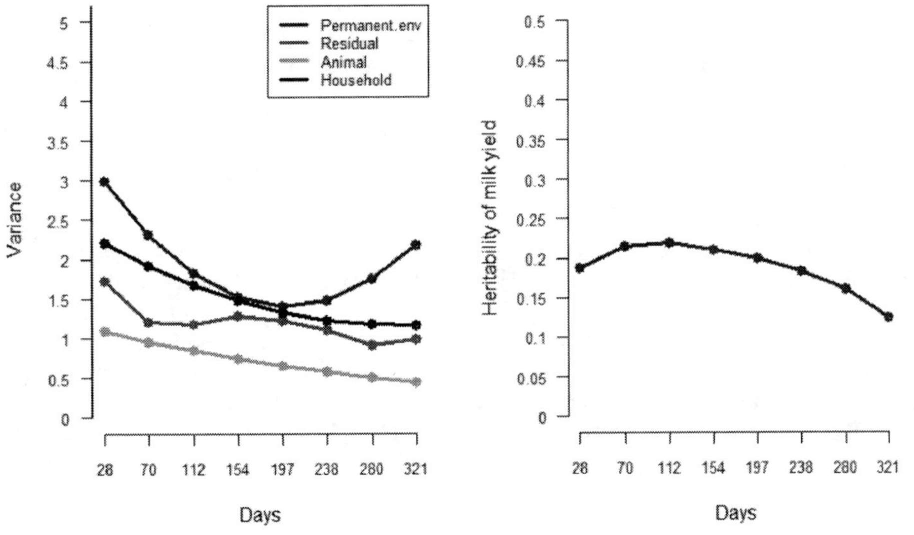

Figure 3. heritability estimates of milk yield over 28, 70, 112, 154, 197, 238, 280, and 321 days of lactation. The heritability was estimated as the proportion of additive genetic variance relative to the sum of the additive genetic, permanent environment, and residual variances

Heritability estimates of milk yield (Figure 3) ranged from 0.14 to 0.22 throughout the lactation period with an average value of 0.19. The highest heritability was observed at around three to four months of lactation. GEBVs for the genotyped animals including the bulls and cows from the BAIF bull stud and the cows from the smallholders were estimated. The range of GEBV of smallholder crossbred cows for average daily yield ranged from +1.9 to -1.35 kg/day, whereas the range of GEBV of bull stud animals ranged from +1.4 to -1.3kg/day. Since the average daily milk yield in this population is 7.5 kg/day, this indicates that there is good opportunity for substantial genetic gain through selection based on GEBV for crossbred performance.

ACKNOWLEDGEMENT

We acknowledge the Bill and Melinda Gates Foundation grant OPP1112185. The contributions of J. Khadse, V. Podtar, Y. Gaundare, K. Bhave, S. Jhadav, S. Khade, V. Dhanikachalam & P. Deshpande in collecting and processing the samples is gratefully acknowledged, as is the support and guidance of G. Sohani and A. Pande.

REFERENCES

Meyer K. (2007) *J Zhejiang Univ Sci B.* **8**:815.

Sargolzaei M., Chesnais JP. and , Schenkel FS. (2014) *BMC Genomics.* **15**:478.

MITOCHONDRIAL GENE EXPRESSION IS ASSOCIATED WITH ORGAN AND TISSUE METABOLISM IN DAIRY CATTLE

J. Dorji[1,2], I.M. MacLeod[1], C.J. Vander Jagt[2], A.J. Chamberlain[2] and H.D. Daetwyler[1,2]

[1]School of Applied Systems Biology, La Trobe University, Bundoora, Victoria, 3083 Australia
[2]AgriBio, Centre for AgriBioscience, Agriculture Victoria, Victoria, 3083 Australia

SUMMARY

Mitochondria and their genes (MGs) are central to cellular energy metabolism in eukaryotes. Mitochondrial DNA mutations are associated with mitochondrial diseases and affected productivity in animals. The variation in mitochondrial phenotypes are plausibly affected through differential expression of MGs, but currently, MG expression across tissues types are unknown. We profiled MG expression using RNAseq of 16 tissues of dairy cattle. Our results show MGs are differentially expressed among tissues. Specifically, upregulated in heart, leg muscles, tongue and downregulated in leukocytes, thymus, lymph node, and lung. Besides, all DE MGs were regulated in a single direction within a tissue.

INTRODUCTION

Mitochondria are autonomous organelles having their own maternally inherited haploid genome. The mitochondrial genome in cattle small (~16.6 kb) encoding 37 genes (13 proteins, 22 tRNAs and 2 rRNAs) and non-coding region (Anderson *et al.* 1982). Mitochondria are functionally responsible for cellular energy metabolism, and the proteins encoded by a mitochondrial genome are integral components of the complex electron transport chain (ETC). Energy (ATP) is generated by the transfer of electron through ETC in the process known as oxidative phosphorylation (OXIPHOS) and caters to varying energy demands across the tissue types in the body (Wang *et al.* 2012).

Mitochondrial mutations are associated to mitochondrial dysfunction and diseases, including diabetes, obesity, and aging (Taylor and Turnbull 2005) and are increasingly studied in humans to understand the underlying biology and to develop therapies for mitochondrial diseases. In livestock, mitochondrial mutations are indicated in productivity (Schutz *et al.* 1994), but little is known regarding the potential causal effects or underlying biology of mitochondrial mutations.These mitochondrial phenotypes are plausibly the result of differential expression of MGs. However, MG expression across the tissue in dairy cattle is less known.

Therefore, we characterized MG expression across 16 tissues in dairy cow using RNA sequencing technology to gain insights into the relationship between cellular energy metabolism and potential biological effects of mitochondrial mutations.

MATERIALS AND METHODS

We sampled 16 tissues from two adult lactating cows (2181 and 6819) aged 8 years from the Agriculture Victoria Research dairy herd at Ellinbank after euthanasia. Cows were born from different sire and dam. Cow 2181 was 208 days in milk and in the fifth lactation while cow 6819 was 173 days in milk and in the seventh lactation. Blood was collected by venipuncture of the coccygeal vein and processed according to the standard protocol in the RiboPure™ blood kit (Ambion by Life Technologies). Other tissue samples were dissected into 1cm squares, sealed in a 5ml tube and flash-frozen in liquid nitrogen and stored at -80 °C. RNA was extracted from leukocytes using RiboPure Blood Kit (Ambion) following the manufacturer's instructions. RNA was extracted from ~30 mg of ground tissue using Trizol (Invitrogen) according to standard protocol. RNAseq libraries were prepared from all samples with a RIN > 6 using the SureSelect Strand Specific RNA Library Prep Kit (Agilent) according to manufacturer's instructions. Each library was uniquely barcoded, randomly assigned to one of two pools and sequenced on a HiSeq™ 3000 (Illumina) in a 150-cycle paired-end run. One hundred and fifty bases paired-end reads were called with bcltofastq and output in fastq format.

Each library of paired-end reads was aligned to Ensembl bovine genome UMD3.1 using STAR version 2.5.3ab (Dobin *et al.* 2013), and quality checked for alignment. We used the SAM output file to generate gene expression counts in a tissue using R package featureCounts (Liao *et al.* 2014). The gene counts were filtered for lowly expressed reads. The gene counts were normalized (counts per million) and analyzed for differential expression using the R package edgeR (Robinson *et al.* 2010). A design matrix with the overall mean of gene expression across all tissues was used as the intercept and compared to gene expression in each sample. Specifically, we used glmQLTest method to identify DE genes (Log_2Fold Change (LFC) > |0.6|, p<0.01) and direction of regulation was determined by the sign of LFC values (i.e. +ve LFC as upregulated and -ve as downregulated).

RESULTS AND DISCUSSION

Differential expression of genes. We compared nuclear and MGs expression profiles across all tissue types. Sixteen MGs (13 protein-coding genes, 2 rRNAs and 1 tRNA) were DE. The highest number of upregulated MG was in heart followed by the tongue, leg muscle, spleen, and kidney cortex, while the highest number of downregulated genes were in leukocytes, thymus, lymph node and lung (Figure 1). The MGs were not DE for the rest of the tissues. Interestingly, all DE MGs were regulated in a single direction (i.e. regulated either all up or all down), while the overall nuclear gene expression was regulated in both directions.

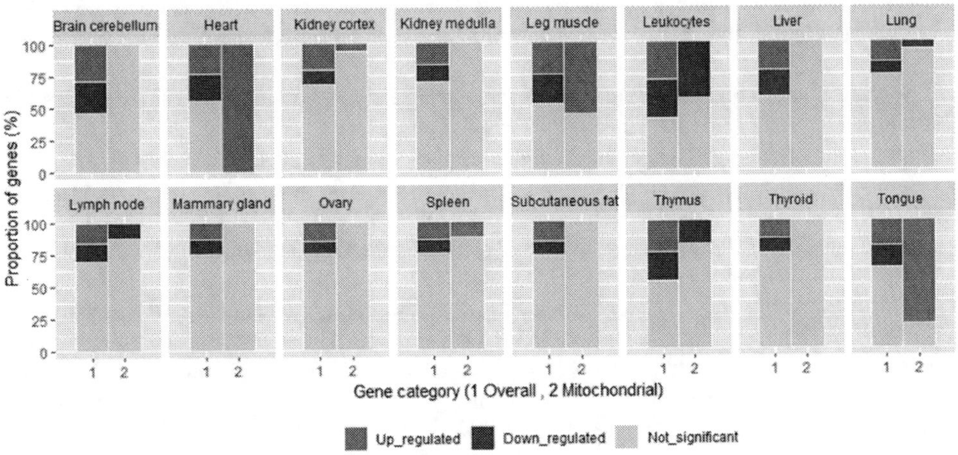

Figure 1. Proportion of differentially expressed overall gene and mitochondrial genes 16 tissue types of dairy cattle

This phenomenon of regulation of MGs in a single direction in tissue may be explained from the mechanism of transcription of mitochondrial DNA. The mitochondrial DNA is transcribed as a near-complete polycistronic unit (Shokolenko and Alexeyev 2017). The polycistronic transcription could result in an almost equal number of all MG transcripts, and thereby regulation of the genes in one direction compared to the mean.

The highest fold changes of MG expression across tissues were in heart, leg muscle, tongue, kidney and liver, and relatively lower fold changes in lung and adipose tissues (Figure 2). The consistency of MG expression in tissue is highlighted by clustering together of most tissues across two animals.

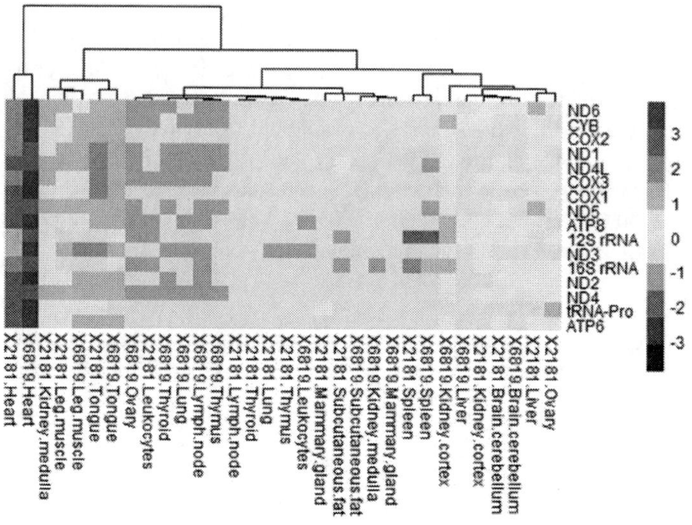

Figure 2. Heatmap showing mitochondrial gene expression across the tissues of two cows (X6819 and X2181) highlighting higher expression in heart, skeletal muscles and tongue

Specific organs and tissues. Heart tissue had the highest number of DE MG as well as the highest fold change compared to the mean expression of all 16 tissues. This coincides with the specific metabolic rate of the organ, which is the highest among tissues in mammals (Table 1). The overall DE genes in the heart were significantly enriched for OXIPHOS pathway (adj $p < 1.2e\text{-}52$). This is logical because a heart requires a continuous and reliable energy supply for the contraction of cardiac muscle to pump blood through the body. Almost 95% of the energy demand of the heart is met from OXIPHOS in mitochondria (Stanley and Chandler 2002).

Tongue (muscular organ) and leg muscle followed heart in the number of DE MGs as well as fold changes with significant enrichment for OXIPHOS pathways. The skeletal muscle has a low basal metabolic rate but has the capacity to increase depending on the activity (Glaister 2005) by as high as 1000-fold during exercise (Spriet 1992). The muscular strength and movement of the tongue are vital for harvesting forage, chewing and regurgitation: typically, in use for up to 20 hours per day.

Kidney, liver and brain despite relatively higher specific metabolic rates compared to the residual tissues neither show higher fold MG expression nor DE MGs (except kidney cortex). This indicates that the OXIPHOS may not be a primary energy generation process in these tissues.

Table 1. Specific metabolic rates of organs and tissues across species (kcal/kg/day)

Species	Heart	Kidney	Brain	Liver	Skeletal	Adipose	Residuals	Source
Human	440	440	240	200	13	4.5	12.0	(Wang *et al.* 2010)
Cattle	429	412	185	130	-	-	10.7	(Wang *et al.* 2012)
Sheep	588	496	255	200	-	-	15.5	(Wang *et al.* 2012)

It follows that leukocytes were the tissue with the most downregulated MG because the role of mitochondria in leukocytes are primarily for non-energy production roles (Kramer *et al.* 2014). The leukocytes mostly derive their energy from glycolysis.

Differential expression of mitochondrial genes and mitochondrial copy number across tissues. The number of mitochondria per cell and the amount of Mitochondrial DNA per cell are closely regulated within a given cell type but differ widely between cell types (Robin and Wong 1988). Currently, the effect of mitochondrial content on the variation of MG expression across the tissue remains poorly known. Tissues with high MG expression in this study are concurrently reported to have high mitochondrial content and vice versa. For example, heart, skeletal muscle, omental fat and breast cells contained 6970, 3650, 400-600 and 25 copies respectively (Miller *et al.* 2003; Lindinger *et al.* 2010; Yu *et al.* 2007).

Differential expression of MGs across tissues and association with energy metabolism indicate gene expression of mitochondrial protein genes (from both nuclear and mitochondrial genomes) can be extended to identify candidate genes involved in energy use efficiency and related traits (e.g. feed efficiency). The genetic markers identified for a trait can be used in genomic selections.

CONCLUSIONS

This study shows that mitochondrial genes are differentially expressed across the tissues in dairy cattle. Mitochondrial gene expression is upregulated in tissues with high energy demand and vice versa. Within a tissue, the mitochondrial genes are regulated in a single direction. Because mitochondrial gene expression was associated with tissues metabolic rates, we are now using RNA sequencing to determine if there are associations between mitochondrial protein gene expression and specific cattle phenotypes.

REFERENCES

Anderson S., de Bruijn M.H.L., Coulson A.R., Eperon I.C., Sanger F. and Young I.G. (1982) *J. Mol. Biol.* **156**: 683

Dobin A., Davis C.A., Schlesinger F., Drenkow J., Zaleski C. and Jha S. (2013) *Bioinformatics,* **29**.

Glaister M. (2005) *Sports Med.* **35:** 757.

Kramer P.A., Ravi S., Chacko B., Johnson M.S. and Darley-Usmar V.M. (2014) *Redox Biol.* **2:** 206.

Liao Y., Smyth G.K. and Shi W. (2014) *Bioinformatics.***30:** 923.

Lindinger A., Peterli R., Peters T., Kern B., Von Flue M., Calame M., Hoch M., Eberle A.N. and Lindinger, P. W. (2010) *Obe. Surg.* **20:** 84.

Miller F.J., Rosenfeldt F.L., Zhang C., Linnane A.W. and Nagley P. (2003) *Nucleic Acids Res.* **31:** e61.

Robin E.D. and Wong R. (1988) *J. Cell. Physiol.* **136:** 507.

Robinson, M.D., Mccarthy D.J. and Smyth G. K. (2010) *Bioinformatics.* **26:** 139.

Schutz M.M, Freeman A.F., Lindberg G.L., Koehler C.M and Beitz D.C. (1994) *Livest. Prod. Sci.* **37**: 283.

Shokolenko I.N. and Alexeyev M.F. (2017) *Front Biosci (Landmark Ed).* **22:** 835.

Spriet L.L. (1992) *Can. J. Physiol. Pharm.***70:** 157.

Stanley W.C. and Chandler M.P. (2002) *Heart Fail. Rev.***7:** 115.

Taylor R.W. and Turnbull D.M. (2005) *Nat. Rev. Genet.* **6:** 389.

Wang Z., Ying Z., Bosy-Westphal A., Zhang J., Schautz B., Later W., Heymsfield S.B. and Müller M.J. (2010) *Am. J. Clin. Nut.* **92:** 1369.

Wang Z., Zhang J., Ying Z. and Heymsfield S.B. (2012). *ISRN Zoology.***9**.

Yu, M., Zhou, Y., Shi, Y., Ning, L., Yang, Y., Wei, X., Zhang, N., Hao, X. and Niu, R. (2007) *IUBMB Life.* **59:**450.

ACCURACY OF GENOMIC PREDICTIONS FOR MILK PRODUCTION TRAITS IN PHILIPPINE DAIRY BUFFALOES

J.R.V. Herrera[1,3], E.B. Flores[2], N. Duijvesteijn[3], N. Moghaddar[3] and J.H.J. van der Werf[3]

[1]Philippine Carabao Center-University of the Philippines Los Banos, College, Laguna, 4031 Philippines
[2]Philippine Carabao Center National Headquarters, Muñoz Nueva Ecija, 3119 Philippines
[3]School of Environmental and Rural Science, University of New England, Armidale, NSW, 2351 Australia

SUMMARY

A three-fold cross-validation scheme was used to compare the accuracies of genomic prediction for milk yield (MY), fat yield (FY) and protein yield (PY) from Philippine dairy buffaloes using Genomic Best Linear Unbiased Prediction (GBLUP) and single-step GBLUP (ssGBLUP) with the accuracies based on pedigree BLUP (pBLUP). To also assess the bias of the prediction, the regression coefficient (slope) of the adjusted phenotypes on the predicted breeding values was also calculated. Results showed that genomic methods (GBLUP and ssGBLUP) provide more accurate predictions compared to pBLUP. Average GBLUP and ssGBLUP accuracies were 0.24 and 0.29, respectively, while average pBLUP accuracies (for GENO and ALL data) were 0.21 and 0.22, respectively. Slopes of the two genomic methods were also closer to one, indicating lesser bias, compared to pBLUP. Average GBLUP and ssGBLUP slopes were 0.89 and 0.84, respectively while average pBLUP (for GENO and ALL data) slopes were 0.80 and 0.54, respectively. With the higher accuracy of prediction and lesser bias, ssGBLUP will be used by PCC to identify replacement breeding bulls at a younger age.

INTRODUCTION

The Philippine Carabao Center (PCC) has put in place a genetic improvement program that includes a system of evaluating genetically superior individual animals for milk and milk component traits; maintenance of nucleus herds of dairy buffaloes (mostly Bulgarian Murrahs) as source of breeding animals and provision of frozen semen from the best riverine buffalo germplasm for artificial insemination. However, present constraints are: only ~ 1200 cows are in the nucleus herds, currently testing only 12 bulls per year, accuracies of progeny test bulls are low due to small number of daughters with lactation records, and generation interval is long for AI sires, ~ 8 yrs (Flores 2014).

The availability of the Affymetrix 90K Buffalo Genotyping Array (Affymetrix, Inc., Sta. Clara, CA) in 2013 makes it now possible to do genomic studies in the bubaline species. When the trait of interest cannot be recorded on the selection candidate, genomic selection schemes are very attractive even when the number of phenotypic records is limited, because traditional breeding requires progeny testing schemes with long generation intervals (Schaeffer 2006).

Genomic prediction studies in dairy buffaloes are very limited and were based on small data sets. Tonhati *et al.* (2016) used single-step GBLUP (ssGBLUP) to estimate the Predicted Transmitting Ability (PTA) accuracies for seven milk traits on 452 Brazilian buffaloes. Using a 5-fold cross validation, Liu et al. (2018) evaluated the reliability of genomic Estimated Breeding Values (GEBVs) and their correlation with EBVs for six milk production traits from 412 Italian Mediterranean buffaloes. The objective of this study was to determine the accuracy of genomic prediction and bias for milk yield (MY), fat yield (FY) and protein yield (PY) from Philippine dairy buffaloes using Genomic Best Linear Unbiased Prediction (GBLUP) and ssGBLUP compared to prediction accuracy and bias based on pedigree BLUP (pBLUP).

MATERIALS AND METHODS

The 305-day-MY,-FY,-PY data and blood samples used in this study were obtained from the PCC. All animals are housed in nucleus/institutional farms and cooperatives managed by PCC. Data collection and storage is managed by the Center's Animal Breeding and Genomics Section (ABGS). Descriptive statistics of the phenotypic data are presented in Tables 1 and 2. Number of animals with one, two and three lactation records are shown in Tables 3 and 4.

Two data sets were analysed. One only contains female buffaloes that have both phenotypes and genotypes (hereby referred to as GENO) (Table 1). Analyses done on this data were pBLUP and GBLUP. The other data set (hereby referred to as ALL) (Table 2) contains all the above animals, plus additional females with phenotype but are not genotyped. Analyses done on this data were pBLUP and ssGBLUP. A pedigree file containing 2642 animals was used for pBLUP and ssGBLUP.

Table 1. Descriptive statistics of GENO data to be used for pBLUP and GBLUP analyses

Trait*	No. of animals	No. of records	No. genotyped	Mean (kg)	Min (kg)	Max (kg)	SD (kg)
MY	904	1773	904	1573.2	103.1	3054.5	505.9
FY	856	1384	856	119.0	30.2	206.9	27.7
PY	856	1384	856	70.7	22.5	127.9	16.0

*MY-milk yield, FY-fat yield, PY- protein yield

Table 2. Descriptive statistics of ALL data to be used for pBLUP and ssGBLUP analyses

Trait*	No. of animals	No. of records	No. genotyped	Mean (kg)	Min (kg)	Max (kg)	SD (kg)
MY	1975	3821	904	1466.3	103.1	3150.9	518.0
FY	1918	3405	856	111.9	29.3	210.1	29.1
PY	1918	3405	856	66.3	19.9	128.8	17.3

*MY-milk yield, FY-fat yield, PY- protein yield

Animals were genotyped using the Axiom 90k Buffalo genotyping array. Polymorphic markers were identified using the Axiom Analysis Suite set on default settings. Additional quality control measures applied include: a SNP was removed if its minor allele frequency (MAF) is less than 0.05, is out of Hardy-Weinberg equilibrium (P-value less than 1×10^{-15}), has no genome location and is not found in the autosomes. After quality control, only 60,827 SNPs in 29 autosomes were used.

A 3-fold cross-validation scheme was used to compare accuracy of prediction and bias using GBLUP and ssGBLUP with those of pBLUP. Animals were assigned to one of 3 test sets: with 1 lactation record, 2 lactation records, and 3 lactation records (Tables 3 and 4). Phenotypes of animals in the test sets were masked, and breeding values were then estimated for each set either by pBLUP and GBLUP for the GENO data or pBLUP and ssGBLUP for ALL data using univariate repeatability animal models. All models included breed, lactation and herd-year-season (HYS) as fixed effects; and animal and permanent environmental effects as random effects. pBLUP uses a numerator relationship matrix (also known as A-matix) based on the pedigree. The creation of the genomic relationship matrix (also known as a G-matrix) used in GBLUP and ssGBLUP is based on Van Raden (2008). The ssGLUP (Legarra *et. al.* 2014) uses a H-matrix where the G-matrix replaced the A_{22} matrix (weight of 1 on the G-matrix). Breeding values (BVs) were generated using ASReml 4.1 (Gilmour *et al.* 2008).

Accuracy of prediction was calculated as the correlation between the predicted BVs of the test set and adjusted phenotypes, which were corrected for fixed effects, divided by the square root of the heritability of the trait, corrected for the number of lactations used in the test set. To assess the bias of the prediction, the regression coefficient (slope) of the adjusted phenotypes on the predicted breeding values was also calculated, with slopes of approximately one showing zero bias. Slopes greater than or lesser than one indicates underestimation and overestimation, respectively, of BVs.

Table 3. Number of animals (number of records) for test and training sets for MY

test set	training set	
	GENO	ALL
329[a] (329)	575 (1444)	1646 (3492)
281[b] (562)	623 (1211)	1694 (3259)
294[c] (882)	610 (891)	1681 (2939)

*a,b,c-number of animals with 1,2,3 lactation records, respectively

Table 4. Number of animals (number of records) for test and training sets for FY and PY

test set	training set	
	GENO	ALL
441[a] (441)	415 (943)	1477 (2964)
302[b] (604)	554 (780)	1616 (2801)
113[c] (339)	743 (1045)	1805 (3066)

*a,b,c-number of animals with 1,2,3 lactation records, respectively

RESULTS AND DISCUSSION

Results showed that genomic methods (GBLUP and ssGBLUP) provide more accurate predictions compared to pBLUP (Table 5). Average pBLUP (for GENO and ALL data) accuracies for the three traits were 0.21 and 0.22, respectively, while average GBLUP and ssGBLUP accuracies were 0.24 and 0.29, respectively. GBLUP and ssGBLUP accuracies were on average 0.03 and 0.07 higher, respectively, compared to pBLUP accuracies.

In the case of prediction bias, slopes for all methods were below one, indicating overestimation of breeding values (Table 6). However, slopes of the two genomic methods are closer to one, indicating lesser bias, compared to pBLUP slopes. Average pBLUP (for GENO and ALL data) slopes for the three traits were 0.80 and 0.54, respectively, while GBLUP and ssGBLUP slopes were 0.89 and 0.84, respectively.

CONCLUSIONS

These results are preliminary and additional animals will be genotyped to increase the sample size to have a more reliable conclusion. As of the moment, results showed that GBLUP and ssGBLUP provide more accurate predictions compared to pBLUP. Moreover, prediction bias of the two genomic methods are lesser compared to pBLUP. With the higher accuracy of prediction and lesser bias, ssGBLUP will be used by PCC to identify replacement breeding bulls at a younger age.

Table 5. Average prediction accuracy for pBLUP, GBLUP and ssGBLUP estimated from 3-fold cross-validation scheme

Trait	GENO			ALL		
	pBLUP	GBLUP	Increase in accuracy	pBLUP	ssGBLUP	Increase in accuracy
MY	0.20 ±0.04	0.28 ±0.06	0.08	0.17 ± 0.02	0.30 ± 0.04	0.13
FY	0.23 ±0.04	0.24 ±0.05	0.01	0.26 ±0.14	0.30 ± 0.01	0.04
PY	0.20 ±0.05	0.20 ±0.05	0	0.23 ±0.14	0.26 ± 0.02	0.03
Average	0.21	0.24	0.03	0.22	0.29	0.07

*MY-milk yield, FY-fat yield, PY- protein yield

Table 6. Estimated slopes calculated from breeding values from pBLUP, GBLUP and ssGBLUP

Trait	GENO		ALL	
	pBLUP	GBLUP	pBLUP	ssGBLUP
MY	0.69 ± 0.39	0.85 ± 0.28	0.42±0.07	0.85±0.16
FY	0.94 ± 0.17	0.99 ± 0.22	0.62±0.36	0.88±0.04
PY	0.76 ± 0.11	0.83 + 0.34	0.57±0.38	0.79±0.10
Average	0.80	0.89	0.54	0.84

*MY-milk yield, FY-fat yield, PY- protein yield

ACKNOWLEDGEMENTS

The authors would like to acknowledge: i) PCAARRD-DOST for funding the genotyping project, ii) various coordinators, herd managers and directors of various PCC centers for making data and samples available, iii) staff of PCC's ABGS for preparing the blood samples for genotyping. JRVH would like to acknowledge PCC-SEARCA for his PhD scholarship.

REFERENCES

Flores EB (2014) PhD Thesis. University of New England.

Gilmour A.R., Gogel B.J., Cullis B.R., Welham S.J. and Thompson R. (2009) ASReml User Guide Release 4.0. 2015. Hemel Hempstead, UK: VSN International Ltd.

Legarra, A., Christensen O.F., Aguilar I. and Misztal I. (2014) *Livest. Sci.* **166**: 54.

Liu J.J., Liang A.X., Campanile G., Plastow G., Zhang C., Wang Z., Salzano A., Gasparrini B., Cassandro M. and Yang L.G. (2018) *J. Dairy Sci.* **101**: 433.

Tonhati H., Cardoso D.F., Jordan D. and Santos A. (2016). 11th World Buffalo Congress paper presentation.

Schaeffer L. R. (2006). *J Anim. Breed. Genet.* **123**:218.

VanRaden P.M. 2008. *J Dairy Sci* **91**:4414.

MEDIUM DENSITY BEADCHIP GENOTYPE DATA REVEALS GENOMIC STRUCTURE OF SOUTH AFRICAN MERINO-BASED BREEDS

E.F. Dzomba[1] and F.C. Muchadeyi[2]

[1]Discipline of Genetics, School of Life Sciences, University of KwaZulu-Natal, Private Bag X01, Scottsville, 3209 South Africa
[2]Agricultural Research Council, Biotechnology Platform, Private Bag X5, Onderstepoort, 0110 South Africa

SUMMARY

In South Africa, the Merino sheep was used to develop a number of Merino-derived breeds with the goal of improving productivity and robustness to challenging production systems. The current study evaluated population genetic structure and breed relationships of Merino and Merino-based sheep breeds of South Africa using the Ovine50K beadchip. Five breeds evaluated consisted of the Dohne Merino (n=50), South African Merino (n=10), South African Mutton Merino (n=10), Meatmaster (n= 48) and the Afrino (51). The Dohne Merino and Afrino were observed to have the highest genetic diversity of H_E=0.3699 and H_E=0.3642, respectively. PCA clustered Meatmaster divergently from other breeds forming an outgroup with no indications of admixture or gene flow into other breeds. Highest differentiation was observed between Meatmaster and South African Mutton Merino (F_{ST}=0.1699). Lowest F_{ST} values were found between Afrino and Dohne Merino F_{ST}=0.01746. Meatmaster differed from Dohne Merino and South African Merino on chromosomes 19; 11 and 9. The SNP *s10035.1* on chromosome 11 had an F_{ST} of 0.6361 and is associated with *SEPT9,* which encodes septin-9 protein while that on chromosome 9 (s71002.1; F_{ST} = 0.6198) is within 1MB region of *SOX17* gene. A SNP that differentiated the subpopulation of Afrino vs Dohne Merino and South African Merino on chromosome 12 (SNP *s62606.1*, F_{ST} = 0.5690) was observed to be associated with *THRSP* gene that plays a role in lipogenesis regulation, more especially in lactating mammary gland. Results obtained in this study indicated breed relations but also genetic structuring in South African Merino based sheep breeds.

INTRODUCTION

The Merino sheep breed is regarded as one of the oldest and most economically influential breeds in the world (Al-Atiyat *et al.* 2016). These sheep are known for their fine and soft wool (DAFF 2015). In South Africa, the breed was introduced in the 1780's from Spain (Buduram 2004; DAD-IS 2009) and has become adapted to South African climatic conditions (Snyman 2014). It is suggested that South African Merino carry a variety of genes that confer adaptation to a range of production environments (Peters *et al.* 2010). The South African Merino is believed to be a composite breed between Spanish, Saxony, Rambouillet, American and Australian sheep breeds (Mason 1996). From the South African Merino, several Merino based breeds have been developed (Hlophe 2011) for either wool, mutton or as dual-purpose breeds such as the Afrino, Meatmaster and Dohne Merino breeds in South Africa (Snyman 2014).

The Dohne Merino is regarded as a synthetic breed (Buduram 2004), developed from crossing German Mutton Merino rams, commonly known as South African Mutton Merino with the South African Merino ewes (Kotze 1951; Jordan 2013). This breed was developed to improve productivity and robustness to survive in a parasite infested region of the Eastern Cape province of South Africa. Afrino sheep originated from a cross between Ronderib Afrikaner, Merino and South African Merino (Snyman 2014) and are reared predominantly for meat production. Meatmaster was developed from multiple breeds of South African Mutton Merino, Van Roy and Ile deFrance, Damara, Wiltshire Horn,

Dormer and other breeds with the objective of improving meat qualities of the fat- tailed sheep breeds (Peters *et al.* 2010; Snyman 2014).

This study investigated population genetic structure and breed relationships between Merino and Merino-based sheep breeds of South Africa. Based on the crossbreeding history and the divergent breeding goals amongst breeds, it was hypothesised that there is genetic diversity and population substructure among these Merino-type breeds.

MATERIALS AND METHODS

The study used SNP genotype data from a total of 169 animals obtained from five different Merino and Merino-based sheep populations consisting of South African Mutton Merino (n=10), Dohne Merino (n=50), South African Merino (n=10), Afrino (n=51) and Meatmaster (n=48). The samples were obtained from different farms in the Eastern Cape region close to Middelburg.

The Illumina OvineSNP 50K Bead Chip array was used to genotype a total of 54,241 SNPs on 169 individuals. The genotype data was subjected to quality control using PLINK V1.07 (Purcell 2009) to remove individuals with more than 5% missing genotypes, SNPs with a call rate less than 95% and a minor allele frequency (MAF) less than 5% as well as individual animals that deviated from Hardy Weinberg equilibrium ($P< 0.0001$) from further downstream analysis.

Within breed diversity estimates of observed heterozygosity (H_O); expected heterozygosity (H_E) and inbreeding co-efficient (F_{IS}) where estimated using Plink v1.07 (Purcel, 2009).

To evaluate population structure, Principle Component Analysis (PCA) was carried out using Golden Helix SNP and Variation Suite (SVS) v8.1(Golden Helix Inc 2012). The eigen values for the principle components (PC) 1 and 2 were used in plotting the clustering of individuals.

An analysis of molecular variation (AMOVA) was done to understand the partitioning of diversity in the five sheep breeds using ARLEQUIN V3.1 (Excoffier and Lischer 2010). AMOVA was analysed for (i) all five breeds and (ii) within and between the developed (Afrino; Meatmaster and Dohne Merino) versus pure breeds (SA Merino and SA Mutton Merino).

The per population and per marker F_{ST} based outlier loci method was used to determine genetic differentiation between pairs of populations of the five sheep breeds and the affected SNPs using Golden Helix SNP and Variation Suite (SVS) V8.1 (Golden Helix Inc. 2012). Associated genes were mapped within 1MB region of significant SNPs (F_{ST}>0.50) using ENSEMBLE genome browser and NCBI (NCBI; www.ncbi.nlm.nih.gov) *ENSEMBL* Ovine (*Ovis Aries*) genome build AOR4.

RESULTS AND DISCUSSION

Highest genetic diversity values were observed in Dohne Merino and Afrino with the expected heterozygosity (H_E) of 0.3699±0.1286 and 0.3642±0.1294, respectively (Table 1). Across all populations, the observed heterozygosity values (H_O) were higher than the expected heterozygosity with an average inbreeding coefficient (H_{IS}) of 0.0567. Low inbreeding values were obtained ranging from -0.0897 in South African Mutton Merino to -0.0191 observed for Dohne Merino.

Table 1. Expected and observed heterozygosity in the five Merino sheep breeds of South Africa

Breed	No. of animals	H_O±SD	H_E±SD	F_{IS}
Afrino	51	0.3746±0.1439	0.3642±0.1294	-0.0287
Meatmaster	48	0.3664±0.1520	0.3537±0.1356	-0.0359
SA Mutton Merino	10	0.3846±0.1960	0.3529±0.1327	-0.0897
SA Merino	10	0.3861±0.1857	0.3363±0.1290	-0.0634
Dohne Merino	50	0.3769±0.1425	0.3699±0.1286	-0.0191
All breeds	166	0.3672±0.1077	0.3893±0.1094	0.0567

Principle component analysis resulted in four main clusters of (i) Dohne Merino and SA Merino; (ii) SA Mutton Merino (iii) Afrino and (iv) Meatmaster (Figure 1). The first principle component (PC1) explained 50.48% of the variation and clustered the Meatmaster divergently from other breeds forming an outgroup. Clustering of Meatmaster separately from other breeds can be explained by the history of the breed. The breed was developed from fat-tailed hair sheep and its clustering is possibly reflecting the composite blood lines of South African Mutton Merino and other breeds with 50% of its blood lines coming from Damara (Snyman 2014). The Afrino, which is also a composite breed with Merino and other none Merino sheep of the Afrikaner breed formed a unique cluster albeit closer to the SA mutton Merino. The Dohne Merino on the other hand is a result of selection and crossing of Merino breeds, which explains its clustering close to the SA Merino sheep.

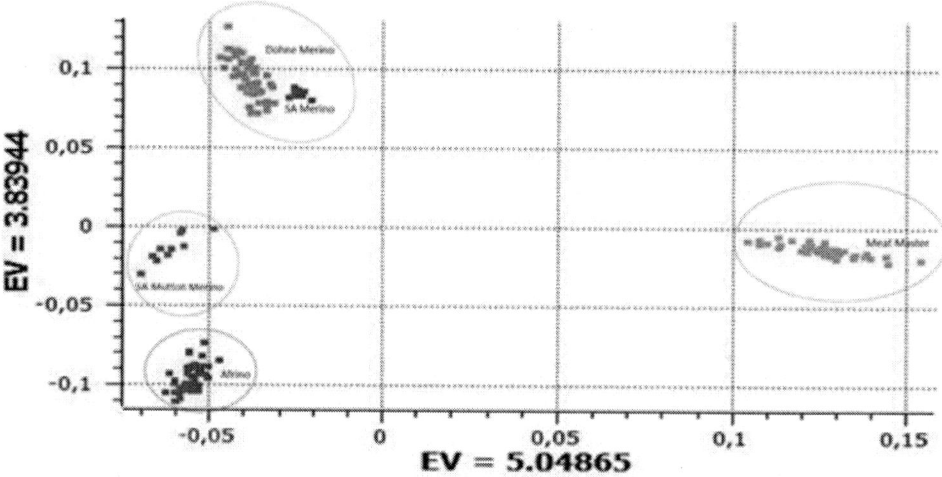

Figure 1. Principal component analysis for the 5 South African Merino-type breeds

AMOVA results (Table 2) suggested some level of population sub-structuring that exist in the South African merino based sheep breeds under study. Slightly over 10% of the variation was between breeds whilst variation amongst the individuals within breed was over 80%. A small percentage (1.25%) was observed between pure breeds and composite breeds developed from Merino and other sheep breeds. Overall, ±10% of the variation was observed between breeds, demonstrating existence of some divergence between breeds albeit small.

Table 2. Analysis of Molecular Variance (AMOVA) for the South African Merino-type breeds

Dataset	Percentage of variation (%)		
	Between population	Between populations within group	Within populations
All 5 breeds	10.62	-	91.38
Developed and pure breeds	1.25	10.74	88.01

Pairwise population F_{ST} ranged from 0.0175 – 0.1699. The Meatmaster diverged significantly from other sheep breeds with F_{ST} ranging from F_{ST} = 0.1100 between Meatmaster and Dohne Merino to F_{ST}=0.1699 between Meatmaster and SA Mutton Merino. Low F_{ST} values was observed between Dohne Merino and Afrino (F_{ST} = 0.01746). This suggest that the breeds have not been kept in isolation from each other for a long period with a possibility for gene flow between breeds. This is regarded as outcrossing and appears to shape the genetic structure of the breeds. Genetic differentiations between Meatmaster from Dohne Merino and South African Merino were observed on chromosome 19 (SNP *OAR19_45719557.1*; F_{ST} = 0.6475); 11 (SNP *s10035.1*, F_{ST} = 0.6361 and SNP *s56939.1*, F_{ST} = 0.6011 and chromosome 9 *SNP s71002.1*, F_{ST} = 0.6198) as illustrated in Table 3. SNP *s10035.1* on chromosome 11 is found to be within 1MB region of *SEPT9* (which encodes septin-9 protein) and while that on chromosome 9 (*s71002.1*) is within 1MB region of *SOX17* gene. A SNP that differentiated the subpopulation of Afrino vs Dohne Merino and South African Merino from chromosome 12 (SNP *s62606.1*, F_{ST} = 0.5690) is associated with *THRSP* gene (a gene that plays a role in lipogenesis regulation, more especially in lactating mammary gland).

Table 3. Breed differentiating SNPs (F_{ST} > 0.5) and the associated ENSEMBLE genes

Subpopulation	F_{ST}	SNP ID	CHR	Related gene
MMA VS DME & SAM	0,6475	*OAR19_45719557_X,1*	19	No gene
	0.6361	*s10035.1*	11	*SEPT9*
	0.6011	*s56939.1*	11	*EXOC7,FOXJ1*
	0.6198	*s71002.1*	09	*SOX17*
AFR VS DME & SAM	0.6197	*OAR5_51499408.1*	5	*ARHGAP26*
	0.5690	s62606.1	12	THRSP

CONCLUSIONS

In conclusion, results showed that the variation within the breeds was greater compared to between breed diversity. Significant genetic variation is still maintained in South African Merino-type sheep breeds. These results are in agreement with the origins, breeding practices as well as the geographical regions from which these breeds are found. The low but present genetic variation between Merino and merino derived breeds could be beneficial in the exchange of genes through outbreeding. Unfavourable traits observed in a certain group can be improved by outcrossing to increase genetic diversity against a population structure.

REFERENCES

Al-Atiyat R., Flood W., Franklin I., Kinghorn B. and Ruvinsky A. (2016) *Small Ruminant Res.* **136**: 137.

Buduram P.O. (2004) MSc Dissertation, University of Free State.

DAD-IS (2009) http://www.fao.org/dad-is/en/

DAFF (2015) https://www.daff.gov.za/daffweb3/branches/economic-development-trade-marketing/marketing/annual-publications

Excoffier L. and Lischer H.E.L. (2010) *Mol. Ecol. Resour.* **10**: 564.

Hlophe S.R. (2011) Doctoral dissertation, University of Zululand.

Kotzé, J.J.J. (1951) *Farming in South Africa* **28**: 110.

Mason I.L. (1996) A World Dictionary of Livestock Breeds, Types and Varieties, 4th ed. CAB International, Wallingford, UK.

Peters F.W., Kotze A., Van der Bank F.H., Soma P. and Grobler J.P. (2010) *Small Ruminant Res.* **90**: 101.

Snyman M.A. (2014) Grootfontein Agricultural Development Institute. http://gadi.agric.za/InfoPacks/2014013%20South%20African%20Sheep%20breeds%20-%20Afrino.pdf

EVALUATION OF BODY MORPHOLOGY AND SHAPE OF BLACK TIGER PRAWN (*PENAEUS MONODON*) BY MORPHOMETRIC ANALYSIS

M.M. Hasan[1,2,§], N.M. Wade[1], C. Bajema[1], P.T. Thomson[1,3], D.R. Jerry[1], H.W. Raadsma[1,2] and M.S. Khatkar[1,2]

[1]ARC Research Hub for Advanced Prawn Breeding
[2]Sydney School of Veterinary Science, Faculty of Science, The University of Sydney, Camden, NSW, Australia
[3]School of Life and Environmental Sciences, Faculty of Science, The University of Sydney, Camden, NSW, Australia

SUMMARY

Morphometric analysis of body shape is an important tool for assessing shape variation among a wide variety of organisms. We carried out a morphometric analysis of prawns (n=8,164) by using land-marked points on images. Principal component (PCA) analysis revealed that the data were clustered in three major landmark derived distance groups, and the first principal component (PC1) as a representation of overall size of the prawn primarily related to overall length, and accounted for 94.9% of the variation, whereas PC2 explained 1.5% of the total variation mainly associated with the measure of the body thickness. The preliminary results suggest that morphometric techniques can be used to reveal shape variation in shrimp species.

INTRODUCTION

Morphometrics is largely defined as an approach to evaluate variation and co-variation of shapes with other traits. Traditional morphometric analysis is based on univariate/multivariate analysis of variables such as distance measures, angels, and/or ratios, to capture and describe complex two- and three-dimensional shapes. Over the last two decades, morphometric analysis of body shape has gradually become an important tool for assessing and studying shape variation among a wide variety of organisms. This technique has been applied in various fields including phylogenetic analysis, species identification and classification, examining extent of phenotypic plasticity, defining and explaining of ancestral morphotype (Adams *et. al.* 2013). Although morphometric approaches are successful in objectively describing complex shapes, these methods are lacking in the application of genetic analyses and genomic selection studies. Most aquaculture studies have been restricted to several fish species, only a limited number of studies are available regarding shrimp species using small number of animals (Claverie and Wainwright 2014; Adams 2013; Aktas *et al.* 2006). To address this deficit, we carried out a morphometric analysis of a large number of shrimp by using land-mark points derived from images.

Landmarks are fixed points of reference on an object that can be matched between and within samples. Derived distances between landmarks are employed to capture two- and three-dimensional shapes using sophisticated data collection tools. The first step of any geometric morphometric data analysis is the superimposition of the landmark configurations. Superimposition is the process of merging two or more landmarks data positions to remove any differences between data collection times and scales all the samples to a common unit. After superimposition, coordinates of each sample are represented by a single point in a common shape and space, and shapes are defined based on the combination of a number of landmarks. To reduce the high-dimensionality of the landmarks, ordination methods are generally employed to summarize the variation of shape space. The most commonly employed ordination technique is principal component analysis (PCA), a method that is used to perform an eigen decomposition of the covariate matrix and evaluate the distribution of samples along each of the resulting eigenvectors or principal component (PC) axes (Slice 2007). Therefore, the resulting PC axes are linear groupings of the original data that are statistically independent of each other. Standard multi-variate statistical analysis techniques (e.g., MANOVA) can be used for exploring statistically significant shape differences among groups related to categorical variables and fixed effects (e.g., sex, geographical location, experimental treatments etc.) (Webster *et. al.* 2011). The purpose of this study was to develop an efficient morphometric approach to characterize shrimp morphometric traits and investigate factors that affect the variability of the traits across sex, environments and other treatments. The characterized morphometric shape traits will further be utilized in the *P. monodon* selection breeding program by estimating genetic parameters and genomic selection.

MATERIALS AND METHODS

Despite the lack of morphometric analyses in shrimp, there is a consensus on the traits that are of industry/commercial significance. These traits include; total length, body length, carapace length, length width ratios. Typically, measurements are made between one easily distinguishable landmark to another and that distance is ascribed with a trait name. The morphological measurements proposed in this study have been generated using an amalgamation of methods used in other studies as well as what measurements are possible within the limitations of the available images (Figure 1 and Table 1 and 2) (Aktas *et. al.* 2006; Hung *et. al.* 2014).

Table 1. Lists of landmark description used in our analysis

Landmark number	Landmark Description
1	Placed on the most anterior point of the antennal scale
2	Placed on the most anterior point of the tail
3	Placed on the most posterior point of the tail
4	Placed on the junction of the carapace and abdomen placed at the most dorsal point
5	Placed on the midway along the carapace on the ventral side of the prawn
6	Placed on the junction of the carapace and abdomen placed at the most ventral point
7	Placed dorsally on the midpoint of the first abdominal segment
8	Placed ventrally on the midpoint of the first abdominal segment
9	Placed dorsally on the midpoint of the third abdominal segment
10	Placed ventrally on the midpoint of the third abdominal segment
11	Placed dorsally on the midpoint of the last abdominal segment
12	Placed ventrally on the midpoint of the last abdominal segment

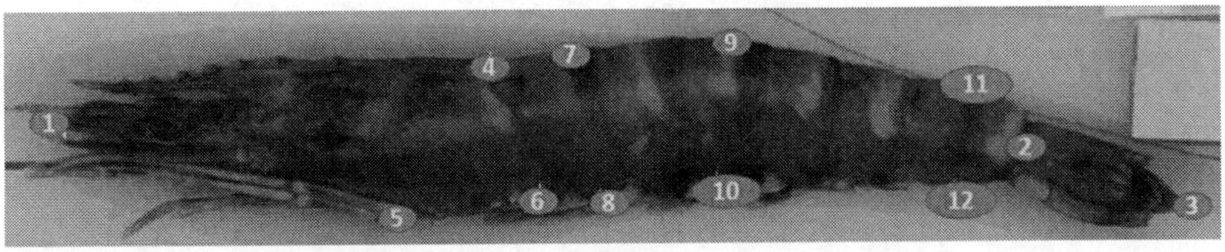

Figure 1. Scored morphometric traits of *P. monodon*

Table 2. Lists of traits derived from landmark data

Trait	Landmark Coordinates	Trait Description
Total length	1-3	From the most anterior point of the antennal scale to most posterior point of the tail
Body length	1-2	From the most anterior point of the antennal scale to most anterior point of the tail
First abdominal segment height	7-8	From the dorsal midpoint to the ventral midpoint on the first abdominal segment
Third abdominal segment height	9-10	From the dorsal midpoint to the ventral midpoint on the third abdominal segment
Last abdominal segment height	11-12	From the dorsal midpoint to the ventral midpoint on the last abdominal segment

For the morphometric shape analysis, 8164 photographed specimens were used to estimate 66 morphometric distances derived from twelve landmarks (Figure 1). All the images were annotated for these twelve landmarks using ImageJ software. These unambiguous landmarks were representative of prawn size and shape from various aspect ratios. To evaluate the reliability of landmark data, repeated measurement of 300 image landmarks were done by the same operator and between two independent operators. The overall repeatability of the measurements was high ($r = 0.98$).

RESULTS AND DISCUSSION

PCA analysis revealed that the first three PCA accounted for 97.5% of the variability of the data. The data were clustered in three major landmark derived distance groups (Figure 1). The first principal component (PC1) accounted for 94.9% of the recorded variation in the landmark distance data (Table 1). PC1 was taken as a representation of overall size of the prawn, since all PC1 loadings were positives and PC1 scores were significantly correlated with d_1_3 (which corresponds total length). Moreover, variation in d_1_3 contributed the highest variability in the PCA. For instances, a comparison between the image of highest and lowest PC1 scores revealed very contrasting sized prawns (data not shown). The higher variance accounted by PC1 may be due to the presence of a wide range of size in the samples, and often observed when samples in different sizes were studied (Stauffer 1991; Claverie and Wainwright 2014). On the other hand, the second principal component (PC2) explained 1.5 % of the total variation among landmarks, which is the measure of the proportionate body width/shape of this species. For example, extremes on this axis are very thin/elongated body shape at the low end and the thick/fatty shape at the high end.

Table 3. Variability explained by different PCs

	PC1	PC2	PC3	PC4	PC5	PC6
Standard deviation	192.21	24.54	20.19	16.16	12.62	10.46
Proportion of Variance	0.949	0.015	0.010	0.006	0.004	0.002
Cumulative Proportion	0.949	0.964	0.975	0.981	0.985	0.988

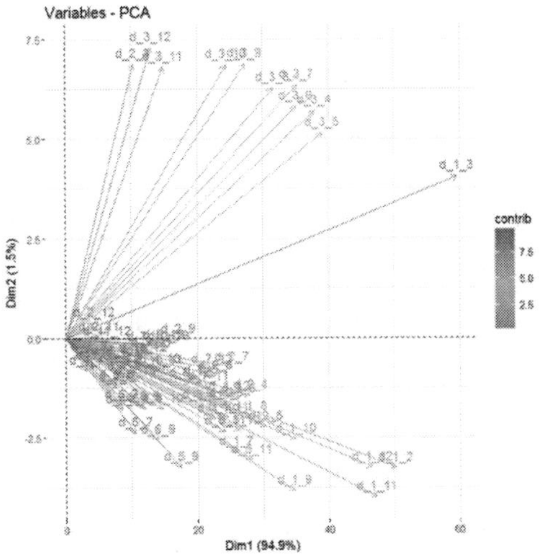

Figure 2. PC analysis showing contribution of individual variables (*i.e.* distances) towards variability in shape among shrimp

Further analyses will be carried out to explore environmental (e.g., pond) and genotype (e.g., family) specific variability of *P. monodon* shape phenotypes. We will estimate additive, dominance and epistatic variances of body morphological traits of *P. monodon* using relationship matrices constructed from genome-wide dense SNP markers. The morphometric analyses will also include the accuracy of genomic predictions using genomic best linear unbiased prediction models with or without including non-additive genetic effects. We will further conduct genomic prediction studies of *P. monodon* by integrating GXE information and environmental variables including prediction of performance in new environments using machine learning models (Leiboff *et. al.* 2015). Furthermore, landmark derived information could also be used as inputs for training Deep Learning methods in particular those which use Convoluted Neural Networks to obtain predicted landmark distances direct from the images. This would increase efficiency and throughput of large-scale applications of morphometrics in semi-computerised digital systems.

CONCLUSIONS

This study shows that highly variable morphometric shapes can be characterized by multivariate quantitative analysis of landmark data, and there is potential to use this shape variability information as selection criteria in shrimp breeding programmes.

REFERENCES

Adams D. C., Rohlf F. J. and Slice, D. E. (2013) *Hystrix* **24:** 7.
Aktas M., Turan C. and Bozkurt A. (2006) *J. Anim. Vet. Adv.*
Claverie T. and Wainwright P. C. (2014) *PLoS One* **9:** e112732.
Hung D., Nguyen N. H., Hurwood D. A. and Mather P. B. (2014) **65:** 198.
Leiboff S., Li X., Hu H. C., Todt N., Yang J., Li X., and Schnable P. S. (2015) *Nature Comm* **6:** 8974.
Slice D. E. (2007) *Annu. Rev. Anthropol.* **36:** 261.
Stauffer Jr J. R. (1991) *Copeia*, 141.
Webster M. M., Atton N., Hart P. J. and Ward A. J. (2011) *PLoS One* **6:**: e21060.

GENOTYPE X ENVIRONMENT INTERACTION IN SHRIMP BREEDING: A REVIEW AND PERSPECTIVES

M.M. Hasan[1,2,§], P.C. Thomson[1,3], H.W. Raadsma[1,2] and M.S. Khatkar[1,2]

[1]ARC Research Hub for Advanced Prawn Breeding
[2]Sydney School of Veterinary Science, Faculty of Science, The University of Sydney, Camden, NSW, Australia
[3]School of Life and Environmental Sciences, Faculty of Science, The University of Sydney, Camden, NSW, Australia

SUMMARY

Genotype by environment interactions (GxE) can effectively suppress the response to selection in shrimp breeding programs. In order to understand the magnitude of GxE, we conducted a systematic review of literature on GxE in shrimp aquaculture. A departure from unity in the genetic correlation between the same trait measured in different environments indicated a GxE. Significant GxE was detected for survival and growth traits. The mean genetic correlation estimates for survival (n=22) and growth (n=26) traits were 0.53 (ranged -0.99 to 0.99) and 0.65 (range 0.47 to 0.99) across environments, indicating genetic improvement in one environment may not be fully realized in other environments.

INTRODUCTION

Global shrimp aquaculture has increased from 1 million metric ton in 1995 to almost 5 million metric ton in 2018, with further prediction of growing demand (Global Aquaculture Alliance 2018). Sustainable increased production of shrimp requires the development of genetically improved high-quality seed stock. Globally, breeding programs have been implemented for a number of shrimp species in particular for *Penaeus vannamei*. Despite the development of such breeding programs, realised genetic gains for different traits have been limited (Nguyen 2016). This may be due to several factors that limit response to selection, including genotype x environment interactions (GxE). GxE is the difference in phenotypic expression of traits in different environments by genetically similar organisms that result in inconsistent performance of superior genotypes across different environments. Evidence of GxE is observed by a re-ranking of breeding values of individuals for each environment. Moreover, as compared to plant and livestock production such phenomena are more obvious in aquaculture breeding programmes, where the majority of the commercial grow-out conditions can be very diverse e.g., ponds, outdoor net and cages, raceways compared to the breeding and selection environments under which brood stock are raised and selected. Given such large variability in production environments, a critical understanding of the importance of GxE is vital for the design of optimal breeding programmes for aquaculture species, including shrimp. Evidence of substantial GxE may require shrimp breeders to develop environment-specific strains. Furthermore, in the era of high throughput phenomics and genomic selection, novel selection strategies may present themselves to mitigate the impact of GxE by making environment specific selection decisions at a central location (i.e., measure performance under diverse commercial environments and link back through genomics to central seed stock providers). The purpose of this study is to review and characterise the magnitude of GxE for important health and production traits within shrimp species, and to discuss the possible knowledge gaps and the implications of these findings in shrimp breeding programmes.

MATERIALS AND METHODS

Publically available GxE studies of shrimp species were selected using the 'PubMed' search engine with the 'topic' search terms 'genotype', 'environment', 'interaction', 'shrimp/prawn' and 'aquaculture'. This search gave 20 published articles, which were selected for this review. For each selected GxE study, we recorded the publication date, species name, breeding design, pedigree, number of individuals/families sampled, type of study (e.g., marker/pedigree based), traits, environment type and presence of GxE. Information about the publication year, country of study, commonly used species, pedigree information and extent of GxE were assembled. Various traits and environmental conditions were categorized into three broad trait groups and three environmental groups, (Table 1). The extent of GxE among the selected studies was determined using two approaches. Firstly, estimates of genetic correlation between the same trait across different environments (r_g) was taken as a measure of GxE. Among the selected studies on GxE interaction, a total of 15 studies reported such genetic correlation between environments. Secondly, the proportion of variance accounted for by interactions between families (genotype) and environments, as reported in six different studies, were compiled. All data were analysed in the R statistical environment (version: 3.5.0) (R Core Team 2013).

Table 1. Definition of trait groups and environments

Traits	Definition
Growth	Body weight, growth rate, body length
Survival	Overall or in a challenge test
Body composition	Protein, lipid, carbohydrate contents
Environment	**Definition**
Habitat	Pond vs. pond (location), pond vs. tank or cage
Feed	Feed type and feeding regimes.
Stress	Salinity, ammonia concentration, temperature, density (high vs. low), pathogens.

RESULTS AND DISCUSSION

The most commonly reported species in the GxE studies was *P. vannamei* followed by *P. japonicus*. Growth and survival were the most commonly studied traits in family-based breeding populations. The earliest GxE study on shrimp was conducted in 2002. There was no specific trend found in terms of the number of studies per year, although highest number of studies were reported in 2018, and most of the studies were conducted in Australia. We recorded a total of 62 genetic correlation estimates primarily focussed on growth, whilst some studies also reported on body composition and feeding efficiency. Habitat type was the most common environmental factor studied reported in the GxE studies for shrimp.

By analysing six studies on variance attributed to GxE, we detected that habitat type and stress had significant interaction for the shrimp growth and survival traits at a family level (genotype). On an average, the habitat x family interaction for growth accounted for 7% of the variation in the model, and the stress environment accounted for 0.5% and 9% of the variation for the growth and survival traits. A higher level of variability in GxE interaction was observed for survival traits compared to growth traits (Figure 1).

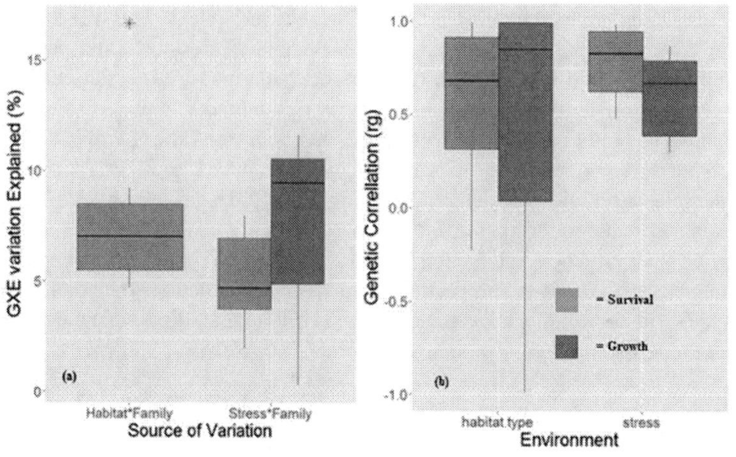

Figure 1. (a) Percent of GxE variation accounted for in ANOVA models; and (b) mean rg values across traits by environments

Growth traits were most commonly reported across all the studies. This might be due to the fact that these traits are directly correlated with economic return in shrimp farming. Moreover, these traits are also easier to record as compared to other traits. The mean r_g value for growth trait was 0.65 with a range of 0.47 to 0.99 (Figure 1b). This positive r_g value implies that selection for improved growth performance in one environment is likely to lead to a genetic response in the same direction in other environments but will result in re-ranking of genotypes. Survival as another economically important trait in shrimp aquaculture, was the second most reported traits for GxE interaction studies in shrimp. The mean r_g value for survival was 0.53 for all the studies with an extreme range of -0.99 to 0.99. This indicates that the re-ranking of candidates available for selection for survival is potentially of great importance pending the environmental conditions in which their progeny are expected to perform. This extreme level of re-ranking for survival traits can be as a result of different causative mortality factors in different environments. Furthermore, there might be limited additive genetic variation for resilience against these factors and making changes in co-variance across environments relatively more important.

Opportunities for genomic selection in overcoming GxE. Traditionally, GxE is viewed as a problem in breeding programs. Where commercial animals are raised in a different environment compared to that where the original selection of breeding animals is carried out, which may result in GxE. Consequently, genetic gain made in the nucleus breeding environment may not be fully realized in the production environment. This scenario is most likely in shrimp and other aquaculture species. Firstly, since aquatic animals are immersed in their water environments which greatly affect physiological activities and overall production can vary significantly across different locations/farms. Secondly, like in poultry and pigs, selection and breeding of broodstock in shrimp species is usually conducted in the hatchery/nucleus which is often pathogen free, but the actual breeding objective usually is to improve the performance of family lines (or their crosses) in commercial environments. Compared to traditional selection breeding program, which rely on pedigree information of family line-bred animals, genomic selection with reference populations of animals grown in several environments can be far more effective. Such multi-environment reference populations could be combined with high-throughput genomics and phenomics using machine learning. For instance, high-throughput image-based phenotyping of very large numbers has been reported for several genomic selection studies with plants and animals (Leiboff 2015; Zenger *et al.* 2017). The large datasets from high-throughput phenotyping can be utilized to improve resilience of animals across environments. Multi-environment reference population could facilitate the computation of more accurate genomic breeding values (GEBV) for all genotyped and pedigreed selection candidates in their ranking for performance and resilience in specific environments. This also opens the possibility of creating specialized lines by genomic selection. Creating such specialized lines would also facilitate maintenance of broader genetic diversity across lines (Mulder 2017).

Developing breeding programmes by incorporating environmental information. Variable environmental factors, including disease and pathogen challenge, salinity, temperature and occurrence of severe weather events, are potentially major constraints to profitable shrimp farming and may limit sustainable and uniform production. Moreover, in the face of climate change, these factors may have profound impact depending on climatic zones, geographic locations and type of aquaculture systems (e.g., intensive vs. extensive). In such situations, animals bred for one specific set of conditions may not perform well in other environment. Moreover, climatic instability may facilitate physiological stress, as well as outbreaks of existing pathogens or may promote dispersal of new diseases. However, combined spatial modelling of all risk factors and selective breeding strategies can be applied to tackle these issues. Environment specific genetic lines can be established by identification of genotypes that perform optimally in specific locations or specific conditions. For example, using detailed environmental data, prevalence of both short and long-term pathogen attack can be predicted (Singh *et al.* 2006). Based on current weather forecast data, such predictions can potentially be carried out for coming year or season, in combination with historical data on disease, pathogens and climatic factors. It is important to develop genotypes resilient to disease and pathogens and suited for a range of environments to minimize the effect of GxE. Although no experiments have been carried out with shrimp, studies with wheat and barley have shown that specific disease outbreaks can be successfully predicted and adapted resistant cultivars used to suit local conditions (Singh *et al.* 2006; Gillberg *et al.* 2017). Similar approaches can be applied to shrimp breeding and production by developing locally adapted and resilient shrimp lines for specific environments, providing a possible means to achieve yield improvement.

CONCLUSIONS

This review shows substantial GxE of importance for a number of economic traits of commercially farmed shrimp species and that such GxE effects warrant consideration in the development of shrimp breeding programmes. Furthermore, it also suggested the potential for genomic selection combined with large scale phenomics across diverse environments to overcome limitations imposed by GxE in advanced shrimp breeding programmes.

REFERENCES

Gillberg J., Marttinen P., Mamitsuka H. and Kaski, S. (2017) *BioRxiv.* **213**:231.
Leiboff S., Li X., Hu H. C. and Schnable, P. S. (2015) *Nat. Comm.* **6**: 8974.
Mulder H. A. (2017) *J. Anim. Breed. Genet.* **134:** 435.
Nguyen N. H. (2016) *Fish & Fisheries* **17:** 483.
Singh R. P. Hodson, D. P. and Ward, R. W. (2006) *CAB Reviews.* **1:** 1.
Team R. C. (2013). R: A language and environment for statistical computing.

BIOECONOMIC MODELLING OF AUSTRALIAN BLACK TIGER PRAWN *PENAEUS MONODON* UNDER INTENSIVE POND CULTURE

M.C. Marín-Riffo[1,4], H.W. Raadsma[1,4], D.R. Jerry[2,4,5], G.J. Coman[3,4] and M.S. Khatkar[1,4]

[4]ARC Research Hub for Advanced Prawn Breeding
[1]Sydney School of Veterinary Science, University of Sydney, Camden, NSW, Australia
[2]Centre for Sustainable Tropical Fisheries and Aquaculture, James Cook University
[3]Aquaculture Program, CSIRO Agriculture and Food
[4]Queensland Bioscience Precinct, St. Lucia, QLD, 4067 Australia
[5]Tropical Futures Institute, James Cook University, Singapore

SUMMARY

We developed a bio-economic model for the black tiger prawn *Penaeus monodon* for intensive pond culture in Australia. Hatchery, grow-out, and combined hatchery-growout scenarios were evaluated by simulating a varying number of ponds in operation. Annual, six-monthly, quarterly and monthly stockings were included in each business scenario. Net present value (NPV) was the main economic variable used for comparison. Highest profitability in the combined business was obtained in a monthly stocking cycle and largest number of ponds included. Feed cost was the major cost in all scenarios studied. Sensitivity analyses showed feed and survival parameters were critical for profitable business models.

INTRODUCTION

In Australia, consumption of marine prawns is a growing market, and most prawns are sourced from wild-caught stocks (~21,000 tonnes in 2016-17; ABARE 2017), or from imported fisheries and aquaculture product (~31,000 tonnes in 2016-17; ABARE 2017). Seasonal variability in supply of wild-caught prawns, along with potential for domestic replacement of imported product, provide significant market opportunities for locally farmed prawns (Treadwell *et al.* 1991). Despite these opportunities, the Australian prawn farming industry has not grown to dominate the domestic market (~5,000 tonnes in 2016-17; ABARE 2017). Limiting factors are the high-risk nature of the prawn industry, site-specific limitations, capital-intensive investment needed, high reliance on technology and technical expertise (Robertson 2006). The high-risk nature of aquaculture businesses can be reduced by controlling the biological and technological variables, and determining critical thresholds for cost and returns in a profitable business.

P. monodon is the main species of marine prawn farmed in Australia (Government of Queensland, 2018). The advantage and attractiveness for growers of farming this species is its rapid growth rate, large size compared to other species, lower food input costs, and its demand in the market. Prawn production can be divided into post larvae (PL) production hatchery and grow-out animals in ponds. In hatcheries, two critical factors driving productivity are optimum utilization of tank facilities, and larval production from domesticated breeding stock or wild broodstock. Grow-out business operations purchase larvae at day 15 (PL15) from commercial hatcheries to stock their ponds. Some big aquaculture enterprises undertake both hatchery and grow-out operations as a combined business. The commercial viability of grow-out farms has been analyzed in Australia by (Treadwell *et al.* 1991), who demonstrated that small scale farms (30 ha) are profitable under stocking scenarios of either one or two annual production cycles. (Hardman *et al.* 1990) showed that small size farms (6 and 20 ha) had a high uncertainty of returns and risk involved at low and high density stocking rates. Farm sizes over 30 ha have not been modelled yet, and there is an opportunity to increase efficiency of large-scale production systems using multiple stocking events per year and scaling up operational size to capture increased efficiency.

This study presents a bio-economic model of *P. monodon* in Australia under intensive farming systems for three business models (hatchery, grow-out and combined) comparing both scale and variable stocking per year.

MATERIALS AND METHODS

A model of an aquaculture production program of *P. monodon* was developed using the parameters of a typical prawn farming operation in Australia. Information from three different sub models-namely a biological model, and technical model and an economic model were combined into a single model with options to change parameters interactively. The models were initially developed in MS Excel and finally in the form of R scripts. A total of 298 different input parameters based on published studies or experts' opinion were used (data not shown). Breeding and production designs were generated for production over a business period of 10 years. Three different business were simulated for comparing economic performance: 1) Hatchery business: Production size (number of PLs) depended on number of females used for reproduction per spawning event. Small, medium and large production, consisted of 50, 250 and 500 females per spawning cycle, were evaluated. 2) Grow-out business: This included acquisition of PL15 from a commercial hatchery in each production cycle, growing animals in ponds until the harvest size, and sale ex-factory. Each production size depended on the number installed ponds (fixed capacity). A total of 50, 250 and 500 ponds under production were evaluated. 3) Combined business: Both hatchery and grow-out were combined in one business using 50, 250 and 500 fixed ponds, and number of females required for stocking event. Different stocking events per year were added in each business viz. annual, six-monthly, quarterly and monthly. Each stocking type was repeated for 10 years of the business period. All the available ponds were used in annual and six-monthly stocking cycles, whereas in quarterly and monthly cycles, the ponds were used to minimize empty capacity at any given time as shown in Table 1.

In the biological sub-model, a 28-day cycle covered production of eggs, nauplius, protozoea, mysis, PL1 to PL15 in hatcheries. PL15 were transferred to grow-out ponds under intensive culture with a grow out period from 29 to 168 day to obtain an average body weight of 34.45 g at harvest after 140 days. In the technological sub-model, resources required were based on the biological variables for each production phase. In the hatchery, each input was directly linked to the number of larvae generated and tanks used. In the grow-out phase, each input was linked to biomass and number of available ponds. In the economic sub-model, income, investment, depreciation, and cost variables were evaluated on a continuous time scale *i.e.* analyzed on per day basis. Profit of each business was determined by net present value of cash flow (NPV) in USD (Ross *et al.* 2008).

RESULTS AND DISCUSSION

This study presents the first systematic bioeconomic modeling of hatchery and grow-out operations of *P. monodon* for the Australian Industry using a variety of input variables used in Australia and other countries.

The hatchery business had positive NPV of cash flow when serving at least 50 ponds monthly (with a farm size of 250 ponds) (Table 1). For all the other scenarios the hatchery business was not profitable. Monthly stocking events reduced the effect of fixed costs but proportionally increased the cost of feed and breeders, the later was offset by the proportional increase in the sales of PLs. When customers are stocking annually, six monthly or quarterly, hatchery facilities were underutilized which increased the proportion of fixed cost in the hatchery business. In this study we have used fixed input parameters for production and reproductive performance, however, in practice, there can be a large variation in reproductive performance in hatchery (Uddin and Rahman 2015), any such variation in breeding behavior can change production costs. A decrease in reproductive performance is potentially a high-risk variable which can affect profitability in large-scale hatcheries.

Table 1. NPV of cash flow (in millions of USD) in hatchery, grow-out, combined business, as per the number of ponds installed (50, 250, 500) and multiple stocking per year (annually, six-monthly, quarterly, monthly), analyzed over a business period of 10 years

Ponds available	Ponds harvested annually	Stocking event	Business		
			Hatchery	Grow-out	Combined
50	50	Annually	-1.95	0.83	-0.40
	100	Six-monthly	-1.62	10.8	9.94
	100	Quarterly	-0.88	11.8	11.7
	120	Monthly	-0.45	16.2	16.5
250	250	Annually	-6.64	7.30	1.35
	500	Six-monthly	-4.90	57.4	53.2
	500	Quarterly	-1.09	62.7	62.3
	600	Monthly	1.69	85.1	87.5
500	500	Annually	-12.5	15.6	3.76
	1000	Six-monthly	-8.99	116.1	107.8
	1000	Quarterly	-1.38	126.3	125.6
	1200	Monthly	4.37	171.3	176.3

In the grow-out business, all the scenarios modelled with 50 to 500 ponds stocked annually, six-monthly, quarterly, and monthly production cycles were profitable. The highest profitability was obtained with monthly stocking in farm operations where 500 ponds were used (171.26 million USD), which was over 200 times more profitable than 50 ponds stocked annually (0.83 million USD). When farms with 500 available ponds were stocked on a monthly basis, 100 ponds were used simultaneously such that a total 1200 ponds were harvested over a period of twelve months (Table 1). In monthly production cycles, there was a minimal time gap to prepare and use each pond, this maximized use of facilities and productivity. In all farm sizes, stocking ponds annually significantly underutilized available resources and this is shown in the proportionally heavily reduced profitability compared to stocking at higher frequency (Table 1).

The combined business (incorporating a hatchery with grow-out business) produced higher profit than grow-out business only when the hatchery business was profitable i.e. when at least 250 ponds were used under a monthly stocking cycle. A negative NPV of cash flow was obtained when 50 ponds were stocked and harvested annually, since under the combined business, the negative NPV of cash flow of the hatchery had to be absorbed (Table 1). However, in a combined business the farm is independent from commercial suppliers of PLs but this comes at a significant cost.

Sensitivity analyses of key biological and economical parameters were used to analyze impact on NPV in each business. In the base model survival in the ponds was set at 66%; when survival decreased to 50%, NPV of cash flow decreased in grow-out and combined business. Conversely when survival increased from 66% to 70% NPV of all business models increased due to the increase in income. Labor cost was increased from 100% in base model to 110% to examine the impact on NPV cash flow, which was marginal on each business, as the labor requirements were proportional to the biomass produced and number of animals farmed and harvested. The main cost in each business was feed. Here a fixed biological variable of Feed Conversion Ratio (FCR) was used in the model. FCR can have a high impact on NPV of cash flow in each business. Increment in feed cost of 20% with respect to base model had a high impact on profitability in each business (the results not shown).

In this study the number of broodstock used per stocking event was fixed using an average reproductive performance. However, in practice, reproductive performance can be highly variable (Coman *et al.* 2006) which may require the culture of larger numbers of broodstock, which in turn affects profitability. An industrial scale production of *P. monodon* culture requires a strategic plan for consistent supply of broodstock which can be insured by well domesticated stocks preferably incorporating selective breeding programmes. We have used scenarios without any genetic improvement. The efficiency of prawn farming can be significantly increased by incorporating advanced genetic improvement programmes using genomic information (Khatkar *et al.* 2017; Zenger *et al.* 2018), however, bioeconomic modeling of such breeding operations is critical for evaluating their profitability. The bio-economic modelling developed and presented in the present study can be extended further to evaluate different breeding designs.

In this study, NPV was analyzed in time continuous *i.e.* per day basis and was very responsive to changes in biological parameters of high effect (*i.e.* survival and feed). Analysis of NPV on a continuous time basis produced a much better assessment of economic impact which is impossible to capture under simple models or models comparing profitability on an annual basis (Treadwell *et al.* 1991; Hardman *et al.* 1990).

CONCLUSIONS

Under Australian conditions grow-out business for *P. monodon* appears to be profitable for a range of farm sizes (50 to 500 ponds). Hatchery business are only profitable in large-scale operations supplying customers on monthly basis. Combining hatchery and grow-out is only warranted for large-scale operations under intensive stocking. Overall, survival and feed parameters had greatest impact on NPV for the scenarios modelled here. This is the first and most comprehensive bio-economic model for *P. monodon* under intensive culture conditions using NPV on a continuous time basis.

REFERENCES

ABARE. (2017). Australian fisheries and aquaculture statistic 2017 http://www.agriculture.gov.au/abares

Coman G.J., Arnold S.J., Peixoto S., Crocos P.J., Coman F.E., and Preston N.P. (2006). *Aquaculture* **252**: 372.

Government of Queesland. (2018) Black tiger prawn aquaculture https://www.business.qld.gov.au.

Hardman J.R.P., Treadwell R., and Maguire G.B. (1990) *Memoirs of the Queensland Museum* **31**: 421.

Khatkar M.S., Zenger K.R., Jones D.B., Prochaska. J., van der Steen H.A.M, Jerry D.R. and Raadsma H.W. (2017) *Proc. Assoc. Advmt. Anim. Breed. Genet.* **22**:121.

Robertson, C. (2006) "Australian prawn farming manual: health management for profit". DPI&F, Queensland.

Ross S.A., Westerfield R. and Jordan B.D. (2008) "Fundamentals of corporate finance", 9th ed. McGraw-Hill Irwin. New York.

Treadwell R.F., Maguire G., and McKelvie L. (1991) Economic aspects of aquaculture in North Queensland. ABARE, Camberra.

Uddin S., and Rahman M. (2015) *Journal of the Bangladesh Agricultural University* **13**: 315

Zenger K. R., Khatkar M. S., Jones D. B., Khalilisamani N., Jerry D. R., and Raadsma H. W. (2019) *Front. Genet.* **9**:693.

Poster

ECONOMIC WEIGHING OF TRAITS IN A PRELIMINARY SELECTION INDEX FOR OSTRICHES IN SOUTH AFRICA

A. Engelbrecht[1], S.W.P. Cloete[1,2] and P.R. Amer[3]

[1]Directorate Animal Sciences, Western Cape Department of Agriculture, Elsenburg, 7607 South Africa
[2]Department of Animal Sciences, Stellenbosch University, Matieland, 7602 South Africa
[3] AbacusBio Limited, Dunedin, 9058 New Zealand

SUMMARY

A selection index to identify the best males and females from the Oudtshoorn research flock to supply the ostrich industry with performance-recorded breeding material was built using a cost approach. Traits of economic importance that were included were growth, adult weight, chick production and survival, feather weight and quality, nodule development and the presence of hair follicles. Economic values were weighted by discounted genetic expressions coefficients to determine economic weights for these traits and combined with genetic standard deviations to indicate the relative emphasis that is put on each trait in the resulting selection index.

INTRODUCTION

The South African ostrich industry has a lack of pedigreed flocks due to the wide use of large breeding colonies where individual reproduction cannot be recorded. The government owned Oudtshoorn Research Farm is one of the only institutions with a breeding structure (pair breeding) and recording system that can generate data suitable for genetic evaluation (Cloete *et al.* 2008b). Data on quantitative and qualitative production traits are also routinely recorded on the farm and can be used to estimate genetic parameters for traits of economic importance.

Quantification of the economic value of the relevant genetic traits is needed to optimise genetic selection strategies (Amer *et al.* 1999). The ostrich industry is basically a slaughter industry, with most income being derived from the products, namely meat, leather and feathers. Live weight or growth to slaughter would therefore be an important trait to consider, since live weight is highly correlated to both carcass weight and skin size (Engelbrecht *et al.* 2009; 2013). Survival and reproduction rate should also be considered in a breeding plan for the ostrich industry (Cloete *et al.* 2008b). Genetic and economic information for these traits can be combined into a multiple trait selection index, which gives proper weight to each trait to estimate genetic merit and ensure more efficient selection than when selecting for one trait at a time (Hazel 1943). Account must also be taken of the frequency of expression of each trait in an individual and its descendants, as well as discounting to account for time delays in the expression of some traits relative to others (McClintock & Cunningham 1974; Amer 1999).

This paper provides a description of simple economic weight calculations for the development of a maternal and a paternal index, which will allow identification of the best males and females from the progeny of Research Farm flock for replacement purposes, as well as for selling to commercial farmers.

MATERIALS AND METHODS

Traits of economic importance were identified by making use of a bio-economic model prepared by Van Zyl (2001), adapted according to current industry cost structures. Once traits of economic importance were decided upon, economic values for the various traits were calculated and economic weights were subsequently derived.

Input cost variables and revenue streams were computed to determine the economic value of each trait, following methods described by Byrne *et al.* (2016). The economic value of a trait was defined as the amount by which profit may be expected to change for each unit of improvement in the trait concerned (Hazel 1943). These economic values were transformed into economic weights by accounting for the different rates of transmission and timing of expression (Amer 1999).

"Days to slaughter" was one of the first traits of economic importance chosen to be used in the selection index. For skin quality, nodule size and the presence of hair follicles was chosen as the most important traits influencing skin income (see Engelbrecht 2013). Feather weight and feather quality was chosen as predictor traits of feather income. Other traits chosen for use in the selection index were the number of chicks per female per season and adult live weight.

Calculations of the economic values for these traits were based on various assumptions and industry averages (Table 1), as well as current industry price structures (April 2019).

Table 1. Base input parameters used for the calculation of economic values

Description	Value
Annual investment/cost of farm	5 000 000 ZAR
Cost of one labourer per annum	50 000 ZAR
Number of labourers	4
Number of birds slaughtered per year	1000
Interest / mortgage minus inflation	0.07%
Maintenance/depreciation	0.05%
Number of chicks per female	20
Survival to slaughter	60%
Mature breeder weight	130kg
Breeder feed intake per day	2.6kg
Juvenile bird (7-month-old) feed intake per day	1.9kg
Price/kg breeder diet	4.50 ZAR
Price/kg maintenance diet	3.50 ZAR
Price/kg grower diet	4.00 ZAR
Length of breeding season	224 days
Average live weight gain per day at slaughter age	300g/day
Maintenance requirements per day	116.5g/day

Economic value of chick production. The economic value of number of chicks per season was calculated based on increasing the average number of chicks per female per production year (20) by one. This translated to needing 4.76% (1-20/21) fewer breeder birds to produce the same number of chicks as before, which would represent a saving of 4.76% on the cost of breeder birds. The annual cost of a breeder pair was calculated to be 7807.80 ZAR, so this translates to an economic value of 371.80 ZAR.

Economic value of survival to 7 months of age. If one assumes that chicks die on average at 2.5 months of age, the costs incurred to rear a chick from hatch to 2.5 months of age need to be added to the value of a day-old chick to determine the economic value. This equates to 570.26 ZAR.

Economic value of weight at 7 months of age. It was assumed that farmers will slaughter animals earlier if they grow faster. Target slaughter weight was assumed to be 100kg. Due to the high genetic correlation between slaughter weight and weight at 7 months of age (Engelbrecht et al. 2011), and because preliminary selection decisions must be done before slaughter, it was decided to use a target weight of 70kg at 7 months of age as a selection trait instead. An increase of 1kg at 7 months, would translate to a 1.43kg heavier bird at slaughter age (100/70), which translates into a cost saving of 4.76 days (1.43kg/0.3kg average live weight gain/day) because the bird can be slaughtered earlier. The feed saving was calculated only for 3.33 days (4.76 x 0.7) though, since the birds were assumed to be a little heavier over their life. The saving on maintenance feed, labour, interest and depreciation amounted to an economic value of 35.77 ZAR per kg of 7 month weight.

Economic value of adult weight. If we assume that maintenance requirement equates to (live weight)$^{0.75}$, and average mature weight is 130kg, then increasing mature weight by 1kg will increase maintenance feed requirements by 0.57%, which translates to a negative economic value of 22.55 ZAR per adult bird per season.

Economic value of nodule size. Using the weighted average price of all skins and the distribution of nodule size scores across all skins, the economic value of an improvement of one unit in nodule size score was estimated at 137.31 ZAR.

Economic value of hair follicle prevalence. The economic value for hair follicles was calculated by considering the effect of improving hair follicle score (scored on a scale of 1 to 9) by one. Taking into account the distribution of hair follicle scores across all skins evaluated on the research farm, it was calculated that the number of skins being downgraded as a result of hair follicles would decrease by 12.9% as a result. Using the weighted average price difference between grades and the current distribution of grades, the economic value of an improvement in hair follicle score was estimated to be 104.67 ZAR.

Economic value of feather weight. The economic value of feather weight was calculated from an average price of 1562 ZAR for 1.17kg of slaughter feathers. The effect of an increase of 100g in total feather weight therefore equates to 133.50 ZAR.

Economic value of feather quality. The distribution of feather grades for white plumes and other wing feathers, the pricing for the different feather grades and the average feather weights for the different feather categories were used to derive an economic value based on the effect of an improvement of 1 grade in quality on the weighted average price. This amounted to an economic value of 162.54 ZAR for feather quality.

Accounting for number of expressions. Transmission factors (allowance of 0.5) for traits expressed by chicks instead of by the actual parent, were multiplied by the number of animals expressing each trait to convert units of economic values to a constant expression scale to give the final economic weights.

Percent emphasis of traits. Genetic parameters obtained from previous analysis of data from the pair-bred ostrich flock at the Oudtshoorn Research Farm (Cloete *et al.*, 2008a; Engelbrecht, 2013) were then used to compute percent emphasis for each trait. Estimates of the genetic standard deviation and variability for feather weight and quality were obtained by analysing weight and length records (as indicator of quality) for mature white wing feathers (n=3092).

RESULTS AND DISCUSSION

Table 2 describes the specific trait weights for a female line index. Traits with high positive economic weights were chick production, nodule size score and feather quality, while hair follicle score had a high negative economic weight.

The relative weights of the traits reflect the current industry situation, where feather prices are disproportionally elevated, while meat and skin prices are under pressure. It also reflects the increasing importance of hair follicles for determining skin quality. Similar to other livestock industries, reproduction (chick production) remains the most important trait of economic importance. For the male line index the corresponding relative trait emphasis were 17.36 for survival, 25.05 for 7-month weight, 3.49 for adult weight, 9.93 for nodule size score, 13.26 for hair follicle score, 13.09 for feather weight and 17.82 for feather quality.

Table 2. Economic values, economic weights, genetic standard deviations and the relative emphasis of different ostrich traits in a female line index per mating season

Trait	EV (ZAR)	Trans-mission	Number of times expressed / parent	Economic weight	Genetic s.d.	Relative emphasis
Chick production	371.80	1	1	371.80	6.52	32.26
Survival to 7 months	570.26	0.5	20	5702.63	0.15	11.76
7-month weight	35.77	0.5	12	214.62	5.94	16.97
Adult weight	22.31	1	1	-22.31	7.96	2.36
Nodule size score	137.31	0.5	12	823.87	0.61	6.72
Hair follicle score	104.67	0.5	12	-628.03	1.07	8.98
Feather weight	133.50	1	1	267.01	2.50	8.87
Feather quality	162.54	1	1	325.07	2.79	12.07

CONCLUSION

Economic weights were determined for a wide range of economically important ostrich traits to provide a basis from which a robust comparison of selection candidates can be made. However, according to Hazel (1943) an index constructed from data taken on a flock in one locality may not be widely applicable, due to differences in genetic constitution, environmental factors and management practices. Optimizing genetic improvement in the ostrich industry will therefore only be possible once commercial ostrich farms improve their breeding structures to allow access to data from different flocks and environments. Once an exhaustive set of genetic parameters are available for all the traits, these should also be factored into the selection index. Furthermore, 5-year averages for economic costings should potentially be used to account for the variability in market conditions applicable to the industry to increase the accuracy of the index over time.

ACKNOWLEDGEMENTS

The South African ostrich industry is thanked for providing information for calculating economic values. Financial support of the Western Cape Department of Agriculture and the Western Cape Agricultural Research Trust is also gratefully acknowledged.

REFERENCES

Amer P.R. (1999) *NZ J. Agric. Res.* **42**: 325.

Amer P.R., McEwan J.C., Dodds K.G. and Davis G.H. (1999) *Livestock Prod. Sci.* **58**: 75.

Byrne T.J., Santos B.F.S., Amer P.R., Martin-Collado D., Pryce J.E. and Axford M. (2016) *J. Dairy Sci.* **99**: 8146.

Cloete S.W.P., Brand Z., Bunter K.L. and Malecki I.A. (2008a) *Austr. J. Exp. Agric.* **48**: 1314.

Cloete S.W.P., Engelbrecht A., Olivier J.J. and Bunter K.L. (2008b) *Austr. J. Exp. Agric.* **48**: 1247.

Engelbrecht A. (2013) PhD dissertation, Stellenbosch University, South Africa.

Engelbrecht A., Cloete S.W.P., Bunter K.L. & van Wyk J.B. (2009) *Proc. Assoc. Advmt. Anim. Breed. Genet.* **18**: 548.

Engelbrecht A., Cloete S.W.P., Bunter K.L. and van Wyk J.B. (2011) *Proc. Assoc. Advmt. Anim. Breed. Genet.* **19**: 183.

Engelbrecht A., Cloete S.W.P., Bunter K.L., van Wyk J.B. and Hoffman L.C. (2013) *Proc. Assoc. Advmt. Anim. Breed. Genet.* **20**: 381.

Hazel L.N. (1943) *Gen.* **28**: 476.

McClintock A. and Cunningham E. (1974) *Anim. Sci.* **18**: 237.

Van Zyl P. (2001) MSc (Agric) Thesis, Stellenbosch University, South Africa.

DISCOVERY OF SIGNATURES OF SELECTION IN BEEF AND DAIRY CATTLE USING ULTRA HIGH-DENSITY SNP GENOTYPES

I.A.S. Randhawa[1,3], M.S. Khatkar[1], P.C. Thomson[2], R.D. Schnabel[4], J.F. Taylor[4] and H.W. Raadsma[1]

[1]Sydney School of Veterinary Science, University of Sydney, Camden, NSW, Australia
[2]School of Life and Environmental Sciences, University of Sydney, Camden, NSW, Australia
[3]School of Veterinary Sciences, University of Queensland, Gatton, QLD, Australia
[4]Division of Animal Sciences, University of Missouri, Columbia, MO, 65211 USA

SUMMARY

Molecular data can provide insights into historical natural or artificial selection events and the genetic architecture underlying breed-specific traits. We examined the impact of phase of ancestral and derived alleles, marker density and composition of reference population panels to validate known and discovered novel genomic regions under selection in Angus and Holstein cattle. Using a composite selection signals method which combines multiple tests into a single score, 57 regions in Angus and 55 regions in Holstein cattle were detected by using ultra-high-density genotypes (2.5M SNPs) compared to four regions in Angus and five regions in Holstein Friesian using a low-density 50K SNP genotyping strategy. The detected regions include many regions known to harbour variants associated with selected beef and dairy traits, as well as several novel putative selection signatures. We conclude that both marker density and composition of the reference panel affects the power to detect selection signatures in domestic cattle.

INTRODUCTION

Discovery of genetic variants and genes has been intensified during the past two decades to understand the biological control of agricultural and health traits in livestock species (Kemper and Goddard 2012). Selective breeding has improved trait performance by increasing the frequency of beneficial alleles throughout the genome. Genomic regions under selection are likely to be of functional importance and recurrent selection has left distinct imprints throughout the genome, called signatures of selection, by producing deviations in allele frequencies, reduced local nucleotide variability and increased linkage disequilibrium (LD) within long haplotypes (Randhawa *et al.* 2014). Angus and Holstein cattle, in particular, have been subjected to long-term selection for beef and dairy production, respectively, and have previously been studied to detect signatures of selection to identify the genomic regions that harbour functional variants influencing beef and dairy traits (Randhawa *et al.* 2016). Genomic investigations are frequently resource-intensive, however, detection of selection signatures can provide insights into the genetic architecture underlying breed-specific traits (Gibbs *et al.* 2009) in a relatively cost-effective manner.

Detection of selection signatures is strongly influenced by significance thresholds and the power of the selection tests, genotypic marker density, sample size, minor allele frequency and origin of ancestral alleles, structure of candidate populations, and the composition of the reference population (Randhawa *et al.* 2016). With the advent of ultra-high density genotyping platforms and whole-genome resequencing, the information per screened individual has risen exponentially, however often at a higher cost per tested sample, resulting in relatively few individuals per population being screened. As such, studies of selection signatures can be confounded by SNP density, sample size of candidate and reference populations. This study examined the impact of phase of ancestral and derived alleles, marker density and composition of reference population panels to validate known and to discover novel genomic regions under selection in Angus and Holstein cattle.

MATERIALS AND METHODS

A composite selection signals method (CSS), which here combined three tests into a single score (Randhawa *et al.* 2014), was used for the analysis of selection signatures. Firstly, the CSS method was evaluated by substituting one of the constituent tests, viz., ΔDAF (change in derived allele frequency), which requires ancestral and derived allele phase to be known, with the ΔSAF (change in selected allele frequency) test, which does not require known allele phases (Randhawa *et al.* 2014) by using phase-known low-density SNP data produced using the Illumina BovineSNP50 BeadChip assay ("50K SNPs") to assay 1630 samples representing 57 breeds of European cattle (Table 1) (Randhawa *et al.* 2014). Next, the CSS method was evaluated for the confounding effects of using different reference panel compositions in selection tests across breeds, by using ultra-high-density SNPs genotyped with a pre-screening assay comprising almost 3 million validated SNPs in collaboration with Affymetrix (Santa Clara, CA) to design the Axiom Genome-Wide BOS 1 Array Plate ("2.5M SNPs") comprising 105 samples (Table 1) from seven breeds: Angus, Holstein, Brahman, Hanwoo, Murray Grey, Simmental and Wagyu (Rothammer *et al.* 2013). For both datasets, imputation of missing genotypes and haplotype phasing was performed with BEAGLE 3.3 (Browning and Browning 2009). CSS scores were smoothed by averaging $-\log_{10}(p)$ values for SNPs within 1 Mb or 50 kb overlapping windows. The top 0.1% of smoothed CSS scores were used for significance thresholds and the boundaries of selection regions were defined by the contiguous SNPs within the top 1% and 0.5%, respectively for 50K and 2.5M datasets. Genome-wide significant regions under selection were detected in Angus and Holstein by using the 2.5M SNPs and these were compared with regions detected using the 50K dataset and the meta-assembly reported by Randhawa *et al.* (2016). It is noteworthy that the individuals from each of the breeds genotyped with the two panels were different. Therefore, this study provides a comparison of independent breed samples genotyped using two SNP densities.

Table 1. DNA samples, their geographic origin, country of sampling and genotypes data

Breeds	Geographic origin	Country of sampling	Two datasets	
			50K	2.5M
Angus	Scotland	Australia, New Zealand, USA	128	29
Holstein	Netherlands	Australia, France, New Zealand, USA	160	40
Other breeds' DNA	Worldwide	Worldwide	1342	36
Total DNA samples	-	-	1630	105
Reference samples*	-	-	165	45
SNPs (N) genotyped	-	-	54,609	2,575,339
SNPs (N) after QC	-	-	35,284	1,583,288

* Reference population comprised randomly selected equal numbers of samples from each breed with 50K (n = 165) or 2.5M (n = 45) genotypes.

RESULTS AND DISCUSSION

After quality control (call rate ≥ 95%), a total of 35,284 and 1,583,288 autosomal SNPs from the 50K and 2.5M datasets, with a mean inter-marker interval of 70.845 kb and 1.584 kb, respectively, and with MAF > 0.01 were used in the analyses (Table 1). Analyses of the 50K SNP dataset suggested that using either of the allele frequency-based tests of selection, ΔSAF and ΔDAF, detected the same regions under selection in both breeds within the top 0.5% of CSS scores. Correlations of scores between the two approaches were 0.92 in Angus and 0.91 in Holstein (Figure 1). Notably, using ΔSAF for computing CSS resulted in the detection of additional regions relative to using ΔDAF, because the latter is limited to derived alleles, while the former can detect selection affecting both ancestral and derived alleles. Overall, 4 regions (BTAs: 4, 13, 18, 21) and 5 regions (BTAs: 8, 10, 13, 20, 26) containing selection signatures were detected in Angus and Holstein, respectively, using 50K SNPs. Of these 9 regions, all but one Holstein region (BTA 13) were detected in the analysis of the 2.5M SNPs, suggesting the robustness of the CSS approach to SNP density using independent samples.

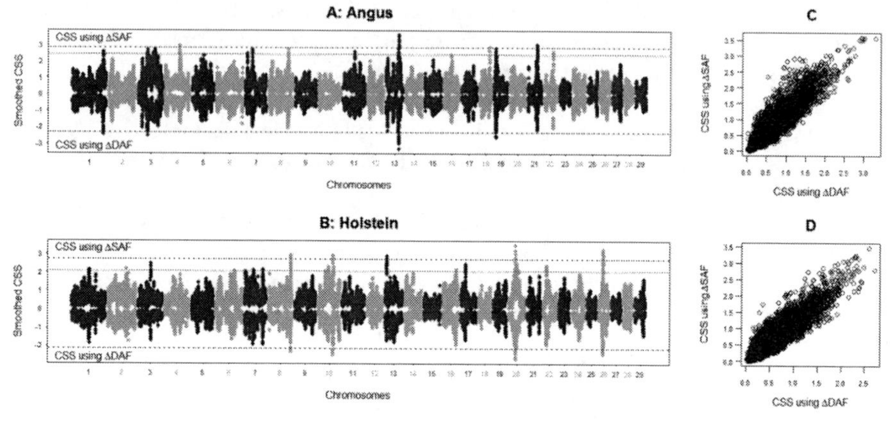

Figure 1. Comparison between the genome-wide smoothed CSS scores computed using ΔSAF and ΔDAF based on the 50K SNP dataset in Angus (A) and Holstein (B). Dashed (red and blue) and dotted (green) lines indicate the top 0.1% and 0.5% thresholds. Scatter plots of smoothed CSS scores computed using ΔSAF and ΔDAF in Angus (C) and Holstein (D)

Evaluation of three different reference panel compositions (1: single-breed, by Holstein vs Angus, 2: multi-breed, using all samples per breed, 3: multi-breed, with equal numbers of samples per breed) using three control regions on BTA6 (*ABCG2*, *KIT* and the casein cluster) in Holstein suggested that option 3 was optimal (results not shown). Hence, the final CSS analyses were performed using the optimal reference population strategy for Angus and Holstein and the ultra-high-density 2.5M SNPs and identified a total of 57 and 55 genomic regions, respectively (Figure 2).

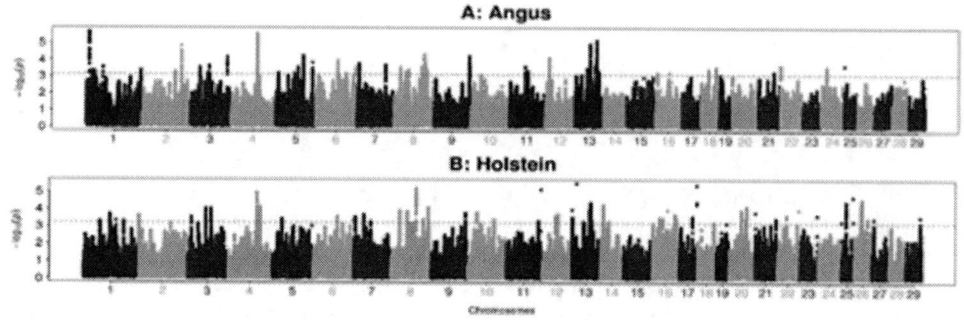

Figure 2. Manhattan plots showing smoothed CSS ($-\log_{10}(p)$) for Angus (A) and Holstein (B) using the 2.5M dataset. Horizontal dashed (red) lines indicate top 0.1% SNP thresholds

The significant CSS regions found in Angus and Holstein were compared against meta selection score (MSS) data, previously computed using a collection of published signatures of selection from 14 and 22 studies, respectively (Randhawa *et al.* 2016). Most of the significant CSS regions in both breeds coincided with regions with MSS $\leq$ 3. However, the distribution of MSS > 3, which indicates that a signature of selection has been validated in more than one study, was more frequently co-located with significant CSS regions in Holstein than in Angus. We also investigated the localization of CSS peaks for the presence of genes, because the bovine genome has variable gene density with an average of approximately nine genes per Mb. Gene locations in the UMD3.1 assembly indicate that 7.5% of 1 Mb genomic regions contain no genes (gene-sparse), whereas, 5% of 1 Mb regions contain 30-78 genes (gene-dense). Selection signatures in Angus and Holstein were mostly present in regions with variable gene-density and only ~7% (4 out of 57) and ~ 2% (1 out of 55), respectively, of significant CSS regions did not contain annotated genes. Given that the breeds are specialized for beef and dairy production, only five regions were found in common (BTAs: 1, 13, 19, 21 and 22).

This study detected several novel regions in both breeds; however, gene-dense regions can make it difficult to predict the underlying functional mutation. The identified signatures of selection show that selective forces have operated on the genetic architecture controlling growth and body size of beef cattle, and the physiological and anatomical structure of mammary glands, and quantity and quality of various milk components in dairy cattle. For example, *ABCG2* has been found to be involved in milk yield and composition and is a lactation regulator. *ABCG2*, along with the *NCAPG-LCORL* genes, has been found to be associated with stature in Holstein (Randhawa *et al.* 2016). *GHR* (Rothammer *et al.* 2013) is also a strong candidate gene with a major effect on milk yield and composition and is linked to many QTLs and is located in a region in which strong selection signatures have been identified in multiple cattle breeds (Khatkar *et al.* 2014). Additional genes underlying selective sweeps detected by CSS in this study, such as *SAR1B, AGTRAP* and *KIF1B* (Flori *et al.* 2009) are involved in the functioning of mammary glands, milk production and disease resistance in high producing dairy cows. The non-dairy related genes include *KIT* for white-spotting coat colour. Moreover, *PDGFRA* and *KDR* (Flori *et al.* 2009; Gautier and Naves 2011; Randhawa *et al.* 2014), *MGAT1*, *SPOCK1* (Gibbs *et al.* 2009) and *SIGLEC* genes (Khatkar *et al.* 2014) have been implicated with reproduction traits, due to their roles in fertilization, embryonic development and growth.

REFERENCES

Browning B.L. and Browning S.R. (2009) *The American Journal of Human Genetics* **84**: 210.

Flori L., Fritz S., Jaffrézic F., Boussaha M., Gut I., Heath S., Foulley J.-L. and Gautier M. (2009) *PLoS One* **4**: e6595.

Gautier M. and Naves M. (2011) *Mol. Ecol.* **20**: 3128.

Gibbs R.A., Taylor J.F., Van Tassell C.P., Barendse W., Eversole K.A., Gill C.A., Green R.D., Hamernik D.L., Kappes S.M., *et al.* (2009) *Science* **324**: 528.

Kemper K.E. and Goddard M.E. (2012) *Hum. Mol. Genet.* **21**: R45.

Khatkar M.S., Randhawa I.A.S. and Raadsma H.W. (2014) *Livestock Science* **166**: 144.

Randhawa I.A.S., Khatkar M.S., Thomson P.C. and Raadsma H.W. (2014) *BMC Genet.* **15**: 34.

Randhawa I.A.S., Khatkar M.S., Thomson P.C. and Raadsma H.W. (2016) *PLoS One* **11**: e0153013.

Rothammer S., Seichter D., Forster M. and Medugorac I. (2013) *BMC Genomics* **14**: 908.

COMPARING THE CARBON DIOXIDE AND METHANE EMISSIONS OF HOLSTEIN AND JERSEY COWS IN A KIKUYU PASTURE-BASED SYSTEM

N.M. Bangani[1], K. Dzama[1], C.J.C. Muller[1], F.V. Nherera-Chokuda[2] and C.W. Cruywagen[1]

[1]Department of Animal Science, Stellenbosch University, Stellenbosch, 7600 South Africa
[2]National Emergent Red Meat Producers' Organisation, Pretoria, South Africa

SUMMARY

The aim of this study was to estimate enteric carbon dioxide (CO_2) and methane (CH_4) production in Holstein and Jersey cows and compare the enteric gas emissions of the two breeds by parity and lactation stage. Data were test-day records of 122 Holstein and 99 Jersey cows from October 2005 to September 2014. Cows were kept on kikuyu pasture and received on as-fed basis 7 kg of a concentrate mixture containing 17% crude protein (CP) fed in two equal portions after each milking throughout the lactation period. The DMI of cows was estimated using the National Research Council method (NRC, 2001) and the pasture intake as the difference between the DMI and concentrate DMI. The Cornell Net Carbohydrate and Protein System equations (CNCPS) were used to estimate CH_4 (g/d) and CO_2 (kg/day) production. Holsteins produced more CO_2/day and CH_4/day than Jerseys. The production of the two enteric gases increased with lactation stage and parity in both breeds. When CO_2 and CH_4 were expressed as proportions of dry matter intake (DMI) or 100 kg body weight (BW), Holsteins had lower emissions both by parity and lactation stage. With CO_2/kg energy corrected milk (ECM), breeds did not differ; however, Jerseys produced lower CH_4/kg ECM. It is concluded that enteric CO_2 and CH_4 emission is affected by breed, lactation stage and parity. It is therefore recommended that the two production stages be accounted for when estimating the methane emission factor (MEF, CH_4/head/year).

INTRODUCTION

Carbon dioxide and CH_4 are natural by-products of enteric fermentation (Hook *et al.* 2010). Although the production of CH_4 (methanogenesis) is an essential process, it constitutes energetic inefficiency. Depending on feed composition and quality, methanogenesis represent a loss of about 2 to 12 % of dietary gross energy consumed by the host animal (Hook *et al.* 2010) with high-producing lactating animals losing at least 6% (Qiao *et al.* 2014). Enteric gases also contribute to greenhouse gas (GHG) concentrations in the atmosphere that are linked to global climate change (Broucek 2014). Methane has a global warming potential 25 times that of CO_2 (Broucek 2014). Although the CO_2 emitted is considered neutral as it arises from metabolism of plant-derived feeds (Smith *et al.* 2014), the inclusion of CO_2 emissions when balancing for carbon flows in the farm is essential to ensure that all sources of carbon emission are accounted for (Chianese *et al.* 2009).

As methanogenesis is inevitable and essential for rumen functioning, more studies are required to find ways to reduce CH_4 emissions. This would improve feed efficiency in ruminants as well as assist in mitigating their effects on global climate change. Breed comparison on enteric CH_4 emissions per unit of DMI or kg product produced is one strategy to be considered. In studies comparing Holstein and Jersey cows (Münger & Kreuzer 2006; Capper & Caddy 2012; Olijhoek *et al.* 2018), conflicting results on breed differences were obtained. Furthermore, the Intergovernmental Panel on Climate Change (IPCC 2006) is encouraging the development of country-specific MEF for different animal categories to enable close estimation of the country's emissions. Estimating emissions by sub-categories such as production stage, e.g., lactation stage and parity, will bring better accuracy as herd population vary throughout the year. According to Mangino *et al.* (2003), overlooking the effects of the production stages assumes that individual animal characteristics remain constant throughout a given year. No literature could be found on the effect of production stages on enteric gas emissions. The aim of this study is therefore to estimate enteric CO_2 and CH_4 production of Holstein and Jersey cows and compare breeds as affected by parity and lactation stage.

MATERIALS AND METHODS

The CNCPS equations were used to estimate CH_4 (g/day) and CO_2 (kg/day) emissions of cows (Van Amburgh *et al.* 2015). The MEF was calculated as daily CH_4 × 365 days. Data used as input in the models were from test-day records of 122 Holstein and 99 Jersey cows from the Elsenburg Research Farm of the Western Cape Department of Agriculture collected from October 2005 to September 2014. Cows were in a kikuyu pasture-based system and were supplemented with a7 kg/cow/day concentrate mixture containing 17% CP on as-fed basis, fed in two equal portions after each milking throughout the lactation period. The DMI of cows was estimated using the National Research Council method (NRC 2001) and the pasture intake was estimated as the difference between the estimated DMI and concentrate DMI. Cow parity varied from 1 to 7. Milk yield and components data were collected on test dates of approximately 35-days intervals in each lactation period, the same intervals were used for modelling enteric gases emitted. Lactation period was divided into four stages by creating class intervals from the test-day production data as follows: calving to 30 days as post-calving transition period (will just be referred to as transition period); 31 to 100 days as early lactation; 101 to 200 days as mid-lactation and above 201 days as late lactation. Data were analysed using the repeated measure methods of the PROC MIXED procedure of SAS Enterprise Guide version 7.1. The main effects were breed, parity and lactation stage, the interaction effects were breed × lactation stage, and breed × parity. The cow was fitted as a random effect while the response variables (CH_4, CO_2, and their efficiency measures) were repeated measurements within a cow at fixed test-dates of approximately 35 days intervals within parities. The between-breeds, between parity and between lactation stage variations and their interactions were compared using Bonferroni test and were declared different at $P < 0.05$. The following equation was used for statistical analysis:

$$Y_{ijkl} = \mu + B_i + P_j + LS_k + (B \times P)_{ij} + (B \times LS)_{ik} + cow_l(B_i) + \varepsilon_{ijkl}$$

RESULTS AND DISCUSSION

Holstein and Jersey cows had overall average BW of 567±3.5 vs. 411±3.8 kg, DMI of 17.8±0.11 vs.14.4±0.12 kg/day and produced 23.8±6.2 vs.17.9±4.4 kg milk/day, which when corrected for fat and protein contents was 22.7±5.9 vs. 19.4±4.8 kg ECM/day, respectively. The CO_2 produced by cows in this study ranged from 6.54 to 15.8 kg/day in Holstein cows and 6.3 to 14.1 kg/day in Jersey cows. Comparable results on kg CO_2 produced per day were reported but only on Holsteins, by Kinsman *et al.* (1995) using infrared gas analyser; Chianese *et al.* (2009) using the Integrated Farm System Model, and Lee *et al.* (2017) using the sulphur hexafluoride tracer technique.

In this study, Holsteins produced more CO_2 than Jerseys; however, when expressed as CO_2/kg DMI or CO_2/100 kg BW, Holstein had lower CO_2 emissions (Table 1). This is because the DMI/kg BW in this study was lower in Holsteins than Jerseys (3.1% vs. 3.5%). Similarly, Muller and Botha (1998) also reported lower DMI/kg BW (3.4 vs. 4.0% DMI/kg BW) in Holsteins than Jerseys, respectively. The CO_2/kg DMI decreased from primiparous to second lactation, then levelled so that the third lactation and mature cows did not differ. With lactation stage, CO_2/kg DMI was higher in transition stage followed by a decline that levelled from early lactation (Table 1). The CO_2/100 kg BW decreased with parity while it increased with lactation stage reaching a plateau in mid-lactation, followed by a decline in late lactation stage (Table 1). This seems to indicate that the higher DMI and heavier BW of multi-parous and later lactation stage cows had a diluting effect on CO_2 emitted, resulting in lower CO_2/kg DMI and lower CO_2/kg BW.

Table 1. Mean enteric CO_2 and CH_4 emissions and their efficiencies of Holstein and Jersey cows as affected by parity and lactation stage

	Parity										
	Parity 1		Parity 2		Parity 3		Parity 4+ (mature cows)		P-values		Intxn
	H	J	H	J	H	J	H	J	Breed	P	B×P
CO_2/day	$10.2^{d}\pm0.05$	$8.9^{g}\pm0.06$	$11.0^{c}\pm0.06$	$9.4^{g}\pm0.06$	$11.5^{b}\pm0.06$	$9.9^{e}\pm0.06$	$11.8^{a}\pm0.06$	$10.1^{d}\pm0.06$	<.01	<.01	<.01
CO_2/DMI	$0.65^{c}\pm0.001$	$0.69^{a}\pm0.001$	$0.63^{d}\pm0.002$	$0.68^{b}\pm0.002$	$0.62^{e}\pm0.002$	$0.66^{c}\pm0.002$	$0.62^{e}\pm0.002$	$0.66^{c}\pm0.002$	<.01	<.01	0.02
CO_2/BW	$2.01^{e}\pm0.01$	$2.42^{a}\pm0.01$	$1.97^{f}\pm0.01$	$2.34^{b}\pm0.01$	$1.95^{g}\pm0.01$	$2.32^{c}\pm0.02$	$1.92^{h}\pm0.01$	$2.31^{d}\pm0.02$	<.01	<.01	0.01
CO_2/ECM	$0.56^{a}\pm0.01$	$0.56^{a}\pm0.01$	$0.51^{b}\pm0.01$	$0.53^{b}\pm0.01$	$0.50^{c}\pm0.01$	$0.50^{c}\pm0.01$	$0.48^{d}\pm0.01$	$0.49^{d}\pm0.01$	0.38	<.01	0.08
MEF	$124.8^{d}\pm0.78$	$102.5^{g}\pm0.86$	$137.5^{c}\pm0.87$	$112.0^{f}\pm0.93$	$143.6^{b}\pm0.95$	$119.5^{e}\pm0.97$	$147.7^{a}\pm0.97$	$123.8^{d}\pm0.98$	<.01	<.01	<.01
CH_4/day	$342^{d}\pm2.1$	$281^{g}\pm2.4$	$377^{c}\pm2.4$	$307^{f}\pm2.5$	$393^{b}\pm2.6$	$327^{e}\pm2.7$	$405^{a}\pm2.7$	$339^{d}\pm2.7$	<.01	<.01	<.01
CH_4/kg DMI	$21.7^{a}\pm0.03$	$21.6^{b}\pm0.03$	$21.4^{c}\pm0.03$	$21.7^{a}\pm0.03$	$21.2^{d}\pm0.04$	$21.7^{a}\pm0.04$	$21.1^{c}\pm0.04$	$21.6^{b}\pm0.04$	<.01	<.01	<.01
CH_4/kg BW	$67.4^{b}\pm0.4$	$76.3^{a}\pm0.5$	$67.2^{b}\pm0.5$	$76.0^{a}\pm0.5$	$66.8^{bc}\pm0.5$	$76.6^{a}\pm0.5$	$66.0^{c}\pm0.5$	$76.7^{a}\pm0.5$	<.01	0.41	<.01
CH_4/ECM	$18.7^{a}\pm0.1$	$17.6^{b}\pm0.2$	$17.5^{b}\pm0.2$	$17.2^{bc}\pm0.2$	$17.0^{c}\pm0.2$	$16.4^{de}\pm0.2$	$16.6^{cd}\pm0.2$	$16.1^{e}\pm0.2$	0.01	0.01	<.01

	Lactation stage (days in milk)										
	Transition (<30)		Early lactation (31-100)		Mid-lactation (101-200)		Late lactation (201+)				
	H	J	H	J	H	J	H	J	Breed	LS	B×LS
CO_2/day	$9.4^{e}\pm0.07$	$8.3^{f}\pm0.07$	$11.4^{b}\pm0.05$	$9.8^{d}\pm0.06$	$11.9^{a}\pm0.05$	$10.2^{c}\pm0.06$	$11.8^{a}\pm0.05$	$10.1^{c}\pm0.06$	<.01	<.01	<.01
CO_2/DMI	$0.68^{b}\pm0.002$	$0.72^{a}\pm0.002$	$0.62^{e}\pm0.001$	$0.66^{c}\pm0.002$	$0.61^{e}\pm0.001$	$0.65^{d}\pm0.002$	$0.62^{e}\pm0.001$	$0.65^{d}\pm0.002$	<.01	<.01	0.03
CO_2/BW	$1.70^{h}\pm0.02$	$2.06^{f}\pm0.02$	$2.07^{e}\pm0.01$	$2.47^{b}\pm0.01$	$2.09^{d}\pm0.01$	$2.49^{a}\pm0.01$	$1.99^{g}\pm0.01$	$2.38^{c}\pm0.01$	<.01	<.01	0.06
CO_2/ECM	$0.40^{d}\pm0.01$	$0.40^{d}\pm0.01$	$0.48^{c}\pm0.01$	$0.49^{c}\pm0.01$	$0.56^{b}\pm0.01$	$0.56^{b}\pm0.01$	$0.61^{a}\pm0.01$	$0.62^{a}\pm0.01$	0.38	<.01	0.85
MEF	$111.1^{e}\pm1.0$	$89.2^{f}\pm1.1$	$143.5^{b}\pm0.9$	$118.9^{d}\pm0.9$	$150.3^{a}\pm0.8$	$125.4^{c}\pm0.9$	$148.8^{a}\pm0.9$	$124.3^{c}\pm0.9$	<.01	<.01	0.12
CH_4/day	$304^{e}\pm2.8$	$244^{f}\pm3.0$	$393^{b}\pm2.3$	$326^{d}\pm2.5$	$412^{a}\pm2.3$	$344^{c}\pm2.4$	$408^{a}\pm2.3$	$341^{c}\pm2.4$	<.01	<.01	0.12
CH_4/kg DMI	$21.5^{b}\pm0.04$	$20.7^{d}\pm0.05$	$21.5^{b}\pm0.03$	$22.0^{a}\pm0.03$	$21.2^{c}\pm0.03$	$22.0^{a}\pm0.03$	$21.3^{c}\pm0.03$	$22.0^{a}\pm0.03$	<.01	<.01	<.01
CH_4/kg BW	$54.5^{g}\pm0.6$	$59.8^{f}\pm0.6$	$71.4^{d}\pm0.5$	$81.9^{b}\pm0.5$	$72.5^{d}\pm0.4$	$84.0^{a}\pm0.$	$68.9^{e}\pm0.5$	$79.9^{c}\pm0.5$	<.01	<.01	<.01
CH_4/ECM	$12.7^{d}\pm0.2$	$11.4^{e}\pm0.2$	$16.7^{c}\pm0.2$	$16.3^{c}\pm0.2$	$19.4^{b}\pm0.2$	$19.0^{b}\pm0.2$	$21.1^{a}\pm0.2$	$20.7^{a}\pm0.2$	<.01	<.01	0.02

[a-h] Means within rows with different superscripts differ at $P<0.05$

CO_2: carbon dioxide (kg/day), **CH_4:** methane (g/day),**MEF:** methane emission factor (kg/head/year),**DMI:** dry matter intake (kg/day), **BW:** body weight (100 kg), **ECM:** energy corrected milk (kg/day), **Intxn:** interaction

The CO_2/kg ECM increased with lactation stage and decreased with parity, but breed did not have an effect (Table 1). The increase in emitted CO_2/kg ECM can be related to reducing milk production with advancing lactation stages as nutrient partitioning shift in mid and late lactation towards supporting pregnancy and building body reserves in preparation for the next calving, while the decrease by parity is attributed to high milk volumes produced by older cows.

As expected, Holsteins produced more CH_4 than Jerseys both by parity and lactation stage (Table 1). In both breeds, daily CH_4 produced increased as parity and lactation stages progressed, mature Holstein cows produced 15.6% and Jersey cows 17.1% more CH_4 than their primiparous counterparts. With lactation stages, the highest CH_4 emissions were observed during mid-lactation, corresponding with the peak DMI. The increase in daily CH_4 emitted from the transition period to mid-lactation was 26.2% and 29.1% in Holstein and Jersey cows, respectively. This indicates that accounting for parity and lactation stages will bring more accuracy in estimating MEF. Similarly, to CO_2, Jerseys produced moreCH_4/kg DMI and CH_4/kg BW. TheCH_4/kg ECM was, however, lower in Jerseys than Holsteins (Table 1). In agreement, Dalla Riva *et al.* (2014) also found greater CO_2equivalent emissions per unit ECM production in Holsteins compared to Jerseys, while Olijhoek *et al.* (2018) reported no differences between Holstein and Jersey cows in CH_4/kg ECM.

CONCLUSIONS

Carbon dioxide and CH_4 emissions varied by breed, parity and lactation stage. Dry matter intake had a positive relationship with enteric gas emissions. Factors affecting DMI such as breed, parity, stage of lactation, milk production and BW have an indirect effect on enteric gases emissions. Because of major variations in enteric emissions caused by parity and lactation stages, it is concluded that accounting for production stages should result in a higher accuracy in estimating MEF.A study on estimating heritability and repeatability of enteric emissions is recommended to determine the extent of differences observed between and within breeds that can be associated with additive genetic variance.

REFERENCES

Broucek J. (2014) *J. Environ. Protection* **5**: 1482.
Capper J.L. and Cady R.A. (2012) *J. Dairy Sci.* **95**: 165.
Chianese D.S.,Rotz C.A. and Richards T.L. (2009)*Applied Eng. Agric.* **25**: 431
Dalla Riva A., Kristensen T., De Marchi M., *et al.* (2014) *Acta Agraria Kaposváriensis* **18**: 75.
Hook S.E., Wright A.G. and McBride B.W. (2010) *Archaea*, Hindawi Publishing Corp., Canada.
IPCC. 2006. IPCC Guidelines for National Greenhouse Gas Inventories **4**: 10.1.
Johnson D.E. and Ward G.M. (1996) *Environ. Monitoring and Assess.* **42**: 133.
Kinsman R., Sauer F. D., Jackson H. A., and Wolynetz M. S. (1995) *J. Dairy Sci.***78**: 2760.
Lee M.A., Davis A.P., Chagunda M.G.G. and Manning P. (2017) *Biogeosciences***14**: 1403
Mangino J., Jacobs H. and Peterson, K. (2003) *Proc. 12th Int.Emission Inventory Conference, US EPA: San Diego, CA. 11–21.*
Muller C.J.C. and Botha J.A. (1998) *S. Afr. J. Anim. Sci.* **28**: 161
Münger A., and Kreuzer M. (2006) *Int.Congr. Ser.* **1293**: 119.
NRC (2001). Nutrient Requirements of Dairy Cattle. 7th rev. ed.
Olijhoek D.W., Løvendahl P., Lassen J., Hellwing A.L.F., *et al.* (2018) *J. Dairy Sci.* **101**: 9926.
Qiao J., Tan Z. and Wang M. (2014) *Sci. Agric.* **5**: 345.
Smith P., Bustamante M., Ahammad H., Clark H., Dong H., *et al.*(2014).Contribution of Working Group III to the Fifth Assessment Report of the IPCC, Cambridge University Press.
Van Amburgh M.E., Collao-Saenz E.A., Higgs R.J., Ross D.A., Recktenwald E.B., Raffrenato E., Chase L.E., Overton, T.R., Mills J.K. and Foskolos A. (2015) The Cornell Net Carbohydrate and Protein System: Updates to the model and evaluation of version 6.5.

INVESTIGATION INTO THE EFFECTS OF NUMBER OF SNPS AND NUMBER OF REFERENCE INDIVIDUALS ON IMPUTATION ACCURACY

M.H. Ferdosi and N.K. Connors

Animal Genetics & Breeding Unit*, University of New England, Armidale, NSW, 2351 Australia

SUMMARY

Imputation is used in Australia's national beef genetic evaluation, BREEDPLAN, to resolve missing genomic information in the combined panels of genomic data and clarify relationships through building a genomic relationship matrix (GRM). The accuracy of imputation is dependent on many factors such as the size of reference population and the density of markers. Here we demonstrated an improvement in imputation accuracy of Angus genotypes by increasing the numbers of individuals and the density of markers in the reference population. The results show a considerable increase in imputation accuracy by increasing density of markers up to 15k SNPs.

INTRODUCTION

Imputation of missing genotypes or imputation from low to high density genotypes has become common practice before building the GRM for routine breeding value estimation through single step best linear unbiased prediction (ssBLUP). Imputation can be used to make a common Single Nucleotide Polymorphism (SNP) panel from several varying density SNP panels, enabling all genotypes to be analysed together, and build the GRM (VanRaden 2008). However, the accuracy of imputation can vary and needs to be addressed. Most imputation algorithms firstly phase the genotypes and then impute the missing SNPs. Therefore, factors that affect haplotype inference accuracy also influence imputation accuracy. The main factors affecting imputation accuracy are the sample size of the reference population (high density genotypes) and to some extent, the target population (low density genotypes to be imputed), marker densities, genotyping accuracy, relatedness both between and within reference and target populations, allele frequencies of markers, and population structure (Browning and Browning 2011). In this study we evaluate the effects of number of SNPs and size of reference population on imputation accuracy of Angus genotypes.

MATERIALS AND METHODS

Genomic data. Angus genotypes included in the November 2018 TransTasman Angus BREEDPLAN analysis were used to assess the imputation accuracy, after quality control using the BREEDPLAN genomic pipeline (Connors *et al.* 2017) were used to assess the imputation accuracy. A subset of 11,041 individuals with 32,453 SNPs were selected from 56,369 individuals, where they had less than 5% missing SNPs in loci and individuals had less than 1% missing genotype (call rate greater than 99%). The missing rate in the final dataset was very low (0.2 percent).

Subsets of individuals to assess the imputation accuracy. Genotypes were divided into reference population subsets based on the individual's date of birth, resulting in 7 subsets with quantities of 777, 1,717, 2,945, 4,148, 6,063, 7,947 and 9,766 individuals, respectively (shown coloured in Figure 1). The last subset (subset 8) was not used in the reference population. Each reference subset was removed from the 11,041 total individuals, such that the remaining animals formed the target population. For example, in the first subset there were 777 individuals in the reference population and the remaining 10264 individuals formed the target population; the second subset had 1,717 reference and 9,324 target individuals, and so on. Subsets were cumulative, i.e. the second reference subset with 1,717

* A joint venture of NSW Department of Primary Industries and the University of New England

individuals included the first subset's 777 individuals.

Subsets of SNPs to assess the imputation accuracy. All genotypes originally consisted of 32,453 SNPs. Each reference population mentioned above used all 32,453 SNPs to impute the SNPs of the target populations. The target populations had subsets of SNPS retained as original, with the remaining SNPs masked, to be imputed back. Retained SNPs were selected randomly each time (and sampled only once), with subsets of 500, 700, 1000, 1500, 2000, 3000, 4000, 5000, 7000, 10000, 15000, 20000 and 30000 SNPs. The SNPs that were not in these subsets (masked SNPs) were imputed back and compared with the original SNPs to check the imputation accuracy.

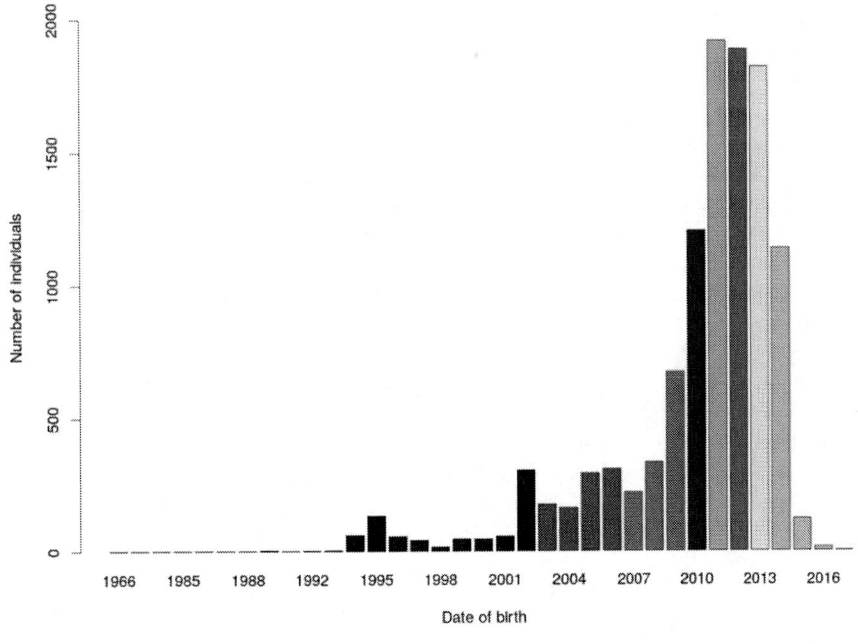

Figure 1. Frequency of number of individuals by their date of birth. Each of the colours represents a different subset used in this study (accumulative, i.e. the second group includes individuals with both red and black colours). In total 8 subsets were used

Imputation. FImpute Version 2.2 and default parameters (Sargolzaei *et al.* 2014) was used to impute missing genotypes without using pedigree information

Imputation Accuracy. The imputation accuracies were calculated for both SNPs and individuals. For individuals, the imputation accuracy was equal to the number of masked SNPs imputed correctly divided by the total number of non-missing SNPs for each individual. A single inaccurately imputed individual can result in multiple changes in relationship between genotyped individuals and reflected in the GRM. As such the minimum imputation accuracy for individuals is used to quantify the effects of changing reference and SNP numbers.

Imputation accuracy of SNPs is calculated by dividing the number of correctly imputed SNPs by the number of SNPs being imputed in that subset. As genomic relationships between individuals depends on all SNPs, the mean imputation accuracy of SNPs was used to quantify effects of changing reference and SNP numbers.

RESULTS AND DISCUSSION

Increasing the number of SNPs and the number of individuals in the reference population resulted in an increase in imputation accuracies of both SNPs and individuals (Figure 2). There was much higher variation in SNP imputation accuracy than that of individual imputation accuracy. The increase in both imputation accuracies was substantial when using up to 15,000 SNPs, at which point the increase in accuracy plateaus. The increasing number of SNPs in each subset decreased the number of SNPs to be imputed, and as such increased potential linkage disequilibrium (LD) between SNPs, even though they were chosen randomly. While Figure 2(B) demonstrates the mean imputation accuracy of SNPs, some SNPs showed much lower accuracy than the mean. For example, some SNPs imputed from the 700 SNPs subset had an imputation accuracy close to 0.1 (data not shown) and some SNPs imputed from the 30,000 SNPs subset had an imputation accuracy close to 0.4. The low imputation accuracy for these SNPs may indicate that these SNPs were not mapped correctly, or they were in regions with high recombination rate (e.g. recombination hotspots). In addition, it may indicate that these SNPs may be in regions where the quality of genotyping is generally low.

The number of individuals in the reference population was also important however with less effect in comparison to the number of SNPs. Figure 2 (A) shows an increase in imputation accuracy for each reference population increasing in individuals. Based on Figure 2 (B) more than 2,000 individuals are required in order to see substantial improvement in the SNPs imputation accuracies. In this study, we reported the imputation accuracy for all the subsets, although there was not considerable difference between subsets. A better approach may be to consider the accuracy for every immediate generation. Only genotype information was considered to evaluate imputation accuracies, however if reliable pedigree is available the imputation accuracy can be increased, especially if less than 15,000 SNPs are used. Here, the subsets of SNPs were selected randomly, rather than selecting SNPs based on their LD or other methods, as well as including SNPs with low minimum allele frequencies (<0.05). Further study is required to identify the effect of pedigree on the imputation accuracy of genotypes, along with SNP selection methods and allele frequency. Importantly, by selecting numbers of individuals based on date of birth, the results may be affected by population structure. Increasing the number of individuals in the reference population also increased relationships between reference and target populations, due to the structure of subset populations. The results showed the accuracy of imputation was affected by density of markers more than the number of animals in the reference population.

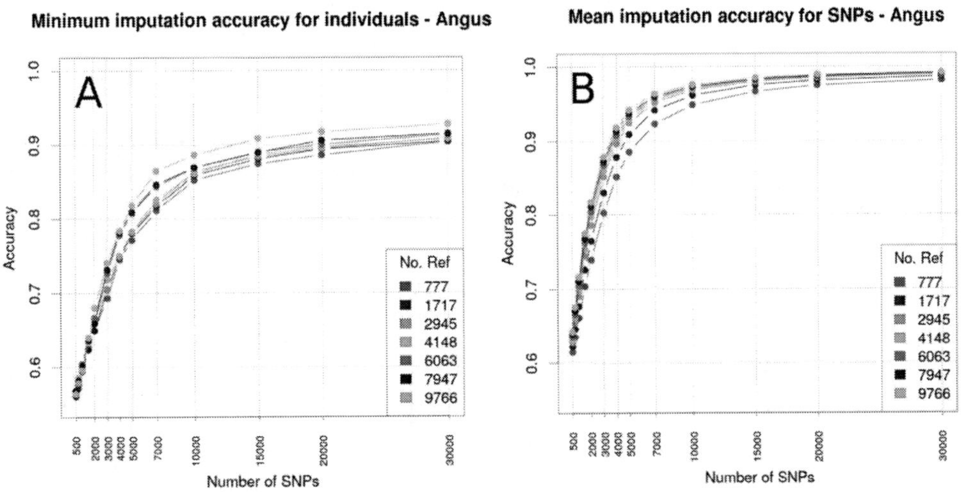

Figure 2. Effect of the number of SNPs on imputation accuracy. (A) Effect of the number of SNPs on the minimum imputation accuracy of individuals. (B) Effect of the number of SNPs on the mean imputation accuracy of SNPs

CONCLUSIONS

In this study we showed how the number of SNPs and the number of individuals in the reference population could affect the imputation accuracy. It should be noted that since the individuals were sorted based on their date of birth, increasing the number of individuals in the reference population also increased the average relationship between reference and target populations. We also showed more than 15,000 SNPs and 2,000 individuals in the reference were required to achieve more than 90 percent imputation accuracy for both individuals and SNPs when the pedigree information was not used and SNPs were selected randomly.

ACKNOWLEDGEMENTS

MHF, NKC and this study was supported by Meat and Livestock Australia project L.GEN.0174. The authors thank Angus Australia and Angus New Zealand for providing data for this study.

REFERENCES

Chadeau-Hyam M., Hoggart C., O'Reilly P., Whittaker J., De Iorio M. and Balding D. (2008) *BMC Bioinformatics* **9**: 364.

Erbe M., Hayes B.J., Matukumalli L.K., Goswami S., Bowman P.J., Reich C.M., Mason B.A. and Goddard M.E. (2012) *J. Dairy Sci.* **95**: 4114.

Browning S.R. and Browning B.L. (2011) *Nat Rev Genet* **12:** 703.

Connors N., Cook J., Girard C., Tier B., Gore K., Johnston D. and Ferdosi M. (2017) *Proc. Assoc. Advmt. Anim. Breed. Genet.* **22:**

Sargolzaei M., Chesnais J.P. and Schenkel F.S. (2014) *BMC Genomics* **15:**

VanRaden P.M. (2008) *J Dairy Sci* **91:** 4414.

FACTORS AFFECTING DEVELOPMENT OF HORNS AND SCURS IN DOMESTIC RUMINANTS

I.A.S. Randhawa,[1] R.E. Lyons,[1] B.J. Hayes,[2] L.R. Porto-Neto[3] and M.R. McGowan[1]

[1]School of Veterinary Science, University of Queensland, Gatton, QLD, 4343 Australia
[2]Centre for Animal Science, QAAFI, University of Queensland, St Lucia, QLD, 4072 Australia
[3]Agriculture and Food, CSIRO, St Lucia, QLD, 4067 Australia

SUMMARY

Most domesticated ruminants were naturally horned. Originally this was to help defend themselves from predators and compete for food resources and mating partners. Genetic control of the presence or absence (polledness) of horns or scurs (unattached horns) has been extensively investigated. However, variation in rate of development, size and shape of horns is also affected by non-genetic factors such as sex, nutrition, age and photoperiod. This study aimed to provide an overview of the impacts of these factors on horn development in different species of ruminants and to empirically investigate the interaction between genetics, sex and horn-status in 7 breeds of Australian beef cattle. This framework suggested that horns evolved through sexual selection, as horn ontogenesis is dependent upon genetics, sex of the individual and variation in testosterone.

INTRODUCTION

Cattle (*Bos taurus* and *Bos indicus*) and sheep (*Ovis aries*) are very important in livestock production systems world-wide because of the multitude of useful products derived from them and, their ability to be raised under highly variable environmental and management conditions. Both species typically grew symmetrical paired horns of various shapes and sizes, and these have become a prominent feature of the phenotypic diversity of modern breeds. Generally, the head-status in these species can be termed as *horned*: keratin coated permanent protrusions attached to the skull itself, *knob*: hard bony lump at horn site (sheep only), *scurred*: rudimentary horns loosely attached to the skin rather than the skull itself, and *polled*: complete absence of horns, scurs, and knobs (Duijvesteijn *et al.* 2018). Morphologically, horns are different than other horn-like structures, e.g., antlers. Horns are comprised of a bony core covered by a hard layer of keratinous tissue. The presence of *horns* and *scurs* is difficult to determine at an early age, because both originate postnatally as free-floating buds and subsequently, only the former fuses into the cranium (Dove 1935; Mariasegaram *et al.* 2010; Wiener *et al.* 2015). Generally, both species grow a pair of horns laterally on the head.

Horns are remarkably diverse in form and size, and were originally likely to help animals' defend themselves, and compete for food resources and mate partners (Kiltie 1985; Lundrigan 1996; Davis *et al.* 2011). However, given the complete absence of horns in females of some Bovids and comparatively smaller size of horns in females of others, it has been suggested that horns originated and evolved primarily as male weaponry (Marshall and Hammond 1914; Kiltie 1985). In various species, diversity in horn shape and size has been associated with different types of agonistic male-male behaviours – namely stabbing, wrestling, fencing and ramming (Geist 1966).

Further, it has been suggested that there may be a link between horn development, dominance and fertility in males (Lundrigan 1996). Thus, differences in horn development between males and females could have been modulated by selective pressure, with larger horns becoming important traits of sexual selection in males, whereas in females less direct selection for horns resulted in considerably lower diversity in horn phenotypes (Bro-Jorgensen 2007; Stankowich and Caro 2009; Davis *et al.* 2011).

Cattle and sheep have been extensively investigated for the role of genetics in horn phenotype. Although a clear understanding of the genes and biological pathways involved is yet to be achieved, gene mapping strategies have led to development of diagnostic tools for early-in-life prediction of horn phenotype (Connors *et al.* 2018; Duijvesteijn *et al.* 2018; Randhawa *et al.* 2019). Evidence suggests that genetics of the individual determines the horn status as a qualitative trait, however, sex of the individual partly controls horn morphology. Studies on nutritional impact on horn development are very limited but has been shown to be important in some species (Monteith *et al.* 2018). Generally postnatal growth of horns up to maturity is linked to body growth, although in some species the horn grows continuously or periodically throughout life (Rughetti and Festa-Bianchet 2011; Wiener *et al.* 2015). Photoperiod has been reported to regulate horn growth in Iberian ibex (*Capra pyrenaica*), Spanish ibex (*Capra pyrenaica hispanica*) and two European mouflon species (*Ovis orientalis musimon* and *Ovis gmelini musimon*). Also in these species variations in testicular function and plasma concentrations of testosterone and prolactin at different ages have been linked to rate of horn growth (Toledano-Diaz *et al.* 2007; Santiago-Moreno *et al.* 2012). Development of horns has shown diverse patterns in beef cattle. The objectives of this study were to review aforementioned factors influencing horn growth in bovids and to investigate the interaction between genetics and sex in Australian beef cattle.

MATERIALS AND METHODS

Phenotypic information about head-status (*horn*, *scur*, *poll*) and sex (male, steer and female) from 6,664 registered animals of 7 breeds in Australia (Table 1) were obtained from the BREEDPLAN database (http://breedplan.une.edu.au/index.php). Accuracy of assignment of head-status phenotypes can be compromised by several issues, including often failing to be able to differentiate between *horn* and *scur* prior to dehorning as a calf, late onset of *scurs*, incorrect assignment of cattle as *poll* when in fact they were dehorned and data transcription errors. To better understand the potential prevalence of these inaccuracies, a resource population of Droughtmaster cattle owned by The University of Queensland (UQ-herd) was phenotyped by two experienced beef cattle research scientists (Table 1). Genotypes on each animal of the UQ-herd (n=84) and samples previously collected (n=6,664) for BREEDPLAN were analysed using the 5-SNP markers in the POLL region applied through the optimized poll testing (OPT) assay (Randhawa *et al.* 2019).

Table 1. Number of samples tested, and cattle by sex and head-status phenotypes

Breeds	Number tested	Sex			Head-status		
		Male	Steer	Female	Horn	Scur	Poll
Brahman	2722	1772	191	759	2138	130	454
Charolais	517	351	-	166	63	9	445
Droughtmaster	488	114	278	96	102	114	272
Hereford	2740	1997	264	479	415	139	2186
Limousin	24	20	-	4	3	2	19
Santa Gertrudis	108	103	1	4	42	11	55
Shorthorn	65	51	-	14	-	-	65
Total	6,664	4,408	734	1,522	2,763	405	3,496
(%)	(100)	(66.2)	(11.0)	(22.8)	(41.5)	(6.0)	(52.5)
UQ-herd	84	9	-	75	15	23	44

RESULTS AND DISCUSSIONS

Table 2 shows the distribution (%) of head-status within each sex for various genotypes determined by OPT. Genotypes represent different combinations of the copies of *horn* (H) and *poll* (Pc & Pf) alleles (Randhawa *et al.* 2019). Pc and Pf are denoted for Celtic and Friesian types, respectively, both sequence variants associated with polledness in cattle (Medugorac *et al.* 2012; Allais-Bonnet *et al.* 2013; Wiedemar *et al.* 2014). Theoretically, single gene-based inheritance suggests that HH animals grow *horns*, HP (HPc or HPf) grow *scurs* and PP (PcPc, PcPf or PfPf) are *polled*. In reality, translation of genotypes to phenotypes is complex for HP. Overall, phenotypic concordances of HH, HP and PP indicated considerable ambiguity in the BREEDPLAN data (Connors *et al.* 2018). HH and PP deviations (less than 5%) are not plausible to any known factor affecting horn development in male and female cattle. However, note that HH steers had the highest frequency of *scur* (8.12%) and *poll* (7.78%) phenotypes, because castration would have caused a cessation or delayed horn growth resulting in misclassification of phenotypes. Genotypes of the UQ-herd were 100% consistent for expected phenotype, except HPc (53%) and HPf (100%) in females were *polled,* when they were expected to be scurred. Interestingly, both data also showed that carriers of Friesian allele (HPf) are more likely to be *polled* than *scurred* as compared to Celtic (HPc) across all sex-classes. Pf has been speculated as more likely to be a causative variant (Wiedemar *et al.* 2014), whereas, gene-edited introgression of Pc has successfully produced *polled* calves (Carlson *et al.* 2016). Different gene pathways have been shown to be involved in horn and scur development (Mariasegaram *et al.* 2010). Previously, horn and scur growth control by two loci (genes), i.e., Poll (P/p) and Scur (Sc/sc), has been debated and the gene-pair interactions postulated that horns are developed by pp (HH) regardless of Sc/sc alleles, however, ScSc cause scurs for PP males and females as well as Pp males can scur with either ScSc or Scsc while Pp female can only scur when ScSc (Long and Gregory 1978). Our results contrast to this model because scurs in PP were rarely observed (5 out of 428), seemingly errors, in limited breeds (4 Droughtmaster and 1 Hereford), which is in line with the findings that only Pp can cause scurs (Wiedemar *et al.* 2014). On the other hand, sex hormones affecting horn agenesis in goats (Pailhoux *et al.* 2001) and morphological alterations in sheep (Marshall and Hammond 1914) suggest sex specific horn regulation , horn dysmorphism or complete agenesis. Our observations warrant further investigation about the perceived involvement of the Sc gene (Tetens *et al.* 2015), and the role of the peripubertal changes in sex-hormones affecting phenotypes for HP (Pp).

Table 2. Distribution of Poll gene tested genotypes for sex-wise phenotypes in cattle breeds

Genotypes	N	Male head-status (%)			Steer head-status (%)			Female head-status (%)		
		Horn	Scur	Poll	Horn	Scur	Poll	Horn	Scur	Poll
HH	2,623	95.6	3.06	1.28	84.1	8.12	7.78	96.2	1.59	2.17
HPc	2,057	15.4	17.8	66.6	10.2	19.1	70.6	8.48	5.05	86.5
HPf	120	2.60	7.80	89.6	-	-	100	3.33	3.33	93.3
PcPc	1,570	0.85	0.08	99.1	0.85	2.56	96.6	0.40	-	99.6
PcPf	267	0.45	0.45	99.1	-	-	100	-	-	100
PfPf	27	-	-	100	-	-	100	-	-	100
HP *	2,177	14.7	17.3	68.0	9.80	18.3	71.9	8.19	4.95	86.9
PP *	1,864	0.77	0.14	99.1	0.76	2.27	96.9	0.33	-	99.7
UQ Herd										
HH	15	100	-	-	-	-	-	100	-	-
HPc	43	-	100	-	-	-	-	-	47	53
HPf	2	-	-	-	-	-	-	-	-	100
PcPc	24	-	-	100	-	-	-	-	-	100

* Combined distributions of heterozygous HP (HPc+HPf) and homozygous PP (PcPc+PcPf+PfPf).

CONCLUSIONS

Form and function of headgear vary in each horned species. However, most ruminant species can be recognized by the unique shape and size of male horns, which are primarily used as anti-predators and to combat for mating partners. Given the evolved involvement of horns in sexual selection, horn growth is dependent upon genetics and sex of the individual. Not only the gender itself, but also variation in the sex related hormones (e.g., lower testosterone in steers) can impact the appearance and growth of horns and scurs. As such, the trait is complex and our empirical data suggested that visual detection of head-status is challenging leading to ambiguous classifications, and this error is minimised when phenotypes were assessed within the UQ-herd by skilled personnel. In future study, it is hoped that heterozygous animals expressing as *poll* and *scur* can be used to elucidate whether the scur phenotype is influenced by sex (hormones) or an entirely separate gene.

ACKNOWLEDGEMENTS

Meat and Livestock Australia provided financial support for this project (L.GEN.1713).

REFERENCES

Allais-Bonnet A., Grohs C., Medugorac I., Krebs S., Djari A., Graf A., Fritz S., Seichter D., Baur A., *et al.* (2013) *PLoS One* **8**: e63512.

Bro-Jorgensen J. (2007) *Evolution* **61**: 1316.

Carlson D.F., Lancto C.A., Zang B., Kim E.-S., Walton M., Oldeschulte D., Seabury C., Sonstegard T.S. and Fahrenkrug S.C. (2016) *Nat. Biotechnol.* **34**: 479.

Connors N.K., Tier B. and Johnston D.J. (2018) *Wld Congr. Genet. Appl. Livest. Prod.* . 344.

Davis E.B., Brakora K.A. and Lee A.H. (2011) *Proceedings. Biological sciences* **278**:.2857.

Dove W.F. (1935) *J. Exp. Zool.* **69**:.347.

Duijvesteijn N., Bolormaa S., Daetwyler H.D. and van der Werf J.H.J. (2018) *Genet. Sel. Evol.* **50**:.28.

Geist V. (1966) *Behaviour* **27**:.175.

Kiltie R.A. (1985) *Biol. J. Linn. Soc.* **24**:.299.

Long C.R. and Gregory K.E. (1978) *J. Hered.* **69**:.395.

Lundrigan B. (1996) *J. Mammal.* **77**:.462.

Mariasegaram M., Reverter A., Barris W., Lehnert S.A., Dalrymple B. and Prayaga K. (2010) *BMC Genomics* **11**:.370.

Marshall F.H. and Hammond J. (1914) *The Journal of Physiology* **48**:.171.

Medugorac I., Seichter D., Graf A., Russ I., Blum H., Göpel K.H., Rothammer S., Förster M. and Krebs S. (2012) *PLoS One* **7**:.e39477.

Monteith K.L., Long R.A., Stephenson T.R., Bleich V.C., Bowyer R.T. and Lasharr T.N. (2018) *The Journal of Wildlife Management* **82**:.67.

Pailhoux E., Vigier B., Chaffaux S., Servel N., Taourit S., Furet J.-P., Fellous M., Grosclaude F., Cribiu E.P., *et al.* (2001) *Nat. Genet.* **29**:.453.

Randhawa I.A.S., McGowan M.R., Porto-Neto L.R., Hayes B.J., Schutt K.M. and Lyons R.E. (2019) *Proc. Assoc. Advmt. Anim. Breed. Genet.* **23**:(in press).

Rughetti M. and Festa-Bianchet M. (2011) *The Journal of Animal Ecology* **80**:.438.

Santiago-Moreno J., Gomez-Brunet A., Toledano-Diaz A., Salas-Vega R., Gomez-Guillamon F. and Lopez-Sebastian A. (2012) *The Journal of endocrinology* **214**:.155.

Stankowich T. and Caro T. (2009) *Proceedings. Biological sciences* **276**:.4329.

Tetens J., Wiedemar N., Menoud A., Thaller G. and Drögemüller C. (2015) *Anim. Genet.* **46**:.224.

Toledano-Diaz A., Santiago-Moreno J., Gomez-Brunet A., Pulido-Pastor A. and Lopez-Sebastian A. (2007) *Anim. Reprod. Sci.* **102**:.300.

Wiedemar N., Tetens J., Jagannathan V., Menoud A., Neuenschwander S., Bruggmann R., Thaller G. and Drögemüller C. (2014) *PLoS One* **9**:.e93435.

Wiener D.J., Wiedemar N., Welle M.M. and Drögemüller C. (2015) *PLoS One* **10**:.e0127691.

LEVELS OF PERFORMANCE RECORDING IN THE AUSTRALIAN BEEF INDUSTRY

B.W. Gudex and C.A. Millen

Southern Beef Technology Services, Agricultural Business Research Institute, Armidale, NSW, 2350, Australia

SUMMARY

The collection of quality performance data underpins genetic evaluation systems, including BREEDPLAN. This paper provides a snapshot of the current levels of performance recording in the Australian beef industry for a range of BREEDPLAN traits, including growth, carcase, birth and fertility. Levels of performance recording were found to vary depending on a number of factors that included trait, herd, location/breed and sex.

INTRODUCTION

The quantity and quality of performance information recorded by Australian beef producers is a key component influencing the accuracy of the Estimated Breeding Values (EBVs) calculated by BREEDPLAN (Nicol *et al.* 1985) and the level of genetic progress made in the industry. BREEDPLAN is the world's most widely used genetic evaluation system for beef cattle with over 80 breed associations from 12 countries utilising the service. Up to 25 EBVs are published in each BREEDPLAN genetic evaluation, using information from pedigree, performance and, in some instances, genotypes (e.g. Millen and Crook 2019, these proceedings). Southern (SBTS; Millen *et al.* 2018) and Tropical Beef Technology Services (TBTS) are partner projects that provide extension services to large sectors of the Australian beef industry and are joint initiatives of the Agricultural Business Research Institute (ABRI), Meat & Livestock Australia (MLA) and 21 breed associations.

This paper reviews the quantity of performance data that is submitted to BREEDPLAN for the purposes of calculating EBVs from breed associations that are stakeholders in the SBTS and TBTS projects.

MATERIALS AND METHODS

Data. The analysis contained in this paper was carried out on data from individual seedstock herds that were members of SBTS and TBTS stakeholder breed associations. According to the Australian Registered Cattle Breeders Association (2017), these breed associations represent approximately 67% of the total Australian seedstock industry. The beef cattle breeds represented in the SBTS and TBTS projects are:

- **SBTS:** Blonde d'Aquitane, Charolais, Devon, Gelbvieh, Hereford, Limousin, Murray Grey, Red Angus, Red Poll, Salers, Shorthorn, Simmental, Speckle Park and Wagyu.
- **TBTS:** Belmont Red, Brahman, Brangus, Droughtmaster, Santa Gertrudis, Senepol and Simbrah. Simbrah are registered with Simmental and thus included in the SBTS analysis.

Analysis. The BREEDPLAN "Completeness of Performance" (CoP) product (Millen *et al.* 2018) summarises the quantity of pedigree and performance information that has been submitted to BREEDPLAN into a report and a star rating. For participating societies, CoP reports are generated for individual herds when they submit data to BREEDPLAN and once a year for all herds enrolled in BREEDPLAN.

The CoP Report can also be produced for individual breed associations and for SBTS and TBTS stakeholder breed associations. In this paper, the average level of performance recording for the birth years 2012 to 2016 is reported as a percentage of the total registrations received. Animals under two years of age when the report was generated (August 2018) are not included as those animals will not have been old enough to have had the opportunity to be fully recorded (e.g. 600d weight). Additionally, the performance recording trends from 2004 to 2016 are reported.

The CoP star rating for each herd is calculated based on the proportion of registered calves that have performance recorded for each trait in a five-year period. For this paper, 895 herds were included in the star rating analysis on the proviso that they had joined BREEDPLAN prior to 2016 and had calves registered in their herd between 2012 and 2016.

RESULTS AND DISCUSSION

Between 2012 and 2016, there were 381,094 and 267,315 cattle born and registered with the SBTS and TBTS stakeholder breed associations, respectively. The levels of performance recording varied due to a number of factors including trait, herd, location (generalised to northern (TBTS) and southern (SBTS)) and sex. Breed was not specifically analysed as part of this paper but is acknowledged as a further source of variation (unpublished results) due to breed specific objectives and differences in which EBVs are reported for each breed.

The absence of particular EBVs for a given breed does not necessarily represent a lack of interest in the EBV but instead is often due to a lack of trait data for BREEDPLAN to analyse. This chicken and egg scenario - which comes first, the data or the EBV - is a common situation for a number of the smaller breeds. While the BREEDPLAN analysis needs data to analyse, breeders see limited value in recording a trait for which they don't receive an EBV. This situation also occurred with the accumulation of genotypes for the Hereford breed. The growth in the number of genotypes was slow leading up to the implementation of Single-Step BREEDPLAN but the number increased 3.8-fold in the following 17 months after implementation (Millen and Crook 2019, these proceedings).

Three performance recording trends are noticeable across both SBTS and TBTS breeds (Figure 1). The first is that there is a reduction in performance recording as animals get older. This is logical given that animals are removed from the herd for various reasons (e.g. death or sale) with age and thus are not available for recording. Additionally, a higher level of birth trait recording is expected as these traits can be submitted by non-BREEDPLAN members at the time of registration with their breed association. The second trend is that there are approximately 15% more females registered than males and this bias has been consistently observed since 2004. This suggests that some males are not being registered with their breed societies and may reflect a reluctance to register males that will be or have been castrated. The third trend is that the remaining males have higher levels of recording than females. For some traits this is unavoidable (scrotal size) but for other traits there are reasons why this is not ideal. For example, data from ultrasound carcase scanning (scan) of heifers can be more valuable for genetic evaluation than that from bulls. As heifers mature and lay down fat earlier than bulls, they typically have more variation in the scan traits which is valuable to the analysis. Secondly, heifers typically represent a better cross section of the herd as they are subject to less selection pressure than bulls. This reduces selection bias in the data submitted to BREEDPLAN. While Corrigan and Parnell (2006) also observed a reduction in performance recording with age, their results from a similar population (8 southern breeds including Angus but no Wagyu or smaller breeds) found a higher level of registration in males and a generally comparable rate of recording in males and females. One similarity was their observation of a 4% higher level of scan in males than females which is approximately half the difference between the sexes observed in this study.

The differentiation between (a) TBTS and (b) SBTS breeds seen in Figure 1 not only reflects breed type (predominately *Bos indicus* versus *Bos taurus* respectively*)*, but also location (Northern versus Southern Australia). In general, levels of performance recording are lower in TBTS breeds but this is not true for every trait, for example Days to Calving (DtC). The higher level of DtC recording in the TBTS breeds reflects the desire to address the lower fertility levels typically observed in *Bos indicus* cattle. Birthweight (Bwt) is the trait where the biggest divergence between SBTS and TBTS breeds occurs. This is likely to be in part due to the extensive nature of the northern beef production system making collection of this trait difficult. The recording levels of some traits can also be evaluated in the context of other traits. For example, mature cow weight (MCW) and the 200-day weight (200d) of the progeny should be recorded at a similar time. Therefore, it is interesting that while SBTS breeds record 27% more 200d than TBTS breeds, the recording of MCW is similar (12% vs. 14% for SBTS and TBTS breeds respectively).

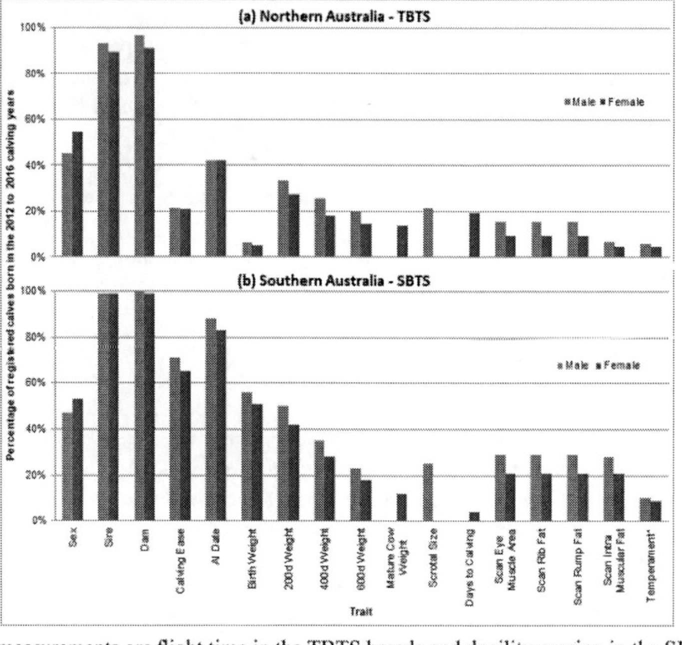

* Temperament measurements are flight time in the TBTS breeds and docility scoring in the SBTS breeds

Figure 1. Levels of performance recording in TBTS and SBTS breeds

Previously, Corrigan and Parnell (2006) reported higher levels of Bwt (+7%), 200d (+16%) and scan (+7%) recording than reported for SBTS breeds in this study. This is potentially explained by differences between the two populations in both breed composition and calving years analysed. Analysis of the SBTS breeds CoP trends in the decade subsequent to Corrigan and Parnell (2006) indicate that calving year is unlikely to be a major factor with both Bwt and scan recording increasing (0.9% and 0.5% per year respectively) while 200d weight decreased (0.4% per year).

Across all SBTS & TBTS breeds since 2004, the traits with the biggest increase in recording levels were calving ease (+1.1% per yr) and scrotal size (+0.7% per year). The biggest decrease was observed in DtC, with the TBTS breeds declining by 7% between 2009 and 2010, and SBTS breeds declining by 10% between 2010 and 2013. This decline corresponded with a change in DtC recording requirements to address incomplete recording observed in this trait at the time. While DtC recording in the TBTS breeds has now recovered to pre-2010 levels, the SBTS breeds have remained at 3% to

4% (Figure 1) since 2013. The recovery observed in the TBTS breeds likely reflects fertility being perceived as a larger issue in *Bos indicus* cattle while the lack of response by SBTS breeds may be partially attributed to an increase in the amount of artificial insemination being utilised (DtC records are only currently analysed for naturally mated cows). Increasing DtC recording remains a key extension focus for both SBTS and TBTS.

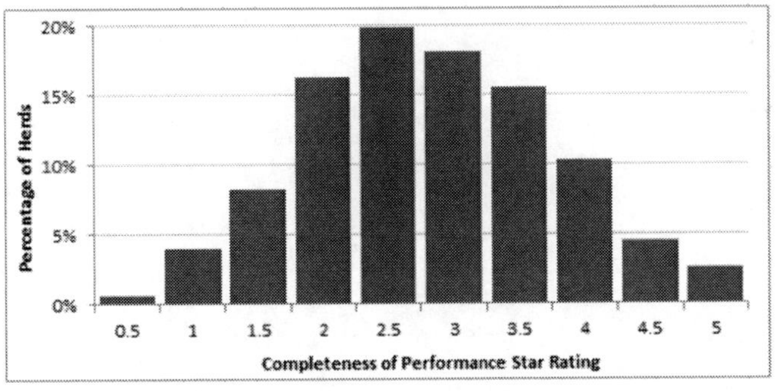

Figure 2. Distribution of individual herds for completeness of performance star rating

The distribution of star ratings (Figure 2) shows the variation in the levels of individual herd recording. While 70% of herds record a moderate level of data (between 2 and 3.5 stars), there are a small percentage of herds at each extreme. Herds do not have to record every trait on every animal to get a 4- or 5-star ratings as this is not always possible. However, these herds are still collecting a considerable amount of performance data across most traits for the majority of their animals. The distribution of star ratings is not breed or region specific; instead, it is observed across all SBTS and TBTS breeds.

CONCLUSIONS

While some herds are doing an exceptional job of collecting performance data, this study has identified that there is considerable room for improvement in the levels of performance recording by the Australian beef industry. The levels of recording were found to vary depending on a number of factors that included trait, herd, location/breed and sex. The recording trends highlighted several important traits, in particular DtC, that should be emphasised in extension messages to breeders.

ACKNOWLEDGEMENTS

The SBTS and TBTS projects are funded by ABRI, MLA and 21 stakeholder Australian-based breed associations. The authors thank these organisations for their support of both projects.

REFERENCES

Australian Registered Cattle Breeders Association. (2017) Annual Report. Armidale.

Corrigan L. and Parnell P.F. (2006) In 'Australian Beef The Leader, The Impact of Science on the Beef Industry', pp. 81-90, CRC for Beef Genetic Technologies, Armidale.

Millen C.A., Gudex B.W., Teseling C. and Skinner S. (2018) *Proc. Wld. Cong. Genet. Appl. Livest. Prod.* **11**:729

Millen C.A. and Crook B.J. (2019) *Proc. Assoc. Advmt. Anim. Breed. Genet.* **23**: *In Press*

Nicol D.C., Graser H-U., Tier B. and Hammond K. (1985) *Proc. Assoc. Advmt. Anim. Breed. Genet.* **5**: 151.

Gap in pagination due to unavailable paper.

Pages 492-495

SHOULD ANGUS BREEDERS LIVE-ANIMAL ULTRASOUND SCAN FOR INTRAMUSCULAR FAT IN THE GENOMICS ERA?

C.J. Duff[1,2], J.H.J. van der Werf[2], P.F. Parnell[1] and S.A. Clark[2]

[1]Angus Australia, 86 Glen Innes Road, Armidale, NSW, 2350 Australia
[2]School of Environmental and Rural Science, University of New England, Armidale, NSW, 2351 Australia

SUMMARY

Accuracy of selection and response to selection for the breeding objective traits of carcase IMF, AUS-MEAT marbling score and MSA marbling score were generated by modelling seventeen different phenotyping and genotyping scenarios using selection index theory. The data used in the study was produced from the Angus Sire Benchmarking Program (ASBP). The study showed that the highest rates of selection accuracy and response (ΔG) were achieved through using a combination of ultrasound scan phenotyping for intramuscular fat (IMF) and genotyping with a reference population including carcase IMF and marbling score phenotypes. For phenotype only scenarios, Central Ultrasound Processing (CUP) showed substantially higher response to selection and accuracy of selection to the Esaote Aquila ultrasound system (PIE) for IMF resulting from the higher heritability and stronger genetic correlations to the direct carcase breeding objective traits. The value of ultrasound scan phenotyping diminishes as the prediction accuracy of the genomic breeding value (GBV) increases. The GBV only scenario surpassed phenotype only scenarios once GBV prediction accuracy was greater than 59% when compared to CUP IMF phenotyping and 46% when compared to the PIE IMF phenotyping.

INTRODUCTION

Improving meat quality is an important breeding objective for many Angus beef producers. Traditionally, meat quality traits measured on the carcase, like intramuscular fat (IMF) and marbling score, have proven expensive and difficult to measure on selection candidates. Consequently, breeders typically use correlated ultrasound scan measurements on the live animal to increase selection accuracy for these traits as outlined by Williams (2002). The most common ultrasound scanning technology used to predict carcase IMF in Australian Angus herds is the Esaote Aquila system produced by Pie Medical (PIE). This technology facilitates crush-side and real-time image capture, interpretation and analysis using inbuilt software and algorithms. An alternative approach for the prediction of carcase IMF is the Central Ultrasound Processing (CUP) system. The CUP system uses proprietary software and algorithms to predict carcase IMF through a centralised image analysis laboratory based on images that are also captured crush-side through ultrasound scanning. Selection for carcase traits can also be enhanced by using genomic information as derived from a reference population with IMF and marbling score phenotypes coupled with genotypes, as described by Goddard *et al.* (2010). This study models and compares different phenotyping and genotyping scenarios focussing on accuracy of selection and response to selection for the breeding objective traits of carcase IMF, AUS-MEAT marbling score and MSA marbling score.

MATERIALS AND METHODS

To estimate parameters for this study, all phenotypes, associated fixed effects and pedigree were generated from the Angus Sire Benchmarking Program (ASBP). The animals in the study (n=1,622) were steer progeny of registered Angus sires (n=173) from 7 different co-operator herds located in

New South Wales and Victoria, Australia. At feedlot intake, at an average age of 502 days (SD 74.1), the steers were ultrasound scanned crush-side at the 12th and 13th rib site by experienced technicians using the Esaote Aquila system (PIE Medical, Maastricht, The Netherlands) equipped with a 3.5-MHz, 18-cm transducer. IMF was predicted using an algorithm within the PIE software providing a real-time and crush-side assessment. At the same time, images using the same ultrasound hardware, from the same location on the animal, were captured using Central Ultrasound Processing (CUP) image capture software and sent to the CUP laboratory in Ames, Iowa, USA for image interpretation for IMF using proprietary software and algorithms. The steers were then killed at an average age of 795 days (SD 70.0) following an average feeding period of 293 days. Their carcases were graded (MSA and AUS-MEAT) and meat samples collected for laboratory assayed IMF measurement.

ASReml software (Gilmour *et al.* 2015) was used to fit an animal model to each trait to estimate parameters based on univariate and bivariate mixed model analysis using three generations of pedigree. Fixed effects fitted in all models included the contemporary group and dam age. Age at measurement was fitted for ultrasound scan traits, while carcase weight was fitted for direct carcase traits. The contemporary group included a concatenation of herd, year of birth, sex, birth type (twin v single), breeder-defined management group, observation date (ultrasound scan or kill date) and management group history (Graser *et al.* 2005). Heritabilities, as well as phenotypic and genetic correlations were estimated from the resulting variance components.

With the resulting phenotypic and genetic parameters, the GenSel program (van der Werf 2019) was used to calculate selection accuracy and response using selection index theory and deterministically optimizing generation intervals by truncation selection across age classes. Seventeen different phenotyping and genotyping scenarios were compared, focussing on the breeding objective traits of carcase IMF (CIMF), AUS-MEAT marbling score (AMBL) and MSA marbling score (MMBL). The herd structure was based on a self-replacing Angus breeding herd with 500 females, mating ratio of 2.5% and a 90% weaning rate. The first two scenarios included phenotype only cases with PIE IMF (scenario 1) or CUP IMF (scenario 2) where all males and females bred in the herd are measured as yearlings with the specified ultrasound scan technology. In scenarios 3 to 7, being genotype only cases, genomic breeding values (GBV) were estimated from a reference population with the traits CIMF, AMBL and MMBL measured. GBV accuracies were 17%, 47%, 70%, 85% or 95%, resulting from the single trait GBV accuracies of 10%, 30%, 50%, 70% and 90%. Scenarios 8 to 17 included combinations of the abovementioned cases by combining one phenotyping with one genotyping scenario.

RESULTS AND DISCUSSION

The characteristics of the traits included in this study are summarised in Table 1. Heritabilities were moderate to high for all traits. The heritability for CUP IMF was significantly higher than PIE IMF at 0.51 and 0.37, respectively, displaying similar standard errors (0.09, 0.08). The genetic and phenotypic correlations with the objective traits CIMF, AMBL and MMBL tended to be stronger for CUP IMF compared to PIE IMF (Table 2). The heritabilities are comparable to those reported by Duff *et al.* (2018), however the genetic correlations are more favourable for CUP IMF to the objective traits in this study.

From the GenSel analysis, comparing phenotyping only scenarios, selection on CUP IMF showed substantially higher response to selection and accuracy of selection compared to using PIE IMF. Selecting on CUP IMF resulted in a 29%, 50% and 54% increase in selection response for CIMF, AMBL and MMBL, respectively (Table 3). The genotyping only scenarios surpassed the phenotype only scenarios once GBV prediction accuracy was greater than 59% when compared to CUP IMF phenotype selection and 46% when compared to PIE IMF (Figure 1).

Table 1. Number of records and the descriptive statistics for carcase weight, IMF and marbling score

Trait[1]	n	Mean	SD[2]	Minimum	Maximum	CV (%)[3]
PIE IMF	1622	4.5	1.2	1.3	7.7	26.5
CUP IMF	1457	4.9	1.8	1.1	10.3	35.8
CIMF	1475	10.1	3.3	3.2	25.1	32.6
AMBL	1473	2.7	1.2	0.0	8.0	46.4
MMBL	1474	514.4	120.2	160.0	1030.0	23.4
CWT	1462	460.2	37.4	334.9	568.6	8.1

[1] PIE IMF: Ultrasound Scan IMF using PIE (%); CUP IMF: Ultrasound scan IMF using CUP (%); CIMF: Carcase Intramuscular Fat by Near Infrared Spectrophotometry; AMBL: AUS-MEAT Marbling Score; MMBL: MSA Marbling Score; CWT: Hot Standard Carcase Weight (kg). [2] SD: Standard Deviation; [3]CV: Coefficient of Variation.

Table 2. Heritabilities, genetic correlations and phenotypic correlations for IMF and carcase marbling traits (standard error in parenthesis)

Trait[1]	PIE IMF	CUP IMF	CIMF	AMBL	MMBL
PIE IMF	**0.37 (0.08)**	0.79 (0.09)	0.64 (0.11)	0.45 (0.14)	0.46 (0.14)
CUP IMF	0.34 (0.03)	**0.51 (0.09)**	0.75 (0.09)	0.59 (0.12)	0.64 (0.12)
CIMF	0.27 (0.03)	0.36 (0.02)	**0.62 (0.09)**	0.97 (0.04)	0.96 (0.03)
AMBL	0.19 (0.03)	0.27 (0.03)	0.56 (0.02)	**0.42 (0.09)**	0.99 (0.01)
MMBL	0.21 (0.03)	0.30 (0.03)	0.62 (0.02)	0.94 (0.01)	0.46 (0.09)

[1] Heritabilities on diagonal, genetic correlations above diagonal, phenotypic correlation below diagonal

Table 3. Accuracy of selection (Acc), and response to selection (ΔG) per year for breeding objective traits of CIMF, AMBL and MMBL

Scenario	Abbreviation	Acc CIMF	ΔG CIMF (%)	Acc AMBL	ΔG AMBL (Score)	Acc MMBL	ΔG MMBL (Score)
1	PIE IMF	0.46	0.75	0.32	0.16	0.33	16.95
2	CUP IMF	0.59	0.97	0.47	0.24	0.51	26.13
3	GBV 17%	0.17	0.28	0.17	0.09	0.17	8.90
4	GBV 47%	0.47	0.79	0.47	0.24	0.47	24.85
5	GBV 70%	0.70	1.16	0.70	0.36	0.70	36.81
6	GBV 85%	0.85	1.42	0.85	0.44	0.85	44.93
7	GBV 95%	0.95	1.59	0.95	0.49	0.95	50.25
8	1+3	0.48	0.78	0.36	0.18	0.36	18.64
9	1+4	0.61	1.00	0.53	0.27	0.53	27.92
10	1+5	0.75	1.25	0.71	0.37	0.72	37.62
11	1+6	0.87	1.46	0.85	0.44	0.86	45.06
12	1+7	0.96	1.60	0.96	0.49	0.96	50.34
13	2+3	0.60	0.99	0.48	0.25	0.52	26.96
14	2+4	0.68	1.12	0.59	0.30	0.62	32.11
15	2+5	0.78	1.30	0.73	0.37	0.75	38.98
16	2+6	0.88	1.47	0.86	0.44	0.87	45.18
17	2+7	0.96	1.60	0.95	0.49	0.96	50.03

Combining information of scanned phenotype and GBV showed the highest selection responses overall. However, the value of phenotyping with CUP IMF or PIE IMF diminished as the GBV accuracy increases (Figure 1). Based on Goddard *et al.* (2011), with the effective number of chromosome segments calculated according to Daetwyler *et al.* (2008) and an assumed effective population size for Angus of 90 (Clark *et. al.* 2019), we expect a GBV accuracy of 48% for CIMF, given the current Angus Australia reference population size of approximately 3,300 animals with carcase IMF phenotypes and genotypes. When the reference population number doubles, the GBV accuracy increases to 61%.

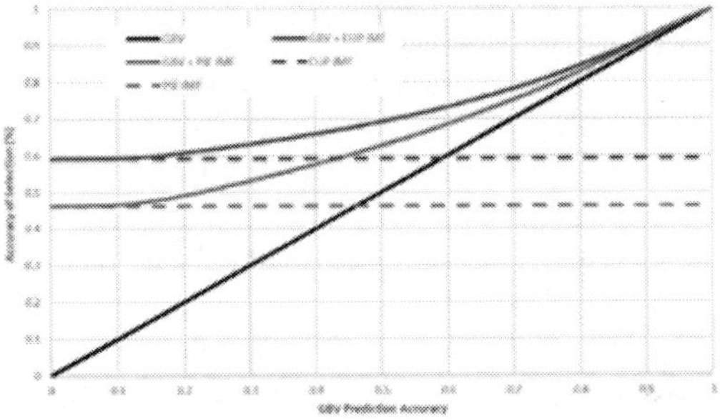

Figure 1. Accuracy of selection of CIMF comparing phenotyping and genotyping scenarios

CONCLUSIONS

The study showed, for Australian Angus breeders, the highest rates of selection accuracy and response (ΔG) will be achieved through using a combination of CUP phenotyping for IMF and genotyping with a reference population of related animals with carcase IMF and marbling score phenotypes. However, the value of ultrasound scan phenotyping diminishes as the GBV prediction accuracy increases, which is mainly a function of the reference population size.

This study also confirmed the potential benefit of utilising the CUP ultrasound scan technology for genetic analysis and selection for carcase IMF. This will be achieved through higher selection accuracy resulting from higher heritability and stronger genetic correlations to the direct carcase breeding objective traits.

ACKNOWLEDGMENTS

The authors thank Angus Australia for their financial support of this research and access to the data from the Angus Sire Benchmarking Program (ASBP). Meat and Livestock Australia (MLA) have provided co-funding for the ASBP (PSH.0528).

REFERENCES

Clark, S.A., Granleese, T. and Parnell, P.F. (2019) *Proc. Assoc. Advmt. Anim. Breed. Genet.* **23**:

Daetwyler, H.D., Villanueva, B. and Woolliams, J.A. (2008) *PLOS ONE* 3, e3395

Duff C.J., van der Werf J.H.J. and Clark S. A. (2018) *Proc. 11th World Congr. Genet. Appl. Livest. Prod.* **Electronic Poster Session - Species - Bovine (beef) 1**: 262.

Gilmour, A.R., Gogel, B.J., Cullis, B.R., Welham, S.J. and Thompson, R. (2015). ASReml User Guide Release 4.1 Functional Specification, VSN International Ltd, Hemel Hempstead, HP1 1ES, UK.

Goddard M.E., Hayes B.J. and Meuwissen, T.H.E. (2010) *Genet. Res., Camb.* **92**: 413.

Goddard M.E., Hayes B.J. and Meuwissen, T.H.E. (2011) *J. Anim. Breed. Genet.* **128**: 409.

Graser H-U., Tier B., Johnston D.J. and Barwick S.A. (2005) *Aus. J. Exp. Agric.* **45**: 913.

van der Werf, J. (2019). Teaching Software used for Quantitative Genetics at UNE. Available from: http://www-personal.une.edu.au/~jvanderw/software.htm (verified 1 April 2019).

Williams A.R. (2002) *J .Anim. Sci.* **80**: E183.

BULL DISCOVERY POWERED BY GENOMICS – A PRACTICAL CASE STUDY

M.J. Reynolds, A.I. Byrne, C.J. Duff and P.F. Parnell

Angus Australia, 86 Glen Innes Rd, Armidale, NSW, 2350 Australia

SUMMARY

The benefits of genomic testing are widely understood but the initial cost of investing in the technology can be daunting for many producers. This case study examines the impact of genomics in a real beef herd by comparing the Estimated Breeding Values (EBVs) of animals with and without genomic information included. The study demonstrated that the ranking of some selection candidates will change with the inclusion of genomic information, which could significantly affect selection decisions and rates of potential genetic gain. On average across the reported traits 5.33 new animals, based on Genomic Breeding Values (GEBV), replaced existing bulls in the top 15 when based on EBVs.

INTRODUCTION

Having led the Australian beef industry with the adoption of numerous reproductive and genetic improvement technologies, Australian Angus breeders are now embracing genomic technology in increasing numbers. When combined with pedigree and performance information, genomic information adds an additional source of information in the Angus BREEDPLAN analysis resulting in the generation of EBVs with additional accuracy, and ultimately enabling more accurate selection decisions. (Johnston *et al.* 2018).

The incorporation of genomic information is of most value when an animal's existing EBV has low accuracy (Dehnavi *et al.* 2018). For example, (i) when an animal is very young; (ii) for traits that are hard to measure, or traits that cannot be measured prior to an animal entering the breeding herd; (iii) for traits that have a low heritability; (iv) in situations where collecting effective performance information is problematic, such as in small herds, or when an animal has been removed from its contemporary group; and (v) in situations where little information is recorded with Angus BREEDPLAN for the animal, such as recently imported overseas sires.

The study will examine the impact of the incorporation of genomic information into the calculation of EBVs for a group of Angus bulls to determine the potential impact to selection decisions.

MATERIALS AND METHODS

The Angus BREEDPLAN genetic evaluation includes pedigree, performance and genomic information from the Angus Australia and New Zealand Angus Association databases to evaluate the genetics of Angus and Angus-influenced animals across Australia and New Zealand, using analytical software developed by the Animal Genetics and Breeding Unit (AGBU) (Graser *et al.* 2005). Over 17,500 animals were genomically tested through Angus Australia during 2018, and over 50,000 Australian Angus animals now have genomic information analysed within the Angus BREEDPLAN genetic evaluation.

To evaluate the effect that the inclusion of genomic information has on the calculation of EBVs, the mid-September 2018 Angus BREEDPLAN analysis was conducted twice, initially with all available pedigree, performance and genomic information included (GEBV), as per the EBVs published to industry, and secondly, with only the available pedigree and performance information included, but with the available genomic information removed (EBV).

From within the results of the two analyses, GEBVs and EBVs for a group of selection candidates were extracted for analysis, being 138 bulls born in 2016 from the herd of a current Angus Australia member who has a long history of performance recording for genetic analysis. The 138 bulls represented the entire group of male calves present in the herd at the time of DNA sample collection, which is 93% of the 149 male calves born in the herd in that birth year.

All 138 bulls were genotyped with the Zoetis i50K for Angus product, containing 29,834 SNPs after genotype QA has been applied.

RESULTS AND DISCUSSION

Summary of Data. A summary of the performance information analysed, in the GEBV and EBV analyses, for the 138 bulls is included in Table 1

Table 1. Count of performance measurements for the 138 analysed bulls

Traits Observed	BWT (kg)	200 (kg)	400 (kg)	600 (kg)	SC (cm)	Scan EMA (cm2)	Scan RIB (mm)	Scan P8 (mm)	Scan IMF (%)
Bulls	135	134	130	106	121	119	119	115	119

BWT: Birth weight, 200: 200-day weight, 400: 400-day weight, 600: 600-day weight, MCW: Mature cow weight, SC: Scrotal circumference, Scan EMA: Scanning measurement of Eye muscle area, Scan RIB: Scanning measurement of rib fat, Scan P8: Scanning measurement of fat at the P8 site, Scan IMF: Scanning measurement of the intramuscular fat

Change in Average EBVs. Across the 138 bulls, little movement in the average EBVs for the selection candidates was observed. The inclusion of genomic information increased the average accuracy of the EBVs for the bulls, with the magnitude of the increase varying for different traits (Table 2).

Table 2. Average EBV and accuracy change in 138 bull with the incorporation of genomic information

	BWT (kg)	200 (kg)	400 (kg)	600 (kg)	MCW (kg)	SS (cm)	CWT (kg)	EMA (cm2)	RIB (mm)	P8 (mm)	RBY (%)	IMF (%)
EBV	3.9	48	85	111	86	1.7	62	5.8	-0.1	-0.4	0.4	2.1
EBV Accuracy	64	60	60	62	58	55	52	50	52	52	48	47
GEBV	4.1	49	87	114	88	1.6	64	5.8	-0.1	-0.4	0.4	2.1
GEBV Accuracy	71	69	67	69	64	64	60	58	62	59	54	57
BV Difference	+0.2	+1	+2	+3	+2	-0.1	+2	0	0	0	0	0
Accuracy Difference	+7%	+9%	+7%	+7%	+6%	+9%	+12%	+8%	+10%	+7%	+6%	+10%

BWT: Birth weight, 200: 200-day weight, 400: 400-day weight, 600: 600-day weight, MCW: Mature cow weight, SS: Scrotal size, CWT: Carcase Weight, EMA: Eye muscle area, RIB: Rib fat, P8: Rump fat, RBY: Retail beef yield, IMF: Intramuscular fat

Change in EBV Spread. While the average EBVs for the bulls did not change substantially, there was an increase in the spread of EBVs for all traits with the inclusion of genomic information. (Table 3). The increase in EBV spread observed from the incorporation of genomic information aids in better identifying genetic differences between individuals.

Change in Ranking of Individuals. Likewise, while the average EBVs for the bulls did not change substantially, there was a considerable re-ranking observed in EBVs when genomic information was included. The correlation between EBVs and GEBVs is listed in Table 4. The reported change is the result of an increase in the genetic linkage between individual bull and the reference population and the replacement of assumptions around pedigree relationships with genomic information.

Table 3. Standard deviation of EBVs in 138 bulls with the incorporation of genomic information

	BWT (kg)	200 (kg)	400 (kg)	600 (kg)	MCW (kg)	SS (cm)	CWT (kg)	EMA (cm2)	RIB (mm)	P8 (mm)	RBY (%)	IMF (%)
EBV	1.4	5.4	8.3	13.1	16.5	0.7	7.6	1.2	1.1	1.4	0.8	0.5
GEBV	2.0	6.9	10.7	17.4	23.6	1.0	10.8	2.1	1.5	1.9	1.1	0.7
Difference	43%	28%	29%	33%	43%	43%	42%	75%	36%	36%	38%	40%

Refer to Table 2 for breeding value descriptions

Table 4. Correlation between EBVs and GEBVs for 138 bulls with the incorporation of genomic information

	BWT	200	400	600	MCW	SS	CWT	EMA	RIB	P8	RBY	IMF
Correlation	0.85	0.85	0.84	0.88	0.87	0.87	0.86	0.71	0.84	0.85	0.79	0.78

Refer to Table 2 for breeding value descriptions

At a practical level, the implementation of selection within a commercial beef breeding program requires the identification of selection candidates that best meets the breeding objective. Whilst movement in individual breeding value may be significant enough to affect whether an animal meets the required breeding objective thresholds, the individual re-ranking has a far more significant effect on an individual's selection. Therefore, it was more significant for the within herd selection of an individual, where they rank within contemporary the group of selection candidates, rather than their reported breeding value.

Table 5 displays bulls, in green, not ranked in the top 15 prior to inclusion of genomic information. Genomic testing has enabled these bulls to be ranked higher and considered for selection. Without genomics they ranked lower and outside the top 15. On average across the reported traits 5.33 new animals, based on GEBV, replaced existing bulls in the top 15 when based on EBV. Bulls marked in red, are those individuals which were replaced within the top 15 and if selected could have resulted in a loss in potential genetic gain.

The paper highlights the value genomic information delivers to the EBV calculation, in addition to the potential risk breeders take implementing selection decisions without the incorporation. The significant movement and potential losses in genetic gain represent expected movement, as a result genomic testing. This feature validates the growing trend seen within registered Angus cattle in Australia around the increased utilisation of genomic technologies.

CONCLUSION

This case study has provided a real beef herd example illustrating the effect of genomic testing on the EBVs for a group of selection candidates. The inclusion of genomic information did not substantially change the EBVs of the selection candidates on average, but a considerable increase in the spread of EBVs and considerable re-ranking of the selection candidates was observed. The accuracy of the EBVs for the selection candidates increased on average, but the magnitude of accuracy increase varied by trait. Most importantly, in a practical context, the inclusion of genomic information changed the candidates selected for use within the breeding program. It must be noted that the observations in this case study may vary with a different group of selection candidates.

Table 5. Ranking top 15 bulls for each trait, with and without the inclusion of genomic information (Red=dropped outside top 15 with genomic inclusion, Yellow= changed rank order but remained in the top 15, Green=moved into the top 15 with genomic inclusion)

						EBV						
Rank	**BWT**	**200**	**400**	**600**	**MCW**	**SS**	**CWT**	**EMA**	**RIB**	**P8**	**RBY**	**IMF**
1	M63	M223	M149	M81	M81	M174	M81	M149	M42	M249	M300	M197
2	M289	M198	M223	M223	M287	M307	M177	M242	M8	M53	M131	M167
3	M8	M81	M198	M333	M250	M109	M149	M153	M29	M238	M271	M188
4	M212	M225	M225	M198	M46	M263	M103	M131	M308	M307	M214	M90
5	M20	M148	M148	M149	M333	M298	M225	M177	M249	M42	M202	M289
6	M162	M149	M118	M225	M223	M65	M148	M163	M65	M38	M133	M82
7	M165	M102	M81	M245	M225	M223	M248	M214	M20	M308	M149	M225
8	M53	M118	M248	M46	M198	M321	M83	M111	M176	M120	M89	M102
9	M169	M186	M237	M148	M186	M315	M46	M202	M45	M29	M328	M80
10	M89	M237	M102	M83	M83	M142	M237	M300	M145	M20	M242	M117
11	M140	M248	M186	M287	M156	M328	M8	M38	M238	M145	M156	M140
12	M141	M240	M103	M102	M323	M275	M223	M248	M307	M118	M275	M240
13	M14	M257	M240	M257	M220	M333	M165	M99	M53	M117	M248	M63
14	M155	M197	M257	M156	M6	M57	M257	M81	M120	M315	M64	M198
15	M179	M188	M290	M118	M257	M259	M98	M148	M54	M198	M174	M223
						GEBV						
Rank	**BWT**	**200**	**400**	**600**	**MCW**	**SS**	**CWT**	**EMA**	**RIB**	**P8**	**RBY**	**IMF**
1	M63	M225	M148	M245	M287	M307	M81	M300	M65	M249	M300	M104
2	M8	M81	M225	M81	M245	M54	M103	M214	M29	M145	M64	M167
3	M329	M102	M245	M225	M250	M275	M83	M242	M35	M315	M214	M197
4	M289	M240	M102	M46	M81	M65	M149	M274	M308	M307	M280	M188
5	M89	M245	M81	M287	M46	M177	M225	M153	M8	M29	M242	M90
6	M212	M223	M103	M333	M83	M315	M148	M111	M42	M212	M237	M53
7	M179	M103	M248	M83	M333	M333	M177	M156	M120	M53	M275	M82
8	M155	M148	M46	M156	M156	M32	M46	M163	M249	M123	M99	M117
9	M14	M198	M118	M148	M234	M194	M245	M64	M145	M308	M131	M236
10	M106	M149	M223	M102	M6	M321	M156	M21	M53	M8	M89	M156
11	M321	M248	M149	M38	M220	M142	M248	M148	M176	M38	M285	M323
12	M141	M333	M240	M103	M80	M109	M86	M202	M45	M238	M217	M80
13	M169	M83	M198	M6	M280	M263	M287	M99	M162	M65	M274	M125
14	M99	M287	M83	M223	M320	M298	M21	M42	M179	M162	M170	M243
15	M311	M118	M142	M149	M186	M89	M142	M321	M63	M179	M272	M35

Refer to Table 2 for breeding value descriptions

ACKNOWLEDGEMENTS

The authors gratefully acknowledge the contribution of staff at the Agricultural Business Research Institute (ABRI), and in particular Dr Brad Crook, who conducted the genetic evaluations required for this case study.

REFERENCES

Dehnavi E., Ansari Mahyari S., Schenkel F.S. and Sargolzaei M. (2018) *J. Dairy. Sci.* **101**: 6.

Graser H-U., Tier B., Johnston D.J. and Barwick S.A. (2005) *Aust. J. Exp. Agric.* **45**: 913.

Johnston D.J., Ferdosi M.H., Connors N.K., Boerner V., Cook J., Girard C.J., Swan A.A. and Tier B. (2018) *World Congress on Genetics Applied to Livestock Production.* **11**: 269

BENEFITS OF GENOMIC INFORMATION IN THE ANGUS INDUSTRY – THE RENNYLEA EXPERIENCE

S. Dominik[1], L. Porto-Neto[2], A. Reverter[2] and L. Corrigan[3]

[1]CSIRO Agriculture and Food, Armidale, NSW, 2350 Australia
[2]CSIRO Agriculture and Food, St Lucia, QLD, 4067 Australia
[3]Rennylea Angus Stud, Bowna, NSW, 2644 Australia

SUMMARY

Performance data, pedigree and genotype information from Rennylea Angus stud was used to estimate pedigree-based estimated breeding values (EBV) and single-step genomic breeding values (GEBV). Pedigree-based EBV and GEBV and their accuracies were compared. For closely related animals, GEBV are more powerful in differentiating animals' genetic potential compared to EBV. The greatest benefit of GEBV for providing an estimate of the underlying genetics was for young sires, even more so, when only a small proportion of animals are being selected for breeding.

INTRODUCTION

Genomic selection was rapidly adopted in the dairy compared to the beef industry, due to advantages, such as implementation in a single breed and a large reference population (Goddard 2012). The use of genomic information is gaining momentum in the Australian Angus industry and Angus Australia is establishing a well-recorded reference population with the Angus Sire Benchmarking Program, which is essential for effective industry application (Angus Australia 2017). Some leading breeders, including Rennylea Angus stud, have been whole-herd genotyping with high density genomic profiles for several years. In this study we explored the Rennylea data with the aim to demonstrate the potential benefit of genomic information for Rennylea Angus, based on a within-herd analysis.

MATERIALS AND METHODS

Data. Data from Rennylea Angus stud were obtained from the Angus Australia database. Data included records of animals born between 1987 and 2017 with 7,758 animals in the pedigree. Three traits were analysed for this study: weaning weight (WW, in kg, 7,325 records), intra-muscular fat from ultrasound scanning at 500 days (IMF, in mm, 5,239 records) and scrotal circumference at 400 days (SS, in cm, 2,086 records).

Rennylea commenced genotyping animals born in 2014. We found 337 sires with 22.6 progeny genotyped on average and progeny numbers per sires ranged from 1 to 338 progeny. For the comparison of EBV to GEBV we will focus on the 74 large sires with at least 25 progeny.

Genotype information was generated with a number of different genotyping platforms with varying SNP densities, ranging from 18K to 778K. Most animals were genotyped with a SNP density of 29,834. However, only 15,388 SNP were in common to the SNP chip with the second largest number of animals (18,815 SNP with 1,679 animals). These SNP were kept for further analyses. An imputation step was carried out, to impute scattered missing genotypes to the set of 15,388 common SNP. After merging all SNP platforms, the imputed genotype file contains 15,388 SNP genotypes for 3,954 animals. The frequency of the first allele averages 0.486 and ranges from 0.0 to 1.0 and 92.5 % of SNP have a minor allele frequency > 1 %.

Statistical analysis. Variance components and heritabilities were estimated in linear mixed animal models from pedigree using ASReml (Gilmour *et al.* 2009). Fixed effects included sex (male / female), birth year (1987 – 2017), birth month (February-April and July – September), artificial insemination (yes/no), embryo transfer (yes / no), recording dates and management groups (35 groups). Significant effects were retained in the model. Animals from multiple births and management groups with less than 10 animals were excluded. Using the same models, EBV and GEBV were estimated using QXPAK (Perez-Enciso and Misztal 2004). The accuracies were calculated as:

accuracy (ACC) = $\sqrt{1-\frac{s_i^2}{V_g^2}}$, with s_i= standard error of the estimated breeding value for the ith individual and $V_g^{\,2}$ = genetic variance. Correlations between EBV and GEBV were calculated for all animals, all sires and young sires only to establish the concordance of the ranking. The proportion of agreement between animals selected at top 1% or 10% on EBV versus GEBV from all animals, sires and young sires were established.

RESULTS AND DISCUSSION

Trait heritabilities from pedigree were low to moderate (Table 1). Compared with heritabilities published by Angus Australia the h^2=0.58 for SS was high and the other heritabilities were within the range of the published estimates (Angus Australia 2019).

Table 1. Mean (± standard deviation (stddev)), minimum (Min) and maximum (Max), phenotypic and genetic variances (V_p and V_g) and heritability h^2 (± standard error (stderr)) for weaning weight (WW, kg), scrotal circumference (SS, cm) and intramuscular fat (IMF, %)

Trait	Mean ± stddev	Min	Max	V_p	V_g	h^2 ± stderr
WW	229.50 ± 43.54	201.00	421.00	674.64	164.28	0.27 ± 0.03
SS	37.05 ± 2.67	26.00	47.00	6.22	3.58	0.58 ± 0.07
IMF	4.39 ± 1.55	0.20	8.40	1.75	0.52	0.30 + 0.03

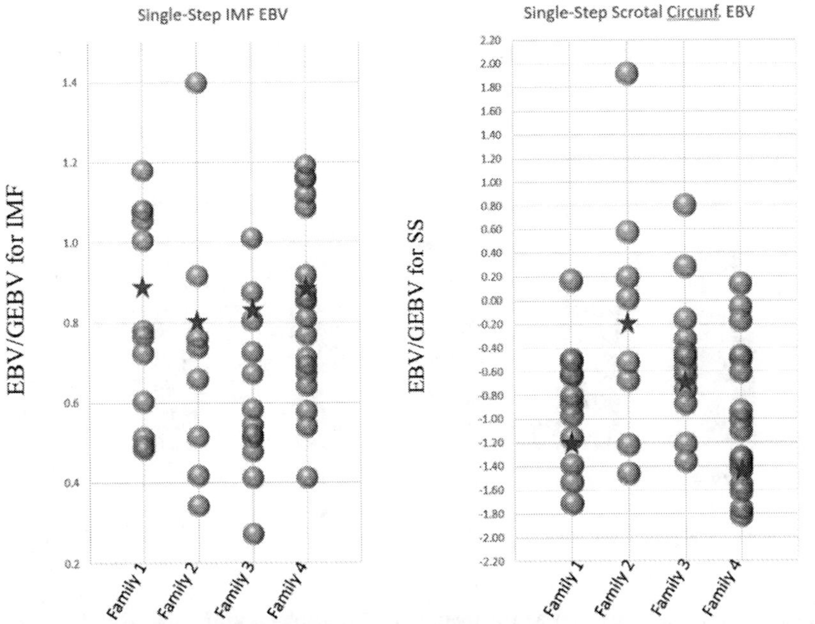

Figure 1. Distribution of GEBV (bubbles) for IMF (left) and SS (right) for the four largest full-sib families. Marked with a star is the value for EBV of full-sibs from the same family

The comparison of EBV and GEBV for the four largest full-sib families for IMF and SS demonstrates that GEBV are different for the members of the same family (Figure 1), because the genomic relationship matrix accounts for different relationships than 0.5 for members of full-sib families. For closely related animals, GEBV are more powerful in differentiating animals' genetic potential, compared to EBV and enables better within-family selection.

A comparison of pedigree-based EBV and GEBV for all 7,460 animals, all 3,020 sires and all 1,450 young bulls showed high concordance of ranking, expressed as correlation (Table 2). The correlations for young bulls were higher for WW, a trait that the animals themselves have a phenotype for, compared to IMF. This means that, for young bulls, GEBV provide a better estimate of the underlying genetics.

Table 2. Correlations between estimated breeding values (EBV) and genomic breeding values from single step (GEBV) for all 7,460 animals, all 3,020 sires and 1,450 young 2016-17 bulls

	All animals	All sires	Young bulls
WW	0.88	0.85	0.80
SS	0.89	0.90	0.66
IMF	0.91	0.85	0.70

The concordance between animals selected on EBV versus GEBV from all animals, all sire or young bulls was higher at 10% selected compared to 1% (Figure 2). We previously demonstrated that GEBV provide a better estimate of an animal's genetic potential (Figure 1), consequently, the lower the concordance the higher the benefit from selecting on GEBV. Figure 2 shows the benefit of selecting on GEBV, particularly for higher selection intensities (1%) and for the WW trait.

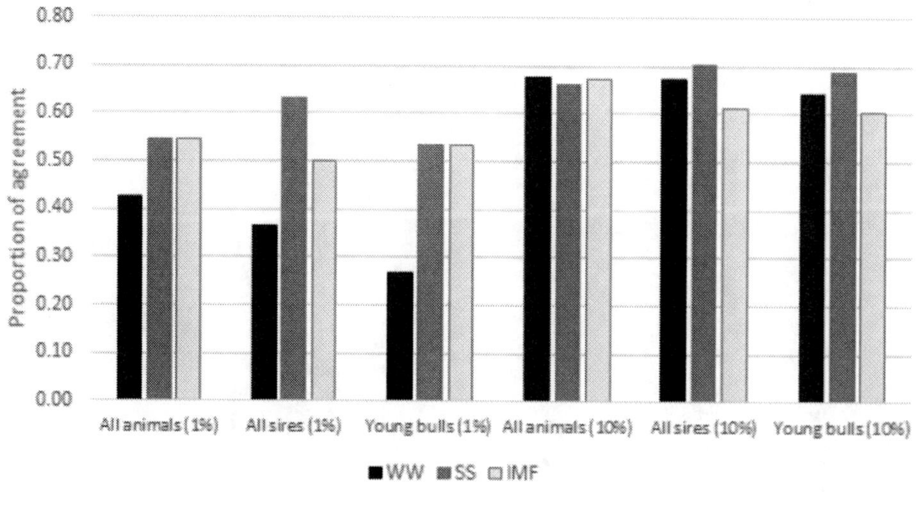

Figure 2. Concordance, expressed as the proportion of agreement, of all animals, all sires and young bulls if selected on GEBV and EBV at 1% (left to middle) and 10% (middle to right)

The plots in Figure 3 show the close alignment of the accuracies of EBV and GEBV. It also demonstrates that the younger bulls (red) are getting the most benefit of higher accuracies from GEBV compared to EBV with the "cloud" of accuracies for the GEBV of young bulls being distinctly higher. This benefit is least obvious for WW.

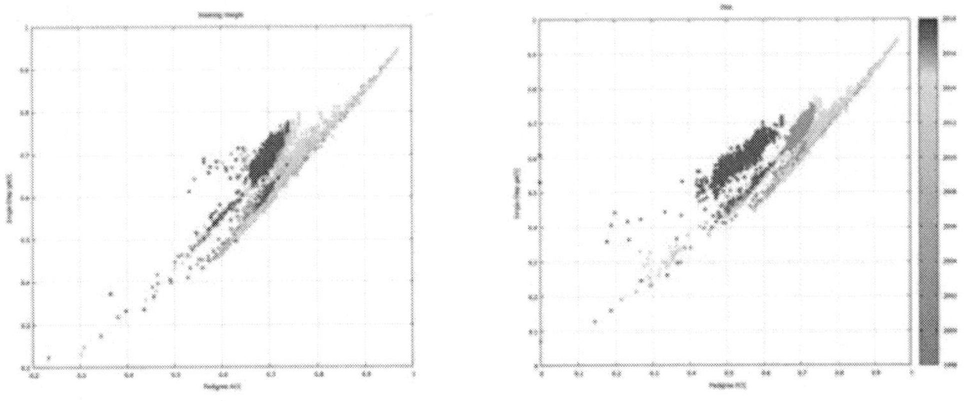

Figure 3. Plots of accuracies for EBV (x-axis) and GEBV (y-axis) for WW (left) and IMF (right). Scale on the right shows colour by age with younger animals are coloured in red and older animal in green

CONCLUSION

The within-herd analysis of the Rennylea data demonstrates the benefit of using genomic information for more accurate estimates of individual animal breeding values only for this particular data set. To generate whole industry benefit from genomic information well recorded reference populations such as the Angus Sire Benchmarking Program are required to deliver the best tools for industry application.

ACKNOWLEDGEMENT

The authors would like to gratefully acknowledge the funding for this study by Rennylea Angus stud and the Australian Federal government's Innovation Connections program.

REFERENCES

Angus Australia (2017) https://www.angusaustralia.com.au/harnessing-the-benefits-of-genomics-in-an-angus-seedstock-enterprise/ . (last viewed March 2019).

Angus Australia (2019) https://www.angusaustralia.com.au/breedplan/resources/heritability -of-traits/. (last viewed March 2019).

Gilmour A.R., Gogel B.J., Cullis B.R. and Thompson R. (2009) ASReml User Guide Release 3.0. VSN International Ltd, Hemel Hempstead, HP1 1ES, UK.

Goddard M.E. (2012). *Anim. Prod. Sci.* **52**: 73.

Pérez-Enciso M. and Misztal I. (2004) *Bioinformatics* **20**: 2792.

A SURVEY APPROACH TO EXPLORE INDUSTRY PRIORITIES FOR NOVEL TRAITS IN AUSTRALIAN ANGUS

A.M. Bell[1], A.I. Byrne[2], C.J. Duff[2] and S. Dominik[1]

[1]CSIRO Agriculture and Food, Armidale, NSW, Australia
[2]Angus Australia, Armidale, NSW, Australia

SUMMARY

This study used a survey approach to investigate preferences in the Australian Angus industry for novel traits (immune competence, structural soundness, feed efficiency and temperament) over classical production traits. One hundred and thirty five people responded and ranked 11 attributes. The pool of respondents represented a cross section of the industry. Preferences were high for novel traits, with the exception of feed efficiency. Calving and reproduction traits had the lowest preference, possibly indicating satisfaction with current performance for these traits. Rankings of attributes were the same when responses were split by demographic information. Although differences in attribute rankings were observed, the interpretation of the results is limited to the respondents of the survey.

INTRODUCTION

Beef cattle breeding programs have focussed on production and reproduction traits, the primary profit drivers for the industry (Angus Australia 2014). Novel traits such as structural soundness, temperament and immune competence offer the opportunity to maintain a healthy, sustainable and productive herd into the future (Angus Australia 2017a, b; Angus Media 2017). However, when a novel trait is introduced into an existing breeding program, selection emphasis on current breeding objective traits is reduced. Every industry participant will value these trade-offs differently, depending on production environments, target markets and their own individual challenges and goals for their herd. Survey approaches are useful to explore these differences and *1000Minds* (Hansen and Ombler 2009) has previously been used to evaluate plant and sheep breeding programs (Smith and Fennessy 2011; Byrne *et al.* 2012). The aim of this study is to explore preferences in the Australian Angus industry for novel traits in breeding programs using a *1000Minds* survey (Hansen and Ombler 2009).

MATERIALS AND METHODS

Survey design. The *1000Minds* software was used to implement a choice experiment survey. The software generates "part-worth utilities" which describe the relative importance that participants place on a number of attributes (Hansen and Ombler 2009). The attributes were based on the traits included in the current Australian Angus Selection Indexes (Angus Australia 2014) plus a number of novel traits, including feed efficiency, structural soundness and immune competence, which is the ability of the immune system to respond to a disease challenge (Colditz and Hine 2016). Two or three levels of possible response were specified for each attribute and these were entered into the software in their logical ranking (Table 1).

Participants responded to a number of questions where they chose between two hypothetical scenarios with differing levels of response in pairs of attributes. An example of a question was "Do you prefer Option A (-20kg change in Carcase weight and +2% change in Saleable meat yield) or Option B (+20kg change in Carcase weight and no change in Saleable meat yield)?" and this is illustrated in Figure 1. Participants also had the opportunity to rate the options as equally appealing or rate a combination as impossible. The *1000Minds* software applies the PAPRIKA method that automatically drops impossible choice combinations based on the previously expressed preferences (Hansen and

Ombler, 2009). The method calculates the part-worth utilities based on individual survey responses.

A Wufoo multiple choice survey (Hale *et al.* 2006) was used to collect demographic information. Questions included participant's role in the industry (e.g. breeder, producer), their location, the average number of females in their breeding herd, which production environment (temperate or subtropical) and which production system (grain or grass fed) they were targeting.

Table 1. Attributes included in the survey

Attribute	Level 1	Level 2	Level 3
Immune competence	More expected disease incidence	Less expected disease incidence	
Calving ease	No change	5% fewer calving difficulties	
Marbling score	-1	+1	
Docility	10% less docile	10% more docile	
Mature cow weight	+30kg	No change	-30kg
Sale live weight	-10kg	+10kg	
Carcase weight	-20kg	+20kg	
Saleable meat	No change	+1%	+2%
Structural soundness	10% less structurally sound	10% more structurally sound	
Weaning rate	No change	5% more calves weaned	
Feed efficiency	No change	5% improvement	

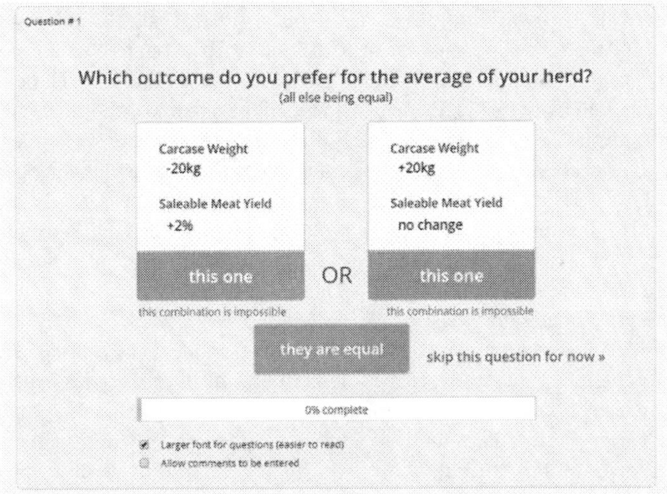

Figure 1. Example of a survey question to determine part-worth utilities

Survey participants. The survey was sent out through Angus Australia to their members, which included around 4,500 breeders and producers, as well as via various rural media publications. Participation was voluntary and this study was approved by the CSIRO Human Ethics Committee (Approval number 145/18).

RESULTS AND DISCUSSION

A total of 135 respondents (Table 2) completed the survey and represent a cross-section of commercial producers, seed stock producers and service providers (e.g. consultants), with distribution across production environments. For analysis purposes of this paper four respondents who identified their location as 'Other' were removed from the dataset. The relatively low response rate may potentially have been due to survey fatigue or to the unfamiliar format of the survey where participants were tasked with choosing between scenarios that may not reflect their personal operation, but the choice provides an informative utility value. The survey was promoted through Angus Australia as a call for members to have their say on traits in current indexes, but more information regarding the survey format and its potential application might have been beneficial in the lead up to the survey period. In addition, this survey technique has previously been used with smaller numbers of focus groups, for example Byrne *et al.* 2012 surveyed 24 sheep industry experts and 25 sheep farmers in Ireland and in comparison the number of respondents in this survey was significantly larger.

Table 2. Role and location of respondents

Role/Location	NSW	VIC/TAS	SA	QLD	WA	Other	Total
Commercial cow-calf producer	10	24	4	1	3	0	42
Seedstock and commercial cow-calf producer	17	9	1	3	5	1	36
Seedstock producer	29	16	1	4	3	2	55
Service provider (eg consultant)	0	1	0	0	0	1	2
Total	56	50	6	8	11	4	135

The total sum of all part-worth utilities always sums to 100%. If all 11 attributes would be ranked the same, they would all have a mean part-worth utility of 9.10%, but as a result of the survey mean part-worth utilities across all traits ranged from 6.50% – 11.86% (Table 3). Three of the four novel traits, with the exception being feed efficiency, had the highest part-worth utilities with values above 10%. For weight and carcase traits, with the exception of mature cow weight, values ranged between 8.40% and 9.50%. The calving and reproduction traits, feed efficiency and mature cow weight ranged between 6.50%-8.07%. The observable difference in preferences for attributes depend on the number of attributes and also their importance to the survey participants (e.g. if all attributes are somewhat important, the observable differences are smaller). Rankings for the attributes across the demographics captured were the same (results not shown). These results indicate that the preferences observed applied regardless of the demographics.

The part-worth utilities demonstrate that amongst the respondents novel traits are desirable characteristics to improve in their herd. Relatively lower part-worth utilities indicate that participants placed lowest emphasis on calving and reproductive traits, possibly indicating that participants are comfortable with current performance for these attributes. The part-worth utility for mature cow weight was 7.16%, but choices for this attribute showed the highest variation, reflecting the trait being a rich topic of industry discussion. Feed efficiency was ranked lowest of the novel traits.

It is interesting to note that immune competence, a trait which is not currently routinely measured, and which is arguably most relevant to the feedlot phase of the animals was ranked more highly than feed efficiency, and by respondents from all sectors of the industry. The more favourable response to the immune competence choice could reflect a desire to make progress towards addressing currently intractable industry issues such as the incidence of bovine respiratory disease in feedlots. On the other hand, the description of the alternate immune competence scenarios (more/less disease incidence) may have played a role, for example when compared with the description of the feed efficiency choice (feed efficiency no change/5 % improvement).

Table 3. Mean, Standard deviation (StdDev), Mininum (Min) and Maximum (Max) part-worth utility values across all participants

Attribute	Mean Part-worth utility value (%)	StdDev	Min	Max
Structural soundness	11.86	3.07	1.02	22.34
Immune competence	10.84	3.03	3.13	18.64
Docility	10.62	3.58	0.81	19.49
Carcase weight	9.53	3.29	1.82	16.22
Marbling score	9.10	4.03	0.80	21.43
Sale live weight	9.00	3.65	0.78	20.41
Saleable meat yield	8.93	3.28	1.75	21.28
Weaning rate	8.39	3.85	0.85	17.17
Feed efficiency	8.07	3.48	0.96	15.60
Cow weight	7.16	5.61	1.71	25.00
Calving ease	6.50	3.93	0.79	14.04

For future studies it may be interesting to capture participant's motivation for the preferences they expressed. For example, was their motivation purely economic in nature, or were there other motivations such as animal welfare concerns, animal management improvements or other strong reasons to desire change. This information could be captured along with the demographic information in a complimentary survey attached to the preference survey.

CONCLUSION

Choice experiment surveys are a useful tool to explore industry preferences. The interpretation of the result is limited to the pool of respondents and the more people respond, the more representative the results are of the wider industry. The results suggest there is interest in potentially exploring novel traits such as immune competence and their application in the Australian Angus industry, while there is a level of satisfaction with other traits such as the more traditional classical production traits such as carcase weight.

ACKNOWLEDGEMENT

We would like to acknowledge the time and effort spent by the breeders and producers who responded to the survey. We are grateful for their contribution.

REFERENCES

Australian Angus Selection Indexes (2014) http://breedplan.une.edu.au/tips/Interpreting%20Australian%20Angus%20Selection%20Indexes.pdf (last viewed April 2019).

Angus Australia (2017a) https://www.angusaustralia.com.au/improving-structural-soundness-in-a-beef-breeding-herd/ (last viewed April 2019).

Angus Australia (2017b) https://www.angusaustralia.com.au/breeding-resilient-angus-cattle/ (last viewed April 2019).

Angus Media (2017) https://www.angus.org/Media/News/FullArticle.aspx?asTitle=Temperament-and-Reproduction&aiId=753 (last viewed April 2019).

Byrne T.J., Amer P.R., Fennessy P.F., Hansen P. and Wickham B.W. (2012) *Animal* **6**: 778.

Colditz I.G. and Hine B.C. (2016) *Anim. Prod. Sci.* **56**: 1961.

Hale K., Campbell R. and Campbell C. (2006). https://www.wufoo.com/ (last accessed April 2019)

Hansen P. and Ombler F. (2009) *J.Multi-Criteria Decision Anal.* **15**: 87.

Smith K.F. and Fennessy P.F. (2011) *Crop and Pasture Science* **62**: 355.

DESIGN AND PURPOSE OF THE MERINO LIFETIME PRODUCTIVITY PROJECT

A.M.M. Ramsay[1], A.A. Swan[2], and B.C. Swain[1]

[1]Australian Merino Sire Evaluation Association, 1534 Prairies Rd, Gunnedah, NSW, 2380
[2]Animal Genetics & Breeding Unit*, University of New England, Armidale, NSW, 2351 Australia

SUMMARY

The Merino Lifetime Productivity project is a major research initiative undertaken by industry with funding support from Australian Wool Innovation, aiming to develop Merino breeding and selection strategies that will maximise the lifetime productivity of Merino ewe enterprises. The project involves mating 134 diverse industry sires across five sites around Australia to generate up to 5,500 Merino ewe progeny that will be annually evaluated through life for a wide range of wool, growth, carcase, reproduction and disease resistance traits. It is planned that all ewe progeny will be genotyped and will add to the genomic reference population for the Merino industry.

INTRODUCTION

The Merino Lifetime Productivity (MLP) project was established in 2015 to generate a unique Merino database that will enable the enhancement and validation of existing Merino ewe breeding and selection approaches in order to increase the long-term productivity of Australian woolgrowers. The project is a ten-year partnership between Australian Wool Innovation, the Australian Merino Sire Evaluation Association, Merino ram breeders, RD&E organisations, and five sire evaluation sites and their hosts around Australia.

The project was largely initiated to answer a range of industry questions relating to lifetime productivity in the Merino ewe enterprise. The practice of selecting animals at increasingly younger ages had raised concerns that industry was selecting for animals that performed well in their early years to the detriment of their lifetime productivity. Further, the diversification of the Merino ewe enterprise to include carcase traits, disease resistance attributes, reproduction along with wool production, led to concerns that current selection approaches might not adequately account for lifetime relationships between all these production elements.

The MLP dataset will be of significant value to the Merino industry for longer term phenotypic, genetic and economic analysis in order to answer a diverse range of commercial producer and ram breeder industry questions. Some of the short-term value of the project will be the submission of data to MERINOSELECT (Brown *et al.* 2007) and the resulting enhanced accuracy of Australian Sheep Breeding Values (ASBVs) for traits which are not routinely measured such as reproduction, body wrinkle and adult age assessments of key production traits.

OBJECTIVES

The immediate objective of the project is to create an extensive dataset that captures the lifetime performance of 5,500 diverse F_1 Merino ewes across five environments. A range of research and development activities will be undertaken using the completed dataset. These will include the precise estimation of genetic parameters for lifetime production across traits and between ages that will help to address the range of industry questions that underpin the project.

Additional objectives include the identification of the best number of visual and quantitative assessments of wool, reproduction, disease resistance and carcase production required to maximise lifetime productivity, and to find the most cost-effective means to achieve flock breeding objectives

* A joint venture of NSW Department of Primary Industries and the University of New England

and lift rates of genetic gain.

The project aims to provide a major stimulus to the collection of reproduction data and to raise the profile and increase collection of adult growth, body composition and wool performance data.

DESIGN AND IMPLEMENTATION

The sites and management. The sites are operated by five sire evaluation groups affiliated with AMSEA and site hosts with details of these key parties outlined in Table 1. Each site has a specific set of protocols that are reinforced by standard sire evaluation requirements which include the management of ewe progeny of the sires (known as F_1 ewes) as much as possible as one management group. In the event that animals need to be split into smaller groups, ewes are stratified across these groups, the exception is the management of ewes during pregnancy whereby ewes are able to be run based on their pregnancy status from pregnancy scanning to lamb tagging as long as each group is genetically linked.

The sire evaluation sites were selected to represent geographically diverse environments and genetically different ewe bases (see further details in Table 1) to ensure that the outcomes of the project are relevant to all Merino breeders. Sites have strong local site committees to integrate industry into the project oversight and support. All sites were selected to have extensive experience in electronic identification, good sheep handling facilities, prior experience in sire evaluation, and proven sheep management skills.

Table 1. MLP site name, number of industry sires joined, details of site hosts and site associations, ewe base details (microns and type), annual rainfall and rain fall distribution

Site	Number of Sires	Host	Ewe Base	Rainfall
Balmoral, Harrow, Vic	50	Balmoral Breeders and Tuloona Pastoral	17.2um fine wool base	470mm winter
MerinoLink, Temora NSW	26	MerinoLink, Moses and Sons, Blue Chip Livestock	18um, previous sire evaluation progeny, and studs with ASBVs	500mm winter
Pingelly, WA	30	Yardstick Sire Evaluation Association, Murdoch University, the University of Western Australia	19.5um, meat focused Merino ewes	440mm winter
New England, Armidale, NSW	30	New England Merino Sire Evaluation Association, CSIRO	17um, ultra-fine flock	800mm summer
Macquarie, Trangie, NSW	32	Macquarie Sire Evaluation Association and NSW Department of Primary Industries	19-21um, two skin types	500mm even

Ewe base joining. The ewe base at each site was generally sourced from a single flock that had been inspected and approved as an even line of ewes by the site committee. There were two exceptions to this; at the MerinoLink site where the ewe base was derived from five performance recorded flocks (and ewe progeny from a previous sire evaluation), and the Macquarie site where the ewe base was made up of an equal split of ewes sourced from two studs that have differing approaches to selection and divergent skin wrinkle.

A total of 90 ewes were randomly allocated to each sire (stratified for age, ewe source, body weight and condition score) and mated via AI with the aim of generating 30 live ewe progeny per sire for lifetime assessment. At several sites the AI program was split over two weeks to manage potential risk of climatic extremes on embryonic or neonatal loss (date of birth was recorded). The target of 30 progeny per sire was considered a balance between involving as many industry sires as possible whilst generating enough sire progeny to produce sufficiently accurate estimates of breeding value for lowly heritable traits such as reproduction. Each site joined to industry sires via AI for two years with the first AI program undertaken in March 2015 and the last in April 2018.

F_1 ewes and wethers. The core MLP sheep are the F_1 female progeny of industry sires that will be assessed throughout their life with ewes only removed for welfare purposes (e.g. severe injury, deformity or disease). The male F_1 progeny have been castrated and exited the core MLP project post weaning. In several instances these F_1 wethers have become the subject of separate projects. Lambs of the F_1 ewes, (F_2 progeny), leave the project following a DNA sample and weaning weight record.

Sires and genetic linkage. Across the five sites there were 166 sire joining groups, and after accounting for designed linkage across sites and years, a total of 134 unique sires from 95 different ram breeding flocks are represented in the project. All MLP sites are genetically linked by use of repeat sires and to the Australian national genetic evaluation service, MERINOSELECT. To achieve adequate linkage across MLP sites and years for traits with low heritability, approximately one in five sires provide linkage.

A sire selection protocol was developed to ensure that the project included sires that were both industry relevant and representative, and that offered extremes and combinations of performance that will help to explore the drivers of Merino lifetime productivity.

To achieve the sire selection goals the sires selected were a mixture of horn and poll, representative of different selection approaches and Merino strains, rams with progeny evaluated for production and without, rams from flocks who were and were not members of MERINOSELECT, a diverse range in performance, those that have been used extensively in industry or have made an industry impact (e.g. have sold sons for record prices or had exceptional show winning performance) and, those that were predicted to increase, decrease or maintain fleece value over time (based on performance information collected prior to the project).

Assessments recorded. A summary of the traits recorded on all F_1 ewes is given in Table 2. These traits are measured from birth through to 6-7 years of age, to capture performance over the productive life of Merino sheep. To complement the measurement program, it is planned that all F_1 ewes will be genotyped on a high-density SNP based chip platform, so that the project will be a valuable resource in adding to the genomic reference population for hard to measure Merino lifetime productivity traits.

Pedigree, reproduction and early environmental effects. Dam and sire pedigree of the F_1 progeny was allocated via DNA with exception of the New England site where daily lambing rounds were conducted to record date of birth, dam pedigree, and number of lambs born (sire pedigree was assumed from AI records). For the remaining four sites, the F_1 date of birth was estimated from the day of AI, birth type was inferred from pregnancy status estimated by ultrasound pregnancy scanning (at or between 40 and 70 days post AI), and rearing type was inferred based on the number of lambs at weaning.

The F_1 ewes are naturally joined as one group to a syndicate of Merino rams annually from 18 months of age, with the older drop of ewes at each site to be joined for 5 years and the younger drop to be joined for 4 years. Generally, each site is naturally joining for a period of five weeks using sires that have ASBVs available and have passed fitness and health requirements. The joining percentage at each site varies slightly from site to site based on local practices. Female reproduction traits on the F_1 ewes are derived from information recorded at joining, pregnancy scanning, in combination with DNA parentage tests on the F_2 lambs to identify maternal pedigree.

Table 2. Visual and measured assessment program per year across all MLP sites

Wool Measurements	Fleece weight, yield, fibre diameter (FD), FD standard deviation, FD coefficient of variation, staple strength, staple length, comfort factor and curvature
Growth and Carcase	Body weight, eye muscle, fat depth and adult ewe size
Health and Welfare	Worm egg count, faecal consistency, dag, urine stain, breech cover, crutch cover, breech wrinkle, and weaner and adult survival
Visual Wool Traits	Fleece rot, wool colour, wool character, dust penetration, staple weathering, staple structure, fibre and non-fibre pigmentation, recessive black, random spot
Visual Conformation Traits	Face cover, jaw, legs/feet, shoulder/back, body wrinkle
Classing	Two classings, flock classing (AMSEA), stud classing (Professional)
Reproduction	Sire and dam pedigree, pregnancy scanning, number of lambs weaned (conception, litter size, ewe rearing ability) body weight and condition score (at pre-joining, pregnancy scanning, pre lambing, weaning)

Analysis. During the sire evaluation phase each site produces an AMSEA accredited standard site report which includes progeny sire results presented within site and year of birth as average visual scores, classing grades, adjusted least square sire means, flock breeding values and indexes. In response to industry demand a separate MLP report is also produced that presents sire results within site within year and packaged as raw data, classing grades, visual scores, adjusted least square sire means, flock breeding values, research breeding values (for reproduction component traits) and indexes.

The details of future analyses are still in development and will include the estimation of genetic parameters for lifetime production across traits and between ages. An analysis to identify measurement and assessment strategies for recording of wool, reproduction, disease resistance and carcase production traits that are required to improve lifetime productivity, and achieve flock breeding objectives, is also planned.

A dedicated analysis committee will direct genetic and economic analysis priorities relevant to both the Merino ram breeder and ram buyer.

CONCLUSIONS

The MLP project will provide a platform for Merino genetics research and development and extension through a range of analyses to test current selection tools across different genotypes and environments and refine and evolve these selection methodologies to deliver greater lifetime productivity.

ACKNOWLEDGEMENTS

We gratefully acknowledge Australian Wool Innovation, nominating Merino breeders, site committees (Balmoral Breeders, MerinoLink, Yardstick Sire Evaluation, New England Merino Sire Evaluation Association, Macquarie Sire Evaluation Association) and site hosts (Tuloona Pastoral, Moses and Son, Murdoch University, the University of Western Australia, CSIRO and NSW Department of Primary Industries) for supporting and funding this work. A special acknowledgement is also made to the Australian Government who support research, development and marketing of Australian wool.

REFERENCE

Brown D.J., Huisman A.E., Swan A.A., Graser H.U., Banks R.B., Ball A.J., Atkins, K.D. and Woolaston, R.R. (2007) *Proc. Assoc. Advmt. Anim. Breed. Genet* **17**: 187.

MERINO LIFETIME PRODUCTIVITY - ECONOMIC VALUE OF MEAT AND WOOL FROM WETHERS AT YEARLING AND ADULT AGE

B.E. Clarke[1], J.M. Young[2], S. Hancock[1] and A.N. Thompson[1]

[1]Murdoch University, Murdoch, WA, 6150 Australia
[2]Farming Systems Analysis Service, 476 Tindale Rd, Denmark, WA, 6333 Australia

SUMMARY

The Merino Lifetime Productivity project is evaluating the lifetime productivity of the ewe progeny of industry sires at five sites around Australia. Additional work at the Pingelly site is evaluating the productivity of the wether progeny. Liveweight, carcase traits and wool production data for wether progeny born in 2016 and 2017 have been collected. This paper presents a preliminary analysis of the value of production of the 2016 and 2017 drop wethers at yearling and adult age. The difference in value of production to 18-22 months of age between the top and bottom ranking sire groups was approximately $50 per head, excluding differences in mortality between sire groups. In future analyses the wether data will be compared to the productivity of the ewes as well as combined with ewe data to calculate flock profitability so comparisons can be drawn between sires in feed intake and whole-body energy.

INTRODUCTION

The Merino Lifetime Productivity (MLP) project is evaluating the productivity of the ewe progeny of industry sires at several sites around Australia (Ramsay *et al.* 2019). At the Pingelly site an additional project is also evaluating the productivity of the wether progeny. Liveweight, carcase traits and wool production data for wether progeny born in 2016 and 2017 have been collected. This paper presents a preliminary analysis of the value of production of the 2016 and 2017 drop wethers at both yearling (10-12 months) and adult (18-22 months) age.

MATERIALS AND METHODS

Animal measurements. Data were collected on F1 wether progeny born in 2016 (n=363 at weaning) and 2017 (n=600 at weaning) from 29 different Merino sires at the Pingelly MLP site (Ramsay *et al.* 2019). There were 15 sires joined by artificial insemination (AI) each year, with one sire used as a link sire across years. The F1 wethers were assessed for fleece, growth and carcase traits until first adult shearing (2016 drop n=341; 2017 drop n=550).

Analysis. Liveweight (WT), condition score (CS), fleece and carcase traits were analysed for the 2016 and 2017 drop progeny separately using the restricted maximum likelihood method (REML, Genstat, 2018. Birth -type-rear-type (BTRT) and sire, as well as their interaction, were fitted as fixed effects. AI date, dam source (nested within AI date), Dam ID (nested within dam source), and Group (prelambing management group for the 2016 drop) were fitted as random terms. The interaction between BTRT and sire was not significant so was excluded from the model.

Economic analysis. The net value of production per animal was calculated to include value of wool produced and sale value of the animal net of selling costs. Mortality was not included in this analysis. Wool value was calculated for the whole progeny group as if sold as one line of fleece wool. Total wool income per head was calculated using average fibre diameter and other wool quality traits of the individuals in a sire group along with the average weight of wool produced. Total wool weight per animal was estimated from measured fleece weight by allowing for the components measured and assuming fixed proportions of fleece (79%), pieces (8%), bellies (8%), stains (1%) and locks

(4%, half board and half table). Nine to ten month measured fleece weight (which excluded belly wool, stains & board locks) was scaled by 112% and adult shearing (which excluded stains and board locks) was scaled by 103%. Prices were calculated to represent the whole clip, "net sweep the floor" for all wool from a flock. The price is net of the cost of selling, wool tax and freight to store. Fleece prices (17µ 2775c/kg clean, 18µ 2450c/kg, 19µ 2305c/kg) and staple strength discounts were based on Fremantle Sale 50 (15 June 2018).

Individual animal value at 10 months of age assumed the top group of animals were sold as prime lamb, the next group as lamb sold into the airfreight market (bag lamb) and the remaining animals as store lambs. Each group was sold on a c/kg dressed weight basis assuming that the dressing percentage was 46%. The top 40% of the entire drop of lambs were valued as prime lambs ($6.27/kg), 30% as air freight lambs ($5.91/kg) and the remaining 30% as store lambs ($4.29/kg). Measured WT, CS and ultrasound fat depth (FAT) was used to allocate individual lambs in each sire group into each category. The prime lamb was the animals with the combination of high WT and FAT. The store lamb was those with the lowest CS. Dividing the entire drop into the 3 groups results in a different average sale price per kg for each sire group based on WT, CS and FAT. Meat prices were based on MLA Prices & Markets Statistics for the week ending Tuesday, June 12, 2018 for WA. Merino lamb (16-22kg) was used to price the animals classed as Prime Lamb, Light Lamb (12-18kg) was used to price the air freight lamb and restocker (0-18kg) was used to price store lamb. The animal value at older ages was based on estimates by Nathan King (pers. comm.) for live shipping trade wethers. The lightest 2% of the entire wether group were classed as unsuitable for export and then the export price grid was applied to all remaining animals. Both wool & meat values are based on spot prices for the market conditions in mid 2018, in subsequent analyses evaluating prices for a broader range of scenarios is justified.

RESULTS AND DISCUSSION

Adjusted sire means for wool and weight traits are given in Tables 1 and 2. There are significant sire differences for all of these traits at both ages.

Table 1. Adjusted sire means for clean fleece weight (CFW), fibre diameter (FD), staple strength (SS), liveweight (WT) and condition score (CS) at postweaning (PW) yearling (Y) and adult (A) age of 2016 born wethers at Pingelly (highlighted cells are trait leaders)

Sire	PWCFW (kg)	ACFW (kg)	PWFD (um)	AFD (um)	PWSS (N/ktex)	ASS N/Ktex)	YWT (kg)	AWT (kg)	YCS	ACS
Sire 1	2.0	3.9	16.9	16.8	36.3	23.3	43.3	55.0	2.77	2.91
Sire 2	2.0	3.9	16.2	15.7	30.9	20.1	41.7	55.4	2.67	2.89
Sire 3	1.9	3.6	17.7	17.6	35.7	22	42.3	52.0	2.93	2.91
Sire 4	2.2	4.1	16.7	16.7	35.5	22.5	44.2	57.1	2.90	3.07
Sire 5	2.0	4.0	17.1	17.8	41.3	28.6	46.0	60.0	3.08	3.20
Sire 6	1.9	3.7	17.3	17.3	36.1	22.4	43.1	56.0	2.87	2.99
Sire 7	2.0	3.6	16.6	16.2	39.8	23.5	43.9	54.6	2.82	2.92
Sire 8	2.3	4.2	18.8	19.2	44.8	29.5	45.4	58.4	2.85	2.99
Sire 9	2.1	3.9	16.8	16.6	37.7	25	42.5	52.9	2.75	2.83
Sire 10	2.2	4.2	17.0	16.9	41.5	21.2	42.6	56.4	3.03	3.24
Sire 11	1.7	3.5	16.2	16.2	38	23.1	42.3	54.5	3.00	3.16
Sire 12	2.0	3.7	16.1	15.9	33.3	18.7	43.7	57.4	2.82	3.03
Sire 13	2.1	4.2	16.8	17.2	36	24.1	43.9	58.0	2.89	3.07
Sire 14	2.0	4.0	16.8	16.9	33	21.2	43.7	58.1	2.69	2.93
Sire 15	2.0	4.0	17.3	17.5	38	27.5	45.1	60.4	2.80	3.01
P value	*<0.001*	*<0.001*	*<0.001*	*<0.001*	*<0.001*	*<0.001*	*<0.001*	*<0.001*	*<0.001*	*<0.01*
LSD	0.19	0.37	0.72	0.8	2.18	2.27	2.77	3.57	0.16	0.20

Table 2. Adjusted sire means for clean fleece weight (CFW), fibre diameter (FD), staple strength (SS), liveweight (WT) and condition score (CS) at yearling (Y) hogget (H) and adult (A) age of 2017 born wethers at Pingelly (highlighted cells are trait leaders)

Sire	YCFW (kg)	HCFW (kg)	YFD (um)	HFD (um)	YSS (N/ Ktex)	HSS (N/ Ktex)	YWT	AWT	YCS	ACS
Sire 1	2.3	2.9	15.6	17.8	27.9	33.4	38.3	55.8	2.95	3.10
Sire 2	2.3	2.6	17.1	19.4	30.5	41.6	44.6	62.6	3.12	3.25
Sire 3	2.3	2.8	16.5	18.8	27.3	35.2	40.1	54.0	3.03	3.07
Sire 4	2.4	2.9	16.3	18.7	24.4	37.3	41.4	54.6	2.97	2.97
Sire 5	2.3	2.9	15.8	17.9	27.5	39.7	44.7	59.9	2.99	3.13
Sire 6	2.4	2.7	16.6	18.7	25.9	37.2	42.3	55.9	3.21	3.22
Sire 7	2.3	2.6	16.2	18.6	24.7	37.6	42.0	57.1	3.16	3.24
Sire 8	2.2	2.4	16.4	18.6	26.1	33.8	43.8	59.3	3.18	3.19
Sire 9	2.3	2.7	16.3	18.3	27.2	38.3	40.1	55.1	2.92	2.95
Sire 10	2.4	2.9	16.3	18.7	31.2	39.4	39.1	55.5	2.80	2.98
Sire 11	2.3	2.7	16.2	18.6	27.3	36.8	39.5	53.2	3.00	2.98
Sire 12	2.3	2.7	16.5	19.1	29.0	37.6	42.8	60.5	3.04	3.18
Sire 13	2.4	2.7	16.7	18.6	26.8	38.7	42.8	57.2	3.00	3.06
Sire 14	2.5	2.9	15.5	18.3	24.9	40.4	41.6	55.8	3.13	3.02
Sire 15	2.3	2.7	16.1	18.6	29.7	43.0	41.9	57.2	3.02	3.11
P value	*<0.01*	*<0.001*	*<0.001*	*<0.001*	*<0.01*	*<0.001*	*<0.001*	*<0.001*	*<0.001*	*<0.001*
LSD	0.2	0.23	0.94	1.16	4.76	5.37	2.68	3.48	0.13	0.13

Economic analysis of the production data (Table 3) shows that, for the 2016 drop wethers, the variation in net value of production was $43.59 at 10 months, and $39.75 at 22 months. Similarly, the variation in net value of production for the 2017 drop (Table 4) was $50.22 at 10 months and $32.27 at 18 months. The re-ranking of sires at 10 versus 18-22 months in combined value of production justifies further analysis of impact of flock structure (turn off age of wethers) on profit.

Table 3. Value of production of 2016 born wethers if sold at 10 and 18 months (highlighted cells are trait leaders). Wool value at 22 months includes wool value at 10mo and 22mo

	Wethers sold at 10 months						Wethers sold at 22 months		
Sire	Wool Value	Sale Value	Proportion in category (%)			Combined	Wool Value[1]	Sale Value	Combined[1]
	$/hd	$/hd	Prime	Air Freight	Store	$/hd	$/hd	$/hd	$/hd
Sire 1	54.24	85.37	0%	28%	72%	139.61	151.30	117.72	269.02
Sire 2	58.54	74.38	0%	4%	96%	132.92	168.18	117.72	285.89
Sire 3	46.95	106.14	30%	70%	0%	153.08	128.43	117.72	246.14
Sire 4	61.05	115.46	74%	26%	0%	176.51	164.25	117.72	281.96
Sire 5	53.25	120.73	91%	9%	0%	173.98	142.19	117.72	259.91
Sire 6	49.22	108.52	40%	56%	4%	157.74	136.02	117.72	253.73
Sire 7	56.37	108.32	54%	31%	14%	164.69	152.45	107.97	260.42
Sire 8	52.95	116.16	83%	11%	6%	169.11	138.98	117.72	256.70
Sire 9	57.71	78.69	10%	0%	90%	136.40	157.25	117.72	274.97
Sire 10	59.23	107.71	33%	67%	0%	166.94	162.30	117.72	280.02
Sire 11	50.06	108.15	48%	52%	0%	158.21	143.44	117.72	261.15
Sire 12	59.31	104.45	43%	30%	26%	163.76	160.99	117.72	278.70
Sire 13	57.62	115.91	80%	20%	0%	173.53	157.44	117.72	275.16
Sire 14	54.73	83.45	0%	17%	83%	138.18	152.89	117.72	270.61
Sire 15	51.90	100.03	23%	35%	42%	151.94	144.03	117.72	261.74
Average	54.87	102.23	41%	30%	29%	157.11	150.68	117.07	267.74

Table 4. Value of production of 2017 born wethers if sold at 10 and 18 months (highlighted cells are trait leaders)

Sire	Animals sold at 10 months						Animals sold at 18 months		
	Wool Value	Sale Value	Proportion in category (%)			Combined	Wool Value[1]	Sale Value	Combined[1]
	$/hd	$/hd	Prime	Air Freight	Store	$/hd	$/hd	$/hd	$/hd
Sire 1	72.99	73.98	0%	0%	100%	146.97	138.48	117.49	255.98
Sire 2	60.32	120.53	100%	0%	0%	180.86	115.01	117.72	232.72
Sire 3	64.64	100.54	0%	100%	0%	165.18	122.58	114.52	237.10
Sire 4	69.62	86.43	37%	0%	63%	156.05	129.77	116.71	246.48
Sire 5	69.32	116.76	95%	5%	0%	186.08	132.57	117.72	250.28
Sire 6	68.21	111.85	76%	24%	0%	180.06	125.84	117.59	243.44
Sire 7	66.51	108.90	57%	43%	0%	175.41	120.65	117.43	238.08
Sire 8	61.91	117.25	100%	0%	0%	179.16	112.54	117.72	230.25
Sire 9	70.08	71.91	0%	0%	100%	141.98	130.77	114.67	245.45
Sire 10	66.62	69.77	0%	0%	100%	136.39	125.24	116.66	241.89
Sire 11	67.89	85.14	0%	53%	47%	153.03	125.92	102.56	228.48
Sire 12	64.58	107.75	50%	50%	0%	172.33	121.33	117.72	239.05
Sire 13	65.57	112.84	92%	0%	8%	178.41	123.36	117.70	241.06
Sire 14	79.70	106.91	24%	76%	0%	186.61	143.52	117.23	260.75
Sire 15	66.44	107.74	69%	20%	11%	174.18	124.34	117.46	241.81
Average	67.63	99.89	47%	25%	29%	167.51	126.13	116.06	242.19

[1] Value includes wool value at 10mo and 18mo.

CONCLUSIONS

This preliminary analysis of MLP wether data, showing significant variation in production value, will contribute to a whole-flock profit per hectare analysis of MLP groups at the Pingelly site. The accuracy of this evaluation will be improved from using measured data from the wethers rather than inferred from ewe production. Together with data on ewe production and reproductive performance, and knowledge of sire group differences in mortality, feed intake and whole-body energy from additional projects linked to MLP Pingelly, this data will contribute to a more consistent and technically rigorous system for calculating profit per hectare for MLP and other bloodline comparisons.

ACKNOWLEDGMENTS

This project is funded by Murdoch University and Australian Wool Innovation. Australian Wool Innovation gratefully acknowledges the funds provided by the Australian government to support research, development and marketing of Australian wool. We acknowledge AMSEA and the Pingelly Breeder Site Committee for their role in the Merino Lifetime Productivity Project.

REFERENCES

Genstat, 19th Edition (*2018*) *VSN International, UK.* www.vsni.co.uk

Ramsay, A.M.M., Swan A.A. and Swain B.C. (2019) *Proc. Assoc. Advmt. Anim. Breed. Genet.* **23**:

ACCOUNTING FOR EWE SOURCE EFFECTS IN GENETIC EVALUATION OF MERINO FLEECE TRAITS

K.L. Egerton-Warburton[1], S.I. Mortimer[2] and A.A. Swan[3]

[1]NSW Department of Primary Industries, Orange, NSW, 2800 Australia
[2]NSW Department of Primary Industries, Armidale, NSW, 2351 Australia
[3]Animal Genetics & Breeding Unit*, University of New England, Armidale, NSW, 2351 Australia

SUMMARY

The significance of ewe bloodline sources and their interactions with sire effects were examined for Merino fleece traits recorded on progeny of Macquarie Merino Sire Evaluation and Merino Lifetime Productivity sites. Ewe source effects significantly influenced fleece traits expressed at post weaning, hogget and adult ages. Sire x ewe genotype interactions on fleece traits across ages were generally unimportant i.e. consistent sire rankings, accounting for small amounts of the phenotypic variation (less than 2%) in the fleece traits. These results support the methods to account for these effects that are used routinely in MERINOSELECT genetic evaluations.

INTRODUCTION

Models used in MERINOSELECT genetic evaluations account for possible sire by ewe genotype effects by fitting a sire by flock-year interaction (Li *et al.* 2015). That is where the flock-year component encompasses both flock environment and ewe genotype effects (G by E and G by G respectively). The procedure removes bias from across-flock genetic evaluations and means that MERINOSELECT breeding values (ASBVs) reflect average genetic merit of sires across all environments and ewe genotypes across which they are used. Mortimer and Casey (2015) reported that sire by ewe bloodline interactions were important for some post weaning (PW) fleece traits of Merinos. However, they highlighted the need for further studies using suitable data (without confounding of sire and other effects). This paper further investigates the importance of sire by ewe bloodline source effects on the genetic evaluation of sires in fleece traits expressed at several ages. Analyses are extended to incorporate later age assessments from the Macquarie Merino Sire Evaluation (MSE) site (hogget, H, and adult, A, traits), together with PW fleece data from the 2017 and 2018 drops of Merino Lifetime Productivity (MLP) progeny from the Macquarie site.

MATERIALS AND METHODS

A detailed description of the design of the Macquarie MSE and the data collection procedures for the progeny were given by Mortimer and Casey (2015). Briefly, progeny were born in 2012 and 2013 following AI of dams drawn from 3 bloodline sources, designated as MSE1, MSE2 and MSE3. Post weaning traits were assessed at shearing at average ages of 8 (2012 drop) and 8.6 months (2013 drop). Later assessments on the 2012 drop were conducted when the sheep were 16 (H) and 25 months of age (A), with 9 months wool growth at both assessments. For the 2013 drop, H traits were assessed at 17.5 months (8 months wool growth) while A traits were assessed at 26.5 months (9 months wool growth). Traits included greasy fleece weight (GFW, kg), clean wool yield (YLD, %), clean fleece weight (CFW, kg), mean fibre diameter (FD, μm), coefficient of variation of FD (FDCV, %), fibre curvature (CURV, degrees/mm), staple length (SL, mm) and staple strength (SS, N/ktex).

Data were also available from the 2017 and 2018 drop progeny of the Macquarie MLP site (PW assessments), which is also run according to the requirements of the Australian Merino Sire Evaluation

* A joint venture of NSW Department of Primary Industries and the University of New England

Association (AMSEA) (AMSEA 2018). Briefly, the progeny were born following AI mating of 15 industry sires in late 2016 and of 16 sires in late 2017 to foundation ewes sourced from commercial flocks. Two flock sources, designated as MLP1 and MLP2, represented a bloodline selected for increased wool production and body size. Three flock sources, designated as MLP3, MLP4 and MLP5, represented a dual-purpose bloodline, where selection aimed to improve wool, fertility and growth traits. Approximately 90 ewes were inseminated to each sire. All ewes were at their second or later lambing opportunity and were allocated to each sire so they were balanced across the ewe flock sources. Foundation ewes were pregnancy scanned and run in separate management groups for single and multiple bearers. Before lambing, ewes were drafted into lambing groups based on AI week and litter size (but not lambed in sire groups). Lambs were tagged approximately 14 days after the end of lambing. DNA samples, derived from ear tissue samples, were collected for determination of parentage. Lambs were weaned at an average age of 119 (2017 drop) and 99 days (2018 drop). Birth type and rearing type were inferred from dam pregnancy scan results and the DNA parentage results, which were obtained through the Australian Innovation Company. Animals were shorn for PW assessment of GFW, YLD, CFW, FD, FDCV, CURV, SL and SS at an average age of 9 months (9 months wool).

Data were available for PW fleece traits from 988 and 996 animals of the MSE for GFW and FD respectively, while 2,017 and 2,021 animals contributed data for GFW and FD respectively from the MLP project (Table 1). Similar record numbers were available at later MSE assessments.

Table 1. Descriptive statistics for fleece traits in Macquarie MSE and MLP data

Trait	Mean	SD	Range	Trait	Mean	SD	Range
GFW (kg)				*YLD (%)*			
PW MSE	3.4	0.73	1.4 - 6.0	PW MSE	68.7	4.67	52.6 - 81.0
PW MLP	3.7	0.73	1.6 - 6.1	PW MLP	63.5	7.09	38.9 - 81.8
H MSE	6.0	0.85	3.7 – 9.0	H MSE	62.8	7.20	39.8 - 78.8
A MSE	6.3	0.95	3.1 - 9.3	A MSE	73.1	4.35	52.6 - 84.0
CFW (kg)				*FD (μm)*			
PW MSE	2.3	0.53	0.9 - 4.2	PW MSE	17.1	1.25	13.6 - 23.0
PW MLP	2.4	0.51	1.0 - 4.2	PW MLP	17.2	1.38	13.1 - 22.2
H MSE	3.8	0.70	1.9 - 6.2	H MSE	19.2	1.40	14.9 - 24.5
A MSE	4.6	0.77	2.2 - 7.1	A MSE	20.0	1.39	15.8 - 25.2
FDCV (%)				*CURV (°/mm)*			
PW MSE	18.1	1.89	13.0 - 25.7	PW MSE	90.4	7.62	65.7 - 117.5
PW MLP	18.8	2.33	12.2 - 28.7	PW MLP	61.1	9.47	32.0 - 111.0
H MSE	17.1	1.93	12.0 - 25.0	H MSE	89.0	7.65	57.8 - 126.8
A MSE	16.7	1.90	11.2 - 25.5	A MSE	89.3	8.00	61.0 - 114.6
SL (mm)				*SS (N/ktex)*			
PW MSE	66.6	9.15	40.0 -96.0	PW MSE	35.4	14.09	10.1 - 73.5
PW MLP	80.0	10.08	49.0 - 120.0	PW MLP	25.0	9.08	4.0 - 61.0
H MSE	78.8	8.82	55.0 - 106.0	H MSE	61.7	13.39	10.0 - 88.9
A MSE	87.0	7.64	63.0 - 112.0	A MSE	42.2	19.70	8.3 - 88.9

For both data sources, analyses were performed using ASReml (Gilmour *et al.* 2015). Significance of fixed effects was first tested in models that fitted a random effect of sire. For the MSE data, fixed effects were as described by Mortimer and Casey (2015), while for the MLP data the fixed effects examined included ewe source within bloodline (5 sources), birth type (single, twin, triplet), rearing type (single, twin, triplet) and dam age (3, 4, 5, 6 and 7 year old at mating), as well as a management group effect. The importance of the interaction of sire with ewe source within bloodline was tested by fitting it as an additional random effect in a second model for each trait. The interaction effect was considered significant ($P<0.05$) if its inclusion increased significantly the log-likelihood between models.

RESULTS AND DISCUSSION

The ewe source effect influenced significantly ($P<0.05$) both fleece weights, FD and CURV at PW ages in both data sets (Table 2 for MSE results, Table 3 for MLP results). The ewe source effect also influenced ($P < 0.001$) H and A records of the MSE progeny for these traits, as well as SS. Larger ranges in fleece weights across ewe bloodline sources at PW assessment were observed for the MLP progeny (0.70 kg for GFW; 0.31 kg for CFW) than for the MSE progeny (0.17 kg for GFW; 0.06 kg for CFW). For FD at this youngest age, the ranges across estimates were similar in both data sets. The range across estimates of ewe bloodline sources for the MSE progeny were similar at H (0.73 kg for GFW; 0.33 kg for CFW) and A (0.78 kg for GFW; 0.37 kg for CFW) assessments, but the range for adult FD was smaller than for hogget FD (1.05 versus 2.08 µm).

Table 2. Significance of ewe source effects[1] and their estimates[2] for post weaning, hogget and adult fleece traits in the Macquarie Merino Sire Evaluation site and percentage of phenotypic variation accounted for by sire x ewe bloodline interaction

	GFW	YLD	CFW	FD	FDCV	CURV	SL	SS
Post weaning traits (Source: Mortimer and Casey 2015)								
Ewe source	*	n.s.	*	***	*	***	***	n.s.
MSE2	-0.09 (0.03)		-0.06 (0.02)	0.35 (0.08)	-0.32 (0.12)	-3.10 (0.51)	3.15 (0.73)	
MSE3	0.08 (0.09)		-0.02 (0.07)	-1.25 (0.21)	-0.19 (0.34)	0.43 (1.42)	-2.44 (1.99)	
Sire x ewe bloodline (%)	0 n.s.	0 n.s.	0 n.s.	4.0 *	2.8 *	1.6 n.s.	2.5 *	2.9 *
Hogget traits								
Ewe source	***	***	***	***	***	***	***	n.s.
MSE2	-0.46 (0.07)	-0.78 (0.32)	-0.33 (005)	0.57 (0.09)	-0.65 (0.13)	-2.28 (0.50)	0.61 (0.75)	
MSE3	0.27 (0.18)	-2.63 (0.91)	-0.07 (0.13)	-1.51 (0.24)	-0.47 (036)	-0.23 (1.43)	0.33 (2.08)	
Sire x ewe bloodline (%)	1.8 n.s.	0 n.s.	0.4 n.s.	0.9 n.s.	3.3 *	0.9 n.s.	4.1 n.s.	0 n.s.
Adult traits								
Ewe source	***	***	***	***	***	***	***	n.s.
MSE2	-0.41 (0.06)	-1.09 (0.26)	-0.37 (0.04)	0.54 (0.09)	-0.65 (0.12)	-1.96 (0.52)	1.90 (0.54)	
MSE3	-0.78 (0.14)	-1.23 (0.64)	-0.12 (0.11)	-0.51 (0.21)	-1.42 (0.30)	0.57 (1.25)	4.34 (1.31)	
Sire x ewe bloodline (%)	0.3 n.s.	0 n.s.	0.5 n.s.	0 n.s.	1.8 n.s.	3.5 *	0.4 n.s.	1.0 n.s.

[1] *, $P < 0.05$; **, $P < 0.01$; ***, $P < 0.001$; n.s., not significant.

[2] Estimates expressed as the deviation from the MSE1 ewe source estimate for each of the MSE2 and MSE3 ewe sources. Standard errors are in brackets.

For the wool production traits at all assessment ages, sire x ewe bloodline source interactions were generally not significant and accounted for less than 2% of the phenotypic variance across both data sets (Tables 2, 3). In contrast to the findings of Mortimer and Casey (2015), the interaction was not significant for wool quality traits at PW assessments in the MLP data, where the interaction accounted for less than 2% of the phenotypic variance for these traits. At later assessments in the MSE data, sire x ewe bloodline source interaction was significant only for FDCV at H assessment and CURV at A assessment, accounting for 3.3 to 3.5% of the phenotypic variation in these traits. Using data on yearling Merino fleece traits from the Information Nucleus flock, Swan *et al.* (2016) reported that sire by site interaction effects were small (1% of the phenotypic variation in CURV) to moderate (6% of the phenotypic variation in CFW).

Table 3. Significance of ewe source effects[1] and their estimates[2] for post weaning fleece traits in the Macquarie Merino Lifetime Productivity progeny and percentage of phenotypic variation accounted for by sire x ewe bloodline interaction

	GFW	YLD	CFW	FD	FDCV	CURV	SL	SS
Post weaning traits								
Ewe source	***	n.s.	***	***	***	*	n.s.	*
MLP2	0.70 (0.12)		0.31 (0.05)	0.69 (0.08)	0.20 (0.17)	1.34 (0.69)		0.92 (0.52)
MLP3	0.70 (0.12)		0.18 (0.05)	-0.73 (0.08)	-1.08 (0.16)	0.21 (0.67)		1.21 (0.51)
MLP4	0.22 (0.17)		0.15 (0.06)	-0.85 (0.09)	-0.91 (0.18)	-0.49 (0.76)		0.57 (0.57)
MLP5	0.68 (0.18)		0.29 (0.08)	-0.44 (0.11)	-0.25 (0.22)	2.14 (0.92)		-0.76 (0.69)
Sire x ewe bloodline (%)	1.5 n.s.	0.3 n.s.	1.8 *	0.1 n.s.	0 n.s.	0.4 n.s.	0.8 n.s.	1.8 n.s.

[1] *, $P < 0.05$; **, $P < 0.01$; ***, $P < 0.001$; n.s., not significant.

[2] Estimates expressed as the deviation from the MLP1 ewe source estimate for each of the MLP2, MLP3, MLP4 and MLP5 ewe sources. Standard errors are in brackets.

CONCLUSIONS

Sire x ewe genotype interactions were generally unimportant and accounted for minor amounts of phenotypic variation in Merino fleece traits across age expressions, indicating that rankings of the sires for fleece traits would be consistent for evaluations conducted across differing ewe genotypes. This study supports the methods adopted by MERINOSELECT genetic evaluations that routinely fit this interaction and account for ewe source effects (Li *et al.* 2015).

ACKNOWLEDGMENTS

Australian Wool Innovation Limited funded this study. We acknowledge generous inputs from: the Macquarie Sire Evaluation Association and its committee; ram breeders who entered rams in the 2012 and 2013 MSE and the MLP project conducted at the Trangie site; and AMSEA. Fleece data collection was ably conducted by staff of NSW Department of Primary Industries.

REFERENCES

AMSEA (2018) 'Merino Sire Evaluation Site Requirements'. Version 180831. =

Gilmour, A.R., Gogel, B.J., Cullis, B.R., Welham, S.J. and Thompson, R. (2015) 'ASReml User Guide Release 4.1 Functional Specification'. VSN International Ltd, Hemel Hempstead, UK.

Li L., Swan A.A., Brown D.J. and van der Werf J.H.J. (2015) *Proc. Assoc. Advmt. Anim. Breed. Genet.* **21**: 386.

Mortimer, S.I. and Casey, A.E. (2015) *Proc. Assoc. Advmt. Anim. Breed. Genet.* **21**: 322.

Swan, A.A., Brown, D.J. and van der Werf, J.H.J. (2016) *Anim. Prod.Sci.* **56**: 87.

INDICATOR TRAITS RECORDED EARLY IN LIFE WILL BE USEFUL SELECTION CRITERIA FOR BREECH FLYSTRIKE RESISTANCE IN WEANER MERINO SHEEP

J.L. Smith[1], S. Dominik[1], S. Lehnert[2], T. Reverter[2] and I.W. Purvis[1]

[1]CSIRO Agriculture and Food, Armidale, NSW, 2350 Australia
[2]CSIRO Agriculture and Food, St Lucia, QLD, 4067 Australia

SUMMARY

This study demonstrated that breechstrike indicator traits such as wrinkle and wool cover are sufficiently variable, heritable and correlated with breech strike incidence to weaner age to be useful indirect selection criteria for breech strike resistance. Further, the indicator traits recorded at birth would be more suitable selection criteria than the same traits recorded at marking or post-weaning. However, birth traits are far more problematic to record, and therefore less likely to be adopted by industry than those recorded at marking, or even later at post-weaning or yearling ages. The relationships between indictor traits and key production traits for Merinos are yet to be evaluated.

INTRODUCTION

Interest in developing alternatives to mulesing in Merino sheep has led to investigations aimed to evaluate the potential of selection for breech flystrike resistance (Greeff *et al.* 2014; Smith *et al.* 2009). For a disease such as flystrike, it is undesirable for economic and animal welfare reasons to challenge animals directly to evaluate their level of resistance. Hence, indirect selection criteria that are variable, heritable, correlated with flystrike and easily recorded in industry flocks early in life are needed. Here, early life indicator traits were explored in the CSIRO Armidale breech flystrike genetic resource flock as potential indirect selection criteria for breech flystrike at weaner age.

MATERIALS AND METHODS

Sheep and data collection. Progeny of the CSIRO breech flystrike genetic resource flock in Armidale, NSW, born between 2005 and 2014 inclusive, were the subject of this study (n = 4,656). The flock structure up to 2010 is detailed in Smith *et al.* (2009). No mating was conducted in 2010 and the design was changed in 2011 to represent a finewool type flock suited to the summer rainfall environment, and incidence of breech flystrike itself became the primary selection criterion. Lambs were born in September and weaner breech flystrike (wBRSTR) was recorded as the count of flystrikes observed during the first challenge season. Struck sheep were treated to kill the infestation and animals could experience more than one strike per season, yielding the count of strikes for the season. The first challenge season occurred approximately from October to April inclusive, when weaners were approximately one to seven months of age. Up to and including 2009, half the lambs from each sire were mulesed and half remained unmulesed. From 2011 onwards, no mulesing was conducted. Neck wrinkle (NWR), body wrinkle (BDWR), breech wrinkle (BRWR), breech wool cover (BCOV) and crutch wool cover (wool coverage in the inguinal area, CCOV) were recorded at birth (b) up to and including 2009. Traits recorded at marking (m, approximately six weeks of age) and post-weaning (p, approximately six months of age), were those recorded at birth (described above) as well as breech bare width (BW, mm), breech bare depth (BD, mm), and tail length (TL, mm, marking only). Here, correlations between indicator traits at birth, marking and post-weaning with wBRSTR were explored to identify breechstrike indicators that can be recorded early in life.

Statistical analysis. Univariate mixed animal models were applied using ASReml (Gilmour *et al.* 2009). No data transformations were performed on any trait data as with the exception of wBRSTR most were sufficient approximations of statistical normality. Fixed effects tested for individual traits were selection line nested within the two design phases (originally 3 levels prior to 2011; 2 levels post 2011); contemporary group (combined birth year, lambing management flock, and property-of-origin (latter for the 2005 drop only)); mules group (2 levels; mulesed and non-mulesed); sex (2 levels; castrated male and female); birth type (for traits recorded at birth, 2 levels; single and multiple); birth-rearing type (for traits recorded from marking onward, 3 levels; born and reared single (SS), born multiple and reared single (MS), born and reared multiple (MM)); age of dam (2 levels; maiden (2yo at first lambing) and adult); operator (up to 4 levels; scorer or measurer of trait). Age, and age specific cannon bone length (CBL) as an indicator of frame and bodyweight (WT) were fitted as covariates where appropriate. Only significant effects were retained in statistical models. Maternal genetic effects were tested by fitting dam as a random effect and significance determined using likelihood ratio testing. Where the inclusion of the maternal effect was non-significant, it was excluded from the final model. In 2005, foundation ewes purchased from industry were assigned to one of the three genetic groups described by MerinoSelect (Banks and Brown 2009). From then onward, all progeny born into the flock were fully pedigreed. Correlations among breech traits were estimated from multivariate mixed animal models.

RESULTS AND DISCUSSION

Weaner breechstrike ranged from 0 to 6 within the first challenge season. Phenotypic and genetic parameters for indicator traits of breechstrike recorded at birth and their correlations with wBRSTR are shown in Table 1. Breechstrike at weaner age was of low heritability. The heritabilities of indicator traits recorded at birth were of low to moderate heritability. Genetic correlations between wBRSTR and indicator traits range from r_g = 0.27–0.59. Genetic correlations between wrinkle traits recorded on different parts of the body (bNWR, bBDWR and bBRWR) were high at birth. Low negative correlations were found between wrinkle traits and CCOV. Phenotypic correlations were low between wrinkle and wool cover traits. In summary, the heritabilities and correlations for indicator traits recorded at birth would be useful for indirect selection for reducing wBRSTR.

Table 1. Arithmetic mean and standard deviation (sd), phenotypic variance (Vp), heritability (bold, diagonal), phenotypic correlations (above diagonal) and genetic correlations (below diagonal) for wrinkle and wool cover traits recorded at birth (b), and weaner (w) breech flystrike (standard error in parentheses)

Trait	bNWR	bBDWR	bBRWR	bCCOV	bBCOV	wBRSTR
Mean, sd	2.57, 0.77	2.18, 0.85	2.30, 0.90	3.55, 0.70	4.49, 0.62	0.181, 0.482
Vp	0.49 (0.02)	0.58 (0.02)	0.72 (0.02)	0.45 (0.02)	0.29 (0.02)	0.21 (0.01)
bNWR	**0.42 (0.06)**	0.65 (0.02)	0.56 (0.02)	-0.02 (0.03)	0.06 (0.06)	0.05 (0.03)
bBDWR	0.91 (0.04)	**0.36 (0.06)**	0.66 (0.02)	-0.03 (0.03)	0.04 (0.03)	0.07 (0.03)
bBRWR	0.82 (0.06)	0.90 (0.4)	**0.37 (0.06)**	-0.05 (0.03)	0.02 (0.03)	0.08 (0.03)
bCCOV	0.04 (0.14)	-0.03 (0.13)	-0.12 (0.13)	**0.38 (0.06)**	0.20 (0.03)	0.01 (0.03)
bBCOV	0.28 (0.17)	0.26 (0.16)	0.11 (0.17)	0.43 (0.15)	**0.22 (0.06)**	0.03 (0.03)
wBRSTR	0.42 (0.15)	0.41 (0.15)	0.47 (0.14)	0.27 (0.14)	0.59 (0.18)	0.18 (0.03)

Genetic and phenotypic parameters for indicator traits recorded at marking and their correlations with wBRSTR are summarised in Table 2. Heritabilities for indicator traits recorded at marking ranged from h^2=0.20-0.43 and were similar to the estimates for the same traits recorded at birth. Tail length at marking had a high heritability. However, the genetic correlations of wBRSTR indicator traits at marking were lower than with the same traits recorded at birth. Favourable genetic correlations between mBW and mBD and wBRSTR were observed. Phenotypic correlations between wBRSTR and indicator traits recorded at marking were low. Genetic correlations between wool cover, wrinkle, and tail length were also mostly low. The exceptions were high favourable genetic correlations between BCOV and BW, and between BCOV and BD. The primary reason for recording BW and BD was validation of the visual sheep scores for BCOV (Australian Wool Innovation Ltd and Meat and Livestock Australia 2007). The strong genetic relationships between these traits confirm that the BCOV score is indeed a suitable industry method for assessing breech cover. In summary, the genetic correlations between indicator traits and wBRSTR recorded at marking were not as favourable as for the same traits recorded at birth.

Table 2. Arithmetic mean and standard deviation (sd), phenotypic variance (Vp), heritability (bold, diagonal), phenotypic correlations (above diagonal) and genetic correlations (below diagonal) for wrinkle and wool cover traits recorded at marking (m), and weaner (w) breech flystrike (standard errors in parentheses)

Trait	mNWR	mBDWR	mBRWR	mCCOV	mBCOV	mBW	mBD	TL	wBRSTR
Mean, sd	2.72, 0.89	2.37, 0.95	2.38, 1.02	3.50, 0.74	4.39, 0.66	26, 10	36, 19	237, 30	0.18, 0.48
Vp	0.54 (0.02)	0.63 (0.02)	0.73 (0.02)	0.42 (0.01)	0.34 (0.01)	53.4 (1.5)	146.0 (4.3)	601.5 (22.4)	0.21 (0.01)
mNWR	**0.39 (0.04)**	0.71 (0.01)	0.67 (0.01)	0.06 (0.02)	0.02 (0.02)	0.00 (0.02)	0.00 (0.02)	0.09 (0.02)	0.09 (0.02)
mBDWR	0.93 (0.02)	**0.36 (0.03)**	0.72 (0.01)	0.06 (0.02)	0.02 (0.02)	-0.01 (0.02)	-0.01 (0.02)	0.04 (0.02)	0.10 (0.02)
mBRWR	0.93 (0.02)	0.91 (0.02)	**0.43 (0.04)**	0.05 (0.02)	0.02 (0.02)	0.01 (0.02)	-0.01 (0.02)	0.05 (0.02)	0.10 (0.02)
mCCOV	0.04 (0.08)	0.08 (0.08)	0.04 (0.08)	**0.35 (0.03)**	0.20 (0.02)	-0.14 (0.02)	-0.17 (0.02)	-0.10 (0.02)	0.06 (0.02)
mBCOV	-0.03 (0.1)	0.09 (0.1)	-0.04 (0.1)	0.46 (0.09)	**0.20 (0.03)**	-0.37 (0.02)	-0.44 (0.02)	-0.12 (0.02)	0.03 (0.02)
mBW	-0.12 (0.1)	-0.13 (0.1)	-0.02 (0.10)	-0.44 (0.09)	-0.86 (0.06)	**0.23 (0.04)**	0.31 (0.02)	0.16 (0.02)	-0.06 (0.02)
mBD	-0.05 (0.1)	-0.16 (0.1)	-0.03 (0.09)	-0.51 (0.08)	-0.80 (0.06)	0.52 (0.10)	**0.25 (0.04)**	0.17 (0.02)	-0.00 (0.02)
TL	0.08 (0.07)	0.03 (0.07)	0.02 (0.07)	-0.25 (0.07)	-0.37 (0.09)	0.41 (0.09)	0.32 (0.09)	**0.62 (0.04)**	0.03 (0.02)
wBRSTR	0.13 (0.11)	0.23 (0.1)	0.26 (0.10)	0.20 (0.11)	0.09 (0.13)	-0.08 (0.13)	-0.25 (0.13)	0.12 (0.10)	0.18 (0.03)

Table 3 shows heritabilities for traits recorded post-weaning and their correlations with wBRSTR. Heritabilities were lower for all traits post-weaning compared to heritabilities for traits recorded at marking and birth. Genetic correlations of indicator traits post-weaning with wBRSTR were low, but higher than with traits recorded at marking. Genetic correlations between post-weaning wrinkle traits were lower than observed at previous recording times. The traits pCCOV and pBCOV had higher genetic correlations than were observed at other times and the correlation for breech measurements with wool cover traits remained negative and high. The genetic parameters for traits recorded post-weaning are generally more favourable than at marking time, but genetic parameters at birth and their relationship with wBRSTR are still the most favourable. It is acknowledged however, that birth traits are far more problematic to record, and therefore less likely to be adopted by industry than those recorded at marking, or even later at post-weaning or yearling ages. Although the heritability of BRWR was lower at post-weaning than marking, the genetic correlation between BRWR and wBRSTR at 0.62 was higher, which makes pBRWR a good alternative to bBRWR.

Table 3. Arithmetic mean and standard deviation (sd), phenotypic variance (Vp), heritability (bold, diagonal), phenotypic correlations (above diagonal) and genetic correlations (below diagonal) for wrinkle and wool cover traits recorded at post-weaning (p), and weaner (w) breech flystrike (standard errors in parentheses)

Trait	pNWR	pBDWR	pBRWR	pCCOV	pBCOV	pBW	pBD	wBRSTR
Mean, sd	2.39, 0.99	2.00, 0.86	2.75, 1.11	3.47, 0.75	3.96, 0.85	39, 14	43, 16	0.18, 0.48
Vp	0.53 (0.01)	0.41 (0.01)	0.63 (0.02)	0.36 (0.01)	0.46 (0.01)	87.4 (2.2)	147.7 (4.0)	0.21 (0.01)
pNWR	**0.24 (0.04)**	0.54 (0.01)	0.38 (0.02)	0.08 (0.02)	0.01 (0.02)	0.00 (0.02)	0.01 (0.02)	0.07 (0.02)
pBDWR	0.82 (0.07)	**0.17 (0.03)**	0.41 (0.02)	0.09 (0.02)	0.02 (0.02)	-0.03 (0.02)	-0.02 (0.02)	0.08 (0.02)
pBRWR	0.66 (0.08)	0.75 (0.08)	**0.30 (0.03)**	0.12 (0.02)	0.05 (0.02)	-0.03 (0.02)	-0.01 (0.02)	0.20 (0.02)
pCCOV	-0.01 (0.09)	0.04 (0.10)	0.15 (0.08)	**0.43 (0.04)**	0.29 (0.02)	-0.24 (0.02)	-0.24 (0.02)	0.09 (0.02)
pBCOV	-0.09 (0.10)	0.10 (0.11)	0.10 (0.09)	0.60 (0.06)	**0.30 (0.03)**	-0.54 (0.01)	-0.64 (0.01)	0.03 (0.02)
pBW	0.05 (0.10)	-0.17 (0.11)	-0.13 (0.09)	-0.55 (0.06)	-0.90 (0.04)	**0.28 (0.03)**	0.43 (0.01)	-0.03 (0.02)
pBD	0.06 (0.10)	-0.11 (0.11)	-0.11 (0.09)	-0.50 (0.06)	-0.94 (0.03)	0.81 (0.05)	**0.30 (0.03)**	0.02 (0.02)
wBRSTR	0.35 (0.13)	0.39 (0.13)	0.62 (0.09)	0.32 (0.10)	0.35 (0.11)	-0.35 (0.10)	-0.34 (0.10)	0.18 (0.03)

CONCLUSION

Traits that can be used early in life and which are correlated strongly with breeding objective traits can help accelerate genetic progress. In this study potential indicator traits for weaner breechstrike resistance measured at early ages, such as wrinkle and wool cover scores, were evaluated and showed potential to be used in the context of breeding for enhanced breechstrike resistance. However, it is acknowledged that trait recording at lambing time is more difficult than during routine husbandry procedures later in life. Once the necessary relationships have been established between these indicator traits and measures of breechstrike at adult ages, and with the other traits of economic importance, incorporation into industry breeding programs can be fully evaluated.

ACKNOWLEDGEMENT

CSIRO acknowledge the support of Australian Wool Innovation Ltd (AWI). AWI gratefully acknowledges the funds provided by woolgrowers and the Australian government to support research, development and marketing of Australian wool. The efforts CSIRO technical staff in looking after the health and welfare of the sheep, is gratefully acknowledged.

REFERENCES

Australian Wool Innovation Ltd and Meat and Livestock Australia (2007) Visual sheep scores.

Banks R.G. and Brown D.J. (2009) *Anim. Genet. Res. Info.* **45**: 29.

Gilmour A.R., Gogel B.J., Cullis B.R. and Thompson R. (2009) ASReml User Guide Release 3.0. VSN International Ltd, Hemel Hempstead, HP1 1ES, UK.

Greeff J.C., Karlsson L.J.E. and Schlink A.C. (2014) *Anim. Prod. Sci.* **54**:125.

Smith J.L., Brewer H.G. and Dyall T. (2009) *Proc. Assoc. Advmt. Anim. Breed. Genet.* **18**: 334.

BREEDING FOR REDUCED BREECH FLYSTRIKE AS PART OF MULTI-TRAIT SELECTION

F.D. Brien[1], S.F. Walkom[2], A.A. Swan[2] and D.J. Brown[2]

[1]Davies Research Centre, University of Adelaide, Roseworthy, SA, 5371 Australia
[2]Animal Genetics & Breeding Unit*, University of New England, Armidale, NSW, 2351 Australia

SUMMARY

We predicted genetic gains for reducing breech flystrike incidence (FSI) based on selection using modified MERINOSELECT indexes. Predicted genetic reductions in FSI after 10 years of selection were larger when the heritability of the trait was moderate (as in a summer rainfall environment) than when a low heritability was assumed (as in a winter rainfall environment). The relative gains between flystrike incidence and fleece weight, fibre diameter and reproductive rate and their implications for Merino breeding programs are discussed.

INTRODUCTION

General consensus within the Australian wool industry is that breeding more resistant sheep will be a critical component of programs to control breech and tail strike in non-mulesed flocks as well as reducing the risk of strike in mulesed sheep. Since 2005, a major research program has been funded to identify optimal breeding strategies for reducing breech strike resistance. Selection lines for breech flystrike were established at two research stations, Mt Barker in WA, in a winter rainfall environment (Greeff and Karlsson 2009) and at CSIRO's Chiswick property, near Armidale, NSW, in a summer rainfall environment (Smith *et al.* 2009). Results have confirmed the presence of significant genetic variability amongst sheep in susceptibility to breech strike and also identified key indirect selection criteria for improving resistance, in particular scores of breech wrinkle, dag, breech cover and urine stain (Smith *et al.* 2009; Greeff *et al.* 2014). Earlier studies (Brown *et al.* 2010; Richards and Atkins 2010) predicted genetic gains from index selection with breech wrinkle score included in the breeding objective, as a proxy for breech flystrike incidence. This paper extends those earlier findings, by predicting genetic gains in breech flystrike incidence from selection using all available breech indicator traits of breech wrinkle, dag and breech cover scores for three different breech objectives within three different environments.

MATERIALS AND METHODS

Predictions of genetic gain from within flock selection were undertaken based on the 'MTINDEX' modelling program (van der Werf 2019), using a batch code written in R. Three breeding objectives were examined, by modifying the Dual Purpose (DP+), Fibre Production (FP+) and Merino Production (MP+) indexes available from MERINOSELECT (Sheep Genetics 2018). The modifications consisted of adding Flystrike incidence (FSI, strikes/ewe/year) as a formal trait to the breeding objective associated with each index. The modified indexes, DP+FSI, FP+FSI and MP+FSI target, respectively, medium wool/dual purpose, superfine/fine wool and fine/medium wool production systems, along with reduction in FSI. The effects of using a range of FSI economic values were examined, from \$0 to -\$240/strike/ewe/year, to include the likely large range in the associated costs of FSI to the sheep industry across Australia. Breeding program assumptions included a flock structure of 5 ewe and 3 ram age groups, a first joining at 18 months of age, a ratio of 1 ram mated to 50 ewes, 90% weaning rate, 48% of ewe hoggets retained for breeding, a death rate in mature ewes of 4% p.a., a 30% loss

* A joint venture of NSW Department of Primary Industries and the University of New England

of variance under selection (Bulmer 1971) and that 65% of selection emphasis was placed on index scores. Genetic parameters used in predictions were from Sheep Genetics, supplemented by values from Hatcher and Preston (2017, 2018) and AWI project reports (Greeff *et al.* 2016; Smith 2016).

Full records of productivity traits and pedigrees were assumed as selection criteria, as well as records of breech wrinkle, dag and breech cover scores. Predictions were conducted for 3 different scenarios (i) moderate heritability (0.20) for FSI (ii) low heritability for FSI (0.10) and (iii) low heritability for FSI (0.10) but high heritability (0.30) for Dag Score.

RESULTS AND DISCUSSION

Genetic gains in FSI (Figure 1). Under conditions where breech flystrike has a moderate heritability (scenario i), predicted gains for FSI after 10 years of selection range from -0.02 to -0.03 (2% to 3%) to a maximum of -0.19 to -0.21 (19% to 21%) strikes/ewe/year, when the economic value for FSI is increased from 0 to -$240. These predicted gains are 3 to 5 times as large as those where heritability for FSI is low (scenario ii). Gains are slightly less for the FP+FSI index and slightly more for the MP+FSI index compared to gains from using the DP+FSI index. At the highest economic value, most (80%+) selection emphasis is placed on reducing FSI (details not shown). Predicted gains scenario iii (h^2 for FSI is low, but the h^2 for Dag Score is high) are no different to the predicted gains for scenario ii, so are not discussed further.

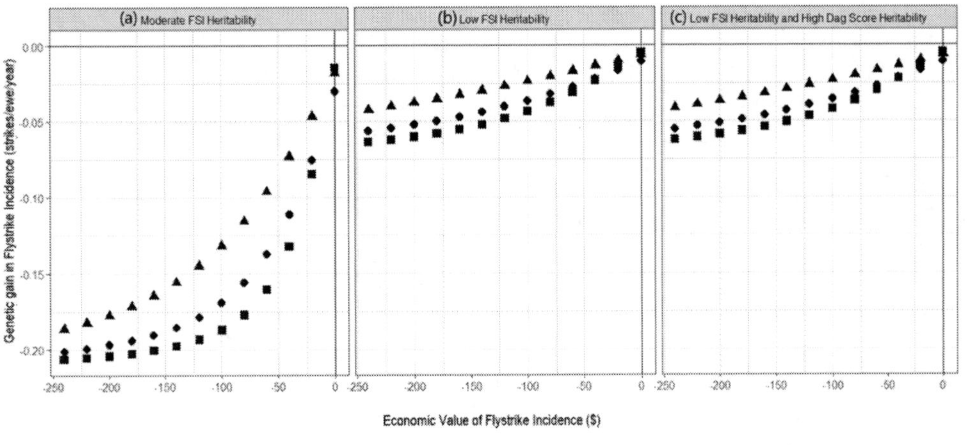

Figure 1. Predicted 10-year genetic gains in Flystrike Incidence (FSI, strikes/ewe/year) assuming (a) moderate FSI heritability (b) low FSI heritability and (c) low FSI heritability and high Dag Score heritability, with economic values from 0 to -$240 for FSI. Results are shown for 3 indexes, Dual Purpose+FSI (DP+FSI ●), Fibre Production+FSI (FP+FSI ▲) and Merino Production+FSI (MP+FSI ■)

When compared to breech flystrike incidence in unmulesed, crutched young sheep in a winter rainfall environment at Mt Barker, WA (from 4% to 9.5% or 0.04 to 0.095 strikes/ewe/year, Greeff *et al.* 2016) and at Chiswick, near Armidale, NSW in a summer rainfall environment (average of 18.1% or 0.181 in weaners and 7.6% or 0.076 in yearlings, Smith 2016), our predicted genetic gains suggest that after 10-15 years of index selection, FSI could be reduced in unmulesed sheep (of average genetic merit for FSI) to low levels (<0.02 or 2%) in average years, given sufficient selection emphasis is given to the trait. Reducing FSI to zero is almost impossible in all years as environmental (and possibly non-additive genetic) variation remains, despite selection effects to reduce incidence. Further, it is important to balance potential gains in reducing FSI with the predicted impact on genetic gains for other important traits, shown below.

Genetic gains in key production traits. Predicted genetic gains for Clean Fleece Weight (CFW), Fibre Diameter (FD) and the Number of Lambs Weaned/Ewe Joined (NLW) are listed in Tables 1 to 3, respectively, listed by economic value for FSI. With increasing economic value for FSI, gains in CFW decrease gradually when the heritability of FSI is low, but decrease more rapidly when the heritability of FSI is moderate, particularly for the DP+FSI and MP+FSI indexes. However, even for the largest economic value for FSI examined (-$240), genetic gains for CFW remain positive.

Table 1. Predicted genetic gain in Clean Fleece Weight (%) after 10 years of index selection, by economic value for Flystrike Incidence (FSI)

h^2 FSI/Index	Economic Value for FSI ($/strike/ewe/year)						
	0	-40	-80	-120	-160	-200	-240
Low h^2							
DP+FSI	2.69	2.50	2.27	2.03	1.80	1.59	1.39
FP+FSI	2.69	2.59	2.48	2.30	2.20	2.06	1.92
MP+FSI	6.04	5.58	4.92	4.22	3.59	3.06	2.63
Moderate h^2							
DP+FSI	2.69	2.07	1.46	1.01	0.70	0.48	0.32
FP+FSI	2.69	2.37	1.97	1.59	1.26	1.00	0.80
MP+FSI	6.04	4.34	2.77	1.84	1.29	0.92	0.67

Table 2. Predicted genetic gain in Fibre Diameter (μ) after 10 years of index selection, by economic value for Flystrike Incidence (FSI)

h^2 FSI/Index	Economic Value for FSI ($/strike/ewe/year)						
	0	-40	-80	-120	-160	-200	-240
Low h^2							
DP+FSI	0.01	-0.03	-0.07	-0.11	-0.14	-0.16	-0.18
FP+FSI	-0.78	-0.80	-0.81	-0.81	-0.81	-0.80	-0.79
MP+FSI	-0.36	-0.42	-0.44	-0.45	-0.44	-0.44	-0.43
Moderate h^2							
DP+FSI	0.01	-0.10	-0.17	-0.21	-0.23	-0.25	-0.26
FP+FSI	-0.78	-0.81	-0.79	-0.75	-0.70	-0.66	-0.62
MP+FSI	-0.36	-0.45	-0.43	-0.40	-0.38	-0.37	-0.36

When the heritability of FSI is moderate, there are small genetic reductions predicted in FD with increasing FSI economic value using the DP+FSI index of up to 0.26μ after 10 years of selection. There are also slightly lower genetic gains in FD with increasing FSI economic value using the FP+FSI index of up to 0.16μ. However, in general there is little impact predicted on genetic gains for FD when also selecting for reduced FSI as part of index selection. Predicted genetic gains for NLW when also selecting for reduced FSI are impacted much less than those for CFW. Relative to gains when FSI has no economic value, there are modest declines of up to a maximum of 31% in NLW gains using the MP+FSI index and up to 45% using the DP+FSI, at the largest economic value for FSI examined (when h^2 for FSI is moderate).

Table 3. Predicted genetic gain in the Number of Lambs Weaned/ewe after 10 years of index selection, listed by economic value for Flystrike Incidence (FSI)

h^2 FSI/Index		Economic Value for FSI ($/strike/ewe/year)						
		0	-40	-80	-120	-160	-200	-240
Low h^2	DP+FSI	0.08	0.08	0.08	0.07	0.07	0.07	0.06
	FP+FSI	0.02	0.02	0.02	0.02	0.03	0.03	0.03
	MP+FSI	0.04	0.04	0.04	0.04	0.04	0.04	0.04
Moderate h^2	DP+FSI	0.08	0.07	0.07	0.06	0.05	0.05	0.04
	FP+FSI	0.02	0.02	0.03	0.03	0.03	0.03	0.03
	MP+FSI	0.04	0.04	0.04	0.03	0.03	0.03	0.03

CONCLUSIONS

Useful reductions in breech flystrike from 0.05 to 0.08 strikes/ewe/year up to 0.19 to 0.21 strikes/ewe/year in a summer rainfall environment were predicted after 10 years of index selection for breeding objectives relevant to the Australian wool industry. For a winter rainfall environment in WA, with a lower heritability for FSI, smaller gains were predicted, from 0.01 to 0.02 up to 0.04 to 0.06 strikes/ewe/year. Predicted gains using a FP+FSI index were less than gains from using MP+FSI or DP+FSI indexes. It is unnecessary to use very high economic values for FSI to achieve significant reductions in the trait. For example, when the heritability for FSI is moderate and the trait is given an economic value of -$80, 62% of the highest genetic gain for FSI predicted (when the economic value is -$240) is obtained with the FP+FSI index, with higher gains of 78% and 86% being obtained with the DP+FSI and MP+FSI indexes, respectively. At that economic value for FSI (-$80), between 46% to 73% of the genetic gains for CFW and between 84% and 125% of the genetic gain for NLW are being retained, with no significant impact on gains in FD. In the absence of formally-derived economic values, -$80 for FSI appears to be a reasonable upper limit to use in Merino breeding programs to achieve a balance between genetically lowering flystrike incidence whilst obtaining competitive genetic gains in productivity traits as outlined above.

ACKNOWLEDGEMENTS

We acknowledge the support of Australian Wool Innovation Ltd (AWI). AWI gratefully acknowledges the funds provided by the Australian government to support research, development and marketing of Australian wool.

REFERENCES

Brown D.J., Swan A.A. and Gill J.S. (2010) *Anim. Prod. Sci.* **50:** 1060.

Bulmer M.G. (1971) *The American Naturalist* **105:** 201.

Greeff J.C. and Karlsson L.J.E. (2009) *Proc. Assoc. Advmt. Anim. Breed. Genet.* **18:** 272.

Greeff J.C., Karlsson L.J.E. and Schlink A.C. (2014) *Anim. Prod. Sci.* **54:** 125.

Greeff J.C., Karlsson L.J.E., Schlink A.C., Stanwyck N., O'Neal R., Windsor A. and Bell S. (2016) Final report to AWI. Project ON00169, AWI, Sydney.

Hatcher S. and Preston J.W.V. (2017) *Small Ruminant Research* **157:** 47.

Hatcher S. and Preston J.W.V. (2018) *Small Ruminant Research* **164:** 48.

Richards J.S. and Atkins K.D. (2010) *Anim. Prod. Sci.* **50:** 1053.

Sheep Genetics (2018) http://www.sheepgenetics.org.au/files/f22de9c3-9a27-4118-bdcf-a39400c74d79/Understanding-the-MERINOSELECT-indexes-April-20142%20(7).pdf

Smith J., Brewer H. and Dyall T. (2009) *Proc. Assoc. Advmt. Anim. Breed. Genet.* **18:** 334.

Smith J. (2016). Final Report to AWI for Project WP639. AWI, Sydney.

Van der Werf, J.H.J. (2019) http://www-personal.une.edu.au/~jvanderw/software.htm.

DEMONSTRATING BREEDPLAN ESTIMATED BREEDING VALUES IN NEW ZEALAND COMMERCIAL BEEF HERDS

M. Tweedie[1], J. Archer[2] and L. Proctor[2]

[1]Beef + Lamb New Zealand Genetics, Level 2, Queens Gardens Court, 3 Crawford Street, Dunedin, 9058 New Zealand
[2]AbacusBio Ltd, 442 Moray Place, Dunedin, 9058 New Zealand

SUMMARY

Demonstration of BREEDPLAN Estimated Breeding Values (EBVs) is important to build confidence in the value of genetic improvement for commercial farming businesses and for user trust in genetic evaluation. Work of this kind has not been completed before in a New Zealand setting. The way that this validation is presented to users is also important in order to maintain confidence. Across the 9 BREEDPLAN traits assessed to date, the responses in calf performance ranged from 25% to 151% of predicted response based on sire EBVs – giving confidence to users that selection on EBVs is translating into gains in performance under New Zealand commercial beef cattle systems.

INTRODUCTION

The uptake of Estimated Breeding Values (EBVs) and other genetic tools has been mixed in New Zealand beef breeding to date. In sheep breeding, the uptake is relatively advanced and many rams sold are first grouped on $ Index to be valued for sale. Given a vast majority of farms run both sheep and beef on the same land – this is an interesting difference.

Beef+Lamb New Zealand Genetics established the Beef Progeny Test (BPT) with a strong focus on adoption. Typically, progeny tests are formed for the purpose of evaluating sires and providing research data primarily. While the BPT also has evaluation and research goals, a strong adoption focus is a defining aspect of the BPT. Proving that the tools work and that the investment pays off from using them is a central theme for the test – it has become a platform for adoption in NZ beef genetics. The use of significant sized commercial farms is a key factor in ensuring that NZ farmers believe and relate to the outputs of the BPT.

MATERIALS AND METHODS

The BPT is an industry and government funded project that tests the genetic merit of beef sires through the performance of their progeny. The project completed its 5th cohort of joining via artificial insemination (A.I) in 2018. Approximately 2200 cows in five large commercial herds (4 herds with Angus cows and 1 herd with Herefords) are artificially inseminated each year using fixed time oestrus synchronisation to approximately 50 sires representing Angus, Charolais, Hereford, Simmental, and Stabilizer breeds. Herds were selected on a range of criteria including number of cows, ability to execute the project, credibility with commercial farmers, geographical spread, ownership/governance/ management, among other factors.

Sires are used across 5 properties; Mendip Hills Station- Cheviot, Tautane Station- Hawkes Bay, Whangara Farms- Gisborne, Caberfeidh Station- Hakataramea Valley and Rangitaiki Station- Taupo. Not every sire is used on every property, however there is sufficient genetic linkage to compare the performance of sires across these wide ranging environments with most sires (except Stabilizers) used across at least two properties and many used across four. Maternal sires were used with the intention of achieving a rate of 25 effective progeny per sire (30 in later years), and terminal sires at a rate to achieve 12.5 effective progeny (15 in later years) giving sufficient accuracy of sire evaluation for most traits - with all progeny being DNA parentage verified to sire and dam. International sires from other global progeny tests have been used to provide benchmarks for New Zealand beef sire performance and to allow future collaboration. Progeny of sires are assessed for over 30 traits- including growth, carcass quality (using live ultra-sound at approximately 18 months of age, and actual carcase assessments based on the Meat Standards Australia measurement system), structural assessments, maternal ability, fertility, cow size and body condition. Heifers from Angus, Hereford and Stabiliser sires are mated naturally to calve at 2 and 3 years of age, with the intention of retaining them into the cow herd. All steers and heifers from Simmental and Charolais sires were slaughtered and carcase measurements collected. Animals for slaughter were drafted into groups based on weight, generally several months prior to slaughter, and all animals within the pre-allocated groups were killed on the same day.

Statistical analysis. An analysis of how calf performance related to sire EBVs was undertaken to provide a demonstration of the utility of EBVs for improving performance in commercial environments. Data was restricted to calves sired by Angus, Hereford and Simmental sires with Breedplan EBVs available. Two cohorts of calves (born 2014 and 2015) were available for growth and ultra-sound scanning traits, with only one cohort available for carcase traits. For each trait a linear model was fitted using R, which included effects of sex, calf age within Herd-year (estimated from conception date scanning and dam-calf DNA match), age of dam (years) and contemporary group (Herd-sex-mob), Sire EBV and Sire Breed. This model assumes the same regression slope on Sire EBV between breeds, with different breed intercepts to allow for different breed EBV bases. A preliminary analysis found that EBV x Breed interaction was not significant, suggesting that there was no evidence within the data analysed for different slope relationships between breeds. The EBVs used in the analysis were from Angus, Simmental and Hereford Group BREEDPLAN analyses, where the BPT data was not included in the analysis (and so were independent of this data).

RESULTS AND DISCUSSION

Figure 1 is an example of the graphs which were produced to summarise the analysis in a visual way. The graph demonstrates that for every kg increase in sire W200 EBV, the progeny performance for weaning weight increased by 0.49 kg, which is 99% of the expected performance in progeny. The major goal of these graphs is to give farmers confidence that a response to the different EBVs is observed in a commercial beef production setting in New Zealand, and that EBVs created largely on data collected in seedstock herds are relevant to commercial beef systems.

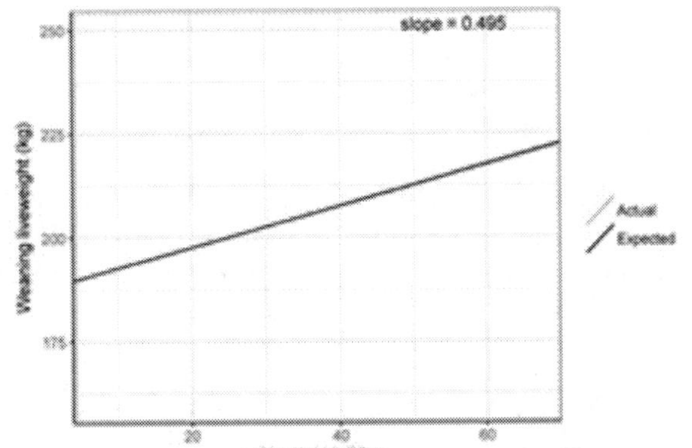

Figure 1. Demonstrating 200 Day Weight EBV

Table 1 gives the slope observed, and the percentage of this slope relative to what might be expected (generally expectation is 0.5 where the trait mirrors the EBV definition exactly). The impact of error in the EBVs on the expected slope is not accounted for in the calculation. Robinson (2005) showed that when the assumption that the independent variable is measured without error is violated, the expected slope is reduced and this can be predicted based on known errors. While accounting for this would be more appropriate from a scientific viewpoint to assess the utility of EBVs, communicating this to a lay audience is difficult, and it is also a "real world" factor which influences the realisation of commercial gains from use of EBVs. Consequently, for extension purposes the expected response was not adjusted to account for imperfect accuracy in sire EBVs.

Table 1. Demonstrating EBVs across traits

Trait	Observed slope	% of expectation
200 Day Weight EBV	0.49	99%
400 Day Weight EBV	0.41	82%
600 Day Weight EBV	0.45	90%
Days to Calving EBV*	0.50	100%
Rib Fat EBV	0.75 (abattoir)	151%
Eye Muscle Area EBV	0.33 (abattoir)	67%
Intra Muscular Fat EBV**	8.0 (abattoir)	25%
Carcass Weight EBV	0.15 (abattoir)	32%
Rib Fat EBV	0.20 (ultrasound scanning)	40%
Rump Fat EBV	0.32 (ultrasound scanning)	65%
Eye Muscle Area EBV	0.33 (ultrasound scanning)	66%
Intra Muscular Fat EBV	0.27 (ultrasound scanning)	54%
		73%

* Conception date as recorded in the BPT is calculated similarly to DTC but doesn't include Gestation length and is based off conception date information from ultra-sound scanning with foetal age called in 5 day increments.
**MSA marble score has been scaled to relate to IMF%. using values from Bindon (2001).

As a general rule, the relationships between calf performance and EBVs which were closest to expectation were for traits where EBVs are generally the most accurate (e.g. growth traits). The responses in calf performance for growth and fertility traits were generally close to expectation from Sire EBV. Responses in carcase ultra-sound measurements were generally lower than expectation (ranging from 40% to 66% of expectation) but still strongly positive.

For actual carcase measurements, the relationships between progeny performance and sire EBV for fat depth, marble score and carcase weight the relationship were positive but substantially below theoretical expectation. For these traits, the accuracy of the EBVs is affected by both the lower level of recording, and that the majority of recording is based on ultra-sound carcase predictions rather than actual carcase data. This is particularly important for marbling predictions, where ultra-sound %IMF is imperfectly correlated to actual marble scores which likely contributes to the lower than expected correlation (where the calculation of expectation does not take this into account). Responses in carcase eye muscle area were similar to those for ultra-sound scan eye muscle area, while for carcase fat depth the responses exceeded that predicted by Sire EBVs. Responses in carcase weight being much lower than expected can be explained by the impact of drafting strategy, where animals were pre-selected into groups to be killed together based on liveweight (generally several months before actual slaughter). Reverter *et al.* (2000) found measurements on the same cattle ultrasound scanned as yearlings and then again at the abattoir were moderate to strongly positive, suggesting that selection using yearling ultrasound measurements of seedstock cattle should result in predictable genetic improvement for abattoir carcass characteristics. This has been demonstrated anecdotally around the world with good levels of genetic gain in carcass traits using ultrasound scanning.

Across the 9 BREEDPLAN traits assessed 73% of sires EBVs turned into calf performance. This was further demonstrated by Angus Australia (2018) which showed an excellent relationship between expected difference between sires and actual differences in their progeny. The outcome was comparable to this work where the percentage of predicted difference and the actual difference averaged 120%. This was reinforced by Thrift and Thrift (2006) where the expected differences between high and low EBV sires were similar to the realised differences.

Our experience in showing these relationships to commercial farmers, with the aim of enhancing uptake of EBVs, is that the main focus of farmers tends to be on the fact that there is a positive relationship, rather than the proportion of the expected relationship actually achieved. Thus while the relationships for some traits might be lower than expected from a scientific perspective, the main message taken from these relationships from an extension viewpoint is that the EBVs are predictive of calf performance, and therefore selection based on EBVs will lead to gains within commercial beef production systems.

CONCLUSION

This work has demonstrated strong positive relationships of Sire EBVs with the performance of their calves. Confidence of commercial beef producers in the application of EBVs is enhanced when they can observe these relationships demonstrated in a local context and in farm systems that similar to their own.

ACKNOWLEDGEMENTS

The B+LNZ Genetics Beef Progeny Test would like to acknowledge the contributions of Simmental NZ, Focus Genetics and in-kind support from Angus NZ, Hereford NZ, Charolais NZ and the 5 site herds is also gratefully acknowledged.

REFERENCES

Angus Australia. (2019, April 29). *EBVs-No-Bull*. Retrieved from https://www.angusaustralia.com.au/content/uploads/2017/02/Ash_POC_EBVs-No-Bull_web2.pdf.

Bindon B.M. (2001) in: Marbling Symposium: Proceedings of a CRC Conference. Coffs Harbour, NSW, October 2001.

Reverter A., Johnston D.J., Graser H.-U., Wolcott M.L. and Upton W.H. (2000) *J. Anim. Sci.* ***78**:* 1786.

Robinson D.L. (2005) In 'Livestock Production Science', pp. 155-161, editor E.H. von Borell, Halle, Salle, Germany.

Thrift F.A. and Thrift T.A. (2006) In 'Professional Animal Scientist', pp. 413-423, editor D. Beede, Michigan State University, East Lansing, Michigan, USA.

TWO YEARS IN: LESSONS FROM THE INTRODUCTION OF HEREFORD SINGLE-STEP BREEDPLAN

C.A. Millen[1,2] and B.J. Crook[2]

[1]Southern Beef Technology Services, University of New England, Armidale, NSW, 2351 Australia
[2]Agricultural Business Research Institute, University of New England, Armidale, NSW, 2351 Australia

SUMMARY

The first multi-country multi-trait single-step GBLUP (ssGBLUP) BREEDPLAN analysis was released for the Hereford breed in October 2017, combining the pedigree, performance and genomic data from Australia, New Zealand and Namibia. Nearly 18 months later, Hereford breeders have embraced genotyping with a 3.8-fold increase in number of animals genotyped. The inclusion of genomic information has influenced the Estimated Breeding Values (EBVs), although this varies by trait and animal. Similarly, exclusion of hard-to-measure phenotypes from the analysis confirms the importance of recording phenotypes for the estimation of accurate EBVs, even when utilising ssGBLUP. The collection of both genotypes and phenotypes, particularly for hard-to-measure traits, remains vital in allowing beef producers to best utilise ssGBLUP.

INTRODUCTION

Since the release of the first genetic evaluation for beef cattle in Australia in the 1980s, the BREEDPLAN model has been continually updated to incorporate new methodology and traits (Graser *et al.* 2005). More recently, several breeds have moved to include genomic information in their BREEDPLAN analyses using ssGBLUP methodology (Johnston *et al.* 2018). This includes the Hereford Single-Step BREEDPLAN (Hereford ssBP) analysis, released in October 2017 and the first multi-country multi-trait ssGBLUP BREEDPLAN analysis. Data from three Hereford associations, being Herefords Australia (HAL), the New Zealand Hereford Association (NZHA) and the Hereford Cattle Breeders Society of Namibia (HCBSN), is combined in a single BREEPDLAN analysis.

Seventeen months on since the release of Hereford ssBP, this paper reviews the size and composition of the Hereford G-Matrix and how this has changed over time. The effect of including both genomic information and phenotypes for hard-to-measure traits on BREEDPLAN EBVs and accuracy is also explored.

MATERIALS AND METHODS

The Hereford ssBP analysis combines pedigree, performance and genomic information from three Hereford breed societies (HAL, NZHA and HCBSN) in a complete multi-trait BREEDPLAN analysis of birth, growth, fertility, carcase and feed intake traits. Calving ease and docility traits are analysed separately using a threshold model analysis (Graser *et al.* 2005) and do not directly incorporate genomic information at present. Twelve standard production analyses are run each year, with 19 trait EBVs reported. For the purposes of the current study, G-Matrix data was extracted from the October 2017, October 2018 and March 2019 Hereford ssBP analyses.

The impact of genomics on EBVs was examined by comparing results from the March 2019 Hereford ssBP analysis with results from the same analysis run without the G-Matrix. Data was analysed for three sets of animals, being (a) sires with an EBV accuracy of ≥75%, (b) young bulls with an EBV accuracy of ≥50% and (c) non-parent genotyped animals with an EBV accuracy of ≥5%.

Likewise, the impact of hard-to-measure trait phenotypes on EBVs and accuracy was examined by comparing the March 2019 results (which included hard-to-measure trait phenotypes) with results from the same analysis where hard-to-measure trait phenotypes were excluded (HtM- trial run).

RESULTS AND DISCUSSION

Hereford G-Matrix composition. In comparing G-Matrices, the number of animals genotyped has increased by 3.8-fold in 17 months, from 3651 animals in the October 2017 analysis to 13764 animals in the March 2019 analysis. The majority of genotypes were supplied by HAL, although there has been a steady increase in the percentage of the NZHA genotypes (0.8% versus 12.6%) since Hereford ssBP was released. While some HCBSN animals have been genotyped, these are not yet available in the format required for inclusion in the Hereford ssBP analysis. Although HAL genotypes currently dominate the G-Matrix, the benefits arising from Hereford ssBP would extend to members in all three Hereford societies subject to linkage between individual herds and the reference population, regardless of which society provides the genotypes.

In October 2017 the G-Matrix was comprised of animals born 1961 to 2016, with 29% born prior to 2010 and 39% in 2015 or later. Just 16% of the animals in the October 2017 G-Matrix were female. Given that the approach employed to build the Hereford reference population included strategic genotyping of influential sires and genotyping of well-recorded animals in the HAL beef information nucleus (BIN; Banks 2011), it is not surprising that the October 2017 Hereford G-Matrix was comprised of many older animals, mostly male. These strategies have been employed by other beef breeds and the Australian sheep industry (Brown *et al.* 2018) as they moved toward the implementation of Single-Step genomics in their genetic evaluations.

However, composition of the G-Matrix shows marked change in March 2019. Of animals in this G-Matrix, 12% were born prior to 2010 and 67% in 2015 or later. While most animals in the reference population were male, female animals now comprise 29% of those genotyped. The rapid growth in the size of the reference population suggests that Hereford breeders are embracing genotyping technology. In particular, the growth in both the percentage of younger animals genotyped and the percentage of female animals genotyped indicates that in the past 17 months Hereford producers have genotyped not only older sires but are also genotyping the younger animals (e.g. young bulls and replacement heifers) in their herds. This is to be expected as the greatest benefit of genomics will be seen in these animals, with increased EBV accuracy at earlier ages. In addition, changing requirements for animal registration (e.g. parent verification; compulsory SNP genotyping for certain classes of animals) combined with extension messages on the benefits of genotyping are likely to have contributed to this rapid uptake.

Rapid uptake of genotyping following the release of genomic analyses is not unprecedented. The Australian dairy industry first released genomic EBVs in 2010 (Nieuwhof *et al.* 2011) and since then, the number of animals being genotyped commercially has grown considerably, from just over 500 animals genotyped in the 2012/2013 financial year to over 19000 in the 2017/2018 financial year (Williams *pers. comm.* 2019). Much of this growth has been in the genotyping of females. While beef producers are unlikely to embrace genotyping of females to the same extent as their dairy counterparts, the release of a low-density panel for Hereford females in mid-2019 may help drive an increase in the genotyping of Hereford females.

Effect of genomics on Hereford BREEDPLAN. Although the inclusion of genomic information did not change the Hereford breed average EBVs (based on 2017 drop), the EBVs of individual animals have changed considerably, some increasing and others decreasing. EBV movements were more pronounced in genotyped non-parent animals than observed for sires, although EBV movements were evident for both groups. This trend was observed across traits, and likely reflects EBV accuracy prior to the inclusion of genomic information.

Hereford ssBP has also led to increased EBV variance for most traits. For genotyped non-parent animals, EBV standard deviation is higher for most BREEDPLAN traits. Increased EBV variance is particularly useful for beef producers, as identification of genetically superior and inferior animals informs selection decisions to drive genetic gain.

Lastly, Hereford ssBP has led to an increase in EBV accuracy. This increase has been greatest for animals with a low EBV accuracy prior to the inclusion of genomic data; a trend that holds across traits (Figure 1). However, the magnitude of this increase varies across traits (Figure 1). For 200 Day Weight (200d), the average increase in EBV accuracy for animals with a prior EBV accuracy of ≤20% was 30%; for Days to Calving (DtC) it was 5.5%. Similar results were observed by Johnston *et al.* (2018) following the implementation of ssGBLUP for Australian Angus and Brahman. They concluded that the difference in magnitude of EBV accuracy increase across traits likely varied due to both heritability and the size of the reference population for each trait. Both are likely to be the case here; 200d is moderately heritable while DtC is lowly heritable. Furthermore, over 85% of animals in the March 2019 G-Matrix have a phenotype for 200d, but less than 5% have DtC phenotypes. This is not surprising given the low level of performance recording observed for DtC in the Australian beef industry (Gudex and Millen 2019, these proceedings). Collection of additional phenotypes is likely to drive larger increases in EBV accuracy than currently observed (in turn leading to increased genetic gain), and thus remains a key extension message for industry.

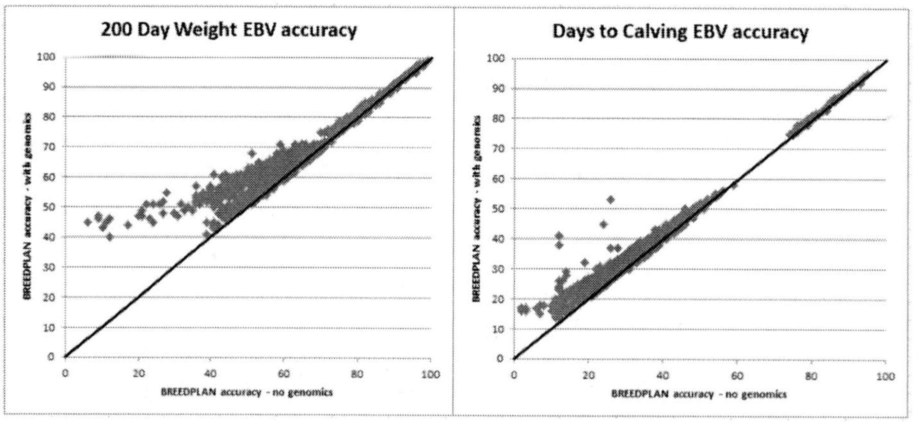

Figure 1. Changes in EBV accuracy for 200 Day Weight and Days to Calving when genomic information is included in the Hereford BREEDPLAN analysis

Effect of hard-to-measure phenotypes on Hereford BREEDPLAN. The removal of DtC and abattoir carcase data had greatest impact on DtC and carcase trait EBVs and accuracy, but little to no effect on EBVs and accuracy for other BREEDPLAN traits. The impact varied by trait, with the biggest impact being evident for DtC. The correlation between the DtC EBV in the ssBP run and the HtM- trial run was 0.35, with the latter results regressed back toward the base (from -2.1 days on average back to -1.7 days) and with reduced standard deviation (from 3.3 days back to 1.6 days). While the average change in DtC EBV was small, substantial changes in DtC EBV were observed for individual animals, the most extreme being changes of ±9.0 days. The exclusion of DtC data also led to a decrease in DtC EBV accuracy, from an average of 80% in the standard production run to an average of 61% in the HtM- trial run.

Less extreme changes were observed for the carcase traits. Correlations between the standard production run and the HtM- trial run varied from 0.95 (for each of Eye Muscle Area (EMA) and Rump Fat (P8)) to 0.99 (Carcase Weight (CWt)). While the exclusion of abattoir carcase data led to small changes in average EBV for the carcase traits, the carcase EBVs did not regress towards the base for all traits and only three of the six carcase traits showed reduced standard deviation in the HtM- trial run (CWt, EMA and P8). Despite this, the exclusion of abattoir carcase data did lead to large changes in carcase EBVs for individual animals. For CWt, changes in EBV ranged from -16.3 to +17.3 kg, and for EMA EBVs, from -5.4 cm^2 to +4.7 cm^2. Small reductions in average EBV accuracy (-1%) were observed for carcase traits in the HtM- trial run.

The exclusion of DtC data had a more extreme impact on DtC EBVs than the exclusion of abattoir carcase data had on carcase trait EBVs. This is most likely due to the multi-trait nature of the Hereford ssBP analysis, where information used in the calculation of carcase EBVs included both abattoir carcase data and correlated ultrasound scan data. Given that ultrasound scan data was not removed, it is likely to have buffered some of the effect of removing abattoir carcase data on the carcase EBVs. The same cannot be said for DtC, where the removal of DtC data led to DtC EBVs regressing towards zero, and with less variation and less EBV accuracy.

This research confirms the importance of a well-recorded reference population for the accurate estimation of breeding values, especially in relation to hard-to-measure traits. In the absence of direct records for these hard-to-measure traits and the absence of significantly correlated trait data, the availability of genotypes and ssBP alone is unlikely to provide sufficient EBV accuracy, limiting the potential for genetic improvements in these traits.

CONCLUSIONS

Since the release of Hereford ssBP, breeders have embraced genomic technology with a 3.8-fold increase in the number of animals genotyped. The inclusion of genomic information does result in EBV movements for individual animals, plus increased EBV variance for most traits and increased EBV accuracy, although this will vary by trait and by animal. This study has also confirmed the value of recording hard-to-measure phenotypes for the accurate estimation of breeding values. To maximise long-term genetic gains, the on-going collection by breeders of both genotypes and phenotypes, particularly for hard-to-measure traits, remains vital.

ACKNOWLEDGEMENTS

The authors wish to thank HAL, NZHA and HCBSN for supporting this work and providing access to data from their societies. The authors also thank the Agricultural Business Research Institute (ABRI), Meat and Livestock Australia (MLA) and 14 Australian-based stakeholder breed associations for their continued support of the SBTS project.

REFERENCES

Banks R. (2011) *Proc. Assoc. Advmt. Anim. Breed. Genet.* **19:** 399.

Brown, D.J., Swan, A.A., Boerner, V., Li, L., Gurman, P.M., McMillan, A.J., van der Werf, J.H.J., Chandler, H.R., Tier, B. and Banks, R.G. (2018) *Proc. World Cong. Genet. Appl. Livest. Prod.* **11:** 460.

Graser H-U., Tier B., Johnston D.J. and Barwick S.A. (2005) *Aust. J. Exp. Agric.* **45:** 913.

Gudex, B.W. and Millen, C.A. (2019) *Proc. Assoc. Advmt. Anim. Breed. Genet.* **23:** in press.

Johnston, D., Ferdosi, M., Connors, N., Boerner, V., Cook, J., Girard, C., Swan, A. and Tier, B. (2018) *Proc. World Cong. Genet. Appl. Livest. Prod.* **11:** 269.

Nieuwhof, G.J., Haile-Mariam, M., Beard, K.T., Konstantinov, K.V. and Hayes, B.J. (2011) *Proc. Assoc. Advmt. Anim. Breed. Genet.* **19:** 363.

PEDIGROMICS: A NETWORK-INSPIRED APPROACH TO VISUALISE AND ANALYSE PEDIGREE STRUCTURES

A. Reverter[1], S. Dominik[2], J.B.S. Ferraz[3], L. Corrigan[4] and L.R. Porto-Neto[1]

[1]CSIRO Agriculture & Food, St. Lucia, Brisbane, QLD, 4067 Australia
[2]CSIRO Agriculture & Food, Armidale, NSW 2350 Australia
[3]Universidade de São Paulo (USP), Pirassununga, SP, Brazil
[4]Rennylea Angus, NSW, Australia

SUMMARY

We exploit the power and visual appeal of network theory to visualise and analyse pedigree structures by bringing into the visualization schema elements of the breeding history for a given population. We name the approach "Pedigromics" and illustrate its potential with three datasets from beef cattle populations of varying characteristics. These include tropical cattle from the Beef CRC Legacy Database, a highly curated pedigree from an Angus seedstock producer in Australia, and a large pedigree from Nelore cattle in Brazil. While conceptually very simplistic, we highlight the ability of Pedigromics to identify emerging features of family structures with a wide range of applications from the understanding of the breeding history of a population, to the identification of influential ancestors and to inform mating allocation alternatives for the design of breeding schemes.

INTRODUCTION

With the advent of cheaply available "omics" technologies, the last decade has seen an explosion of molecular data. Whether at the level of genome sequence variants, transcript expression or protein abundance, the wide adoption of these technologies has allowed us to collect tens of thousands of data points from individual samples. As a consequence, new analytical and computational approaches have been developed in order to interrogate the large datasets in an efficient and effective manner. Paramount among these approaches are the numerical algorithms for the inference of gene networks. Two such approaches include WGCNA (Langfelder and Horvath 2008) and PCIT (Reverter and Chan 2008). Once a gene network has been constructed, its optimality should be established by mathematical and biological criteria. The former includes whether or not the connections follow a non-random scale-free distribution by which most genes have a few connections while a few genes, termed hubs, have lots of connections. Biological criteria include the ability of the network to (1) recover known gene-gene interactions and (2) reveal clusters of genes annotated to a biological process of relevance to the subject matter under scrutiny.

Pedigree information and associated metadata (i.e. sex, year of birth, farm of origin, etc.) is ideally suited to be explored using what have now become standard approaches to visualise and analyse gene networks. Individual animals are represented as nodes in the network, and two animals are connected by an edge when a particular kinship exists, e.g. Parent – Offspring.

We name the approach "Pedigromics" and illustrate its potential with three beef cattle datasets: (1) A Brahman and Tropical Composite population from the Beef CRC Legacy Database; (2) A highly curated pedigree from Rennylea, an Angus producer in Australia; and (3) a large pedigree from Brazilian Nelore cattle.

MATERIALS AND METHODS

Beef CRC Dataset. We made use of the Brahman (BR) and Tropical Composite (TC) populations extensively described in the literature by our group and most recently by Raidan *et al.* (2018). For the present study, the original 2,111 BB and 2,550 TC cattle were further edited to only include animals satisfying the following 3 conditions: (1) having both sire and dam recorded; (2) from the top 20 sires based on number of progeny; and (3) from contemporary groups (CG, including sex, month of birth and property) with at least 10 records and at least two sires represented. These editing criteria resulted in 720 BB and 883 TC cattle used for Pedigromics analyses.

Angus Dataset. We used a pedigree file of 7,551 Angus cattle (4,429 females and 3,122 males) from Rennylea spanning 29 birth years (1989 to 2017). The file contained information from 305 sires averaging 24.5 progeny and ranging from 1 to 338 progeny. Further information used in the Pedigromics visualisation scheme was BreedPlan EBV for weaning weight and the 3,955 animals with SNP genotypes were highlighted.

Nelore Dataset. The dataset contained 51,265 Nelore cattle (25,324 females and 25,941 males) spanning over 50 years to 2017 and with 910 sires averaging 56.3 progeny (range = 1 to 4,613). For the Pedigromics map we captured the progeny (N=31) from the most prolific sire in 2017, namely D4685. We trace the dams of these progenies and all of the male ancestors, i.e. maternal and paternal grandsires, great-grandsires, and so on until exhausting the available genealogy.

In all three cases, we used Cytoscape (available at http://cytoscape.org; Shannon *et al.* 2003), a widely used software program for the visualization and exploration of the networks.

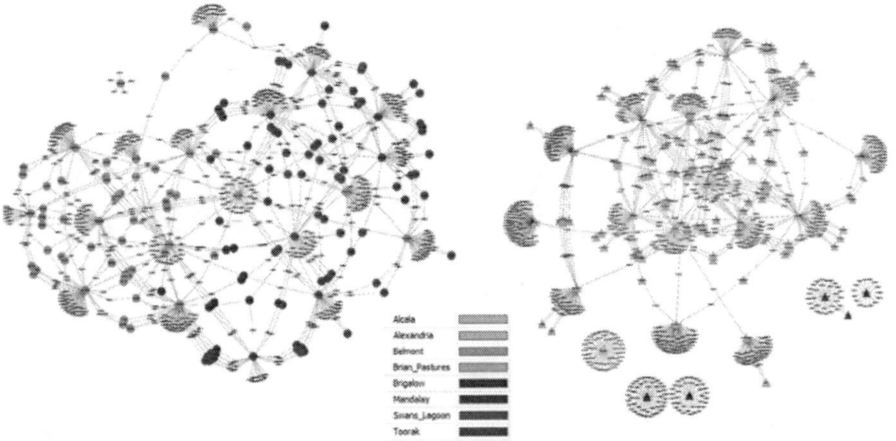

Figure 1. Pedigromics view of the Beef CRC dataset comprised of 1,872 cattle (nodes) connected by 2,108 edges. Two clear clusters emerge corresponding to Brahman (circles, left) and Tropical Composite cattle (triangles, right). Nodes are coloured according to property of origin (see insert). Connections correspond to sire-offspring (black edges) or dam-offspring (red edges). Big and small nodes correspond to parents and offspring, respectively

RESULTS AND DISCUSSION

The Pedigromics view of the Beef CRC (Figure 1), Angus (Figure 2) and Nelore datasets (Figure 3) have been designed to highlight different aspects of each dataset by resorting to a diverse range of criteria for the visualisation scheme. With the Beef CRC dataset (Figure 1), we highlight the power of Pedigromics to quickly identify the presence of two unrelated populations (Brahman and Tropical Composite) while providing a birds eye view of the of cattle across the eight properties.

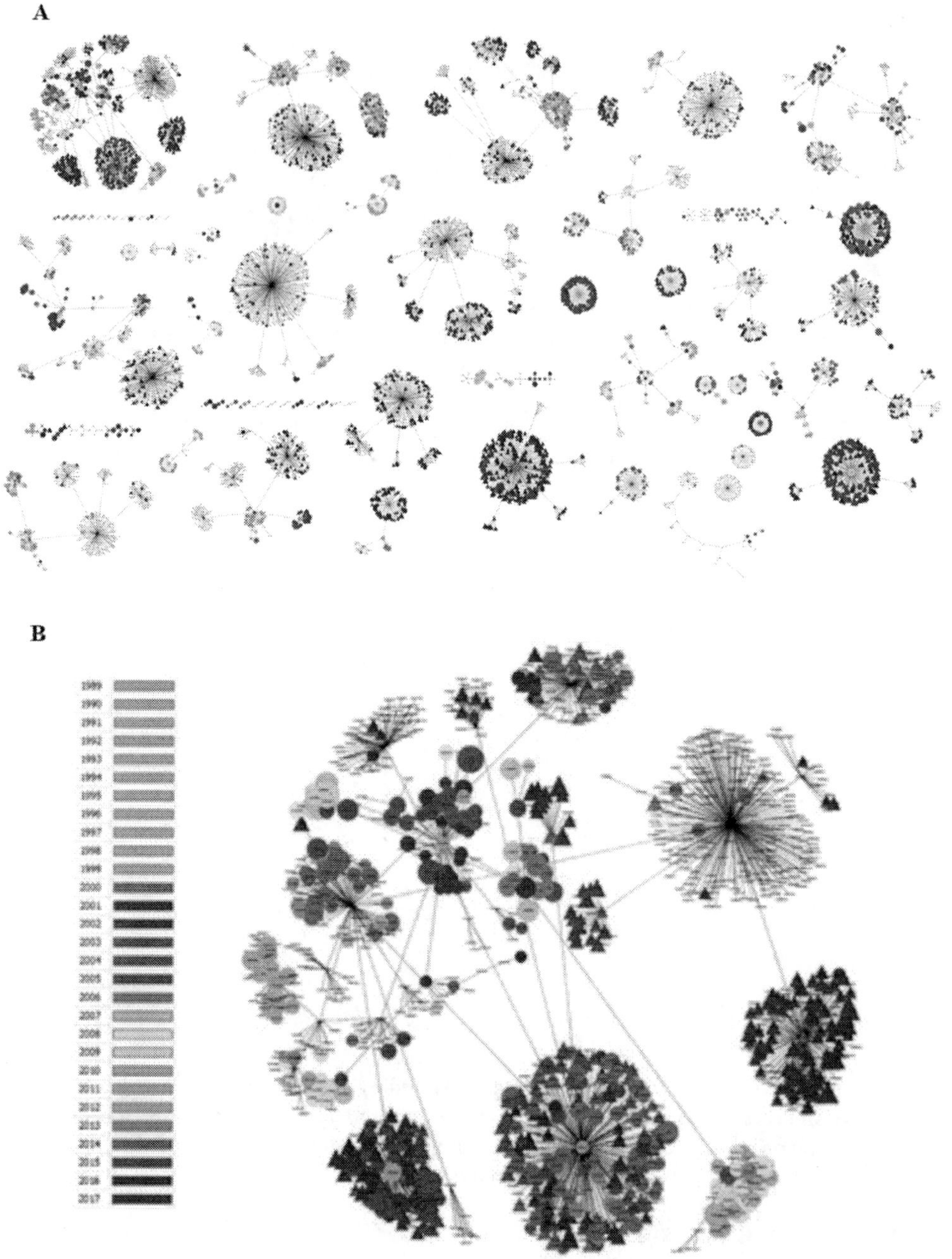

Figure 2. Pedigromics view of the Rennylea Angus dataset for the entire set of 7,551 individuals from 305 sire families (A) and a close up of a single large sire family spanning several years (B). Non-genotyped and genotyped individuals are represented by circles and triangles, respectively. Also, the size of the node has been mapped to the weaning weight EBV

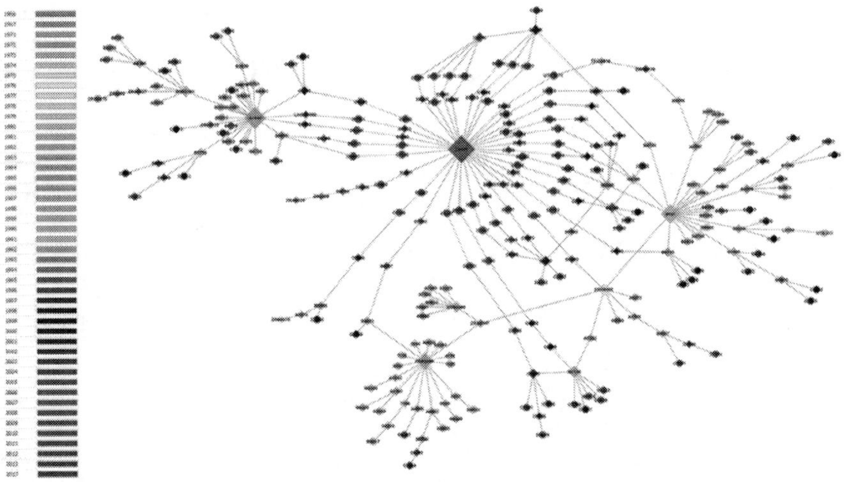

Figure 3. Pedigromics view of the Brazilian Nelore dataset focussing on the ancestors of the 31 progeny of sire D4685 in 2017

With the Angus dataset, we show how the 305 sire families cluster in super-families providing a look similar to constellations in the cosmos. We then zoom into one of these sire super-families so that we can easily trace individual sire families back in time. In addition, by imposing the weaning weight EBV to the size of the nodes, we can immediately see what could be termed as 'heritability at work' by which progeny of a small EBV sire are also small, and vice versa.

Finally, with the Nelore dataset we show how by focusing on the progeny of an influential sire Pedigromics can easily reveal the identity of other influential sires back in time.

CONCLUSIONS

We have presented Pedigromics, a tool to take advantage of network theory and visualisation techniques to analyse pedigree files and reveal features of the population that were not immediately apparent from the flat files. Applied to three distinct datasets allowed us explore different aspects of interest to the history and potential of a breeding population. Further applications of Pedigromics are being explored including assessing conflicts between pedigree and genomic information.

ACKNOWLEDGEMENTS

This work was performed using the legacy database belonging to the Cooperative Research Centre for Beef Genetic Technologies and their core partners including Meat and Livestock Australia. The authors are grateful to Sigrid Lehnert for reviewing this manuscript.

REFERENCES

Langfelder P. and Horvath S. (2008) *BMC Bioinformatics* **9**: 1.

Raidan F.S.S., Porto-Neto L.R., Li Y., Lehnert S.A., Vitezica Z.G. and Reverter A. (2018) *J. Anim. Sci.* **96**: 4028.

Reverter A. and Chan EK. (2008) *Bioinformatics* **24**: 2491.

Shannon P., Markiel A., Ozier O., Baliga N.S., Wang J.T., Ramage D., Amin S., Schwikowski B. and Ideker T. (2003) *Genome Res.* **13**: 2498.

GENETICS OF FEMALE FERTILITY IN WAGYU CATTLE USING FIELD DATA

K. Reid, G. Moser and M. Kelly

Australian Agricultural Company Limited, Brisbane, QLD, 4006 Australia

SUMMARY

Wagyu breeders mainly focus on carcase characteristics when making breeding and selection decisions. The objective of this study was to estimate heritabilities based on pregnancy test records as these are collected as part of standard management practice. Traits included pregnancy rate heifer and cow, body condition score (heifer and cow) and age at first pregnancy, with heritabilities of 0.06, 0.11, 0.07, 0.07 and 0.09 respectively. In this study, pregnancy rates were continuously high (mean >86%). The heritability of pregnancy rate traits were low. As the expected increase in fertility in this population by genetic approaches will be slow, opportunites for reproductive improvement on these particular traits will be limited.

INTRODUCTION

Little attention has been paid to the genetics of fertility in Australian Wagyu for two reasons, (1) Wagyu females are generally highly fertile and (2) males are predominately used as terminal sires. Literature suggests that fertility traits of most beef breeds, including Wagyu, have low to moderate heritability (Oyama *et al.* 2011; Buddenberg *et al.* 1989; Toelle and Robinson 1985). Improving male and female fertility will improve herd productivity. Immediate gains in herd performance can be made by culling non-pregnant cows; however, this is inefficient due to the low selection differential and the low accuracy of selecting on fertility phenotypes. Higher selection intensity is possible through the sire pathway (Baeza-Rodríguez *et al.* 2018). Heritability of fertility traits reported in the literature are population-specific and highly dependent on the actual component of fertility under investigation. In Australia, breeding programs use traits such as days to calving or calving interval to assess fertility in temperate production systems which have low to moderate heritability (Gutiérrez 2002; Johnston and Bunter 1996), while age at puberty and *post partum* anoestrus are highly heritable (Engle *et al.* 2018; Johnston *et al.* 2009). Pregnancy testing is routinely used to make management decisions and can provide information much earlier than days to calving or calving interval. In this study, we estimated heritability for fertility traits using pregnancy test information as a first step to assess their value for standard genetic evaluation.

MATERIALS AND METHODS

Pregnancy test data from 4,278 heifers and 6,047 cows were collected between 2011 and 2019. All animals were pedigree recorded and raised in an extensive environment in Queensland.

Data collection and editing. Heifer Pregnancy Rate (HPR) reflects a female's ability to become pregnant during her first breeding season as a yearling. Age At Pregnancy (AAP) is the age at which she conceives. Both of these are calculated from their first pregnancy test. Cow Pregnancy Rate (CPR) is the resultant pregnancy rate of subsequent parities. At the time of pregnancy testing, heifers and cows are scored for Body Condition Score (BCS) on a scale from 1-5. In this population, the BCS ranged from 2-4 with an average of 3. The number of records used in the analysis, heifer and cow, are shown in Table 1.

Table 1. Number of records and summary of phenotypic data recorded for heifers and cows

Heifers				Cows		
Year brand	N	Mean HPR	Mean AAP	Year brand	N	Mean CPR
2013	552	0.90	524	2011	1785	0.89
2014	911	0.88	508	2012	1248	0.91
2015	1068	0.86	492	2013	1155	0.93
2016	1035	0.92	533	2014	1114	0.95
2017	712	0.90	499	2015	745	0.95

Statistical analyses. Analyses were performed using univariate sire models for HPR & CPR and an animal model for BCS and AAP. The R software package MCMCglmm was used for all models (Hadfield 2010). The MCMC chain was run for 200k iterations with the first 20k used for burn-in. Due to the binary nature of pregnancy rate, observations were converted to an underlying normal distribution using a probit link function which restricts the residual variance to 1, all other traits were analysed as continuous outcomes. Fixed effects in the mixed effects included year and month of birth (BYM), measurement date (MD) and the age of the animal in years (AGE). The general form of the equations in the model was as follows:

$$y=Xb + Za + e,$$

where y is a vector of observations for the trait, X and Z are incidence matrices relating observations to in **y** to levels of fixed effects in **b**, and random solutions in **a**, respectively, and **e** is a vector of residual effect solutions. Variances and mean of the random effects were assumed to be as follows:

$$\mathrm{var}(e) = I\boldsymbol{\sigma}2e,\ \ \mathrm{var}(a) = A\boldsymbol{\sigma}2a$$

where **A** is the numerator relationship matrix, and E(**a**) = $E(\boldsymbol{u})$ = 0. The models used were:

HPR, CPR & AAP: Y = BYM + body condition score + MD + sire

BCS: Y = BYM + body condition score + MD + AGE + animal

RESULTS AND DISCUSSION

Heritability estimates of HPR and CPR, BCS, and AAP are shown below in Table 2.

Table 2. Heritability estimates for Wagyu female fertility traits

	Heifer		**Cow**	
	h^2	95% CI	h^2	95% CI
HPR/ CPR	0.06	0.00, 0.26	0.11	0.03, 0.22
AAP	0.09	0.04, 0.14		
BCS	0.07	0.04, 0.11	0.07	0.00, 0.15

Heritability estimates for all traits were low, and therefore agree with estimates from other studies (Oyama *et al.* 2011; Buddenberg *et al.* 1989; Toelle and Robinson 1985). Heritability of age at first calving was low in studies in Japanese Black Cattle and other breeds such as Herefords and Angus, whereas in tropical cattle, age at first calving and age at puberty have been recorded as moderately to highly heritable (Johnston *et al.* 2009; Gutiérrez *et al.* 2002; Oyama *et al.* 2002; Toelle and Robinson 1985). The differences in heritability are likely due to influences of the environment and breed differences. For example, heritability estimates for age at calving range from 0.8 in temperate environments to 0.3-0.5 under tropical conditions. The pedigreed Wagyu herd studied was managed so to optimise reproductive performance, including nutrition, which may have impacted upon BCS and potentially masked the genetic variation in Wagyu HPR/ CPR. As a result, very little phenotypic variance or genetic variance was observed in this trait.

Pregnancy rate and age at first pregnancy have limited value in selecting for increased fertility in this population. To increase fertility multiple avenues could be explored, some would require changes in management practice; others would require additional measurements to be taken. One approach would be to switch to a trait such as days to calving. This would require recording bull in and out dates as well as a more consistent approach to management and disposal of pregnant females. Alternatively, traits such as age at puberty and post-partum anoestrus could be explored as these could be measured before pregnancy testing and would therefore not be impacted by culling or animal transfers (Johnston *et al.* 2009).

Body condition score is as an indicator of the nutritional status of an animal. Several researchers have reported positive phenotypic correlations between conception rates and high BCS (Selk *et al.* 1988). The distribution of body condition scores for both heifers and cows are shown below in Figure 1.

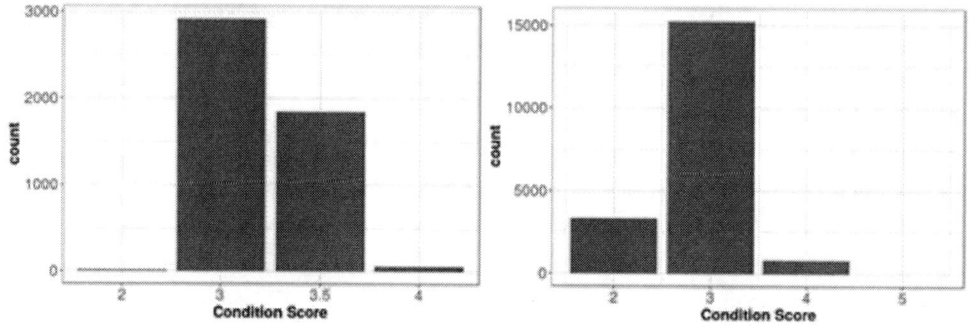

Heifers Cows

Figure 1. Distribution of BCS across all heifers (left) and cows (right) at pregnancy test

However, in this study we found a small correlation between BCS and HPR/ CPR, which could be due to the small variation of BCS within contemporary groups, or the management practice of not mating young females with low BCS.

CONCLUSIONS

Heritabilities of Wagyu female reproductive traits were estimated from pregnancy test data in a stud herd managed to optimise reproductive rate. Heritability estimates were low, and at the lower end of those in the literature. This is likely to be due to the lack of phenotypic variance potentially masking genetic differences in fertility. To successfully implement a genetic improvement program for fertility in this population would require development of alternative fertility traits. Traits such as days to calving and utilising ultrasound measurement to more precisely identify age at puberty or *post partum* anoestrus may be beneficial.

REFERENCES

Baeza-Rodríguez J., Montaño-Bermúdez M., Vega-Murillo V. and Arechavaleta-Velasco. (2018) *J. Appl. Anim. Res.* **46:** 534.

Buddenberg B., Brown C., Johnson Z., Dunn J. and Peterson H. (1989) *J. Anim. Sci.* **67:** 2589.

Engle B., Corbet. N., Allen J., Laing A., Fordyce G., McGowan M., Burns B., Lyons R. and Hayes B. (2018) *J. Anim.Sci.* **97:** 90.

Gutiérrez J., Alvarez I., Fernández I., Royo L., Díez J. and Goyache F. (2002) *Livest. Prod. Sci.* **78:** 215.

Hadfield J (2010). *J. Stat. Softw.* **33**: 1.

Johnston D., Barwick S., Corbet N., Fordyce G., Holroyd R., Williams P and Burrow H. (2009) *Anim. Prod. Sci. 4*: 399.

Johnston D. and Bunter K. (1996) *Livest. Prod. Sci.* **45:** 13.

Oyama K. (2011) *J. Anim.Sci.* **82:** 367.

Oyama K., Katsuta T., Anada K., and Mukai F. (2002). *Asian-Aust. J. Anim. Sci.* **15:** 1680.

Selk G., Wettermann R., Lusby K., Oltjen J., Mobley S., Rasby R. and Garmendia J. (1988) *J. Anim. Sci.* **66:** 3153.

Toelle V. and Robinson O. (1985) *J. Anim. Sci.* **60:** 89.

POPPLEWELL COMPOSITES: ADDING VALUE TO OUR CUSTOMERS' TROPICAL BEEF SUPPLY CHAINS THROUGH GENOMIC EVALUATION AND SELECTION

G.I. Popplewell[1], R.G. Tearle[2], R.A. McEwin[2], M.L. Facy[1] and W.S. Pitchford[2]

[1]Popplewell Composites, 33 Tom Schmidt Court, Mount Samson, QLD, 4520 Australia
[2]Davies Research Centre, School of Animal and Veterinary Sciences, University of Adelaide, Roseworthy, SA, 5371 Australia

SUMMARY

Popplewell Composites has been supplying objectively evaluated Tropical Composite bulls to commercial breeders since 2008. All active and historic animals in its nucleus breeding program were genotyped and Genomic Evaluation (G-BLUP) first conducted in 2016. Since first presenting at AAABG in early 2017, the GenoRater™ genomic evaluation program, built in collaboration with the University of Adelaide, has been used to evaluate more phenotypes, genotypes and traits. Now emerging customer multiplier tier herds are being added to the evaluation. This paper provides an update on the program and data, demonstrating the collaborative growth and connection the breeding program has achieved through application of scientific principles.

INTRODUCTION

Popplewell Composites' nucleus herd was established using animals from Angus, Belmont Red and Bonsmara, Senepol and Brahman breeds, objectively selected on their genetic makeup. The objectives of the program are to deliver meat quality and female fertility improvement through replacement of traditional *Bos indicus* dominated herds with Taurus/Sanga/Indicus Tropically Adapted Composites (Burrow *et al.* 2003). In addition, the aim is to provide continuous additive improvement through quantitative trait selection and introgression of favourable qualitative alleles (e.g. Poll genes for no horns).

The company's value proposition, direct approach to customer and research collaboration and how it conducts its genetic evaluation are all not typical of an Australian beef cattle seed-stock operation. In these aspects, it has more similarities to the operations of some modern Poultry, Pork and Aquaculture breeding companies. Popplewell Composites bulls do not fit into a purebred description or 'breed standard', but instead they are hybrid and multi-breed composite line bulls. The company uses an in-house genomic evaluation tool (GenoRater™) which focuses on being nimble, innovative and combining nucleus and customer herd data.

Customers with very large commercial herds (>10,000 breeding females) are being serviced via development of 'turn-key' bull multiplier herds that are run on the customers' own properties with Popplewell Composites supplying the sires, technology, sample collectors, ear tags, performance recording planning and support. Popplewell Composites also have a loyal client base of progressive family businesses that purchase bulls direct from the nucleus herd.

MATERIALS AND METHODS

Herd management. The Popplewell Composites nucleus cow herd is run in coastal South East Queensland, rotationally grazed on tropical species-based pastures and exposed to tropical parasites. The herd is phenotyped for fertility, birth weight, growth, mature weight, flight speed, tick resistance, coat score, and live-ultrasound carcass traits. Semen tested yearling bulls are sold to commercial and bull multiplier herds in Tropical, Subtropical and Arid regions of Australia. All heifers born into the program are first mated as yearlings which is not typical of tropical breed seed-stock herds.

Multiplier tier herds are located on customers' own properties in North West Queensland and the Pilbara, Western Australia. These herds are phenotyped for fertility, growth, mature weight, flight speed, and live-ultrasound carcass traits. Birth weight and birth date are not recorded in these herds due to the extensive nature of their operations, however an estimated conception date, captured using pregnancy foetal aging techniques, is used to measure female fertility and adjust other traits for age. Birth Weight genomic estimated breeding values are calculated for multiplier animals using data from Nucleus genomic relatives in combined herd analysis.

Genotyping and processing of marker data. At the last AAABG Pitchford *et al.* (2017) presented the results of analysing all of our current and historic animals records and DNA samples (1,104 head combined). These included all animals with phenotypes genotyped on the Illumina GGP Bovine LD chip (versions 3 and 4) and sires genotyped on the Illumina BovineHD chip.

Since our last paper, 579 more animals have have been added from the Popplewell nucleus. All have been genotyped and many have been phenotyped. Also 1,125 animals with genotypes some associated phenotypes have been added from customers' bull multiplier tier herds. Processing of genotypes of calves born in late 2018 in now underway and will further expand the analysis, along with more genotypes and phenotypes from the multiplier herds this year.

The GGP Bovine LD chip is no longer available, so the program has started using a 50K SNP chip from Weatherbys Scientific Australia. This initially presented some difficulties, overcome by imputing up to the original GGP Bovine LD chip.

To create a Genomic Relationship Matrix (GRM) for the statistical analysis, monomorphic SNPs, those with a call rate less than 0.05 and a minor allele frequency (MAF) less than 0.01 are removed, leaving 29,433 SNPS to be used in the analysis. Heterozygosity for each animal is calculated by summing the number of heterozygous genotypes as a proportion of all called genotypes. A standardised matrix of counts for each SNP is generated by subtracting its mean and dividing by its standard deviation. Missing values are replaced by the standardised mean (0). This starting matrix is multiplied by its transpose and divided by the number of SNPs to generate a relationship matrix which is then inverted, ready for trait analysis.

Statistical analysis. As per Pitchford *et al.* (2017); data is analysed using a linear mixed model in ASREML-R (Gilmour *et al.* 2009). Fixed effects are birth year (2008-2017), sex (male or female), dam age (heifer or mature), age (by fitting birth date as a covariate within year), and heterozygosity (% , labelled Het%).

Contemporary groups are defined as management groups within birth year and sex. Management groups for later ages are comprised of current management group and previous management groups as described by Graser *et al.* (2005). Ultrasound traits include day of measurement in the contemporary group and weight as a covariate within contemporary group. Scrotal size included a covariate of age within contemporary group. Lastly, the random animal effects are fitted as the inverse of the GRM.

The carcass, growth, temperament, adaption and carcass traits analysed are; birth weight, weights at 200, 400 and 600 days (kg), tick resistance, log flight time, ultrasound loin eye muscle area (cm^2), P8 fat depth, rib fat depth (mm) and intramuscular fat content (%).

Female fertility is analysed only on naturally mated females as days from joining to calving (Days to Calving or DC) with yearling heifers (HDC) separate from those joined at 2 years or older (mature, MDC). Those that failed to calve have a 32-day penalty added to the maximum DC value in their management group. Dam age and heifer age effects are included in the analysis of for HDC but not MDC, but both include heterozygosity. Mature weight is analysed on lactating cows only, using fixed effects of age in years and heterozygosity.

RESULTS AND DISCUSSION

Significant programming has gone into GenoRater™ to streamline data handling, accommodate for new traits and also new SNP Chip formats since our last report. The analysis continues to perform well at estimating both within and between breed group differences using G-BLUP methods. Using Het% in the analysis continues to be particularly important for female fertility traits, highlighting the importance of the non-additive benefits (hybrid vigour) of composite breeding the program is providing customers, whose breeding objectives' place large economic values on female fertility.

Table 1. Nucleus Traits and Data Volumes as at 15/9/2018

Trait	# Records	Heritability	Standard Error
Birth Weight (kg)	1352	0.51	0.06
200 Day Weight (kg)	1420	0.30	0.06
400 Day Weight (kg)	1380	0.38	0.06
600 WT (kg)	518	0.37	0.12
U Eye Muscle Area (cm)	1361	0.40	0.06
U_P8 Fat (mm)	1359	0.48	0.06
U_Rib Fat (mm)	1361	0.38	0.06
U_Intramuscular Fat (%)	1361	0.26	0.06
Scrotal Circumference (cm)	696	0.47	0.09
Flight Time (Log Sec)	1268	0.15	0.06
Tick Resistance (Score)	589	0.37	0.10
Coat Type (Score)	991	0.72	0.05
Heifer (Yearling Mated)_DC	374	0.32	0.18
Mature DC	1083	0.16	0.04
Mature Cow WT	688	0.58	0.06
Total Animals/Genotypes	1,683		

Although standard errors for most traits are now low, it is expected those higher standard errors for traits Heifer Days to Calving and 600 Day weight will reduce with extra data expected soon.

Genotype and Phenotype volumes have increased significantly from within the nucleus, which is growing in numbers. Pregnancy records, calves and associated data is now starting to flow from the multiplier herds. Multiplier herds will contribute significantly more phenotypes for more traits over the next 12 months.

CONCLUSIONS

The increase in annual numbers of herds, phenotypes and genotypes entering the analysis can be largely attributed to an increase in customer satisfaction and industry demand for the nucleus bred bulls being described and bred objectively, fuelling and funding nucleus herd growth.

Implementing these genomic approaches is allowing the rapid start-up of the integrated multiplier tier herds, as animal pedigrees (in this case for starting base multiplier cows) are not required for (Meuwissen *et al.* 2001; Hayes and Goddard 2011). This was foreshadowed for this program by Pitchford *et al.* (2017) and is now being exploited.

Planned future research includes exploration of using a dominance matrix in the analysis instead of Het% and further incorporation of back-solving to estimate the effect of individual SNPs on traits. There are also plans for new traits and the inclusion of carcass grading data and corresponding genotypes from bull customers' commercial tier herds.

REFERENCES

Burrow H.M., Griffith G,R.,Barwick S.A. and Holmes W.E. (2003) *Proc. Assoc. Advmt. Anim. Breed. Genet.* 294-297

Gilmour A.R., Gogel, B.J., Cullis, B.R. and Thompson, R. (2009) *ASReml User Guide*

Graser H-U., Tier B., Johnston, D.J. and Barwick S.A. (2005) *Aust. J. Exp. Agric.* **45**: 913.

Hayes B.J. and Goddard M.E. (2011) *J. Anim. Sci.* **86**: 2089.

Meuwissen T.H.E., Hayes B.J. and Goddard M.E. (2001) *Genetics* **157**: 1819.

Pitchford W.S., Popplewell G.I. and Tearle R.G. (2017) *Proc. Assoc. Advmt. Anim. Breed. Genet.* **22**: 179.

Quaas R.L. and Pollack E.J. (1980) *J. Anim. Sci.* **51**: 1277.

MERINOLINK/UNE DNA STIMULATION PROJECT: DOUBLING THE RATE OF GENETIC GAIN - WHERE ARE WE AT IN YEAR 2?

S.J. Martin[1,3] and T.Granleese[2,3]

[1]Sally Martin Consulting Pty Ltd, Young, NSW, 2594 Australia
[2]University of New England, Armidale, NSW, 2351 Australia
[3]Meat and Livestock Australia, Armidale, NSW, 2351 Australia

SUMMARY

The MerinoLink/UNE DNA Stimulation Project is a major genetics adoption program in Merino sheep breeding enterprises running from January 2018 to June 2022. The collaborative approach between MLA Donor Company, University of New England, MerinoLink Limited and the project participants has already started to meet the project target of doubling the rate of genetic gain by 2022. The project focusses on working with the project participants to strategically use the genetic and genomic tools currently available. A major component of the adoption strategy has seen the total financial contribution from all participants equating to upward of $1.7 million. In a time of poor seasons recently, this is a significant commitment for participating seed stock breeders to improve their genetic gain.

DNA Parentage and low density genomic testing has allowed Merino seed stock breeders to capture relationships between animals to enhance accuracy of the estimated breeding value predictions, hence improving the rate of genetic gain. In the first year of the project participants have increased the rate of genetic gain within the group by 2.4 Index and are 5 MP+ index points higher than the rest of the MERINOSELECT database when comparing average index improvement.

Flock Profile and RamSelect's Ram Team Manager are tools commercial breeders are using to benchmark the genetic merit of their Merino flocks. This information is aiding participants to make better ram selection and buying decisions to increase genetic merit of their flocks.

INTRODUCTION

On average genetic gain in Merinos is currently low with very large variations across the industry (Granleese *et al.* 2018). Underlying influences are caused by a multitude of factors including inaccurate breeding values (Stephen *et al.* 2018) and/or a lack of understanding of selection theory. The DNA Stimulation Project aims to build understanding and implementation of genetic selection tools. The project focusses on capacity building and working collaboratively at all levels across the industry to communicate how to more effectively use software tools for assisting in the design of breeding programs to increase the rate of genetic gain for participants.

The Project has 30 ram breeders and 18 commercial breeders participating who breed their own rams, 52 commercial breeders who purchase rams and a number of service providers. The project participants are located across New South Wales, Western Australia and Victoria. In total the ram breeders sell 8,500 rams and 19,000 doses of semen annually to over 1,600 ram buying clients. If it is assumed that each ram sold is used for an average of 2.5 mating years and sire 75 lambs per mating year, this means the rams produced by the participant studs will sire ~1.6 million commercial Merino sheep annually. Semen and stud sires used at nucleus level will also influence the broader Merino population directly and indirectly through daughters. Furthermore, commercial participants of the project run a combined number of 575,000 Merino sheep annually.

The aim of the DNA Stimulation Project is to double the annual rate of genetic gain by 2022 (starting in 2018) in line with the National Livestock Genetic Consortium's aims. This paper outlines methods the project uses and provides a progress report of how participants are working towards this goal.

MATERIALS AND METHODS

Project participants sign an agreement to be part of the DNA Project. The project expectations are agreed, and the funding is 58 percent from the project participants and 42 percent from the Meat and Livestock Donor Company.

Key tools used with project participants include DNA parentage, low density genomic tests, Flock Profile genomic tests, Australian Sheep Breeding Values (ASBVs), Sheep Genetics Ramping Up Genetic Gain report (RUGG), Flock Profile (Swan *et al.* 2018), Ram Team Manager (RamSelect.com.au), Rampower within flock indexes and percentile band tables.

The extension and adoption process include a combination of workshops, intensive one-on-one meetings, webinars, phone calls, personal emails, e-newsletters and group email updates. Input for all of these include key personnel within University of New England, Sheep Genetics, Meat and Livestock Australia and participants in the project, including breeders and genetics service providers. Each project participant works through the process of setting an achievable breeding objective which is revisited each workshop or meeting with the project coordinators or service provider. Key questions are covered – are they on track, what is working well (or not), what can be improved, and how are project participants measuring progress?

To measure progress the DNA project uses tools, such as rates of genetic gain, as generated by Sheep Genetics, and workshop feedback, to track the effectiveness of the project. In this project we use the Merino Production Plus (MP+) index as the genetic progress benchmark. "Doubling the rate of genetic gain" will be measured by comparing the five-year genetic gain average compared to the project five year genetic gain average.

RESULTS AND DISCUSSION

A key outcome of the project to date, includes the use of DNA testing technologies to increase the number of ram breeders' participants submitting data to Sheep Genetics with full pedigree. There has been an increase in the number of project participants submitting full pedigree from 25% to 34% between January 2018 and January 2019 shown in Table 1.

Between the start of the DNA Project in late 2017 and the end of April 2019, the project participants have utilised 43,831 DNA parentage tests, 60 Flock Profiles and 11,779 low density genomic tests. The majority of the ram breeders involved the project have now tested their ram breeding nucleus (dams, replacements and sires) with either DNA parentage or low density genomic information. Breeders who have not used DNA parentage testing will maintain their pre-project mothering programs.

Table 1. Average proportion pedigree submission to Sheep Genetics for project participants

	Full Pedigree	Sire Only Pedigree	Dam Only Pedigree	Syndicate Pedigree	No Pedigree
1-Jan-18	0.25	0.41	NA	0.25	0.09
1-Jan-19	0.34	0.38	0.01	0.21	0.07

There is a large range of genetic merit between the ram breeder participants in the DNA Project, with the lowest average MP+ index 110 and the highest 167 MP+ index points. This diversity in the group is being used to share and swap experience to allow informal mentoring within the group.

Overall the collective ram breeder participant index trend from the 2017 to 2018 lamb drop have increased by 2.4 index points. However, the project was not in place when the 2018 drop was conceived, but improved data collection methods were. Hence the project can only claim some of the credit for the improved rate of genetic gain. The 2018 average index is 5.0 points higher than the

Merino industry average (Figure 1). The 2019 drop, which is yet to be measured and submitted to Sheep Genetics will be the first drop of lambs where the project has had full involvement of mating and performance recording.

Ten percent of the ram breeder participants first used Matesel (Kinghorn 2011) for mating allocations for the 2019 joining. The biggest limitation for many of the project participants to utilise Matesel in the past has been the lack of full pedigree data.

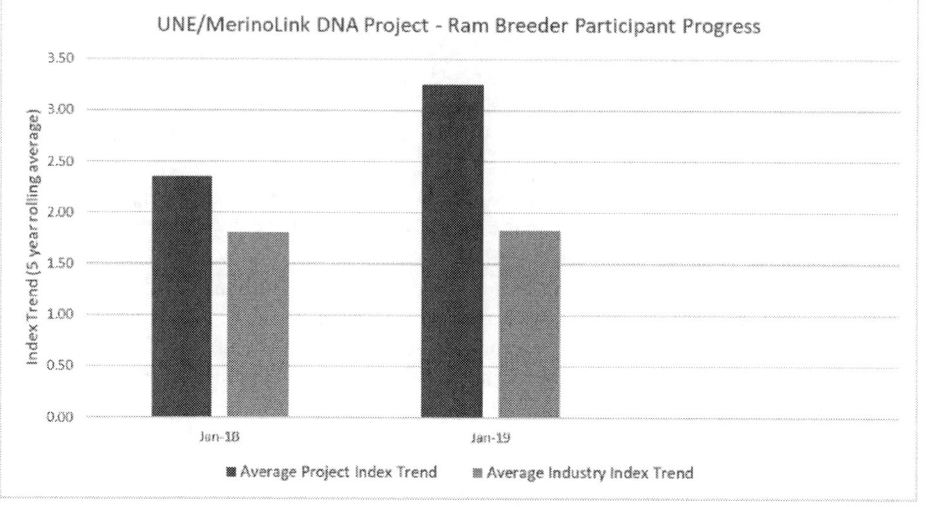

Figure 1. Average index for project participants (grey) and MERINOSELECT members (blue)

Key times identified in the project to provide information, data and advice to assist in decision making are when breeders are setting and reviewing their breeding objectives and developing their mating programs. The majority of the one on-one-ram breeder meetings are planned to coincide with strategic times in their calendar of operations. The one-on-one meetings allow for detailed investigation of the Sheep Genetics' Ramping Up Genetic Gain (RUGG) reports, Sheep Genetics Genetic Trend reports and other information associated with seasonal conditions (constraints).

An annual workshop is run with the project participants. The aim of the ram breeder workshops is to disseminate genetic extension material, update the breeders with their current genetic reports (RUGG) from Sheep Genetics and network with service providers and peers. To date there have been two workshops held for the ram breeders, with 86% able to attend. The overall satisfaction and relevance of the material and content of the workshop has been rated 9.2 and 8.8 out of 10, respectively.

All ram breeder workshop participants were asked what the major obstacles were in implementing changes to enhance progress towards their breeding objective and improve genetic gain. Time and labour ranked first; costs associated with data collection and seasonal conditions ranked second and genomic turnaround time, database integrity, communication and risk vs reward (birth traits) ranked third.

Commercial breeder workshops have also been held, 6 in total to date. These workshops aim to disseminate genetic extension material, review Flock Profile results, update Ram Team Manager and review breeding objectives in order to set actions for the coming 6 to 12 months. Currently 85% of commercial ram breeder participants that attended workshops have at least one ram team registered in RamSelect.

Table 2. Averaged responses to all breeder workshops conducted in 2018

Pre-Workshop Question	Score (Pre)	Post-Workshop Question	Score (Post)
How would you rate your current use of ASBVs when selecting rams for your ram breeding operation?	6.8	After today's workshop how do you rate the importance of using ASBVs in ram selection to reach your breeding objective?	8.8
How would you rate your current use of ASBVs when selecting ewes for your ram breeding operation?	5.4	After today's workshop how do you rate the importance of using ASBVs in ewe selection to reach your breeding objective?	8.4
Are you using the Sheep Genetics website/database to search and sort animals to purchase and/or source semen?	5.5	How confident are you following today's workshop in being able to improve your utilisation of MERINOSELECT and the Sheep Genetics website?	8.5
How would you rate your current knowledge of Matesel?	4.0	After today's workshop, how important do you see the use of Matesel in making the best possible mating decisions?	7.5
How would you rate your current use of Matesel?	3.0	After today's workshop, how confident do you feel to use Matesel in making the best possible mating decisions?	7.2

CONCLUSIONS

Progress is being made, although it is too early to call the project a success. There are positive signs with breeder satisfaction in training, extension and adoption. As more data continues to flow in annually, we will be able to more accurately describe the impact of this unique project.

ACKNOWLEDGEMENTS

The MerinoLink/UNE DNA stimulation project is funded by the Meat and Livestock Australia Donor Company (MDC) [P.PSH.0961], University of New England, MerinoLink and all the project participants. Daniel Brown of AGBU/Sheep Genetics is thanked for providing some data for this paper and his ongoing technical support to the project.

REFERENCES

Chadeau-Hyam M., Hoggart C., O'Reilly P., Whittaker J., De Iorio M. and Balding D. (2008) *BMC Bioinformatics* **9**: 364.

Erbe M., Hayes B.J., Matukumalli L.K., Goswami S., Bowman P.J., Reich C.M., Mason B.A. and Goddard M.E. (2012) *J. Dairy Sci.* **95**: 4114.

Granleese T., Clark S.A., Duijvesteijn N., Bradley P.E., and van der Werf J.H.J. (2018) *Anim. Prod. Sci.* **Published online**: https://doi.org/10.1071/AN17720.

Kinghorn B.P. (2011) *Genet. Sel. Evol.* **43**.

Stephen L., Brown D., Jones C. and Collinson C. (2018) *WCGALP* **11**: 433.

Swan A.A., Gurman P.M., Boerner V., Brown D.J., Clark S.A., Gore K., Granleese T. and van der Werf J.H.J. (2018) *WCGALP* **11**: 346.

AUSTRALIA YOU HAVE FOOTROT, TIME TO START BREEDING AGAINST IT!

S.F. Walkom[1], K.L. Bunter[1], H. Raadsma[2], D.J. Brown[1] and M.B. Ferguson[3]

[1]Animal Genetics & Breeding Unit*, University of New England, Armidale, NSW, 2351 Australia
[2]School of Veterinary Science, University of Sydney, Camden, NSW, 2570 Australia
[3]NeXtgen Agri, Christchurch, New Zealand

SUMMARY

Footrot affects all aspects of sheep production and has substantial welfare and economic impacts in the New Zealand (NZ) and Australian sheep industries. Merino sheep managed in high rainfall environments are particularly susceptible. Funding and support from the NZ fine wool industry has enabled development of a breeding value for footrot susceptibility. Currently, breeding values with a reportable accuracy are available for approximately 30,000 MERINOSELECT animals. However, only 97 of these animals are from Australian flocks. The following manuscript challenges Australian sheep breeders to take advantage of the resource developed by the NZ fine-wool industry and consider their own input into improving the outcomes for Australian breeders looking to reduce footrot susceptibility within their flocks.

INTRODUCTION

Footrot is a highly contagious and difficult to manage hoof disease in sheep and other ungulates that begins with interdigital dermatitis and progresses to separation of the hard horn from the foot (Mulvaney 2013). Both infection and the progression of footrot within the flock are heavily influenced by the prevailing weather conditions and the virulence of the essential infective bacterium *Dichelobacter nodosus* (Egerton and Raadsma 1991). Virulent footrot costs the Australian sheep industry $32M pa (Dhungyel *et al.* 2017) and remains a notifiable disease in many states at the time of this publication. The governmental attitudes to footrot differ between Australia and NZ, and to date the NZ industry (or a section of it) has taken a more proactive approach to reducing footrot susceptibility.

Estimates of heritability for footrot resistance range from 0.10 to 0.30 across populations of Romney (Skerman *et al.* 1988), Australian Merino (Raadsma *et al.* 1994), Scottish Black Face and Mule breeds (Nieuwhof *et al.* 2008) and NZ fine wool Merino (Walkom *et al.* 2018). Accordingly, an opportunity exists to select for reduced susceptibility to footrot. Preliminary research into the development of footrot breeding values for the NZ fine wool industry, and by extension the Australian Merino industry, has been reported (Walkom *et al.* 2018). This paper outlines progression of the development of a footrot breeding value, for Sheep Genetics MERINOSELECT, and the capacity and role of Australian Merino breeders in reducing footrot susceptibility.

GENETIC EVALUATION OF FOOTROT IN NEW ZEALAND MERINO SHEEP

Walkom *et al.* (2018) previously reported the development of research breeding values (RBV: breeding values that are still under development) for the NZ fine wool industry. Briefly, the genetic evaluation of footrot was developed around an initial Central Progeny Test (CPT), commencing in 2013, where approximately 2,000 commercial fine wool NZ Merino ewes are mated to 20-40 Merino type rams annually. Following a standardised protocol for field challenges, resulting wether progeny (and ewes since the 2017 drop) were challenged as hoggets under conditions of virulent footrot. Each foot was then scored on a 5 point scale by trained NZ Merino Company Limited staff, with 0 being not affected, and scores 1 to 5 representing increasing degrees of severity of foot lesions /damage, from

* A joint venture of NSW Department of Primary Industries and the University of New England

water maceration (1) to chronic footrot (5) (Mulvaney 2013). To date the CPT challenge has provided 3,727 footrot phenotypes, and the CPT progeny have been DNA sampled and genotyped (15k Illumina ovine panel). In conjunction with the CPT and in response to the release of sire breeding values in 2016 from the challenged 2013-2014 drops, 16 industry flocks, consisting of Merino, Poll Merino, Polwarth and quarter-bred Merino sheep, have conducted field challenges for yearlings/hoggets and in turn provided a further 8,069 additional phenotypes for genetic evaluation.

The footrot breeding value has evolved since the Walkom *et al.* (2018) publication to a weighted average of the footrot scores, with scores of zero and one grouped with a weighting of 0.5 and score two given a weighting of 1. Thus, an animal with no underrun feet cannot produce a phenotype worse than an animal with a single underrun foot (Score 3+). The phenotypes are also transitioned pre-analysis to a minimum of 40% underrunning, if the challenge strength was below this threshold, as per Walkom *et al.* (2018). Adjusting for differences in disease incidence and transitioning the data to a similar incidence based on the biological progression through footrot scores helps to standardise means and variance across the CPT and industry challenges.

Genetic parameters and breeding values have been estimated from a single-step genomic BLUP (Legarra *et al.* 2014) model incorporating genomic and pedigree information with an equal weighting on the pedigree and genomic information, using WOMBAT (Meyer 2007). The population is predominantly purebred fine wool Merino, with Corriedale and Polwarth influences in some flocks. To account for the effects of breed composition on allele frequencies, the genomic relationship matrix has been adjusted, as per Gurman *et al.* (these proceedings), by utilising a combined NZ Merino and Sheep Genetics genotype reference population and BreedComp (Boerner 2018) to describe an individual animals' breed composition.

To date the analysis produces RBVs, at reportable accuracy of greater than 43% ($\sqrt{h^2} * 0.80$), for just over 30k sheep. The footrot RBVs range from –0.99 to 1.19, with lower breeding values associated with an animal less susceptible to footrot under a severe challenge. Consequently, an RBV of -0.24 (one sd. superior to the mean of 0.00) is associated with a 7% decline in the proportion of progeny expected to present with under-running in at least one foot and 5% decline in progeny expressing at least one chronic foot under severe footrot conditions than the average sire (Figure 1).

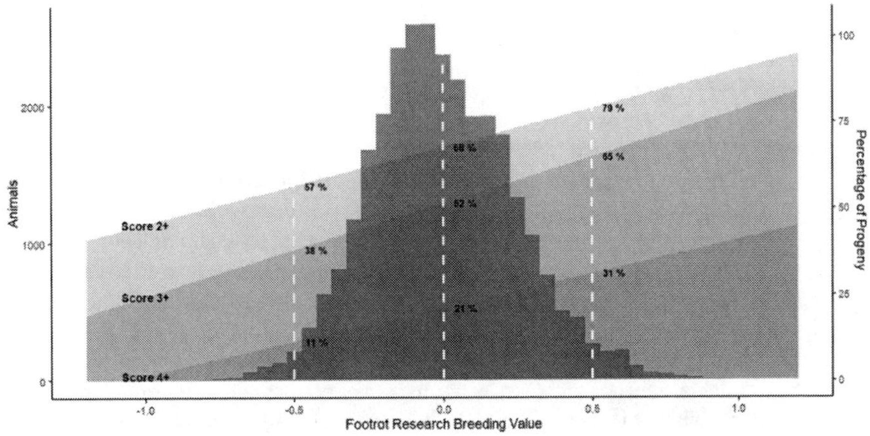

Figure 1. Relationship between footrot RBVs and the expected proportion of progeny to show a minimum of one foot affected (score 2+), underrun (score 3+) or chronic (score 4+) under a severe footrot challenge, as expressed under Central Progeny Test challenge conditions

GENERATING FOOTROT BREEDING VALUES FOR AUSTRALIAN SHEEP

Relying on New Zealand. At the time of publishing, Australian sheep will only get a footrot RBV if an Australian ram has been used in the CPT (32 sires to date), or if a NZ stud has sourced Australian germplasm. Currently, 97 Australian animals (from 27 Australian flocks) have reportable footrot RBVs. The diversity in RBVs for the Australian sires is relatively large and shows no obvious relationship with the major MERINOSELECT indexes: fibre production plus, merino production plus and dual purpose plus (Figure 2). Breeding values reliant on having sires in the progeny test are limited by cost and sires deemed to be relevant to the NZ fine wool industry. An alternative is to sell rams to NZ flocks which are phenotyping their flock for footrot. However, Australian sheep breeders are limited as the trade of live animals between the two countries is currently one-way (from Aus to NZ), which prevents importing evaluated NZ germplasm. Consequently, if they wish to progeny test a sire and then use it on farm they are limited to selling semen into NZ.

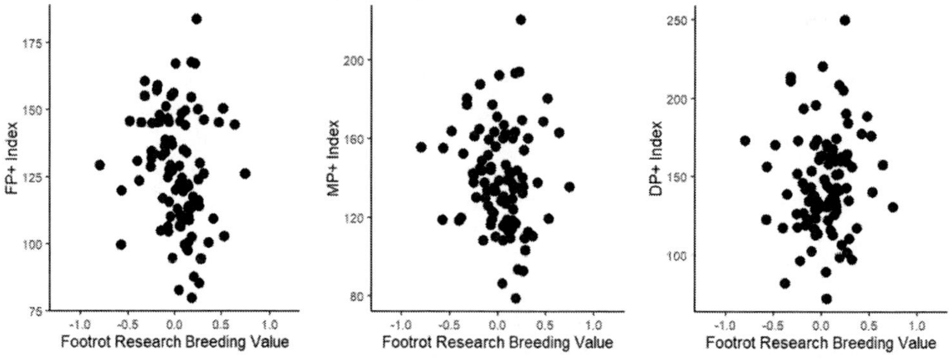

Figure 2. Footrot RBV plotted against index values for fibre production plus (FP+), merino production plus (MP+), and dual purpose plus (DP+) for Australian Merino sires with publishable footrot RBVs

Structured Australian reference flock. The establishment or addition of footrot recording to an existing reference flock of a current reference population, similar to the CPT in NZ, is an appealing option for Australian sheep breeders. This would increase the number of animals with phenotypes collected under common environmental footrot challenge, without the need for outbreaks of footrot on their own property. However, while projects based on large reference populations typically supply good quality data for hard to measure or novel traits, they are limited in the number of records that can be produced and are expensive on a per animal basis. Breeders will also have limited capacity to progeny-test all sires of interest since choice of sire is primarily based on industry good, not their own. Progeny testing provides highly accurate comparisons of sires but require scale and careful design to provide appropriate levels of genetic linkage. Organisers of current resource flocks may also be reluctant or unable to incorporate footrot phenotypes into existing programs. Challenging progeny with footrot is likely to influence the expression of genetic variation in production traits and due to the notifiable status of the disease, animals may not be able to return to breeding flocks. Challenging animals with a disease deliberately is a potential welfare issue and will require ethics approval, something that can become difficult for industry flocks.

Industry recording in Australia. In a single-step genetic evaluation, any animal that has both a genotype and phenotype forms part of the reference population. The phenotypes for such animals are often being collected independently of the need for a reference population and as such should be cost effective if the phenotype is part of standard management / recording practices. However, the notifiable status of the disease in some states and the social stigma of having footrot (something NZ breeders have moved beyond) means that implementing a footrot recording program on-farm in Australia will not be undertaken lightly, even by early adopters. Similar to recording phenotypes for worm egg count, the ability to record the footrot phenotype is challenging as it depends on both suitable environmental conditions and the presence of appropriate strains of *Dichelobacter nodosus*. However, by recording in diverse industry cohorts, multiple strains of the bacterium will be captured within the challenges. As has been shown by NZ breeders, the recording protocols are achievable and reliable footrot phenotypes can be captured by industry for genetic evaluation. The advantage of industry recording is that a phenotype and in turn a highly accurate breeding value can be achieved at a young age i.e. before the animals become selection candidates.

Genomic Prediction. A suitable reference population where large numbers of animals are phenotyped and genotyped underpins the accuracy and utility of any genomic enhanced prediction. Currently, under the single-step model, industry animals that have a genomic relationship with the reference population (from non-phenotyping flocks) are obtaining RBV accuracies ranging from 0.09 to 0.70, with a mean accuracy of 0.22. Low accuracies are expected for most Australian flocks if phenotyping only occurs in NZ, reflecting relatively low genetic relationships between animals in the two countries. As the reference population grows, hopefully including Australian data, the mean accuracy of genomic predictions will improve, and accordingly the potential contribution of genomics to industry breeding programs.

CONCLUSIONS

NZ fine-wool breeders have lead the way in championing the development of a footrot RBV and its likely future publishing as an ASBV in MERINOSELECT. Australian sheep breeders will need to move beyond the social stigma associated with footrot and work together to develop a genomic reference of footrot phenotyped animals within Australia if they wish to take advantage of the current research. If this is achieved, Australian sheep breeders will be able to join their contemporaries in NZ in selecting effectively to reduce footrot susceptibility.

ACKNOWLEDGEMENTS

The authors wish to acknowledge the contribution of New Zealand Merino Company, New Zealand Merino Incorporated and Ministry for Primary Industries (MPI) for their funding contribution at different stages of the project. The authors also wish to acknowledge the New Zealand fine-wool flocks for their contribution of phenotypes, genotypes and pedigree information and their support for the CPT.

REFERENCES

Dhungyel O., McPherson A. and Whittington R. (2017) MLA Report B.AHE.0224

Gurman P.M., Bunter K.L., Boerner V., Swan A.A. and Brown D.J. (2019) *Proc. Assoc. Adv. Anim. Breed. Genet.* **23**: these proceedings.

Mulvaney C. (2013) 'A Guide to Management of Footrot in Sheep' 2nd ed. New Zealand Merino, Christchurch, New Zealand.

Nieuwhof G.J., Conington J., Bűnger L., Haresign W. and Bishop S.C., (2008) *Anim.* **2**: 1289.

Raadsma H.W., Egerton J.R., Wood D., Kristo C. and Nicholas F.W. (1994) *J. Anim. Breed. Genet.* **111**: 367.

Skerman T.M., Johnson D.L., Kane D.W. and Clarke J.N., (1988) *Aust. J. Agric. Res.* **39**: 907.

Walkom S.F., Bunter K.L., Raadsma H., Brown D.J., Gibson W., Swan A.A., Borner V. and Ferguson M.B. (2018) *Proc. Wld. Congress Gene. App. Livestock Prod.* 261-266

EVOLUTION OF GENETIC EVALUATION FOR EWE REPRODUCTIVE PERFORMANCE

K.L. Bunter[1], A.A. Swan[1], P.M. Gurman[1], V. Boerner[1], A.J. McMillan[1] and D.J. Brown[1,2]

[1]Animal Genetics & Breeding Unit*, University of New England, Armidale, NSW, 2351 Australia
[2]Meat and Livestock Australia, Armidale, NSW, 2350 Australia

SUMMARY

This paper describes characteristics of the new analysis for reproductive traits of ewes, as applied to data from maternal breeds. Components of reproduction are treated as different traits for yearling and adult ewes. The new procedures produce breeding values for conception, litter size, rearing ability of ewes, maternal behaviour and condition scores from a multi-breed, multi-trait analysis, using industry and information nucleus phenotypes, genotypes and single-step methodology. Breeding values will be released as research (interim) breeding values by Sheep Genetics while further refinements are made. Future implementation is targeted for Merinos.

INTRODUCTION

Prior to 2007, the phenotype for reproductive performance of a ewe analysed by Sheep Genetics was a single record reflecting the number of lambs weaned per lambing opportunity. This phenotype was inferred from lamb pedigree. Only ewes joined to produce pedigreed lambs over their lifetime obtained an accurate record. In addition, the trait definition was imprecise, as it did not allow for important systematic and temporal effects influencing annual outcomes, it did not capture lamb losses, and contemporary comparisons were also limiting. Since 2007, reproductive performance of ewes has been evaluated using annual phenotypes for the number of lambs born (NLB) or weaned (NLW) (Brown *et al.* 2007). These are also composite traits which combine the components of conception (CON) and litter size (LS), along with the ability of the ewe to successfully rear lamb(s) (ERA). That is, NLB is a function of CON+LS, while NLW is a function of CON+(LS×ERA). Despite better trait definition and models, data for NLB and NLW were frequently censored due to incorrect recording for one or more of the above component traits. Information regarding dry ewes, lambs born dead and an update to individual lamb rearing outcomes were required to correctly identify all components contributing to NLB and NLW.

To improve identification of dry ewes and litter size when lambs were not individually recorded, pregnancy scan data became a secondary data source in 2017. However, inaccurate identification of lambs born dead, lamb survival outcomes and incorrect contemporary grouping across multiple data sources (e.g. lambs vs scans) and component traits (e.g. joining vs lambing contemporary groups) remained problematical. In this paper, we describe new procedures for genetic evaluation of the component traits contributing to ewe reproductive performance. These procedures provide better use of data recorded by breeders subject to both data limitations and opportunities and provide breeding values for the above component (and other) traits in a multi-breed, multi-trait format, using phenotypes, genotypes and single step methodology.

MATERIALS AND METHODS

Inventory-based recording commenced in 2009, giving breeders an opportunity to provide additional detail on joining dates, service sire(s), pregnancy scan outcomes, and management groups (mgp) specified for joining and lambing separately. Component traits were defined annually for joined ewes

* A joint venture of NSW Department of Primary Industries and the University of New England

as CON: 0=failed to conceive, 1=conceived; while LS (1 to n lambs born) and ERA (lambs surviving/ lambs born) were defined for ewes which lambed. For ewes without lambs recorded, CON and LS could be assigned using pregnancy scan results (Bunter *et al.* 2016). After further consultation with breeders, values for CON were altered to 0 if conception occurred more than 60 days after joining commenced, to better reflect commercial targets.

For the reproductive traits, records with obvious data errors and in small contemporary groups were discarded. Additional data recorded for ewes included maternal behaviour score (MBS: from 1: good to 5: poor), as described by Brown *et al.* (2016), and pre-joining weight (WT) and condition score (CS) recorded within the 30 days prior to joining. Data further describing body composition and development, thought to be relevant to reproductive outcomes, include scanned post-weaning carcase fat (PFAT) and eye muscle depth (PEMD), along with post-weaning (PSC) or yearling (YSC) scrotal circumferences. Data for these four traits were extracted from the Sheep Genetics database only for those flocks included in analyses of reproductive performance for maternal breeds. Data from the information nucleus research flocks (van der Werf *et al.* 2010) was also used to commence a genomic reference population for reproductive traits, augmented by more recent genotyping of industry animals with phenotypes.

Based on previous analyses (Bunter and Brown 2013), yearling and adult performances of CON, LS and ERA are treated as separate traits. The basic model for non-genetic effects influencing reproductive traits includes cgp + age, where cgp refers to joining (CON, LS) or lambing (ERA) contemporary groups (based on site-year-timegp-mgp details) and age represents the ewe's age in years (adult ewes). Timegp is assigned based on dates to accommodate evidence of gaps between separate but unspecified joining events. Contemporary groups were further refined (Bunter *et al.* 2017) to include: 1) month of birth and dam status (yearling or adult) in the cgp for yearling traits, and 2) previous status (no lamb, lambed and lost or weaned, unknown) in the cgp for 2yo traits, enabling flock specific effects with respect to these factors. Litter size group (1, 2, 3 or more, unknown) at birth was fitted as an additional model term for ERA, since litter size alters the rearing challenge for ewes (Bunter *et al.* 2017). Additional model terms included birth-rear type group for yearling but not adult reproductive traits. Regressions on age, but not weight, were applied to raw data for PFAT, PEMD, PSC and YSC where significant ($p<0.05$), modelled using contemporary groups previously defined for these traits (Brown *et al.* 2007).

Following a series of univariate and bivariate analyses to estimate genetic parameters for each trait and trait combination, covariance matrices were adjusted for consistency. After preliminary investigation, the genomic relationship matrix (G) was also partitioned into 15 breed sub-groups to improve consistency of genomic with pedigree-based relationships (described by the numerator relationship matrix, A), after correcting obvious anomalies (e.g. unidentified or incorrect pedigree parents) (Gurman *et al.* 2019). Single step procedures were implemented constructing the H-matrix according to Aguilar *et al.* (2010). The research breeding values (RBVs) were subsequently estimated using RACEMULE software (AGBU), which enables data to be analysed fitting desired model terms directly to the data, without pre-adjustment. The predictive capacity of the resulting RBVs was assessed by regressing adjusted phenotypes (post 2013) on sire breeding values estimated from data recorded before 2013. Estimates of variances due to genetic groups are described in Bunter *et al.* (2019). The reportable traits from these analyses are confined to RBVs for conception, litter size, rearing ability, maternal behaviour and condition scores, since ASBVs for all other traits are already reported for a larger sample of flocks through Sheep Genetics.

RESULTS AND DISCUSSION

The new procedures dramatically increase the number of animals with reproductive phenotypes, most commonly litter size (N>700K), compared to a comparable export of data used for analyses of NLB and/or NLW (N<350K). The relative magnitude and direction of the estimated genetic and phenotypic correlations between traits for maternal breeds is shown in Figure 1. Traits correlated with ewe conception (PSC, YSC) and ERA (MBS) enhance the accuracy of prediction for these traits through low to moderate genetic correlations. Traits reflecting individual variation in post-weaning body composition (PFAT and PEMD) were moderately correlated with pre-joining weight and condition score for ewes, and their inclusion improves accuracy of breeding values for YCS/CS. However, after accounting for the model terms noted earlier (e.g. age, dam age, birth-rear type status, and previous status), all body composition traits showed modest to negligible genetic correlations with reproductive traits in this data. Therefore, body composition traits are not strongly predictive of genetic merit for component traits in maternal breeds and recording reproductive outcomes directly is important for improving accuracy of evaluation for reproductive performance.

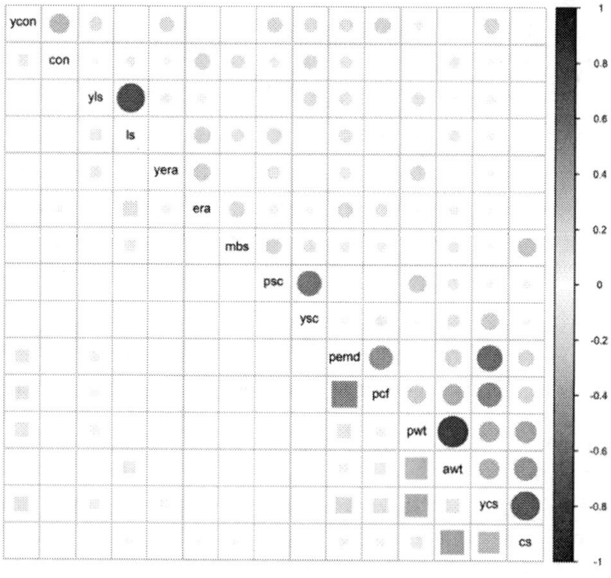

Figure 1. Depiction of genetic (circles) and phenotypic (squares) correlations between traits for MATL analyses. Size and colour of icons reflects the magnitude and direction of correlations (white=negligible correlation)

The distribution of RBVs for reportable traits from a recent analysis for 2017 drop animals is illustrated in Table 1. Even though reproductive traits tend to have low heritability, genetic variation is quite large and there were some useful differences amongst young selection candidates with respect to reproductive component traits. This demonstrates that opportunities exist to select amongst recent candidates for different components of reproduction. Regression coefficients for daughter performance on sire EBVs (Table 1) were generally positive, as expected, and approaching the expected value of 0.5 when backed by significant data contributing to both the sire RBVs and post-2013 validation data (e.g. YLS). Regression coefficients for binary (YCON, CON) traits analysed with linear models are frequently under-predicted (Zhao *et al.* 2001).

Table 1. The distribution of RBVs for animals born in 2017 for reportable traits (with units below) from the MATL analyses, along with the coefficient for the regression of daughter performance (N) on sire RBVs

	Trait RBVs at percentile thresholds								
Percentile	YCS	CS	YCON	CON	YLS	LS	YERA	ERA	MBS
	Score	score	%	%	lambs	lambs	%	%	score
Top	0.85	0.69	30.8	10.1	0.39	0.40	11.8	8.91	-0.45
1%	0.33	0.16	14.3	3.9	0.16	0.18	4.87	3.30	-0.11
5%	0.26	0.11	11.0	3.1	0.13	0.14	3.83	2.58	-0.06
10%	0.20	0.09	8.3	2.5	0.10	0.12	2.95	2.08	-0.04
20%	0.14	0.07	6.0	2.0	0.08	0.09	2.09	1.67	0.00
40%	0.05	0.04	2.1	0.9	0.03	0.03	0.64	0.78	0.04
50%	0.02	0.02	0.5	0.3	0.01	0.00	-0.09	0.27	0.06
60%	-0.03	0.00	-1.0	-0.4	0.00	-0.03	-1.08	-0.24	0.09
80%	-0.12	-0.08	-4.6	-2.0	-0.04	-0.10	-2.90	-1.51	0.15
90%	-0.21	-0.16	-10.4	-3.1	-0.07	-0.14	-3.98	-2.61	0.20
Bottom	-0.51	-0.39	-17.8	-6.5	-0.16	-0.24	-7.36	-4.74	0.46
N	1586	7399	5776	3900	16678	51603	12524	29337	7533
Coefficient	-0.01	0.38	0.07	0.38	0.54	0.37	0.26	0.29	0.16

CONCLUSIONS

The data pipeline and software have been developed to improve evaluation of genetic merit for reproductive outcomes of ewes, allowing a substantially larger proportion of the maternal sheep population to receive more accurate breeding values for reproductive performance. Moreover, the new breeding values for component traits will enable breeders to make more precise selection decisions (e.g. litter size versus lambs surviving). Where data were sufficient, regressions of offspring performance on sire EBVs for reproductive traits generally supported predictive capacity for reportable traits from the new analyses. Accurate recording of reproductive data will underpin the procedures for genetic evaluation of reproductive performance. Similar improvements will be implemented for other breeds (e.g. Merinos) in the near future.

REFERENCES

Aguilar I., Misztal I., Johnson D.L., Legarra A., Tsuruta S. and Lawlor T.J. (2010) *J. Dairy Sci.* **93**: 743.

Brown D.J., Huisman A.E., Swan A.A., Graser H.U., Woolaston R.R., Ball A.J., Atkins K.D. and Banks R.G. (2007) *Proc. Assoc. Advmt. Anim. Breed. Genet.* **17**: 268.

Brown D.J., Fogarty N.M., Iker C.L., Ferguson D.M., Blache D. and Gaunt G.M. (2016) *Anim. Prod. Sci.* **56**: 767.

Bunter K.L. and Brown D.J. (2013). *Proc. Assoc. Advmt. Anim. Breed. Genet.* **20**: 82.

Bunter K.L., Newton J.E. and Brown D.J. (2017). *Proc. Assoc. Advmt. Anim. Breed. Genet.* **22**: 565.

Bunter K.L., Swan A.A., Brown D.J. (2019). *Proc. Assoc. Advmt. Anim. Breed. Genet.* **23**: (in press)

Bunter K.L., Swan A.A., Brown D.J., Brien F.D. and Smith J. (2018) *Anim. Prod. Sci.* **58**: 791.

Bunter K.L., Swan A.A., Purvis I.W. and Brown D.J. (2016) *Anim. Prod. Sci.* **56**: 679.

Gurman P.M., Bunter K.L., Boerner V. and Swan A.A. (2019) *Proc. Assoc. Advmt. Anim. Breed. Genet.* **23:** (in press).

Van der Werf J.H.J., Kinghorn B.P. and Banks R.G. (2010) *Anim. Prod. Sci.* **50**: 998.

Zhao L., Chen Y. and Schaffner D.W. (2001). *Appl. Environ. Microbiol.* **67**: 2129.

THE INFLUENCE FEED COST HAS ON CHANGING BEEF CATTLE GREEN-HOUSE GAS EMISSIONS

B.J. Walmsley, A.L. Henzell and S.A. Barwick

Animal Genetics & Breeding Unit*, University of New England, Armidale, NSW, 2351 Australia

SUMMARY

Genetic trends are presented for estimated greenhouse gas (GHG) emissions of young Angus animals at pasture and in the feedlot, and of Angus cows at pasture for a self-replacing, 100d-finished production system. GHG emissions are predicted to have increased over time, accompanying genetic gains in productivity traits and feed intake. The trends support the need for multiple trait selection that appropriately considers feed intake and the whole production chain. The results show the cost of feed used in the breeding objective impacts on the GHG emissions reductions that can be achieved with selection. Small reductions in GHG emissions can be achieved when feed is expensive, e.g. $130/t, and carbon is priced at $0/t. When feed is inexpensive GHG emissions increase and an $80/t carbon price is needed to make GHG emission changes negligible.

INTRODUCTION

The need to reduce greenhouse gas (GHG) emissions from livestock, especially beef cattle, is the subject of increased societal pressure due to concern about climate change. In the absence of feed efficiency improvement, Walmsley *et al.* (2017) showed that productivity gains in Australian Angus cattle that have occurred through breeding are associated with increases in feed intake. The association between feed intake and GHG emissions was shown by Blaxter and Clapperton (1965). Many studies have examined reducing GHG emissions genetically, but often they inadequately describe feed intake, only examine responses for a particular trait, or do not focus on the whole production chain. This paper examines genetic trends in estimated GHG emissions in Angus cattle and shows how multiple-trait selection strategies can reduce GHG emissions associated with feed intake for the whole production chain.

METHODS

Breeding objectives. Breeding objectives for net return per cow were derived with BreedObject (Barwick *et al.* 2005) for a 100d feedlot-finished (self-replacing cow herd at pasture, steers finished at 640 kg live at 22 months) production system. Traits in the breeding objective were sale weight, dressing %, saleable meat %, rump fat depth, marbling score, feedlot entry weight, weaning weight (direct & maternal), calving ease (direct and maternal), mature cow weight, cow weaning rate, residual feed intake when pasture availability is limited, residual feed intake when pasture availability is greater than required, residual feed intake in the feedlot, and cow condition score. When GHG emissions were part of the breeding objective, the cost of emissions was included in the economic value of each breeding objective trait. Residual GHG emission traits were not included in this paper as EBVs for these are not currently available and little is known about their association with other traits. The general form of the economic value calculated for traits was Δ returns – Δ feed costs – Δ non-feed management costs. The feed requirement associated with a unit change in each objective trait was estimated using the equation systems described by Freer *et al.* (2007).

Trends in methane production. EBVs for the breeding objective traits were predicted from BREEDPLAN EBVs published in January 2017 for 1,895,481 Angus animals born from 1985 through

* A joint venture of NSW Department of Primary Industries and the University of New England

2015, and summarised by year of birth. Predictions use the relation $\hat{\mathbf{g}} = \hat{\mu}\,\mathbf{G}_{11}^{-1}\mathbf{G}_{12}$, where $\hat{\mathbf{g}}$ and $\hat{\mu}$ are breeding objective trait EBVs and BREEDPLAN EBVs, respectively, and $\mathbf{G}_{11}$ and $\mathbf{G}_{12}$ are genetic covariances among BREEDPLAN EBVs and between these and the breeding objective traits. Genetic parameters used were derived from industry and literature estimates and are those used for developing Angus indexes in Australia. Genetic trends in estimated total feed intake (excluding any period of surplus feed) per animal were as presented by Walmsley *et al.* (2017). The trends in methane production are derived using the phenotypic relationships between feed intake and methane production of Charmley *et al.* (2017) when animals are at pasture, equation 7 of Johnson *et al.* (2017) when animals are in the feedlot, and the recommended global warming constant for methane (28; Edenhofer *et al.* 2014) for determining kg of CO_2-equivalent.

Production system responses to selection. When included in the breeding objective the cost of feed has been shown to have important impacts on index rankings of beef cattle (Walmsley *et al.* 2018). In this paper, feed was costed at \$130/t (expensive) and lowered by 30% (\$91/t; inexpensive) to demonstrate the impact cost of feed has on GHG emissions. The carbon price used in the breeding objective varied with feed price. When feed price was expensive the carbon prices examined were \$0, \$20 and \$40/t. When feed price was inexpensive the carbon prices examined varied between \$0 and \$80/t. More detail can be found in Barwick *et al.* (2019).

Changes in individual traits after a single generation of selection were calculated assuming a selection intensity of $i = 1$, using the MTIndex software of J. van der Werf. These changes were used to determine the changes predicted to occur in feed intake and CO_2-equiv. Changes in total GHG emissions and net return were summed over the breeding objective traits for the young animal at pasture, young animal in the feedlot and the cow. These were also summed to give changes for the whole production system. GHG emissions were also expressed per unit of feed intake and per unit of product for the whole production system.

RESULTS AND DISCUSSION

Figure 1a demonstrates genetic changes that have occurred in the breeding objective liveweight traits (sale weight, feedlot entry weight and weaning weight) based on Australian Angus EBVs. Genetic merit for all these traits increased between 1985 and 2015 with sale weight having the highest rate of increase. Figure 1b shows that predicted GHG emissions have also increased, both at pasture and in the feedlot, from increases in feed intake associated with growth potential increases. The total GHG emissions at pasture are higher because the period of time at pasture is longer than the period in the feedlot. Figure 1c shows that young animal GHG emissions per day are greater in the feedlot than in young animals at pasture. Figure 1d shows the trends in GHG emissions per day for the cow alone and for the cow and calf prior to weaning.

Figure 2a shows that when feed is \$130/t and carbon has a price of \$0/t in the breeding objective a small reduction in total GHG emissions is predicted when selecting for multiple-trait merit. As carbon price increases further reductions in GHG emissions are predicted. In contrast, when feed is inexpensive and carbon is \$0/t in the breeding objective, increases in GHG emissions are predicted as a response to gains in live weight traits. GHG emissions do not become negligible when feed cost is inexpensive until the carbon price reaches \$80/t. Figures 2b, c and d illustrate that GHG emissions per unit feed, per unit product, and net return when the feed cost is inexpensive do not reach levels comparable to that when the feed cost is expensive until the carbon price is \$60-\$80/t.

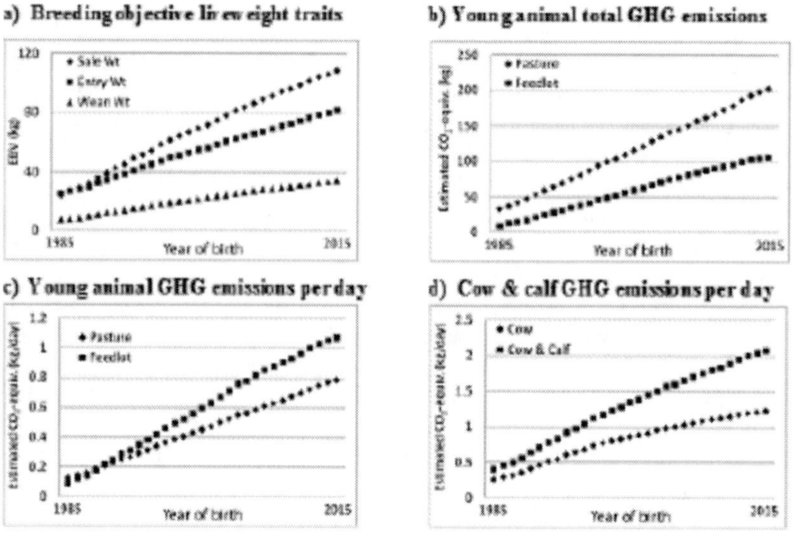

Figure 1. Genetic trends in liveweight traits of the breeding objective, young animal overall and daily estimated GHG emissions at pasture and in the feedlot, and cow and cow/calf emissions at pasture for Angus cattle for a self-replacing cow herd with steers 100-d feedlot finished. Trends exclude any period of surplus feed

The trends in Figure 1 illustrate that greenhouse gas emissions have increased as a result of genetic change in productivity traits and feed intake. Compared to 1985, the young animal at pasture in 2015 produces an extra 0.66 kg CO_2-e per day which equates to ~170 kg CO_2-e over the pasture period. The animal in the feedlot is producing an extra 0.98 kg CO_2-e per day or 98 kg CO_2-e during the 100 days in the feedlot. The cow in 2015 is producing 0.97 kg CO_2-e per day extra and when combined with her calf to weaning they produce an extra 1.67 kg CO_2-e per day. When viewed in association with the unfavourable trend in feed efficiency shown by Walmsley *et al.* (2017) these GHG emissions trends illustrate that there is a need to focus on multiple trait selection to improve feed efficiency and GHG emissions along with productivity. The association between feed intake and GHG emissions means that either trait can be recorded and this will be expected to provide useful information on the other trait that can be used in selection.

CONCLUSION

The production system responses in Figure 2 illustrate that feed price and carbon price have important impacts on GHG emissions changes under multiple-trait selection. When feed is \$130/t, decreases in GHG emissions are predicted even with a \$0/t carbon price. When feed is considered cheap and carbon is not costed, increases in GHG emissions are predicted to occur in response to liveweight and feed intake increases. GHG emission reductions are not predicted unless carbon is priced at ~\$80/t. These results further support that animal genetic improvement and pasture stocking rate management need to be considered jointly. When stocking rates are low and genetic improvement has the extra benefit of improving pasture utilisation by increasing growth potential, GHG emissions are likely to increase. At high stocking rates, when pasture utilisation and the cost of feed are inherently higher, multiple-trait selection can drive genetic change and limit feed intake and GHG emissions increases across the whole production system.

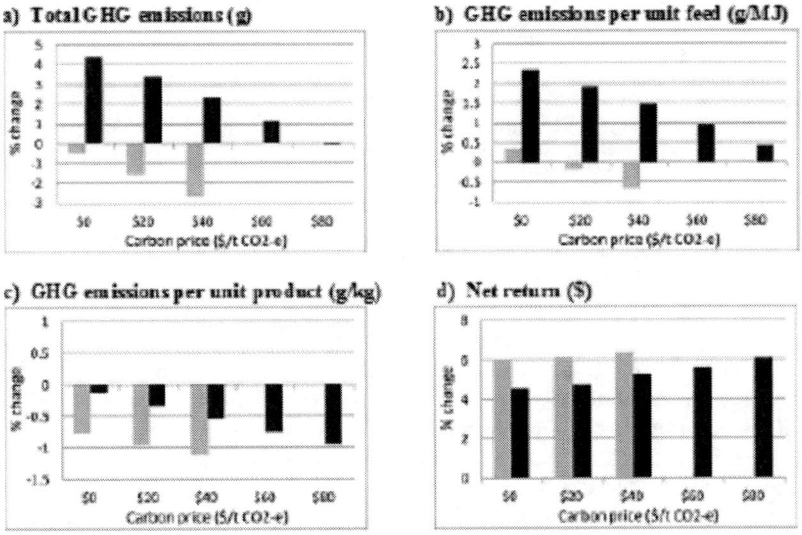

Figure 2. The effect of carbon price on production system responses to selection when feed price is expensive ($130/t; grey bars) or inexpensive (30% lower; solid bars) for Angus cattle from a self-replacing cow herd with steers 100-d feedlot finished

ACKNOWLEDGEMENTS

We thank the Angus Society of Australia for data access, and NSW Dept. of Primary Industries and Meat & Livestock Australia for financial support.

REFERENCES

Barwick S.A. and Henzell A.L. (2005) *Aust. J. Exp. Agric.* **45**: 923.

Barwick S.A., Henzell A.L., Herd R.M., Walmsley B.J. and Arthur P.F. (2019) *Genet. Select. Evol.* **51:** 18.

Blaxter K.L. and Clapperton J.L. (1965) *Br. J. Nutr.* **19**: 511.

Charmley E., Williams S.R.O., Moate P.J., Hegarty R.S., Herd, R.M., Oddy V.H., Reyenga P., Staunton K.M., Anderson A. and Hannah M.C. (2016) *Anim. Prod. Sci.* **56**: 169.

Edenhofer O., Pichs-Madruga R., Sokona Y., Kadner S., Minx J.C. and Brunner S. (2014) In 'Climate change 2014: Mitigation of climate change', Cambridge University Press, New York.

Freer M., Dove H. and Nolan J.V. (Eds) (2007) 'Nutrient requirements of domesticated ruminants'. CSIRO Publishing, Collingwood Vic.

Johnson D.E., Hill T.M., Ward G.M., Johnson K.A., Branine M.E., Carmean B.R. and Lodman D.W. (1993) In 'Atmospheric methane', pp 199-229, editor M.A.K. Khalil, Springer-Verlag.

Walmsley B.J. and Barwick S.A. (2018) *Proc. World Congr. Genet. App. Livest. Breed.* **11**: 866.

Walmsley B.J., Henzell A.L. and Barwick S.A. (2017) *Proc. Assoc. Advmt. Anim. Breed. Genet.* **22**: 61.

COSTS OF AND RETURNS FROM PERFORMANCE RECORDING IN BEEF AND SHEEP STUDS

R.G. Banks[1], Yue Zhang[2], Y. Zhang[1] and J. Shovelton[3]

[1]Animal Genetics & Breeding Unit*, University of New England, Armidale, NSW, 2351 Australia
[2]UNE Business School, University of New England, Armidale, NSW, 2351 Australia
[3]Meridian Agriculture, 96 Harbours Road, Yendon, Victoria, 2253 Australia

SUMMARY

Genetic improvement of the Australian beef herd and sheep flock depends heavily on investment by stud producers in performance recording. A survey was conducted to estimate levels of investment and profitability in beef and sheep studs in Australia, and to assess whether these differed between studs using BREEDPLAN or Sheep Genetics and those not. A voluntary survey was developed and invitation to participate widely circulated. The participation rate was low, but participants included a sample of the largest users of BREEDPLAN and Sheep Genetics. Results suggest that BREEDPLAN and Sheep Genetics users have larger businesses in both scale and level of investment, and those businesses are more profitable, than non-users. Knowledge of operating margins and return on investment is valuable in understanding seedstock sector capacity to increase investment, particularly in hard-to-measure traits and genotyping. On the basis of the results here, such capacity appears likely to be constrained. Obtaining more comprehensive survey results would be valuable, but it is not obvious how to achieve that.

INTRODUCTION

Commercial sheepmeat and/or grassfed beef producers derive their long-term genetic improvement from ram and bull breeders, and the higher the rate of genetic progress those breeders can achieve, the better the commercial producers are able to meet or beat the cost-price squeeze. An essential ingredient of achieving high rates of genetic gain in the ram- and bull-breeding enterprises is the investment in performance recording both on- and off-farm. How much the ram- and bull-breeder are prepared to invest must depend in part on their operating margins – the costs of recording have to be covered by sales of bulls and rams.

Particularly as breeding moves more fully into the genomics era, the demand for high quality recording will rise. This is because of the dependence for genomic prediction on the existence of a reference population, which should be recorded for all traits in the breeding objective. Archer *et al.* (2004) estimated performance recording costs in beef seedstock businesses in Australia, but no comprehensive data on the costs of recording, of operating margins for breeders, and the variation in those costs and margins, is available. This is important information for the industry, since the levels of these parameters will influence the capacity of the ram- and bull-breeding sectors to scale up investment in recording, particularly for hard-to-measure traits. If that capacity is limited, this will place limits on the accuracy of genomic selection that can be achieved.

This project aimed to inform industry about the average of and variation in investment levels in recording costs and in operating margins. This knowledge could potentially assist in developing strategies for increasing adoption of genomic technologies. If it is the case that there is limited scope for increasing volume and quality of recording, this will limit the benefits that can be delivered to the commercial production sector, and accordingly, novel mechanisms to facilitate increased recording may need to be considered.

---

* A joint venture of NSW Department of Primary Industries and the University of New England

MATERIALS AND METHODS

A survey was developed with inputs from ABRI and Sheep Genetics staff, and an invitation to participate in the survey distributed widely through breed societies in sheep and beef cattle, through ABRI outlets, via Sheep Genetics and via sheep Breed Societies. The survey questions covered:

- Scale of business, including number of breeding females and numbers of rams and ewes sold
- Returns from sales, both per unit of genetic material sold and in total
- Operating costs including both variable costs and overheads including plant and equipment
- Level of recording and associated costs
- Participation or not in BREEDPLAN and Sheep Genetics

The survey included up to 82 items of information (sheep) and 42 items of information (beef), and 18 questions directly relating to income or costs.

The initial aim was to obtain data from 300 studs in each of beef and sheep, including at least 30 in each species not using BREEDPLAN or Sheep Genetics. With the aim of increasing participation rate, direct calls were made to 1,450 sheep studs (605 Merino, 539 Dorset and Poll Dorset, 134 Suffolk, 178 Border Leicester) and 1,670 cattle studs (197 Hereford, 307 Angus, 143 Simmental, 215 Charolais, 157 Santa Gertrudis, 2,633 Brahman, 157 Droughtmaster, 189 Brangus and 52 Braford), and the survey sent directly to over 350 sheep studs and 200 cattle studs.

RESULTS AND DISCUSSION

Survey participation. Response to the survey was not as high as hoped. Invitations to participate were distributed twice, based on limited response to the first call. In total, 68 responses were received from beef studs, and 81 from sheep studs, including 26 and 51 from non-BREEDPLAN and non-Sheep Genetics users respectively. The number of respondents not using BREEDPLAN or Sheep Genetics were close to or more than the targets for these groups. Numbers of respondents from BREEDPLAN or Sheep Genetics users were much smaller than the target.

These overall response rates (around 4% of BREEDPLAN and Sheep Genetics users) are lower than the levels quoted as typical for surveys (10-15% for external surveys of values, political opinions etc). This may reflect the amount and/or nature of the questions. Despite the low response rate, the survey results provide some basis for estimating investment levels and return on investment for the larger users of BREEDPLAN and Sheep Genetics. The herd and flock size distributions among participants are similar to those of the largest 10-15% of all BREEDPLAN herds and 25-30% of all Sheep Genetics users.

Characteristics of seedstock businesses. Table 1 includes data on scale and financial parameters of the 4 categories of businesses.

In both species, the most obvious feature is that businesses that utilise the industry genetic evaluation systems are significantly larger than those that don't, and that levels of investment are higher, and value of sales are higher. Together these translate into clear differences in levels of profitability. In both species, the most obvious feature is that businesses that utilise the industry genetic evaluation systems are significantly larger than those that don't.

Table 1. Key scale and financial parameters of businesses surveyed

Parameter	Beef		Sheep	
	In BREEDPLAN	Not in BREEDPLAN	In Sheep Genetics	Not in Sheep Genetics
Females per herd or flock	270	40	559	381
Bulls or rams sold per year	81	10	151	63
Bulls or rams sold per breeding female per year	0.30	0.26	0.27	0.17
Average total operating costs	$242,669	$67,795	$100,977	$56,121
Total performance recording and genetic evaluation costs	$65,530	$1,468	$4,351	$704
Average sale value of sales of bulls or rams	$5,103	$2,452	$1,089	$729
Total value of sales (includes semen and females)	$511,034	$36,896	$214,192	$60,094
Gross margin per business	$268,365	-$30,899	$113,215	$3,973
Return on Investment	21%	-20%	24%	2%

All aspects of herd or flock operating costs are higher for businesses using BREEDPLAN or Sheep Genetics (the only exception is cost of bulls used – which is potentially impacted by the breeder assessment of the value of bulls).

The direct investments in recording and genetic evaluation are higher for businesses using BREEDPLAN or Sheep Genetics, and all aspects of herd or flock income are higher for businesses using BREEDPLAN or Sheep Genetics. Note that the costs associated with recording and genetic evaluation included a category for "other measurement costs not specified" which was substantial for businesses using BREEDPLAN or Sheep Genetics, and which includes genotyping costs. Gross margins and estimated return on investment were calculated. The estimate of return on investment did not include an owner-operator allowance, which would likely vary with the scale of the enterprise. Both gross margin per business and estimated return on investment were substantially larger for studs using BREEDPLAN or Sheep Genetics than not.

None of the gross margins per stud calculated are likely to be large enough to support a family business, with the possible exception of that for beef studs using BREEDPLAN. This implies that for any of the enterprises surveyed, any significant increase in investment would be challenging.

CONCLUSIONS

The key findings from the survey are that among the businesses surveyed:

- businesses that used BREEDPLAN or Sheep Genetics were larger, and had greater value of sales, level of operating and non-owner overhead costs, and return on investment than business that did not use the industry genetic evaluation systems.
- in both species, the distributions of these parameters were all skewed, meaning that for example the average return on investment is markedly higher than the median.
- businesses that did not use BREEDPLAN or Sheep Genetics were not only smaller, but had lower sales value, lower operating costs and overheads, and lower return on investment. In the beef studs not using BREEDPLAN, average estimated return on investment was

negative even before consideration of owner-operator allowance.
- there was wide variation in all aspects of financial performance, with coefficients of variation averaged across parameters of 176% and 154% for Sheep Genetics users and non-users respectively. There was a positive relationship between business size and gross margin per cow in beef, although there is no clear relationship between these parameters in sheep.

Three general observations about the seedstock sector can be made:
- larger herds or flocks contribute the majority of the performance data (this is not based on this survey, but on data provided by ABRI and Sheep Genetics). In beef studs using BREEDPLAN, the median herd size ranges from 25-50 depending on breed, meaning that contemporary groups of 8-10 animals must be widespread.
- users and non-users of BREEDPLAN or Sheep Genetics are distinct groups, in terms of scale, and level of investment, meaning that different communication or extension approaches and messages are likely to be needed for effective engagement with the groups.
- changing from non-adoption to adoption is financially a significant decision, particularly in beef. From this, it can be hypothesised that that "conversion" of non-adopters to adoption will only happen at a low rate, and that significant increases in the level of performance recording in studs are unlikely.

Overall, the survey data and analysis suggests that there is some capacity for increased investment in recording in studs using BREEDPLAN and Sheep Genetics. However, this conclusion is tentative, reflecting:
- the owner-operator allowance will reduce surplus funds available for investment.
- we have no way of assessing the level of uncertainty around the value of future sales – which must be a consideration for stud breeders when considering their level of investment in recording and other items of breeding business expense.
- while the BREEDPLAN and Sheep Genetics users achieve higher average and total sales value, they have to – because their investment in the business are significantly higher than for non-users.

Having a sound information base on the economics of the seedstock sector could provide a basis for assessing likely uptake of new recording and evaluation technologies. More extensive survey data would provide a stronger base for such assessments, and consideration should be given to how to achieve higher response rates than achieved in this study.

ACKNOWLEDGEMENTS

This study was supported by Meat and Livestock Australia project L.GEN.1802. The assistance provided by ABRI, Sheep Genetics and Angus Australia is greatly appreciated.

REFERENCES

Archer J. A., Barwick S. A. and Graser, H.-U. (2004) *Aust. J. Exp. Agric.* **44**: 393.

IMPACT OF KEY MESSAGES ON ACCURACY OF SHEEP BREEDING PROGRAMS

L.M Stephen[1] and D.J Brown[2]

[1]NSW Department of Primary Industries, Armidale, NSW, 2350 Australia
[2]Animal Genetics & Breeding Unit*, University of New England, Armidale, NSW, 2351 Australia

SUMMARY

Sheep Genetics has been providing genetic evaluation services to the Australian Sheep Industry for Terminal, Maternal and Merino breeders since 2005. This paper looks at the average performance of flocks for genetic gain and other data quality metrics over two five year periods for the flocks that are represented in both time periods. These averages were compared using a t-test for significance. All three analyses have made significant improvements in the amount of pedigree being contributed to the analyses. Merino and maternal analyses have had significance differences in index gain. The terminal analyses were the only analysis that showed a significant improvement in the proportion of effective progeny, while no analyses improved the index accuracy over the study.

INTRODUCTION

Sheep Genetics is the national genetic evaluation for the Australian Sheep Industry, offering across flock services for the Maternal, Merino and Terminal breeds since 2005 (Brown *et al.* 2007). The two major pieces of information that they provide are Australian Sheep Breeding Values (ASBVs) for individual traits, and selection indexes that represent the standard breeding objectives. Sheep Genetics activities to increase adoption and use of genetic tools at the member level include:

- Individual breeder consultation on farm
- Regional forums for groups of geographically close clients
- Leading Breeding forums for breeders wishing to take their business further, and
- Service provider training aiming to upskill the people dealing with the client base (Collison *et al.* 2018).

Two key aims of these activities are: 1) to assist producers and service providers to increase their the understanding in how to use genetic technologies in their breeding programs and 2) to improve the quality of information submitted to Sheep Genetics by the breeder. It has been noted that the variation of data quality within the analysis is high (Collison *et al.* 2018).

But have the messages been received? Banks (1990) noted that the lamb industry was slow to adopt recommendations and methods based on animal breeding theory. Swan *et al.* (2009) demonstrated that the rates of genetic gain being achieved across the sheep industry is sub optimum across the Merino and Maternal segments. Stephen *et al.* (2018) identified significant differences in key performance recording factors between the flocks that are in the top 20% of genetic gain and those in the bottom 20%.

This study will determine if the collective education activities undertaken provided by Sheep Genetics and the sheep industry has been able to not only significantly lift the rates of gain of the industry, but also the quality of the information that is contributing to the ASBVs provided to industry over its first ten years.

---

* A joint venture of NSW Department of Primary Industries and the University of New England

MATERIALS AND METHODS

Rolling five year averages were taken for the first two time periods since Sheep Genetics was launched (2006-2010) and (2011-2015) for each flock. The LAMBPLAN maternal and terminal analyses along with the MERINOSELECT analysis were used for flocks that had a minimum of 3 years data in both time periods. Data from research flocks and Central Test Sire Evaluation have been excluded as their primary objectives are not genetic gain.

The diagnostic traits included in this study were the overall flock performance through index trend (indextrend, average annual change in a the Sheep Genetics index), the quality of information recorded included through effective progeny as stated by Brown *et al. (2000)* (eff, averaged across years and key traits) and average pedigree known (avpedknown, completeness of pedigree know from last 3 generations). The amount of data recorded was reflected by the accuracy of the index that best represents the analyses (indexacc). The indexes used were the Merino Production plus for Merinos, Maternal Carcase Production plus for Maternal's, and Lamb2020 for Terminals. For all data the mean, maximum and minimum flock performance and standard deviation was calculated. A summary of the data is shown in Table 1

Table 1. Number of flocks and average flock size of flocks included in analysis

		2005-2010	2011-2015
Merino	No of flocks	120	120
	Average flock size	614	779
Maternal	No of flocks	116	116
	Average flock size	530	586
Terminal	No of flocks	331	331
	Average flock size	313	347

The means were compared using a two-sided T-test to test for significance differences across the time periods $P<0.05$. Values that are significant have been bolded in Table 2 below.

RESULTS AND DISCUSSION

Examination of the differences in index trend between the two time periods it is of note that the reductions in standard deviations for the maternal and terminal index trends indicates that the flocks participating in these analyses are tightening around the average index trend. It is of concern that the flocks that are at the minimum on rates of gain are falling further behind the industry with a decrease in the rate of gain.

There are at least 2 potential reasons why flocks are further behind;

1. The breeding objective of the flock does not match the index that has been used in this study
2. The flock in question is currently not using ASBVs in their own selection decisions and only relying on the ASBVs for their clients to use on ram sale day.

In examining the reasons it is thought that the second point is more common in the industry. While not every flocks breeding objective will be a direct match to the index, the key profit drivers are generally similar to those in the indexes used in this study. This generally sees a high correlation between their objective and the index that is most common to the analyses. However there is a sub-section of breeders who believe that the function of ASBVs and indexes is a marketing tool for use by the commercial ram buying sector and not in their breeding program selections.

Table 2. Summary of data from two time periods from the Sheep Genetics analysis[1]

		Avpedknown		Index trend		Index acc		Eff	
		2005-2010	2011-2015	2005-2010	2011-2015	2005-2010	2011-2015	2005-2010	2011-2015
Merino	Mean	**22.2**	**31.2**	**0.7**	**1.7**	32.2	34.1	61.2	64.1
	Max	91.8	93	3.4	5.9	52.3	53.2	90.9	89.7
	Min	0.0	0.0	-4.8	-4.5	7.7	8.9	0.0	0.0
	SD	18.4	22.7	1.4	1.7	8.3	8.4	22.1	20.2
Maternal	Mean	**74.3**	**82.2**	**1.7**	**1.3**	42	42	67.8	70.6
	Max	99.4	99.8	5.2	4.3	52.6	52.4	92.7	91
	Min	10.8	23.3	-0.4	-0.8	25.4	28.2	0.0	3.0
	SD	21.2	15.8	1.2	1	5.5	5.6	17.7	14.9
Terminal	Mean	77.2	84.1	**0.4**	**0.5**	47.3	47.5	**58.6**	**62.4**
	Max	99.9	99.9	1.3	1.2	66.3	66.7	90.2	90.1
	Min	3.2	0.0	-0.2	-0.5	24	25.6	0.0	0.0
	SD	21.5	16.8	0.3	0.3	5.5	5.7	19.5	18.3

[1]$P<0.05$ significant differences are bolded

When examining Table 2, it is noted that there has been an increase in the average pedigree known across all three analyses. This could be due to a range of factors including the increase in pedigree recording technologies that are now available. These technologies include DNA parentage testing and pedigree association tags. The latest drop from the Merino analysis that was not included in this study had 44% of data being from animals with full pedigree (Collison *et all* 2018). It is worth noting that this percentage is significantly higher than the work done by Casey and Hygate (1992), whose survey work showed only 6% of Merino flocks had full pedigree recorded. As a result of extensive awareness programs there has also been an increase in the perceived value that pedigree contribute to the analysis. This is encouraging as Stephen *et al.* (2018) showed that the Merino flocks that were making the most genetic gain had higher percentages of full pedigree recorded and pedigree known.

Table 2 shows there are significant differences in index trends across the Merino and Maternal analyses. The declining rate of gain in the maternal analysis could be due to the differing breeding objectives of breeds within the maternal analysis not aligning with the standard industry index. A contributing factor also being that prior to 2012 the flocks participating in maternal analysis were analysed within breed only. This result is consistent with the work of Swan (2017) in showing that the maternal breeds had a decrease in rates of gain.

In looking at the index trend for the Terminal genetic evaluation traits not included in the index would be contributing. Genomically predicted ASBVs for lean meat yield (LMY) and intramuscular fat (IMF) were first available for breeders to use for the 2012 drop of animals. Breeders have been able to use these breeding values in selection decisions however these were outside of the standard index selection. This results in a lower selection efficiency achieved against the index as animals that are highest on index may not be selected if the eating quality ASBVs are not acceptable to the breeders objectives. Swan (2017) has since developed indexes for the Terminal analysis that now include LMY and EQ ASBVs, these were deployed to industry in May 2017.

The Merino rates of gain more than doubled during this period which is significant and important for industry given the merino's large contribution to the Australian sheep flock. This is encouraging as Brown *et al.* (2001) highlighted the lack of use of genetic technologies by Merino breeders, while Swan (2009) highlighted the sub optimum gains achieved in the industry. This increase in the rate of gain has coincided with a significant increase in the number of flocks participating in MERINOSELECT since 2010.

The Terminal analysis was the only analysis that has had a significant change in effective progeny although both the Maternal and Merino analyses have seen small non-significant increases. Table 1 illustrates that Terminal breeding programs are genuinely lower in numbers compared to Maternal and Merinos analyses. In addition to flock size being smaller the measurements in a terminal breeding program are taken earlier in life.

In examining index accuracy there hasn't been a significant increase across any of the three analyses. The Merino analysis has seen a slight increase in average index accuracy across the two time periods that could be attributed to the increase in pedigree. The data recorded for these indexes have generally remained the same number of key traits that breeders will record prior to ram sales, with little future trait recording on ewes. But with more recent innovations around recording dam pedigree and mating data included in commercial software packages there is no reason why more of the traits in the objectives cannot be measured in Merino and Maternal breeds.

CONCLUSIONS

This study shows that there are certain aspects of the key messages around improving rates of gain that have been adopted by breeders in their breeding programs. Across all three analyses there has been a concerted effort to increase the average pedigree known in their flocks across the time periods. Rates of gain for the Merino industry have managed to increase significantly which is also encouraging for the Australian sheep industry given the Merinos large contribution to the Australian sheep flock. However there is still more opportunities for breeders in improving the accuracy of the information through either recording more traits in the index or adding more historical pedigree. The future of genetic evaluation will be aided by technologies to improve pedigree and performance recording which will help improve the accuracy of the information. While the tools are now available in industry to identify flocks who are sub optimum for either genetic gain or data quality and provide targeted assistance to improve these metrics. It is envisaged with these tools breeders can be more proactive in inreasing the accuracy of the information they are recording. This in turn with using the ASBVs and indexes to select the genetically superior animals will lead to an increase in genetic gain for the flocks and the Australian Sheep Industry.

REFERENCES

Banks R.G. (1990) *Proc. Assoc. Advmt. Anim. Breed. Genet.* **8:** 237.

Brown D.J., Tier. B. and Banks R.G. (2000) *Proc. Assoc. Advmt. Anim. Breed. Genet.* **14:** 441.

Brown D.J., Huisman A.E., Swan A.A., Graser H.-U., Woolaston R.R., Ball A.J., Atkins K.D. and Banks R.G. (2007) *Proc Assoc Advmt. Anim Breed Genet.* **17**: 187.

Casey A.E. and Hygate, L.C. (1992) NSW Agriculture Animal Industries Report 2,28pp

Collison C.E., Brown D.J., Gill J.S., Chandler H.R., Apps R., Swan A.A. and Banks R.G. (2018) *Proc. 8th WCGALP* **11:** 661.

Stephen L.M, Brown D.J, Jones C.E. and Collison C.E. (2018) *Proc. 8th WCGALP* **11**: 433.

Swan A.A. Brown. D.J. and Banks R.G. (2009) *Proc. Assoc. Advmt. Anim. Breed. Genet* **18**: 326.

Swan A.A. Pleasants T. and Pethick D. (2017) *Proc Assoc. Advmt. Anim. Breed. Genet* **21**:29.

FARMER APP ADOPTION IS INFLUENCED BY AGE, USE OF ADVISORS AND FARMER NETWORKS

P.J. Schulz[1,2], J. Prior[1], L. Kahn[1,2] and G. Hinch[1]

[1]University of New England, Armidale, NSW, 2351 Australia
[2]Sheep Cooperative Research Centre, Armidale, NSW 2351 Australia

SUMMARY

No peer-reviewed information is available on the adoption of mobile applications (apps) by Australian farmers, despite the increase in number and usage. An online survey of 400 sheep industry participants found that age, use of advisors and farmer network participation had a significant influence on the level of farmer app adoption. However, when surveyed, farmers ranked networks and advisors of low importance when deciding to use an agricultural app. Older age brackets had lower levels of app adoption but the actual level of adoption was high compared with common perceptions. Age should not be viewed as a barrier to adoption or overlooked in R&D and marketing.

INTRODUCTION

Mobile apps (computer applications for mobile devices) have the potential to revolutionise the way farmers undertake training, extension, and support more informed decision making. Despite the increased number and use of agricultural apps, there is no peer-reviewed information on the adoption of apps by Australian farmers and their role in extension, training and decision making. Therefore, online surveys were conducted to examine the extent to which existing technology adoption models are relevant for the adoption of app technology for the sheep industry, as well as identify factors of importance to farmers for app adoption.

MATERIALS AND METHODS

An initial online survey was conducted in 2017 utilising the Sheep CRC communications database, which consisted of predominantly sheep farmers but also included service providers, researchers and others. The survey was completed by 400 respondents with 79% being sheep farmers and the majority of the remainder being industry service providers. A subsequent survey was conducted using the Sheep CRC RamSelect app user database (n=54) to examine factors affecting adoption from a targeted app user group. RamSelect, (viz. RamSelect Plus), is a web-based application, www.ramselect.com.au, viewable on mobile devices rather than downloaded, allowing ram buyers to find and rank sale rams based on Australian Sheep Breeding Values (ASBVs) for their flock.

Both surveys were voluntary participation and conducted via SurveyMonkey, an online survey builder website, via industry email invite. Participants included those that did and did not use apps. Survey questions were based upon aspects of the technology adoption model Diffusion of Innovations (DOI) theory (Rogers 2003), as well as derivatives from the Technology Acceptance Model (TAM) (Davis *et al.* 1989), Unified Theory of Acceptance and Use of Technology (UTAUT) (Venkatesh *et al.* 2003), and the ADOPT tool (Kuehne *et al.* 2011). These models encompass widely accepted principles for technology adoption (e.g. DOI), plus new considerations from more modern adoption theories such as ADOPT.

Survey participants ranked the importance of app attributes in relation to their actual use of apps and indicated if the attributes were very important, important, somewhat important or not important in deciding whether they were going to use an agricultural app. These rankings were then scored 4 (very important) to 1 (not important). Within the survey, participants were also asked if they owned a smartphone or tablet, if they used mobile apps (in general) and if they used agricultural apps. They indicated their age bracket, whether they participated in a farmer group or network and whether they used farm advisors.

Statistical analysis. App attribute importance ranking was analysed using a repeated measures ANOVA, with a Greenhouse-Geisser correction, determining if the mean scores for app attribute importance were significantly different. Post hoc tests were conducted, using Bonferroni correction, to reveal if any attribute importance score means were significantly different from each other through a pairwise comparison. A Chi-square test was used to investigate if there was a relationship between age and levels of smartphone/tablet ownership, mobile app use and agricultural app use. The Chi-square test was also used to test the same adoption relationship with use of advisors and farmer network participation. A one-way ANOVA with Tukey post hoc test was also conducted to determine if the level of adoption for each technology was different between each age brackets but full results not reported here. All test assumptions were checked and validated. Survey data was analysed using IBM SPSS version 25.

RESULTS AND DISCUSSION

Of the respondents to the Sheep CRC database survey, 91.9% owned a smartphone or tablet, 83.4% used mobile applications and 63.3% used agricultural applications. Similarly, of the respondents to the RamSelect app user survey, 92.6% owned a smartphone or tablet, 85.2% used mobile apps and 70.4% used agricultural apps. Note that RamSelect can be accessed without a mobile device and hence some users would define its use not to be that of an app.

App attribute importance. The three most important app attributes identified from the Sheep CRC database survey were usefulness, ease-of-use, and providing a better way of doing something the user already does. The three least important factors related to farmer networks, fellow farmer use and advisor recommendations, which all ranked not important to somewhat important. Attributes of compatibility, interest in the subject, quickness to see benefits, complexity, trialability and price, were identified as a group of medium importance factors influencing farmer app use. Survey data analysis determined that the mean scores for app attribute importance were significantly different ($F(7.866, 2352.003) = 279.879$, $p<0.0005$). Post hoc tests revealed that several attribute importance score means were significantly different from each other, which led to the segregation of the groups of attributes with high, medium and low importance.

It is evident that the fundamental principles affecting technology adoption, such as usefulness and ease-of-use, remain vital to technology adoption, including apps. Providing a better way of doing something the user already does was also of high importance. This should be of particular interest to developers working on tools that provide information or methods the user has not previously considered, as they may have an additional barrier to adoption. The three lower importance factors had been hypothesised by the authors to be of higher importance. Therefore, further analysis of the survey data was undertaken, revealing that those who were part of a farmer network or used advisors had significantly higher levels of agricultural app adoption.

Farmer networks. Respondents from the Sheep CRC database survey who were part of a farmer group or network had significantly higher levels of app use (Table 1). Figure 1 illustrates a significant finding that is the proportion of survey participants who used agricultural apps ranged from 70.4% for those involved in farmer networks to 44.6% for those who were not.

Table 1. Percentage of smartphone/tablet ownership, mobile app and agricultural app use within age bracket, farmer network and advisor use categories in the Sheep CRC database survey (percentage and number of respondents in each category are shown)

	Smartphone/tablet			Mobile app use			Agricultural app use		
	n	Yes	*p*-value*	n	Yes	*p*-value*	n	Yes	*p*-value*
Age bracket									
18-40 years	64	98.4%^		64	96.9%^		63	85.7%	
41-55 years	138	94.9%^	<.001	138	88.4%^	<.001	135	68.9%	<.001
56-70 years	119	90.8%^		119	78.2%		116	51.7%^	
≥ 71 years	32	68.8%		31	51.6%		31	35.5%^	
Farmer network									
Yes	237	92.6%	.564	220	85.9%	.076	178	70.4%	<.001
No	88	90.7%		75	78.1%		41	44.6%	
Use of advisors									
Yes	88	96.6%	.053	88	90.9%	.022	88	73.9%	.012
No	260	90.0%		259	80.3%		252	58.7%	

* Significance level *p*-value ≤ 0.05. ^ Post hoc tests were conducted finding these age brackets not being significantly different from each other.

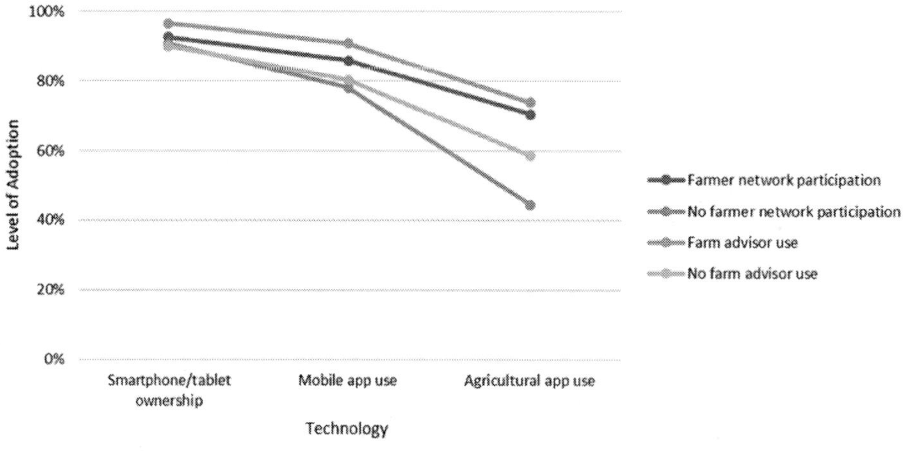

Figure 1. Percentage of technology adoption between farmer network participation (n=354) and farm advisor use (n=348) from Sheep CRC database survey respondents

A similar association between farmer network involvement and app adoption was evident in the RamSelect app user survey. The proportion of participants who used mobile apps ranged from 90.9% for those involved in farmer networks to 55.6% for those who were not with the difference being highly significant ($\chi2(1) = 7.287$, $p = .007$). Furthermore, the proportion of participants who used agricultural apps ranged from 81.4% for those involved in farmer networks or groups to 33.3% for those who were not, with the difference being significant ($\chi^2(2) = 8.821$, $p = .012$).

Use of advisors. In the Sheep CRC database survey, the highest proportion of mobile app use was in those who used advisors but the more significant association was between advisor use and agricultural app use with 73.9% using agricultural apps compared with 58.7% (Table 1, Figure 1).

Rather than being a contradiction to attributes of importance for app adoption, this inconsistency between farmers' ranking of what is important to their agricultural app adoption and the clear difference in app adoption between farmer networks and advisor use, it is likely due to farmer adopter category (e.g. early adopter). Therefore, those who participate in farmer networks or use advisors, are potentially an excellent target market for agricultural app developers due to their high level of adoption.

Age. Results from the Sheep CRC database survey showed that age had a significant influence on smartphone or tablet ownership but only when compared with the oldest age bracket of 71 years or older (Table 1). Age also had a significant influence on mobile app use when comparing those 55 years and younger with those in two older age brackets. The most significant influence of age was on agricultural app adoption where there was a significant difference between the main three age brackets where 85.7% of 18-40 year olds used agricultural apps, 68.9% in the 41-55 years bracket used agricultural apps but this dropped to 51.7% in the 56-70 years group, although this is still over half of that age group (Table 1). No significant difference was found between those 56-70 years to those 71 or older for agricultural app use.

A similar significant effect of age was found in the RamSelect app user survey where smartphone or tablet ownership ranged from 100% to 83.3% ($\chi^2(2) = 6.183$, $p = .045$) the use of mobile apps ranged from 100% to 63.2% ($\chi^2(2) = 11.461$, $p = .003$). There was a large range between age groups for the use of agricultural apps (81.3% to 55.6%) but the variation was not found to be significant ($\chi^2(4) = 4.187$, $p = .381$) possibly due to smaller sample size.

From this we can deduce that there is a significant effect of age on the level of smartphone or tablet ownership, mobile app and agricultural app use. However, it is valuable for app developers and marketers to note that the level of smartphone ownership and app use was still high across most age brackets compared to anecdotal evidence that few "older" people are using apps.

CONCLUSIONS

This study shows that while some principles from existing technology adoption models are relevant for the adoption of app technology, there are some that are not so important and farmers have identified their factors of importance to app adoption. The level of agricultural app adoption was significantly higher amongst respondents using advisors and farmer networks, yet farmers did not rank advisors or networks as important to their app uptake. Therefore, an alternative model may be needed specific to the adoption of apps for agriculture. The level of general app use was high amongst survey respondents but the level of agricultural app use was lower suggesting there is a lack of relevant or useful apps in the marketplace and/or inadequate marketing. While age did have a significant effect on app adoption, the level of app use amongst most age brackets was high and age should not be viewed as a barrier to adoption.

ACKNOWLEDGEMENTS

Recognition of statistics consulting support from Dr Brenda Vo and Dr Ian Reeves (University of New England) and to the Sheep CRC communications and RamSelect team for assistance in facilitating the survey work. The Sheep CRC contributed financial support for this work.

REFERENCES

Davis F., Bagozzi R. and Warshaw, P. (1989) *Manage Sci.* **35:** 982.

Kuehne G., Llewellyn R., Pannell D., Wilkinson R., Dolling P. and Ewing M. (2011) "ADOPT: a tool for predicting adoption of agricultural innovations", *presented at the 55th Annual National Conference of Australia Agricultural Economics Society*, Melbourne, Victoria.

Rogers, E. (2003) *Diffusion of Innovations* (5th ed.), New York, NY: Free Press.

Venkatesh V., Morris M., Davis G. and Davis F. (2003) *MIS Quarterly* **27:** 425.

ANGUSSELECT™

M.J. Reynolds, A.I. Byrne, C.J. Duff and P.F. Parnell

Angus Australia, 86 Glen Innes Rd, Armidale, NSW, 2350 Australia

SUMMARY

A key practice for producers looking to increase their rate of genetic gain is the consistent application of a breeding objective to selection decisions. AngusSELECT™ enables users of Angus genetics to identify the animals that are most aligned with their breeding objective through an intuitive suite of decision support tools. This paper will showcase the functionality of the tools and provide analytics to demonstrate their use to identify Angus genetics that meet the breeding objectives of individual cattle breeders.

INTRODUCTION

Angus Australia is focussed on enhancing the value and profitability of Angus genetics throughout the beef supply chain by ensuring that its members, and their commercial customers, have access to world leading genetic evaluation technologies and associated tools for genetic improvement. Consistent with this focus, Angus Australia has developed decision support tools specifically tailored to the needs of its members.

AngusSELECT™

AngusSELECT™ (https://www.angusaustralia.com.au/angusselect) composes a suite of decision support tools which support breeders in identifying desirable genetics to include within their breeding program and improve the profitability of Angus genetics within the beef supply chain. These tools include; (i) SaleSELECT – search and sort the registered Angus animals that are currently available for sale and identify those animals that are most aligned with users' breeding goals and objectives; (ii) SemenSELECT – search and sort the registered Angus bulls for which semen is available and identify the most suitable genetics for use within users' artificial breeding programs; (iii) ASBPSELECT – search and sort the Angus bulls that have been entered in the Angus Sire Benchmarking Program and identify bulls carrying genetics that are suitable for use within users' breeding programs.

Entering selection criteria. The AngusSELECT™ decision support tools assist users in the application of breeding goals and objectives to selection candidates. The two-step selection tool assists users to select relevant catalogues and then apply selection criteria to identify animals which meet their requirements.

Selecting catalogues. Users have access to a multitude of ways to select catalogues within AngusSELECT™, which hosts sale, semen, embryo and ASBP sire catalogues, which users are able to refine with a number of intuitive search features. These features include: (i) selecting specific catalogues; (ii) specifying by region (only available in SaleSELECT); (iii) specifying within distance from a location (only available in SaleSELECT); (iv) specifying within postcode (only available in SaleSELECT); and, (v) specifying with the interactive map utilising the power of Google Maps and interfacing with locations of currently listed sale catalogues. This provides users a modern and convenient approach to identifying selection candidates in relevant geographic locations (only available in SaleSELECT).

Selecting candidates. Once users have identified the catalogue(s) to be filtered, the second step supports them in the application of selection criteria to the candidates that are available. Users can define selection criteria to filter the catalogues through a range of tabs including: (i) Animal details – sex, register, calving year, colour, has photo, has video, etc.; (ii) Breeding – Sire ID, exclude sire

ID, dam ID, etc.; (iii) Genetic Tests – parent verified, red gene, arthrogryposis multiplex (AM), contractual arachnodactyly (CA), developmental duplications (DD), neuropathic hydrocephalus (NH), etc.; (iv) EBVs and Indexes – birth weight (BWT), 400 day weight (400), carcase weight (CWT), intramuscular Fat (IMF), Angus Breeding Index (ABI) etc.

When users apply criteria relating to EBVs they have the option to set minimum or maximum breeding value thresholds or alternatively apply percentile band cut offs using an interactive slider bar. Regardless the level of selection pressure users choose to apply, the AngusSELECT suite of decision support tools will support them in their genetic selection decision.

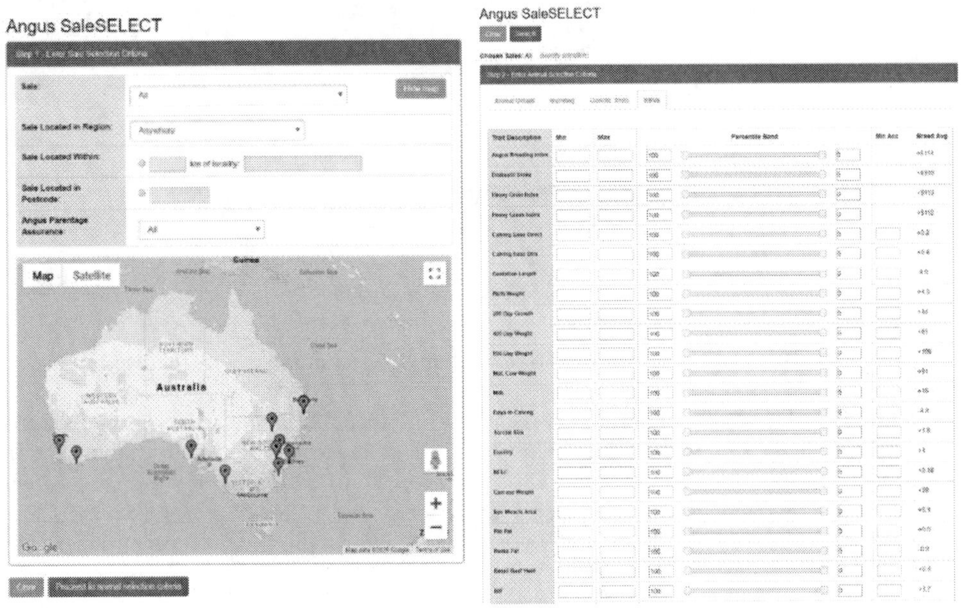

Figure 1. Interface for selecting catalogues of interest and displaying the EBV selection criteria users can apply to searches in SaleSELECT

Viewing candidate information. A comprehensive array of information is provided on animals meeting the search requirements. The information is displayed over a number of tabs specific to the listed animal, including: (i) Summary – a quick guide to the information stored on the Angus Australia database for the animal; (ii) Basic details – includes details as to the ID, name and tattoo of the animal, birth date, gender, mating type and registration status; (iii) Sale details - includes details around the sale of the animal or its semen; (iv) Ownership – includes details of the current ownership and ownership history of the animal; (v) Pedigree – an interactive tab enabling users to navigate through the pedigree of the animal; (vi) Progeny – includes details of any progeny of the animal; (vii) EBVs – includes the current EBVs for the animal in tabular form; (viii) EBV chart – displays the current EBVs for the animal in graphical form; (ix) Genetic results – includes the DNA testing results for the animal, including parentage verification and status for genetic conditions and traits; (x) Photos/ Videos – includes photos and videos of the animal.

A number of new concepts have been implemented to simplify the display of genetic information, including the display of percentile band information below each EBV, and a revised, more intuitive graphical display of EBV information via the EBV chart (Figure 2).

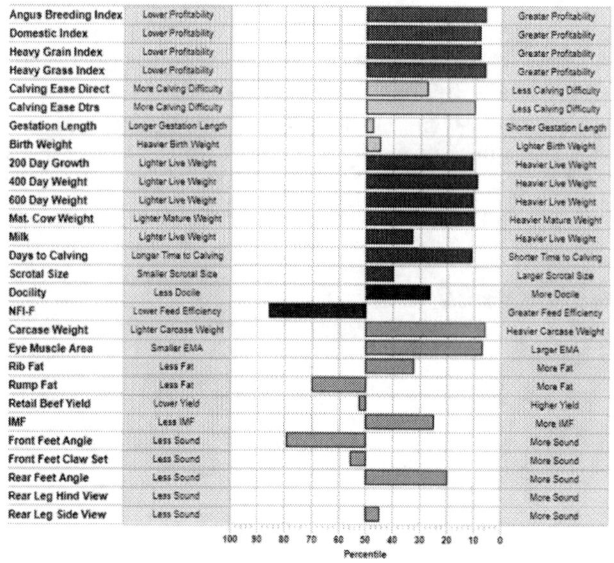

Figure 2. The EBV chart displays EBV information in an intuitive graphical display

Report Centre. In addition to viewing candidate information, an interactive report centre provides users with the ability to generate reports and analyse the results of any search that they have conducted. Specifically, users can: (i) download pdf reports and .csv files of search results in a format that is suitable for printing or importing into software programs such as Microsoft Excel; (ii) download pdf reports for individual animals in a print-friendly format; (iii) generate reports summarising the average, minimum and maximum EBVs for animals returned in a search; (iv) generate reports summarising the sires represented within the results of a particular search; (v) generate an interactive graph comparing the EBVs of animals returned in a search for two specific traits or selection indexes; (vi) generate an interactive graph summarising the distribution of EBVs of animals returned in a search for a specific EBV or selection index (Figure 3).

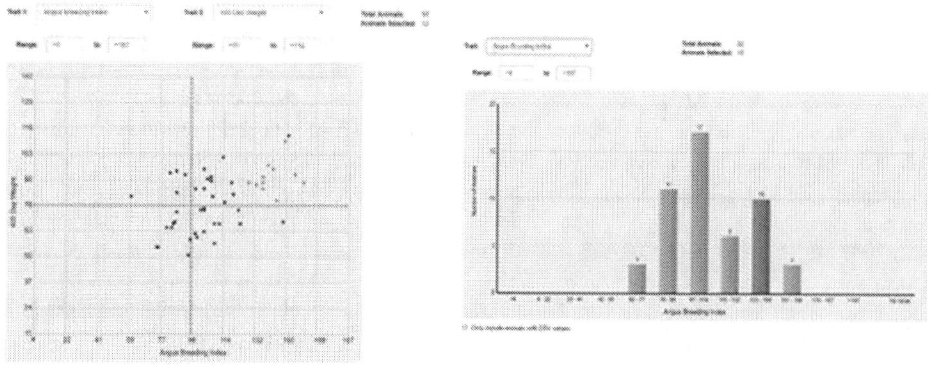

Figure 3. Two of the reports (trait comparison and trait distribution reports) available in the report centre of AngusSELECT™

USAGE OF AngusSELECT™

The initial 10 months following full implementation of AngusSELECT™ in November 2018 has highlighted a number of interesting trends around user queries. Usage of the tools has been high with an average of 15,794 catalogue searches a month, which cumulatively equated to 157,944 searches for the initial 10-month period (Figure 4).

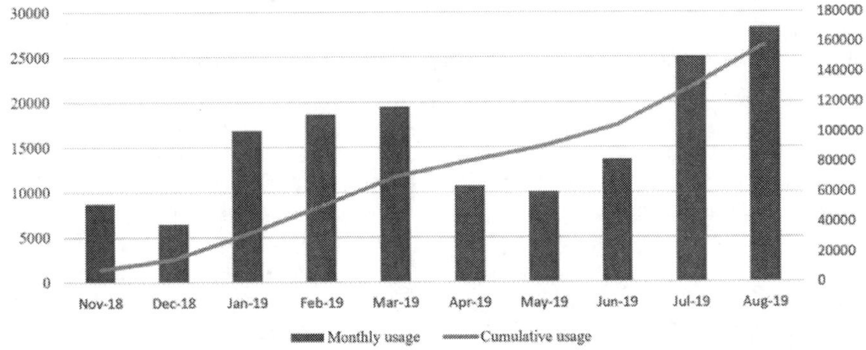

Figure 4. AngusSELECT™ catalogue searches by month and cumulative usage between November 2018 and August 2019

Angus.Tech is the software currently used to manage Angus Australia's extensive pedigree, performance and genomic database and encompasses AngusSELECT™ tools. With regards to the selection criteria applied to candidates, the IMF EBV has been the most searched EBV criteria across the Angus.Tech software, reporting over 1,800 more queries than the second and third most searched in birth weight and eye muscle area respectively (Table 1). The Angus Breeding Index was the most searched of the four available indexes.

Table 1. EBV Criteria used for searching within Angus.Tech between 1st November 2018 and 31st of August 2019

Criteria used for animal searches	Number of queries
Intramuscular Fat EBV	7,035
Birth weight EBV	5,211
Eye muscle area EBV	4,018
Angus breeding index	3,781
400 day weight EBV	3,672
200 day growth EBV	3,204
Mature cow weight EBV	3,185
600 day weight EBV	3,020
Heavy grain index	2,637
Calving ease direct EBV	2,547

CONCLUSION

Angus Australia's software development initiative delivers users a modern and intuitive range of decision support tools, in AngusSELECT™, to support them to increase their rate of genetic gain in the herds. Usage analytics indicate that the suite of AngusSELECT™ tools are being used to identify Angus genetics to meet breeding objectives of individual cattle breeders.

ENHANCED DATA CAPTURE IN AUSTRALIAN RED MEAT SUPPLY CHAINS FOR THE GENETIC IMPROVEMENT OF EATING QUALITY AND CARCASE YIELD

S.Z.Y. Guy[1,2] and D.J. Brown[1,2]

[1]Animal Genetics & Breeding Unit*, University of New England, Armidale, NSW, 2351 Australia
[2]Advanced Livestock Measurement Technologies Project, Meat & Livestock Australia, North Sydney, NSW, 2060 Australia

SUMMARY

Australian beef cattle and sheep breeders have made substantial and sustained genetic gains in growth and carcase yield. This has been achieved through the breeding objective traits of saleable meat yield in beef cattle, and percentage of carcase lean (also known as lean meat yield) in sheep. However, the need to also simultaneously select for eating quality is increasingly being acknowledged. While it is possible to counteract the antagonistic genetic relationships using index selection, data on both eating quality and carcase yield are required. Genomic reference populations play an essential role in obtaining these data. The aim of this paper is to summarise the measurement program of carcase and eating quality traits for Australian beef and sheep genomic reference populations, and the challenges involved in the process of data collection. The paper then discusses enhanced data capture through online objective carcase measurement technologies that will further contribute to genetic gains in eating quality and yield in red meat supply chains. The incorporation of technologies in harsh processing environments is challenging. However, if such technologies are appropriately incorporated in the processing chain, and also validated for the measurement of eating quality and yield for their respective species, this may decrease the cost and increase the efficiency of carcase data collection for genetic improvement. Combined with enhanced processor systems that allow tracking of individual carcases, as well as on-farm information collection and transmission systems, the aim is to move towards better collection and flow of accurate phenotypes to genetic evaluation systems. Enhanced data capture of disease and defect phenotypes are also important. Ultimately, a whole supply chain approach is required to increase collection of accurate carcase measures, which will lead to higher accuracies in estimated breeding values and greater genetic gains in eating quality and carcase yield.

INTRODUCTION

Australian beef cattle and sheep breeders have made substantial and sustained genetic gains in growth and carcase yield (Johnston 2007; Swan *et al.* 2009). This has been achieved through the breeding objective traits of saleable meat yield in beef cattle, and percentage of carcase lean (also known as lean meat yield) in sheep. However, consumer-eating experience of red meat is becoming increasingly important, and care must be taken to avoid selection for carcase yield at the detriment of eating quality. While it is possible to counteract this antagonistic relationship in breeding programs using index selection, it is difficult to select simultaneously for both eating quality and carcase yield due to the lack of data on these traits.

Genetic gains in eating quality and carcase yield rely heavily on genomic reference populations, which are measured for growth, carcase and eating quality traits on individual animals. The aim of this paper is to summarise the measurement program of carcase and eating quality traits for Australian genomic reference populations, and the challenges involved in the process of data collection. The paper then discusses enhanced data capture through online objective carcase measurement technologies that will further contribute to genetic gains in eating quality and yield in red meat supply chains.

* A joint venture of NSW Department of Primary Industries and the University of New England

AUSTRALIAN BEEF AND SHEEP GENOMIC REFERENCE POPULATIONS

Genomic reference populations are constructed as a representation of the genetic diversity present in the industry. In Australia, the genomic reference populations established include breed-specific Beef Information Nucleus (Banks 2001), the Information Nucleus/MLA Resource Flock for sheep (van der Werf *et al.* 2010), as well as well-recorded industry herds or flocks. These reference populations are genotyped and extensively phenotyped for a range of traits, including production and carcase traits. By understanding the genetic relationships between (correlated) traits, and use of genomic technologies to quantify their genetic relationship with reference populations, estimated breeding values (EBVs) for industry animals can be derived for hard-to-measure traits. A major future challenge will be the continued collection of phenotypes to derive accurate genomic predictions (Swan *et al.* 2011).

Data capture on eating quality and carcase yield. The eating quality and carcase yield traits collected on Australian beef and lamb genomic reference populations are outlined in Table 1. The genetic analyses of these carcase traits require additional data about each animal, including pedigree, date of birth (to calculate age at slaughter), birth type, rear type and management group. These data are essential for the genetic analysis of carcase traits to appropriately separate genotype from environmental effects. There are also requirements for how these animals are managed on-farm and specific protocol for measurements. For example, animals within a defined contemporary group must be kept in the same group across time, and there should not to be any systematic selection of animals to be measured based on one or more criteria (i.e. no harvesting based on weight). These requirements are in place to ensure the data structure integrity for accurate genetic parameter estimation and robust estimation of breeding values.

Table 1. Eating quality and yield carcase traits collected on Australian beef and lamb genomic reference populations, and used in genetic evaluation systems

Species	Eating Quality	Carcase yield
Beef	• Intramuscular fat % • Shear force • Marble score*	• Hot standard carcase weight • Retail beef yield (subset only) • Eye muscle area* • Rib Fat* • P8 (rump) fat
Sheep	• Intramuscular fat % • Shear force • Consumer testing (subset only) – tenderness, juiciness, liking of flavour, overall liking	• Computed Tomography (subset only) – lean %, fat %, bone % • Hot standard carcase weight • Primal bone out (subset only) • Fat (c-site, GR-site) • Eye muscle depth and width (for calculation of eye muscle area)

*part of Meat Standards Australia (MSA) beef grading

Challenges in carcase data collection. The collection of these phenotypes from reference sheep/ beef carcases is a highly coordinated exercise. It is essential to negotiate with the relevant processor on the slaughter process and sample collection prior to slaughter. A team of research staff must be present at the kill to match up farm identification numbers with body numbers. Each carcase must be followed during processing to ensure that identity is not lost, and that the same traits are being collected in the same way in different abattoirs e.g. no excessive trimming. The operating environment is harsh due to moisture, heat, and fast chain speeds. Depending on the processing facilities, data are either collected from the slaughter floor (e.g. rump fat in beef), in the chillers (e.g. fat depth in lamb), and/

or meat samples are collected and transported to meat science laboratories for further analysis (e.g. intramuscular fat). Connecting different data points on individual animals requires integrated systems within the processing plant, and also alignment with on-farm and meat laboratory data.

OPPORTUNITIES FOR ENHANCED DATA CAPTURE

There is currently no technology available to measure eating quality and yield traits in real-time. Enhanced data capture and information flow have been identified as essential keys to increase producer confidence in carcase grading and pricing, transparency, competition, productivity and growth in the red meat supply chains (ACCC 2017). As a result, the development of technologies for independent and accurate descriptions of carcase performance has been prioritised by industry. Objective and predictive measures on carcase yield and eating quality, provided as feedback, will allow producers to improve decision-making to increase compliance and profitability.

The Advanced Livestock Measurement Technologies (ALMTech) project, 2015-2019. The ALMTech project for globally competitive Australian meat is funded by the Australian Government, in partnership with Research and Development Corporations, commercial companies, state departments and universities. This project involves many collaborations within multiple supply chains to maximise effective decision making, reduce risk and optimise profitability for all partners in the meat supply chain.

ALMTech has the following objectives:

- Accurate measurement (objective and predictive) systems of live animals, carcases and cuts in major beef, lamb and pork supply chains
- Valued producer feedback to effect decision making on compliance and profitability
- Enhanced supply chain information systems to extract maximum value of products

Figure 1 outlines the ALMTech activities and examples of the technologies currently being investigated for the enhanced data carcase composition, eating quality and health during processing.

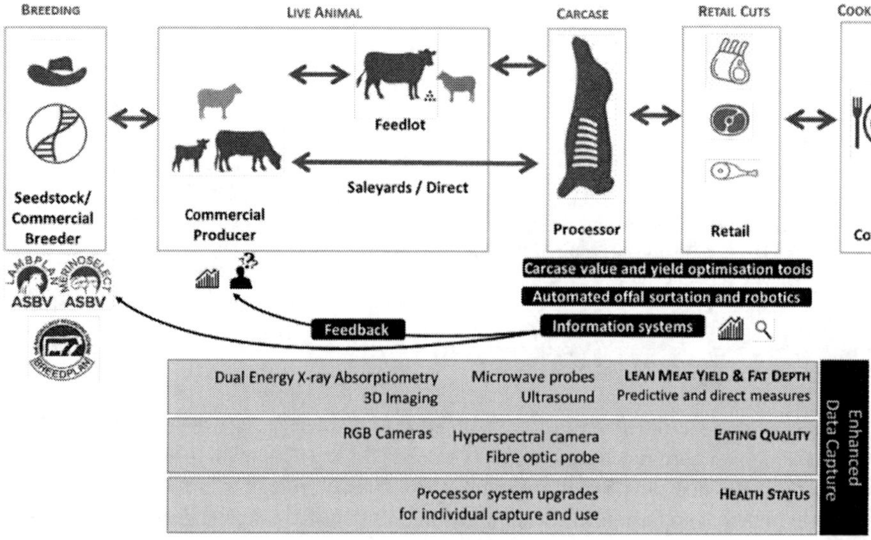

Figure 1. A snapshot of activities and technologies from the Advanced Livestock Measurement Technologies (ALMTech) project (2015-2019), aiming to enhance data capture and information flow along the red meat supply chain

Implications for animal breeding. In order to make effective selection decisions in the seedstock sector, a certain level of understanding of the process further down the supply chain is required. The supply chain is in fact a value chain, where there is a full range of activities required at different stages to create the red meat product. In addition to this, a whole chain approach is required to ensure the flow of data on accurate carcase measures.

Data collection on genomic reference populations have been aligned with, or identical to, the traits that impact value, especially in the processing and retail sectors. The measurement of reference populations with objective carcase measurement technologies will also assist with preparations for future market signals that may arise from the installation of technologies in commercial processors.

A validated technology appropriately incorporated in a processing chain will decrease the cost and increase the efficiency of objective carcase data collection on genetically informed animals. This also provides seedstock breeders more opportunities to collect carcase data through progeny testing, or testing of surplus (cull) breeding animals. These will help breeders to balance the selection based on live animal measures (such as weights, eye muscle area and fat depth) with carcase measures, and also allow them to overcome the unfavourable genetic correlations between yield and eating quality. Increased collection of accurate carcase measures ultimately will lead to higher EBV accuracies and therefore greater potential genetic gains in yield and eating quality.

There are also opportunities to select for good animal health using meat inspection data, which has primarily been investigated in pig production (Mathur *et al.* 2018; Horst *et al.* 2019). Although there are challenges with the accurate collection of such data on an individual basis, and the constraint of low and variable prevalence, preliminary genetic parameters demonstrate the potential to use this data for the selection of health conditions that may not be observable in the live animal.

CONCLUSIONS

Genetic gains in eating quality and carcase yield currently rely heavily on genomic reference populations. While there are opportunities to improve the efficiency and accuracy of data capture of carcase traits on reference animals, the incorporation of technologies in the harsh processing environments is challenging. However, combined with enhanced processor animal and carcase tracking systems that allow tracking of individual carcasses, as well as on-farm information flow systems, the aim is to move towards better collection and flow of accurate phenotypes to genetic evaluation systems. An entire supply chain approach is required to increase collection of accurate carcase measures, which will lead to higher EBV accuracies and therefore greater genetic gains in eating quality and carcase yield.

ACKNOWLEDGEMENTS

The Advanced measurement technologies for globally competitive Australian meat (ALMTech) project is supported by funding from the Australian Government Department of Agriculture and Water Resources as part of its Rural R&D for Profit programme, in partnership with Research and Development Corporations, commercial companies, state departments and universities.

REFERENCES

ACCC (2017) Cattle and beef market study – Final Report.
Banks R.G. (2001) *Proc. Assoc. Advmt. Breed. Genet.* **19**: 399.
Horst A., Gertz M. and Krieter J. (2019) *Livest. Sci.* **211**: 155.
Johnston D.J. (2007) *Proc. Assoc. Advmt. Breed. Genet.* **17**: 8.
Swan A.A., Brown D.J. and R.G. Banks. (2009) *Proc. Assoc. Advmt. Breed. Genet.* **18**: 326.
Mathur P.K., Vogelzang R., Mulder H.A. and Knol E.F. (2018) *Animals* **8**: 16
van der Werf J., Kinghorn B. and Banks R.G.. (2010) *Anim. Prod. Sci.* **50**: 998.

PHENOTYPIC VARIATION IN RETAIL BEEF YIELD IN ANGUS CATTLE

K.A. Donoghue[1], L.M. Cafe[2], B.J. Walmsley[3] and C. Duff[4]

[1]NSW Department of Primary Industries, Agricultural Research Centre, Trangie, NSW, 2823
[2]NSW Department of Primary Industries, Livestock Industries Centre, Armidale, NSW, 2351
[3]Animal Genetics & Breeding Unit*, University of New England, Armidale, NSW, 2351 Australia
[4]Angus Australia, 86 Glen Innes Road, Armidale, NSW, 2350

SUMMARY

This paper reports preliminary estimates of phenotypic variation and relationships for carcase traits using records from 386 Angus steers. Carcase traits recorded included carcase weight (CWT), retail beef yield (RBY), carcase intramuscular fat (CIMF), carcase P8 fat (CP8), carcase rib fat (CRIB) and carcase eye muscle area (CEMA). Carcase trait averages for animals used in this study were higher than for those used for the previous estimation of RBY parameters, and the differences evident between summary statistics for historic and contemporary carcase traits records highlights the value of continually recording carcase traits, as well as the value of re-examining relationships between carcase traits. Phenotypic correlations between carcase traits indicated that animals with higher RBY also had, phenotypically, heavier CWT ($r_p = 0.12$), larger CEMA ($r_p = 0.31$), lower CRIB and CP8 ($r_p = -0.28$ and -0.10 respectively) and lower CIMF ($r_p = -0.16$). For every 1-kg increase in RBY EBV, recorded CWT, RBY and CEMA were greater by 2.08kg, 0.24% and 2.77cm^2, respectively. In contrast, for every 1-kg increase in RBY EBV, recorded CIMF, CRIB and CP8 were lower by 0.39%, 1.31mm and 1.81mm, respectively. These preliminary results highlight the importance of continually recording carcase traits as well as the value of periodically re-examining relationships between carcase traits.

INTRODUCTION

Retail Beef Yield (RBY) is a key driver of profitability in the Australian beef industry, and the RBY EBV underpins BreedObject $Indices which allow selection for profitability in the major beef breeds. However, in current genetic evaluations, the trait of RBY is estimated almost exclusively from correlations with live weight, live animal scanned eye muscle area and fat depths. This is due to very limited recording of abattoir carcase data and specifically, RBY phenotypes. RBY phenotypes are not generated routinely because accurate data is time-consuming and expensive to collect. Previous estimates of phenotypic and genetic parameters for RBY were undertaken on a relatively small number of animals (n=1,930) born between 1994 and 2000 (Reverter *et al.* 2000; Reverter *et al.* 2003). Genetic improvement in carcase traits in the 20 years since the last RBY records were collected has led to significant improvements in carcase traits (Johnston 2007).

The purpose of this project is to generate RBY phenotypes on 1,000 fully pedigreed, genotyped and genetically described Angus steers. These will be used to re-estimate BREEDPLAN RBY parameters for incorporation into BREEDPLAN analyses to provide RBY EBVs. Animals in the project will be measured for a comprehensive suite of traits from birth until slaughter, including live weight and scan data taken from weaning until feedlot exit, and these traits will be assessed for their relationship with RBY. This will provide information to determine the effectiveness of exploiting early in life measurements as genetic indicators for RBY which may reduce the need to record actual RBY. The collection of RBY data will provide a reference population on which genomic information will be applied to provide RBY EBVs. Data used in this study comprises animals from the first two cohorts (of four) of the RBY project and provides preliminary estimates of phenotypic relationships between RBY and other carcase traits.

* A joint venture of NSW Department of Primary Industries and the University of New England

MATERIALS AND METHODS

The animals used in this project were steers from the NSW DPI Trangie and Glen Innes research herds born in 2015 (n=150) and 2016 (n=236). All steers were fully pedigreed and, in collaboration with the Angus Sire Benchmarking Program had a comprehensive suite of traits measured from birth, based on BREEDPLAN collection protocols, and genomic testing (GGP-LD or Angus GS profile) conducted. There were 109 sires represented in the dataset, with the number of progeny for each sire ranging from 1 to 7. Fifty three of the 109 sires had progeny represented in both herds (Trangie and Glen Innes), and these 53 sires accounted for 58% (224/386) of the data.

Following weaning on the birth property, all steers were backgrounded on irrigated ryegrass /oats pasture and other pastures at Grafton Primary Industries Institute and were grain finished at UNE Tullimba Research Feedlot for 150 days to fit the John Dee, Warwick Angus Gold Grid. The steers were slaughtered in lots of approximately 50 animals at John Dee, Warwick, to suit the number of sides which can be boned for out for RBY at the facility in one shift. The cohorts were split into kill groups to ensure the integrity of the contemporary groups and based on balance for sire and live weight.

Following commercial AUS-MEAT carcase preparation (Anon. 2007), carcasses were weighed and hot P8 fat depth recorded. Fat trim following hot standard carcase weight measurement was restricted over the 12/13 rib to allow meaningful MSA rib fat data collection, and standard excess fat trim was conducted on the remainder of the carcase. Carcasses were tagged and chilled overnight, and Meat Standards Australia (MSA) carcase grading data (Anon. 2008) was collected by registered MSA graders on the right hand sides the following morning prior to boneout. A sample of the cube roll was collected for lab analysis to determine IMF% by the Meat Science Department at the University of New England following the near infrared spectrophotometry (NIR) method described by Perry *et al.* (2001).

For boneout a standard set of AUS-MEAT boneless primals, standard trim of 10mm fat, and estimated chemical lean of trim of 65% was conducted. The primal cuts included in the measure of RBY were: cube roll, chuck, chuck tender, blade, brisket, intercostals, FQ shin/shank, rib end meat, inside skirt, FQ trim, topside/inside, thick flank, rump, striploin, tenderloin, thin flank, silverside, HQ shin/shank, HQ trim. Carcase components of FQ fat trim, FQ bone, HQ fat trim and HQ bone did not contribute to RBY. Boning room pre-trim was weighed for completeness, though this was usually only 50 to 200g of fat and added to the weight of fat trim. The cold weight of the sides was measured prior to quartering and entering the boning room. This cold weight was used as a measure of reliability of the RBY data, as the recovery of this weight should be close to 100% and should be consistent across sides.

Data available for the analyses included carcase weight (CWT), eye muscle area (CEMA), retail beef yield percentage (RBY), intramuscular fat percentage (CIMF), subcutaneous fat depths at the P8 rump (CP8) and the 12th and 13th rib (CRIB) sites.

Model of analysis. Contemporary group for carcase records included effects of herd of origin and kill date. Variance and covariance components were estimated with an animal model using ASReml (Gilmour *et al.* 2015). A fixed effect of contemporary group, random direct genetic effects, and residual effects were included in the standard model. Univariate analyses were undertaken to obtain phenotypic variances for each trait, while pairwise bivariate analyses were undertaken to obtain phenotypic correlations. Regression of recorded carcase traits on the corresponding BREEDPLAN carcase EBVs were undertaken using PROC REG (SAS). Carcase traits were adjusted as per standard BREEDPLAN analyses (CWT adjusted to 750 days and all other traits adjusted to 400 kg carcase weight) for the regression analyses. BREEDPLAN carcase EBVs on animals were obtained prior to the inclusion of carcase data from the project in the genetic evaluation.

RESULTS AND DISCUSSION

Summary statistics for the carcase traits are shown in Table 1. Most carcase trait averages for animals used in the previous estimation of RBY parameters were markedly lower than those presented in Table 1 (From Reverter *et al.* 2003: Slaughter age = 630±128 days; CWT= 269±55 kg; RBY=67.03±3.69%; CIMF=4.64±2.23%; CP8=10.19±4.57 mm; CRIB=8.21±4.45 mm; CEMA=81.59±15.9 cm²). The differences evident between summary statistics for historic and contemporary carcase trait records highlight the value of recording carcase traits in modern cattle populations, as well as the value of re-examining relationships between carcase traits in this population.

Table 1. Descriptive statistics for carcase traits

Trait	No. records	Average (SD)	Minimum	Maximum	σ^2_p (SE)
Age at slaughter (days)	386	596.4 (21.8)	536.0	635.0	-
CWT (kg)	386	344.7 (30.1)	269.0	428.5	889.10 (72.96)
RBY (%)	386	73.84 (2.02)	68.69	78.92	1.59 (0.12)
CIMF (%)	386	6.29 (2.04)	2.01	13.99	2.91 (0.22)
CP8 (mm)	386	16.24 (4.18)	10.00	30.00	13.73 (1.04)
CRIB (mm)	386	12.58 (4.45)	4.00	28.00	11.09 (0.81)
CEMA (cm²)	386	81.48 (8.44)	57.00	108.0	37.27 (2.89)

Phenotypic correlations (r_p) between carcase traits were similar to those reported in the literature for the same traits in temperate breeds (Reverter *et al.* 2003) for the majority of traits (Table 2). Of those that were different to previous estimates, the majority of r_p were in the same direction but of a different magnitude; r_p between RBY and CIMF (-0.16 vs -0.31 in literature); r_p between and RBY and CP8 (-0.10 vs -0.40 in literature); r_p between RBY and CEMA (0.31 vs 0.14 in literature). The remaining estimates that indicated relationships in a different direction to previous estimates are most likely a reflection of the relatively low numbers analysed so far in this project; r_p between CWT and CIMF (0.12 vs -0.19 in literature); r_p between CWT and CP8 (0.15 vs -0.19 in literature); r_p between CWT and CRIB (0.19 vs -0.19 in literature).

Table 2. Phenotypic correlations (SE) between carcase traits

Trait	RBY	CIMF	CP8	CRIB	CEMA
CWT	0.12 (0.05)	0.12 (0.06)	0.15 (0.05)	0.19 (0.05)	0.38 (0.05)
RBY		-0.16 (0.05)	-0.10 (0.05)	-0.28 (0.05)	0.31 (0.05)
CIMF			0.07 (0.05)	0.06 (0.05)	-0.14 (0.05)
CP8				0.25 (0.05)	-0.09 (0.05)
CRIB					-0.17 (0.05)

Phenotypic correlations between carcase traits would indicate that animals with higher RBY also had, phenotypically, heavier CWT (r_p = 0.12), larger CEMA (r_p = 0.31), lower CRIB and CP8 (r_p = -0.28 and -0.10 respectively) and lower CIMF (r_p = -0.16). In summary, results so far would indicate that phenotypic relationships between recorded carcase traits follow the pattern expected, given previous estimates, and make sense biologically. This confirms the quality of the data that is being collected and highlights the value of the data to estimate updated genetic parameters for these traits once more records have been obtained.

Regression coefficients from regression of recorded carcase traits on corresponding BREEDPLAN carcase EBVs are presented in Table 3. These results indicate that, on average, for every 1-kg increase in RBY EBV, recorded CWT, RBY and CEMA were greater by 2.08kg, 0.24% and 2.77cm^2, respectively. In contrast, for every 1-kg increase in RBY EBV, recorded CIMF, CRIB and CP8 were lower by 0.39%, 1.31mm and 1.81mm, respectively.

Table 3. Regression coefficients from regression of recorded carcase traits on corresponding BREEDPLAN carcase EBVs

Trait	CWT EBV	RBY EBV	CIMF EBV	CP8 EBV	CRIB EBV	CEMA EBV
CWT	2.11**	2.08	7.22	-5.26**	-5.44**	-0.25
RBY	-0.01*	0.24**	-0.08	-0.05	-0.07	0.13**
CIMF	0.006	-0.39**	1.31**	0.08	0.10	-0.07
CP8	-0.05**	-1.81**	-0.22	1.17**	1.27**	-0.26**
CRIB	-0.03	-1.31**	-0.10	0.70**	1.00**	-0.33**
CEMA	-0.04	2.77**	-1.67**	-0.52**	-0.71**	1.31**

* Significance at P<0.05; ** Significance at P<0.001

CONCLUSIONS

In the majority of cases, phenotypic correlations and regression coefficients were such that relationships between carcase traits were in the expected direction and similar in magnitude to literature estimates. As the number of carcase records increase as this project continues it is anticipated that genetic parameters for carcase traits will be estimated. This will allow more detailed investigations into the appropriateness of the current BREEPLAN carcase genetic parameters for contemporary Angus cattle.

ACKNOWLEDGEMENTS

This work was funded by NSW DPI and Angus Australia through the MLA Donor Company program. The capable assistance of staff at Glen Innes Agricultural Research and Advisory Station, Trangie Agricultural Research Centre, Grafton Primary Industries Institute and Angus Australia is gratefully acknowledged. The management and staff at John Dee, Warwick abattoir provided much capable assistance during collection of the carcase data.

REFERENCES

Anon. (2007) 'AUS-MEAT National Accreditation Standards' 2007 Edition (AUS-MEAT Ltd, Brisbane).

Anon. (2008) *Aust. J. Exp. Agr.* **48**: 1360.

Gilmour A.R., Gogel B.J., Cullis B.R., Welham S.J. and Thompson R. (2015) 'ASReml User Guide Release 4.1'. VSN International Ltd, Hemel Hempstead, UK.

Perry D., Shorthhose W.R., Ferguson D.M. and Thompson J.M. (2001) *Aust. J. Exp. Agr.* **41**: 953.

Johnston D.J. (2007) *Proc. Assoc. Advmt. Anim. Breed. Genet.* **17**: 8.

Reverter A., Johnston D.J., Perry D., Goddard M.E. and Burrow H.M. (2003) *Aust. J. Agric. Res.* **54**: 119.

Reverter A., Johnston D.J., Graser H.-U., Wolcott M.L. and Upton W.H. (2000) *J. Anim. Sci.* **78**: 1786.

DESIREBULL, A DECISION SUPPORT TOOL TO SIMPLIFY GENETIC INFORMATION FOR EFFECTIVE USE BY COMMERCIAL BEEF PRODUCERS

L.A. Penrose[1], M. Suarez[1] and B.J. Walmsley[1, 2]

[1]NSW Department of Primary Industries, Livestock Industry Centre, Armidale, NSW, 2350 Australia
[2]Animal Genetics & Breeding Unit*, University of New England, Armidale, NSW, 2351 Australia

SUMMARY

DeSireBull™ is a new decision support tool for commercial beef producers currently being developed by the NSW Department of Primary Industries, co-funded by Meat and Livestock Australia. It has been acknowledged that selecting a bull(s) can be a complex and overwhelming task for producers, when faced with large amounts of information on which to make their decision. Both producers and industry have developed ways to try and simplify the selection process, but this often results in sub-optimal decisions, negatively impacting the genetic progress of herds. DeSireBull™ aims to incorporate several strategies to simplify genetic information and aid producers in finding the right bull for their breeding objective.

INTRODUCTION

When faced with the task of purchasing bulls for use in the breeding herd, commercial beef producers can often be overwhelmed by the amount of information provided to them, be it genetic, or observable animal characteristics (Martin-Collado *et al.* 2018). In any given sale catalogue, there can be up to thousands of pieces of information, presented in numerical or graphical form. Given that bull purchases are an infrequent event, interpreting and digesting all of this information can be a somewhat complex, confusing and time consuming task.

Due to the complexity of the bull selection process, producers tend to limit the number of selection criteria and often do not realise the impact that this decision may have on other traits or the overall breeding objective. As a result, the genetic performance of herds is not progressing as fast and effectively as it could. This impacts the potential productivity and profitability of producers, the competitiveness of beef against other proteins, and the competitiveness of the Australian beef industry against its overseas competitors. This has been recognised by the NSW Department of Primary Industries, which is developing a decision support tool aimed at simplifying the use of genetic information to aid the bull selection process. This paper will revisit the complexity of the breeding selection process, discuss DeSireBull™ and its aims of simplifying the selection process for commercial beef cattle producers.

COMPLEXITY OF INFORMATION & THE CONSEQUENCES OF DECISION MAKING

The complexity of decision making when it comes to selecting a bull has been recognised as a somewhat cumbersome task for commercial beef cattle producers (Ipsos, 2016). The commercial producer has many factors to consider when running their beef enterprise of which, EBVs and genetic information often form a very small fraction (Figure 1). They must also select the right breeding animals to incorporate into the breeding program whilst considering the influence of other factors including current herd performance, costs, environmental constraints and commercial target market (Ispos 2016; BreedObject 2016).

---

* A joint venture of NSW Department of Primary Industries and the University of New England

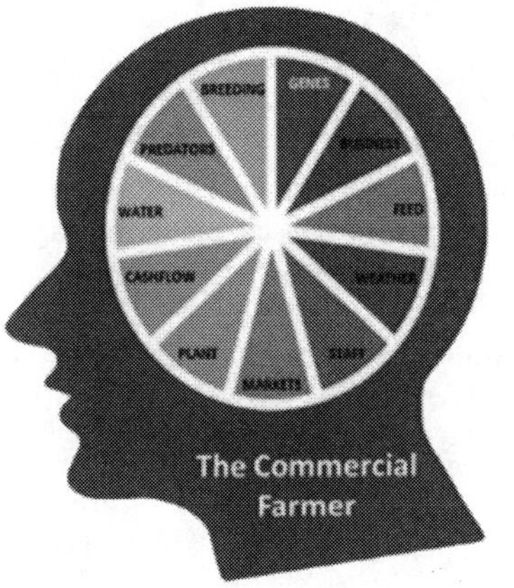

Figure 1. Considerations of the commercial farmer (Ipsos 2016)

Printed sale catalogues and online tools are the main information sources for producers when selecting bulls to buy. Martin-Collado *et al.* (2018) have identified some key sources of decision complexity and acknowledge that when making selection decisions, producers may be overwhelmed by one or in most cases, a combination of the following:

1. The number of animals available for sale,
2. The amount of information presented to them for each animal,
3. Trade-offs between traits of interest e.g. production and functional traits,
4. The existence of animals with unique features e.g. genetic conditions or polled animals,
5. The different trait units,
6. The format in which the information is presented.

The compounding effect is producers using their own simplifying strategies (heuristics), when selecting breeding animals, often results in sub-optimal decisions (Martin-Collado *et al.* 2018). When trying to avoid "negative" traits by using minimal selection criteria, producers often do not realise the implications this might have on other related traits. It is an optimal combination of all traits that contributes to the high genetic merit and suitability of an animal to a commercial production system and enables genetic progress. The over-simplified selection process used by many producers is impacting the potential productivity and profitability of their enterprise, as well as the beef industry in its entirety.

CURRENT TOOLS USED TO SIMPLIFY INFORMATION

There are a number of strategies currently used within the beef industry, both by farmers and industry leaders to try and reduce the complexity of information available to commercial producers to make selection decisions. Table 1 provides an indication of the ease of employing a range of selection strategies developed by industry, and the producers' use of heuristics, highlighting how effective they are in reducing decision complexity and whether or not they are based on genetic information (Martin-Collado *et al.* 2018). Whilst producers' own simplifying strategies are usually easy to implement and greatly simplify the selection process, it is worth noting that these methods are rarely based on genetic criteria alone, result in sub-optimal selection and hinder the long-term productivity and profitability of the enterprise.

Strategies used by industry leaders range in their difficulty to implement and their ease of interpretation. EBVs and selection indexes are most commonly used by industry leaders as they are based on genetic information. Selection indexes have been available since the late 1980's to aid selection decisions by combining all EBVs with bio-economic modelling to create a single value (Barwick & Henzell, 2005). Whilst selection indexes take account of the complexity and uncertainty of both current and future production systems they are generally developed by breed societies and poorly understood by producers (Martin-Collado *et al.* 2018). Furthermore, selection indexes were designed to be used by producers as a first option in the selection process. However, more than one index is often presented alongside EBVs thereby increasing the complexity of decision making.

Table 1. Some of the most common breeding industry and farmers' strategies and tools consciously used to deal with complexity of the selection decision

Actor	Strategy heuristic (H), tool (T)	Ease of implementation	Extent of complexity reduction	Based on genetic criteria
Farmers	(H) Truncated selection/ culling	Medium	Medium-High	Not only
	(H) Trait non-attendance			
	(H) Information non-attendance	Easy	Medium-High	Not only
	(H) Follow breeders, breeding company or other famers	Easy	Medium-High	Not only
	advice	Easy	High	Not only
BC/IO /B&FO#	(T) EBVs	Difficult	Low	Yes
	(T) Selection indexes and sub-indexes	Difficult	^Medium-High	Yes
	(T) Star systems	Easy	Medium	Yes
	(T) Charts	Easy	Medium	Yes
	(T) Animal of the month	Easy	High	Not only
	(T) Pre-selection of animals offered to farmers	Medium	Medium-High	Not only

Modified from "Complexity of animal breeding choice making", by Martin-Collado, D. *et al.* (2018)
^ Depends if they are used just as another source of information to add to all the existing ones
# Breeding companies/Industry organisations/Breeder and farmer organisations

DESIREBULL™ – SIMPLIFYING GENETIC INFORMATION

The overall aim of DeSireBull™ is to aid the bull selection process, increasing the number of commercial producers effectively using genetic information and ultimately improving the productivity and profitability of herds. DeSireBull™ is intended to be a part of the BREEDPLAN toolkit, and provide an online platform for bull vendors to upload all relevant genetic information for their bulls, as well as additional, non-genetic information. DeSireBull™ will combine existing genetic tools, including BREEDPLAN data, in an innovative way so the genetic information is supplied in a more easily understood format. In addition to assisting bull purchasers, DeSireBull™ will be used by sellers and agents, and act as a channel for feedback to inform purchase decisions based on genetic information.

Combining simplifying strategies. The aim of DeSireBull™ is to combine a number of selection strategies in order to assist commercial producers in finding relevant animals and present all animal information in a format that is easy to process and understand. The three methods outlined in Table 1 of particular interest to this project are the use of selection indexes, trait subgroups (components of the index), and graphical representation e.g. star systems and charts, all of which are based on genetic criteria. Furthermore, producers will also be able to filter on a particular EBV subgroup of interest. Also under investigation is the use of an algorithm developed by Kinghorn (2013), which will demonstrate the consequences selection on one EBV subgroup will have on other subgroups. By incorporating these strategies into one tool, DeSireBull™ aims to simplify the presentation of and increase the use of genetic information by commercial producers within the Australian beef industry.

Additional features and future uses. DeSireBull™ will use a condensed version of the BreedObject questionnaire and bio-economic models to help producers define their breeding objective to develop their own selection index. Using a customised index as search criteria, producers will be able to find animals that best meet their breeding objectives. As well as providing feedback to seedstock breeders on customer requirements, additional information could also be harvested which would be extremely useful to industry as a way to examine buying behaviour and its potential consequences. This could be used to establish/review extension and education programs targeting commercial producers' buying behaviour. Such a tool could also be used as a channel to enable more direct feedback that is currently difficult to access, such as MSA carcass, processor and feedlot information, to inform current and future breeding decisions. Ideally, the successful development of DeSireBull™ will facilitate the streamlining of on-farm use of feedback for future breeding decisions based on current realised performance.

CONCLUSIONS

When it comes to selecting bulls to purchase, commercial beef producers are faced with the task of processing large amounts of information on each animal, all with different units of measurement and presented in many different formats. This amounts to a complex, cumbersome and time consuming task that has been shown to be quite overwhelming for producers. To cope with this information, producers often come up with their own ways to simplify the decision making process, resulting in sub-optimal decisions that hinder genetic progress, and impact their productivity and return. There are a number of tools currently available within industry that aim to simplify genetic information, however they are seemingly adding to the complexity of the decision process. DeSireBull™ is a tool that aims to incorporate these tools and further streamline genetic information so that it is easier to process, understand and compare. This project is well underway and so far has had positive feedback from a vast number of commercial beef producers who see value in a tool that can simplify genetic information, allow the comparison of all animals available for sale and provide feedback to both the buyer and seller.

ACKNOWLEDGEMENTS

This project is an initiative co-funded by NSW Department of Primary Industries (DPI) and Meat and Livestock Australia (MLA) Donor Company. It would not be possible to develop this tool without the additional cooperation and support of the Australian Business Research Institute (ABRI), Animal Genetics and Breeding Unit (AGBU), University of New England (UNE) and Beef and Lamb Genetics New Zealand.

REFERENCES

Barwick S.A. and Henzell A.L. (2005) *Aust. J. Exp. Agric.* **45**: 923.

BreedObject. (2016) *Developing your own index.* Retrieved from http://www.breedobject.com/page/About_BreedObject.html

Fennessy P., Byrne T., Amer P. and Martin G. (2014) Evaluating the impact of animal genetics and genomics RE&E investment. *MLA Final report.* Retrieved from https://www.mla.com.au/

Ipsos. (2016) Understanding the usage and perceptions of genetics and genomics in the Australian beef and sheep sectors. Retrieved from https://www.mla.com.au/

Kinghorn, B. (2013) DESIRE Target your genetic gains. Retrieved from https://bkinghor.une.edu.au/desire.htm.

Martin-Collado D., Byrne T., Diaz C. and Amer, P.R. (2018) *J. Anim Breed Genet.* **135:** 395.

THE LIVESTOCK PHENOTYPE REVOLUTION: ENABLING A STEP-CHANGE IN FARM MANAGEMENT AND SCIENTIFIC DISCOVERY

C.E.F. Clark[1], S.C. Garcia[1], K. Marshall[2], J.E. Pryce[3,4], P. Greenwood[5,6] and S. Lomax[1]

[1]School of Life and Environmental Sciences, Faculty of Science, University of Sydney, Camden, NSW, 2570 Australia
[2]School of Environmental and Rural Science, University of New England, Armidale, NSW, 2351 Australia
[3]Agriculture Victoria Research, AgriBio, Bundoora, Victoria, 3083 Australia
[4]School of Applied Systems Biology, La Trobe University, Bundoora, Victoria, 3083 Australia
[5]New South Wales Department of Primary Industries, Armidale Livestock Industries Centre, University of New England, Armidale NSW, 2351 Australia
[6]CSIRO Agriculture and Food, FD McMaster Laboratory Chiswick, Armidale, NSW, 2350 Australia

SUMMARY

Risks and challenges for the global agricultural industry are arising alongside well defined societal and environmental trends. Whilst genomic selection has historically enabled rapid genetic progress, new phenotypes are now required to fully unlock the power of genomics enabling the creation of robust, high welfare, agricultural systems. Here we reveal why technology is being adopted by industry to drive on-farm value, how these data are now generating new phenotypes and what phenotypes are being created now for the future. We propose to leverage these data to revolutionise current surveillance techniques enabling early warning of potential pest and disease outbreaks whilst also focusing industry extension to improve animal health. How the diversity in new phenotypes can be used for scientific discovery will also be revealed.

INTRODUCTION

Consumers now base their consumption decisions on a range of factors with a significant emphasis on provenance and animal welfare (Taylor and Signal 2009). Alongside these societal changes has been a shift towards larger, more intensive production systems. Technology is well placed to focus farmer attention to individual animals requiring intervention earlier than by visual diagnosis. Technology now not only has the capability to 'learn' from humans to 'observe' discrete behaviours but to remotely monitor individual plant and animal physiology and morphology. In this paper we describe how livestock phenotypes have evolved through time alongside those that are required for the sustainability and competitiveness of the Australian agricultural industry. Such phenotypes will revolutionise biosecurity surveillance and on-farm management in both developed and developing nations, and when collected and collated, enable significant scientific advancement and the creation of new advanced traits through genetic/genomic selection.

CONSUMER, LIVESTOCK INDUSTRY AND ENVIRONMENTAL TRENDS

'Resilient, profitable and competitive agricultural and rural industries need to actively consider what is coming and be well placed to respond' (Hajcowitz and Eady 2015). This review of the trends likely to impact rural industries over the next 20 years introduces the megatrends impacting Australian rural industries, namely the growing demand for food and fibre, climate and environmental change and the associated risk profile, increasing middle class wealth and shift in diet towards animal proteins, information empowered consumers and expectations and the rapid advances in digital technology and genetic science. Our global agricultural industry is thus tasked with increasing agricultural output by 70% to 2050 on a reduced land area due to urbanisation and desertification (Alexandratos and Bruisnma 2012). In value terms, the Australian Agriculture industry, through the National Farmers Federation, has set itself a target of increasing whole industry gross value from the current ~$60 billion to over $100 billion by 2030 (EY 2019). The same industry is consolidating into larger entities whilst rapidly moving towards a highly interconnected, globalised, production environment between consumer and producer. As such, the red meat industry has calculated a 3.5 billion risk related to the loss of society freedom to operate (RMAC 2017).

Alongside these risks and challenges will be the continued development of digital technologies enabling the detailed monitoring of livestock 24h per day (Molfino *et al.* 2017) and such technology will continue to rapidly reduce in cost. As an example, the generation of a watt through solar photovoltaics has reduced from $1,910 USD in 1956 to $0.82 in 2013 with a continued halving of cost every 5 years (Farmer and Lafond 2016). Such advancements in technology will be required to meet the complex, intersecting challenges posed above and at the forefront of these challenges will be the requirement for robust, high welfare, agricultural systems. These systems will be enabled through the creation and use of new phenotypes.

PHENOTYPES ENABLING A STEP CHANGE IN FARM MANAGEMENT

A phenotype is a set of observable characteristics of an individual resulting from the interaction of its genotype with the environment; which relates well to the origins of the word being to 'show type'. However, we have come from a time where the description of such phenotypes were derived from visual observation through to a time now where advancements in science and technology are now automating those processes and allowing new phenotypic data to be collected (Clark and Garcia 2016). New phenotypic data can be collected as technology now monitors the structure and function of the individual over durations that exceed human capability. For instance, small, frequent changes in a variable, changes occurring over considerable time periods or the internal function of the animal. Thus, we need to rethink the definition of phenotype. 'Observable', from physics, is a physical quantity that can be measured, so here we describe a phenotype as the set of physical quantities that characterise an individual resulting from the interaction of its genotype with the environment. In other words, data describing the structure and/or function of a body.

Such technology derived data required to generate new phenotypes is already being collected on farm. Lyons *et al.* (2016) highlights the vast differences in the type of technology adopted with technologies that collect and collate data and automatically implement actions such as automatic cup removers were more prevalent than those that generate data with no clear action, outcome or decision support system directly linked. These findings together suggest that it is not the technology *per se* that is limiting adoption by famers but the lack of decision support systems around them and the value that is derived from this technology on farm.

In the next 10 years we will also see new technologies implemented on farm in line with the advances made using machine learning and other techniques to reveal valuable new phenotypes. Gardenier *et al.* (2018) recently presented a robust, cost-effective perception system suitable for automatic lameness detection in dairy cattle. This system detected and localised performance for carpal/tarsal joints and hooves, along with tracking performance for hooves showing the capability of such techniques to autonomously record the structure and function related to the movement of animals.

Whilst this paper has so far focused on the value of new phenotypes using data from developed countries, phenotyping tools that are cheap, reliable and easy to use are required for developing countries. The use of such technologies would enable a step-change in farm management in developing countries shifting decades of agricultural research to such farms overnight. Wearable devices for remote recording of livestock health, movement and reproductive status are being tested in African systems but these are currently unaffordable by most farmers (Marshall *et al.* 2019). Increasing animal protein price due to increased population demand and wealth coupled with a rapidly declining technology price will create the opportunity to dramatically advance developing country livestock systems in the near future. Irrespective of whether the country be developed or developing, management value will continue to drive technology adoption on individual farms. A dedicated programme to collect and collate these data will enable the creation of new phenotypes not only for genetic selection but also for scientific discovery and biosecurity.

SCIENTIFIC DISCOVERY

The collection and collation of data to create new phenotypes presents a new approach to scientific discovery. Differences in phenotypes between animals has, and will continue to be, discovered through building the processes of an animal from the ground up; for instance, understanding metabolic and cell processes and interactions leading to differences in structure and function between individuals. Garner *et al.* (2016) provides an example of a different approach, where phenotypic diversity is firstly determined through the collection and collation of data. Extremes in phenotype are then sourced to enable break-through metabolic and physiological discoveries as to why these differences occur. The aforementioned study used the rate of decline in milk production with increasing temperature humidity index to create a heat tolerance phenotype and then with these data genomics predictions for the trait (Nguyen *et al.* 2016). These genomics predictions were then used to source cattle with trait extremes that were then exposed to heat-stress conditions in controlled-climate chambers to discover why the cause of differences. In this regard, Garner *et al.* (2016) revealed superior thermoregulatory ability enabled through improved vasodilation, with further work proposed determine differences in thermal physiology. Using a similar approach, Molfino (2019) used data generated by automatic milking systems to create groups of cows associated with efficiency in the context of voluntary cow traffic. The more 'efficient' cows produced 9% more milk with 5% less milkings per day than 'inefficient' cows, which in practice could allow more cows per robot unit and therefore improve whole system performance. Furthermore, in a subsequent study, the same author showed that efficient cows significantly spent 40 min more per day grazing ($P = 0.04$), 42 min more per day ruminating ($P < 0.001$) and 55 min less resting ($P = 0.004$) in comparison to inefficient cows, providing strong evidence of phenotypic differences between animals. Although the implications are yet to be investigated, this is an example of a 'phenotype' that was neither relevant nor detectable previously, but that with the advancement of technology could be detected and used to not only improve farm efficiency but also for scientific discovery. We propose that the approach of Garner *et al.* (2016) and Molfino (2019) will form the basis for scientific discovery as to the structure and function of livestock into the future and rapidly enable the selection of more robust animals for our future environment.

BIOSECURITY

Of the 91% of Australian dairy farmers that report unusual health events to their veterinarian, only 6% contact their state department of primary industries (Dairy Australia 2012). It is clear that silos of information related to the potential spread of pests and diseases are maintained across Australia, which is a concern given our increasingly interconnected, global production environment. In this regard, Newell *et al.* (2017) reported 83% of veterinarians to hold data within their practice without sharing with state organisations or other veterinary practices and almost all veterinarians (89%) identified cases of reduced health or abnormal appearance that remained undiagnosed. This presents a large risk to the ongoing health of Australia's livestock industry. Despite this, 71% of respondents were willing to share data with a centralised database provided a streamlined process was in place. The collection and collation of these data into a centralised system would enable the creation of phenotypes associated with pests and diseases as per the creation of the aforementioned heat stress trait (Nguyen *et al.* 2016). This would revolutionise current surveillance techniques enabling early warning of potential outbreaks whilst also focusing industry extension

CONCLUSION

Risks and challenges for the global agricultural industry are arising alongside well defined societal and environmental trends. Whilst genomic selection has historically enabled rapid genetic progress, new phenotypes are now required to fully unlock the power of genomics enabling the creation of robust, high welfare, agricultural systems. Technology that collects and collates data and automatically implement actions is adopted by industry as these technologies typically drive on-farm profit. In the next 10 years we will also see new technologies implemented on farm in line with the advances made using machine learning and other techniques to reveal valuable new phenotypes. Increasing animal protein price due to increased population demand and wealth coupled with a rapidly declining technology price, will create the opportunity to dramatically advance developing country livestock systems in the near future. We propose to leverage these phenotypic data to revolutionise current surveillance techniques enabling early warning of potential pest and disease outbreaks whilst also focusing industry extension to improve animal health. A new approach is also proposed for scientific discovery whereby the causes of livestock structure and function are determined by animal population phenotypic screening to determine the causes of diversity.

REFERENCES

Alexandratos N. and Bruinsma J. (2012) 'World agriculture towards 2030/2050: the 2012 revision.' (FAO: Rome).

Baquet A., Hambleton R. and Jose D. (1997). 'Introduction to risk management.' (1st ed.) Washington: USDA.

Clark C.E.F. and Garcia S.C. (2016) 67th Annual Meeting of the European Association for Anim. Prod. Belfast, Ireland.

Farmer J. and Lafond F. (2016) *Res. Pol.* **45**:647.

Dairy Australia. (2012). 'Dairy 2012: Situation and Outlook' (pp. 67). Victoria, Australia.

EY (2019). 'Summary Report.' Ernst and Young, March 2019; 41p.

Garner, J.B., Douglas, M.L., Williams, S.R.O., Wales, W.J. Marett, L.C. Nguyen, T.T.T., Reich C.M. and Hayes B.J. (2016) *Sci. Rep.* **6**:34114

Gardenier J., Underwood J. and Clark C.E.F (2018). International Conference on Robotics and Automation. Brisbane, Australia.

Hajkowicz, S. and Eady, S. (2015). 'Publication No. 15/065'. Rural Industries Research and Development Corporation

Lyons N.A., Eastwood C.R., Dela Rue B.T. and Kerrisk K.L. (2016). Precision Dairy Conference. Netherlands.

Marshall K., Gibson J.P., Mwai O. Mwacharo J.M., Haile A., Getachew T., Mrode R. and Kemp S.J. (2019) *Front. Genet.* **10**:297.

Molfino, J. (2019). PhD Thesis. The University of Sydney. 147 p.

Newell E., Lyons N., Ryan D., Ward M., Lomax S. and Clark C. (2017) 8th European Conference on Precision Livestock Farming. Nantes, France.

Nguyen T.T., Bowman P.J., Mekonnen H.M., Pryce J.E. and Hayes B.J. (2016) *J. Dairy Sci.* **99**:2849.

RMAC (2017). 'Meat Industry Strategic Plan.' Canberra: Meat Industry Advisory Council.

Talukder, S., Thomson, P. C., Kerrisk, K. L., Clark, C. E. F., and Celi, P. (2015). *Therio.* **83**:739.

Taylor N. and Signal T.D. (2009) *J. Appl. Anim. Welf. Sci.* **12**:345.